D0145906

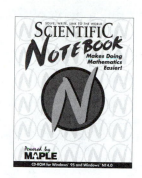

ORDER FORM

_____Yes! Send me a copy of ***Scientific Notebook*™ *for Windows®95 and Windows NT® 4.0*** (ISBN: 0-534-34864-5)

_____Copies x $59.95* =__________

Residents of: AL, AZ, CA, CT, CO, FL, GA, IL, IN, KS, KY, LA, MA, MD, MI, MN, MO, NC, NJ, NY, OH, PA, RI, SC, TN, TX, UT, VA, WA, WI must add appropriate state sales tax.

Subtotal __________
Tax __________
Handling __$4.00__
Total Due __________

Payment Options

_______ Check or money order enclosed

Bill my ____VISA ____MasterCard ____American Express

Card Number: ______________________________

Expiration Date: ______________________________

Signature: ______________________________

Please ship my order to: *(Credit card billing and shipping addresses must be the same)*

Name ______________________________

Institution ______________________________

____

Street Address______________________________

____

City ____________________ State __________ Zip+4__________

Telephone ( )____________________ e-mail ____________________

Your credit card will not be billed until your order is shipped. Prices subject to change without notice. We will refund payment for unshipped out-of-stock titles after 120 days and for not-yet-published titles after 180 days unless an earlier date is requested in writing from you.

Mail to:

Brooks/Cole Publishing Company
Source Code 8BCTC020
511 Forest Lodge Road
Pacific Grove, California 93950-5098
Phone: (408) 373-0728; Fax: (408) 375-6414
e-mail: info@brookscole.com

* Call after 12/1/97 for current prices: 1-800-487-3575

10/98

Differential Equations with Computer Lab Experiments

Second Edition

Dennis G. Zill

Loyola Marymount University
Los Angeles, California

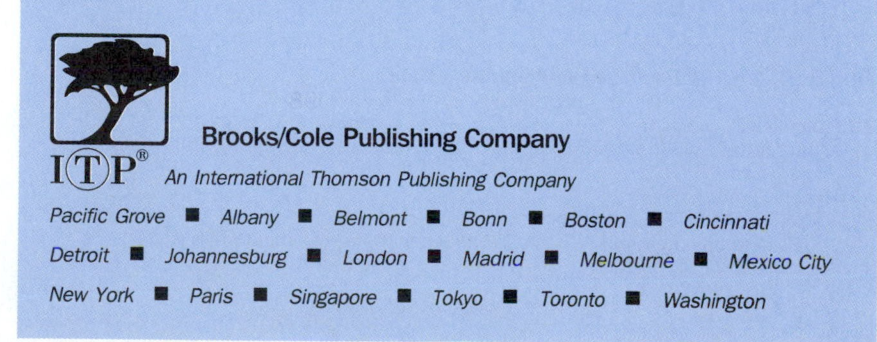

Brooks/Cole Publishing Company
An International Thomson Publishing Company

Pacific Grove ■ Albany ■ Belmont ■ Bonn ■ Boston ■ Cincinnati
Detroit ■ Johannesburg ■ London ■ Madrid ■ Melbourne ■ Mexico City
New York ■ Paris ■ Singapore ■ Tokyo ■ Toronto ■ Washington

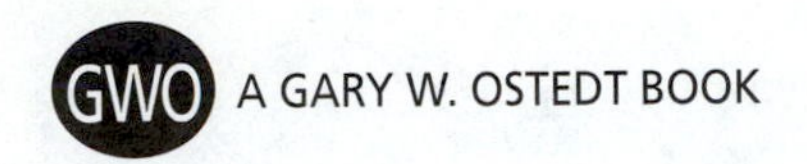

A GARY W. OSTEDT BOOK

Publisher: *Gary W. Ostedt*
Marketing Team: *Caroline Croley, Michele Mootz, Jean Thompson*
Editorial Associate: *Carol Ann Benedict*
Production Editor: *Nancy Velthaus*
Manuscript Editor: *Blair Woodcock*
Interior Design: *Rita Naughton*
Cover Design: *Roy R. Neuhaus*
Cover Photo: *Telegraph Colour Library/FPG International*
Typesetting: *Bi-Comp, Inc.*
Cover Printing: *Phoenix*
Printing and Binding: *Courier*

For more information, contact:

BROOKS/COLE PUBLISHING COMPANY
511 Forest Lodge Road
Pacific Grove, CA 93950 USA

International Thomson Publishing Europe
Berkshire House 168–173
High Holborn
London WC1V 7AA England

Thomas Nelson Australia
102 Dodds Street
South Melbourne, 3205
Victoria, Australia

Nelson Canada
1120 Birchmount Road
Scarborough, Ontario
Canada M1K 5G4

International Thomson Editores
Seneca 53
Col. Polanco
11560 México, D. F., México

International Thomson Publishing GmbH
Königswinterer Strasse 418
53227 Bonn
Germany

International Thomson Publishing Asia
#05-10 Henderson Building
Singapore 0315

International Thomson Publishing Japan
Hirakawacho Kyowa Building, 3F
2-2-1 Hirakawacho
Chiyoda-ku, Tokyo 102
Japan

Printed in the United States of America

10 9 8 7 6 5 4 3 2

Library of Congress Cataloging-in-Publication Data

Zill, Dennis G.,
Differential equations with computer lab experiments / Dennis G. Zill.—2nd ed.
p. cm.
Includes index.
ISBN 0-534-35173-5 (alk. paper)
1. Differential equations. 2. Differential equations—Computer-assisted instruction. I. Title.
QA372.Z53 1998
515'.35—dc21

97-31466
CIP

Contents

Preface

Differential Equations with Computer Lab Experiments was written originally to provide an alternative to the more traditional texts in differential equations. The first and now the second edition of this text reflect, in part, how my own course in differential equations has evolved under the influences of the "reform movement" in differential equations and the movement to incorporate active learning strategies in the classroom. In addition, some goals of this text are to encourage the use of the growing number of computer aids in the study of differential equations, to introduce qualitative and numerical aspects of differential equations early on in the course, to motivate some independent thinking, and to encourage some outside reading and even writing on the part of the student. I have also intentionally chosen to limit the content to allow the instructor the time that is necessary to explore, demonstrate, and utilize collaborative learning strategies.

Changes in This Edition

The most significant change in this new edition is that the separate manual of computer lab experiments that came bundled with the first edition has been eliminated. Apparently this was a source of confusion for both instructors and students. Now computer lab experiments, along with projects and writing assignments, are gathered together in a separate section at the end of each chapter. These problems can be used in a variety of ways: as topics for classroom discussion, as classroom presentations, or as assignments to single individuals or groups. The problems are often open-ended and some are challenging, *and all are intended to be directed by the instructor.*

There is a much greater emphasis on visualization and conceptual problems than in the first edition. Almost all exercise sets end with a separate section called *Discussion Problems.* These problems require the student to synthesize and extend what he or she has learned in the section and as such can also be used for classroom discussions.

In the first several sections of the text, the emphasis is on discerning what a differential equation can tell us without trying to solve it. The concepts of autonomous differential equations, critical points, one- and two-dimensional phase portraits, and stability are presented in a spiral fashion, beginning at a fairly intuitive level. I have always felt that the traditional manner of presenting these concepts, namely, gathered together in a chapter on systems of differential equations, tends to be an overwhelming accumulation of terminology and theory. Since these qualitative aspects of differential equations are discussed at a level commensurate with the typical sophomore's mathematical sophistication, certain aspects of stability theory are not covered. It is left to the instructor to embellish this material in places if desired.

Some additional changes are:

- The text now starts with an essay intended to motivate the subject of differential equations.
- There is an increased emphasis on real-life applications and mathematical modeling.
- A chapter review test has been added at the end of each chapter.
- The solution of differential equations is now spread throughout *each* section in Chapter 5 (The Laplace Transform) rather than confined to just one section as in the first edition.
- The discussion on reduction of order leading to a formula for a second solution of a linear second-order differential equation has been eliminated as a stand-alone section in favor of simply *using* the method of reduction of order at the points where it is needed.

Supplements

- *Complete Solutions Manual,* Warren S. Wright, Loyola Marymount University. This manual for instructors contains worked-out solutions or suggestions to every problem, project problem (writing problems excluded), and computer lab assignment.
- *Student Solutions Manual,* Warren S. Wright, Loyola Marymount University. This manual provides complete solutions to every fourth problem in each exercise set. The answers to discussion problems, project problems, and computer lab assignments are, however, excluded from this manual.
- *Computer Programs,* C. J. Knickerbocker, St. Lawrence University. This disk contains a listing of computer programs for many of the numerical methods considered in the text. Each program is written in three languages: BASIC, FORTRAN, and Pascal. This disk is available in Macintosh and IBM compatible versions.
- *Grapher,* Steve Scarborough, Loyola Marymount University. This program has routines for graphing functions, parametric and polar

equations, series, interpolating polynomials, direction fields, and numerical solutions of differential equations. For the Macintosh only.

- *ODE Solver: Numerical Procedures for Ordinary Differential Equations,* Thomas Kiffe and William Rundell, Texas A&M University, is a software package containing most of the standard numerical routines used to solve initial-value and boundary-value problems for ordinary differential equations. Output from these routines can be viewed in tabular as well as in graphical form. This program is available in Macintosh and IBM compatible versions.

Use of Technology

It is intended that this text be used in conjunction with a certain amount of computer software. A computer algebra system (CAS), such as *Mathematica, Maple, Macsyma, DERIVE,* along with one or two smaller specialized programs, such as *Grapher, ODE Solver, PhasePlane* (all from Brooks/Cole Publishing Co.), *DEQSolve* (Innosoft International, Inc.), *Differential Systems* (U-betcha Publications), and *MATLAB* (The MathWorks, Inc.), are recommended.

Acknowledgments

I would like to recognize and thank the following individuals who had a considerable influence in shaping this revision through their thoughtful reviews of the manuscript: David Dudley, Phoenix College; Bryan Johns, Seattle Central Community College; Renate McLaughlin, University of Michigan–Flint; and Frederick J. Wicklin, SAS Institute, Inc. I would also like to express my deep appreciation to Prof. Mark Michael, Department of Mathematics, Kings College, Wilkes-Barre, Pennsylvania, and Dr. Rick Wicklin, SAS Institute, Inc., for taking the time out of their busy schedules to suggest and develop some of the interesting new projects and computer lab experiments found in this edition.

Finally, work done on this revision was supported in part by NSF grant DUE-9453608. This support consisted of a course development grant from the *Los Angeles Collaborative for Teacher Excellence* (LACTE). I am grateful to Dr. Jacqueline Dewar, coordinator of LACTE on the Loyola Marymount University campus, for her encouragement and suggestions.

Dennis G. Zill
Los Angeles

An Introduction to Differential Equations

The words *differential* and *equations* certainly suggest solving some kind of equation that contains derivatives. So it is, *sort of.* Let's start with a simple example. In calculus you learned that the derivative dy/dt of a function $y = f(t)$ is itself another function $f'(t)$ found by an appropriate rule. If $y = e^{t^2}$, then $dy/dt = 2te^{t^2}$. If we replace e^{t^2} by the symbol y, we obtain

$$\frac{dy}{dt} = 2ty \tag{1}$$

Now imagine that a friend of yours simply hands you the *differential equation* in (1)—and you have no idea where it came from—and the friend asks: What is y? You are now face-to-face with one of the basic problems in the course called *Differential Equations:* How do you solve such an equation for the unknown function $y = f(t)$? The problem is loosely equivalent to the familiar reverse problem of differential calculus: Given a derivative, find an antiderivative.

The first paragraph tells something, but not the complete story, about the course that you are about to begin. There is more to the study of differential equations than just memorizing methods someone has devised to solve them. Over the centuries, differential equations would often spring from the efforts of mathematicians, scientists, and engineers to describe some physical phenomenon or to translate an empirical or experimental law into mathematical terms.

Let's consider another example—this time from physics. In calculus you also learned that a derivative is an instantaneous rate of change; in a physical setting, we can interpret an instantaneous rate of change as velocity or as acceleration.

Now suppose a small heavy object, such as a rock, is dropped from a tall building as illustrated in Figure I.1. The problem consists of finding the height of the rock above ground level. In order to describe a dynamic

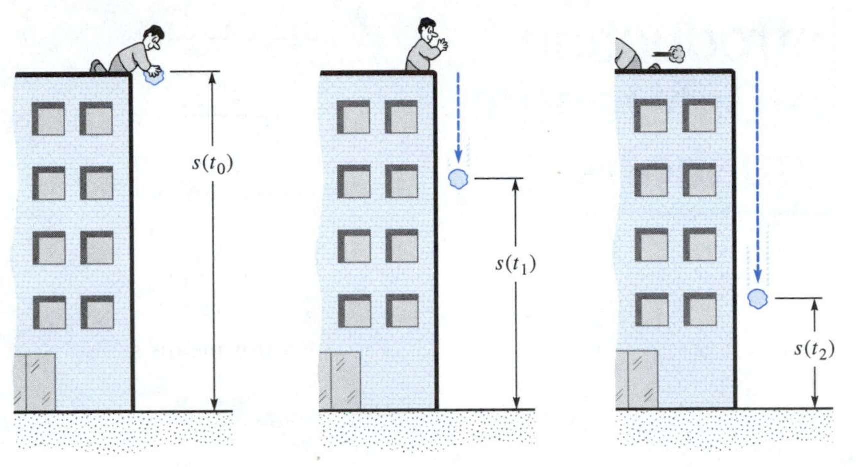

Figure I.1. $t_0 < t_1 < t_2$.

problem such as this in mathematical terms, we use Newton's empirical second law of motion, which states that the resultant or net force $F = \Sigma F_k$ acting on a moving body of mass m is given by $F = ma$, where a is its instantaneous acceleration. Suppose $s(t)$ denotes the rock's position or height above ground at time t, and that upward is the positive direction. Because the acceleration of the rock is the second derivative $a = d^2s/dt^2$, if we assume that the only force acting on the rock is the force due to gravity, then Newton's second law gives

$$m\frac{d^2s}{dt^2} = -mg \tag{2}$$

The minus sign is used on the right side of (2) because the weight $W = mg$ of the rock, the force due to gravity, is directed downward and is therefore opposite to the assumed positive direction. Since the motion is taking place near the surface of the earth, it is also assumed that the acceleration g due to gravity is constant (32 ft/s^2, 9.8 m/s^2, or 980 cm/s^2). So after dividing through by m,

$$\frac{d^2s}{dt^2} = -g \tag{3}$$

Now that we have an equation—a differential equation—that describes the position or height of the rock while it is in the air, we must now solve it; that is, we must find some method or procedure for determining the *solution* $s(t)$. This solution must *satisfy* the equation, which means that it must be a function that is at least twice differentiable.

Here is another example. One of the simplest and one of the most famous problems in physics/mathematics was posed by Johann Bernoulli in 1696. Consider a flexible and smooth wire fixed at two points P_1 and P_2, and suppose a bead B is allowed to slide freely from point P_1 to point P_2

under the influence of gravity. Here is the problem:

> *How should the wire be bent, that is, what should the shape of the wire be, so that the bead slides from P_1 to P_2 in the shortest time?*

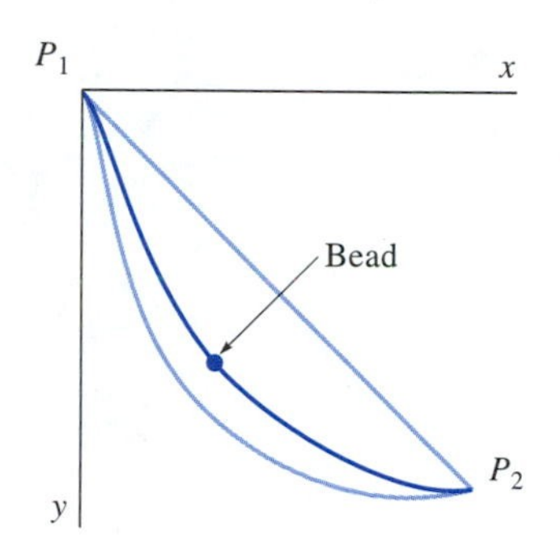

Figure I.2.

In Figure I.2, we have drawn several shapes and have fixed point P_1 at the origin in an xy-plane. One's intuition immediately suggests that, since the shortest distance between two points is a straight line, the shape of the wire should be a straight line. Wrong! Years earlier Galileo conjectured that the shape should be an arc of a circle. A better guess but still wrong! This problem, which came to be called the *brachistochrone problem* (in Greek, *brachistos* means shortest and *chronos* means time) stumped many famous mathematicians, so Bernoulli solved it himself. Although we will not go through the details, needless to say, Bernoulli's analysis led to a differential equation

$$y\left[1 + \left(\frac{dy}{dx}\right)^2\right] = c \tag{4}$$

where c is a constant. Solving this equation showed that the wire had to be bent into the shape of . . .—sorry, you will be asked later to complete the story.

You can't get far into the mastery of engineering, the physical, social, and business sciences without mathematics—and differential equations in particular. People who delve into the construction of mathematical descriptions of things such as animal population growth; the competition of two animal species in an ecosystem; the spread of communicable diseases; molecular reactions; dissemination of drugs in a body; the flow of traffic; the spread of rumors, information, and technology; the theory of learning and memorization; war; the expansion of the universe, and so on, invariably try to answer questions involving the words "How fast?" For example, *How fast does a disease spread through a community? How fast does a person learn?* "How fast" refers to a rate of change; a derivative is rate, and so the mathematical translation or formulation—that is, the *mathematical model*—of experiments, observations, or theories may be a differential equation.

Chapter 1

Getting Started

1.1 Definitions and Terminology

In the motivational introduction to this text, we gave some intuitive ideas about the nature and origins of differential equations. But in order to study, read, and be conversant in a specialized topic, one first has to learn the terminology—the jargon—of that discipline. As we have just seen, $d^2s/dt^2 = -g$ and $y[1 + (dy/dx)^2] = c$ are both differential equations, but it is helpful to be able to distinguish between different kinds of differential equations. Soon you will be able to appreciate the significance in the designations that the first equation is *linear* and the second is *nonlinear.*

The goal of this first section then is to give names to things and to state two definitions.

A Word About Notation Throughout precalculus and most of calculus, the symbol x is traditionally used as an independent variable; a function of a single variable is written $y = f(x)$. Because so many applications in the study of differential equations involve time, it is convenient now to use the symbol t at the outset as an independent variable. This choice also frees the symbol x to be used as a dependent variable. Symbols such as x and y, denoting dependent variables, are understood to be real-valued functions of the independent variable t. Derivatives will be written using either the Leibniz notation dy/dt, dx/dt, d^2y/dt^2, d^2x/dt^2, ... or the prime notation y', x', y'', x'', ... Thus, we can write the differential equations

$$\frac{d^2s}{dt^2} = -g \quad \text{and} \quad y\left[1 + \left(\frac{dy}{dx}\right)^2\right] = c \tag{1}$$

a little more compactly as $s'' = -g$ and $y[1 + (y')^2] = c$.

For the remainder of this section, we are going to formalize some of the concepts presented in the introduction. We start with a definition of a general differential equation.

Definition 1.1 Differential Equation

An equation containing derivatives of one or more unknown functions with respect to one or more independent variables, is said to be a **differential equation** (DE).

Differential equations are classified by **type**, **order**, and **linearity**.

Classification by Type If an equation contains only ordinary derivatives of one or more functions with respect to a single independent variable, it is said to be an **ordinary differential equation** (ODE). For example,

$$\frac{dy}{dt} + 10y = e^t, \quad \frac{d^2y}{dt^2} - \frac{dy}{dt} + 6y = 0, \quad \text{and} \quad \frac{dx}{dt} + \frac{dy}{dt} = 5x - y$$

are ordinary differential equations. An equation involving partial derivatives of one or more functions of two or more independent variables is called a **partial differential equation** (PDE). For example,

$$\frac{\partial u}{\partial y} = -\frac{\partial v}{\partial x} \quad \text{and} \quad \frac{\partial^2 u}{\partial x^2} = \frac{\partial^2 u}{\partial t^2} - 2\frac{\partial u}{\partial t}$$

are partial differential equations.

Classification by Order The **order** of a differential equation (ODE or PDE) is the order of the highest derivative in the equation. For example,

$$\underset{\text{second-order}}{\frac{d^2y}{dt^2}} + 5\underset{\text{first-order}}{\left(\frac{dy}{dt}\right)^3} - 4y = e^t \tag{2}$$

is a second-order ordinary differential equation. The equations in (1) are second order and first order, respectively. First-order differential equations are sometimes written in terms of differentials. For example, by dividing $(y - t)dt + 4t\,dy = 0$ by the differential dt and using $dy = \left(\frac{dy}{dt}\right)dt$, we recognize $4t\frac{dy}{dt} + y = t$ as a first-order differential equation.

In symbols, we can express an nth-order ordinary differential equation by the general form

$$F(t, y, y', \ldots, y^{(n)}) = 0 \tag{3}$$

where F is a real-valued function of $n + 2$ variables: $t, y, y', \ldots, y^{(n)}$, and where $y^{(n)}$ is d^ny/dt^n. We also make the assumption hereafter that it is possible to solve an ordinary differential equation in the form (3) uniquely for the highest derivative $y^{(n)}$ in terms of the remaining $n + 1$ variables. The differential equation

$$\frac{d^ny}{dt^n} = f(t, y, y', \ldots, y^{(n-1)}) \tag{4}$$

where f is a real-valued function, is referred to as the **normal form** of (3). Thus, when it suits our purposes, we use the normal forms

$$\frac{dy}{dt} = f(t, y) \quad \text{and} \quad \frac{d^2y}{dt^2} = f(t, y, y')$$

to represent general first- and second-order ordinary differential equations. See the Remarks.

Classification as Linear or Nonlinear An nth-order ordinary differential equation (3) is said to be **linear** if F is linear in $y, y', \ldots, y^{(n-1)}$. This means that an nth-order ordinary differential equation is linear when equation (3) is $a_n(t)y^{(n)} + a_{n-1}(t)y^{(n-1)} + \cdots + a_1(t)y' + a_0(t)y - g(t) = 0$ or

$$a_n(t)\frac{d^ny}{dt^n} + a_{n-1}(t)\frac{d^{n-1}y}{dt^{n-1}} + \cdots + a_1(t)\frac{dy}{dt} + a_0(t)y = g(t) \tag{5}$$

Note that a linear ordinary differential equation is characterized by two properties: The dependent variable y and all its derivatives are of the first degree, and each coefficient depends only on the independent variable t. The equations, in turn,

$$t\,dy + y\,dt = 0, \quad y'' - 2y' + y = 0, \quad t^3\frac{d^3y}{dt^3} + 3t\frac{d^2y}{dt^2} - 5y = e^t$$

are linear first-, second-, and third-order ordinary differential equations. A **nonlinear** ordinary differential equation is simply one that is not linear. Nonlinear functions of y or its derivatives, such as $e^{y'}$ and $\sin y$, cannot appear in a linear equation. Therefore,

nonlinear term: coefficient depends on y	nonlinear term: nonlinear function of y	nonlinear term: power not 1
↓	↓	↓
$(1 + y)y' + 2y = e^t,$	$\dfrac{d^2y}{dt^2} + \sin y = 0,$	$\dfrac{d^4y}{dt^4} + y^2 = 0$

are examples of nonlinear first-, second-, and fourth-order ordinary differential equations, respectively. The second-order equation in (1) is linear; the first-order equation in (1) and the second-order equation in (2) are nonlinear.

Solutions One of our goals in this course is to *solve* differential equations. In Definition 1.2, we define the all-important notion of a **solution** of an ordinary differential equation (3).

> **Definition 1.2 Solution of a Differential Equation**
>
> Any function ϕ defined on some interval I, which when substituted into a differential equation reduces the equation to an identity, is said to be a **solution** of the equation on the interval.

You cannot think *solution* without simultaneously thinking *interval*. A solution of an ordinary differential equation (3) is a function ϕ that is defined over an interval I and possesses at least n derivatives in I, and

$$F(t, \phi(t), \phi'(t), \ldots, \phi^{(n)}(t)) = 0 \quad \text{for all } t \text{ in } I$$

We say that $y = \phi(t)$ *satisfies* the differential equation. The interval I is called the **interval of definition** or the **interval of existence** of the solution, and can be an open interval (a, b), a closed interval $[a, b]$, an infinite interval (a, ∞), and so on.

■ EXAMPLE 1 Verification of a Solution

Verify that $y = t^4/16$ is a solution of the nonlinear equation $dy/dt = ty^{1/2}$ on the interval $(-\infty, \infty)$.

Solution One way of verifying that the given function is a solution is to see, after substituting, whether each side of the equation is the same for every t in the interval. From

$$\textit{left-hand side:} \quad \frac{dy}{dt} = 4\cdot\frac{t^3}{16} = \frac{t^3}{4}$$

$$\textit{right-hand side:} \quad ty^{1/2} = t\left(\frac{t^4}{16}\right)^{1/2} = t\frac{t^2}{4} = \frac{t^3}{4}$$

we see that each side of the equation is the same for every real number t. Note that $y^{1/2} = t^2/4$ is, by definition, the nonnegative square root of $t^4/16$. ■

■ EXAMPLE 2 Verification of a Solution

The function $y = te^t$ is a solution of the linear equation $y'' - 2y' + y = 0$ on the interval $(-\infty, \infty)$. To see this, we compute $y' = te^t + e^t$ and $y'' = te^t + 2e^t$. After substituting observe that for every real number t,

$$y'' - 2y' + y = (te^t + 2e^t) - 2(te^t + e^t) + te^t = 0$$ ■

The interval I, over which a solution $y = \phi(t)$ exists or is defined, does not need to be the same as the domain of the function ϕ. Example 3 illustrates the difference.

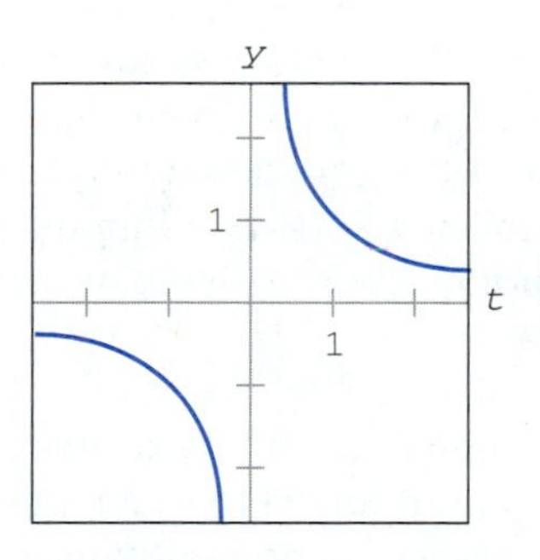

(a) Function $y = 1/t$, $t \neq 0$

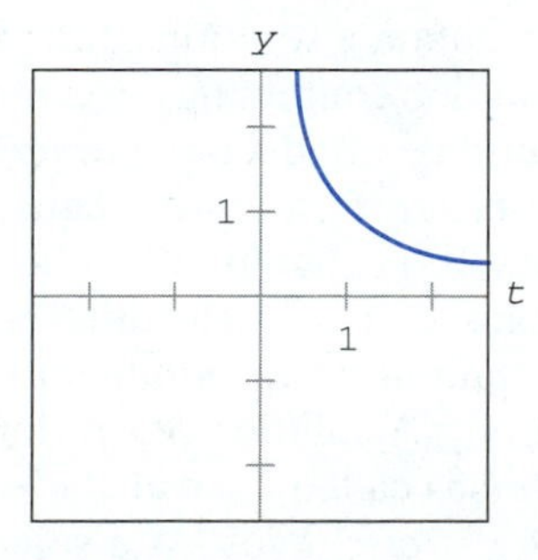

(b) Function $y = 1/t$, $(0, \infty)$

Figure 1.1.

■ EXAMPLE 3 Domain/Interval of Existence

You should verify that $y = 1/t$ satisfies the linear first-order differential equation $ty' + y = 0$. Considered simply as a *function*, the domain of $y = 1/t$ is the set of all real numbers t except 0. As we see in Figure 1.1(a), the y-axis ($t = 0$) is a vertical asymptote for its graph and so the function is not differentiable at $t = 0$. Thus when we say $y = 1/t$ is a *solution* of the differential equation, we mean $y = 1/t$ defined on an interval on which it is *differentiable*, in this case, an interval not containing 0. We can take either $(-\infty, 0)$ or $(0, \infty)$ to be an interval of definition or an interval of existence. See Figure 1.1(b). ■

Explicit and Implicit Solutions You should be familiar with the terms *explicit functions* and *implicit functions* from your study of calculus. Because some methods for solving differential equations lead directly to these two forms, solutions of differential equations can be further distinguished as either explicit solutions or implicit solutions. A solution in which the dependent variable is expressed solely in terms of the independent variable and constants is said to be an **explicit solution**. For our purposes, let us think of an explicit solution as an explicit formula $y = \phi(t)$ that we can manipulate, evaluate, and differentiate. In Example 1, $y = t^4/16$, $-\infty < t < \infty$ is an explicit solution of $dy/dt = ty^{1/2}$. Note, too, that this same differential equation possesses the constant solution $y = 0$, $-\infty < t < \infty$. An explicit solution of a differential equation that is identically zero on an interval I

is called a **trivial solution**. A relation $G(t, y) = 0$ is said to be an **implicit solution** of an ordinary differential equation (3) on an interval I, provided there exists at least one function ϕ that satisfies the relation as well as the differential equation on I. In other words, $G(t, y) = 0$ defines the function ϕ implicitly.

EXAMPLE 4 Verification of an Implicit Solution

The relation $t^2 + y^2 - 25 = 0$ is an implicit solution of the differential equation

$$\frac{dy}{dt} = -\frac{t}{y} \tag{6}$$

on the interval $-5 < t < 5$. By implicit differentiation, we obtain

$$\frac{d}{dt}t^2 + \frac{d}{dt}y^2 - \frac{d}{dt}25 = \frac{d}{dt}0 \quad \text{or} \quad 2t + 2y\frac{dy}{dt} = 0$$

Solving the last equation for the symbol dy/dt gives (6). In addition, you should verify that the functions $y_1 = \sqrt{25 - t^2}$ and $y_2 = -\sqrt{25 - t^2}$ satisfy the relation, in other words, $t^2 + y_1^2 - 25 = 0$ and $t^2 + y_2^2 - 25 = 0$, and are solutions of the differential equation on $-5 < t < 5$. ■

Any relation of the form $t^2 + y^2 - c = 0$ *formally* satisfies (6) for any constant c. However, it is understood that the relation should always make sense in the real number system; thus, for example, we cannot say that $t^2 + y^2 + 25 = 0$ is an implicit solution of the equation. (Why not?)

Because the distinction between an explicit solution and an implicit solution should be intuitively clear, we will not belabor the issue by always saying "here is an explicit (implicit) solution."

Families of Solutions The study of differential equations is similar to that of integral calculus. A solution is sometimes referred to as an **integral** of the equation, and its graph is called an **integral curve** or a **solution curve**. When evaluating an antiderivative or indefinite integral in calculus, we use a single constant c of integration. Analogously, when solving a first-order differential equation $F(t, y, y') = 0$ we *usually* obtain a solution containing a single arbitrary constant or parameter c. A solution containing an arbitrary constant represents a set $G(t, y, c) = 0$ of solutions called a **one-parameter family of solutions**. When solving an nth-order differential equation $F(t, y, y', \ldots, y^{(n)}) = 0$ we seek an **n-parameter family of solutions** $G(t, y, c_1, c_2, \ldots, c_n) = 0$. This simply means that a single differential equation can possess an infinite number of solutions corresponding to the unlimited number of choices for the parameter(s). A solution of a differential equation that is free of arbitrary parameters is called a **particular solution**. For example, the one-parameter family $y = ct - t\cos t$ is a solution of the linear first-order equation $t\,dy/dt - y = t^2 \sin t$ on the interval $(-\infty, \infty)$ (verify). Figure 1.2 shows some of the solution curves of this

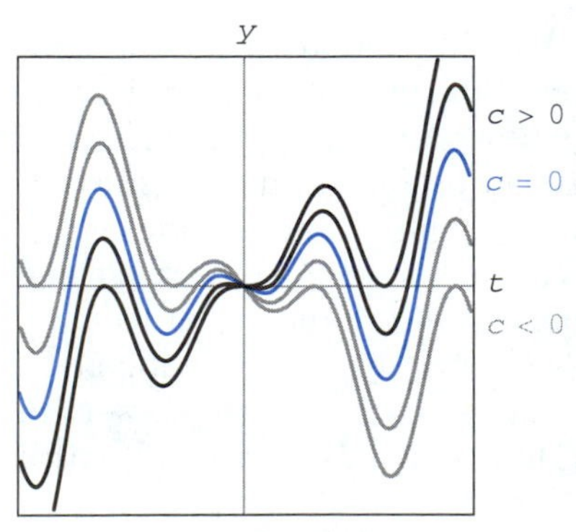

Figure 1.2.

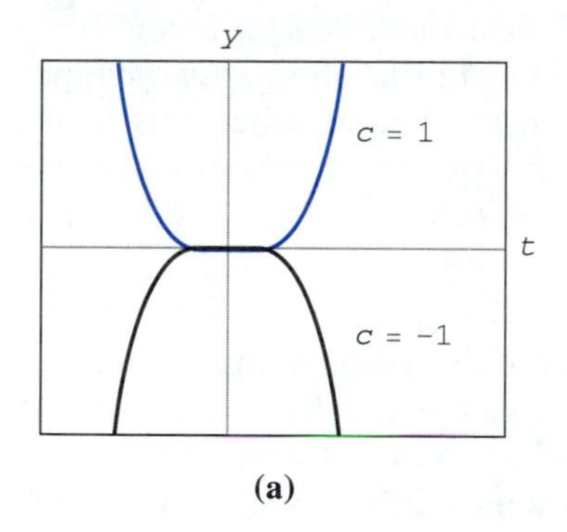

(a)

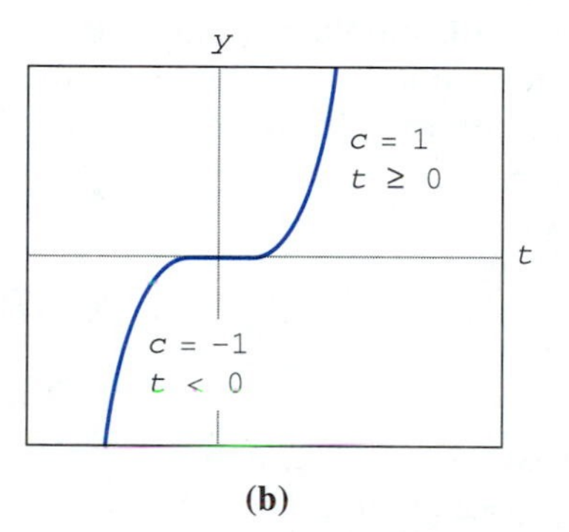

(b)

Figure 1.3.

equation. The solution $y = -t\cos t$, the colored curve in the figure, is a particular solution of the differential equation corresponding to $c = 0$. Similarly, on the interval $(-\infty, \infty)$, $y = c_1e^t + c_2te^t$ is a two-parameter family of solutions of the linear second-order equation $y'' - 2y' + y = 0$ (verify). Some particular solutions of the equation are the trivial solution $y = 0$ ($c_1 = c_2 = 0$), $y = te^t$ ($c_1 = 0$, $c_2 = 1$), $y = 5e^t - 2te^t$ ($c_1 = 5$, $c_2 = -2$), and so on.

The next example shows that a solution of a differential equation can be a piecewise defined function.

EXAMPLE 5 A Piecewise Defined Solution

You should verify that the one-parameter family $y = ct^4$ is a solution of the differential equation $ty' - 4y = 0$ on the interval $(-\infty, \infty)$. See Figure 1.3(a). The piecewise defined differentiable function

$$y = \begin{cases} -t^4, & t < 0 \\ t^4, & t \geq 0 \end{cases}$$

is a particular solution of the equation, but cannot be obtained from the family $y = ct^4$ by a *single* choice of c; the solution is constructed from the family by choosing $c = -1$ for $t < 0$ and $c = 1$ for $t \geq 0$. See Figure 1.3(b). ■

Sometimes a differential equation possesses a solution that cannot be obtained by specializing *any* of the parameters in a family of solutions. Such a solution is called a **singular solution**.

EXAMPLE 6 A Singular Solution

So far we have seen that $y = t^4/16$ and $y = 0$ are solutions of the differential equation $y' = ty^{1/2}$ on $(-\infty, \infty)$. Later on (Section 2.2) we shall demonstrate, by actually solving it, that the differential equation possesses the one-parameter family of solutions $y = (t^2/4 + c)^2$. When $c = 0$, the resulting particular solution is $y = t^4/16$. But notice the trivial solution $y = 0$ is not a member of this family; there is no way of assigning a value to the constant c to obtain $y = 0$. Hence $y = 0$ is a singular solution of the equation. ■

General Solution If *every* solution of an nth-order ordinary differential equation $F(t, y, y', \ldots, y^{(n)}) = 0$ on an interval I can be obtained from an n-parameter family of solutions $G(t, y, c_1, c_2, \ldots, c_n) = 0$ by appropriate choices of the parameters c_i, $i = 1, 2, \ldots, n$, we then say that the family is the **general solution** of the differential equation. In solving linear differential equations, we shall impose relatively simple restrictions on the coefficients of the equation; with these restrictions one can always be assured not only that a solution does exist on an interval but also that a family of solutions does indeed yield all possible solutions. Nonlinear equations, with the exception of some first-order equations, are usually difficult or impossible to solve in terms of familiar elementary functions: finite combinations of integer powers of t, roots, exponential and logarithmic functions, and trigonometric and inverse trigonometric functions. Furthermore, if we happen to have a family of solutions for a nonlinear equation, when this family constitutes a general solution is not obvious. On a practical level then, the designation "general solution" is applied only to linear differential equations.

Systems of Differential Equations Until this point we have been discussing single differential equations containing one unknown function. But often in theory as well as in many applications we must deal with systems of differential equations. A **system of ordinary differential equations** is two or more equations involving the derivatives of two or more unknown functions of a single independent variable. For example, if x and y denote dependent variables, and t denotes the independent variable, then a system of two first-order differential equations is given by

$$\begin{aligned}\frac{dx}{dt} &= f(t, x, y)\\ \frac{dy}{dt} &= g(t, x, y)\end{aligned} \tag{7}$$

A **solution** of a system such as (7) is a pair of differentiable functions $x = \phi_1(t), y = \phi_2(t)$ defined on a common interval I that satisfy each equation of the system on this interval.

Remarks

(*i*) Although you should be able to distinguish between an ordinary and a partial differential equation, we shall consider only the theory and applications of ordinary differential equations in this text. Consequently, from this point on the words *differential equation* will refer to an ordinary differential equation.

(*ii*) It may not seem like a big deal to assume that $F(t, y, y', \ldots, y^{(n)}) = 0$ can be solved for $y^{(n)}$, but one should be careful here. There are exceptions and there certainly are some problems connected with this assumption. See Problems 33 and 34 in Exercises 1.1.

(*iii*) You will sometimes hear instructors of courses in differential equations speak of "closed-form solutions." Translated, this phase usually refers to explicit solutions that are expressible in terms of the *elementary functions* mentioned earlier in this section.

(*iv*) There is another classification of differential equations that we mention now in passing, but which will be important in the subsequent chapters. If the variable of differentiation, that is, the independent variable, does not appear explicitly in a differential equation, the equation is then said to be **autonomous**. If t is the independent variable, normal forms of autonomous first- and second-order differential equations are

$$\frac{dy}{dt} = g(y) \quad \text{and} \quad \frac{d^2y}{dt^2} = g(y, y')$$

respectively. For example, the first-order differential equation $dy/dt = y^2$ is autonomous; the equation $dy/dt = 2ty$ is nonautonomous. The second-order equation in Example 2 is autonomous.

(*v*) Although we will primarily use the letter t to denote the independent variable or variable of differentiation, on occasion other symbols, such as x, will be used. For example, in the differential equations

$$\frac{dy}{dx} + xy = e^x \quad \text{and} \quad \frac{dx}{dt} = x - x^2$$

x represents an independent variable in the first equation, but a dependent variable in the second.

EXERCISES 1.1

Answers to odd-numbered problems begin on page 433.

In Problems 1–10, state whether the given differential is linear or nonlinear.

1. $(1 - t)y'' - 4ty' + 5y = \cos t$

2. $t\dfrac{d^3y}{dt^3} - \left(\dfrac{dy}{dt}\right)^4 + y = 0$

3. $yy' + 2y = 1 + t^2$

4. $t^2\,dy + (y - ty + te^t)\,dt = 0$

5. $t^4y^{(4)} - t^2y'' + 3y = 0$

6. $\dfrac{d^2y}{dt^2} + \dfrac{dy}{dt} + y = \cos(t + y)$

7. $\dfrac{d^2y}{dx^2} = \sqrt{1 + \left(\dfrac{dy}{dx}\right)^2}$

8. $\dfrac{d^2r}{dt^2} = -\dfrac{k}{r^2}$

9. $(\sin\theta)y''' - (\cos\theta)y' = 2$

10. $(1 - y^2)\,dx + x\,dy = 0$

In Problems 11–16, verify that the indicated function is a solution of the given differential equation. Assume an appropriate interval of definition.

11. $2y' + y = 0;\ y = e^{-t/2}$

12. $\dfrac{dy}{dt} + 20y = 24;\ y = \frac{6}{5} - \frac{6}{5}e^{-20t}$

13. $y'' - 6y' + 13y = 0;\ y = e^{3t}\cos 2t$

14. $y'' + y = \tan t;\ y = -(\cos t)\ln(\sec t + \tan t)$

15. $2xy\,dx + (x^2 + 2y)\,dy = 0;\ x^2y + y^2 = 1$

16. $\dfrac{dX}{dt} = (2 - X)(1 - X);\ \ln\dfrac{2 - X}{1 - X} = t$

In Problems 17–20, verify that the indicated function is a family of solutions of the given differential equation. Assume an appropriate interval of definition.

17. $y' = y(1 - y);\ y = \dfrac{1}{1 + c_1e^{-t}}$

18. $y' + 2ty = 1;\ y = e^{-t^2}\displaystyle\int_0^t e^{u^2}\,du + c_1e^{-t^2}$

19. $\dfrac{d^2y}{dt^2} - 4\dfrac{dy}{dt} + 4y = 0;\ y = c_1e^{2t} + c_2te^{2t}$

20. $t^3\dfrac{d^3y}{dt^3} + 2t^2\dfrac{d^2y}{dt^2} - t\dfrac{dy}{dt} + y = 12t^2;$

$$y = c_1t^{-1} + c_2t + c_3t\ln t + 4t^2$$

In Problems 21 and 22, verify that the given pair of functions is a solution of the given system of differential equations on $(-\infty, \infty)$.

21. $\dfrac{dx}{dt} = x + 3y$

$$\frac{dy}{dt} = 5x + 3y;$$

$$x = e^{-2t} + 3e^{6t},\quad y = -e^{-2t} + 5e^{6t}$$

22. $\dfrac{d^2x}{dt^2} = 4y + e^t,$

$\dfrac{d^2y}{dt^2} = 4x - e^t;$

$x = \cos 2t + \sin 2t + \frac{1}{5}e^t,\ y = -\cos 2t - \sin 2t - \frac{1}{5}e^t$

23. Verify that the piecewise defined function

$$y = \begin{cases} -t^2, & t < 0 \\ t^2, & t \geq 0 \end{cases}$$

is a solution of the differential equation $ty' - 2y = 0$ on $(-\infty, \infty)$.

24. In Example 4, we saw that $y_1 = \sqrt{25 - t^2}$ and $y_2 = -\sqrt{25 - t^2}$ are solutions of $dy/dt = -t/y$ on the interval $(-5, 5)$. Explain why the piecewise defined function

$$y = \begin{cases} \sqrt{25 - t^2}, & -5 < t < 0 \\ -\sqrt{25 - t^2}, & 0 \leq t < 5 \end{cases}$$

is not a solution of the differential equation on the interval $(-5, 5)$.

In Problems 25 and 26, the indicated function is a solution of the given differential equation on an interval *I*. Determine at least one such interval.

25. $y' = 25 + y^2;\ y = 5 \tan 5t$

26. $2y' = y^3 \cos t;\ y = (1 - \sin t)^{-1/2}$

Discussion Problems

27. Make up a differential equation that does not possess any real solutions.

28. Make up a differential equation that you feel confident possesses only the trivial solution $y = 0$. Explain your reasoning.

29. What function(s) that you know from calculus is such that its first derivative is itself? Write your answer in the form of a differential equation with a solution.

30. What function(s) that you know from calculus is such that its second derivative is itself? What function(s) is such that its second derivative is the negative of itself? Write each answer in the form of a differential equation with a solution.

31. Suppose the two-parameter family $y(t) = c_1y_1(t) + c_2y_2(t)$ is the general solution of a linear second-order differential equation on an interval I. If $t = 0$ is in I, find a particular solution in the family that satisfies both conditions: $y(0) = 2,\ y'(0) = 0$. State any assumptions that you must make.

32. Discuss, and illustrate with examples, how to solve differential equations of the forms $dy/dt = f(t)$ and $d^2y/dt^2 = f(t)$. Find a two-parameter family of solutions of the second-order differential equation in (1).

33. The differential equation $t(y')^2 - 4y' - 12t^3 = 0$ has the form given in (3). Determine whether the equation can be put into the normal form $dy/dt = f(t, y)$.

34. The normal form (4) of an nth-order differential equation is equivalent to (3) whenever both forms have exactly the same solutions. Make up a first-order differential equation in which the equation $F(t, y, y') = 0$ is *not* equivalent to the normal form $dy/dt = f(t, y)$.

35. Find a linear second-order differential equation for which $y = c_1t + c_2t^3$ is a two-parameter family of solutions. Make sure your equation is free of arbitrary parameters.

36. Find values of m so that the indicated function is a solution of the given differential equation. Explain your reasoning.
(a) $y' + 2y = 0;\ y = e^{mt}$
(b) $y'' - 5y' + 6y = 0;\ y = e^{mt}$
(c) $ty'' + 2y' = 0;\ y = t^m$
(d) $t^2y'' - 7ty' + 15y = 0;\ y = t^m$

37. Examine the general form (5) of a linear differential equation. Discuss: Under what conditions is $y = 0$ a solution of a linear equation?

38. We know that $y' = 0$ if and only if y is a constant function. Use this idea to determine whether the given differential equation possesses constant solutions.
(a) $y' = y^2 + 2y - 3$ **(b)** $(y - 1)y' = 1$
(c) $y'' + 4y' + 6y = 10$

39. Interpret the following statement as a differential equation:

> *The rate at which a body cools is proportional to the difference between the temperature of the body and the temperature of the surrounding medium.*

Information about the nature of a solution of a differential equation can often be obtained from the equation itself. Before working Problems 40–45, review your calculus text on the geometric significance of the derivatives dy/dt and d^2y/dt^2.

40. Discuss: If a solution $y(t)$ of the differential equation $dy/dt = 2y - 8y^2$ has maximum and minimum values, why is it at least plausible that $y_{\min} = 0$ and $y_{\max} = \frac{1}{4}$?

41. Discuss: Is the following proposition true or false?

If $y(t)$ is a solution of the differential equation $y' = f(t, y)$ on an interval I, then $y(t)$ is continuous on I.

42. Suppose $y(t)$ is a solution of $y' = y^2 + 4$ on an interval I. Describe the graph of $y(t)$ on I. For example, can the graph of $y(t)$ have a relative extremum?

43. Given that $y = 5$, $-\infty < t < \infty$ is a constant solution of $y' = 5 - y$. Sketch several plausible solution curves in the regions defined by $y < 5$ and $y > 5$. Explain your reasoning.

44. Discuss: From its normal form, can you predict a plausible interval of existence for a solution of the differential equation $(t - 2)y' + y = 1$?

45. Discuss: Based only on the differential equation, why it is unlikely that $d^2y/dt^2 = ay$, $a > 0$ possesses *periodic* solutions. Sketch several plausible solution curves. Explain your reasoning.

1.2 Initial-Value Problems

Often we impose certain constraints on a solution of a DE even before we attempt to solve the equation. In this section these constraints or side conditions are numerical values specified for the unknown function $y(t)$ or its derivatives at the same point t_0. For example, suppose we want to find a solution of the first-order differential equation $dy/dt = ty^{1/2}$ such that the solution curve passes through a point in the ty-plane, say, (1, 2); that is, we want the solution $y(t)$ to satisfy $y(1) = 2$. This kind of problem is referred to as an *initial-value problem.*

Freely Falling Body In (3) of the introduction we saw that the simple differential equation $d^2s/dt^2 = -g$ described the height $s(t)$ of a small rock dropped from the rooftop of a building, and in Problem 32 in Exercises 1.1 you should have deduced that this equation possesses the two-parameter family of solutions $s(t) = -\frac{1}{2}gt^2 + c_1t + c_2$, where c_1 and c_2 are arbitrary constants. Clearly, we want specific values of the constants, that is, we want a particular solution $s(t)$. When a differential equation is used to describe or model a physical system changing with time t, it is usually accompanied by side conditions that enable us to find specific constants in its solution.

In our rock example, if $t_0 = 0$ denotes the time at which the rock is released, and $t = T$ is the time when it hits the ground, the rock is in the air for a finite period of time, $0 < t < T$. As soon as the rock has left the hands of the person on the rooftop, the person can no longer control it. In other words, what takes place subsequently is determined by what takes place *initially*. Put another way, the *state* of the rock, namely, its *position* above the ground and its *velocity* for $0 < t < T$, as well as the time T, obviously depend on whether the rock is simply dropped or thrown straight down (or up), and how high the building is. So if the building is s_0 units high, then our solution $s(t)$ of the differential equation must satisfy $s(0) = s_0$. Additionally, as shown in Figure 1.4, if we interpret the word *dropped* to mean *released from rest,* then the rock's initial velocity is zero, that is,

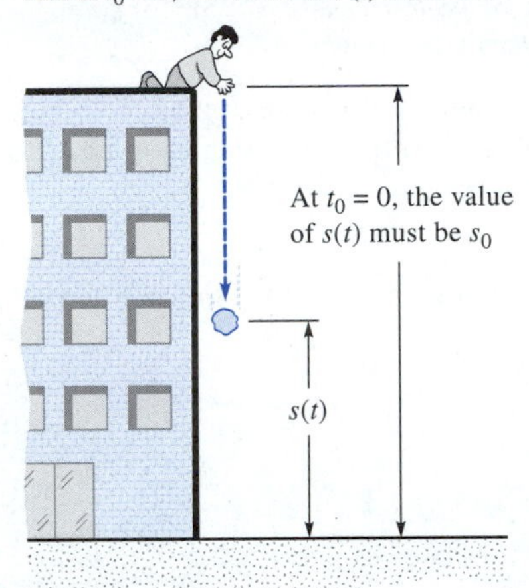
If the rock is dropped or released from rest at $t_0 = 0$, the value of $s'(t)$ must be 0
At $t_0 = 0$, the value of $s(t)$ must be s_0
$s(t)$

Figure 1.4. The state of the rock, its position, and velocity for $t > 0$, are determined by its position and velocity at $t = 0$.

$s'(0) = 0$. The prescribed height and velocity at $t_0 = 0$: $s(0) = s_0$, $s'(0) = 0$ are called **initial conditions**. Were the rock *thrown* rather than dropped from the rooftop then the condition $s'(0) = 0$ is replaced by $s'(0) = v_0$, where the symbol v_0 represents a nonzero initial velocity; v_0 is positive or negative depending on whether the rock is thrown upward or downward, respectively. The collective problem

$$\underbrace{\frac{d^2s}{dt^2} = -g,}_{\text{Find a solution } s(t) \text{ of the DE}} \qquad \underbrace{s(0) = s_0, \qquad s'(0) = v_0}_{\text{so that } s(t) \text{ also satisfies these side conditions}} \tag{1}$$

is an example of an **initial-value problem**, which, in generality, is defined next. You should verify that the well-known formula from physics $s(t) = -\frac{1}{2}gt^2 + v_0t + s_0$ is a solution of (1). The interval over which this solution is defined is $[0, T]$.

Initial-Value Problem On some interval I containing t_0, the problem

$$\textit{Solve:} \quad \frac{d^ny}{dt^n} = f(t, y, y', \ldots, y^{(n-1)})$$

$$\textit{Subject to:} \quad y(t_0) = y_0, y'(t_0) = y_1, \ldots, y^{(n-1)}(t_0) = y_{n-1}$$

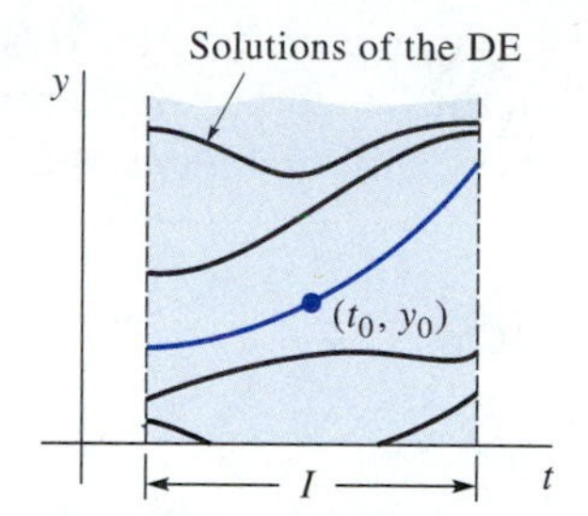

Figure 1.5. First-order IVP.

where $y_0, y_1, \ldots, y_{n-1}$ are arbitrarily chosen real constants, is called an ***n*th-order initial-value problem** (IVP). The given values of the unknown function $y(t)$ and its derivatives at the single point t_0: $y(t_0) = y_0$, $y'(t_0) = y_1, \ldots, y^{(n-1)}(t_0) = y_{n-1}$, are called **initial conditions**. Note that there are precisely n initial conditions associated with an nth-order ordinary differential equation.

Geometric Interpretations It is easy to interpret first- and second-order initial-value problems,

$$\begin{aligned} &\textit{Solve:} && \frac{dy}{dt} = f(t, y) \\ &\textit{Subject to:} && y(t_0) = y_0 \end{aligned} \tag{2}$$

and

$$\begin{aligned} &\textit{Solve:} && \frac{d^2y}{dt^2} = f(t, y, y') \\ &\textit{Subject to:} && y(t_0) = y_0, \quad y'(t_0) = y_1 \end{aligned} \tag{3}$$

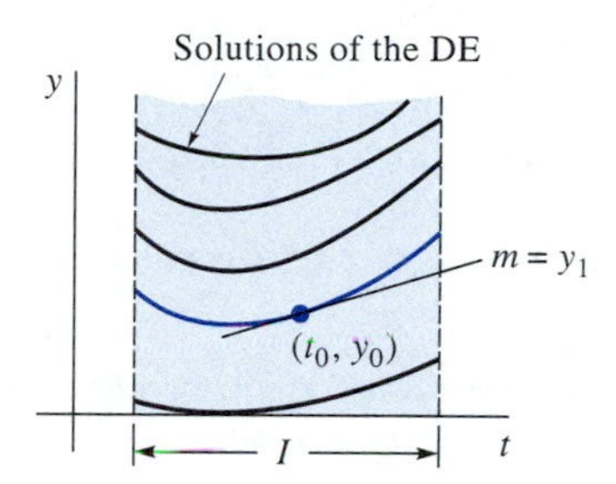

Figure 1.6. Second-order IVP.

in geometric terms. For (2) we seek a solution of the differential equation defined on an interval I containing t_0 so that a solution curve passes through the prescribed point (t_0, y_0). See Figure 1.5. For (3) we not only want a solution curve to pass through the point (t_0, y_0) but we also want the slope of the curve at this point to be the number y_1. See Figure 1.6.

If an n-parameter family of solutions can be found, then solving an nth-order initial-value problem entails, for the most part, using the n initial conditions to find n specialized constants so that the resulting particular solution "fits" everything—differential equation and initial conditions.

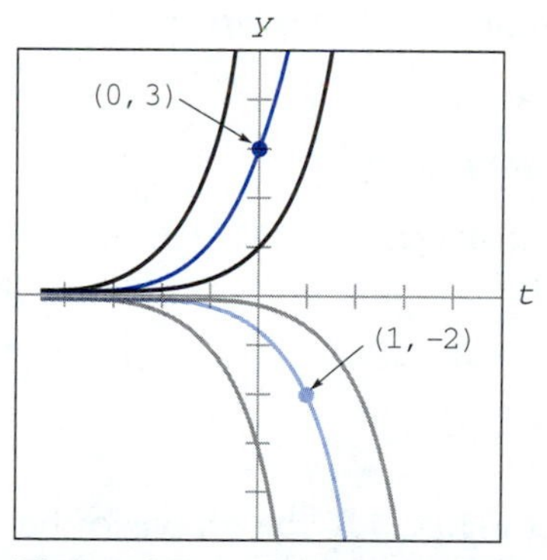

Figure 1.7. Solutions of IVPs.

■ EXAMPLE 1 First-Order IVPs

It is readily verified that $y = ce^t$ is a one-parameter family of solutions of the simple first-order equation $y' = y$ on the interval $(-\infty, \infty)$. If $y(0) = 3$ is the required initial condition, then substituting $t = 0$, $y = 3$ in the family determines the constant $3 = ce^0 = c$. Thus the function $y = 3e^t$ is a solution of the initial-value problem $y' = y$, $y(0) = 3$. If we demand that a solution of the differential equation pass through the point $(1, -2)$ rather than $(0, 3)$, then $y(1) = -2$ would yield $-2 = ce$ or $c = -2e^{-1}$. The function $y = -2e^{t-1}$ is a solution of the initial-value problem $y' = y$, $y(1) = -2$. The graphs of these two functions are shown in color in Figure 1.7. ■

■ EXAMPLE 2 Second-Order IVP

The two-parameter family $x = c_1 \cos 4t + c_2 \sin 4t$ is a solution of the second-order equation $x'' + 16x = 0$. Find a solution of the initial-value problem

$$x'' + 16x = 0, \quad x(\pi/2) = -3, \quad x'(\pi/2) = 1 \tag{4}$$

Solution We first apply $x(\pi/2) = -3$ to the given family of solutions: $c_1 \cos 2\pi + c_2 \sin 2\pi = -3$. Since $\cos 2\pi = 1$ and $\sin 2\pi = 0$, we find that

$c_1 = -3$ and so we now arrive at the one-parameter family $x(t) = -3\cos 4t + c_2 \sin 4t$. We next apply $x'(\pi/2) = 1$ to the last family. Differentiating and then setting $t = \pi/2$ and $x' = 1$ gives $12 \sin 2\pi + 4c_2 \cos 2\pi = 1$, from which we see $c_2 = \frac{1}{4}$. Hence

$$x = -3\cos 4t + \frac{1}{4}\sin 4t$$

is a solution of (4). ■

Existence and Uniqueness Two fundamental questions arise in considering an initial-value problem:

> *Does a solution of the problem exist? If a solution exists, is it the* **only** *solution of the problem?*

For an initial-value problem, such as (2), we ask:

Existence { *Does the differential equation $dy/dt = f(t, y)$ possess solutions?* *Do any of the solution curves pass through the point (t_0, y_0)?*

Uniqueness { *When can we be certain that there is precisely one solution curve passing through the point (t_0, y_0)?*

Note, in Examples 1 and 2, that the phrase "*a* solution" is used rather than "*the* solution" of the problem. The indefinite article "a" was used deliberately to suggest the possibility that there may exist other solutions. At this point it has not been demonstrated that there is a single, or unique, solution of each problem. The next example illustrates an initial-value problem with two solutions.

■ EXAMPLE 3 An IVP Can Have Several Solutions

Each of the functions $y = 0$ and $y = t^4/16$ satisfies the differential equation $dy/dt = ty^{1/2}$ and the initial condition $y(0) = 0$, and so the initial-value problem

$$\frac{dy}{dt} = ty^{1/2}, \quad y(0) = 0$$

has at least two solutions. As illustrated in Figure 1.8, the graphs of both functions pass through the same point (0, 0). ■

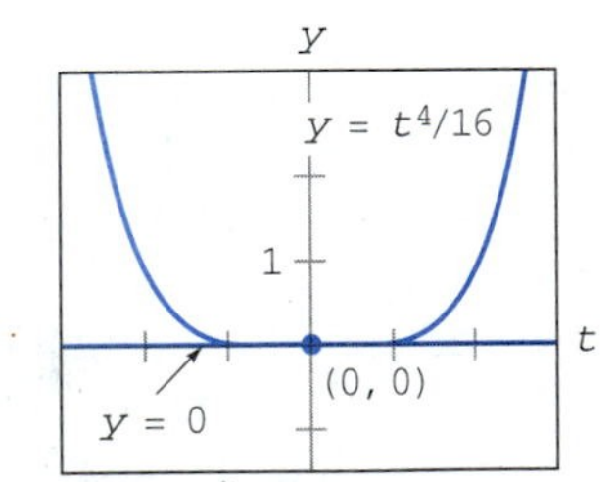

Figure 1.8. Two solutions of the same IVP.

Within the safe confines of a formal course in differential equations, one can be fairly confident that *most* differential equations will have solutions and that solutions of initial-value problems will *probably* be unique. Real life, however, is not so idyllic. Thus, to avoid the possibility of wasting valuable time and energy, it is desirable to know before attempting to solve an initial-value problem whether a solution actually exists and, when it does, whether it is the only solution of the problem. Since we are going to consider first-order differential equations in the next chapter, we state here, without proof, a straightforward theorem that gives conditions sufficient to guarantee the existence and uniqueness of a solution of a first-order initial-value problem (2). We shall wait until Chapter 3 to address the question of existence and uniqueness of a second-order initial-value problem.

Theorem 1.1 Existence of a Unique Solution

Let R be a rectangular region in the ty-plane defined by $a \leq t \leq b$, $c \leq y \leq d$ that contains the point (t_0, y_0) in its interior. If $f(t, y)$ and $\partial f/\partial y$ are continuous on R, then there exist some interval I_0: $t_0 - h < t < t_0 + h, h > 0$, contained in $a \leq t \leq b$, and a unique function $y(t)$ defined on I_0 that is a solution of the initial-value problem (2).

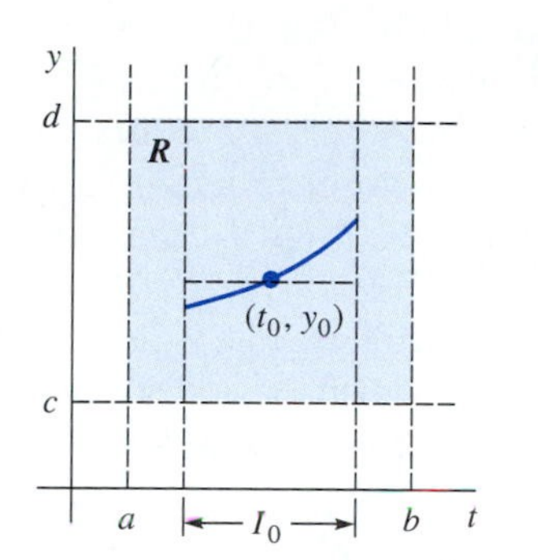

Figure 1.9. Rectangular region R.

The foregoing result is one of the most popular existence and uniqueness theorems for first-order differential equations because the criteria of continuity of $f(t, y)$ and $\partial f/\partial y$ are relatively easy to check. The geometry of Theorem 1.1 is illustrated in Figure 1.9. Observe that the interval I_0, described by $t_0 - h < t < t_0 + h$, is centered at t_0.

■ EXAMPLE 4 **Example 3 Revisited**

We saw in Example 3 that the differential equation $dy/dt = ty^{1/2}$ possesses at least two solutions whose graphs pass through (0, 0). Examination of the two functions

$$f(t, y) = ty^{1/2} \quad \text{and} \quad \frac{\partial f}{\partial y} = \frac{t}{2y^{1/2}}$$

shows that they are continuous in the upper half-plane defined by $y > 0$. Hence Theorem 1.1 guarantees that for any initial condition $y(t_0) = y_0$, as long as $y_0 > 0$, there is only one solution of the differential equation, defined on some interval I_0 centered at t_0, whose graph passes through the point (t_0, y_0). Thus, for example, even without solving it we know that the initial-value problem $dy/dt = ty^{1/2}$, $y(2) = 1$ has a unique solution on some interval $2 - h < t < 2 + h$. ■

In Example 1, Theorem 1.1 guarantees intervals on which there are no other solutions of the initial-value problems $y' = y$, $y(0) = 3$ and $y' = y$, $y(1) = -2$ other than $y = 3e^t$, and $y = -2e^{t-1}$, respectively. This follows from the fact that $f(t, y) = y$ and $\partial f/\partial y = 1$ are continuous throughout the entire ty-plane.

Interval of Existence/Uniqueness Suppose $y(t)$ represents a solution of the initial-value problem (2). The following three sets on the real t-axis may not be the same: the domain of the function $y(t)$, the interval I over which the solution $y(t)$ exists or is defined, and the interval I_0 of existence and uniqueness. In Example 3 of Section 1.1 we illustrated the difference between the domain of a function and the interval I of existence. Now suppose (t_0, y_0) is a point in the interior of the rectangular region R in Theorem 1.1. It turns out that the continuity of the function $f(t, y)$ on R by itself is sufficient to guarantee the existence of at least one solution of $dy/dt = f(t, y)$, $y(t_0) = y_0$ defined on some interval I. The **interval of existence** I of this initial-value problem is the largest interval containing t_0 over which the solution $y(t)$ is defined and differentiable. The interval I depends on both $f(t, y)$ and the initial condition $y(t_0) = y_0$. The extra condition of continuity of the first partial derivative $\partial f/\partial y$ on R enables us to say that not only does a solution exist on some interval I_0 containing t_0, but it is the *only* solution satisfying $y(t_0) = y_0$. However, Theorem 1.1 does not give any indication of the sizes of the intervals I and I_0; *the interval I of existence need not be as wide as the region R and the interval I_0 of existence and uniqueness may not be as large as I.* The number $h > 0$ that defines the interval I_0: $t_0 - h < t < t_0 + h$, could be very small, and so it is best to think that the solution $y(t)$ is *unique in a local sense*, that is, a solution defined *near* the point (t_0, y_0).

ODE Solvers It is sometimes possible to obtain an *approximate* graphical representation of a solution of a differential equation or a system of differential equations without actually obtaining either an explicit or implicit solution. This graphical representation requires computer software with a utility known generically as an **ODE solver**. In the case of a first-order differential equation $dy/dt = f(t, y)$, we need only supply $f(t, y)$ and specify an initial value $y(t_0) = y_0$. For example, in view of Theorem 1.1 we are assured that the differential equation

$$\frac{dy}{dt} = -y + \sin t$$

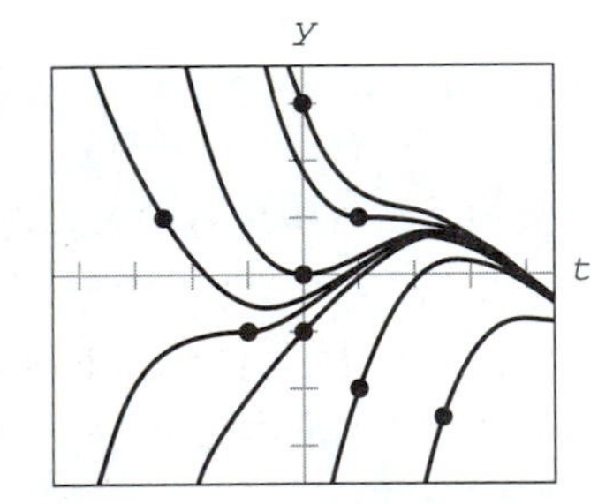

Figure 1.10. Some solutions of $y' = -y + \sin t$.

possesses only one solution passing through each point (t_0, y_0) in the ty-plane. Figure 1.10 shows the solution curves generated by an ODE solver that pass through $(-2.5, 1)$, $(-1, -1)$, $(0, 0)$, $(0, 3)$, $(0, -1)$, $(1, 1)$, $(1, -2)$, and $(2.5, -2.5)$. We will elaborate on ODE solvers in Section 2.5.

Remarks

(i) The conditions in Theorem 1.1 are sufficient but not necessary. When $f(t, y)$ and $\partial f/\partial y$ are continuous on a rectangular region R, it must always follow that a solution of (2) exists and is unique whenever (t_0, y_0) is a point interior to R. However, if the conditions stated in the hypothesis of Theorem 1.1 do not hold, then anything could happen: Problem (2) *may* still have a solution and this solution *may* be unique, or (2) may have several solutions, or it may have no solution at all. A rereading of Example 4 reveals that the hypotheses of Theorem 1.1 do not hold on the line $y = 0$ for the differential equation $dy/dt = ty^{1/2}$ and so it is not

surprising, as we saw in Example 3, that there are two solutions defined on a common interval $-h < t < h$ satisfying $y(0) = 0$. On the other hand, the hypotheses of Theorem 1.1 do not hold on the line $y = 1$ for the differential equation $dy/dt = |y - 1|$. Nevertheless, it can be proved that the solution of the initial-value problem $dy/dt = |y - 1|$, $y(0) = 1$ is unique. Can you guess the solution?

(ii) You are encouraged to read, think about, work, and then keep in mind Problem 31 in Exercises 1.2.

EXERCISES 1.2

Answers to odd-numbered problems begin on page 433.

In Problems 1 and 2, use the fact that $y = 1/(1 + c_1e^{-t})$ is a one-parameter family of solutions of $y' = y - y^2$ to find a solution of the initial-value problem consisting of the differential equation and the given initial condition.

1. $y(0) = -\frac{1}{3}$ **2.** $y(-1) = 2$

In Problems 3–6, use the fact that $x = c_1 \cos t + c_2 \sin t$ is a two-parameter family of solutions of $x'' + x = 0$ to find a solution of the initial-value problem consisting of the differential equation and the given initial conditions.

3. $x(0) = -1,\ x'(0) = 8$ **4.** $x(\pi/2) = 0,\ x'(\pi/2) = 1$

5. $x(\pi/6) = 1/2,\ x'(\pi/6) = 0$

6. $x(\pi/4) = \sqrt{2},\ x'(\pi/4) = 2\sqrt{2}$

In Problems 7–10, use the fact that $y = c_1e^t + c_2e^{-t}$ is a two-parameter family of solutions of $y'' - y = 0$ to find a solution of the initial-value problem consisting of the differential equation and the given initial conditions.

7. $y(0) = 1,\ y'(0) = 2$ **8.** $y(1) = 0,\ y'(1) = e$

9. $y(-1) = 5,\ y'(-1) = -5$ **10.** $y(0) = 0,\ y'(0) = 0$

In Problems 11 and 12, determine by inspection at least two solutions of the given initial-value problem.

11. $y' = 3y^{2/3},\ y(0) = 0$ **12.** $ty' = 2y,\ y(0) = 0$

In Problems 13–20, determine a region of the ty-plane for which the given differential equation will have a unique solution whose graph passes through a point (t_0, y_0) in the region.

13. $\dfrac{dy}{dt} = y^{2/3}$ **14.** $\dfrac{dy}{dt} = \sqrt{ty}$

15. $t\dfrac{dy}{dt} = y$ **16.** $\dfrac{dy}{dt} - y = t$

17. $(4 - y^2)y' = t^2$ **18.** $(1 + y^3)y' = t^2$

19. $(t^2 + y^2)y' = y^2$ **20.** $(y - t)y' = y + t$

In Problems 21–24, determine whether Theorem 1.1 guarantees that the differential equation $y' = \sqrt{y^2 - 9}$ possesses a unique solution through the given point.

21. $(1, 4)$ **22.** $(5, 3)$ **23.** $(2, -3)$ **24.** $(-1, 1)$

25. **(a)** Verify that $y = \tan(t + c)$ is a one-parameter family of solutions of the differential equation $y' = 1 + y^2$.

(b) Since $f(t, y) = 1 + y^2$, and $\partial f/\partial y = 2y$ are continuous everywhere, the region R in Theorem 1.1 can be taken to be the entire ty-plane. Find a solution from the family in part (a) that satisfies $y' = 1 + y^2$, $y(0) = 0$. Explain why the solution is not defined on the interval $-2 < t < 2$.

(c) Determine the largest interval I of existence for the solution of the initial-value problem in part (b).

26. **(a)** Verify that $y = -1/(t + c)$ is a one-parameter family of solutions of the differential equation $y' = y^2$.

(b) Since $f(t, y) = y^2$, and $\partial f/\partial y = 2y$ are continuous everywhere, the region R in Theorem 1.1 can be taken to be the entire ty-plane. Find a solution from the family in part (a) that satisfies $y(0) = 1$. Find a solution from the family in part (a) that satisfies $y(0) = -1$. Determine the largest interval I of existence for the solution of each initial-value problem.

(c) Find a solution from the family in part (a) that satisfies $y' = y^2$, $y(0) = y_0$, $y_0 \neq 0$. Explain why the largest interval I over which this solution exists is either $-\infty < t < 1/y_0$ or $1/y_0 < t < \infty$.

(d) Determine the largest interval I of existence for the solution of the initial-value problem $y' = y^2, y(0) = 0$.

Discussion Problems

27. Figure 1.11 shows the graphs of six members of a family of solutions of a first-order differential equation $dy/dt = f(t, y)$. Match the given initial condition with an appropriate graph.

(a) $y(1) = 1$ **(b)** $y(-2) = -2$

(c) $y(4) = 0$ **(d)** $y(-5) = \sqrt{5}$

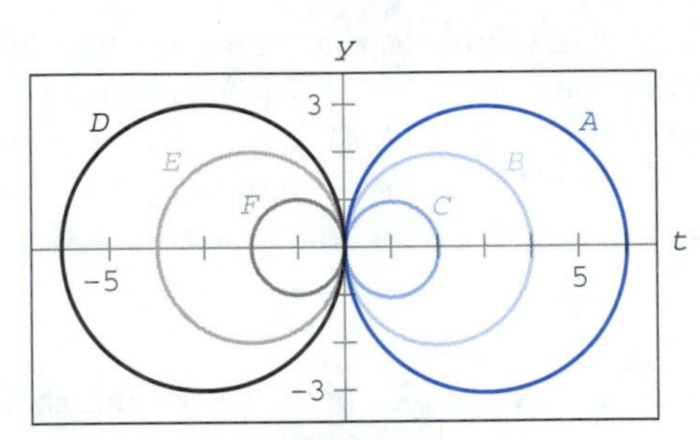

Figure 1.11.

28. Figure 1.12 shows the graphs of four members of a family of solutions of a second-order differential equation $d^2y/dt^2 = f(t, y, y')$. Match each solution curve with appropriate initial conditions.

(a) $y(1) = 1, y'(1) = -2$
(b) $y(-1) = 0, y'(-1) = -4$
(c) $y(1) = 1, y'(1) = 2$ **(d)** $y(0) = -1, y'(0) = 2$
(e) $y(0) = -1, y'(0) = 0$ **(f)** $y(0) = -4, y'(0) = -2$

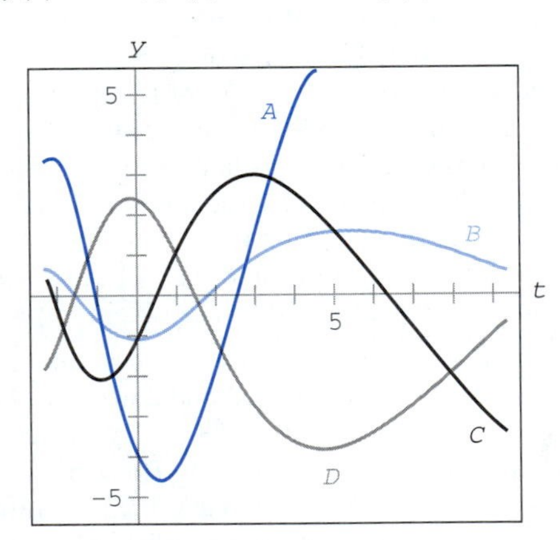

Figure 1.12.

29. Discuss: A first-order initial-value problem, such as $dy/dt = f(t, y), y'(t_0) = y_1$, where y_1 is an arbitrary real number, does not, in general, make sense.

30. (a) Verify that $y = ct, -\infty < t < \infty$ is a one-parameter family of solutions of $ty' = y$.

(b) Observe that all members of the family, $y = 0$ and $y = t$ in particular, satisfy the initial condition $y(0) = 0$. Moreover, the function

$$y = \begin{cases} 0, & t < 0 \\ t, & t \geq 0 \end{cases}$$

satisfies the same initial condition. Discuss: Is this piecewise defined function a solution of the initial-value problem $ty' = y, y(0) = 0$?

31. Consider the differential equation $dy/dt = f(t, y)$, where $f(t, y)$ is a polynomial function in the variables t and y. Discuss: Why can't the members in the family of solution curves intersect as shown in Figure 1.13?

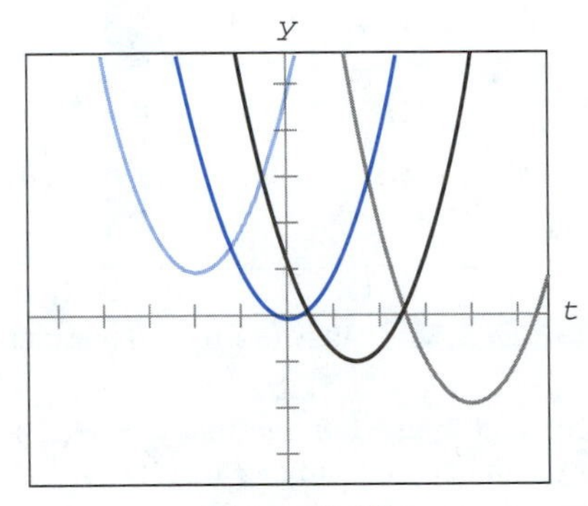

Figure 1.13.

32. The functions

$$y(t) = t^4/16, -\infty < t < \infty \quad \text{and} \quad y(t) = \begin{cases} 0, & t < 0 \\ t^4/16, & t \geq 0 \end{cases}$$

shown in Figures 1.14(a) and 1.14(b), respectively, are clearly different. Both functions are solutions of the differential equation $dy/dt = ty^{1/2}$ on the interval $(-\infty, \infty)$ and both functions also satisfy the initial condition

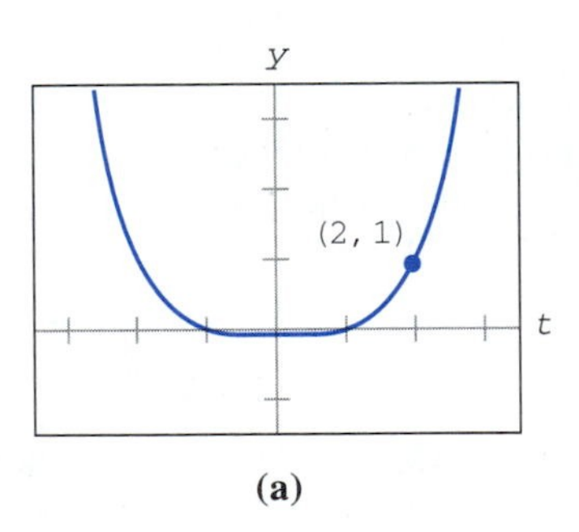

(a)

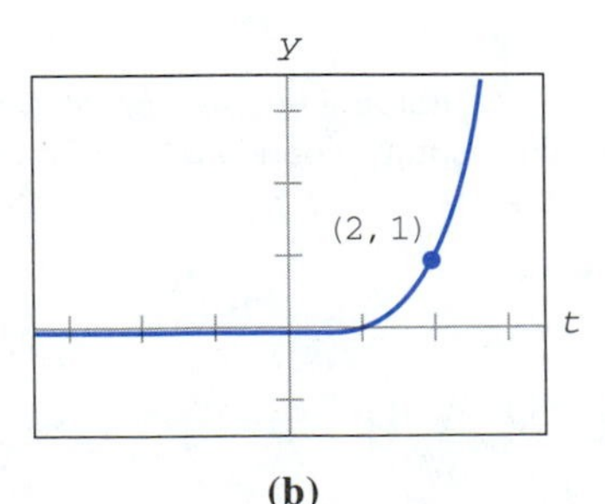

(b)

Figure 1.14. Two solutions of $dy/dt = ty^{1/2}, y(2) = 1$.

$y(2) = 1$. Resolve the apparent contradiction between this fact and the last sentence in Example 4.

33. **(a)** Given that $y = ce^t$ is a one-parameter family of solutions for $y' = y$, and $y = -1/(t + c)$ is a one-parameter family of solutions for $y' = y^2$. Describe, in graphical terms, the differences between the solutions of

$$y' = y, \quad y(0) = y_0 \quad \text{and} \quad y' = y, \quad y(t_0) = y_0, \quad y_0 \neq 0$$

Repeat for

$$y' = y^2, \quad y(0) = y_0 \quad \text{and} \quad y' = y^2, \quad y(t_0) = y_0$$

(b) A first-order differential equation, $dy/dt = g(y)$, which has no explicit dependence on the variable of differentiation, is said to be **autonomous**. The two equations in part (a) are autonomous. Assume that solutions of the general initial-value problems

$$\frac{dy}{dt} = g(y), \ y(0) = y_0 \quad \text{and} \quad \frac{dy}{dt} = g(y), \ y(t_0) = y_0$$

exist. Use your conclusion in part (a) as a basis for conjecturing how the graphs of these two solutions are related. Discuss why solutions of similar initial-value problems involving a nonautonomous differential equation $dy/dt = f(t, y)$ would not share this property.

34. **(a)** Use the family of solutions given in part (a) of Problem 33 to sketch a few representative solutions of the initial-value problems $y' = y^2$, $y(t_0) = y_0$, $y_0 > 0$ and $y' = y^2$, $y(t_0) = y_0$, $y_0 < 0$.

(b) Explain how the shape of the solution curves in part (a) is consistent with the information obtained solely from the differential equation $y' = y^2$. [*Hint*: See Problem 42 in Exercises 1.1.]

(c) Now forget about the known family of solutions, and let $y(t)$ be a solution of the initial-value problem $y' = y^2$, $y(t_0) = y_0$, $y_0 \neq 0$. Explain, using Theorem 1.1, how we can deduce that the graph of $y(t)$ does not cross the t-axis. [*Hint*: See Problems 26(d) and 31.]

35. Consider the differential equation $y' = y - t + 1$. If we differentiate the differential equation and resubstitute y', then $y'' = y' - 1 = (y - t + 1) - 1 = y - t$. Discuss the geometric significance of the information in the following table. Use this information as an aid in sketching a family of solution curves for the differential equation. Also reread Problem 31.

$y' = 0$ for $y = t - 1$	$y'' = 0$ for $y = t$
$y' > 0$ for $y > t - 1$	$y'' > 0$ for $y > t$
$y' < 0$ for $y < t - 1$	$y'' < 0$ for $y < t$

Chapter 1 in Review

Answers to this review are posted at our Web site diffeq.brookscole.com/books.html#zill 98

In Problems 1–4, answer true or false.

1. The differential equation $(y')^3 + y^4 = 1$ is first-order.
2. A solution $y(t)$ of $y' = t^2y^2 + 4$ is an increasing function over an interval on which it is defined.
3. The initial-value problem $y' = f(t, y)$, $y(0) = 0$, where $f(t, y) = y^{1/2}$, has no solution since $\partial f/\partial y$ is discontinuous on the line $y = 0$.
4. There exists an interval centered at 2 on which the constant function $y = 1$ is defined and is the unique solution of the initial-value problem $y' = (y - 1)^3$, $y(2) = 1$.
5. Discuss briefly why the following statement about $y' = 2t(1 + y^2)$ is false:

   Since $f(t, y) = 2t(1 + y^2)$ and $\partial f/\partial y = 4ty$ are continuous throughout the entire ty-plane, there exists a unique solution $y(t)$ defined for all t whose graph passes through any specified initial point (t_0, y_0).

6. Discuss briefly why the following statement about $y' = f(t, y)$ is true:

If f and $\partial f/\partial y$ are continuous in a rectangular region R of the ty-plane, then no two solution curves of the equation can intersect in R.

7. Discuss briefly whether the following statement about $y' = f(t, y)$ is true or false:

If f and $\partial f/\partial y$ are continuous in a rectangular region R of the ty-plane, then two different initial points (t_0, y_0) in R will yield two different solutions.

8. What is the slope of the tangent line to the graph of the solution of $dy/dt - 4\sqrt{y} = 5t^2$ that passes through $(-2, 4)$?

In Problems 9–14, match each of the given differential equations with one or more of the solutions:

(a) $y = 0$ **(b)** $y = 2$ **(c)** $y = 2t$ **(d)** $y = 2t^2$

9. $ty' = 2y$ **10.** $y' = 2$ **11.** $y' = 2y - 4$

12. $ty' = y$ **13.** $y'' + 9y = 18$ **14.** $ty'' - y' = 0$

15. **(a)** Give the domain of the function $y = t^{2/3}$.

(b) Give an interval over which $y = t^{2/3}$ is a solution of $3ty' - 2y = 0$.

16. Fill in the blank $\dfrac{d^2}{dt^2}(c_1 \cosh t + c_2 \sinh t) =$ __________, and then write your result in the form of a differential equation with a solution.

In Problems 17 and 18, interpret each statement as a differential equation.

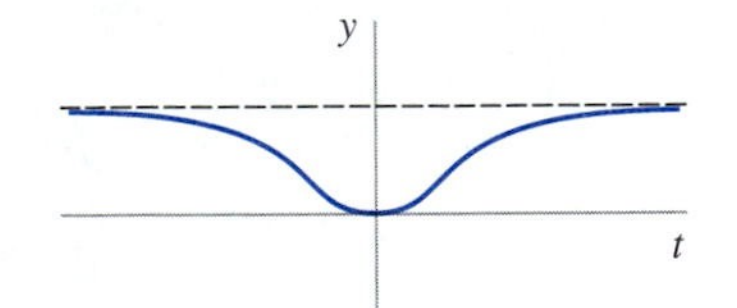

Figure 1.15.

17. The slope of the tangent line to the graph of a differentiable function $y = f(t)$ at any point P is the sum of the squares of the coordinates of P.

18. In a community with a fixed population N, the rate at which a disease spreads is proportional to the product of the number $x(t)$ of people that are currently infected with the number of people that are not currently infected.

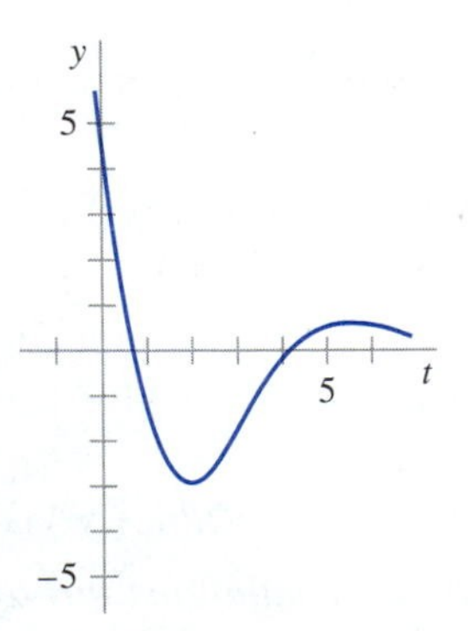

Figure 1.16.

19. Consider the initial-value problem: $y' = e^{-t^2}$, $y(0) = 0$. Since y' gives slope and $y' \to 0$ as $t \to \pm\infty$, it is likely that a solution curve possesses a horizontal asymptote. Differentiate the differential equation and then explain why the graph given in Figure 1.15, is *not* a plausible solution curve. Make any corrections in Figure 1.15 that you deem necessary.

20. The graph of a solution of a second-order initial-value problem $d^2y/dt^2 = f(t, y, y')$, $y(2) = y_0$, $y'(2) = y_1$, is given in Figure 1.16. Estimate y_0 and y_1.

Projects and Computer Lab Experiments for Chapter 1

Section 1.1

1. **Differential Equation of a Family of Curves** Occasionally, instead of seeking an n-parameter family of solutions (curves) for an nth-order differential equation, we have to turn the problem around: Starting with an n-parameter family of curves, can we find an associated nth-order differential equation that describes the family? In parts (a)–(d), first find a Cartesian equation for the family of curves and then find the differential equation of the family whose order matches the number of parameters in the Cartesian equation. Make sure your differential equation does not contain any arbitrary parameters.
   (a) Family of straight lines passing through (2, 5).
   (b) Family of parabolas with vertex on the x-axis.
   (c) Family of circles of radius 1 centered on the x-axis.
   (d) Family of circles of radius 1.
   (e) Inspect the differential equations in parts (a)–(c) and note any singular solutions. Consult a calculus text and supply an interpretation for your answer in part (d).

2. **Integral Solutions** Many of the functions in applied mathematics are defined by means of integrals. Before working this problem, look up Leibniz's rule for differentiating a definite integral under the integral sign. Be sure to consider two cases of this rule: constant limits of integration and variable limits of integration. Verify that the indicated function is a solution of the given differential equation.
   (a) $\dfrac{d^2y}{dt^2} + \omega^2 y = f(t);\qquad y = \dfrac{1}{\omega}\displaystyle\int_0^t f(u)\sin\omega(t-u)\,du$
   (b) $ty'' + y' + ty = 0;\qquad y = \dfrac{1}{\pi}\displaystyle\int_0^\pi \cos(t\sin u)\,du$
   (c) $\dfrac{dy}{dx} = -\dfrac{1}{1+x^2};\qquad y = \displaystyle\int_0^\infty e^{-xt}\frac{\sin t}{t}\,dt,\quad x > 0$

3. **Using a CAS** In the computer algebra systems *Mathematica* and *Maple*, the operation of differentiation of a function f with respect to the independent variable t is carried out by the commands

   D[f, t] and **diff(f, t)**;

   respectively. The nth derivative of f is obtained from

   D[f, {t, n}] and **diff(f, t$n)**;

   Determine the appropriate differentiation commands for the CAS you have on hand.
   (a) Use the CAS to carry out the required differentiations and simplifications to verify that $y = te^{2t}\sin 3t$ is a solution of the DE

$$y^{(4)} - 8y''' + 42y'' - 104y' + 169y = 0$$

(b) Use the CAS to carry out the required differentiations and simplifications to verify that $y = 10\dfrac{\cos(5 \ln t)}{t} - 7\dfrac{\sin(5 \ln t)}{t}$ is a solution of the DE

$$t^3y''' + 2t^2y'' + 20ty' - 78y = 0$$

(c) Use the CAS to carry out the required differentiations and simplifications to verify that

$$x = -3e^t \cos 2t - e^t \sin 2t, \quad y = -2e^{-3t} - 3e^t \cos 2t + 3e^t \sin 2t, \quad z = e^{-3t} + 2e^t \cos 2t + 2e^t \sin 2t$$

constitutes a solution of the system of first-order DEs

$$\frac{dx}{dt} = 2x + y + 2z$$

$$\frac{dy}{dt} = 3x + 6z$$

$$\frac{dz}{dt} = -4x - 3z$$

Section 1.2

4. **Graph of a Solution** Use an ODE solver to graph the solution curve for the given initial-value problem.

(a) $t\dfrac{dy}{dt} = y \ln t, \quad y(2) = 1$ **(b)** $\dfrac{dy}{dt} = \dfrac{2}{t+1} - y^2, \quad y(0) = 2$

(c) $t\dfrac{dy}{dt} + y = t, \quad y(1) = 2$ **(d)** $t^2\dfrac{dy}{dt} - 3ty + 2y = \sin t, \quad y(1) = 0$

5. **Graph of a Family of Solutions** Use an ODE solver in the four parts of this problem.

(a) Graph some members of the family of solutions for $2\dfrac{dy}{dt} = 3 - 4y + y^2$.

(b) Graph some members of the family of curves whose slope at a point (t, y) is given by $1 - 2y$.

(c) Graph some members of the family of solutions for $\dfrac{dy}{dt} = \dfrac{y}{t}$. Use the family of graphs to try to guess an implicit solution $G(t, y, c) = 0$ of the equation.

(d) Graph some members of the family of solutions for $\left(\dfrac{dy}{dt}\right)^2 = y$.

6. **Interval of Existence** Use an ODE solver to estimate the length of the interval of existence of a solution of the differential equation $\dfrac{dy}{dt} = 3y^{4/3} \cos t$ subject to the given initial condition.

(a) $y(\pi) = 0$ **(b)** $y(\pi) = \frac{1}{8}$

(c) $y(\pi) = 8$ **(d)** $y(\pi) = 1$

7. Interval of Existence Again Consider the relation

$$4t^2 + 4ty + 4y^2 = 3$$

(a) Find two explicit functions $y = f_1(t)$ and $y = f_2(t)$ defined by this relation. Give the domains of f_1 and f_2. Use a graphing utility to obtain the graphs of f_1 and f_2 on the same set of coordinate axes.

(b) Verify that the relation given in part (a) is an implicit solution of each of the initial-value problems

$$\sqrt{1-y^2}\,dt + \sqrt{1-t^2}\,dy = 0, \quad y(0) = \frac{\sqrt{3}}{2}$$

and
$$\sqrt{1-y^2}\,dt + \sqrt{1-t^2}\,dy = 0, \quad y(0) = -\frac{\sqrt{3}}{2}$$

[*Hint*: Recall two things: The quadratic formula and that dy/dt gives slope.]

(c) Use an ODE solver to obtain the solution curve of each initial-value problem in part (b). Compare these graphs with the graphs of f_1 and f_2 in part (a).

(d) For each of the initial-value problems in part (b) give an explicit solution and an interval of existence of each solution. [*Hint*: Find the coordinates of the points at the ends of the solution curves in part (c).]

8. Properties of a Solution without the Solution We have seen that qualitative information about solutions of differential equations can often be obtained directly from the equation. Suppose, for the purposes of this problem, that we do not know any explicit solutions of the equation $y' - y = 0$. From Theorem 1.1 we know that the initial-value problem $y' - y = 0$, $y(0) = 1$ has a unique solution—call it $E(t)$—on some interval I_0. Let us assume for the sake of discussion that it has been shown that I_0 is the interval $(-\infty, \infty)$.

(a) Explain why $E(t)$ possesses derivatives of all orders on the interval.

(b) What is the nth derivative $E^{(n)}(t)$?

(c) Verify that $y = cE(t)$ is a one-parameter family of solutions of the differential equation on $(-\infty, \infty)$.

(d) Verify that $y = E(t_1 + t)$, where t_1 is some real number, is a solution of the initial-value problem $y' - y = 0$, $y(0) = E(t_1)$. Use this fact to show that $E(t_1 + t) = E(t_1)E(t)$.

(e) Show that $E(-t) = 1/E(t)$.

(f) Give a plausibility argument to justify the conclusions that on the interval $(0, \infty)$, the function $E(t)$ is strictly increasing and that its graph is concave upward. What is $\lim_{t\to\infty} E(t)$?

(g) Explain why $0 < E(t) < 1$ on $(-\infty, 0)$. What is $\lim_{t\to-\infty} E(t)$?

(h) Obtain an approximation for the values of $E(\frac{1}{2})$ and $E(-\frac{1}{2})$. Sketch a graph of $E(t)$.

9. Picard's Method of Iteration

(a) By integrating both sides of the DE, show that the initial-value problem $y' = f(t, y)$, $y(t_0) = y_0$ and

$$y(t) = y_0 + \int_{t_0}^{t} f(x, y(x))\, dx \tag{1}$$

are equivalent. Now suppose that $y_0(t)$ is an arbitrary continuous function that represents a guess of the solution of the original IVP. Since $f(t, y_0(t))$ is a known function depending solely on t, it can be integrated. With $y(t)$ replaced by $y_0(t)$, the right-hand side of (1) defines another function, which we denote as $y_1(t)$, that is,

$$y_1(t) = y_0 + \int_{t_0}^{t} f(x, y_0(x))\, dx$$

When we repeat the procedure, yet another function is given by

$$y_2(t) = y_0 + \int_{t_0}^{t} f(x, y_1(x))\, dx$$

In this manner we obtain a sequence of functions $y_1(t)$, $y_2(t)$, $y_3(t)$, . . . , whose nth term is defined by

$$y_n(t) = y_0 + \int_{t_0}^{t} f(x, y_{n-1}(x))\, dx \tag{2}$$

for $n = 1, 2, 3, \ldots$. The repetitive use of (2) is known as **Picard's method of iteration**; the functions $y_1(t)$, $y_2(t)$, $y_3(t)$, . . . are called **Picard iterates**. Under certain conditions $\lim_{n\to\infty} y_n(t) = y(t)$, where $y(t)$ is a solution of the original IVP. In the application of (2), it is a common practice to choose the initial function as $y_0(t) = y_0$.

(b) Use Picard's method of iteration to find approximations y_1, y_2, y_3, y_4, and y_5 to the solution of $y' = y - 1$, $y(0) = 2$. Use a graphing utility to graph the iterates on the same pair of coordinate axes. Deduce a solution $y(t)$ of the IVP by determining the value of $\lim_{n\to\infty} y_n(t)$. Graph $y(t)$ and compare with the graphs of the Picard iterates.

Writing Assignments

10. Clairaut's Equation Write a short report on the solution of **Clairaut's differential equation**. Show how a Clairaut equation possesses a singular solution that is defined parametrically. Use a specific example and a figure to illustrate the connection between the family of solutions and the singular solution. If possible, introduce the concept of an envelope of a family of curves.

11. Picard's Method Again Do some outside reading and write a short report on Picard's method of iteration. Address the questions: Under what conditions does $\lim_{n\to\infty} y_n(t)$ converge to a solution of the initial-value problem? What must be done when the integration cannot be carried out symbolically? Is Picard's method a practical method for approximating solutions of first-order IVPs?

Chapter 2

First-Order Equations

2.1 What a First-Order Equation Can Tell Us

Most differential equations cannot be solved. Before demanding your money back for this course, perhaps the last sentence should be balanced by saying that many differential equations possess solutions, but the problem is finding them. When we say that a solution of a DE exists we do not mean that there also exists a method for finding it in the sense of being able to exhibit an exact solution, namely, an implicit or an explicit solution. It may be that the best we can do is to analyze a DE *qualitatively* or *numerically*.

In this section we shall examine two ways of analyzing a first-order DE qualitatively.

Fundamental Questions Let us imagine for the moment that we have in front of us a first-order differential equation $dy/dt = f(t, y)$ for which we can neither find nor devise a method for obtaining an explicit or an implicit solution. This is not as bad nor as unusual a predicament as one might think. Although there are plenty of differential equations that *can* be solved—over the years mathematicians have produced a large bag of ingenious procedures to solve some specialized equations—the focus of research nowadays is not so much on finding functions, formulas, or equations that satisfy differential equations, but rather on gleaning as much information as possible directly from the differential equation itself. In the case a first-order equation $dy/dt = f(t, y)$, we have just seen that whenever $f(t, y)$ and $\partial f/\partial y$ satisfy certain continuity conditions, the qualitative questions about existence and uniqueness of solutions can be answered. We shall see in the second part of this section that other qualitative questions about properties of solutions—for example: How does a solution behave near a certain point? How does a solution behave as $t \to \infty$?—can often be answered when the function f depends solely on the variable y. We begin, however, with a simple concept from calculus: A derivative dy/dt gives slope.

2.1.1 Direction Fields

A First-Order DE Defines Slope Because a solution $y(t)$ of a first-order differential equation $dy/dt = f(t, y)$ is necessarily a differentiable function on its interval I of existence, it must also be continuous on I. Thus the corresponding solution curve on I has no breaks and must possess a tangent line at each point $(t, y(t))$. The slope of the tangent line at $(t, y(t))$ on a solution curve is the value of the first derivative dy/dt at this point, and this we know from the differential equation: $f(t, y(t))$. Now suppose that (t, y) represents any point in the ty-plane at which the function f is defined. The function f assigns a value $f(t, y)$ to the point; the value is the slope of a line. A short line segment, called **lineal element**, is drawn through (t, y)

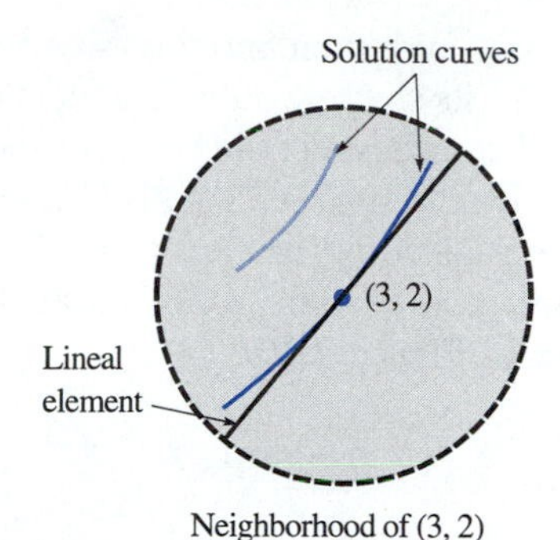

Figure 2.1. Enlargement of a circular neighborhood of (3, 2).

with slope $f(t, y)$. For example, consider the equation $dy/dt = 0.2ty$, where $f(t, y) = 0.2ty$. At the point (3, 2), for example, the slope of a lineal element is $f(3, 2) = 1.2$. As shown in Figure 2.1, a solution curve that passes through (3, 2) does so tangent to the line segment; a different solution curve that passes close to (3, 2) will have a similar shape in a *small* neighborhood of the point.

Now suppose we systematically evaluate $f(t, y)$ over a rectangular grid of points in the ty-plane and draw a lineal element at each point where f is evaluated. The collection of all these lineal elements is called a **direction field** or a **slope field** of the differential equation $dy/dt = f(t, y)$. Visually, the direction field suggests the appearance or shape of a family of solution curves of the differential equation and consequently it may be possible to see certain qualitative aspects of the solutions. A single solution curve that wends its way through the direction field must follow the flow pattern of the field; it is tangent to a lineal element when it intersects a point in the grid.

Solution Curves without the Solution Sketching a direction field by hand is straightforward, but time consuming; it is probably one of those tasks about which an argument can be made for doing it once or twice in a lifetime, but it is overall most efficiently carried out by means of computer software. Figure 2.2(a) was obtained using the direction field application in one of the software supplements to this text, with $dy/dt = 0.2ty$ and a 5×5 grid with points defined by (mh, nh), $h = 1$, m and n integers, $-5 \le m \le 5$, $-5 \le n \le 5$. In Figure 2.2(a), notice that at any point along the t-axis ($y = 0$) and the y-axis ($t = 0$) the slopes are $f(t, 0) = 0$ and $f(0, y) = 0$, respectively, so the lineal elements are horizontal. Moreover, observe that as t and y increase, $f(t, y) = 0.2ty$ increases, and the lineal elements become almost vertical. Reading left to right, imagine a solution curve starting at a point in the second quadrant, moving steeply downward,

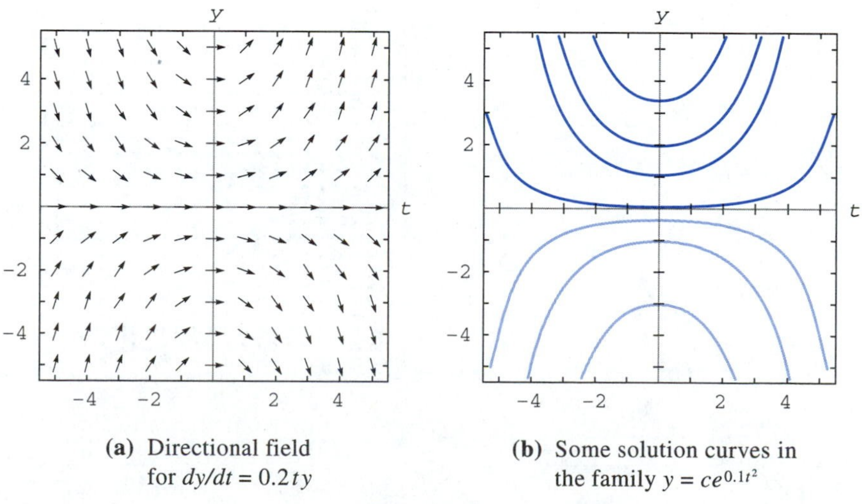

(a) Directional field for $dy/dt = 0.2ty$

(b) Some solution curves in the family $y = ce^{0.1t^2}$

Figure 2.2.

becoming flat as it passes through the y-axis, and then moving steeply upward into the first quadrant—in other words, its shape would be concave upward and would be similiar to a horseshoe. From this it could be surmised that $y \to \infty$ as $t \to \pm\infty$. Now it can be shown that $y = ce^{0.1t^2}$ is a one-parameter family of solutions of the differential equation $dy/dt = 0.2ty$. For the purpose of comparison with Figure 2.2(a), some representative graphs of members of this family are shown in Figure 2.2(b).

EXAMPLE 1 Direction Field

Use a direction field to sketch an approximate solution curve for the initial-value problem $dy/dt = \sin y$, $y(0) = -\frac{3}{2}$.

Solution Before proceeding, recall that from the continuity of $f(t, y) = \sin y$ and $\partial f/\partial y = \cos y$, Theorem 2.1 guarantees the existence of a unique solution curve passing through any specified point (t_0, y_0) in the plane. Now we set our computer software again for a 5×5 rectangular region, and specify (because of the initial condition) points in that region with vertical and horizontal separation of $\frac{1}{2}$ unit, that is, at points (mh, nh), $h = \frac{1}{2}$, m and n integers such that $-10 \leq m \leq 10$, $-10 \leq n \leq 10$. The result is shown in Figure 2.3(a). Since the lineal elements are horizontal at $y = 0$ and $y = -\pi$, it makes sense then in a neighborhood of the initial point $(0, -\frac{3}{2})$ that a solution curve has the shape shown in color in Figure 2.3(a).

Of course, questions concerning a solution of a single initial-value problem can also be investigated by means of an ODE solver. It would appear from Figures 2.3(a) and 2.3(b) that it could be conjectured that y approaches a constant (0 and $-\pi$?) as $t \to -\infty$ and as $t \to \infty$.

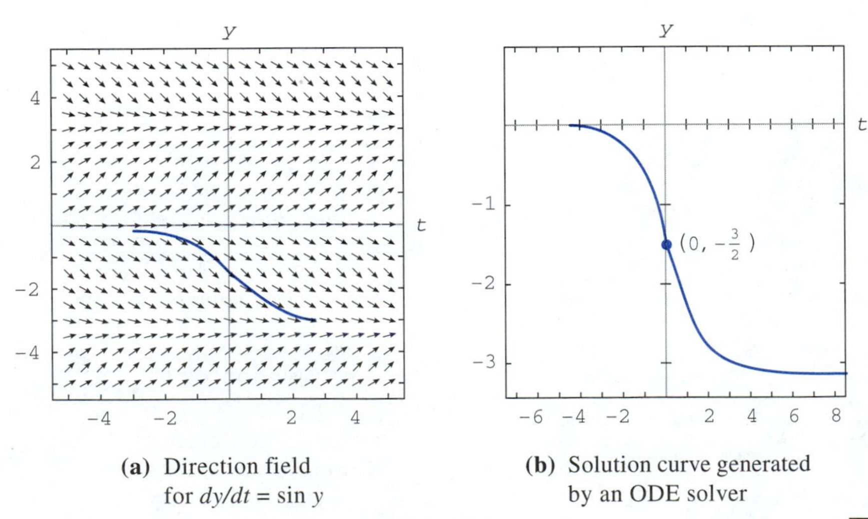

(a) Direction field for $dy/dt = \sin y$

(b) Solution curve generated by an ODE solver

Figure 2.3. ■

2.1.2 Phase Portraits and Stability

Interpretation of the derivative dy/dt as a function that gives slope played the key role in the construction of direction fields. In the discussion that follows, we will employ another telling property of the first derivative, namely, if $y(t)$ is a differentiable function, and if $dy/dt > 0$ (or $dy/dt < 0$) for all t in an interval I, then $y(t)$ is increasing (or decreasing) on I.

Autonomous DEs In the Remarks at the end of Section 1.1, we introduced another kind of classification of ordinary differential equations, a classification that is of particular importance in the qualitative investigation of differential equations. Recall that a first-order equation $F(y, y') = 0$—a differential equation in which the independent variable t does not appear explicitly—is said to be an **autonomous differential equation**.

For purposes of discussing autonomous equations it is a little more convenient to use the symbol x to represent the dependent variable rather than the usual symbol y. Hence an autonomous first-order differential equation $F(x, x') = 0$ is one whose normal form is

$$\frac{dx}{dt} = g(x) \tag{1}$$

We shall assume throughout that g and its derivative g' are continuous functions of x on some interval I. The differential equations

$$\frac{dx}{dt} = \underbrace{1 + x^2}_{g(x)} \quad \text{and} \quad \frac{dx}{dt} = \underbrace{2tx}_{f(t,\,x)}$$

are autonomous and nonautonomous, respectively. We will see later in this chapter that many differential equations encountered in applications are autonomous and of the form (1).

Critical Points The zeros of the function g in (1) are of special interest. We say that a real number x_1 is a **critical point** of the autonomous differential equation (1) if it is a zero of g, that is, $g(x_1) = 0$. Critical points are also called **equilibrium points** and **stationary points**. Moreover, substituting $x(t) = x_1$ into (1) makes both sides of the equation zero, so we see that:

> *If x_1 is critical point of (1) then $x(t) = x_1$ is a constant solution of the autonomous equation.*

A constant solution of (1), such as $x(t) = x_1$, is called an **equilibrium solution**; equilibria are the *only* constant solutions of (1).

As mentioned earlier, we can also tell when a nonconstant solution $x(t)$ is increasing or decreasing by determining the algebraic sign of the derivative dx/dt; this we do by identifying the intervals over which $g(x)$ is positive or negative.

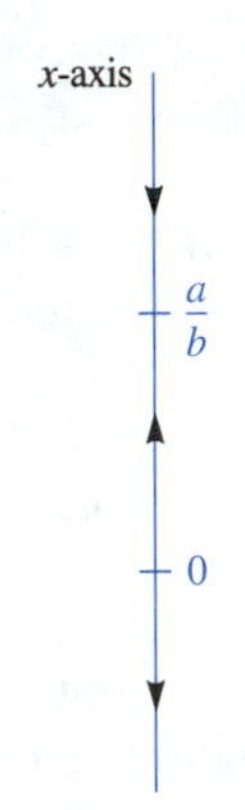

Figure 2.4.

EXAMPLE 2 Autonomous First-Order DE

Inspection of the differential equation

$$\frac{dx}{dt} = x(a - bx) \tag{2}$$

$a > 0$, $b > 0$, shows that it is autonomous. From $g(x) = x(a - bx) = 0$ we also see that 0 and a/b are critical points of the equation. By putting these two points on a vertical line, we divide the line into three intervals: $-\infty < x < 0$, $0 < x < a/b$, $a/b < x < \infty$. The arrows on the line shown in Figure 2.4 indicate the algebraic sign of $g(x) = x(a - bx)$ on these intervals, and whether $x(t)$ is increasing or decreasing. The following table explains the figure.

Interval	Sign of $g(x)$	$x(t)$	Arrow
$(-\infty, 0)$	minus	decreasing	points down
$(0, a/b)$	plus	increasing	points up
$(a/b, \infty)$	minus	decreasing	points down

■

Figure 2.4 is called a **one-dimensional phase portrait**, or simply a **phase portrait**, of the differential equation $dx/dt = x(a - bx)$. The vertical line is called a **phase line**. A phase portrait such as this can also be interpreted in terms of motion of a moving particle. If we imagine that $x(t)$ denotes the position of a particle at time t on a vertical line whose positive x-direction is upward, then dx/dt is the velocity of the particle. Positive velocity indicates motion upward and negative velocity indicates that the particle is moving downward. If a particle is placed at a critical point, then it must remain there for all time. Whence the origin of the alternative name *stationary point.*

Solution Curves without the Solution Without solving an autonomous differential equation, we can usually say a great deal about its solution curves. Relating this back to the first topic of this section, note that a direction field of an autonomous differential equation (1) is independent of t, so at any point on a line parallel to the t-axis all slopes are the same. Thus if $x(t)$ is a solution of (1), then any horizontal translation $x(t - a)$, a a constant, is also a solution. See Problems 33 and 34 in Exercises 1.2. Since the function g in (1) is independent of the variable t, we may consider it defined for $-\infty < t < \infty$ or $0 \le t < \infty$. Also, since g and its derivative g' are continuous functions of x on some interval I, the fundamental results of Theorem 1.1 hold in some horizontal strip or region R in the tx-plane corresponding to I, and so through any point (t_0, x_0) in R there passes only one solution curve of (1). See Figure 2.5(a). For the sake of discussion let us suppose that (1) possesses exactly two critical points x_1 and x_2 and that

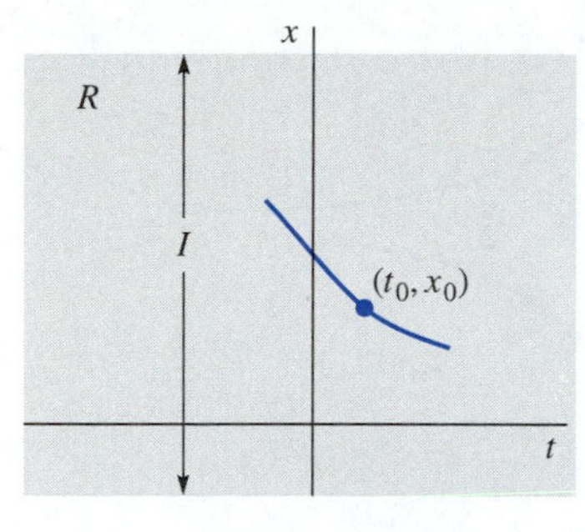

(a) Region R

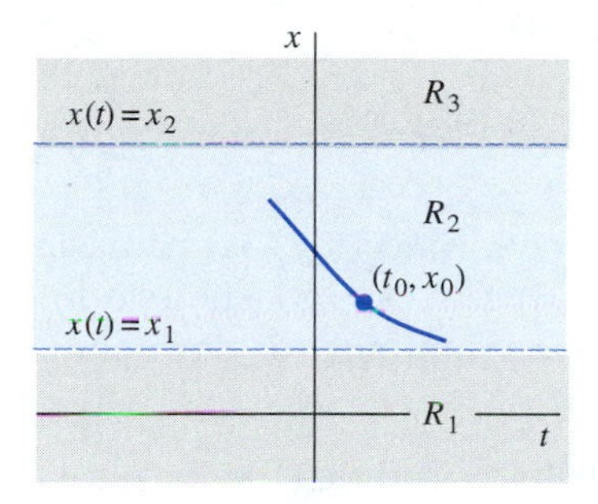

(b) Subregions R_1, R_2, and R_3

Figure 2.5.

$x_1 < x_2$. The graphs of the equilibrium solutions $x(t) = x_1$ and $x(t) = x_2$ are horizontal lines, and these lines partition the region R into three subregions R_1, R_2, and R_3, as illustrated in Figure 2.5(b). Without proof, here are some conclusions that we can draw about a nonconstant solution $x(t)$ of (1):

- If (t_0, x_0) is in a subregion R_i, $i = 1, 2, 3$, and $x(t)$ is a solution whose graph passes through this point, then $x(t)$ remains in the subregion for all t. As illustrated in Figure 2.5(b) the solution $x(t)$ in R_2 is bounded below by x_1 and above by x_2, that is, $x_1 < x(t) < x_2$ for all t. The solution curve stays within R_2 for all t because the graph of a nonconstant solution of (1) cannot cross the graph of an equilibrium solution. See Problem 30 in Exercises 2.1.
- By continuity of g we must then have either $g(x) > 0$ or $g(x) < 0$ for all x in a subregion R_i, $i = 1, 2, 3$. In other words, $g(x)$ cannot change signs in a subregion. See Problem 32 in Exercises 2.1.
- Since $dx/dt = g(x(t))$ is either positive or negative in a subregion, a solution $x(t)$ is strictly monotonic—that is, $x(t)$ is either increasing or decreasing in a subregion R_i, $i = 1, 2, 3$. Therefore $x(t)$ cannot be oscillatory, nor can it have a relative extremum (maximum or minimum). See Problem 33 in Exercises 2.1.
- If $x(t)$ is bounded above by a critical point (as in subregion R_1) or bounded below by a critical point (as in subregion R_3) then $x(t)$ must approach this point either as $t \to \infty$ or as $t \to -\infty$. If $x(t)$ is bounded, that is, bounded above and below by two consecutive critical points (as in subregion R_2), then $x(t)$ must approach both critical points, one as $t \to \infty$ and the other as $t \to -\infty$.

With the foregoing facts in mind let us reexamine the differential equation in Example 2.

EXAMPLE 3 **Example 2 Revisited**

The three intervals determined on the x-axis or phase line by the critical points 0 and a/b now correspond in the tx-plane to three subregions:

$$R_1 : -\infty < x < 0, \quad R_2 : 0 < x < \frac{a}{b}, \quad R_3 : \frac{a}{b} < x < \infty$$

where $-\infty < t < \infty$. The phase portrait in Figure 2.4 tells us that $x(t)$ is decreasing in R_1, increasing in R_2, and decreasing in R_3. If $x(0) = x_0$ is an initial value, then in R_1, R_2, and R_3, we have, respectively:

(i) For $x_0 < 0$, $x(t)$ is bounded above. Since $x(t)$ is decreasing, $x(t)$ decreases without bound for increasing t and $x(t) \to 0$ as $t \to -\infty$. This means the negative t-axis, $x = 0$, is a horizontal asymptote for a solution curve.

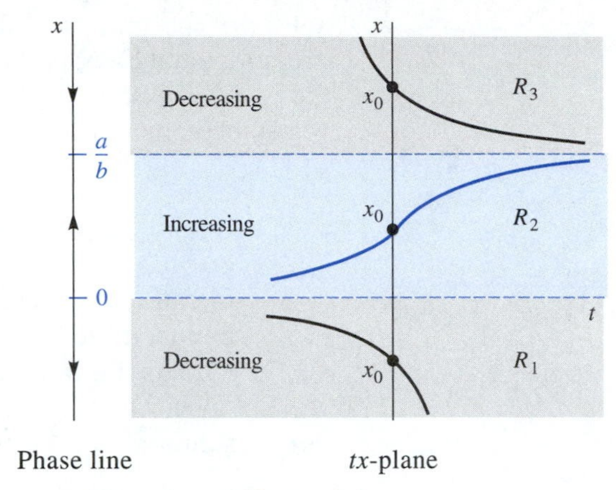

Figure 2.6.

(ii) For $0 < x_0 < a/b$, $x(t)$ is bounded. Since $x(t)$ is increasing, $x(t) \to a/b$ as $t \to \infty$, and $x(t) \to 0$ as $t \to -\infty$. The two lines $x = 0$ and $x = a/b$ are horizontal asymptotes for any solution curve starting in this subregion.

(iii) For $x_0 > a/b$, $x(t)$ is bounded below. Since $x(t)$ is decreasing, $x(t) \to a/b$ as $t \to \infty$. This means $x = a/b$ is a horizontal asymptote for a solution curve.

In Figure 2.6, the original phase portrait is reproduced to the left of the *tx*-plane in which the subregions R_1, R_2, and R_3 are shaded. The graphs of the equilibrium solutions $x = a/b$ and $x = 0$ (the *t*-axis) are shown in Figure 2.6 as colored, dashed lines; the solid graphs represent typical graphs of $x(t)$ illustrating the three cases just discussed. ■

In a subregion such as R_1 in Example 3, where $x(t)$ is decreasing and unbounded below, we must necessarily have $x(t) \to -\infty$. Do *not* interpret this last statement to mean $x(t) \to -\infty$ as $t \to \infty$; we could have $x(t) \to -\infty$ as $t \to T$, where $T > 0$ is a finite number that depends on the initial condition $x(t_0) = x_0$. Thinking in dynamic terms, $x(t)$ could "blow up" in finite time; thinking graphically, $x(t)$ could have a vertical asymptote at $t = T > 0$. A similar remark holds true for the subregion R_3. The next example illustrates these concepts.

■ EXAMPLE 4 Solution Curves

The autonomous equation $dx/dt = (x - 1)^2$ possesses the single critical point 1. From the phase portrait in Figure 2.7(a), we conclude that a solution $x(t)$ is an increasing function in the subregions $-\infty < x < 1$ and $1 < x < \infty$, where $-\infty < t < \infty$. For an initial condition $x(0) = x_0 < 1$, a solution $x(t)$ is increasing and bounded above by 1, so $x(t) \to 1$ as $t \to \infty$; for $x(0) = x_0 > 1$, a solution $x(t)$ is increasing and unbounded.

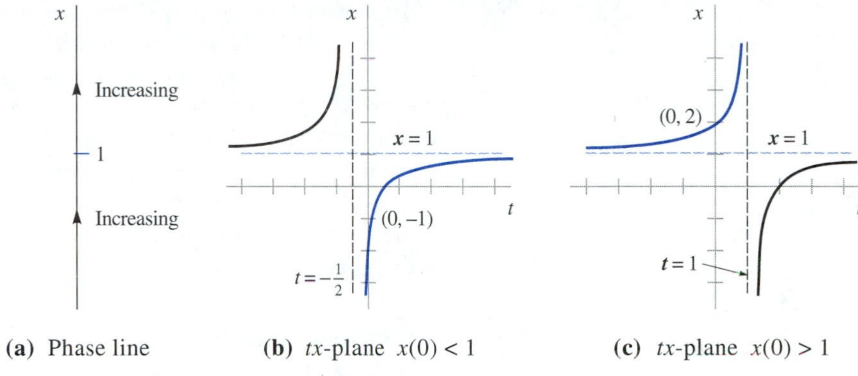

(a) Phase line **(b)** tx-plane $x(0) < 1$ **(c)** tx-plane $x(0) > 1$

Figure 2.7.

Now you should verify that $x(t) = 1 - 1/(t + c)$ is a one-parameter family of solutions of the differential equation. For a given initial condition, for example, $x(0) = -1 < 1$, we find $c = \frac{1}{2}$ and $x(t) = 1 - 1/(t + \frac{1}{2})$. Observe $t = -\frac{1}{2}$ is a vertical asymptote and $x(t) \to -\infty$ as $t \to -\frac{1}{2}$ from the right. See Figure 2.7(b). For an initial condition, say this time, $x(0) = 2 > 1$, we find $c = -1$ and $x(t) = 1 - 1/(t - 1)$. The last function has a vertical asymptote at $t = 1$ and thus $x(t) \to \infty$ as $t \to 1$ from the left. See Figure 2.7(c). ■

Stability An isolated critical point x_1 of a first-order autonomous differential equation can be classified as either asymptotically stable or unstable. Although we shall not go into great detail at this time, we shall try to make the distinction between these two concepts intuitively clear. Let us again imagine a moving particle whose position is given by a solution $x(t)$ of the differential equation. We say that x_1 is **asymptotically stable** provided that, whenever its initial position at t_0 is sufficiently close to x_1, the particle remains close to x_1 for all time $t > t_0$ and, additionally, it approaches x_1 after a long period of time. That is,

$$\lim_{t \to \infty} x(t) = x_1$$

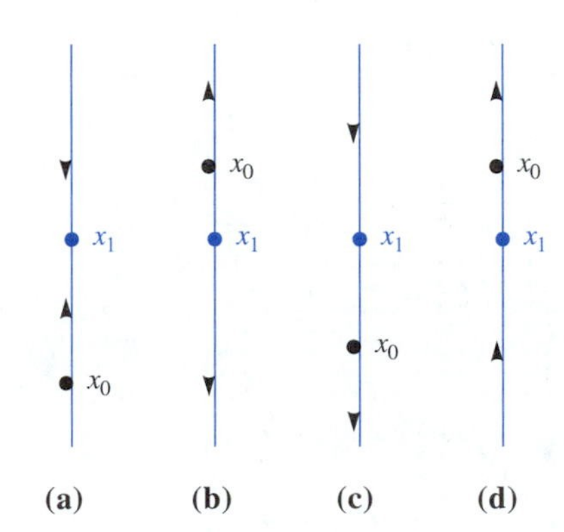

(a) (b) (c) (d)

(a) x_1 is asymptotically stable

(b) x_1 is unstable

(c) x_1 is unstable

(d) x_1 is unstable

Figure 2.8.

Graphically this means that the solution curve passing through an initial point (t_0, x_0), which is sufficiently close to an asymptotically stable critical point x_1, has a horizontal asymptote $x(t) = x_1$. In terms of our moving particle analogy, a critical point is **unstable** if there are slight changes in the position of the particle from x_1 that result in the particle moving away, or receding, from x_1. You should recognize from Figure 2.8 that these two concepts are the basic idea of a phase portrait for an autonomous first-order differential equation. For example, the two arrowheads in Figure 2.4 pointing toward the critical point a/b mean that any solution $x(t)$ satisfying $x(0) = x_0$, $0 < x_0 < a/b$ or $x_0 > a/b$, *must approach* a/b as $t \to \infty$. On the other hand, in the same figure, two arrowheads point away from 0; any solution $x(t)$ satisfying $x(0) = x_0$, $x_0 < 0$, or $0 < x_0 < a/b$, *must move away* from the t-axis. Thus in Example 3 we classify the critical point a/b as asymptotically stable and the critical point 0 as unstable. In Figure 2.7(a), the arrowheads on either side of the number 1 point in the same direction (upward); this means that a solution $x(t)$ satisfying $x(0) = x_0$, $x_0 < 1$, approaches 1 as $t \to \infty$, whereas a solution $x(t)$ satisfying $x(0) = x_0$, $x_0 > 1$, becomes unbounded as t increases. Since at least *some* of the solutions move away from 1, we classify the critical point in Example 4 as unstable.

Attractors and Repellers An asymptotically stable critical point, x_1, for the reasons just explained and illustrated in Figure 2.8(a), is also called an **attractor**. When solutions $x(t)$ behave near an unstable critical point x_1 as shown in Figure 2.8(b), then x_1 is referred to as a **repeller**. Although the critical points in Figures 2.8(c) and 2.8(d) are not given special names, some texts refer to them as **semistable** instead of unstable.

EXERCISES 2.1

Answers to odd-numbered problems begin on page 433.

2.1.1

In Problems 1–4, use the given computer-generated direction field to sketch an approximate solution curve for the indicated differential equation that passes through the given point.

1. $y' = \dfrac{t}{y}$

(a) $y(0) = 5$ **(b)** $y(3) = 3$
(c) $y(4) = 2$ **(d)** $y(-5) = -3$

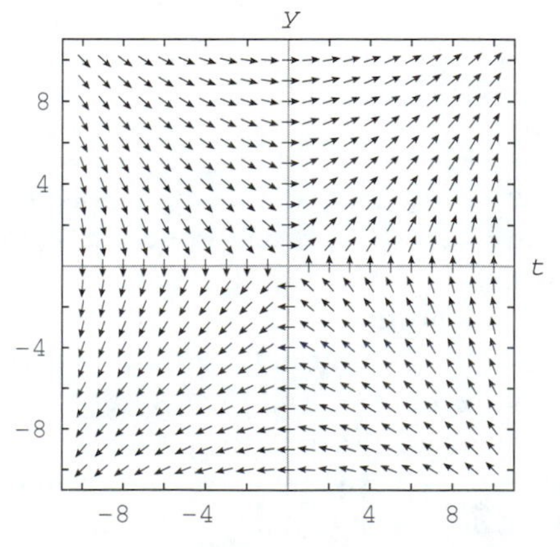

Figure 2.9.

2. $y' = e^{-0.01ty^2}$

(a) $y(-6) = 0$ **(b)** $y(0) = 1$
(c) $y(0) = -4$ **(d)** $y(8) = -4$

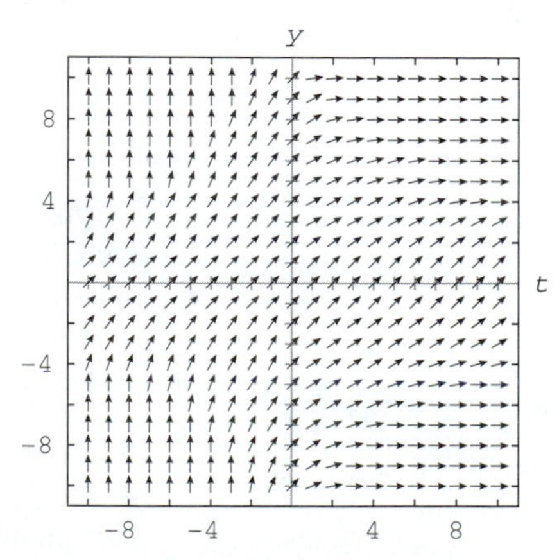

Figure 2.10.

3. $y' = 1 - ty$
(a) $y(0) = 0$ **(b)** $y(-1) = 0$
(c) $y(2) = 2$ **(d)** $y(0) = -4$

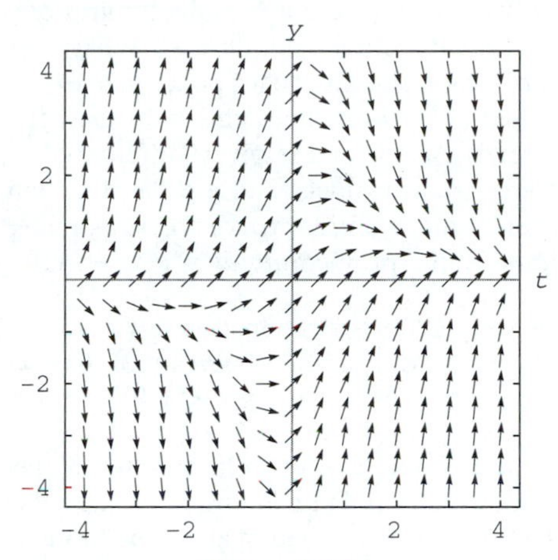

Figure 2.11.

4. $y' = (\sin t) \cos y$
(a) $y(0) = 1$ **(b)** $y(1) = 0$
(c) $y(3) = 3$ **(d)** $y(0) = -\frac{5}{2}$

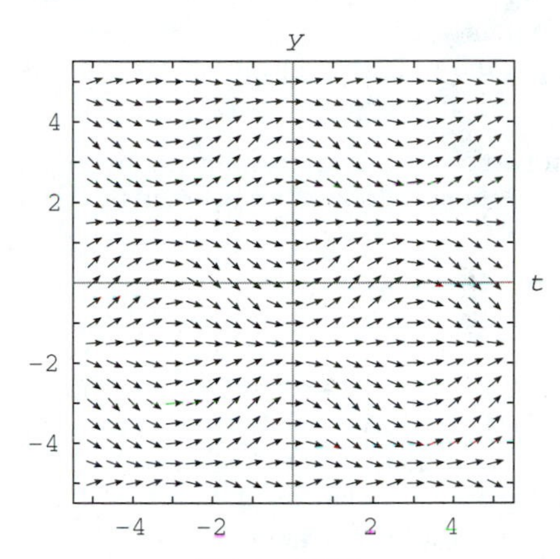

Figure 2.12.

In Problems 5–12, obtain a direction field for the given differential equation. Sketch several approximate solution curves.

5. $y' = t$ **6.** $y' = t + y$

7. $y\frac{dy}{dt} = -t$ **8.** $\frac{dy}{dt} = \frac{1}{y}$

9. $\frac{dy}{dt} = 0.2t^2 + y$ **10.** $\frac{dy}{dt} = te^y$

11. $y' = y - \cos\frac{\pi}{2}t$ **12.** $y' = 1 - \frac{y}{t}$

2.1.2

In Problems 13–20, find the critical points and phase portrait of the given autonomous first-order differential equation. Classify each critical point as asymptotically stable or unstable. Of the unstable critical points, which are repellers?

13. $\frac{dx}{dt} = x^2 - 3x$ **14.** $\frac{dx}{dt} = x^2 - x^3$

15. $\frac{dx}{dt} = (x - 2)^2$ **16.** $\frac{dx}{dt} = 10 + 3x - x^2$

17. $\frac{dx}{dt} = x^2(4 - x^2)$ **18.** $\frac{dx}{dt} = x(2 - x)(4 - x)$

19. $\frac{dx}{dt} = x\ln(x + 2)$ **20.** $\frac{dx}{dt} = \frac{xe^x - 9x}{e^x}$

21. Consider the autonomous first-order differential equation $dx/dt = x - x^3$ and the initial condition $x(0) = x_0$. Sketch the graph of a typical solution $x(t)$ where x_0 has the given values.
(a) $x_0 > 1$ **(b)** $0 < x_0 < 1$
(c) $-1 < x_0 < 0$ **(d)** $x_0 < -1$

22. Consider the autonomous first-order differential equation $dx/dt = x^2 - x^4$ and the initial condition $x(0) = x_0$. Sketch a graph of a typical solution $x(t)$ when x_0 has the given values.
(a) $x_0 > 1$ **(b)** $0 < x_0 < 1$
(c) $-1 < x_0 < 0$ **(d)** $x_0 < -1$

23. The autonomous differential equation

$$m\frac{dv}{dt} = mg - kv$$

where k is a positive constant of proportionality and g is the acceleration due to gravity, governs the velocity

v of a body of mass m falling under the influence of gravity. Because the term $-kv$ represents air resistance, the velocity of a body falling from a great height does not increase indefinitely.

(a) Find the limiting or terminal velocity of the body. Explain your reasoning.

(b) Find the terminal velocity of the body if air resistance is proportional to v^2.

24. When two chemicals are combined, the rate at which a new compound is formed is governed by the differential equation

$$\frac{dX}{dt} = k(\alpha - X)(\beta - X)$$

where $k > 0$ is a constant of proportionality and $\beta > \alpha > 0$. Here $X(t)$ denotes the number of grams of the new compound formed in time t.

(a) Describe the behavior of X as $t \to \infty$.

(b) Consider the case when $\alpha = \beta$. What is the behavior of X as $t \to \infty$ if $X(0) < \alpha$? From the phase portrait of the differential equation, can you predict the behavior of X as $t \to \infty$ if $X(0) > \alpha$?

(c) Verify that an explicit solution of the differential equation in the case when $k = 1$ and $\alpha = \beta$ is $X(t) = \alpha - 1/(t + c)$. Find a solution satisfying $X(0) = \alpha/2$. Find a solution satisfying $X(0) = 2\alpha$. Graph these two solutions. Does the behavior of the solutions as $t \to \infty$ agree with your answers to part (b)?

25. The number 0 is a critical point of the autonomous differential equation $dx/dt = x^n$, where n is a positive integer. For what values of n is 0 asymptotically stable? Unstable? Repeat for the equation $dx/dt = -x^n$.

26. The following **derivative test** enables us to distinguish between asymptotically stable and unstable critical points without the necessity of constructing a phase portrait of the differential equation.

Let x_1 be a critical point of the autonomous first-order differential equation (1). If $g'(x_1) < 0$ then x_1 is asymptotically stable, whereas if $g'(x_1) > 0$ then x_1 is unstable.

Use this derivative test to classify the critical points of the differential equations that follow.

(a) $dx/dt = \sin x$ **(b)** $dx/dt = \sin 2x - \sin x$

(c) $dx/dt = x^2e^{2x} - x$

Discussion Problems

27. (a) The equation $t^2 + y^2 = c$, $c > 0$, represents a family of concentric circles centered at the origin. Sketch the circles corresponding to $c = \frac{1}{4}$, $c = 1$, $c = \frac{9}{4}$, and $c = 4$.

(b) Now suppose we wish to sketch the direction field of $dy/dt = t^2 + y^2$ *by hand*. There is no rule that says we have to use a rectangular grid of points to sketch a direction field. Discuss: What is true about the lineal elements at each point on one of the circles in part (a)? How does this help in sketching the direction field for the given differential equation? Carry out your ideas and use the direction field to sketch the solution curve corresponding to the initial-value problem $dy/dt = t^2 + y^2$, $y(0) = 1$.

28. Consider the differential equation $dy/dt = ty - 3y$. Discuss: What do the solution curves look like within the open circle $(t - 4)^2 + (y - 3)^2 < 0.01$?

29. A critical point x_1 is said to be **isolated** if there exists some open interval that contains x_1 but no other critical point. Discuss: Can there exist an autonomous first-order differential equation of the form given in (1) for which *every* critical point is nonisolated? Do not think profound thoughts.

30. Suppose x_1 is a critical point of the autonomous differential equation (1). Discuss: Why can't the graph of a nonconstant solution of (1) cross the horizontal line $x(t) = x_1$?

31. (a) Using the autonomous equation (1), discuss how it is possible to obtain information about the location of **points of inflection** of a solution curve.

(b) Consider the differential equation $dx/dt = x^2 - x - 6$. Use your ideas from part (a) to find intervals on the x-axis for which solution curves are concave up and to find intervals for which solution curves are concave down. Discuss: Why does *each* solution curve of an initial-value problem of the form $dx/dt = x^2 - x - 6$, $x(0) = x_0$, where $-2 < x_0 < 3$, have a point of inflection with the same x-coordinate? What is the x-coordinate? Carefully sketch the solution curve for which $x(0) = -1$. Repeat for $x(2) = 2$.

32. Consider the function g in (1). Discuss: Why can't $g(x)$ change signs in one of the subregions R_i discussed on page 31?

33. Suppose $x(t)$ is a nonconstant solution of the autonomous differential equation (1). Discuss: Why can't $x(t)$ be oscillatory or have a relative extremum (maximum or minimum)?

34. Suppose an autonomous differential equation (1) possesses no critical points. Discuss: Why can no solution of (1) be bounded? Why does a solution $x(t)$ of the equation assume all real values?

In Problems 35 and 36, consider the autonomous differential equation $dx/dt = g(x)$ where the graph of $g(x)$ is given. Use the graph to locate the critical points of each differential equation. Sketch a phase portrait of each differential equation and give the stability classification of each critical point. Sketch, by hand, a direction field in a region of the tx-plane that you think appropriate based on the graph of $g(x)$.

35. **36.**

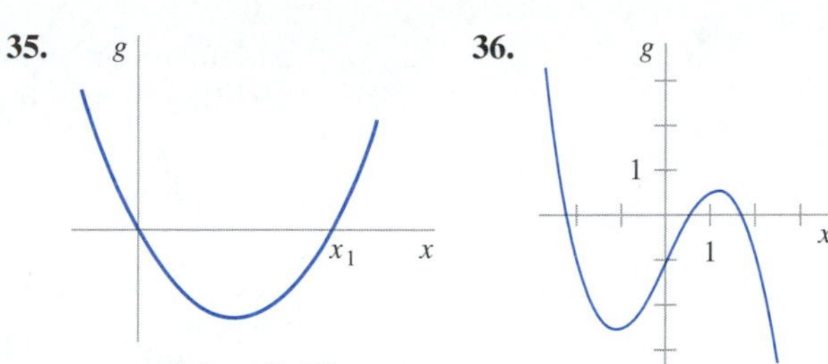

Figure 2.13.

Figure 2.14.

2.2 Separable Variables

We are now ready to get down to the business of actually solving some differential equations.

Certain kinds of first-order DEs can be solved by integration. The basic idea in this and in the next section is to put the DE into an appropriate form and then to integrate both sides of the equality. It may be helpful to review the integration formulas for the standard functions studied in calculus as well as some of the techniques of integration, such as integration by parts and the use of partial fractions.

Solution by Integration We begin our study of the methodology of solving first-order differential equations $dy/dt = f(t, y)$ with the simplest of all equations. When f is independent of the variable y, that is, $f(t, y) = g(t)$, the differential equation

$$\frac{dy}{dt} = g(t) \tag{1}$$

can be solved by integration. If $g(t)$ is a continuous function, then integrating both sides of (1) gives the solution

$$y = \int g(t)\, dt = G(t) + c$$

where $G(t)$ is an antiderivative (indefinite integral) of $g(t)$. For example, if

$$\frac{dy}{dt} = 1 + e^{2t} \quad \text{then} \quad y = \int (1 + e^{2t})\, dt = t + \frac{1}{2}e^{2t} + c$$

Equation (1), as well as its method of solution, is just a special case when f in the normal form $dy/dt = f(t, y)$ is a product of a function of t and a function of y.

Separable Equations A first-order differential equation of the form

$$\frac{dy}{dt} = g(t)\, h(y)$$

is said to be **separable** or to have **separable variables**. Observe that by dividing by the function $h(y)$, a separable equation can be written as

$$p(y)\frac{dy}{dt} = g(t) \tag{2}$$

where, for convenience, we have denoted $1/h(y)$ by $p(y)$. From this last form we see immediately that (2) reduces to (1) when $h(y) = 1$.

Now if $y = \phi(t)$ represents a solution of (2), we must have $p(\phi(t))\phi'(t) = g(t)$, and therefore,

$$\int p(\phi(t))\phi'(t)\,dt = \int g(t)\,dt \tag{3}$$

But $dy = \phi'(t)\,dt$, and so (3) is the same as

$$\int p(y)\,dy = \int g(t)\,dt \quad \text{or} \quad H(y) = G(t) + c \tag{4}$$

where $H(y)$ and $G(t)$ are, in turn, antiderivatives of $p(y) = 1/h(y)$ and $g(t)$.

Method of Solution Equation (4) indicates the procedure for solving separable equations. A one-parameter family of solutions, defined implicitly, is obtained by integrating both sides of $p(y)\,dy = g(t)\,dt$.

Note There is no need to use two constants in the integration of a separable equation, because if we write $H(y) + c_1 = G(t) + c_2$ then the difference $c_2 - c_1$ can be replaced by a single constant c, as in (4). In many instances throughout the chapters that follow, we will relabel constants in a manner convenient to a given equation. For example, multiples of constants or combinations of constants can sometimes be replaced by a single constant.

■ EXAMPLE 1 Solving a Separable DE

Solve $(1 + t)\,dy - y\,dt = 0$.

Solution Dividing by $(1 + t)y$, we can write $dy/y = dt/(1 + t)$, from which it follows that

$$\begin{aligned}
\int \frac{dy}{y} &= \int \frac{dt}{1+t} \\
\ln|y| &= \ln|1 + t| + c_1 \\
y &= e^{\ln|1+t| + c_1} \\
&= e^{\ln|1+t|} \cdot e^{c_1} \qquad \leftarrow \text{laws of exponents} \\
&= |1 + t|e^{c_1} \\
&= \pm e^{c_1}(1 + t) \qquad \leftarrow \begin{cases} |1+t| = 1+t, \; t \geq -1 \\ |1+t| = -(1+t), \; t < -1 \end{cases}
\end{aligned}$$

Relabeling $\pm e^{c_1}$ as c then gives $y = c(1 + t)$. ■

Note If each integral in a problem, or even most of the integrals, results in a logarithm, a judicious choice for the constant of integration is $\ln|c|$ rather than c. In Example 1, if we replace c_1 in the second line of the solution by $\ln|c|$, then

$$\ln|y| = \ln|1 + t| + \ln|c| \quad \text{gives} \quad \ln|y| = \ln|c(1 + t)|$$

In this way we do not have to relabel constants since the last equation is immediately equivalent to $y = c(1 + t)$.

■ EXAMPLE 2 **An Initial-Value Problem**

Solve the initial-value problem

$$\frac{dy}{dt} = -\frac{t}{y}, \quad y(4) = -3$$

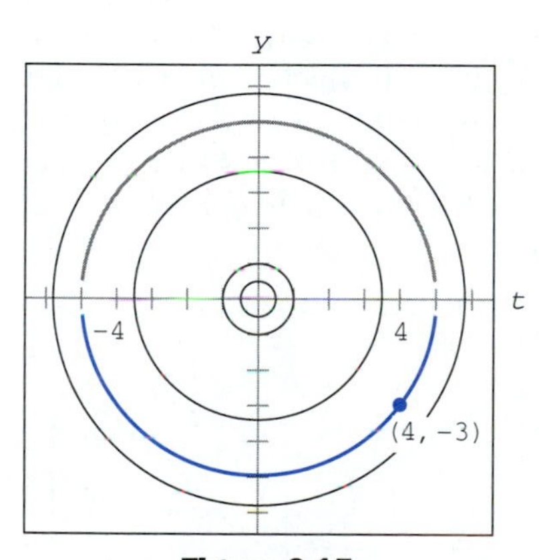

Figure 2.15.

Solution From $y\, dy = -t\, dt$, we get

$$\int y\, dy = -\int t\, dt \quad \text{and} \quad \frac{y^2}{2} = -\frac{t^2}{2} + c_1$$

We can write this solution as $t^2 + y^2 = c^2$ by replacing the constant $2c_1$ by c^2. The relation represents a family of concentric circles.

When $t = 4$, then $y = -3$ so that $16 + 9 = 25 = c^2$. Thus the initial-value problem determines $t^2 + y^2 = 25$. Because of its simplicity, we can solve this last implicit solution for an explicit function or solution that satisfies the initial condition. In Example 4 of Section 1.1, we have seen this function as $y_2(t) = -\sqrt{25 - t^2}$, $-5 < t < 5$. The solution curve is shown in color in Figure 2.15. ■

■ EXAMPLE 3 **An Initial-Value Problem**

Solve the initial-value problem

$$\cos t(e^{2y} - y)\frac{dy}{dt} = e^y \sin 2t, \quad y(0) = 0$$

Solution Dividing the equation by $e^y \cos t$ gives

$$\frac{e^{2y} - y}{e^y}\, dy = \frac{\sin 2t}{\cos t}\, dt$$

By using termwise division and the trigonometric identity $\sin 2t = 2 \sin t \cos t$, we see that the preceding equation is the same as

$$\int (e^y - ye^{-y})\, dy = 2\int \sin t\, dt$$

Integration by parts then gives

$$e^y + ye^{-y} + e^{-y} = -2\cos t + c \tag{5}$$

Substituting $y = 0$ and $t = 0$ in (5) yields $c = 4$ and so a solution of the initial-value problem is

$$e^y + ye^{-y} + e^{-y} = -2\cos t + 4 \tag{6}$$ ■

Use of Computers Unless it is important or convenient, there is no need to try to solve an implicit solution for y in terms of t. Equation (6) shows that this task may present more problems than just the drudgery of symbol pushing—it simply can't be done! Implicit solutions such as (6) are somewhat frustrating; neither the graph of the equation, nor an interval over which a solution satisfying $y(0) = 0$ is defined, are apparent. The problem of "seeing" what an implicit solution looks like can be overcome in some cases by means of technology. Using the contour plot application of a computer algebra system (CAS), we have illustrated in Figure 2.16 some of the level curves of the function $G(t, y) = e^y + ye^{-y} + e^{-y} + 2\cos t$. The family of solutions defined by (5) are the level curves $G(t, y) = c$. Figure 2.17 illustrates, in color, the level curve $G(t, y) = 4$, which is the particular solution (6). The other curve in Figure 2.17 is the level curve $G(t, y) = 2$, which is the member of the family $G(t, y) = c$ that satisfies $y(\pi/2) = 0$. See Problem 7 in *Projects and Computer Lab Experiments for Chapter 2*.

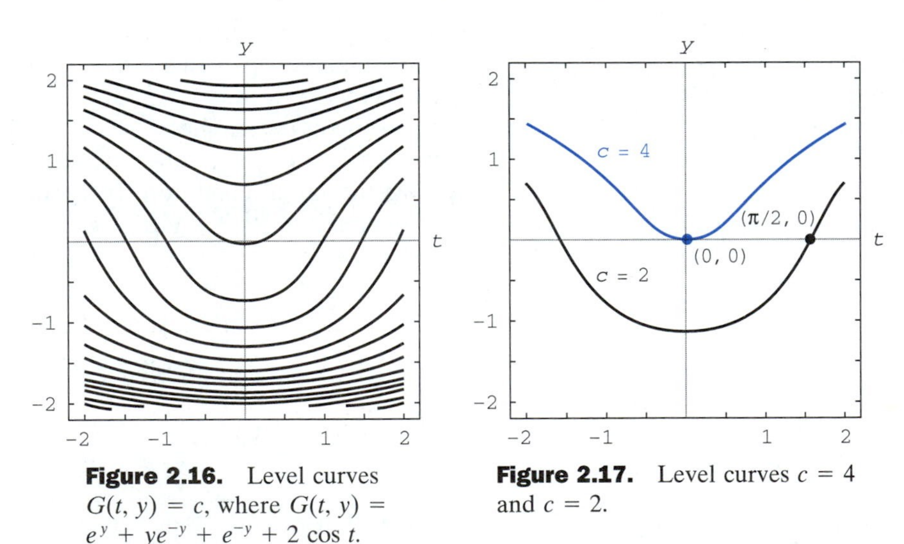

Figure 2.16. Level curves $G(t, y) = c$, where $G(t, y) = e^y + ye^{-y} + e^{-y} + 2\cos t$.

Figure 2.17. Level curves $c = 4$ and $c = 2$.

Some care should be exercised when separating variables since the variable divisors could be zero at a point. As the next example shows, a constant solution may sometimes get lost in the shuffle of solving the problem.

■ EXAMPLE 4 Losing a Solution

Solve the initial-value problem

$$\frac{dy}{dt} = y^2 - 4, \quad y(0) = -2$$

Solution We put the equation in the form

$$\frac{dy}{y^2 - 4} = dt \tag{7}$$

and use partial fractions on the left side. We have

$$\left[\frac{-1/4}{y+2} + \frac{1/4}{y-2}\right] dy = dt \tag{8}$$

so that

$$-\frac{1}{4}\ln|y+2| + \frac{1}{4}\ln|y-2| = t + c_1 \tag{9}$$

In this case we can solve the implicit solution (9) for y in terms of t. Multiplying the equation by 4 and combining logarithms gives

$$\ln\left|\frac{y-2}{y+2}\right| = 4t + c_2 \quad \text{and so} \quad \frac{y-2}{y+2} = ce^{4t}$$

Here we have replaced $4c_1$ by c_2 and e^{c_2} by c. Finally, solving the last equation for y, we get

$$y = 2\frac{1 + ce^{4t}}{1 - ce^{4t}} \tag{10}$$

But substituting $t = 0$ and $y = -2$ in (10) leads to the predicament

$$-2 = 2\frac{1+c}{1-c} \quad \text{or} \quad -1 + c = 1 + c \quad \text{or} \quad -1 = 1$$

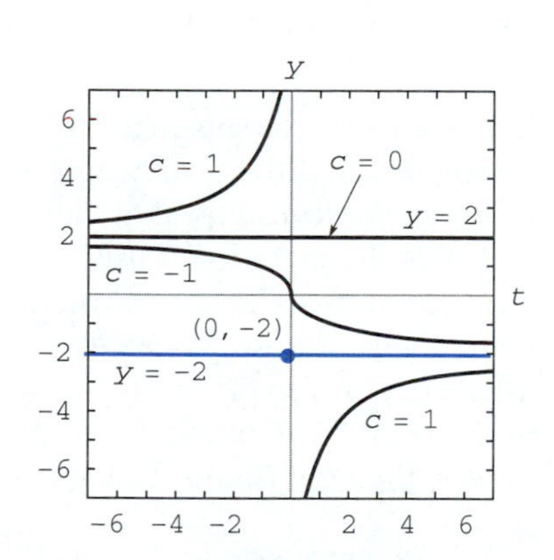

Figure 2.18.

The last equality prompts us to examine the differential equation a little more carefully. The fact is, the equation

$$\frac{dy}{dt} = (y+2)(y-2)$$

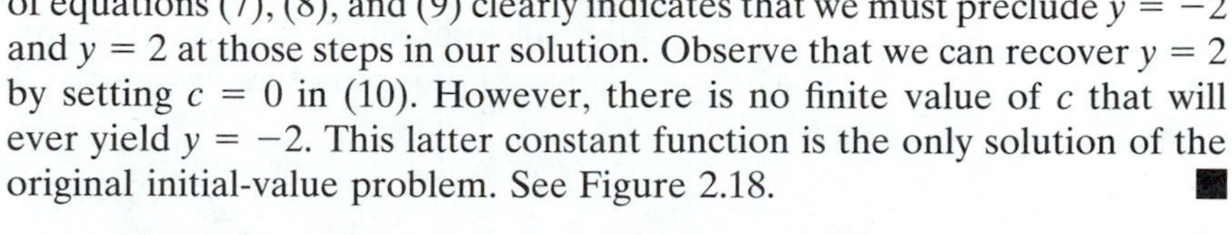
is satisfied by two constant functions, namely, $y = -2$ and $y = 2$. Inspection of equations (7), (8), and (9) clearly indicates that we must preclude $y = -2$ and $y = 2$ at those steps in our solution. Observe that we can recover $y = 2$ by setting $c = 0$ in (10). However, there is no finite value of c that will ever yield $y = -2$. This latter constant function is the only solution of the original initial-value problem. See Figure 2.18. ■

If, in Example 4, we had chosen to use $\ln|c|$ for the constant of integration, the form of the one-parameter family of solutions would be

$$y = 2\frac{c + e^{4t}}{c - e^{4t}} \tag{11}$$

It is interesting to observe that (11) reduces to $y = -2$ when we set $c = 0$, but now there is no finite value of c that will give $y = 2$.

If an initial condition leads to a particular solution by finding a specific value of the parameter c in a family of solutions for a first-order differential equation, it is a natural inclination for most students (and instructors) to relax and be content. However, a solution of an initial-value problem may not be unique. We saw in Example 3 of Section 1.2 that the initial-value problem

$$\frac{dy}{dt} = ty^{1/2}, \quad y(0) = 0 \tag{12}$$

has at least two solutions, $y = 0$ and $y = t^4/16$. We are now in a position to solve the equation. Separating variables and integrating

$$y^{-1/2}\, dy = t\, dt$$

give

$$2y^{1/2} = \frac{t^2}{2} + c_1 \quad \text{or} \quad y = \left(\frac{t^2}{4} + c\right)^2$$

When $t = 0$, then $y = 0$, and so necessarily $c = 0$. Therefore $y = t^4/16$. The trivial solution $y = 0$ was lost by dividing by $y^{1/2}$. In addition, the initial-value problem (12) possesses infinitely many more solutions, since for any choice of the parameter $a \geq 0$ the piecewise defined function

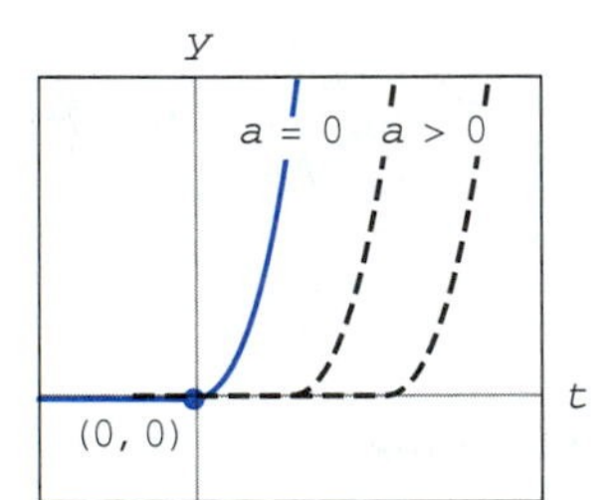

Figure 2.19.

$$y = \begin{cases} 0, & t < a \\ \dfrac{(t^2 - a^2)^2}{16}, & t \geq a \end{cases}$$

satisfies both the differential equation and initial condition. See Figure 2.19.

Substitutions We solve a differential equation by recognizing it as a certain kind of equation—separable, for example—and then carrying out a procedure, consisting of equation-specific mathematical steps, that yields a sufficiently differentiable function that satisfies the equation. Often the first step in solving a given differential equation consists of transforming it into another differential equation by means of a **substitution**. For example, suppose we wish to transform the first-order equation $dy/dt = f(t, y)$ by the substitution $y = \phi(t, u)$, where u is regarded as a function of the variable t. If ϕ possesses first-partial derivatives, then the Chain Rule gives

$$\frac{dy}{dt} = \phi_t(t, u) + \phi_u(t, u)\frac{du}{dt}$$

By replacing dy/dt by $f(t, y)$ and y by $\phi(t, u)$ in the foregoing derivative, we get the new first-order differential equation

$$f(t, \phi(t, u)) = \phi_t(t, u) + \phi_u(t, u)\frac{du}{dt}$$

which, after solving for du/dt, has the form $du/dt = F(t, u)$. If we can determine a solution $u = g(t)$ of this second equation, then a solution of the original differential equation is $y = \phi(t, g(t))$.

Substitutions: Homogeneous Equations If a function f possesses the property

$$f(kt, ky) = k^{\alpha} f(t, y)$$

for some real number α, then f is said to be a **homogeneous function** of degree α. For example, $f(t, y) = t^3 + y^3$ is homogeneous of degree 3 since

$$f(kt, ky) = (kt)^3 + (ky)^3 = k^3(t^3 + y^3) = k^3 f(t, y)$$

whereas $f(t, y) = t^3 + y^3 + 1$ is seen not to be homogeneous.

A differential equation of the form $M(t, y)\, dt + N(t, y)\, dy = 0$ is said to be **homogeneous** if both coefficients M and N are homogeneous functions of the *same* degree. Such an equation can be reduced to separable variables by *either* the substitution $y = ut$ or $t = vy$, where u and v are new dependent variables.

EXAMPLE 5 Solving a Homogeneous DE

Solve $(t^2 + y^2)\, dt + (t^2 - ty)\, dy = 0$.

Solution Inspection of $M(t, y) = t^2 + y^2$ and $N(t, y) = t^2 - ty$ shows that these coefficients are homogeneous functions of degree 2. If we let $y = ut$, then $dy = u dt + t du$ so that, after substituting, the given equation becomes

$$\begin{aligned}
(t^2 + u^2t^2)\, dt + (t^2 - ut^2)[u\, dt + t\, du] &= 0 \\
t^2(1 + u)\, dt + t^3(1 - u)\, du &= 0 \\
\frac{1 - u}{1 + u}\, du + \frac{dt}{t} &= 0 \\
\left[-1 + \frac{2}{1 + u}\right] du + \frac{dt}{t} &= 0 \qquad \leftarrow \text{long division}
\end{aligned}$$

After integration, the last line gives

$$\begin{aligned}
-u + 2\ln|1 + u| + \ln|t| &= \ln|c| \\
-\frac{y}{t} + 2\ln\left|1 + \frac{y}{t}\right| + \ln|t| &= \ln|c| \qquad \leftarrow \text{resubstituting } u = y/t
\end{aligned}$$

Using the properties of logarithms, we can write the preceding solution as

$$\ln\left|\frac{(t + y)^2}{ct}\right| = \frac{y}{t} \quad \text{or} \quad (t + y)^2 = cte^{y/t}$$

■

Remarks

(*i*) Recall from the fundamental theorem of calculus that if g is a function continuous on an interval I containing t_0 then

$$\frac{d}{dt}\int_{t_0}^{t} g(u)\, du = g(t)$$

In other words, $\int_{t_0}^{t} g(u)\,du$ is an antiderivative of the function g. There are times when this latter form is more convenient than (2). For example, if g is continuous on an interval I containing t_0 then a solution of the simple initial-value problem $dy/dt = g(t)$, $y(t_0) = y_0$ that is defined on I is given by

$$y(t) = y_0 + \int_{t_0}^{t} g(u)\,du$$

You should verify this.

(*ii*) In some of the preceding examples, we saw that the constant in the one-parameter family of solutions for a first-order differential equation can be relabeled when convenient. Also, it can easily happen that two individuals solving the same equation correctly arrive at dissimilar expressions for their answers. For example, by separation of variables we can show that one-parameter families of solutions for $(1 + y^2)\,dt + (1 + t^2)\,dy = 0$ are

$$\arctan t + \arctan y = c \quad \text{or} \quad \frac{t + y}{1 - ty} = c$$

As you work your way through the exercises, keep in mind that families of solutions may be equivalent in the sense that one family may be obtained from another either by relabeling the constant or by applying algebra and trigonometry.

(*iii*) Although either of the indicated substitutions can be used for every homogeneous differential equation, in practice we try $t = vy$ whenever the function $M(t, y)$ is simpler than $N(t, y)$. Also it could happen that after using one substitution, we may encounter integrals that are difficult or impossible to evaluate in closed form; switching substitutions may result in an easier problem.

(*iv*) Every autonomous differential equation $dx/dt = g(x)$, or using the symbols t and y, $dy/dt = g(y)$, is separable. Observe in Example 4 that -2 and 2 are critical points of the autonomous equation $dy/dt = y^2 - 4$. It is evident in Figure 2.18 that these two critical points are, in turn, asymptotically stable and unstable. In addition, the fact that we ended up with a constant singular solution of the initial-value problem in Example 4 should be no surprise when you realize that $y(t) = 2$ and $y(t) = -2$ are the equilibrium solutions of the equation and $y(t) = -2$ automatically satisfies the initial condition $y(0) = -2$. With an appeal to Theorem 1.1, we see that $y(t) = -2$ is the only solution defined on an interval I_0 containing $t_0 = 0$.

EXERCISES 2.2

Answers to odd-numbered problems begin on page 435.

In Problems 1–20, solve the given differential equation by separation of variables.

1. $\dfrac{dy}{dt} = \sin 5t$
2. $\dfrac{dy}{dt} = (t+1)^2$
3. $dt + e^{3t}\,dy = 0$
4. $dt - t^2\,dy = 0$
5. $t\dfrac{dy}{dt} = 4y$
6. $\dfrac{dy}{dx} + 2xy = 0$
7. $\dfrac{dy}{dx} = e^{3x+2y}$
8. $e^t y\dfrac{dy}{dt} = e^{-y} + e^{-2t-y}$
9. $y \ln x \dfrac{dx}{dy} = \left(\dfrac{y+1}{x}\right)^2$
10. $\dfrac{dy}{dt} = \left(\dfrac{2y+3}{4t+5}\right)^2$
11. $\sec^2 t\,dy + \csc y\,dt = 0$
12. $\sin 3t\,dt + 2y\cos^3 3t\,dy = 0$
13. $(e^y+1)^2 e^{-y}\,dt + (e^t+1)^3 e^{-t}\,dy = 0$
14. $y dy = t(1+t^2)^{-1/2}(1+y^2)^{1/2}\,dt$
15. $\dfrac{dS}{dr} = kS$
16. $\dfrac{dQ}{dt} = k(Q-70)$
17. $\dfrac{dP}{dt} = P - P^2$
18. $\dfrac{dN}{dt} + N = Nte^{t+2}$
19. $\dfrac{dy}{dx} = \dfrac{xy+3x-y-3}{xy-2x+4y-8}$
20. $\dfrac{dy}{dx} = \dfrac{xy+2y-x-2}{xy-3y+x-3}$

In Problems 21–24, solve the given differential equation subject to the indicated initial condition.

21. $\dfrac{dx}{dt} = 4(x^2+1), \quad x\left(\dfrac{\pi}{4}\right) = 1$
22. $\dfrac{dy}{dx} = \dfrac{y^2-1}{x^2-1}, \quad y(2) = 2$
23. $x^2\dfrac{dy}{dx} = y - xy, \quad y(-1) = -1$
24. $\dfrac{dy}{dt} + 2y = 1, \quad y(0) = \dfrac{5}{2}$

In Problems 25 and 26, find a solution of the given differential equation whose graph passes through the indicated point.

25. $\dfrac{dy}{dt} = y^2 - 9, \quad$ **(a)** $(0, 0)$ **(b)** $(0, 3)$ **(c)** $\left(\dfrac{1}{3}, 1\right)$
26. $t\dfrac{dx}{dt} = x^2 - x, \quad$ **(a)** $(0, 1)$ **(b)** $(0, 0)$ **(c)** $\left(\dfrac{1}{2}, \dfrac{1}{2}\right)$

In Problems 27–32, solve the given homogeneous differential equation by using an appropriate substitution.

27. $(t-y)\,dt + t\,dy = 0$
28. $t\,dt + (x-2t)\,dx = 0$
29. $(y^2+ty)\,dt - t^2\,dy = 0$
30. $(y^2+ty)\,dt + t^2\,dy = 0$
31. $\dfrac{dy}{dx} = \dfrac{y-x}{y+x}$
32. $\dfrac{dy}{dx} = \dfrac{x+3y}{3x+y}$

Discussion Problems

33. Find a function whose square plus the square of its derivative is equal to 1.
34. Show that there are more than 1.65 million digits in the y-coordinate of the point of intersection of the solution curves for the initial-value problems
$$\frac{dy}{dx} = y, \quad y(0) = 1 \quad \text{and} \quad \frac{dy}{dx} = y + \frac{y}{x\ln x}, \quad y(e) = 1$$
35. In Example 2, explain why the solution $y_2(t)$ is not defined on the interval $[-5, 5]$.
36. Consider the differential equation in Problem 17. Find an explicit solution of the equation whose graph passes through $(0, \frac{1}{4})$. Use a graphing utility to obtain this solution curve. Use the graph to estimate the point Q on the solution curve where the tangent line is steepest. Discuss how the differential equation can be used to locate Q. Carry out your ideas and find the point.
37. **(a)** Without solving, explain why the initial-value problem
$$\frac{dy}{dt} = \sqrt{y}, \quad y(t_0) = y_0$$
has no solution for $y_0 < 0$.
(b) Solve the initial-value problem in part (a) for $y_0 > 0$ and find the largest interval I over which the solution exists.
38. Discuss: Are the points $(0, 1)$ and $(2, 1)$ on the graph of the solution of $(t-1)\,dy + 2y\,dt = 0$, $y(0) = 1$?
39. **(a)** Sketch, in the xy-plane, the regions for which the differential equation
$$(1-x^2)\left(\frac{dy}{dx}\right)^2 = 1 - y^2$$
has real solutions.

(b) Find solutions of the differential equation in the various regions found in part (a). Find singular solutions of the equation.

(c) Find a solution of the differential equation whose graph has negative slope at every point and passes through the point $(0, \sqrt{3}/2)$. In view of (*ii*) of the Remarks, express the answer in several entirely different forms.

40. Let F be a function. Discuss how one might solve differential equations of the given form.

(a) $\dfrac{dy}{dt} = F\left(\dfrac{y}{t}\right)$ (b) $y\dfrac{dy}{dt} + t = F(t^2 + y^2)$

2.3 Linear Equations

We continue our quest for solutions of first-order DEs by next examining linear equations. Linear differential equations are an especially "friendly" family of differential equations in that, given a linear equation, whether first-order or a higher-order kin, there is always a good possibility that we can find some sort of solution of the equation that we can look at.

The technique for solving a linear first-order equation, like a separable equation, consists of integration; but integration only after the original equation has been multiplied by a special function called an *integrating factor.*

Standard Form Recall from Section 1.1 that a differential equation that is of the first degree in the dependent variable and all its derivatives is said to be linear. When $n = 1$ in (5) of Section 1.1 we see that a **linear first-order differential equation** is any equation of the form

$$a_1(t)\frac{dy}{dt} + a_0(t)y = g(t) \tag{1}$$

When $g(t) = 0$, the linear equation is said to be **homogeneous**,* otherwise it is **nonhomogeneous**.

By dividing both sides of (1) by the lead coefficient $a_1(t)$, we obtain a more useful form, the **standard form**, of a linear equation:

$$\frac{dy}{dt} + P(t)y = f(t) \tag{2}$$

We seek a solution of (2) on an interval I for which both functions P and f are continuous.

In the discussion that follows, we illustrate a property and a procedure and end up with a formula representing the form that every solution of (2) must have. But more than the formula, the property and the procedure are important, because these two concepts carry over to linear equations of higher order.

*Do not confuse this with the first-order equations $M(t, y)\,dt + N(t, y)\,dy = 0$ studied in the last section, where M and N were homogeneous functions of the same degree.

The Property The differential equation (2) has the property that its solution is the **sum** of the two solutions, $y = y_c + y_p$, where y_c is a solution of the **associated** or **related** homogeneous equation

$$\frac{dy}{dt} + P(t)y = 0 \tag{3}$$

and y_p is a particular solution of the nonhomogeneous equation (2). To see this, observe

$$\frac{d}{dt}[y_c + y_p] + P(t)[y_c + y_p] = \underbrace{\left[\frac{dy_c}{dt} + P(t)y_c\right]}_{0} + \underbrace{\left[\frac{dy_p}{dt} + P(t)y_p\right]}_{f(t)} = f(t)$$

Now the homogeneous linear equation (3) is also separable. This fact enables us to find y_c by writing (3) as

$$\frac{dy}{y} = -P(t)\,dt$$

and integrating. Solving for y gives $y_c = c_1e^{-\int P(t)\,dt}$. For convenience, let us write $y_c = c_1y_1(t)$, where $y_1 = e^{-\int P(t)\,dt}$. The fact that $dy_1/dt + P(t)y_1 = 0$ will be used immediately in finding y_p.

The Procedure We can now find a particular solution of equation (2) by a procedure known as **variation of parameters**. The idea here is to find a function u so that $y_p = u(t)y_1(t)$, y_1 defined above, is a solution of (2). In other words, our assumption for y_p is basically $y_c = c_1y_1(t)$ except c_1 is replaced by the "variable parameter" u.

Now substituting $y_p = uy_1$ into (2) gives

$$\frac{d}{dt}[uy_1] + P(t)uy_1 = f(t)$$

$$u\frac{dy_1}{dt} + y_1\frac{du}{dt} + P(t)uy_1 = f(t)$$

$$u\underbrace{\left[\frac{dy_1}{dt} + P(t)y_1\right]}_{0} + y_1\frac{du}{dt} = f(t)$$

so that

$$y_1\frac{du}{dt} = f(t)$$

By separating variables, the last equation yields

$$du = \frac{f(t)}{y_1(t)}\,dt \quad \text{and} \quad u = \int \frac{f(t)}{y_1(t)}\,dt$$

From the definition of y_1 we see that

$$y_p = uy_1 = e^{-\int P(t)\,dt} \int e^{\int P(t)\,dt} f(t)\,dt$$

and so
$$y = y_c + y_p = c_1 e^{-\int P(t)\,dt} + e^{-\int P(t)\,dt} \int e^{\int P(t)\,dt} f(t)\,dt \tag{4}$$

Thus if (2) has a solution, it must be of the form given in (4). Conversely, it is a straightforward exercise in differentiation to verify that (4) constitutes a one-parameter family of solutions of equation (2).

You should not memorize the formula given in (4). There is an equivalent but easier way of solving (2). If (4) is multiplied by

$$e^{\int P(t)\,dt} \tag{5}$$

and then
$$e^{\int P(t)\,dt} y = c_1 + \int e^{\int P(t)\,dt} f(t)\,dt \tag{6}$$

is differentiated
$$\frac{d}{dt}[e^{\int P(t)\,dt} y] = e^{\int P(t)\,dt} f(t) \tag{7}$$

we see that (6) is

$$e^{\int P(t)\,dt} \frac{dy}{dt} + P(t)\, e^{\int P(t)\,dt} y = e^{\int P(t)\,dt} f(t) \tag{8}$$

Dividing the last result by $e^{\int P(t)\,dt}$ gives (2).

Method of Solution The recommended method of solving (2) actually consists of (6)–(8) worked in *reverse order*. In other words, if (2) is multiplied by (5), we get (8). The left side of (8) is recognized as the derivative of the product of $e^{\int P(t)\,dt}$ and y. This brings us to (7). We then integrate both sides of (7) to get (6), which determines y. Because we can solve (2) by integration after multiplication by $e^{\int P(t)\,dt}$, we call this function an **integrating factor** for the differential equation. For convenience, we summarize these results. We again emphasize that you should not memorize formula (4) but work through the following procedure each time:

Solving a Linear First-Order Equation

(*i*) Put a linear equation of form (1) into standard form (2) and then determine the integrating factor $e^{\int P(t)\,dt}$.

(*ii*) Multiply (2) by the integrating factor. The left side of the resulting equation is automatically the derivative of the integrating factor and y. Write

$$\frac{d}{dt}[e^{\int P(t)\,dt} y] = e^{\int P(t)\,dt} f(t)$$

and then integrate both sides of this equation.

■ **EXAMPLE 1** **Solving a Linear DE**

Solve $\dfrac{dy}{dt} - 3y = 6$.

Solution This linear equation can be solved by separation of variables. Alternatively, since the equation is already in the standard form (2), we see that the integrating factor is $e^{\int(-3)\,dt} = e^{-3t}$. We multiply the equation by this factor and recognize that

$$e^{-3t}\frac{dy}{dt} - 3e^{-3t}y = 6e^{-3t} \quad \text{is the same as} \quad \frac{d}{dt}[e^{-3t}y] = 6e^{-3t}$$

Integrating both sides of the last equation gives $e^{-3t}y = -2e^{-3t} + c$. Thus a solution of the differential equation is $y = -2 + ce^{3t}$, $-\infty < t < \infty$. ■

Constant of Integration Notice in the general discussion and in Example 1 we disregarded a constant of integration in the evaluation of the indefinite integral in the exponent of $e^{\int P(t)\,dt}$. If you think about the laws of exponents and the fact that the integrating factor multiplies both sides of the differential equation, you should be able to answer why writing $\int P(t)\,dt + c$ is unnecessary. See Problem 45 in Exercises 2.3.

When a_1, a_0, and g in (1) are constants, the differential equation is autonomous. In Example 1, you can verify from the form $dy/dt = 3(y + 2)$ that -2 is a critical point and that it is unstable and is a repeller. Thus a solution curve with an initial point either above or below the graph of the equilibrium solution $y = -2$ pushes away from this horizontal line as t increases.

General Solution Suppose again that the functions P and f in (2) are continuous on a common interval I. In the steps leading to (4), we showed that *if* (2) has a solution on I, then it must be of the form given in (4). Conversely, it is a straightforward exercise in differentiation to verify that any function of the form given in (4) is a solution of the differential equation (2) on I. In other words, (4) is a one-parameter family of solutions of equation (1) and *every* solution of (2) defined on I is a member of this family. Consequently we are justified in calling (4) the **general solution** of the differential equation on the interval I. Now by writing (2) in the normal form $y' = F(t, y)$ we can identify $F(t, y) = -P(t)y + f(t)$ and $\partial F/\partial y = -P(t)$. From the continuity of P and f on the interval I, we see that F and $\partial F/\partial y$ are also continuous on I. With Theorem 1.1 as our justification, we conclude that there exists one and only one solution of the initial-value problem

$$\frac{dy}{dt} + P(t)y = f(t), \quad y(t_0) = y_0 \tag{9}$$

defined on *some* interval I_0 containing t_0. But when t_0 is in I, finding a solution of (9) is just a matter of finding an appropriate value of c in (4); that is, for each t_0 in I there corresponds a distinct c. In other words, the interval I_0 of existence and uniqueness in Theorem 1.1 for the initial-value problem (9) is the entire interval I.

EXAMPLE 2 General Solution

Solve $t\dfrac{dy}{dt} - 4y = t^6e^t$.

Solution By dividing by t we get the standard form

$$\frac{dy}{dt} - \frac{4}{t}y = t^5e^t \tag{10}$$

From this form we identify $P(t) = -4/t$ and $f(t) = t^5e^t$ and observe that P and f are continuous for $t > 0$. Hence the integrating factor is

we can use $\ln t$ instead of $\ln |t|$ since $t > 0$

$$e^{-4\int dt/t} = e^{-4\ln t} = e^{\ln t^{-4}} = t^{-4}$$

Here we have used the basic identity $b^{\log_b N} = N$, $N > 0$. Now we multiply (10) by t^{-4},

$$t^{-4}\frac{dy}{dt} - 4t^{-5}y = te^t \quad \text{and obtain} \quad \frac{d}{dt}[t^{-4}y] = te^t$$

It follows from integration by parts that the general solution defined for $t > 0$ is

$$t^{-4}y = te^t - e^t + c \quad \text{or} \quad y = t^5e^t - t^4e^t + ct^4$$

■

EXAMPLE 3 General Solution

Find the general solution of $(t^2 - 9)\dfrac{dy}{dt} + ty = 0$.

Solution We write the equation in standard form

$$\frac{dy}{dt} + \frac{t}{t^2 - 9}y = 0 \tag{11}$$

and identify $P(t) = t/(t^2 - 9)$. Although P is continuous on $(-\infty, -3)$, $(-3, 3)$, and on $(3, \infty)$, we shall solve the equation on the first and third intervals. On these intervals, the integrating factor is

$$e^{\int t\,dt/(t^2-9)} = e^{\frac{1}{2}\int 2t\,dt/(t^2-9)} = e^{\frac{1}{2}\ln(t^2-9)} = \sqrt{t^2 - 9}$$

After multiplying the standard form (11) by this factor, we get

$$\frac{d}{dt}[\sqrt{t^2 - 9}\,y] = 0, \quad \text{and integrating gives} \quad \sqrt{t^2 - 9}\,y = c$$

For $t > 3$ or $t < -3$, the general solution of the equation is $y = \dfrac{c}{\sqrt{t^2 - 9}}$.

■

■ EXAMPLE 4 An Initial-Value Problem

Solve the initial-value problem $\frac{dy}{dt} + y = t,\ y(0) = 4$

Solution The equation is already in standard form, and $P(t) = 1$ and $f(t) = 1$ are continuous on $(-\infty, \infty)$. The integrating factor is $e^{\int dt} = e^t$, and so integrating

$$\frac{d}{dt}[e^t y] = te^t$$

gives $e^t y = te^t - e^t + c$. Solving this last equation for y yields the general solution

$$y = t - 1 + ce^{-t} \tag{12}$$

But from the initial-condition we know that $y = 4$ when $t = 0$. Substituting these values in (12) implies $c = 5$. Hence the solution of the problem is

$$y = t - 1 + 5e^{-t},\ -\infty < t < \infty \tag{13}$$ ■

Recall that the general solution of every linear first-order differential equation is a sum of two special solutions: y_c, the general solution of the homogeneous equation (3), and y_p, a particular solution of the nonhomogeneous equation (2). In (12) of Example 4, we identify $y_c = ce^{-t}$ and $y_p = t - 1$. Figure 2.20, obtained with the aid of a graphing utility, shows the particular solution (13) in color and other representative members of the one-parameter family (12). It is interesting to observe that as t gets large, the graphs of *all* members of the family are close to the graph of $y_p = t - 1$, which is shown in solid black in Figure 2.20. This is because the contribution of $y_c = ce^{-t}$ to the values of a solution becomes negligible for increasing values of t. We say that $y_c = ce^{-t}$ is a **transient term** since $y_c \to 0$ as $t \to \infty$. Although this behavior is not a characteristic of all general

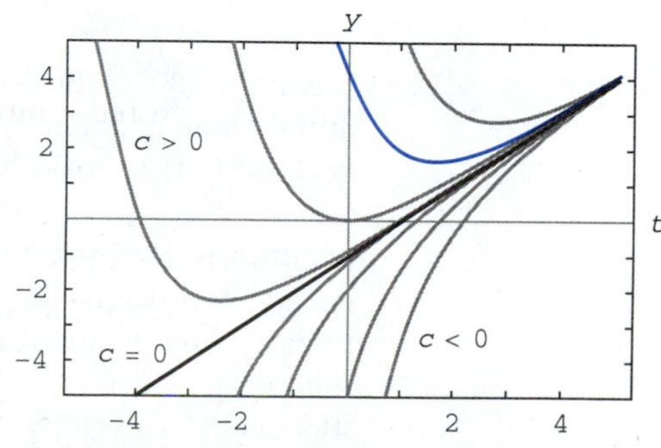

Figure 2.20. Some solutions of $y' + y = t$

solutions of linear equations (see Example 2), the notion of a transient is often important in applied problems.

■ EXAMPLE 5 A Discontinuous $f(t)$

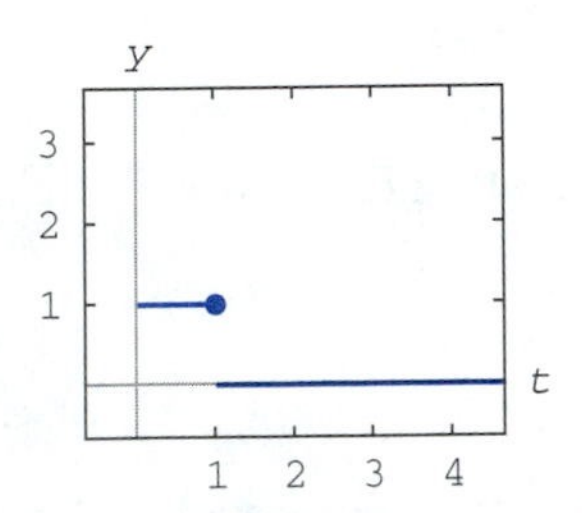

Figure 2.21.

Find a continuous solution satisfying

$$\frac{dy}{dt} + y = f(t), \quad \text{where} \quad f(t) = \begin{cases} 1, & 0 \le t \le 1 \\ 0, & t > 1 \end{cases}$$

and the initial condition $y(0) = 0$.

Solution From Figure 2.21, we see that f is piecewise continuous with a discontinuity at $t = 1$. Consequently, we solve the problem in two parts corresponding to the two intervals over which f is defined. For $0 \le t \le 1$ we have

$$\frac{dy}{dt} + y = 1 \quad \text{or, equivalently,} \quad \frac{d}{dt}[e^t y] = e^t$$

Integrating this last equation and solving for y gives $y = 1 + c_1 e^{-t}$. Since $y(0) = 0$ we must have $c_1 = -1$ and therefore $y = 1 - e^{-t}, 0 \le t \le 1$. Then for $t > 1$, the equation

$$\frac{dy}{dt} + y = 0$$

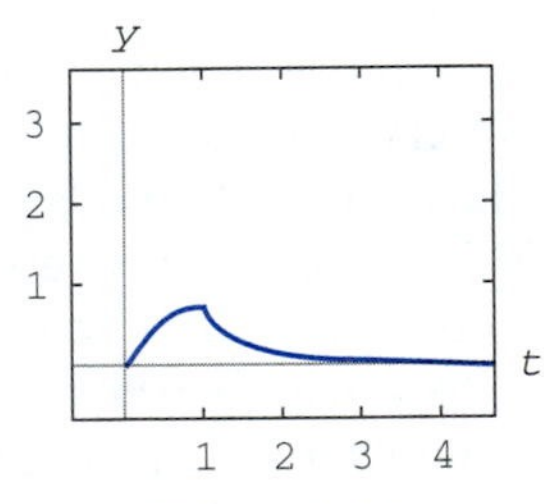

Figure 2.22.

leads to $y = c_2 e^{-t}$. Hence we can write

$$y = \begin{cases} 1 - e^{-t}, & 0 \le t \le 1 \\ c_2 e^{-t}, & t > 1 \end{cases}$$

Now, in order that y be a continuous function at $t = 1$, we must have $\lim_{t \to 1^+} y(t) = y(1)$. This last requirement is equivalent to $c_2 e^{-1} = 1 - e^{-1}$, or $c_2 = e - 1$. As Figure 2.22 shows, the function

$$y = \begin{cases} 1 - e^{-t}, & 0 \le t \le 1 \\ (e-1)e^{-t}, & t > 1 \end{cases} \tag{14}$$

is continuous. Although we shall refer to this function as a solution of the initial-value problem for $0 \le t < \infty$, technically the interval is not correct because (14) is not differentiable at $t = 1$. ■

Functions Defined by Integrals The basic distinction between the so-called "elementary functions" (see page 7) and the "nonelementary functions" is simply a matter of familiarity. In other words, in practice we use functions such as t^n, e^t, $\sin t$, and $\cos t$ more often than we do functions such as $\mathrm{erf}(t)$, $\mathrm{erfc}(t)$, $\mathrm{Si}(t)$, and $\mathrm{Ci}(t)$. Some simple elementary functions do not possess antiderivatives that are again elementary functions; integrals of these kinds of functions are called **nonelementary**. For example, you may have seen in calculus that $\int e^{u^2}\,du$ and $\int \sin u^2\,du$ are nonelementary integrals. In applied mathematics, some important functions are *defined* in terms of nonelementary integrals. Two such functions are the **error function** and the **complementary error function**:

$$\operatorname{erf}(t) = \frac{2}{\sqrt{\pi}}\int_0^t e^{-u^2}\,du, \quad \operatorname{erfc}(t) = \frac{2}{\sqrt{\pi}}\int_t^\infty e^{-u^2}\,du \tag{15}$$

It is known that $(2/\sqrt{\pi})\int_0^\infty e^{-u^2}\,du = 1$, so by adding the two functions in (15) and using the additive interval property of integrals, we see that the complementary error function $\operatorname{erfc}(t)$ is related to $\operatorname{erf}(t)$ by $\operatorname{erf}(t) + \operatorname{erfc}(t) = 1$. Because of its importance in areas such as probability and statistics, the error function has been extensively tabulated. Note however that $\operatorname{erf}(0) = 0$ is one obvious functional value. Values of $\operatorname{erf}(t)$ can also be found using a CAS. See Problem 34 in Exercises 2.3 for the definition of $\operatorname{Si}(t)$.

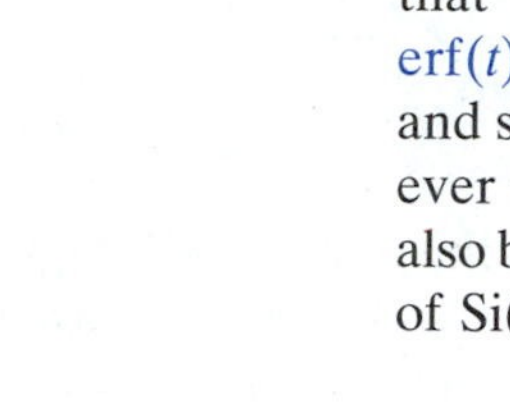

■ EXAMPLE 6 The Error Function

Solve the initial-value problem $\dfrac{dy}{dt} - 2ty = 2$, $y(0) = 1$.

Solution Before working through this solution you should reread part (i) in the Remarks at the end of the preceding section.

Since the equation is already in standard form, we see that the integrating factor is e^{-t^2}, and so from

$$\frac{d}{dt}[e^{-t^2}y] = 2e^{-t^2} \quad \text{we get} \quad y = 2e^{t^2}\int_0^t e^{-u^2}\,du + ce^{t^2} \tag{16}$$

The lower limit of integration in (16) is chosen as 0 because of the initial condition. By multiplying the integral in (16) by $\sqrt{\pi}/\sqrt{\pi}$, this solution can be written in terms of the error function:

$$y = e^{t^2}[c + \sqrt{\pi}\operatorname{erf}(t)] \tag{17}$$

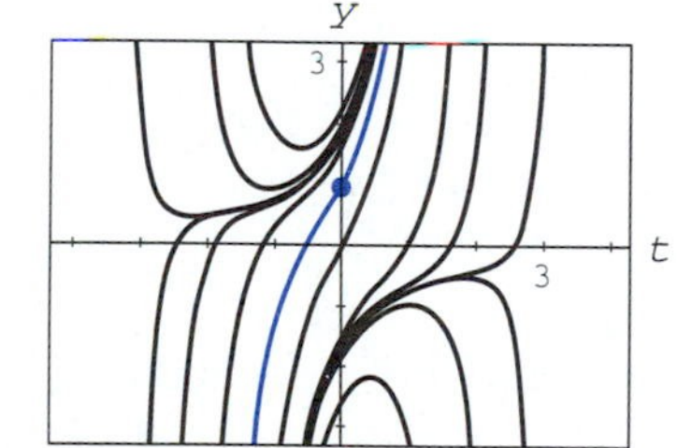

Figure 2.23. Some solutions of $y - 2ty = 2$.

Applying $y(0) = 1$ to this last expression then gives $c = 1$. Hence, the solution of the problem is $y = e^{t^2}[1 + \sqrt{\pi}\operatorname{erf}(t)]$, $-\infty < t < \infty$. The graph of this solution, shown in color in Figure 2.23 among other members of the family defined by (17), was obtained with the aid of a computer algebra system. ■

Use of Computers Some computer algebra systems are capable of producing explicit solutions for some kinds of differential equations. For example, to solve the equation $y' + 2y = t$, the input commands

DSolve[y'[t] + 2 y[t] == t, y[t], t] (in *Mathematica*)

and **dsolve(diff(y(t), t) + 2*y(t) = t, y(t));** (in *Maple*) yield, in turn, the output

$$\text{y[t]} -> -\left(\frac{1}{4}\right) + \frac{\text{t}}{2} + \frac{\text{C[1]}}{\text{E}^{2\,\text{t}}}$$

and y(t) = 1/2 t − 1/4 + exp(−2 t) _C1

Translated to standard symbols this means $y = -\frac{1}{4} + \frac{1}{2}t + ce^{-2t}$.

Substitutions: Bernoulli's Equation The differential equation

$$\frac{dy}{dt} + P(t)y = f(t)\,y^n \tag{18}$$

where n is any real number, is called **Bernoulli's equation**. Note that for $n = 0$ and $n = 1$, equation (18) is linear. For $n \neq 0$ and $n \neq 1$, the substitution $u = y^{1-n}$ reduces any nonlinear equation of form (18) to a linear equation. Example 7 illustrates the solution of a Bernoulli equation.

EXAMPLE 7 Solving a Bernoulli DE

Solve $t\frac{dy}{dt} + y = t^2y^2$.

Solution We first divide by t to rewrite the nonlinear equation in the form given in (18),

$$\frac{dy}{dt} + \frac{1}{t}y = ty^2$$

With $n = 2$, we next substitute

$$y = u^{-1} \quad \text{and} \quad \frac{dy}{dt} = -u^{-2}\frac{du}{dt} \quad \leftarrow \text{Chain Rule}$$

into the last equation and simplify. The result is the linear equation

$$\frac{du}{dt} - \frac{1}{t}u = -t$$

The integrating factor for this equation on, for example, $(0, \infty)$ is

$$e^{-\int dt/t} = e^{-\ln t} = e^{\ln t^{-1}} = t^{-1}$$

Integrating $$\frac{d}{dt}[t^{-1}u] = -1$$

gives $$t^{-1}u = -t + c \quad \text{or} \quad u = -t^2 + ct$$

Since $u = y^{-1}$, we have $y = 1/u$, so a solution of the original equation is

$$u = \frac{1}{-t^2 + ct}$$ ■

Note that we have not obtained the general solution of the original nonlinear differential equation in Example 7 because $y = 0$ is a singular solution of the equation.

Remarks

(*i*) In (1), the values of t for which $a_1(t) = 0$ are called **singular points** of the equation. Singular points are potentially troublesome. In (2), if $P(t) = a_0(t)/a_1(t)$ is discontinuous at a point, the discontinuity *may* carry over to solutions of the differential equation. In Example 3, $P(t) = t/(t^2 - 9)$ is discontinuous at $t = -3$ and $t = 3$ and every function in the general solution $y = c/\sqrt{t^2 - 9}$ is discontinuous at the same points. On the other hand, in Example 2, $P(t) = -4/t$ is discontinuous at $t = 0$, but the general solution $y = t^5e^t - t^4e^t + ct^4$ is noteworthy in that *every* function in this one-parameter family is continuous at $t = 0$ and is defined on $(-\infty, \infty)$, not just $(0, \infty)$ as stated in the solution. However, $t = 0$ can still cause a problem. See Problem 40 in Exercises 2.3.

(*ii*) Occasionally a first-order differential equation is not linear in one variable but is linear in the other variable. For example, the differential equation

$$\frac{dy}{dt} = \frac{1}{t + y^2}$$

is not linear in the variable y. But its reciprocal,

$$\frac{dt}{dy} = t + y^2 \quad \text{or} \quad \frac{dt}{dy} - t = y^2$$

is recognized as linear in the variable t. You should verify that the integrating factor $e^{\int (-1)\,dy} = e^{-y}$ and integration by parts yield an implicit solution of the first equation: $t = -y^2 - 2y - 2 + ce^y$.

(*iii*) Because they thought they were appropriately descriptive, mathematicians "adopted" certain terms from engineering and made them their own. The word *transient*, used earlier, is one of these terms. In future discussions involving linear equations the words **input** and **output** will occasionally pop up. The function f in (2) is called the **input** or **driving function**; a solution of the differential equation for a given input is called the **output** or **response**.

EXERCISES 2.3

Answers to odd-numbered problems begin on page 435.

In Problems 1–22, find the general solution of the given differential equation. Give the largest interval over which the general solution is defined. Determine whether there are any transient terms in the general solution.

1. $\frac{dy}{dt} = 5y$

2. $\frac{dy}{dt} + 2y = 0$

3. $\frac{dy}{dt} + y = e^{3t}$

4. $3\frac{dy}{dt} + 12y = 4$

5. $y' + 3t^2y = t^2$

6. $y' + 2ty = t^3$

7. $t^2y' + ty = 1$

8. $y' = 2y + t^2 + 5$

9. $t\frac{dy}{dt} - y = t^2 \sin t$

10. $t\frac{dy}{dt} + 2y = 3$

11. $t\frac{dy}{dt} + 4y = t^3 - t$

12. $(1 + t)\frac{dy}{dt} - ty = t + t^2$

13. $t^2y' + t(t + 2)y = e^t$

14. $ty' + (1 + t)y = e^{-t} \sin 2t$

15. $y\,dx - 4(x + y^6)dy = 0$

16. $y\,dt = (ye^y - 2t)dy$

17. $\cos t\frac{dy}{dt} + (\sin t)y = 1$

18. $\cos^2 t \sin t\, dy + (y \cos^3 t - 1)dt = 0$

19. $(t + 1)\frac{dx}{dt} + (t + 2)x = 2te^{-t}$

20. $\frac{dP}{dt} + 2tP = P + 4t - 2$

21. $\frac{dr}{d\theta} + r \sec \theta = \cos \theta$

22. $(x + 2)^2\frac{dy}{dx} = 5 - 8y - 4xy$

In Problems 23–28, solve the given differential equation subject to the indicated initial condition. Give the largest interval over which the solution is defined.

23. $ty' + y = e^t, \quad y(1) = 2$

24. $y\frac{dx}{dy} - x = 2y^2, \quad y(1) = 5$

25. $L\frac{di}{dt} + Ri = E, \quad i(0) = i_0$, L, R, and E are constants

26. $\frac{dT}{dt} = k(T - T_m), \quad T(0) = T_0$, k, T_m, and T_0 are constants

27. $(t + 1)\frac{dx}{dt} + x = \ln t, \quad x(1) = 10$

28. $y' + (\tan t)y = \cos^2 t, \quad y(0) = -1$

In Problems 29–32, find a continuous solution satisfying the given differential equation and the indicated initial condition. Graph *f* and the solution.

29. $\frac{dy}{dt} + 2y = f(t), \quad f(t) = \begin{cases} 1, & 0 \le t \le 3 \\ 0, & t > 3 \end{cases}, \quad y(0) = 0$

30. $\frac{dy}{dt} + y = f(t), \quad f(t) = \begin{cases} 1, & 0 \le t \le 1 \\ -1, & t > 1 \end{cases}, \quad y(0) = 1$

31. $\frac{dy}{dt} + 2ty = f(t), \quad f(t) = \begin{cases} t, & 0 \le t \le 1 \\ 0, & t > 1 \end{cases}, \quad y(0) = 2$

32. $(1 + t^2)\frac{dy}{dt} + 2ty = f(t), \quad f(t) = \begin{cases} t, & 0 \le t \le 1 \\ -t, & t > 1 \end{cases}, \quad y(0) = 0$

33. (a) Express the solution of the initial-value problem $y' - 2ty = -1$, $y(0) = \sqrt{\pi}/2$, in terms of $\operatorname{erfc}(t)$.

(b) Use tables or a computer to calculate $y(2)$. Use a CAS to graph the solution on the interval $[0, \infty)$.

34. The sine integral function is defined by

$$\operatorname{Si}(t) = \int_0^t \frac{\sin u}{u}\, du$$

(a) Show that the solution of the initial-value problem $t^3y' + 2t^2y = 10 \sin t$, $y(1) = 0$ is $y = 10t^{-2}[\operatorname{Si}(t) - \operatorname{Si}(1)]$.

(b) Use tables or a computer to calculate $y(2)$. Use a CAS to graph the solution on the interval $(0, \infty)$.

In Problems 35 and 36, solve the given Bernoulli equation.

35. $t\frac{dy}{dt} + y = \frac{1}{y^2}$

36. $\frac{dy}{dt} - y = e^ty^2$

In Problems 37 and 38, solve the given Bernoulli equation subject to the indicated initial condition.

37. $t^2\frac{dy}{dt} - 2ty = 3y^4, \quad y(1) = \frac{1}{2}$

38. $y^{1/2}\frac{dy}{dt} + y^{3/2} = 1, \quad y(0) = 4$

Discussion Problems

39. Reread the discussion following Example 1. Construct a linear first-order differential equation for which all nonconstant solutions approach the horizontal asymptote $y = 4$ as $t \to \infty$.

40. Discuss, with reference to Theorem 1.1, the existence and uniqueness of a solution of the initial-value problem consisting of the differential equation in Example 2 and the given initial condition.
(a) $y(0) = 0$, **(b)** $y(0) = y_0$, $y_0 \neq 0$,
(c) $y(t_0) = y_0$, $t_0 \neq 0$, $y_0 \neq 0$.

41. Find the general solution of the differential equation in Example 3 on the interval $(-3, 3)$.

42. Reread the discussion following Example 4. Construct a linear first-order differential equation for which all solutions are asymptotic to the line $y = 3t - 5$ as $t \to \infty$.

43. Construct a linear first-order differential equation for which $y_c = c/t^3$ and $y_p = t^3$.

44. Solve the initial-value problem:

$$\frac{dy}{dt} + y = \sqrt{1 + t^3}, \quad y(0) = 1$$

Discuss how to go about finding the numerical value of $y(2)$. Carry out your ideas.

45. In determining the integrating factor (5), we did not use a constant of integration in the evaluation of $\int P(t)\,dt$. Explain why using $\int P(t)\,dt + c$ has no effect on the solution of (2).

46. Suppose P is continuous on an interval I and that a is a point in this interval. What is the solution of $y' + P(t)y = 0$, $y(a) = 0$ on I?

47. The following system of differential equations is encountered in the study of a special type of radioactive series of elements:

$$\frac{dx}{dt} = -\lambda_1 x$$

$$\frac{dy}{dt} = \lambda_1 x - \lambda_2 y$$

where λ_1 and λ_2 are constants. Solve the system subject to $x(0) = x_0$, $y(0) = y_0$.

48. **(a)** Suppose $P(t)$ represents the population of some animal species present in an environment at time t. If the symbol $\propto$ means "proportional to," give a physical interpretation of the mathematical statement

$$\frac{dP}{dt} \propto P$$

(b) If k is a constant of proportionality, give a physical interpretation of the differential equation $\dfrac{dP}{dt} = kP$ in the two cases $k > 0$ and $k < 0$.

2.4 Mathematical Models

So far our experience with first-order DEs has been limited to either solving them or analyzing them to see what they "say" about their solutions. But mathematics is a language and a tool. As you undoubtedly remember from algebra and calculus, we translate words into mathematics when solving a "word problem." So too, we can interpret words, empirical laws, observations, or simply assumptions, into mathematical terms. When we try to describe something, let us call it a system, in mathematical terms we are forming a *model* of that system. If something in the system changes with time, say, either growing or decreasing at a certain *rate*—and a rate of change is a derivative—then a mathematical model of the system may be a differential equation.

What Is a Mathematical Model? It is often desirable to express the behavior of some real-life system or a phenomenon, whether physical, sociological, or even economic, in mathematical terms. The mathematical description of a system or a phenomenon is called a **mathematical model**. A mathematical model is constructed with certain goals in mind, for example, we may wish to understand the mechanisms of a certain ecosystem by studying the growth of animal populations in that system, or we may wish to date fossils by means of analyzing the decay of a radioactive substance either in the fossil or in the stratum in which it was discovered.

Construction of a mathematical model of a system starts with

(*i*) *Identifying the variables that are responsible for changing the system. We may choose not to incorporate all these variables into the model at first. In this step we are specifying the* ***level of resolution*** *of the model.*

(*ii*) *We next make a set of reasonable assumptions or hypotheses about the system we are trying to describe. These assumptions will also include any empirical laws that may be applicable to the system.*

Every mathematical model is an *idealization* of reality. For some purposes, it may be within reason to be content with low resolution models. For example, in the model of the motion of a rock dropped from a rooftop of a building, (3) on page xii, we tacitly made several simplifying assumptions. For one thing, we assumed that the acceleration of gravity g was constant; but of course g is not constant but varies with position on the earth as well as with distance from the center of the earth. In addition, we assumed that the *only* force acting on the rock was the force of gravity. The retarding force of air friction was ignored. In a beginning course in physics, the differential equation $d^2s/dt^2 = -g$ is a reasonable model for describing the motion of a hard compact object, such as a rock or a coin, that is thrown straight up or released from a point near the surface of the earth, but would hardly be reasonable for modeling the motion of a falling feather. If you are a scientist whose job it is to predict accurately, say, the flight path of a fast-moving long-range projectile, you must take air resistance and other factors, such as the curvature of the earth, into account.

As mentioned in the introduction, since the assumptions made about a system frequently involve a *rate of change* of one or more of the variables, the mathematical formulation of all these assumptions is one or more equations involving *derivatives*. In other words, the mathematical model is a differential equation or a system of differential equations.

After we have formulated a mathematical model that is either a differential equation or a system of differential equations, we are faced with the not insignificant problem of solving it. Having solved it, we deem the model to be reasonable if its solution—and here we now have to add that "solution" may be "an approximation to the exact solution"—is consistent with either experimental data or known facts about the behavior of the system. If the predictions produced by the solution are poor, we can either increase the level of resolution of the model or make alternative assumptions about the mechanisms for change in the system. The steps of the modeling process are then repeated, as shown in the following diagram:

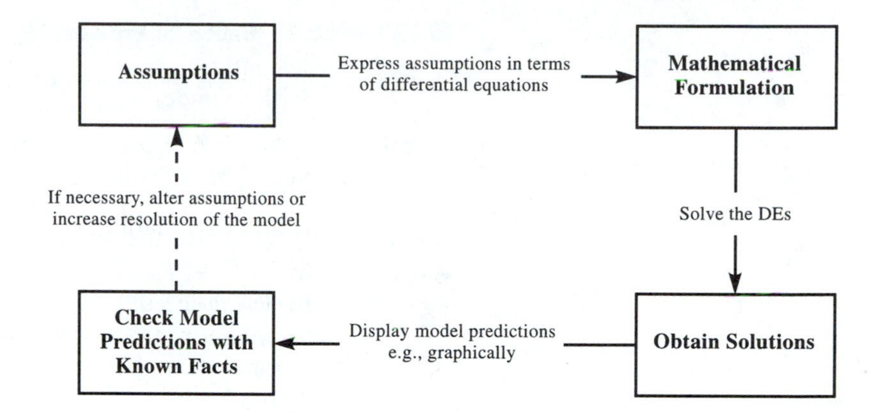

Of course, by increasing the resolution we add to the complexity of the mathematical model and increase the likelihood that we must approximate its solution using some numerical method (see Section 2.5).

A mathematical model of a *physical* system will often involve the variable time t. A solution of the model then gives the **state of the system**—in other words, for appropriate values of t, the values of the dependent variable (or variables) describe the system in the past, present, and future.

2.4.1 Linear Models

Population Dynamics One of the earliest attempts to model human population growth by means of mathematics was by the English economist Thomas Malthus in 1798. Basically, the idea of the malthusian model is the assumption that the rate at which a population of a country grows is proportional to the total population $P(t)$ of the country at time t. In other words, the more people there are at time t, the more there are going to be in the future. In mathematical terms this assumption can be expressed as

$$\frac{dP}{dt} \propto P \quad \text{or} \quad \frac{dP}{dt} = kP \tag{1}$$

where k is a constant of proportionality. This simple model, which fails to take into account many factors (immigration and emigration, for example) that can influence human populations to either grow or decline, nevertheless turned out to be fairly accurate in predicting the population of the United States during the years 1790–1860. The differential equation given in (1) is still often used to model, over short intervals of time, the populations of bacteria or small animals.

The constant of proportionality k in (1) can be determined from the solution of the initial-value problem $dP/dt = kP$, $P(t_0) = P_0$ using a subsequent measurement of P at a time $t_1 > t_0$.

EXAMPLE 1 Bacterial Growth

A culture initially has P_0 number of bacteria. At $t = 1$ hr, the number of bacteria is measured to be $\frac{3}{2}P_0$. If the rate of growth is proportional to the number of bacteria $P(t)$ present at time t, determine the time necessary for the number of bacteria to triple.

Solution We first solve the differential equation in (1) subject to the initial condition $P(0) = P_0$. Then we use the empirical observation that $P(1) = \frac{3}{2}P_0$ to determine the constant of proportionality k.

Now the equation $dP/dt = kP$ is both separable and linear. When it is put in the form

$$\frac{dP}{dt} - kP = 0$$

we can see by inspection that the integrating factor is e^{-kt}. Multiplying both sides of the equation by this term immediately gives

$$\frac{d}{dt}[e^{-kt}P] = 0$$

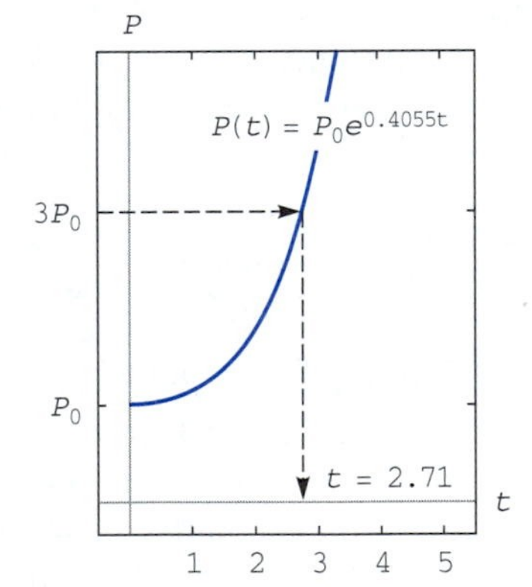

Figure 2.24.

Integrating both sides of the last equation yields

$$e^{-kt}P = c \quad \text{or} \quad P(t) = ce^{kt}$$

At $t = 0$, it follows that $P_0 = ce^0 = c$, so $P(t) = P_0e^{kt}$. At $t = 1$, we have $\frac{3}{2}P_0 = P_0e^k$ or $e^k = \frac{3}{2}$. From the last equation, $k = \ln\frac{3}{2} = 0.4055$. Thus

$$P(t) = P_0e^{0.4055t}$$

To find the time at which the number of bacteria have tripled, we solve $3P_0 = P_0e^{0.4055t}$ for t. It follows that $0.4055t = \ln 3$, so

$$t = \frac{\ln 3}{0.4055} \approx 2.71 \text{ hr}$$

See Figure 2.24. ■

Radioactive Decay The nucleus of an atom consists of combinations of protons and neutrons. Many of these combinations of protons and neutrons are unstable,that is, the atoms decay or transmute into the atoms of another substance. Such nuclei are said to be radioactive. For example, over time the highly radioactive radium, Ra-226, transmutes into the radioactive gas radon, Rn-222. To model the phenomenon of radioactive decay, it is assumed that the rate dA/dt at which the nuclei of a substance decays is proportional to the amount of the substance (more precisely, the number of nuclei) $A(t)$ remaining at time t:

$$\frac{dA}{dt} \propto A \quad \text{or} \quad \frac{dA}{dt} = kA \tag{2}$$

The model (2) for decay also occurs in a biological setting, such as determining the time that it takes for 50% of a drug to be eliminated from a body by excretion or metabolism. The point of (1) and (2) is simply this:

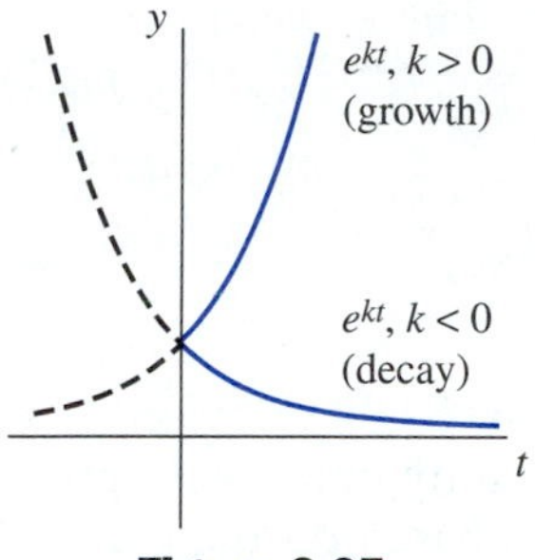

Figure 2.25.

A single differential equation can serve as a mathematical model for many different phenomena.

Of course, since equations (1) and (2) are exactly the same, their general solutions are exactly the same (namely, ce^{kt}); the difference is only in the symbols and their interpretation. As shown in Figure 2.25, the exponential function e^{kt} increases as t increases for $k > 0$ and decreases as t increases for $k < 0$. Thus problems describing growth (whether of animal populations, bacteria, or even capital) are characterized by a positive value of k, whereas problems involving decay yield a negative k value. Accordingly, we say that k is either a **growth constant** ($k > 0$) or a **decay constant** ($k < 0$).

You should also verify that the growth and decay behavior just described is consistent with the fact that the differential equation $dx/dt = kx$ is autonomous and that a solution $x(t)$ moves away from 0 (that is, grows) because 0 is an unstable critical point of the equation for $k > 0$, and a solution $x(t)$ moves closer and closer to 0 as $t \to \infty$ (that is, decays) because 0 is an asymptotically stable critical point of the equation for $k < 0$.

Half-Life In physics the **half-life** is a measure of the stability of a radioactive substance. The half-life is simply the time it takes for one-half of the atoms in an initial amount A_0 to disintegrate, or transmute, into the atoms of another element. The longer the half-life of a substance, the more stable it is. For example, the half-life of highly radioactive radium, Ra-226, is about 1700 years. In 1700 years, one-half of a given quantity of Ra-226 is transmuted into radon, Rn-222. The most commonly occurring uranium isotope, U-238, has a half-life of approximately 4,500,000,000 years. In about 4.5 billion years, one-half of a quantity of U-238 is transmuted into lead, Pb-206.

EXAMPLE 2 Half-Life of Plutonium

A breeder reactor converts relatively stable uranium 238 into the isotope plutonium 239. After 15 years it is determined that 0.043% of the initial amount A_0 of the plutonium has disintegrated. Find the half-life of this isotope if the rate of disintegration is proportional to the amount remaining at time t.

Solution Let $A(t)$ denote the amount of substance remaining at time t. As in Example 1, the solution of the initial-value problem

$$\frac{dA}{dt} = kA, \quad A(0) = A_0$$

is $A(t) = A_0e^{kt}$. If 0.043% of the atoms of A_0 have disintegrated, then 99.957% of the substance remains. To find the decay constant k we use $0.99957A_0 = A(15)$, that is, $0.99957A_0 = A_0e^{15k}$. Solving for k then gives $k = \frac{1}{15}\ln(0.99957) = -0.00002867$. Hence

$$A(t) = A_0e^{-0.00002867t}$$

Now the half-life is the corresponding value of time at which $A(t) = A_0/2$. Solving for t gives $A_0/2 = A_0e^{-0.00002867t}$ or $\frac{1}{2} = e^{-0.00002867t}$. The last equation yields

$$t = \frac{\ln 2}{0.00002867} \approx 24{,}180 \text{ years}$$ ■

Carbon Dating About 1950, the chemist Willard Libby devised a method of using radioactive carbon as a means of determining the approximate ages of fossils. The theory of **carbon dating** is based on the fact that the isotope carbon 14 is produced in the atmosphere by the action of cosmic radiation on nitrogen. The ratio of the amount of C-14 to ordinary carbon in the atmosphere appears to be a constant, and as a consequence, the proportionate amount of the isotope present in all living organisms is the same as that in the atmosphere. When an organism dies, the absorption of C-14, by either breathing or eating, ceases. Thus, by comparing the proportionate amount of C-14 present, say, in a fossil with the constant ratio found in the atmosphere, it is possible to obtain a reasonable estimation of its age. The method is based on the knowledge that the half-life of the radioactive C-14 is approximately 5600 years. For his work, Libby won the Nobel Prize for chemistry in 1960. Libby's method has been used to date wooden furniture in Egyptian tombs and the woven flax wrappings of the Dead Sea scrolls.

■ EXAMPLE 3 **Age of a Fossil**

A fossilized bone is found to contain 1/1000 the original amount of C-14. Determine the age of the fossil.

Solution The starting point is again $A(t) = A_0e^{kt}$. To determine the value of the decay constant k we use the fact that $A_0/2 = A(5600)$, or $A_0/2 = A_0e^{5600k}$. From $5600k = \ln\frac{1}{2} = -\ln 2$ we get $k = -(\ln 2)/5600 = -0.00012378$. Therefore $A(t) = A_0e^{-0.00012378t}$. With $A(t) = A_0/1000$, we have $A_0/1000 = A_0e^{-0.00012378t}$, so that $-0.00012378t = \ln(1/1000) = -\ln 1000$. Thus

$$t = \frac{\ln 1000}{0.00012378} \approx 55{,}800 \text{ years}$$ ■

The date found in Example 3 is really at the border of accuracy for this method. The usual carbon-14 technique is limited to about 9 half-lives of the isotope, or about 50,000 years. One reason is that the chemical analysis needed to obtain an accurate measurement of the remaining C-14 becomes somewhat formidable around the point of $A_0/1000$. Also, this analysis demands the destruction of a rather large sample of the specimen. If this measurement is accomplished indirectly, based on the actual radioactivity of the specimen, then it is difficult to distinguish between the radiation from the fossil and the normal background radiation. But in more recent developments, the use of a particle accelerator has enabled scientists to separate the C-14 from the stable C-12 directly. By computing the precise value of the ratio of C-14 to C-12, the accuracy of this method can be extended to 70,000–100,000 years.

Newton's Law of Cooling According to Newton's empirical law of cooling, the rate at which a body cools is proportional to the difference between the temperature of the body and the temperature of the surrounding medium, the so-called ambient temperature. If we let $T(t)$ represent the temperature of a body at time t, T_m the temperature of the surrounding medium, and dT/dt the rate at which a body cools, then Newton's law of cooling translates into the mathematical statement

$$\frac{dT}{dt} \propto T - T_m \quad \text{or} \quad \frac{dT}{dt} = k(T - T_m) \tag{3}$$

where k is a constant of proportionality. If T_m is constant, and since we have assumed the body is cooling (that is, $T > T_m$), then it stands to reason that $k < 0$. Equation (3), like (1) and (2), is separable as well as linear.

Series Circuits In a series circuit containing only a resistor and an inductor, Kirchhoff's second law states that the sum of the voltage drop across the inductor ($L(di/dt)$) and the voltage drop across the resistor (iR) is the same as the impressed voltage ($E(t)$) on the circuit. See Figure 2.26(a).

Thus we obtain the linear differential equation for the current $i(t)$,

$$L\frac{di}{dt} + Ri = E(t) \tag{4}$$

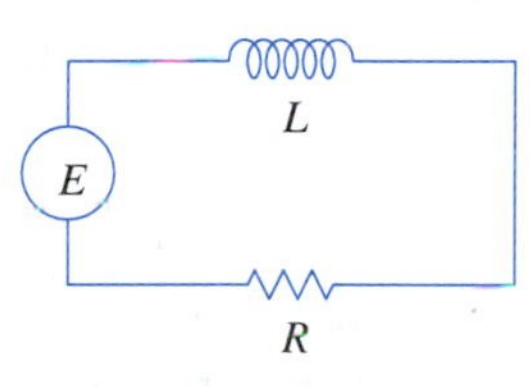

(a) L-R series circuit

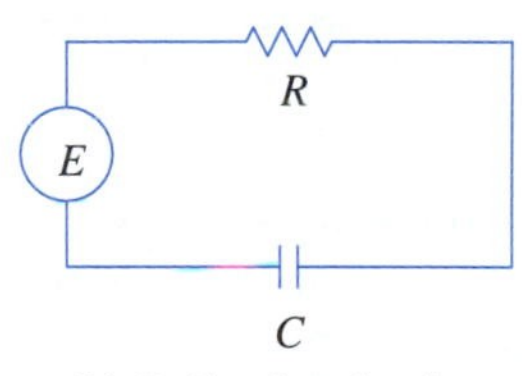

(b) R-C series circuit

Figure 2.26.

where L and R are constants known as the inductance and the resistance, respectively. The current $i(t)$ is called the **response** of the system.

The voltage drop across a capacitor with capacitance C is given by $q(t)/C$, where q is the charge on the capacitor. Hence, for the series circuit shown in Figure 2.26(b), Kirchhoff's second law gives

$$Ri + \frac{1}{C}q = E(t) \tag{5}$$

But current i and charge q are related by $i = dq/dt$, so (5) becomes the linear differential equation

$$R\frac{dq}{dt} + \frac{1}{C}q = E(t) \tag{6}$$

EXAMPLE 4 Series Circuit

A 12-volt battery is connected to a series circuit in which the inductance is $\frac{1}{2}$ henry and the resistance is 10 ohms. Determine the current i if the initial current is zero.

Solution From (4) we see that we must solve

$$\frac{1}{2}\frac{di}{dt} + 10i = 12$$

subject to $i(0) = 0$. First, we multiply the differential equation by 2 and read off the integrating factor e^{20t}. We then obtain

$$\frac{d}{dt}[e^{20t}i] = 24e^{20t}$$

Integrating each side of this equation and solving for i gives

$$i = \frac{6}{5} + ce^{-20t}$$

Now $i(0) = 0$ implies $0 = \frac{6}{5} + c$, or $c = -\frac{6}{5}$. Therefore the response is

$$i(t) = \frac{6}{5} - \frac{6}{5}e^{-20t}$$ ■

From (4) of Section 2.3, we can write a general solution of (4):

$$i(t) = \frac{e^{-(R/L)t}}{L}\int e^{(R/L)t}E(t)\,dt + ce^{-(R/L)t} \tag{7}$$

In particular, when $E(t) = E_0$ is a constant, (7) becomes

$$i(t) = \frac{E_0}{R} + ce^{-(R/L)t} \tag{8}$$

Note that as $t \to \infty$, the second term in equation (8) approaches zero, that is, it is a **transient term**; any remaining terms are called the **steady-state** part of the solution. In this case E_0/R is also called the **steady-state current**; for large values of time it then appears that the current in the circuit is simply governed by Ohm's law ($E = iR$).

Mixture Problem The mixing of two fluids sometimes gives rise to a linear first-order differential equation. In the next example, we consider the mixture of two salt solutions with different concentrations.

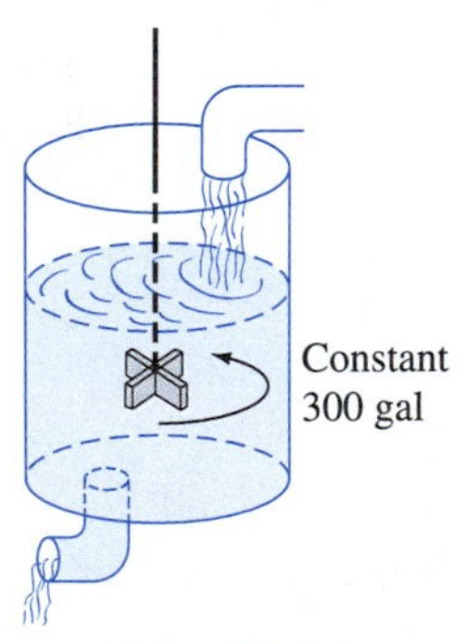

Figure 2.27.

■ EXAMPLE 5 Mixture of Two Salt Solutions

Initially 50 lb of salt are dissolved in a large tank holding 300 gal of water. A brine solution is pumped into the tank at a rate of 3 gal/min, and the well-stirred solution is then pumped out at the same rate. See Figure 2.27. If the concentration of the solution entering is 2 lb/gal, determine the amount of salt in the tank at time t. How much salt is present after 50 min? After a long time?

Solution Let $A(t)$ be the amount of salt (in pounds) in the tank at time t. For problems of this sort, the net rate at which $A(t)$ changes is given by

$$\frac{dA}{dt} = \left(\begin{array}{c}\textit{rate of}\\ \textit{substance entering}\end{array}\right) - \left(\begin{array}{c}\textit{rate of}\\ \textit{substance leaving}\end{array}\right) = R_1 - R_2 \tag{9}$$

Now the rate at which the salt enters the tank is, in lb/min,

$$R_1 = (3 \text{ gal/min}) \cdot (2 \text{ lb/gal}) = 6 \text{ lb/min}$$

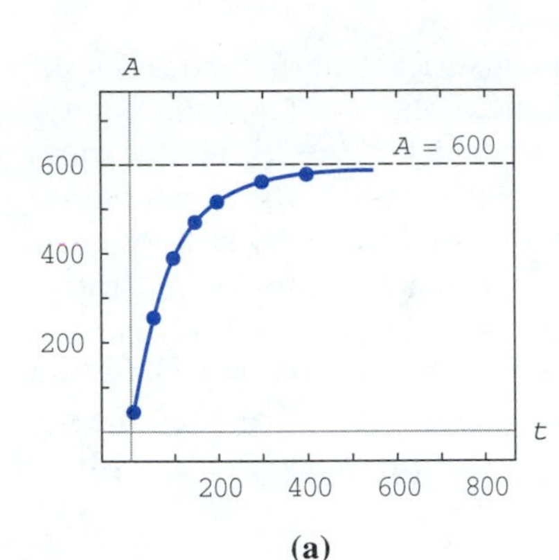

(a)

t (min)	A (lbs)
50	266.41
100	397.67
150	477.27
200	525.62
300	572.62
400	589.93

(b)

Figure 2.28.

whereas the rate at which salt is leaving is

$$R_2 = (3 \text{ gal/min}) \cdot \left(\frac{A}{300} \text{ lb/gal}\right) = \frac{A}{100} \text{ lb/min}$$

Thus equation (9) becomes

$$\frac{dA}{dt} = 6 - \frac{A}{100} \quad \text{or} \quad \frac{dA}{dt} + \frac{1}{100}A = 6 \qquad \textbf{(10)}$$

which we solve subject to the initial condition $A(0) = 50$.

Since the integrating factor is $e^{t/100}$, we can write the last equation in (10) as

$$\frac{d}{dt}[e^{t/100}A] = 6e^{t/100}$$

and therefore

$$e^{t/100}A = 600e^{t/100} + c$$

$$A = 600 + ce^{-t/100}$$

When $t = 0$, $A = 50$, so we find that $c = -550$. Finally, we obtain

$$A(t) = 600 - 550e^{-t/100} \qquad \textbf{(11)}$$

At $t = 50$ we find $A(50) = 266.41$ lb. Also, as $t \to \infty$, it is seen from (11) and Figure 2.28 that $A \to 600$. Of course this is what we would expect; over a long period of time the number of pounds of salt in the solution must be

$$(300 \text{ gal})(2 \text{ lb/gal}) = 600 \text{ lb}$$

This long-term behavior can also be deduced from the fact that 600 is an asymptotically stable critical point of the autonomous differential equation (10). ■

In Example 5, we assumed that the rate at which the solution was pumped in was the same as the rate at which the solution was pumped out. This need not be the case; the mixed brine solution could be pumped out at a rate faster or slower than the rate at which the other solution is pumped in. For example, if the well-stirred solution in Example 5 is pumped out at a slower rate of 2 gal/min, the solution is accumulating at a rate of $(3 - 2)$ gal/min $= 1$ gal/min. After t min there are $300 + 1 \cdot t = 300 + t$ gal of brine in the tank. The rate at which the salt is leaving the tank is then

$$R_2 = (2 \text{ gal/min}) \left(\frac{A}{300 + t} \text{ lb/gal}\right)$$

Hence equation (9) becomes

$$\frac{dA}{dt} = 6 - \frac{2A}{300 + t} \quad \text{or} \quad \frac{dA}{dt} + \frac{2}{300 + t}A = 6$$

You should verify that the solution of this equation subject to $A(0) = 50$ is

$$A(t) = 600 + 2t - (4.95 \times 10^7)(300 + t)^{-2}$$

Newton's Second Law of Motion To construct a mathematical model of the motion of a body moving in a force field, the usual starting point is Newton's second law of motion. Recall, **Newton's first law of motion** states that a body will either remain at rest or will continue to move with a constant velocity unless acted upon by an external force. In each of these two cases, this is equivalent to saying that when the sum of the forces $F = \Sigma F_k$—that is, the net or resultant force—acting on the body is zero, then the acceleration a of the body is zero. **Newton's second law of motion** indicates that when the net force acting on a body is *not* zero, then the net force is proportional to its acceleration a, or more precisely, $F = ma$, where m is the mass of the body.

Falling Bodies and Air Resistance It has been established empirically that when a body moves through a resistive medium such as air, the retarding force due to the medium acts in the direction opposite to that of the motion and is proportional to a power of the body's velocity: kv^{α}. Here k is a constant of proportionality and α is constant in the range $1 \le \alpha \le 2$. Roughly, for slow speeds, $\alpha = 1$ and for higher speeds, $\alpha = 2$. Now suppose a falling body of mass m encounters air resistance proportional to its instantaneous velocity v. If we take, in this circumstance, the positive direction to be oriented downward, then the net force acting on the mass is given by $mg - kv$, where the weight mg of the body is force acting in the positive direction and air resistance is a force acting in the opposite or upward direction. Now since v is related to acceleration a by $dv/dt = a$, Newton's second law becomes $F = ma = m\,dv/dt$. By equating the net force to this form of Newton's second law, we obtain a differential equation for the velocity v of the body at time t,

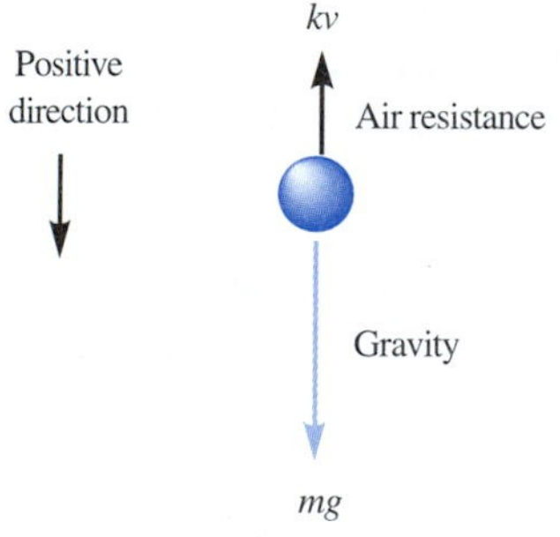

Figure 2.29. Falling body of mass m.

$$m\frac{dv}{dt} = mg - kv \tag{12}$$

Here k is a positive constant of proportionality. See Figure 2.29 and Problem 26 in Exercises 2.4.

2.4.2 Nonlinear Models

Population Dynamics If $P(t)$ denotes the size of a population at time t, the model for exponential growth (1) begins with the assumption that $dP/dt = kP$ for some $k > 0$. In this model the **specific** or **relative growth rate** defined by

$$\frac{dP/dt}{P} \tag{13}$$

is assumed to be a constant k. True cases of exponential growth over long periods of time are hard to find, because the limited resources of the environment will at some time exert restrictions on the growth of a population. Thus (13) can be expected to decrease as P increases in size.

The assumption that the rate at which a population grows (or declines) is dependent only on the number present and not on any time-dependent mechanisms such as seasonal phenomena can be stated as

$$\frac{dP/dt}{P} = f(P) \quad \text{or} \quad \frac{dP}{dt} = Pf(P) \tag{14}$$

The autonomous differential equation in (14), which is widely assumed in models of animal populations, is called the **density-dependent hypothesis**.

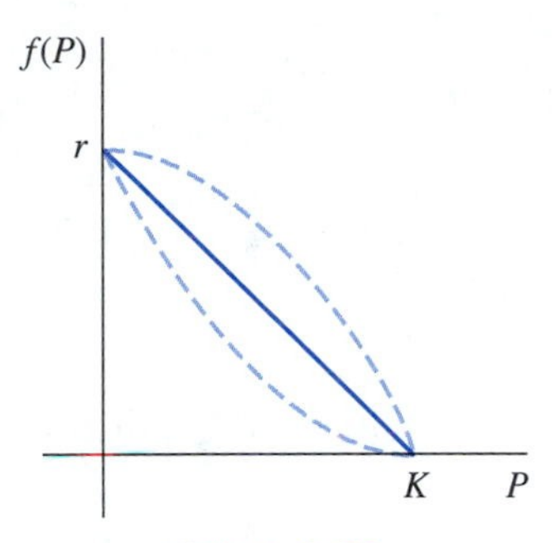

Figure 2.30.

Logistic Equation

Suppose an environment is capable of sustaining no more than a fixed number of K individuals in its population. The quantity K is called the **carrying capacity** of the environment. Hence, for the function f in (14) we have $f(K) = 0$, and we simply let $f(0) = r$. Figure 2.30 shows three decreasing functions f that satisfy these two conditions. The simplest assumption that we can make is that $f(P)$ is linear, that is, $f(P) = c_1P + c_2$. If we use the conditions $f(0) = r$ and $f(K) = 0$, we find, in turn, $c_2 = r$ and $c_1 = -r/K$. Thus f takes on the form $f(P) = r - (r/K)P$. Equation (14) becomes

$$\frac{dP}{dt} = P\left(r - \frac{r}{K}P\right) \tag{15}$$

After relabeling constants, we see that equation (15) is the same as

$$\frac{dP}{dt} = P(a - bP) \tag{16}$$

Around 1840, the Belgian mathematician–biologist P. F. Verhulst was concerned with mathematical models for predicting the human population of various countries. One of the equations he studied was (16), where $a > 0$, $b > 0$. Equation (16) came to be known as the **logistic equation** and its solution is called the **logistic function**. The graph of a logistic function is called a **logistic curve**.

The linear differential equation $dP/dt = kP$ does not provide a very accurate model for population growth when the population itself is large. Overcrowded conditions with the resulting detrimental effects on the environment, such as pollution and excessive and competitive demands for food and fuel, can have an inhibiting effect on population growth. As we shall now see, the solution of the nonlinear equation (16) is bounded as $t \to \infty$. If we rewrite (16) as $dP/dt = aP - bP^2$, the nonlinear term $-bP^2$, $b > 0$, can be interpreted as an "inhibition" or "competition" term. Also, in most applications, the positive constant a is much larger than the constant b.

Logistic curves have proved to be accurate in predicting the growth patterns, in a limited space, of certain types of bacteria, protozoa, water fleas (*Daphnia*), and fruit flies (*Drosophila*).

Solution of the Logistic Equation

Equation (16) can be solved by separation of variables. Decomposing the left side of $\dfrac{dP}{P(a - bP)} = dt$ into partial fractions and integrating give

$$\left[\frac{1/a}{P} + \frac{b/a}{a - bP}\right] dP = dt$$

$$\frac{1}{a}\ln|P| - \frac{1}{a}\ln|a - bP| = t + c$$

$$\ln\left|\frac{P}{a - bP}\right| = at + ac$$

$$\frac{P}{a - bP} = c_1 e^{at}$$

It follows from the last equation that

$$P(t) = \frac{ac_1 e^{at}}{1 + bc_1 e^{at}} = \frac{ac_1}{bc_1 + e^{-at}}$$

If $P(0) = P_0$, $P_0 \neq a/b$, we find $c_1 = P_0/(a - bP_0)$, and so, after substituting and simplifying, the solution becomes

$$P(t) = \frac{aP_0}{bP_0 + (a - bP_0)e^{-at}} \tag{17}$$

Graphs of $P(t)$ We have already examined the graphs of solutions of the autonomous equation (16) for different initial conditions in Figure 2.6. Recall from Section 2.1, 0 is an unstable critical point, and a/b is an asymptotically stable critical point, and that $x(t) = a/b$ is an equilibrium solution of the logistic equation. Thus, for $0 < P_0 < a/b$, the basic graph of $P(t)$ is the same as that given in color in Figure 2.6, and thus we know that the population $P(t) \to a/b$ as $t \to \infty$. The dashed line $P = a/2b$ shown in Figure 2.31 corresponds to the ordinate of a point of inflection of the logistic curve. To see this, we differentiate (16) by the product rule and replace the symbol dP/dt by $P(a - bP)$:

$$\begin{aligned}\frac{d^2P}{dt^2} &= P\left(-b\frac{dP}{dt}\right) + (a - bP)\frac{dP}{dt}\\ &= P(a - bP)(a - 2bP) = 2b^2P\left(P - \frac{a}{b}\right)\left(P - \frac{a}{2b}\right)\end{aligned}$$

From calculus, recall that the points where $d^2P/dt^2 = 0$ are possible points of inflection, but $P = 0$ and $P = a/b$ can obviously be ruled out. Hence $P = a/2b$ is the only possible ordinate value at which the concavity of the graph can change. For $0 < P < a/2b$, it follows that $P'' > 0$, and $a/2b < P < a/b$ implies $P'' < 0$. Thus, as we read from left to right, the graph changes from concave up to concave down at the point corresponding to $P = a/2b$. When the initial value satisfies $0 < P_0 < a/2b$, the graph of $P(t)$ assumes the shape of an S, as we see in Figure 2.31(a). For

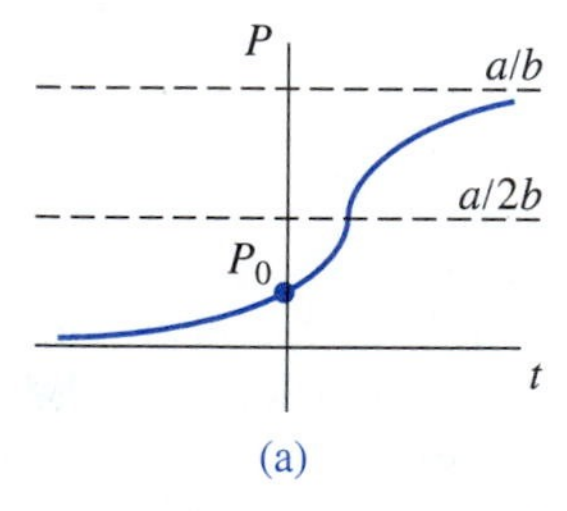

(a)

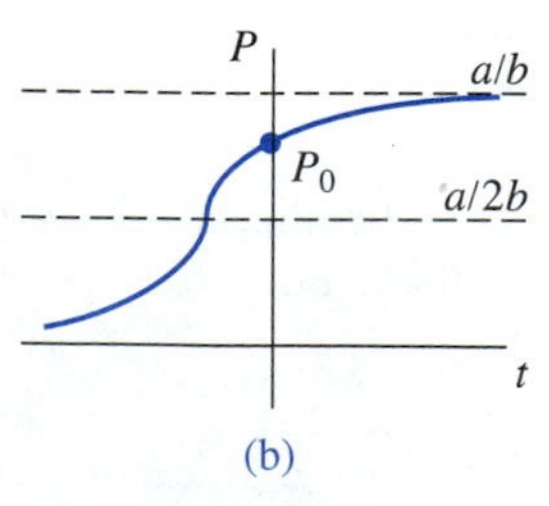

(b)

Figure 2.31.

$a/2b < P_0 < a/b$ the graph is still S-shaped, but the point of inflection occurs at a negative value of t, as shown in Figure 2.31(b).

EXAMPLE 6 Logistic Growth

Suppose a student carrying a flu virus returns to an isolated college campus of 1000 students. If it is assumed that the rate at which the virus spreads is proportional not only to the number x of infected students but also to the number of students not infected, determine the number of infected students after 6 days if it is further observed that after 4 days $x(4) = 50$.

Solution Assuming that no one leaves the campus throughout the duration of the disease, we must solve the initial-value problem

$$\frac{dx}{dt} = kx(1000 - x), \quad x(0) = 1$$

By making the identifications $a = 1000k$ and $b = k$, we have immediately from (17) that

$$x(t) = \frac{1000k}{k + 999ke^{-1000kt}} = \frac{1000}{1 + 999e^{-1000kt}}$$

Now, using the information $x(4) = 50$, we determine k from

$$50 = \frac{1000}{1 + 999e^{-4000k}}$$

We find $-1000k = \frac{1}{4} \ln \frac{19}{999} = -0.9906$. Thus

$$x(t) = \frac{1000}{1 + 999e^{-0.9906t}}$$

Finally $$x(6) = \frac{1000}{1 + 999e^{-5.9436}} = 276 \text{ students}$$

Additional calculated values of $x(t)$ are given in the table in Figure 2.32(b). ■

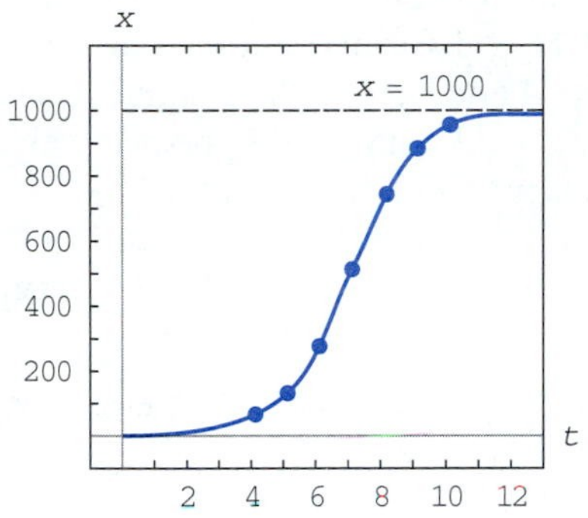

(a)

t (days)	x (number infected)
4	50 (observed)
5	124
6	276
7	507
8	735
9	882
10	953

(b)

Figure 2.32.

Chemical Reactions The disintegration of a radioactive substance, governed by equation (2) of this section, is said to be a **first-order reaction**. In chemistry, a few reactions follow the same empirical law: If the molecules of a substance A decompose into smaller molecules, it is a natural assumption that the rate at which this decomposition takes place is proportional to the amount of the first substance that has not undergone conversion; that is, if $X(t)$ is the amount of substance A remaining at time t, then $dX/dt = kX$, where k is negative, since X is decreasing. An example of a first-order chemical reaction is the conversion of t-butyl chloride into t-butyl alcohol:

$$(CH_3)_3CCl + NaOH \rightarrow (CH_3)_3COH + NaCl$$

Only the concentration of the t-butyl chloride controls the rate of reaction.

Now, in the reaction

$$CH_3Cl + NaOH \rightarrow CH_3OH + NaCl$$

for every molecule of methyl chloride, one molecule of sodium hydroxide is consumed, thus forming one molecule of methyl alcohol and one molecule of sodium chloride. In this case the rate at which the reaction proceeds is proportional to the product of the remaining concentrations of CH_3Cl and of NaOH. If X denotes the amount of CH_3OH formed and α and β are the given amounts of the first two chemicals A and B, then the instantaneous amounts not converted to chemical C are $\alpha - X$ and $\beta - X$, respectively. Hence the rate of formation of C is given by

$$\frac{dX}{dt} = k(\alpha - X)(\beta - X) \qquad \textbf{(18)}$$

where k is a constant of proportionality. A reaction described by equation (18) is said to be of **second-order**.

■ EXAMPLE 7 Second-Order Chemical Reaction

A compound C is formed when two chemicals A and B are combined. The resulting reaction between the two chemicals is such that for each gram of A, 4 g of B are used. It is observed that 30 g of the compound C are formed in 10 min. Determine the amount of C at time t if the rate of the reaction is proportional to the amounts of A and B remaining and if initially there are 50 g of A and 32 g of B. How much of the compound C is present at 15 min? Interpret the solution as $t \rightarrow \infty$.

Solution Let $X(t)$ denote the number of grams of the compound C present at any time t. Clearly $X(0) = 0$ g and $X(10) = 30$ g.

If, for example, there are 2 g of compound C, we must have used, say, a grams of A and b grams of B so that $a + b = 2$ and $b = 4a$. Thus we must use $a = 2/5 = 2(1/5)$ g of chemical A and $b = 8/5 = 2(4/5)$ g of B. In general, for X grams of C we must use

$$\frac{1}{5}X \text{ grams of } A \quad \text{and} \quad \frac{4}{5}X \text{ grams of } B$$

The amounts of A and B remaining at time t are then

$$50 - \frac{1}{5}X \quad \text{and} \quad 32 - \frac{4}{5}X$$

respectively.

Now we know that the rate at which chemical C is formed satisfies

$$\frac{dX}{dt} \propto \left(50 - \frac{1}{5}X\right)\left(32 - \frac{4}{5}X\right)$$

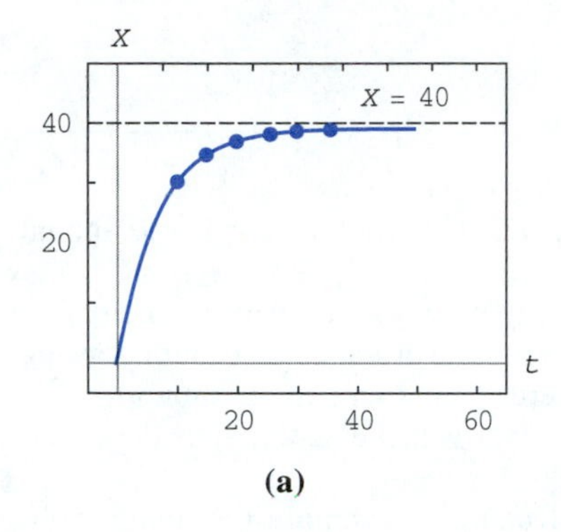

(a)

t (min)	X (g)
10	30 (measured)
15	34.78
20	37.25
25	38.54
30	39.22
35	39.59

(b)

Figure 2.33.

To simplify the subsequent algebra, we factor 1/5 from the first term and 4/5 from the second, and then introduce the constant of proportionality:

$$\frac{dX}{dt} = k(250 - X)(40 - X)$$

By separation of variables and partial fractions, we can write

$$-\frac{1/210}{250 - X}dX + \frac{1/210}{40 - X}dX = k\,dt$$

Integrating gives

$$\ln\left|\frac{250 - X}{40 - X}\right| = 210kt + c_1 \quad \text{or} \quad \frac{250 - X}{40 - X} = c_2 e^{210kt} \tag{19}$$

When $t = 0$, then $X = 0$, and so it follows at this point that $c_2 = \frac{25}{4}$. Using $X = 30$ g at $t = 10$, we find $210k = \frac{1}{10}\ln\frac{88}{25} = 0.1258$. With this information we solve the last equation in (19) for X:

$$X(t) = 1000\frac{1 - e^{-0.1258t}}{25 - 4e^{-0.1258t}} \tag{20}$$

The behavior of X as a function of time t is displayed in Figure 2.33. It is clear from the table accompanying the figure and from equation (20) that $X \to 40$ as $t \to \infty$. This means there are 40 g of compound C formed, leaving

$$50 - \frac{1}{5}(40) = 42 \text{ g of } A \quad \text{and} \quad 32 - \frac{4}{5}(40) = 0 \text{ g of } B$$ ■

Observe that equation (18) is autonomous, and a phase-line analysis of the specific model in Example 7 is consistent with Figure 2.33(a).

Falling Bodies and Air Resistance On page 66 we saw that air resistance of a body moving through air with velocity v is often taken to be kv^{α}, where $1 \le \alpha \le 2$. For high-speed motion—such as the motion of the sky diver shown in Figure 2.34 who is falling before the parachute is opened—it is usually assumed that air resistance is proportional to the square of the instantaneous velocity, in other words, $\alpha = 2$. If the positive direction is again taken to be downward, then a model for the velocity v of a falling body of mass m is given by the nonlinear differential equation

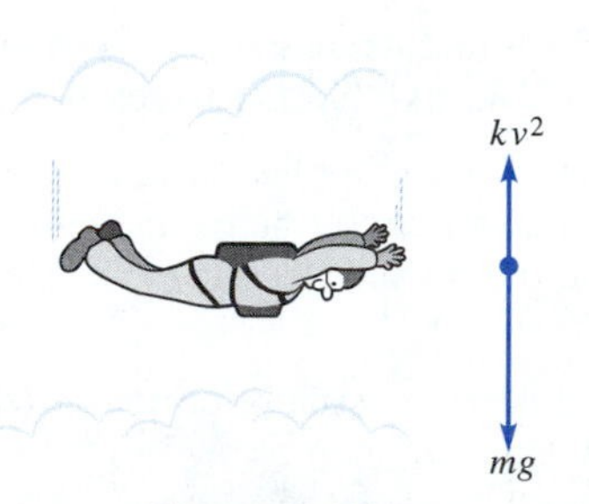

Figure 2.34.

$$m\frac{dv}{dt} = mg - kv^2 \tag{21}$$

where k is a positive constant of proportionality. See Problem 35 in Exercises 2.4.

EXERCISES 2.4

Answers to odd-numbered problems begin on page 436.

2.4.1

1. The population of a certain community is known to increase at a rate proportional to the number of people present at time t. If the population has doubled in 5 yr, how long will it take to triple? to quadruple?

2. Suppose it is known that the population of the community in Problem 1 is 10,000 after 3 yr. What was the initial population? What will be the population in 10 yr?

3. The population of a town grows at a rate proportional to the population at time t. Its initial population of 500 increases by 15% in 10 yr. What will be the population in 30 yr?

4. The population of bacteria in a culture grows at a rate proportional to the number of bacteria present at time t. After 3 hr it is observed that there are 400 bacteria present. After 10 hr there are 2000 bacteria present. What was the initial number of bacteria?

5. The radioactive isotope of lead, Pb-209, decays at a rate proportional to the amount present at time t and has a half-life of 3.3 hr. If 1 g of lead is present initially, how long will it take for 90% of the lead to decay?

6. Initially there were 100 mg of a radioactive substance present. After 6 hr the mass decreased by 3%. If the rate of decay is proportional to the amount of the substance present at time t, find the amount remaining after 24 hr.

7. Determine the half-life of the radioactive substance described in Problem 6.

8. Show that the half-life of a radioactive substance is, in general,

$$t = \frac{(t_2 - t_1)\ln 2}{\ln(A_1/A_2)}$$

where $A_1 = A(t_1)$ and $A_2 = A(t_2)$, $t_1 < t_2$.

9. When a vertical beam of light passes through a transparent substance, the rate at which its intensity I decreases is proportional to $I(t)$, where t represents the thickness of the medium (in feet). In clear seawater, the intensity 3 ft below the surface is 25% of the initial intensity I_0 of the incident beam. What is the intensity of the beam 15 ft below the surface?

10. When interest is compounded continuously, the amount of money increases at a rate proportional to the amount S present at time t: $dS/dt = rS$, where r is the annual rate of interest.

(a) Find the amount of money accrued at the end of 5 yr when \$5000 is deposited in a savings account drawing $5\frac{3}{4}\%$ annual interest compounded continuously.

(b) In how many years will the initial sum deposited be doubled?

(c) Use a hand calculator to compare the number obtained in part (a) with the value

$$S = 5000\left(1 + \frac{0.0575}{4}\right)^{5(4)}$$

This value represents the amount accrued when interest is compounded quarterly.

11. In a piece of burned wood, or charcoal, it was found that 85.5% of the C-14 had decayed. Use the information in Example 3 to determine the approximate age of the wood. (It is precisely these data that archaeologists used to date prehistoric paintings in a cave in Lascaux, France.)

12. A thermometer is taken from an inside room to the outside where the air temperature is 5°F. After 1 min the thermometer reads 55°F, and after 5 min the reading is 30°F. What is the initial temperature of the room?

13. A thermometer is removed from a room where the air temperature is 70°F to the outside where the temperature is 10°F. After $\frac{1}{2}$ min the thermometer reads 50°F. What is the reading at $t = 1$ min? How long will it take for the thermometer to reach 15°F?

14. Equation (3) also holds when an object absorbs heat from the surrounding medium. If a small metal bar whose initial temperature is 20°C is dropped into a container of boiling water, how long will it take for the bar to reach 90°C if it is known that its temperature increased 2° in 1 s? How long will it take the bar to reach 98°C?

15. A 30-volt electromotive force is applied to an L-R series circuit in which the inductance is 0.1 henry and the resistance is 50 ohms. Find the current $i(t)$ if $i(0) = 0$. Determine the current as $t \to \infty$.

16. Solve equation (6) under the assumption that $E(t) = E_0 \sin \omega t$ and $i(0) = i_0$.

17. A 100-volt electromotive force is applied to an R-C series circuit in which the resistance is 200 ohms and the capacitance is 10^{-4} farad. Find the charge $q(t)$ on the capacitor if $q(0) = 0$. Find the current $i(t)$.

18. A 200-volt electromotive force is applied to an R-C series circuit in which the resistance is 1000 ohms and the capacitance is 5×10^{-6} farad. Find the charge $q(t)$ on the capacitor if $i(0) = 0.4$. Determine the charge and current at $t = 0.005$ s. Determine the charge as $t \to \infty$.

19. An electromotive force

$$E(t) = \begin{cases} 120, & 0 \le t \le 20 \\ 0, & t > 20 \end{cases}$$

is applied to an L-R series circuit in which the inductance is 20 henry and the resistance is 2 ohms. Find the current $i(t)$ if $i(0) = 0$.

20. Suppose an R-C series circuit has a variable resistor. If the resistance at time t is given by $R = k_1 + k_2 t$, where $k_1 > 0$ and $k_2 > 0$ are known constants, then (6) becomes

$$(k_1 + k_2 t)\frac{dq}{dt} + \frac{1}{C}q = E(t)$$

Show that if $E(t) = E_0$ and $q(0) = q_0$, then

$$q(t) = E_0 C + (q_0 - E_0 C)\left(\frac{k_1}{k_1 + k_2 t}\right)^{1/Ck_2}$$

21. A tank contains 200 L of fluid in which 30 g of salt is dissolved. Brine containing 1 g of salt per liter is then pumped into the tank at a rate of 4 L/min; the well-mixed solution is pumped out at the same rate. Find the number of grams of salt $A(t)$ in the tank at time t.

22. Solve Problem 21 assuming pure water is pumped into the tank.

23. A large tank is filled with 500 gal of pure water. Brine containing 2 lb of salt per gallon is pumped into the tank at a rate of 5 gal/min. The well-mixed solution is pumped out at the same rate. Find the number of pounds of salt $A(t)$ in the tank at time t.

24. Solve Problem 23 under the assumption that the solution is pumped out at a faster rate of 10 gal/min. When is the tank empty?

25. A large tank is partially filled with 100 gal of fluid in which 10 lb of salt is dissolved. Brine containing $\frac{1}{2}$ lb of salt per gallon is pumped into the tank at a rate of 6 gal/min. The well-mixed solution is then pumped out at a slower rate of 4 gal/min. Find the number of pounds of salt in the tank after 30 min.

26. **(a)** Solve the differential equation (12) subject to $v(0) = v_0$.
(b) Determine the limiting (that is, as $t \to \infty$), or terminal velocity of a falling body of mass m.
(c) Since the positive direction is downward, the position $s(t)$ of a falling body measured from the point of release is related to its velocity $v(t)$ by $ds/dt = v$. Find an explicit expression for $s(t)$.

Miscellaneous Linear Models

27. **(a)** In the theory of learning, the rate at which a subject is memorized is assumed to be proportional to the amount that is left to be memorized. Suppose M denotes the total amount of a subject to be memorized and $A(t)$ is the amount memorized in time t. Determine a differential equation for the amount $A(t)$.
(b) In part (a) assume that the amount of material forgotten is proportional to the amount memorized in time t. Determine a differential equation for $A(t)$ when forgetfulness is taken into account.
(c) Solve the differential equations in parts (a) and (b). Assume $A(0) = 0$. Graph each solution. In each case find the limiting value of A as $t \to \infty$, and interpret the result.

28. The differential equation

$$\frac{dP}{dt} = (k \cos t)\, P$$

where k is a positive constant, is a mathematical model for a population $P(t)$ that undergoes yearly seasonal fluctuations. Solve the equation and graph the solution. Assume $P(0) = P_0$.

2.4.2

29. The number of supermarkets $C(t)$ at time t throughout the country that are using a computerized checkout system is described by the initial-value problem

$$\frac{dC}{dt} = C(1 - 0.0005\,C), \quad C(0) = 1$$

where $t > 0$. How many supermarkets are using the computerized method when $t = 10$? How many companies are estimated to adopt the new procedure over a long period of time?

30. The number of people $N(t)$ at time t in a community who are exposed to a particular advertisement is governed by the logistic equation. Initially $N(0) = 500$, and it is observed that $N(1) = 1000$. If it is predicted that the limiting number of people in the community who will see the advertisement is 50,000, determine $N(t)$.

31. The population $P(t)$ at time t in a suburb of a large city is governed by the initial-value problem

$$\frac{dP}{dt} = P(10^{-1} - 10^{-7} P), \quad P(0) = 5000$$

where t is measured in months. What is the limiting value of the population? At what time will the population be equal to one-half of this limiting value?

32. Two chemicals A and B are combined to form a chemical C. The rate or velocity of the reaction is proportional to the product of the amounts $A(t)$ and $B(t)$ at time t not converted to chemical C. Initially there are 40 g of A and 50 g of B, and for each gram of B, 2 g of A are used. It is observed that 10 g of C are formed in 5 min. How much is formed in 20 min? What is the limiting amount of C after a long time? How much of chemicals A and B remain after a long time?

33. Obtain an explicit solution of the differential equation

$$\frac{dX}{dt} = k(\alpha - X)(\beta - X)$$

governing second-order reactions in the two cases $\alpha \neq \beta$ and $\alpha = \beta$.

34. In a third-order chemical reaction the number of grams X of a compound obtained by combining three chemicals is governed by

$$\frac{dX}{dt} = k(\alpha - X)(\beta - X)(\gamma - X)$$

Solve the equation under the assumption that the constants α, β, and γ are distinct.

35. **(a)** Solve the differential equation (21) subject to $v(0) = v_0$.
(b) Determine the limiting, or terminal, velocity of a falling body of mass m.

Miscellaneous Nonlinear Models

36. **(a)** Determine the differential equation for the velocity $v(t)$ of an object of mass m sinking in water that imparts a resistance proportional to the square of the instantaneous velocity and also exerts an upward buoyant force whose magnitude is equal to the weight of the water displaced (Archimedes' principle). Assume the positive direction is downward.
(b) Solve the differential equation in part (a).
(c) Determine the limiting, or terminal, velocity of the sinking mass.

37. A boat B, starting at the origin, moves in the direction of the positive x-axis, pulling a waterskier along the curve C, called a **tractrix**, as shown in Figure 2.35. The waterskier, initially located on the y-axis at $(0, s)$, is pulled by a rope of constant length s, which is kept taut throughout the motion.

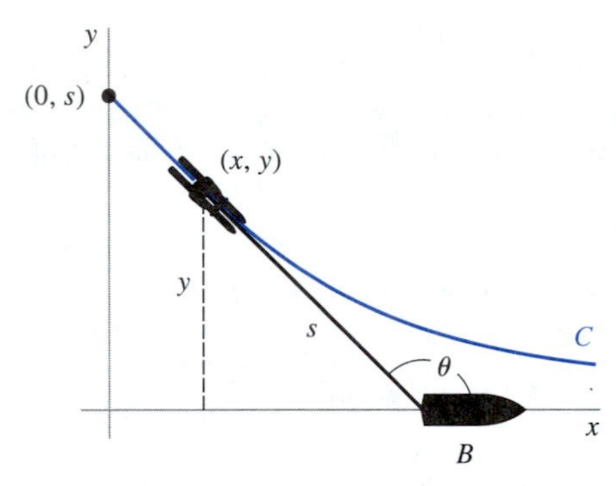

Figure 2.35.

(a) Find the differential equation of the path of motion. [*Hint*: The rope is always tangent to C; consider the angle of inclination θ as shown in the figure.]
(b) Solve the differential equation in part (a). Assume that the initial point on the y-axis is $(0, 10)$ and that the length of the rope is $s = 10$ ft.

38. Suppose a water tank has the form of a right-circular cylinder as shown in Figure 2.36. If water is allowed to drain under the influence of gravity through a hole in the bottom of the tank, then the height h of water at time t is given by

$$\frac{dh}{dt} = -c\frac{A_h}{A_w}\sqrt{2gh}$$

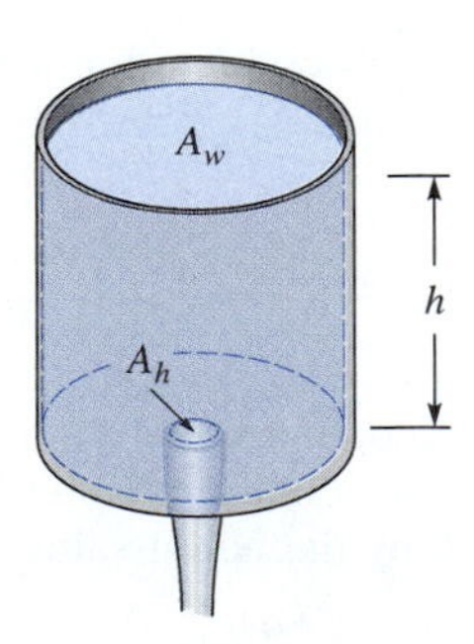

Figure 2.36.

where A_w and A_h are the cross-sectional areas of the water and the hole, respectively, and c is a friction/contraction factor at the hole.

(a) Solve the equation if the initial height of the water is 20 ft and $A_w = 50$ ft^2 and $A_h = \frac{1}{4}$ ft^2.

(b) If $c = 1$, at what time is the tank empty?

(c) How long would it take the tank to empty if the friction/contraction factor is $c = 0.6$?

Discussion Problems

39. **(a)** The model given in (1) fails to take death into consideration, that is, the underlying assumption is that the growth rate equals the birth rate. In another mathematical model of a changing population of a community, it is assumed that the rate at which the population changes is a net rate, that is, the difference between the rate of births and the rate of deaths in the community. Determine a mathematical model for the population if the birth rate as well as the death rate are each proportional to the population $P(t)$ present at time t. Before solving, discuss an interpretation for the solution of this equation; in other words, what kind of population does the differential equation describe? Consider various cases. Verify your conjectures by graphing several solutions.

(b) Determine a model for the population if the birth rate, death rate, immigration rate, and emigration rate are each proportional to the population $P(t)$ present at time t.

40. A cup of coffee cools according to Newton's law of cooling (3). Use data from the graph of the temperature $T(t)$ in Figure 2.37 to estimate T_m, T_0, and k in a model of the form $dT/dt = k(T - T_m)$, $T(0) = T_0$.

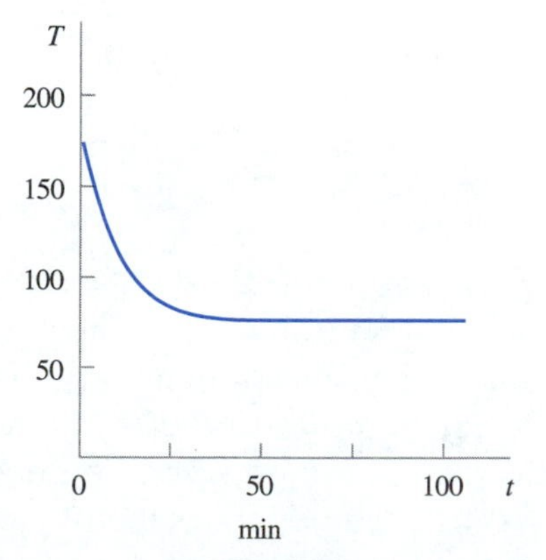

Figure 2.37.

41. The ambient temperature T_m in (3) could be a function of time t. Suppose in an artificially controlled environment, $T_m(t)$ is periodic with period 24 hr as illustrated in Figure 2.38.

(a) Devise a mathematical model for the temperature $T(t)$ of a body within this environment.

(b) Suppose a body with initial temperature $T(0) = 120$ is placed within this environment. Without solving the differential equation in part (a), discuss the behavior of $T(t)$ over a long interval of time.

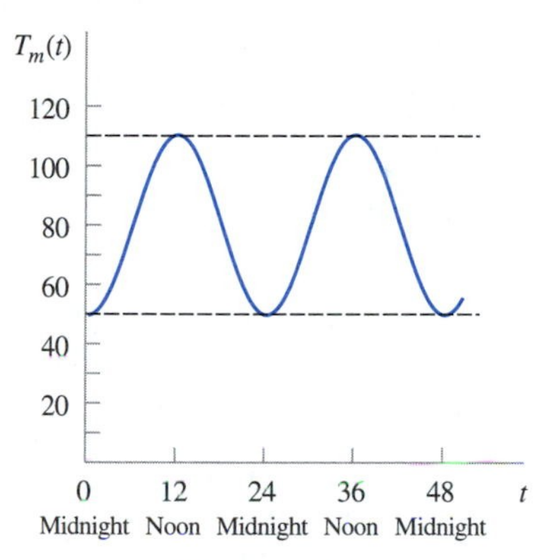

Figure 2.38.

42. A large snowball is shaped into the form of a sphere. Starting at some time, which we can designate as $t = 0$, the snowball begins to melt. Assume for the sake of discussion that the snowball melts in such a manner that its shape remains spherical. Discuss the quantities that change with time as the snowball melts. Discuss an interpretation of "melting" as a rate. If possible, construct a mathematical model that describes the state of the snowball at time $t > 0$. If possible, determine the radius of the snowball as a function of time t. When does the snowball disappear?

43. A homeowner feels that she loses a lot of water from her hot tub over a period of time. She feels that there may be a leak in the tub, but the pool maintenance person thinks that the water loss is normal and is simply due to evaporation. A geometric model for the tub is a frustum of a circular cone with dimensions shown in Figure 2.39. Suppose $h(t)$ is the depth of the water measured in feet and t is time measured in days. Discuss an interpretation of "evaporation" as a rate. If possible, construct a mathematical model that describes the state of the water in the hot tub at time $t > 0$ that is consistent with the data: $h(0) = 4$, $h(3) = 3.4$. What does the model

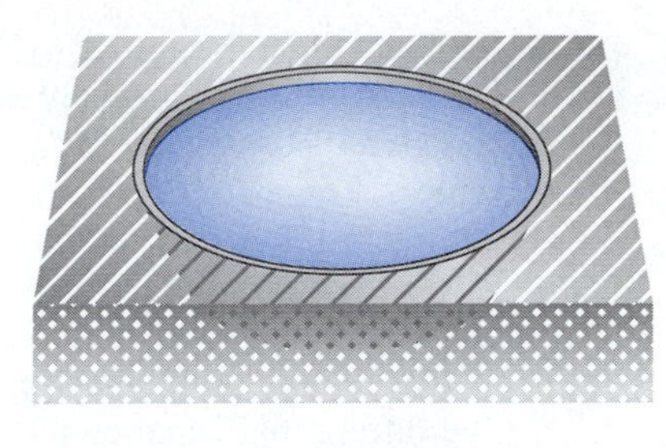

(a) Hot-tub

16
h
4
10

(b) Side view

Figure 2.39.

predict for $h(4)$? If the water is not replenished, when does the model predict that the tub will be empty?

44. Discuss how to solve equation (4) when an L-R series circuit has a variable inductor with inductance defined by

$$L(t) = \begin{cases} 1 - \dfrac{t}{10}, & 0 \le t < 10 \\ 0, & t \ge 10 \end{cases}$$

Carry out your ideas when the resistance is 0.2 ohms, the impressed voltage is $E(t) = 4$, and $i(0) = 0$. Use a graphing utility to obtain the graph of $i(t)$.

45. The models in (12) and (21) represent the velocity of a falling body of mass m when the body is dropped or thrown in the *downward* direction in the two cases when air resistance is proportional to the velocity, and proportional to the square of the velocity, respectively. Discuss: Again consider the two cases of air resistance. Do these models represent the velocity of a body when it is projected straight *upward*? Keep in mind that the constant k of proportionality is assumed to be positive and that, with the positive direction downward, $v > 0$ when the body is falling and $v < 0$ when ascending.

46. Suppose the differential equation

$$\frac{dP}{dt} = kP^{\alpha}$$

where $k > 0$ and $\alpha \ge 1$, is a mathematical model for a population $P(t)$ that exhibits unbounded growth as t increases. Compare the cases $\alpha = 1$ and $\alpha > 1$. Discuss: Why do you think that the model with $\alpha > 1$ is sometimes referred to as a "doomsday equation?" Consider $\alpha = 1.01$ and $\alpha = 2$.

47. Suppose the logistic equation (15) is written as

$$\frac{dP}{dt} = \frac{r}{K}P(K - P)$$

For $0 < P_0 < K$, the population is increasing but cannot exceed the carrying capacity K. For $P_0 > K$, the population cannot sustain itself but does not fall below K. How would you modify the above differential equation to describe a population P that has these same two characteristics but additionally has a **threshold level** m, $0 < m < K$, below which the population cannot sustain itself and becomes extinct. Consider two cases: Extinction in finite time and extinction as $t \to \infty$. [*Hint*: Think phase portrait.]

2.5 Numerical Methods

In the introduction to Section 2.1, we mentioned that a solution of a DE might be known to exist and yet it may be impossible to display it in an explicit or implicit form. This is especially true for nonlinear DEs. We have also seen that we can generate an approximate solution curve by means of an ODE solver and often predict behaviors of solutions by examining equilibrium solutions. But what do we do when we need to know a *numerical* value, say $y(1.5)$, of a solution $y(t)$ that we cannot exhibit?

To answer this question, we must now enter the realm of *numerical analysis*.

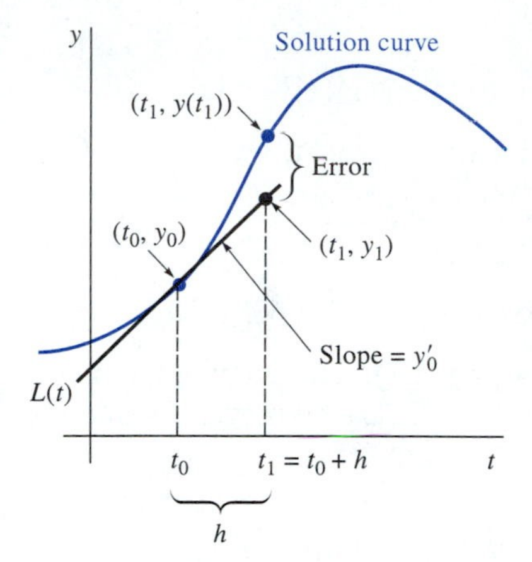

Figure 2.40.

2.5.1 Euler's Methods

Linearization One of the simplest techniques for approximating a solution of a first-order initial-value problem

$$y' = f(t, y), \; y(t_0) = y_0 \tag{1}$$

is known as **Euler's method**, or the method of **tangent lines**. This technique uses the fact that the derivative of a function $y(t)$ at a point t_0 determines a **linearization** of $y(t)$ at $t = t_0$:

$$L(t) = y'(t_0)(t - t_0) + y_0$$

As the name implies, the linearization of $y(t)$ at t_0 is a *linear* function and is simply an equation of the tangent line to the graph of $y = y(t)$ at the point (t_0, y_0). We now let h be a positive increment on the t-axis, as shown in Figure 2.40. Then for $t_1 = t_0 + h$ we have

$$L(t_1) = y'(t_0)(t_0 + h - t_0) + y_0 = y_0 + hy'_0$$

where $y'_0 = y'(t_0) = f(t_0, y_0)$. Letting $y_1 = L(t_1)$ we get

$$y_1 = y_0 + hf(t_0, y_0)$$

The point (t_1, y_1), which is seen in Figure 2.40 to be a point on the tangent line, is an approximation to the point $(t_1, y(t_1))$ on the actual solution curve; that is, $L(t_1) \approx y(t_1)$ or $y_1 \approx y(t_1)$ is a *local linear approximation* of $y(t)$ at t_1. Of course, the accuracy of the approximation depends heavily on the size of the increment h. Usually we must choose this **step size** to be "reasonably small." If we now repeat the process, using (t_1, y_1) as the new starting point, we obtain the approximation

$$y(t_2) = y(t_0 + 2h) = y(t_1 + h) \approx y_2 = y_1 + hf(t_1, y_1)$$

In general it follows that

$$y_{n+1} = y_n + hf(t_n, y_n) \tag{2}$$

where $t_n = t_0 + nh$.

To illustrate Euler's method, we use the iteration scheme (2) on a differential equation for which we know the explicit solution; in this way we can compare the estimated values y_n with the true values $y(t_n)$.

■ EXAMPLE 1 Euler's Method

Consider the initial-value problem

$$y' = 0.2ty, \quad y(1) = 1$$

Use the Euler method to approximate $y(1.5)$ using first $h = 0.1$ and then $h = 0.05$.

Solution We first identify $f(t, y) = 0.2ty$ so that (2) becomes

$$y_{n+1} = y_n + h(0.2t_n y_n)$$

Then for $h = 0.1$ we find

$$y_1 = y_0 + (0.1)(0.2t_0y_0) = 1 + (0.1)[0.2(1)(1)] = 1.02$$

which is an estimate to the value of $y(1.1)$. However, if we use $h = 0.05$, it takes *two* iterations to reach $t = 1.1$. We have

$$y_1 = 1 + (0.05)[0.2(1)(1)] = 1.01$$

$$y_2 = 1.01 + (0.05)[0.2(1.05)(1.01)] = 1.020605$$

Here we note that $y_1 \approx y(1.05)$ and $y_2 \approx y(1.1)$. The remainder of the calculations is summarized in Tables 2.1 and 2.2. Each entry is rounded to four decimal places.

Table 2.1 Euler's Method with $h = 0.1$

t_n	y_n	True Value	Absolute Error	% Relative Error
1.00	1.0000	1.0000	0.0000	0.00
1.10	1.0200	1.0212	0.0012	0.12
1.20	1.0424	1.0450	0.0025	0.24
1.30	1.0675	1.0714	0.0040	0.37
1.40	1.0952	1.1008	0.0055	0.50
1.50	1.1259	1.1331	0.0073	0.64

Table 2.2 Euler's Method with $h = 0.05$

t_n	y_n	True Value	Absolute Error	% Relative Error
1.00	1.0000	1.0000	0.0000	0.00
1.05	1.0100	1.0103	0.0003	0.03
1.10	1.0206	1.0212	0.0006	0.06
1.15	1.0318	1.0328	0.0009	0.09
1.20	1.0437	1.0450	0.0013	0.12
1.25	1.0562	1.0579	0.0016	0.16
1.30	1.0694	1.0714	0.0020	0.19
1.35	1.0833	1.0857	0.0024	0.22
1.40	1.0980	1.1008	0.0028	0.25
1.45	1.1133	1.1166	0.0032	0.29
1.50	1.1295	1.1331	0.0037	0.32

■

In Example 1, the true values were calculated from the known solution $y = e^{0.1(t^2-1)}$. Also, the **absolute error** is defined to be

$$|\textit{true value} - \textit{approximation}|$$

The **relative error** and **percentage relative error** are, in turn,

$$\frac{\textit{absolute error}}{|\textit{true value}|} \quad \text{and} \quad \frac{\textit{absolute error}}{|\textit{true value}|} \times 100$$

It is apparent by comparing Tables 2.1 and 2.2 that the accuracy of the approximations improves as the step size h decreases. Also, we see that even though the percentage relative error is growing, it does not appear to be that bad. But you should not be deceived by one example. Watch what happens in the next example, when we simply change the coefficient of the right side of the differential equation from 0.2 to 2.

■ EXAMPLE 2 **Comparison of Exact/Approximate Values**

Use the Euler method to approximate $y(1.5)$ for the solution of

$$y' = 2ty, \quad y(1) = 1$$

Solution You should verify that the exact or analytic solution is now $y = e^{t^2-1}$. Proceeding as in Example 1, we obtain the results shown in Tables 2.3 and 2.4.

In this case, with a step size $h = 0.1$, a 16% relative error in the calculation of the approximation to $y(1.5)$ is totally unacceptable. At the expense of doubling the number of calculations, a slight improvement in accuracy is obtained by halving the step size to $h = 0.05$.

Table 2.3 Euler's Method with $h = 0.1$

t_n	y_n	True Value	Absolute Error	% Relative Error
1.00	1.0000	1.0000	0.0000	0.00
1.10	1.2000	1.2337	0.0337	2.73
1.20	1.4640	1.5527	0.0887	5.71
1.30	1.8154	1.9937	0.1784	8.95
1.40	2.2874	2.6117	0.3244	12.42
1.50	2.9278	3.4904	0.5625	16.12

Table 2.4 Euler's Method with $h = 0.05$

t_n	y_n	True Value	Absolute Error	% Relative Error
1.00	1.0000	1.0000	0.0000	0.00
1.05	1.1000	1.1079	0.0079	0.72
1.10	1.2155	1.2337	0.0182	1.47
1.15	1.3492	1.3806	0.0314	2.27
1.20	1.5044	1.5527	0.0483	3.11
1.25	1.6849	1.7551	0.0702	4.00
1.30	1.8955	1.9937	0.0982	4.93
1.35	2.1419	2.2762	0.1343	5.90
1.40	2.4311	2.6117	0.1806	6.92
1.45	2.7714	3.0117	0.2403	7.98
1.50	3.1733	3.4904	0.3171	9.08

■

Errors in Numerical Methods In choosing and using a numerical method for the solution of an initial-value problem, we must be aware of the various sources of errors. For some kinds of computation, the accumulation of errors might reduce the accuracy of an approximation to the point of being useless. On the other hand, depending on the use to which a numerical solution may be put, extreme accuracy may not be worth the added expense and complication.

One source of error always present in calculations is **round-off error**. This error results from the fact that any calculator or computer can represent numbers using only a finite number of digits. Suppose for the sake of illustration that we have a calculator that uses base 10 arithmetic and carries four digits, so that $\frac{1}{3}$ is represented in the calculator as 0.3333 and $\frac{1}{9}$ is represented as 0.1111. If we use this calculator to compute $(t^2 - \frac{1}{9})/(t - \frac{1}{3})$ for $t = 0.3334$, we obtain

$$\frac{(0.3334)^2 - 0.1111}{0.3334 - 0.3333} = \frac{0.1112 - 0.1111}{0.3334 - 0.3333} = 1$$

With the help of a little algebra, however, we see that

$$\frac{t^2 - 1/9}{t - 1/3} = \frac{(t - 1/3)(t + 1/3)}{t - 1/3} = t + \frac{1}{3}$$

so that when $t = 0.3334$, $(t^2 - \frac{1}{9})/(t - \frac{1}{3}) \approx 0.3334 + 0.3333 = 0.6667$. This example shows that the effects of round-off error can be quite serious unless some care is taken. One way to reduce the effect of round-off error is to minimize the number of calculations. Another technique on a computer is to use double-precision arithmetic to check the results. In general, round-off error is unpredictable and difficult to analyze, and we will neglect it in the error analysis that follows. We will concentrate on investigating the error made by using a formula or algorithm to approximate the values of the solution.

Truncation Errors for Euler's Method When iterating Euler's formula (2), we obtain a sequence of values $y_1, y_2, y_3, \ldots$. Usually the value y_1 will not agree with $y(t_1)$, the actual solution evaluated at t_1, because the algorithm gives only a straight-line approximation to the solution. See Figure 2.40. The error is called the **local truncation error**, **formula error**, or **discretization error**. It occurs at each step; that is, if we assume that y_n is accurate, then y_{n+1} will contain local truncation error.

To derive a formula for the local truncation error for Euler's method, we use Taylor's formula with remainder. If a function $y(t)$ possesses $k + 1$ derivatives that are continuous on an open interval containing a and t, then

$$y(t) = y(a) + y'(a)\frac{(t-a)}{1!} + \cdots + y^{(k)}(a)\frac{(t-a)^k}{k!} + y^{(k+1)}(c)\frac{(t-a)^{(k+1)}}{(k+1)!} \tag{3}$$

where c is some point between a and t. Setting $k = 1$, $a = t_n$, and $t = t_{n+1} = t_n + h$, we get

$$y(t_{n+1}) = y(t_n) + y'(t_n)\frac{h}{1!} + y''(c)\frac{h^2}{2!} \quad \text{or} \quad y_{n+1} = y_n + hf(t_n, y_n) + y''(c)\frac{h^2}{2!}$$

Euler's method is this formula without the last term; hence, the local truncation error in y_{n+1} is

$$y''(c)\frac{h^2}{2!}, \quad \text{where} \quad t_n < c < t_{n+1}$$

Unfortunately, the value of c is usually unknown (it exists theoretically) and so the exact error cannot be calculated, but an upper bound on the absolute value of the error is

$$M\frac{h^2}{2}, \quad \text{where} \quad M = \max_{t_n < t < t_{n+1}} |y''(t)|$$

When discussing errors arising from the use of numerical methods, it is helpful to use the notation $O(h^n)$. To define this concept we let $e(h)$ denote the error in a numerical calculation depending on h. Then $e(h)$ is said to be of order h^n, denoted by $O(h^n)$, if there exist a constant C and a positive integer n such that $|e(h)| \le Ch^n$ for sufficiently small h. Thus, the local truncation error for the Euler method is $O(h^2)$. We note that in general if a numerical method has order h^n, and h is halved, the new error is approximately $C(h/2)^n = Ch^n/2^n$; that is, the error is reduced by a factor of $\frac{1n}{2}$.

■ **EXAMPLE 3 Bound for Local Truncation Errors**

Find a bound for the local truncation errors for Euler's method applied to

$$y' = 2ty, \quad y(1) = 1$$

Solution This differential equation was studied in Example 2 and its analytic solution is $y(t) = e^{t^2-1}$.

The local truncation error is

$$y''(c)\frac{h^2}{2} = (2 + 4c^2)e^{(c^2-1)}\frac{h^2}{2}$$

where c is between t_n and $t_n + h$. In particular, for $h = 0.1$, we can get an upper bound on the local truncation error for y_1 by replacing c by 1.1:

$$[2 + (4)(1.1)^2]e^{((1.1)^2-1)}\frac{(0.1)^2}{2} = 0.0422$$

From Table 2.3, we see that the error after the first step is 0.0337, less than the value given by the bound.

Similarly, we can get a bound for the local truncation error for any of the five steps given in Table 2.3 by replacing c by 1.5 (this value of c gives the largest value of $y''(c)$ for any of the steps and may be too generous for the first few steps). Doing this gives

$$[2 + (4)(1.5)^2]e^{((1.5)^2-1)}\frac{(0.1)^2}{2} = 0.1920 \tag{4}$$

as an upper bound for the local truncation error in each step. ■

Note in Example 3 that if h is halved to 0.05, then the error bound would be 0.0480, about one-fourth as much as in (4). This is expected because the local truncation error for Euler's method is $O(h^2)$.

In the previous analysis, we assumed that the value of y_n is correct in the calculation of y_{n+1}, but it is not, because it contains local truncation errors from previous steps. The total error in y_{n+1} is an accumulation of the errors in each of the previous steps. This total error is called the **global truncation error**. A complete analysis of the global truncation error is beyond the scope of this text, but it can be shown that the global truncation error for the Euler method is $O(h)$.

We expect that for Euler's method if the step size is halved, then the error would be approximately halved as well. This is borne out in Example 2, where the absolute error at $t = 1.50$ with $h = 0.1$ is 0.5625 and the absolute error with $h = 0.05$ is 0.3171, approximately half as large. See Tables 2.3 and 2.4.

In general, it can be shown that if a method for the numerical solution of a differential equation has local truncation error $O(h^{\alpha+1})$, then the global truncation error is $O(h^{\alpha})$.

Improved Euler's Method Although the Euler formula (2) is attractive for its simplicity, it is seldom used in serious calculations. In the remainder of this section we study methods that give significantly greater accuracy than Euler's method.

The formula

$$y_{n+1} = y_n + h\frac{f(t_n, y_n) + f(t_{n+1}, y^*_{n+1})}{2} \tag{5}$$

where $\quad y^*_{n+1} = y_n + h(t_n, y_n)$

is known as the **improved Euler formula**. Euler's formula is used to obtain the initial estimate y^*_{n+1}. The values $f(t_n, y_n)$ and $f(t_{n+1}, y^*_{n+1})$ are approximations to the slopes of the solution curve at $(t_n, y(t_n))$ and $(t_{n+1}, y(t_{n+1}))$, and consequently, the quotient

$$\frac{f(t_n, y_n) + f(t_{n+1}, y^*_{n+1})}{2}$$

can be interpreted as an average slope on the interval from t_n to t_{n+1}.

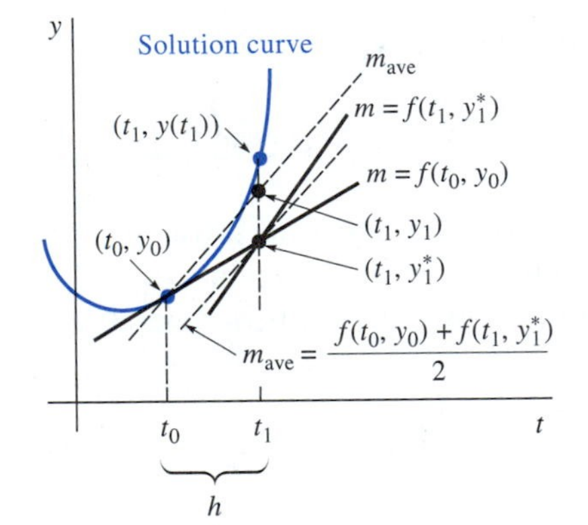

Figure 2.41.

The equations in (5) can be readily visualized. In Figure 2.41 we show the case when $n = 0$. Note that $f(t_0, y_0)$ and $f(t_1, y_1^*)$ are slopes of the indicated straight lines passing through the points (t_0, y_0) and (t_1, y_1^*), respectively. By taking an average of these slopes, we obtain the slope of the dashed skew lines. Rather than advancing along the line with slope $m = f(t_0, y_0)$ to the point with y-coordinate y_1^* obtained by the Euler method, we advance instead along the line through (t_0, y_0) with slope m_{ave} until we reach t_1. It seems plausible from inspection of Figure 2.41 that y_1 is an improvement over y_1^*.

We say that the value of y^*_{n+1} *predicts* a value of $y(t_n)$, whereas the formula in (5) *corrects* this estimate.

EXAMPLE 4 Improved Euler's Method

Use the improved Euler formula to obtain the approximate value of $y(1.5)$ for the solution of the initial-value problem in Example 2. Compare the results for $h = 0.1$ and $h = 0.05$.

Solution For $n = 0$ and $h = 0.1$, we first compute

$$y_1^* = y_0 + (0.1)(2t_0 y_0) = 1.2$$

Then from (5)

$$y_1 = y_0 + (0.1)\frac{2t_0 y_0 + 2t_1 y_1^*}{2} = 1 + (0.1)\frac{2(1)(1) + 2(1.1)(1.2)}{2} = 1.232$$

The comparative values of the calculations for $h = 0.1$ and $h = 0.5$ are given in Tables 2.5 and 2.6, respectively.

Table 2.5 Improved Euler's Method with $h = 0.1$

t_n	y_n	True Value	Absolute Error	% Relative Error
1.00	1.0000	1.0000	0.0000	0.00
1.10	1.2320	1.2337	0.0017	0.14
1.20	1.5479	1.5527	0.0048	0.31
1.30	1.9832	1.9937	0.0106	0.53
1.40	2.5908	2.6117	0.0209	0.80
1.50	3.4509	3.4904	0.0394	1.13

Table 2.6 Improved Euler's Method with $h = 0.05$

t_n	y_n	True Value	Absolute Error	% Relative Error
1.00	1.0000	1.0000	0.0000	0.00
1.05	1.1077	1.1079	0.0002	0.02
1.10	1.2332	1.2337	0.0004	0.04
1.15	1.3798	1.3806	0.0008	0.06
1.20	1.5514	1.5527	0.0013	0.08
1.25	1.7531	1.7551	0.0020	0.11
1.30	1.9909	1.9937	0.0029	0.14
1.35	2.2721	2.2762	0.0041	0.18
1.40	2.6060	2.6117	0.0057	0.22
1.45	3.0038	3.0117	0.0079	0.26
1.50	3.4795	3.4904	0.0108	0.31

■

A brief word of caution is in order here. We cannot compute all the values of y_n^* first and then substitute these values in the first formula of (5). In other words, we cannot use the data in Table 2.3 to help construct the values in Table 2.5. Why not?

Truncation Errors for the Improved Euler Method The local truncation error for the improved Euler method is $O(h^3)$. The derivation of this result is similar to the derivation of the local truncation error for Euler's method and is left for you. See Problem 16. Since the local truncation for the improved Euler method is $O(h^3)$, the global truncation error is $O(h^2)$. This can be seen in Example 4; when the step size is halved from $h = 0.1$ to $h = 0.05$, the absolute error at $t = 1.50$ is reduced from 0.0394 to 0.0108, a reduction of approximately $(\frac{1}{2})^2 = \frac{1}{4}$.

2.5.2 Runge-Kutta Methods

Probably one of the most popular as well as most accurate numerical procedures used in obtaining approximate solutions to the initial-value problem $y' = f(t, y)$, $y(t_0) = y_0$, is the **fourth-order Runge-Kutta method**. As the name suggests, there are Runge-Kutta methods of different orders. These methods are derived using the Taylor series expansion with remainder for $y(t_n + h)$:

$$y(t_{n+1}) = y(t_n + h) = y(t_n) + hy'(t_n) + \frac{h^2}{2!}y''(t_n) + \frac{h^3}{3!}y'''(t_n) + \cdots + \frac{h^{k+1}}{(k+1)!}y^{(k+1)}(c)$$

where c is a number between t_n and $t_n + h$. In the case when $k = 1$, and the remainder $\frac{h^2}{2} y''(c)$ is small, we obtain the familiar iteration formula

$$y_{n+1} = y_n + hy'_n = y_n + hf(t_n, y_n)$$

In other words, the basic Euler method is a **first-order Runge-Kutta** procedure.

We consider now the **second-order Runge-Kutta** procedure. This consists of finding constants a, b, α, and β so that the formula

$$y_{n+1} = y_n + ak_1 + bk_2 \tag{6}$$

where $\quad k_1 = hf(t_n, y_n) \quad$ and $\quad k_2 = hf(t_n + \alpha h, y_n + \beta k_1)$

agrees with a Taylor polynomial of degree 2. It can be shown that this can be done whenever the constants satisfy

$$a + b = 1, b\alpha = \frac{1}{2}, \quad \text{and} \quad b\beta = \frac{1}{2} \tag{7}$$

This is a system of three equations in four unknowns and has infinitely many solutions. Observe that when $a = b = \frac{1}{2}$ and $\alpha = \beta = 1$, (6) reduces to the improved Euler formula. Since the formula agrees with a Taylor polynomial of degree 2, the local truncation error for this method is $O(h^3)$ and the global truncation error is $O(h^2)$.

Notice that the sum $ak_1 + bk_2$ in (6) is a weighted average of k_1 and k_2 since $a + b = 1$. The numbers k_1 and k_2 are multiples of approximations to the slope of the solution curve $y(x)$ at two different points in the interval from t_n to t_{n+1}.

Fourth-Order Runge-Kutta Formula The **fourth-order Runge-Kutta** procedure consists of finding appropriate constants so that the formula

$$y_{n+1} = y_n + ak_1 + bk_2 + ck_3 + dk_4$$

where
$$k_1 = hf(t_n, y_n)$$
$$k_2 = hf(t_n + \alpha_1 h, y_n + \beta_1 k_1)$$
$$k_3 = hf(t_n + \alpha_2 h, y_n + \beta_2 k_1 + \beta_3 k_2)$$
$$k_4 = hf(t_n + \alpha_3 h, y_n + \beta_4 k_1 + \beta_5 k_2 + \beta_6 k_3)$$

agrees with a Taylor polynomial of degree 4. This results in 11 equations in 13 unknowns. The most commonly used set of values for the constants yields the following result:

$$\begin{aligned} y_{n+1} &= y_n + \tfrac{1}{6}(k_1 + 2k_2 + 2k_3 + k_4) \\ k_1 &= hf(t_n, y_n) \\ k_2 &= hf(t_n + \tfrac{1}{2}h, y_n + \tfrac{1}{2}k_1) \\ k_3 &= hf(t_n + \tfrac{1}{2}h, y_n + \tfrac{1}{2}k_2) \\ k_4 &= hf(t_n + h, y_n + k_3) \end{aligned} \tag{8}$$

You are advised to look carefully at the formulas in (8); note that k_2 depends on k_1, k_3 depends on k_2, and k_4 depends on k_3. Also k_2 and k_3 involve approximations to the slope at the midpoint of the interval between t_n and t_{n+1}.

EXAMPLE 5 Runge-Kutta Method

Use the Runge-Kutta method with $h = 0.1$ to approximate $y(1.5)$ for the solution of $y' = 2ty$, $y(1) = 1$.

Solution For the sake of illustration, let us compute the case when $n = 0$. From (8) we find

$$k_1 = (0.1)f(t_0, y_0) = (0.1)(2t_0y_0) = 0.2$$

$$k_2 = (0.1)f(t_0 + \tfrac{1}{2}(0.1), y_0 + \tfrac{1}{2}(0.2)) = (0.1)2(t_0 + \tfrac{1}{2}(0.1))(y_0 + \tfrac{1}{2}(0.2)) = 0.231$$

$$k_3 = (0.1)f(t_0 + \tfrac{1}{2}(0.1), y_0 + \tfrac{1}{2}(0.231)) = (0.1)2(t_0 + \tfrac{1}{2}(0.1))(y_0 + \tfrac{1}{2}(0.231)) = 0.234255$$

$$k_4 = (0.1)f(t_0 + 0.1, y_0 + 0.234255) = (0.1)2(t_0 + 0.1)(y_0 + 0.234255) = 0.2715361$$

$$y_1 = y_0 + \tfrac{1}{6}(k_1 + 2k_2 + 2k_3 + k_4) = 1 + \tfrac{1}{6}(0.2 + 2(0.231) + 2(0.234255) + 0.2715361) = 1.23367435$$

The remaining calculations are summarized in Table 2.7 whose entries are rounded to four decimal places.

Table 2.7 Runge-Kutta Method with $h = 0.1$

t_n	y_n	True Value	Absolute Error	% Relative Error
1.00	1.0000	1.0000	0.0000	0.00
1.10	1.2337	1.2337	0.0000	0.00
1.20	1.5527	1.5527	0.0000	0.00
1.30	1.9937	1.9937	0.0000	0.00
1.40	2.6116	2.6117	0.0001	0.00
1.50	3.4902	3.4904	0.0001	0.00

Inspection of Table 2.7 shows why the fourth-order Runge-Kutta method is so popular. If four-decimal-place accuracy is all that we desire, there is no need to use a smaller step size. Table 2.8 lists a comparison of the results of applying the Euler, improved Euler, and fourth-order Runge-Kutta methods to the initial-value problem $y' = 2ty$, $y(1) = 1$.

Truncation Errors for the Runge-Kutta Method Since the first equation in (3) agrees with a Taylor polynomial of degree 4, the local truncation error for this method is

$$y^{(5)}(c)\frac{h^5}{5!} \quad \text{or} \quad O(h^5)$$

Table 2.8 $y' = 2ty$, $y(1) = 1$

Comparison of Numerical Methods with $h = 0.1$

t_n	Euler	Improved Euler	Runge-Kutta	True Value
1.00	1.0000	1.0000	1.0000	1.0000
1.10	1.2000	1.2320	1.2337	1.2337
1.20	1.4640	1.5479	1.5527	1.5527
1.30	1.8154	1.9832	1.9937	1.9937
1.40	2.2874	2.5908	2.6116	2.6117
1.50	2.9278	3.4509	3.4902	3.4904

Comparison of Numerical Methods with $h = 0.05$

t_n	Euler	Improved Euler	Runge-Kutta	True Value
1.00	1.0000	1.0000	1.0000	1.0000
1.05	1.1000	1.1077	1.1079	1.1079
1.10	1.2155	1.2332	1.2337	1.2337
1.15	1.3492	1.3798	1.3806	1.3806
1.20	1.5044	1.5514	1.5527	1.5527
1.25	1.6849	1.7531	1.7551	1.7551
1.30	1.8955	1.9909	1.9937	1.9937
1.35	2.1419	2.2721	2.2762	2.2762
1.40	2.4311	2.6060	2.6117	2.6117
1.45	2.7714	3.0038	3.0117	3.0117
1.50	3.1733	3.4795	3.4903	3.4904

and the global truncation error is thus $O(h^4)$. It is now obvious why this is called the *fourth-order* Runge-Kutta method.

■ EXAMPLE 6 Bound for Local and Global Truncation Errors

Analyze the local and global truncation errors for the fourth-order Runge-Kutta method applied to $y' = 2ty$, $y(1) = 1$.

Solution By differentiating the known solution $y(x) = e^{t^2-1}$, we get

$$y^{(5)}(c)\frac{h^5}{5!} = (120c + 160c^3 + 32c^5)e^{c^2-1}\frac{h^5}{5!} \tag{9}$$

Thus with $c = 1.5$, (9) yields a bound of 0.00028 on the local truncation error for each of the five steps when $h = 0.1$. Note that in Table 2.7, the actual error in y_1 is much less than this bound.

Table 2.9 lists the approximations to the solution of the initial-value problem at $t = 1.5$, which are obtained from the fourth-order Runge-Kutta method. By computing the value of the exact solution at $t = 1.5$, we can find the error in these approximations. Because the method is so accurate, many decimal places must be used in the numerical solution to see the effect of halving the step size. Note that when h is halved, from $h = 0.1$ to $h = 0.05$, the error is divided by a factor of about $2^4 = 16$, as expected.

Table 2.9 Runge-Kutta Method

h	Approximation	Error
0.1	3.49021064	$1.323210889 \times 10^{-4}$
0.05	3.49033382	$9.137760898 \times 10^{-6}$

■

ODE Solvers Regardless of whether we can actually find an explicit or implicit solution, if a solution of a differential equation exists it represents a smooth curve in the Cartesian plane. The numerical methods in this section are just three of many different ways a solution of a differential equation can be approximated. The basic idea behind *any* numerical method is to somehow approximate the y-values of a solution for preselected values of t. We start at a specified initial point (t_0, y_0) on a solution curve and proceed to calculate a sequence of points $(t_1, y_1), (t_2, y_2), \ldots, (t_n, y_n)$ whose y-coordinates y_i approximate the y-coordinates $y(t_i)$ of points $(t_1, y(t_1))$, $(t_2, y(t_2)), \ldots, (t_n, y(t_n))$ that lie on the usually unknown solution curve. By taking the t-coordinates close together (that is, for small values of h), and by joining the points $(t_1, y_1), (t_2, y_2), \ldots, (t_n, y_n)$ with short line segments, we obtain a polygonal curve that appears smooth and whose qualitative characteristics we hope are close to those of an actual solution curve. Drawing curves is something well-suited to a computer. A routine written to implement a numerical method or to render a visual representation of an approximate solution curve fitting the numerical data produced by this method is sometimes referred to as an **ODE solver** or simply a **solver**. There are many different ODE solvers available, either embedded in a larger software package such as a CAS (*Mathematica*, *Maple*, DERIVE, *Macsyma*), or as a stand-alone package (*Differential Systems, DEQSolve, PhasePlane, Phaser*). Some software packages, such as the supplement to this text by Kiffe and Rundel, generate both numerical data as well as the corresponding approximate solution curves. By way of illustration of the connect-the-dots nature of an ODE solver, the colored polygonal graph in Figure 2.42 is the "solution curve" for the initial-value problem $y' = 0.2ty$, $y(0) = 1$ on the interval [0, 5] using the Euler method with a ridiculously large step size $h = 1$. The smooth black curve is a more believable "solution curve" obtained using the Runge-Kutta method with $h = 0.1$. With $h = 0.1$, it takes 50 steps to get from $t = 0$ to $t = 5$. Only every tenth value of y_n is printed out in Figure 2.42.

t_n	Euler $h = 1$	Runge-Kutta $h = 0.1$	Exact
0.00	1.0000	1.0000	0.9048
1.00	1.0000	1.1052	1.0000
2.00	1.2000	1.4918	1.3499
3.00	1.6800	2.4596	2.2255
4.00	2.6880	4.9530	4.4817
5.00	4.8384	12.1825	11.0232

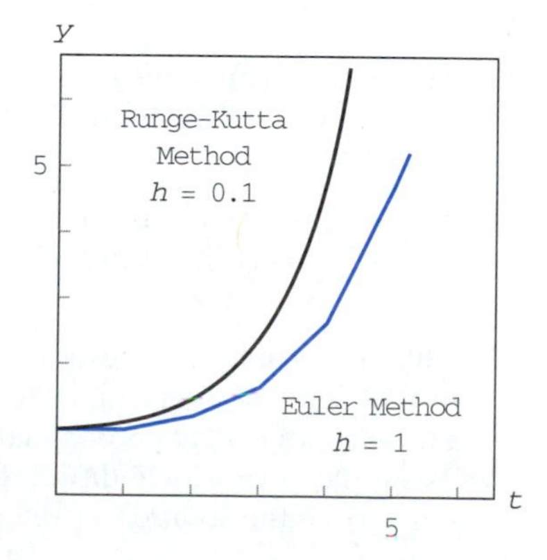

Figure 2.42.

EXERCISES 2.5 Answers to odd-numbered problems begin on page 437.

2.5.1

1. Consider the initial-value problem

$$y' = (t + y - 1)^2, \quad y(0) = 2$$

(a) Solve the initial-value problem in terms of elementary functions. [*Hint*: Let $u = t + y - 1$.]
(b) Use the Euler formula (2) with $h = 0.1$ and $h = 0.05$ to obtain approximate values of the solution of the initial-value problem at $t = 0.5$. Compare the approximate values with the exact values computed using the solution from part (a).

2. Repeat the calculations of Problem 1(b) using the improved Euler formula.

Given the initial-value problems in Problems 3–12, use the Euler formula (2) to obtain a four-decimal approximation to the indicated value. First use (a) $h = 0.1$ and then (b) $h = 0.05$.

3. $y' = 2t - 3y + 1$, $y(1) = 5$; $y(1.5)$

4. $y' = 4t - 2y$, $y(0) = 2$; $y(0.5)$

5. $y' = 1 + y^2$, $y(0) = 0$; $y(0.5)$

6. $y' = t^2 + y^2$, $y(0) = 1$; $y(0.5)$

7. $y' = e^{-y}$, $y(0) = 0$; $y(0.5)$

8. $y' = t + y^2$, $y(0) = 0$; $y(0.5)$

9. $y' = (t - y)^2$, $y(0) = 0.5$; $y(0.5)$

10. $y' = ty + \sqrt{y}$, $y(0) = 1$; $y(0.5)$

11. $y' = ty^2 - \frac{y}{t}$, $y(1) = 1$; $y(1.5)$

12. $y' = y - y^2$, $y(0) = 0.5$; $y(0.5)$

13. As parts (a)–(e) of this problem, repeat the calculations of Problems 3, 5, 7, 9, and 11 using the improved Euler formula (5).

14. As parts (a)–(e) of this problem, repeat the calculations of Problems 4, 6, 8, 10, and 12 using the improved Euler formula (5).

15. Although it may not be obvious from the differential equation, its solution could "behave badly" near a point t at which we wish to approximate $y(t)$. Numerical procedures may give widely differing results near this point. Let $y(t)$ be the solution of the initial-value problem

$$y' = t^2 + y^3, \quad y(1) = 1$$

(a) Use an ODE solver to obtain a graph of the solution on the interval [1, 1.4].
(b) Using the step size $h = 0.1$, compare the results obtained from the Euler formula (2) with the results from the improved Euler formula (5) in the approximation of $y(1.4)$.

16. In this problem we show that the local truncation error for the improved Euler method is $O(h^3)$.
(a) Use Taylor's formula with remainder to show that

$$y''(t_n) = \frac{y'(t_{n+1}) - y'(t_n)}{h} - \frac{1}{2}hy'''(c)$$

[*Hint*: Set $k = 2$ and differentiate the Taylor polynomial (3).]
(b) Use the Taylor polynomial with $k = 2$ to show that

$$y(t_{n+1}) = y(t_n) + h\frac{y'(t_n) + y'(t_{n+1})}{2} + O(h^3).$$

17. Consider the initial-value problem $y' = 2y$, $y(0) = 1$. The analytic solution is $y(t) = e^{2t}$.
(a) Approximate $y(0.1)$ using one step and the Euler method.
(b) Find a bound for the local truncation error in y_1.
(c) Compare the actual error in y_1 with your error bound.
(d) Approximate $y(0.1)$ using two steps and the Euler method.
(e) Verify that the global truncation error for Euler's method is $O(h)$ by comparing the errors in parts (a) and (d).

18. Repeat Problem 17 using the improved Euler method. Its global truncation error is $O(h^2)$.

19. Repeat Problem 17 using the initial-value problem $y' = -2y + t$, $y(0) = 1$. The analytic solution is $y(t) = \frac{1}{2}t - \frac{1}{4} + \frac{5}{4}e^{-2t}$.

20. Repeat Problem 19 using the improved Euler method. Its global truncation error is $O(h^2)$.

21. Consider the initial-value problem $y' = 2t - 3y + 1$, $y(1) = 5$. The analytic solution is $y(t) = \frac{1}{9} + \frac{2}{3}t + \frac{38}{9}e^{-3(t-1)}$.
(a) Find a formula involving c and h for the local truncation error in the nth step if the Euler method is used.
(b) Find a bound for the local truncation error in each step if $h = 0.1$ is used to approximate $y(1.5)$.

(c) Approximate $y(1.5)$ using $h = 0.1$ and $h = 0.05$ with the Euler method. See Problem 3.

(d) Calculate the errors in part (c) and verify that the global truncation error of the Euler method is $O(h)$.

22. Repeat Problem 21 using the improved Euler method, which has global truncation error $O(h^2)$. See Problem 13(a). You may need to keep more than four decimal places to see the effect of reducing the order of error.

23. Repeat Problem 21 for the initial-value problem $y' = e^{-y}$, $y(0) = 0$. The analytic solution is $y(t) = \ln(t + 1)$. Approximate $y(0.5)$. See Problem 7.

24. Repeat Problem 23 using the improved Euler method, which has global truncation error $O(h^2)$. See Problem 13(a). You may need to keep more than four decimal places to see the effect of reducing the order of error.

2.5.2

25. Use the fourth-order Runge-Kutta method with $h = 0.1$ to obtain a four-decimal-place approximation to the solution of the initial-value problem

$$y' = (t + y - 1)^2, \quad y(0) = 2$$

at $t = 0.5$. Compare the approximate values with the exact values obtained in Problem 1.

26. Solve the equations in (7) using the assumption $a = \frac{1}{4}$. Use the resulting second-order Runge-Kutta method to obtain a four-decimal-place approximation to the solution of the initial-value problem

$$y' = (t + y - 1)^2, \quad y(0) = 2$$

at $t = 0.5$. Compare the approximate values with the values obtained in Problem 2.

Given the initial-value problems in Problems 27–36, use the Runge-Kutta method with $h = 0.1$ to obtain a four-decimal-place approximation to the indicated value.

27. $y' = 2t - 3y + 1$, $y(1) = 5$; $y(1.5)$

28. $y' = 4t - 2y$, $y(0) = 2$; $y(0.5)$

29. $y' = 1 + y^2$, $y(0) = 0$; $y(0.5)$

30. $y' = t^2 + y^2$, $y(0) = 1$; $y(0.5)$

31. $y' = e^{-y}$, $y(0) = 0$; $y(0.5)$

32. $y' = t + y^2$, $y(0) = 0$; $y(0.5)$

33. $y' = (t - y)^2$, $y(0) = 0.5$; $y(0.5)$

34. $y' = ty + \sqrt{y}$, $y(0) = 1$; $y(0.5)$

35. $y' = ty^2 - \dfrac{y}{t}$, $y(1) = 1$; $y(1.5)$

36. $y' = y - y^2$, $y(0) = 0.5$; $y(0.5)$

37. If air resistance is proportional to the square of the instantaneous velocity, then the velocity v of a mass m dropped from a height h is determined from

$$m\frac{dv}{dt} = mg - kv^2, \quad k > 0$$

Let $v(0) = 0$, $k = 0.125$, $m = 5$ slugs, and $g = 32$ ft/s^2.

(a) Use the Runge-Kutta method with $h = 1$ to find an approximation to the velocity of the falling mass at $t = 5$ s.

(b) Use an ODE solver to graph the solution of the initial-value problem.

(c) Use separation of variables to solve the initial-value problem and find the true value $v(5)$.

38. A mathematical model for the area A (in cm^2) that a colony of bacteria (*B. dendroides*) occupies is given by

$$\frac{dA}{dt} = A(2.128 - 0.0432A)$$

Suppose that the initial area is 0.24 cm^2.

(a) Use the Runge-Kutta method with $h = 0.5$ to complete the following table.

t (days)	1	2	3	4	5
A (observed)	2.78	13.53	36.30	47.50	49.40
A (approximated)					

(b) Use an ODE solver to graph the solution of the initial-value problem. Estimate the values $A(1)$, $A(2)$, $A(3)$, $A(4)$, and $A(5)$ from the graph.

(c) Use separation of variables to solve the initial-value problem and compute the values $A(1)$, $A(2)$, $A(3)$, $A(4)$, and $A(5)$.

39. Consider the initial-value problem

$$y' = t^2 + y^3, \quad y(1) = 1$$

See Problem 15.

(a) Compare the results obtained from using the Runge-Kutta formula over the interval $[1, 1.4]$ with step sizes $h = 0.1$ and $h = 0.05$.

(b) Use an ODE solver to obtain a graph of the solution on the interval $[1, 1.4]$.

40. Consider the initial-value problem $y' = 2y$, $y(0) = 1$. The analytic solution is $y(t) = e^{2t}$.

(a) Approximate $y(0.1)$ using one step and the fourth-order Runge-Kutta method.

(b) Find a bound for the local truncation error in y_1.

(c) Compare the actual error in y_1 with your error bound.

(d) Approximate $y(0.1)$ using two steps and the fourth-order Runge-Kutta method.

(e) Verify that the global truncation error for the fourth-order Runge-Kutta method is $O(h^4)$ by comparing the errors in parts (a) and (d).

41. Repeat Problem 40 using the initial-value problem $y' = -2y + t$, $y(0) = 1$. The analytic solution is $y(t) = \frac{1}{2}t - \frac{1}{4} + \frac{5}{4}e^{-2t}$.

42. Consider the initial-value problem $y' = 2t - 3y + 1$, $y(1) = 5$. The analytic solution is $y(t) = \frac{1}{9} + \frac{2}{3}t + \frac{38}{9}e^{-3(t-1)}$.

(a) Find a formula involving c and h for the local truncation error in the nth step if the fourth-order Runge-Kutta method is used.

(b) Find a bound for the local truncation error in each step if $h = 0.1$ is used to approximate $y(1.5)$.

(c) Approximate $y(1.5)$ using the fourth-order Runge-Kutta method with $h = 0.1$ and $h = 0.05$. See Problem 27. You will need to carry more than six decimal places to see the effect of reducing the step size.

43. Repeat Problem 42 for the initial-value problem $y' = e^{-y}$, $y(0) = 0$. The analytic solution is $y(t) = \ln(t + 1)$. Approximate $y(0.5)$. See Problem 31.

Discussion Problems

44. A count of the number of evaluations of the function f used in solving the initial-value problem (1) is used as a measure of the computational complexity of a numerical method. Discuss: What are the number of evaluations of f required for each step of the Euler, improved Euler, and Runge-Kutta methods? By considering some specific examples, discuss the accuracy of these methods when used with comparable computational complexities.

45. In the examples in this section, we checked the accuracy of a numerical method by comparing the results obtained from it with the actual solution of the initial-value problem. You might wonder how we can verify the accuracy of a numerical technique when an exact solution cannot be found. To examine this question, we begin by considering a problem whose exact solution can be determined.

(a) Solve the initial-value problem $y' = 15y - 20t$, $y(0) = 1$.

(b) Use the fourth-order Runge-Kutta method to approximate the solution of the problem in part (a). Compare the results.

(c) If you did not know the solution of the problem in part (a), discuss how you might discover that the numerical method is not giving accurate results.

(d) Approximate the solutions of the following initial-value problems and discuss the accuracy of your solution:

$$y' = t^2 + y^3, \quad y(0) = 1 \quad \text{at} \quad t = 0.5;$$

$$y' = t + \sin y, \quad y(0) = 1 \quad \text{at} \quad t = 1$$

Chapter 2 in Review

Answers to this review are posted at our Web site diffeq.brookscole.com/books.html#zill 98

In Problems 1–4, fill in the blanks with the answers.

1. The equation __________ is an example of a DE that is both linear and separable.

2. The function $y = \phi(t)$, where $\phi(t) =$ __________, has the properties that its graph passes through (0, 1) and the slope of the tangent line at a point (t, y) on its graph is t/y.

3. The DE, $y' - ky = A$, where k and A are constants, is autonomous. The critical point __________ of the equation is a(n) __________ (attractor or repeller) for $k > 0$ and a(n) __________ (attractor or repeller) for $k < 0$.

4. The initial-value problem $t\dfrac{dy}{dt} - 4y = 0,\ y(0) = k$, has an infinite number of solutions for $k =$ ____________ and no solution for $k =$ ____________.

In Problems 5–8, examine each direction field and then match with one of the differential equations:

(a) $\dfrac{dy}{dt} = t - y$ **(b)** $\dfrac{dy}{dt} = t^2 - y^2$ **(c)** $\dfrac{dy}{dt} = \dfrac{t^2 - y^2}{t^2 + y^2}$ **(d)** $\dfrac{dy}{dt} = \dfrac{t^2}{y^2}$

Explain the reason for your choice in one sentence. On the appropriate direction field, sketch the solution curve for each differential equation that satisfies $y(-2) = 1$.

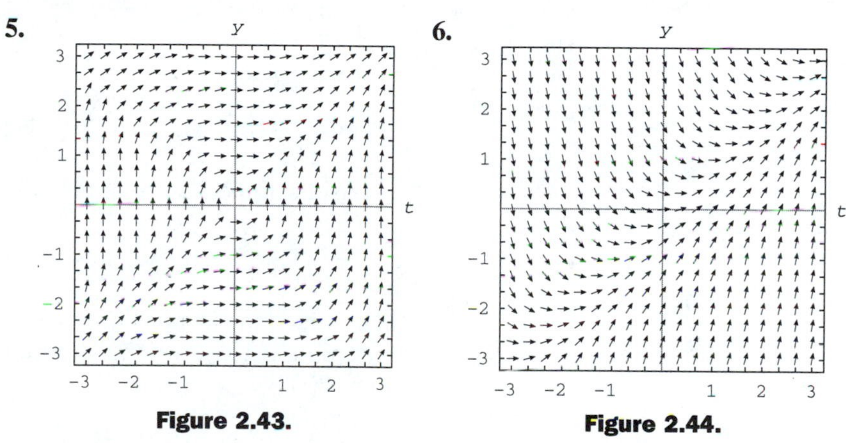

5. Figure 2.43. **6.** Figure 2.44.

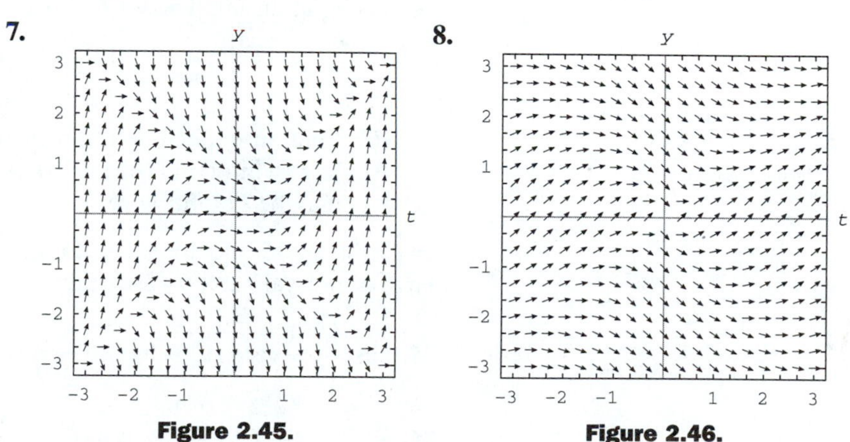

7. Figure 2.45. **8.** Figure 2.46.

9. Write a paragraph that explains and illustrates the basic differences in the direction fields for the differential equations:

(a) $\dfrac{dy}{dt} = f(t, y)$ **(b)** $\dfrac{dy}{dt} = f(y)$ **(c)** $\dfrac{dy}{dt} = f(t)$

10. The one-parameter family $y = c/t$ represents a solution of $dy/dt = f(t, y)$. Give a rough hand sketch of the direction field of the differential equation.

In Problems 11 and 12, construct an autonomous first-order differential equation $dx/dt = g(x)$ whose phase portrait is consistent with the given figure.

11. **12.**

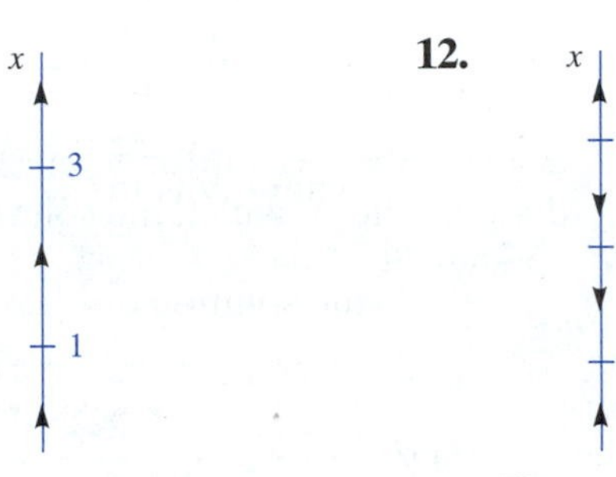

Figure 2.47. **Figure 2.48.**

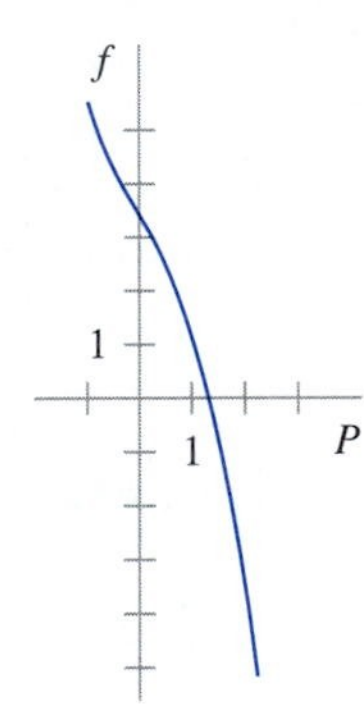

Figure 2.49.

13. Consider the differential equation

$$\frac{dP}{dt} = f(P), \quad \text{where} \quad f(P) = -0.5P^3 - 1.7P + 3.4$$

The function $f(P)$ has one real zero as shown in Figure 2.49. Without attempting to solve the differential equation, estimate the value of $\lim_{t \to \infty} P(t)$.

14. **(a)** Find an implicit solution of the initial-value problem

$$\frac{dy}{dt} = \frac{y^2 - t^2}{ty}, \quad y(1) = -\sqrt{2}$$

(b) Find an explicit solution of the problem in part (a) and give the largest interval I over which the solution exists. A graphic calculator may be helpful here.

In Problems 15 and 16, solve the given initial-value problem and give the largest interval I over which the solution exists.

15. $\sin t \dfrac{dy}{dt} + (\cos t)y = 0, \quad y\left(\dfrac{7\pi}{6}\right) = -2$

16. $\dfrac{dy}{dt} + 2(t + 1)y^2 = 0, \quad y(0) = -\dfrac{1}{8}$

17. In March 1976, the world population reached 4 billion. At that time, a popular news magazine predicted that with an average yearly growth rate of 1.8%, the world population would be 8 billion in 45 years. How does this value compare with that predicted by the model that says the rate of increase is proportional to the population at time t?

18. Air containing 0.06% carbon dioxide is pumped into a room whose volume is 8000 ft^3. The rate at which the air is pumped in is 2000 ft^3/min, and the circulated air is then pumped out at the same rate. If there is an initial concentration of 0.2% carbon dioxide, determine the subsequent amount in the room at time t. What is the concentration at 10 min? What is the steady-state concentration of carbon dioxide?

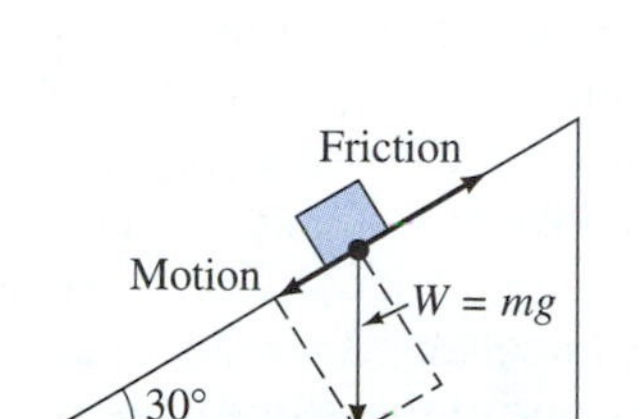

Figure 2.50.

19. A mass weighing 96 lb slides down an incline that makes a 30° angle with the horizontal. See Figure 2.50. If the coefficient of sliding friction is μ, determine a mathematical model for the velocity $v(t)$ of the mass at time t. Use the fact that the force of friction opposing the motion is μN, where N is the normal component of the weight.

20. Use Euler's method with step size $h = 0.1$ to approximate $y(1.2)$ where $y(t)$ is the solution of the initial-value problem $y' = 1 + t\sqrt{y}$, $y(1) = 9$.

Projects and Computer Lab Experiments for Chapter 2

Section 2.1

1. Direction Field

(a) Use a computer to reproduce the direction field given in Figure 2.3(a), but this time use a 7×7 rectangular grid with points (mh, nh), $h = \frac{1}{2}$, $-14 \le m \le 14$, $-14 \le n \le 14$.

(b) On the direction field, sketch by hand a solution curve that corresponds to each initial condition: $y(0) = 4$; $y(-2) = 2$; $y(0) = -3$; $y(-4) = -4$.

(c) Based on the direction field and the solution curves, form a conjecture about the behavior of each of the four solutions $y(t)$ in part (b) as $t \to \pm\infty$.

2. Direction Field

(a) Use a computer to obtain the direction field for $y' = e^{-t} - 3y$ using a 3×3 rectangular grid with points (mh, nh), $h = 0.25$, $-12 \le m \le 12$, $-12 \le n \le 12$.

(b) On the direction field, sketch by hand a solution curve that corresponds to each initial condition: $y(0) = 1$; $y(-2) = 0$; $y(-1) = -2$.

(c) Based on the direction field and the solution curves, form a conjecture about the behavior of *all* solutions $y(t)$ as $t \to \pm\infty$.

3. Equilibrium Solutions

(a) Use a CAS or a graphing utility to find all equilibrium solutions of the given autonomous differential equation.

(*i*) $\dfrac{dx}{dt} = 2x - e^{-x}$ (*ii*) $\dfrac{dx}{dt} = x^4 + 3x^2 + x - 7$

(*iii*) $\dfrac{dx}{dt} = x^2 - x - \cos x$ (*iv*) $\dfrac{dx}{dt} = e^{-x^2} - x^4$

(b) Find all equilibrium solutions of $dx/dt = \sin x$. Use an ODE solver to obtain some representative solution curves. Explain why none of the solution curves are oscillatory, even though $g(x) = \sin x$ is a periodic function of x.

4. **Isoclines and Nullclines** It is sometimes useful to identify the points in a direction field where the lineal elements have the same slope. For the DE $dy/dt = f(t, y)$, any member of the one-parameter family $f(t, y) = c$ is called an **isocline**. For a given value of c, lineal elements passing through points on this curve do so with the same slope.
   (a) On the direction field for part (a) of Problem 2, superimpose the graphs of the isoclines corresponding to $c = -3$, $c = 3$, $c = -6$, and $c = 6$. Use a colored pencil to circle the lineal elements that appear to be bisected by each curve. What is the slope of these lineal elements?
   (b) An isocline corresponding to $c = 0$ is called a **nullcline**. Superimpose the nullcline over the direction field in part (a) and circle the lineal elements that appear to be bisected by this curve. What is the slope of these elements?
   (c) Use a computer to reproduce a direction field for $dy/dt = (y + t)/(y - t)$. Identify the isoclines and nullcline for this DE and superimpose their graphs on the direction field.
   (d) Describe, and then illustrate by example, the isoclines and nullclines for an autonomous differential equation.

5. **Polynomial Approximation** This problem illustrates how solutions of differential equations can be approximated using polynomials.

   The simple nonlinear first-order initial-value problem $y' = t^2 + y^2$, $y(0) = 1$ has no elementary solution. However, let us assume that $y(t)$ can be expanded in a Taylor series centered at the initial point $t = 0$:

$$y(t) = y(0) + \frac{y'(0)}{1!}t + \frac{y''(0)}{2!}t^2 + \frac{y'''(0)}{3!}t^3 + \frac{y^{(4)}(0)}{4!}t^4 + \frac{y^{(5)}(0)}{5!}t^5 + \cdots$$

   Taylor polynomials of degree 0 and 1 are, in turn, $P_0(t) = y(0)$, $P_1(t) = y(0) + \dfrac{y'(0)}{1!}t$. From the initial condition, $P_0(t) = y(0) = 1$; from the differential equation, $y'(0) = 0^2 + y(0)^2 = 1$ so that $P_1(t) = 1 + t$. Taylor polynomials of higher degrees can be obtained by successively differentiating the differential equation, and then using previously calculated values of the lower-order derivatives of y.
   (a) Compute the Taylor polynomials $P_2(t), P_3(t), \ldots, P_7(t)$.
   (b) Use an ODE solver to graph the solution curve for the given initial-value problem.
   (c) Use a graphing utility and graph the solution curve and $P_1(t)$ on the same set of coordinate axes. Repeat for the solution curve and $P_2(t)$, and so on, until you have seven sets of graphs.
   (d) One pitfall of this method is that it may be impossible to obtain the interval of convergence for the Taylor series centered at the initial point. Use the graphs to conjecture a possible interval of convergence.

Section 2.2

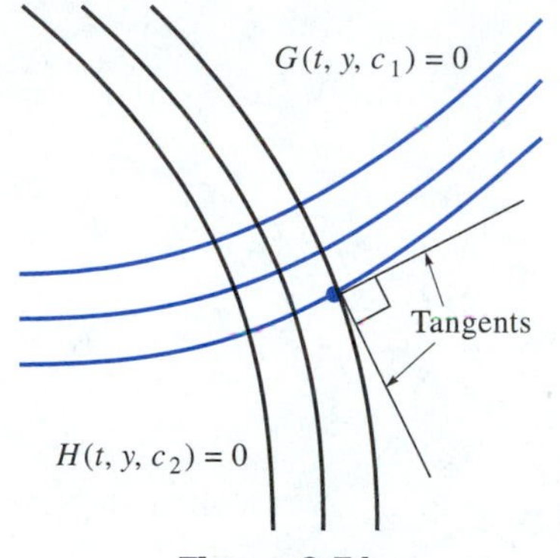

Figure 2.51.

6. **Orthogonal Trajectories** When all curves in a family $G(t, y, c_1) = 0$ intersect orthogonally all curves in another family $H(t, y, c_2) = 0$, the families are said to be **orthogonal trajectories** of each other. See Figure 2.51. If $dy/dt = f(t, y)$ is the differential equation of one family, then the differential equation for the orthogonal trajectories of this family is $dy/dt = -1/f(t, y)$.
   - **(a)** Find the orthogonal trajectories for the family $y = 1/(t + c_1)$. Use a graphing utility to graph both families on the same axes.
   - **(b)** Find the orthogonal trajectories for the family $t^2 - ty + y^2 = c_1$. Use a CAS to plot some of the level curves for both families on the same axes for $t > 0$, $y > 0$.
   - **(c)** Find several examples where orthogonal trajectories occur in the sciences.

7. **Level Curves**
   - **(a)** Find a relation $G(t, y) = c$ that is an implicit solution of the differential equation $(y^3 + y - 2)y' = 4 - t$. Use a CAS to plot some of the level curves defined by this solution for $-15 \leq t \leq 15$ and $-5 \leq y \leq 5$.
   - **(b)** Find an implicit solution of the equation whose graph passes through $(-1, 0)$. Repeat for the graph that passes through $(2, 2)$. Use a CAS to plot these level curves. Use a colored pencil to trace out a *function* $y_1(t)$ that satisfies the differential equation, the defining relation, and the condition $y(-1) = 0$. Repeat for a function $y_2(t)$ and the condition $y(2) = 2$.
   - **(c)** Use the differential equation in the form $dy/dt = (4 - t)/(y^3 + y - 2)$ and the notion of slope of a tangent line to a graph as a means of finding intervals on which the functions $y_1(t)$ and $y_2(t)$ are defined. [*Hint*: See Problem 35 in Exercises 2.2.]
   - **(d)** Use a CAS or a graphic calculator as an aid in approximating the values of $y_1(4)$ and $y_2(4)$.

8. **Solution Curves** Consider the nonlinear first-order differential equation

$$\left(\frac{dy}{dt}\right)^4 - 8\left(\frac{dy}{dt}\right)^2 + 16 - y^4 = 0$$

   - **(a)** Use an ODE solver to discover the number of different sets of solutions and illustrate representative solution curves in each set.
   - **(b)** Use separation of variables to find exact solutions of the differential equations that define each set of solutions.

9. **Sawing a Log** In cutting with a saw through an object such as a round log, if the friction between the sides of the saw blade and the wood through which the blade has already cut is ignored, then it can be assumed that the rate at which the saw blade moves through the log

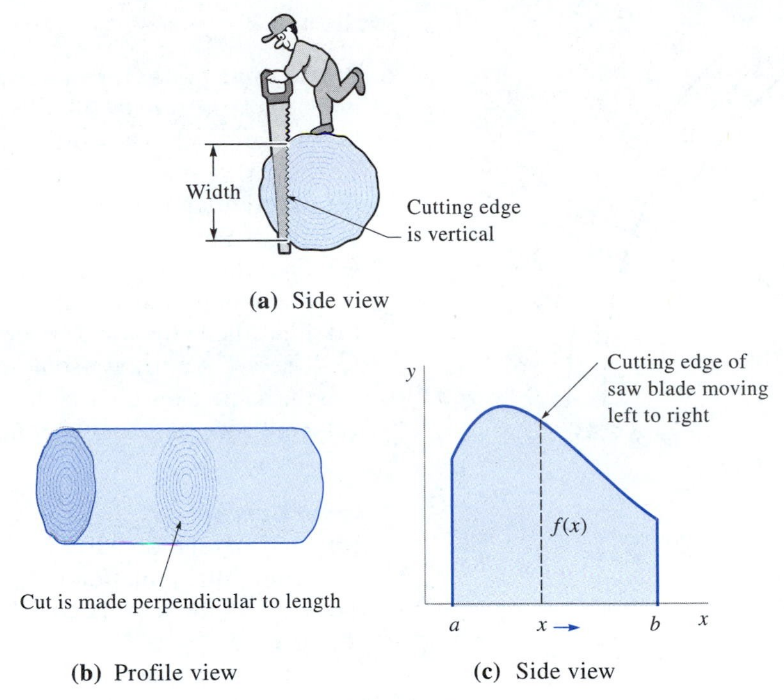

Figure 2.52.

is inversely proportional to the width of the wood in contact with its cutting edge. See Figure 2.52(a). Suppose now that a long uniform (cross sections are the same) piece of wood is cut perpendicular to its length by a vertical saw blade. Let the region bounded by the graph of a continuous positive function f and the x-axis on the interval $a \leq x \leq b$ represent a cross section of the wood. The vertical cut made by the blade can then be represented by a vertical line at x and its width is the height $f(x)$. See Figure 2.52(c). Thus the rate dx/dt at which the saw blade moves to the right is related to $f(x)$ by

$$f(x)\frac{dx}{dt} = K$$

where K represents a rate, namely, the amount of material (measured in square units) removed by the saw blade per unit time. Now imagine that a computerized saw can be programmed so that K is constant. This can be done by specifying the position x of the saw blade as a function of time t. The position $x(t)$ is obtained by solving the DE for a given $f(x)$.

In the parts (a)–(c), suppose that the amount of wood removed by the saw is $K = 1$ square unit per unit time. In each part, determine an appropriate initial condition $x(0) = x_0$ based on the fact that the saw blade enters the region from the left. Then solve the differential equation and state the interval over which the solution is defined.

(a) The cross section of the piece of wood cut by the saw is a triangular region bounded by the graphs of $y = x$, $x = 1$, and the x-axis.

(b) The cross section is a region congruent to the region in part (a), but oriented so that the endpoints of its hypotenuse are located at the points $(0, 0)$ and $(\sqrt{2}, 0)$. [*Hint*: The solution $x(t)$ will be piecewise defined.]

(c) The cross section is a region bounded by a circle of radius 2 centered at $(0, 0)$. Use a CAS if necessary to perform the integration.

(d) In part (c), it is impossible to obtain an explicit solution $x(t)$. Solve the implicit solution for time t as a function of x. Use a graphing utility and graph this function $t(x)$. From your graph, determine the intervals over which x and t are defined.

(e) Use a CAS to compute the derivative $t'(x)$ and verify that it is positive over the open x-interval obtained in part (d) and is zero at the endpoints. What does this fact tell us about the rate of change of the inverse function $x(t)$ with respect to t at the endpoints of the t-interval obtained in part (d)? What is the physical interpretation of this in terms of the motion of the saw blade?

(f) To obtain a discrete sample of the inverse function $x(t)$, use a CAS and your explicit formula for $t(x)$ to construct a table of ordered pairs (t, x) as x runs over the x-interval obtained in part (d) in steps of 0.1.

(g) To check the validity of the implicit solution $x(t)$, we can use the data points in part (f) and a CAS to compute the average rate at which material is being removed at different times. This can be done by computing the area under the graph of $y = f(x) = 2\sqrt{4 - x^2}$ from x_0 to $x_0 + 0.1$ and dividing by the time duration $t(x_0 + 0.1) - t(x_0)$ for various values of x_0. What value should we expect for these average rates of material removal?

Section 2.3

10. Graph of a Family of Solutions Proceed as in Example 6 in Section 2.3 and express the general solution of the linear equation in terms of an integral. Use a CAS to graph some members of the family of solutions.

(a) $\dfrac{dy}{dt} + 2ty = 1$ **(b)** $\dfrac{dy}{dt} + y = \sin t^2$

Section 2.4 Linear Models

11. A Population Model The U.S. census population figures (in millions) for the years 1790, 1800, 1810, . . . , 1950 are given in the table on the left.

Year	Population
1790	3.929
1800	5.308
1810	7.240
1820	9.638
1830	12.866
1840	17.069
1850	23.192
1860	31.433
1870	38.558
1880	50.156
1890	62.948
1900	75.996
1910	91.972
1920	105.711
1930	122.775
1940	131.669
1950	150.697

(a) Use these data to construct a population model of the form

$$\frac{dP}{dt} = kP, \quad P(0) = P_0$$

(b) Construct a table comparing the population predicted by the model in part (a) with the actual census population. Compute the error and percentage error for each entry pair.

(c) What does the model predict for the population in 1960? in 1970? Look up the actual population for each of those years and compare it with the predicted population. Compute the percentage error in each case.

Year	Area Occupied (km^2)
1939	32,800
1944	55,800
1949	73,600
1954	138,000
1959	202,000
1964	257,000
1969	301,000
1974	584,000

Cumulative geographical range of *Bufo marinus* in Queensland, Australia

12. Invasion of the Toads* In 1935, the poisonous American marine toad (*Bufo marinus*) was introduced, against the advice of ecologists, into some of the coastal sugar cane districts in Queensland, Australia, as a means of controlling sugar cane beetles. Due to lack of natural predators and the existence of an abundant food supply, the toad population grew and spread into regions far from the original districts. The survey data given in the accompanying table indicate how the toads expanded their territorial bounds within a 40-year period. As in Problem 11, our goal is to find a population model of the form $dP/dt = kP$, $P(0) = P_0$, but this time we want to construct the model that *best* fits the given data. Note that the data are not given as *number of toads* at 5-year intervals since, for the toads in question, this kind of information would be virtually impossible to obtain.

(a) Discuss:

- What assumptions could one make in order to convert the given data into population data? How realistic are these assumptions?
- Suppose you are writing a grant proposal to study the population of this toad. You want to ask for money to hire two biologists to gather additional data for three months. What additional data would you want to obtain that would give you additional insight into the total toad population?

(b) For ease of computation, we will assume that, on the average, there is one toad per square kilometer. We will also count the toads in units of thousands and measure time in years with $t = 0$ corresponding to 1939. One way to model the data in the table is to use the initial condition $P_0 = 32.8$ and to search for a value of k so that the graph of $P_0 e^{kt}$ appears to fit the data points. Experiment, using a graphic calculator or a CAS, by varying the birth-rate k until the graph of $P_0 e^{kt}$ appears to fit the data well over the time period $0 \leq t \leq 35$.

Alternatively, it is also possible to solve analytically for a value of k that will guarantee that the curve passes through exactly two of the data points. Find a value of k so that $P(5) = 55.8$. Find a different value of k so that $P(35) = 584$.

(c) In practice, a mathematical model rarely passes through every experimentally obtained data point, and so statistical methods must be used to find values of the model's parameters that best fit experi-

*This problem is based on the article "Teaching Differential Equations with a Dynamical Systems Viewpoint" by Paul Blanchard, *The College Mathematics Journal* 25 (1994) 385–395.

mental data. Specifically, we will use **linear regression** to find a value of k that describes the given data points:

- Use the table to obtain a new data set of the form $(t_i, \ln P(t_i))$, where $P(t_i)$ is the given population at $t_1 = 0$, $t_2 = 5, \ldots$.
- Most graphic calculators have a built-in routine to find the line of least squares that fits this data. The routine gives an equation of the form $\ln P(t) = mt + b$, where m and b are the slope and intercept corresponding to the line of best fit. (Most calculators also give the value of the correlation coefficient that indicates how well the data is approximated by a line; a correlation coefficient of 1 or -1 means perfect correlation. A correlation coefficient near 0 may mean that the data do not appear to be fit by an exponential model.)
- Solving $\ln P(t) = mt + b$ gives $P(t) = e^{mt+b}$ or $P(t) = e^b e^{mt}$. Matching the last form with $P_0 e^{kt}$, we see that e^b is an approximate initial population, and m is the value of the birth rate that best fits the given data.

(d) So far you have produced four different values of the birth rate. Do your four values of k agree closely with each other? Should they? Which of the four values do you think is the best model for the growth of the toad population during the years for which we have data? Use this birth rate to predict the toad's range in the year 2039. Given that the area of Australia is 7,619,000 km^2, how confident are you of this prediction? Explain your reasoning.

(e) In part (b) of this problem, we made the assumption that there was an average of one toad per square kilometer. But suppose we are wrong and there were actually an average of two toads per square kilometer. As before, solve analytically for a value of k that will guarantee that the curve passes through exactly two of the data points. In particular, if we now assume that $P(0) = 65.6$, find a value of k so that $P(5) = 111.6$, and a different value of k so that $P(35) = 1168$. How do these values of k compare with the values you found previously? What does this tell us? Discuss the importance of knowing the exact average density of the toad population.

13. Now They Darken, Now They Don't We've all seen them: eyeglasses that are clear while worn indoors, but darken when worn outside on a bright day. Those of us who wear these glasses know that it takes a while for the glass to *activate* (darken) when exposed to sunlight and to *recover* (lighten or bleach) when moving indoors. In this problem, we model the darkening and bleaching of photochromatic glass as a function of time t.

Although there are several different varieties of photochromatic optical glass, we will concentrate on Photogray™ lenses, which are made of silver halide glass. In the presence of light, incident photons provide the energy needed to separate chemically a silver halide ($AgCl$)

crystal. This chemical reaction forms the basis of photography: the silver forms the photographic image, whereas the halide diffuses away in the surrounding emulsion. Suppose, however, that silver halide is combined with molten glass and allowed to solidify. Exposure to light still causes the halide to separate, and the glass to darken, but now the halide cannot diffuse away, so when the light source is removed, the halide recombines with the silver, and the glass reverts toward its initial state.

Because it is hard to quantify "lightness" and "darkness," let's formulate the problem in terms of the amount of light that the glass transmits to the eye: darkened glass transmits less light than bleached glass. Experiments have been conducted to measure this transmission so we will be able to compare our model to actual data.*

Let M be the spectral transmittance (in percent) of the glass. The transmittance measures the fraction of incident light energy that is transmitted through the glass; $M = 0$ means that no light penetrates the material, but M nearly equal to 1 means that most light penetrates the material without being absorbed. A typical value for clear glass is $M \approx 0.8$, whereas sunglasses are typically between 20% and 50% spectral transmittance. Let us assume that a pair of eyeglasses has a spectral transmittance of M_b when it is completely bleached, and transmittance of $M_d < M_b$ when completely darkened. Let us also assume that the darkening occurs in light of intensity I.

t	$M(t)$	t	$M(t)$
0	0.70	65	0.575
5	0.56	70	0.62
10	0.51	75	0.64
15	0.485	80	0.665
20	0.475	90	0.69
30	0.465	100	0.715
40	0.46	110	0.735
50	0.455	120	0.755
60	0.45		

Transmittance as a function of time t (in minutes) for a piece of Photogray glass. The glass was exposed to sunlight for 60 min at 25°C, then brought indoors for 60 min.

(a) Graph the data in the accompanying table.

(b) The data in the table, along with other experimental data, suggest that a simple model for the darkening of the glass might be

$$\frac{dM}{dt} = k_d I(M_d - M) \tag{1}$$

whereas the bleaching process may be modeled by

$$\frac{dM}{dt} = k_b(M_b - M) \tag{2}$$

(c) Physically, equation (1) says that the rate of change of the transmittance is proportional to the intensity of the incident light and the difference between the final (darkened) transmittance and the transmittance at time t. The constant of proportionality k_d must be determined experimentally for each type of glass and for each temperature. According to equation (1), under what conditions will photochromatic glass darken fastest:
- A sunny day (intense light) or a cloudy day (less intensity)?
- When bleached ($M \approx M_b$) or nearly darkened ($M \approx M_d$)?

(d) Why doesn't light intensity appear in equation (2)?

(e) Based on the table given in part (a), determine reasonable values of M_d and M_b for the sample of photochromatic glass used in the experiment. Explain how you arrived at your conclusions.

*See G. P. Smith, *J. Photo. Sci.* 18 (1970) and J. Harris and V. Zanetti, *Phys. Teacher* 28 (1990).

(f) The first half of the data in the table in part (a) was gathered in sunlight. Assume that the average intensity of sunlight was 200 W/m^2. Estimate a value of k_d by doing one of the following:

- If you did Problem 12 on fitting an exponential model to experimental data, estimate the value of k_d by using the procedure described in that problem. How well is this data modeled by an exponential function? [*Hint*: Define $P = M - M_d$ and then show that $dP/dt = -k_d IP$.]
- If you did not do Problem 12, but have access to computational software to solve DEs numerically, then vary the numerical values of k_d until you have produced a solution of equation (1) that satisfies the initial condition $M(0) = 0.7$ and that seems to fit the data.
- If you do not have access to technology, find an explicit solution of equation (1). Then find a value of k_d so that $M(0) = 0.7$ and $M(20) = 0.475$.

(g) We now determine possible values of the parameter k_b. Assume $M_b = 0.78$.

- If you have access to computational software to solve DEs numerically, then vary the numerical values of k_d until you have produced a solution of equation (2) that seems to fit the data for $60 \leq t \leq 120$ and satisfies $M(60) = 0.45$.
- If you do not have access to technology, show that an explicit solution of equation (2) is $M(t) = M_b + (M_{60} - M_b)e^{-k_b(t-60)}$ where $M(60) = M_{60}$. With $M_{60} = 0.45$, use a graphing utility to plot this solution on the interval $60 \leq t \leq 120$ until you find a value of k_b that fits the data.
- In either case, discuss how well these solutions model the data in the table in part (a). Are they better or worse than the data for $0 \leq t \leq 60$?

(h) A moviegoer wearing photosensitive glasses (initial transmittance $M(0) = 0.6$) stands outside a theater on a sunny day ($I =$ 200 W/m^2) for 10 min. She then enters the dark theater, watches a movie for 2 hr, and walks home in the twilight of the approaching evening ($I = 150$ W/m^2).

- Without doing any computations, make a qualitative graph of the transmittance of her glasses as a function of time t. Mark the beginning and the end of the movie on your graph.
- If you have access to computational utilities, use them and your answers to part (f) to help you generate the graph of the transmittance as a function of time that is predicted by equations (1) and (2). What is the transmittance of her glasses when the woman finally returns home?

14. Time of Death Suppose a medical examiner, after arriving at the scene of a homicide, found that the temperature of the body was 82°F. Make up additional, but plausible, data necessary to determine an approximate time of death of the victim using Newton's law of cooling (3) of Section 2.4.

15. **The Hotter Cup of Coffee** Mr. Jones puts two cups of coffee on the breakfast table at the same time. Mr. Jones immediately pours cream into his coffee from a pitcher that was sitting on the table for a long time. He then reads the morning paper for 5 min before taking his first sip. Mrs. Jones arrives at the table 5 min after the cups were set down, adds cream to her coffee, and takes a sip. Determine who drinks the hotter cup of coffee. Assume that both Mr. and Mrs. Jones add exactly the same amount of cream to their cups.

16. **My Tank Runneth Over** In Example 5 of Section 2.5, the size of the tank containing the salt mixture was not given. Suppose, as discussed on page 65, that the rate at which brine is pumped into the tank is the same but that the mixed solution is pumped out a rate of 2 gal/min. It stands to reason, since brine is accumulating in the tank at the rate of 1 gal/min, that any finite tank must eventually overflow. Now suppose that the tank has an open top and has a total capacity of 400 gal.
    (a) When will the tank overflow?
    (b) What is the number of pounds of salt in the tank at the instant it overflows?
    (c) Assume that the tank is overflowing and that brine solution continues to be pumped in at a rate of 3 gal/min and the well-stirred solution continues to be pumped out at a rate of 2 gal/min. Devise a method for determining the number of pounds of salt in the tank at $t = 150$ min.
    (d) Determine the number of pounds of salt in the tank as $t \to \infty$. Does your answer agree with your intuition?
    (e) Use a graphing utility to obtain the graph $A(t)$ on the interval $[0, \infty)$.

17. **Variable Ambient Temperature** In Problem 41 of Exercises 2.4, you were asked to find the function $T_m(t)$ that describes the variable periodic temperature given in Figure 2.38. The resulting linear differential equation can be solved by hand, but it is tedious to do so.
    (a) If directed by your instructor, use a CAS to solve the initial-value problem $dT/dt = k(T - T_m(t))$, $T(0) = 120$.
    (b) Use a root-finding application in a CAS or a graphic calculator to find k if it is known that $T(6) = 80$.
    (c) Use a graphing utility to graph the solution $T(t)$. Does the graph agree with your conjecture about the long-range behavior of $T(t)$ in part (b) of Problem 41 of Exercises 2.4? Try to explain any differences.

18. **What Goes Up Must Come Down**
    (a) A projectile, let us suppose that it is a small cannonball as shown in Figure 2.53, is shot vertically upward from ground level. Assume that the weight of the cannonball is 16 lb. If we ignore air resistance,

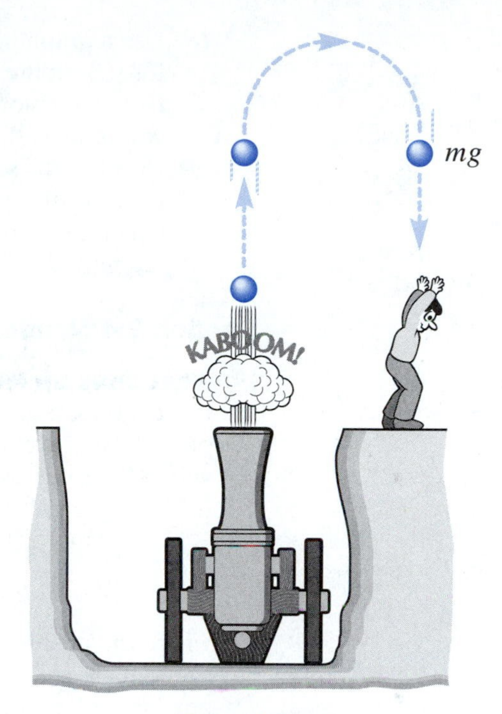

Figure 2.53.

a model of the motion of the cannonball is given by (3) on page xii,

$$\frac{d^2s}{dt^2} = -g, \quad s(0) = 0, \quad s'(0) = v_0$$

where we have assumed that the positive direction is upward, $s(t)$ is the height of the cannonball above ground level, $g = 32$ ft/s^2, and v_0 is its initial velocity. Assume for this entire problem that $v_0 = 300$ ft/s. Solve for $s(t)$ and use this solution to find the velocity $v(t)$. Analytically, determine the maximum height $s_{\max_1}$ attained by the cannonball, show that the time t_a of ascent to $s_{\max_1}$ is the same as the time t_d of descent from a height of $s_{\max_1}$ to the ground, and, finally, show that the impact velocity is simply the negative of the initial velocity.

(b) Now let us suppose that the force of air resistance is proportional to the instantaneous velocity. The model for the velocity of the cannonball in part (a) is given by

$$m\frac{dv}{dt} = -mg - kv, \quad v(0) = v_0$$

Suppose that $k = 0.0025$. Solve for $v(t)$ and use this solution to find the position $s(t)$. See Problem 26 in Exercises 2.4.

(c) Use a graphing utility to obtain the graphs of the two expressions for $s(t)$ on the same coordinate axes. Repeat for the graphs of the two expressions for $v(t)$. Write down all comparisons between the two models that you can draw from these graphs.

(d) Now use the solutions $v(t)$ and $s(t)$ in part (b) to determine *analytically* the time t_a of ascent to the maximum height $s_{\max_2}$, the maximum height $s_{\max_2}$, the time t_d of descent from $s_{\max_2}$ to the ground, and the impact velocity. Does $t_a = t_d$ as in part (a)?

Section 2.4 Nonlinear Models

19. What Goes Up Must Come Down—Another Model This problem is a continuation of Problem 18, but the assumption is that the force of air resistance is proportional to the square of the instantaneous velocity. In this case, the model for the velocity of the cannonball as it is rising is given by

$$m\frac{dv}{dt} = -mg - kv^2, \quad v(0) = v_0$$

where the positive direction is upward. To describe the velocity of the cannonball as it falls back to the ground from its maximum height, we now assume that the positive direction is downward

$$m\frac{dv}{dt} = mg - kv^2, \quad v(0) = 0$$

With $g = 32$ ft/s^2, $v_0 = 300$ ft/s, $k = 0.0003$, solve for $v(t)$ and use this solution to find the position $s(t)$. Use the solutions $v(t)$ and $s(t)$ of the first model to determine the time t_a of ascent to the maximum height $s_{\max_3}$ and the maximum height $s_{\max_3}$. Use the solutions $v(t)$ and $s(t)$ of the second model to determine the time t_d of descent from $s_{\max_3}$ to the ground, and the impact velocity. Fill in the following table comparing the results of Problem 18 and the current results.

Air Resistance	t_a	s_{max}	t_d	Impact Velocity
None				
$0.0025\,v$				
$0.0003\,v^2$				

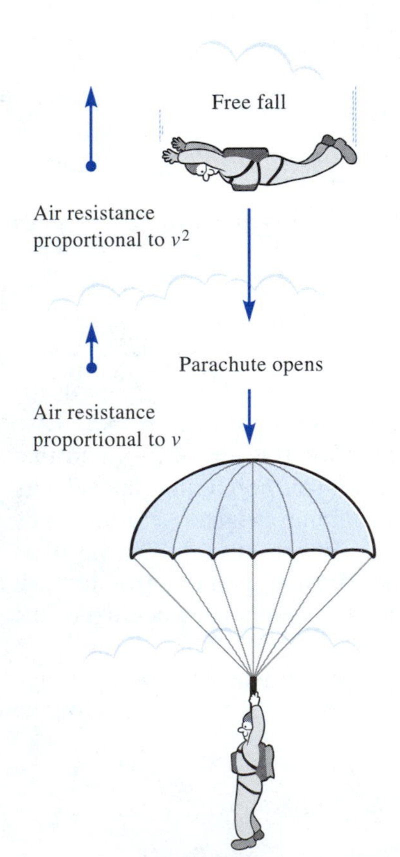

Figure 2.54.

20. Skydiving A skydiver weighing a total of 160 lb steps out the door of an airplane that is flying at an altitude of 12,000 ft. After freely falling for 15 s, as illustrated in Figure 2.54, the parachute is opened. Assume that air resistance is proportional to v^2 while the parachute is unopened and is proportional to the velocity v after the parachute is opened. For a person of this weight, typical values for k in the models (12) and (21) of Section 2.4 are, respectively, $k = 7.857$ and $k = 0.0053$. Determine the time that it takes the skydiver to reach the ground. What is the skydiver's impact velocity?

21. **An Absent-Minded Skydiver** It has happened that a skydiver has stepped out of an airplane without a parachute. Imagine yourself in such a predicament starting at an altitude of 12,000 feet. You can take some consolation in the fact that because of air resistance your velocity will not exceed a limiting value v_t, called (unfortunately) the **terminal velocity**. If air resistance is proportional to v^2, then $v_t = \sqrt{mg/k}$. See Problem 23 in Exercises 2.1 and Problem 35 in Exercises 2.4.

(a) Having nothing better to do on the way down, you decide to compute v_t but, alas, you do not know the empirical constant k. Try this: Use $v_t = \sqrt{mg/k}$ to eliminate k from equation (21) of Section 2.4 and solve the resulting DE subject to $v(0) = 0$. [*Hint*: You will save valuable time if you employ inverse hyperbolic functions.]

(b) What is your distance $s(t)$ from the point you exited the plane? See Figure 2.55.

Figure 2.55.

(c) As you pass a mountain top, you see that its altitude is exactly 10,678 ft and you look at your watch and notice that exactly 10 s have elapsed since exiting the plane. You then reach into your briefcase and withdraw a calculator. Use the root-finding application of your calculator with the foregoing data to approximate v_t (in mi/hr). How do you feel about this number?

(d) How many *more* seconds do you have before you . . . before you reach the ground?

22. A Population Model

(a) Use the census data from 1790, 1850, and 1910 from the table in Problem 11 to construct a population model of the form

$$\frac{dP}{dt} = P(a - bP), \quad P(0) = P_0$$

(b) Construct a table comparing the population predicted by the model in part (a) with the actual population. Compute the error and percentage error for each entry pair.

(c) What does the model predict for the population in 1960? in 1970? Compare the actual population for each of these years with the predicted population. Compute the percentage error in each case. Compare these results with part (c) of Problem 11.

23. Pursuit Curve In a naval exercise, a destroyer S_1 pursues a submerged submarine S_2. Suppose that S_1 at (9, 0) on the x-axis detects S_2 at (0, 0) and that S_2 simultaneously detects S_1. The captain of the destroyer S_1 assumes that the submarine will take immediate evasive action and conjectures its likely new course is the straight line L, indicated in Figure 2.56. When S_1 is at (3, 0), the destroyer changes from its straight-line course toward the origin to a pursuit curve C. Assume that the speed of the destroyer is, at all times, a constant 30 mi/hr and the submarine's speed is a constant 15 mi/hr.

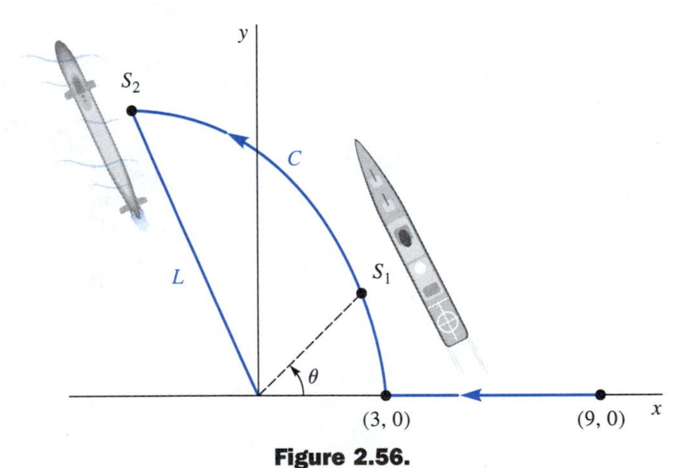

Figure 2.56.

(a) Explain why the captain waits until S_1 reaches (3, 0) before ordering a course change to C.

(b) Using polar coordinates, find an equation $r = f(\theta)$ of the curve C.

(c) Explain why the time, measured from the initial detection, at which the destroyer intercepts the submarine must be less than $(1 + e^{2\pi/\sqrt{3}})/5$.

24. Bifurcation The theory of differential equations with parameters, such as the first-order equation $dx/dt = g(\alpha, x)$, is important in applied problems because the qualitative nature of solutions *may* change drastically for even the slightest change in the parameter α. As the parameter varies, a differential equation may go from having no critical points to having one critical point, then to having several critical points with differing stability characteristics. The splitting of a critical point into two or more critical points is called a **bifurcation**. The word *bifurcation* means literally "to divide or split into (two) parts or branches." The value of α at which such a splitting takes place is called a **bifurcation value**.

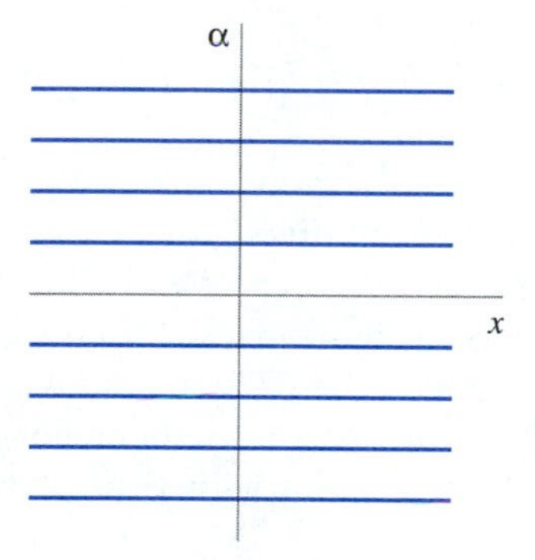

Figure 2.57.

(a) Consider the simple differential equation

$$\frac{dx}{dt} = -x^2 + \alpha$$

where α represents a real number. An analysis of the equation can be done by means of a bifurcation diagram. As shown in Figure 2.57, suppose we draw a set of coordinate axes in the $x\alpha$-plane with the usual sign conventions: positive x-values to the right, positive α-values upward, the intersection of the two axes corresponding to $x = 0$, $\alpha = 0$, and so on. The x-axis is a phase line and its equation is $\alpha = 0$. The other horizontal lines in the figure are additional phase lines with equations given by α = constant. By considering all possibilities for α, $-\infty < \alpha < \infty$, indicate on the horizontal lines with the usual arrows where $g(\alpha, x) = -x^2 + \alpha$ is positive and negative, mark the position of critical points by dots, and classify these points as asymptotically stable or unstable. Is there a bifurcation value? Do you see anything that resembles a "splitting?" Suppose you connect the dots that represent the critical points by a continuous curve. What is the equation of this curve?

(b) Construct a bifurcation diagram for the differential equation,

$$\frac{dx}{dt} = x(\alpha - x^2)$$

and determine the bifurcation value. Use two curves to connect appropriate critical points and give their equations. Discuss why this bifurcation is called a **pitchfork bifurcation**.

25. Harvesting If a constant number of animals h is removed or harvested per unit time, then the population $P(t)$ of animals is modeled by the initial-value problem

$$\frac{dP}{dt} = P(a - bP) - h, \quad P(0) = P_0$$

where a, b, h, and P_0 are positive constants.

(a) Consider the case where $P(t)$ represents a fish population in a specific fishery (P measured in units of millions, time t measured in tens of years), $a = 5$, $b = 1$, and $h = 3.5$. Find the critical points of the differential equation; denote these points by P_1 and P_2, where $P_1 \leq P_2$. Use a phase line to predict the long-term status of the population. Compare this with the long-term status of the population when there is no harvesting ($h = 0$).

(b) Find an explicit solution of

$$\frac{dP}{dt} = P(5 - P) - 3.5, \quad P(0) = P_0$$

If $P_0 = 1$, what is the population in the fishery at $t = 1$ (10 yr)?

(c) Use a graphing utility to obtain representative graphs of $P(t)$ in part (b) in each of the cases $P_0 > P_2$, $P_1 < P_0 < P_2$, and $0 < P_0 < P_1$. Use the same set of coordinate axes. Discuss the effect of P_0 on the fish population in each case. Which case would you label "overfishing?" Verify that the long-term behavior of the population in each of the cases is consistent with your phase-line analysis.

(d) If the population becomes extinct in finite time, find that time in terms of P_0. Calculate the extinction times for specific values of P_0.

(e) Investigate (phase line, graphs, and so on) the effect on the fish population $P(t)$ if the harvesting is proportional to the population. Use $a = 5$, $b = 1$, and $h = 0.3P$.

(f) Consider the differential equation in part (b), with the number 3.5 replaced by a parameter h:

$$\frac{dP}{dt} = P(5 - P) - h$$

Allow h to take on all real values. Read or work Problem 24 and then construct a bifurcation diagram in the Ph-plane for this equation. See Figure 2.57. Use the bifurcation diagram as an aid in answering the following question: Does there exist a harvest rate h_{max}, beyond which the fish population will become extinct for *every* initial population P_0?

26. Water Clock The *clepsydra*, or water clock, was a device used by the ancient Egyptians, Greeks, Romans, and Chinese to measure the passage of an interval of time by observing the change in the height of water that was permitted to flow out a small hole in the bottom of a container.

(a) Suppose the container is made of glass and has the shape of a right-circular cylinder. In Problem 38 in Exercises 2.4 we saw that the height h of water that is draining through a hole in a tank in the form of a vertical right-circular cylinder is given by the differential equation

$$\frac{dh}{dt} = -k\sqrt{h}$$

where $k = cA_h\sqrt{2g}/A_w$, A_w and A_h are the cross-sectional areas of the water and the hole, respectively, c is a friction/contraction factor at the hole, and g is the acceleration due to gravity. Find an explicit solution of this equation that satisfies $h(0) = 2$ ft.

(b) Suppose that $h(0) = 2$ ft corresponds to water filled to the top of the glass tank, the radius of the tank is 1 ft, the hole in the bottom is circular with radius $\frac{1}{32}$ in., and that $c = 0.6$. How far up from the bottom of the tank should a mark be made on its side, as shown in Figure 2.58(a), that corresponds to the passage of 1 hr? Continue and determine where to place the marks corresponding to the passage of 2 hr, 3 hr, ..., 12 hr.

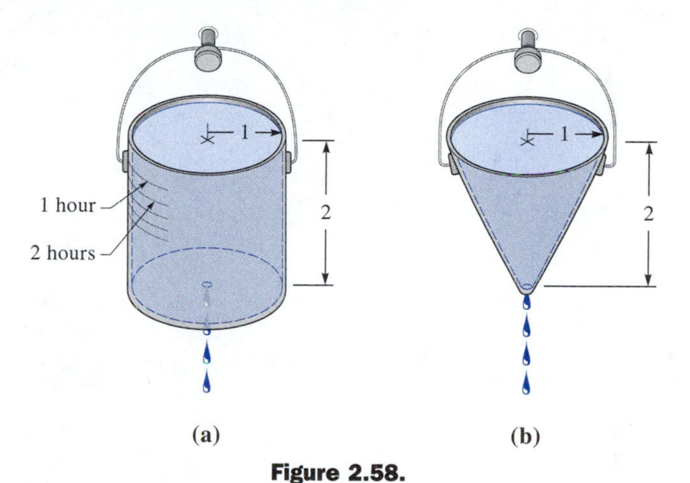

Figure 2.58.

(c) The marks determined in part (b) are not evenly spaced. Why?

(d) The differential equation in part (a) remains valid even when the surface area A_w is not constant. In this case, we express A_w as a function of h, that is, $A_w = A(h)$. Suppose that the tank is conical with circular cross sections as shown in Figure 2.58(b), and suppose again that $h(0) = 2$ ft corresponds to water filled to the top of the tank, that the hole in the bottom is circular with radius $\frac{1}{32}$ in., and that $c = 0.6$. Can this type of water clock measure 12 time intervals of length equal to 1 hr? Explain using sound mathematics.

(e) Suppose that $r = f(h)$ defines the shape of a water clock for which the time marks are equally spaced. Use the differential equation in part (a) to find $f(h)$ and sketch a typical graph of h as a function of r. Assume A_h is constant. [*Hint*: In this situation, dh/dt must be a negative constant. Why?]

27. Solar Furnace In the design of a solar furnace—or for that matter the mirror of a reflecting telescope—it is necessary that a reflecting surface collect all incoming light rays at one point. Suppose, as illustrated in Figure 2.59, that light rays strike a plane curve C in such a manner that all rays L parallel to the x-axis are reflected to a single point O. Assuming that the angle of incidence is equal to the angle of reflection, show that the differential equation that describes the shape

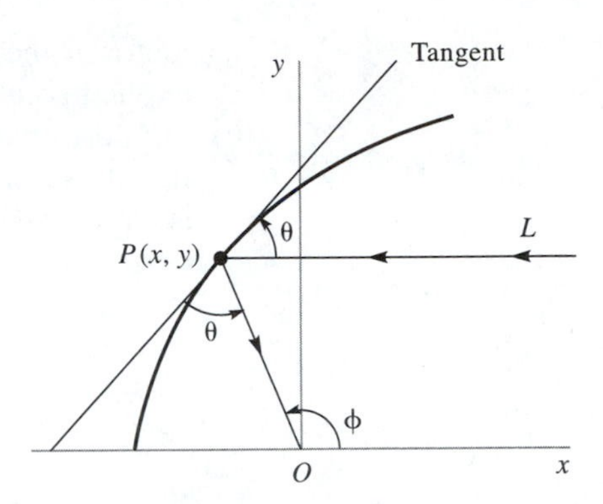

Figure 2.59.

of the curve C is

$$\frac{dy}{dx} = \frac{-x + \sqrt{x^2 + y^2}}{y}$$

[*Hint*: Trigonometry and the fact that $\tan \theta = dy/dx$ will help.]
There are several ways of solving this equation.

(a) First, verify that the differential equation is homogeneous (see Section 2.2). Show that the substitution $y = ux$ yields

$$\frac{u du}{\sqrt{1 + u^2}(1 - \sqrt{1 + u^2})} = \frac{dx}{x}$$

Use a CAS (or another judicious substitution) to integrate the left side of the equation. Identify the curve C.

(b) Now show that the first differential equation can also be written as $y = 2xy' + y(y')^2$. Let $w = y^2$ and then look up Clairaut's differential equation and find out how to solve it. (See Problem 10, *Projects and Computer Lab Experiments for Chapter 1*.) Reconcile any difference between this answer and that obtained in part (a).

(c) Finally, show that the first differential equation can also be solved by means of the substitution $u = x^2 + y^2$.

Section 2.5

28. Euler's Method Use an ODE solver that can implement Euler's method to graph the solution curve of the given initial-value problem with the specified step sizes.

(a) $y' = \cos t$, $0 \leq t \leq 10$; $y(0) = -4$ $(h = 2)$; $y(0) = -2$ $(h = 1)$; $y(0) = 0$ $(h = 0.5)$; $y(0) = 2$ $(h = 0.1)$; $y(0) = 4$ (h is screen resolution, if possible)

(b) $y' + y = 0$, $0 \leq t \leq 5$; $y(0) = -4$ $(h = 2.5)$; $y(0) = -2$ $(h = 1)$; $y(0) = 0$ $(h = 0.5)$; $y(0) = 2$ $(h = 0.1)$; $y(0) = 4$ (h is screen resolution, if possible)

29. Difference Quotients We can obtain variations of Euler's method by replacing the first derivative y' with special difference quotients. Throughout this problem, we consider the initial-value problem $y' = y$, $y(0) = 1$.

(a) Use the **forward-difference quotient**

$$\frac{y(t+h) - y(t)}{h}$$

$h > 0$, to replace y' in the given IVP to show that

$$y(t + h) = (1 + h)y(t)$$

Note that this is the basic Euler's formula. Solve the problem on $[0, 1]$, using $h = \frac{1}{8}$.

(b) Use the **backward-difference quotient**

$$\frac{y(t) - y(t-h)}{h}$$

$h > 0$, to derive

$$y(t+h) = \frac{1}{1-h}y(t)$$

Use this formula to solve the problem on $[0, 1]$ using $h = \frac{1}{8}$.

(c) Find the exact solution of the given IVP. How do the magnitudes of the errors using the methods in parts (a) and (b) compare? How do the signs of the errors compare?

(d) The **central-difference quotient** is the average of the forward- and backward-difference formulas. Average the formulas derived in parts (a) and (b) to show that

$$y(t+h) = \frac{2-h^2}{2(1-h)}y(t)$$

Use this formula to solve the problem on $[0, 1]$ using $h = \frac{1}{8}$.

(e) Create a table comparing the numerical values obtained from the formulas in parts (a), (b), and (d). Which formula gives the best results? Why?

30. Stiff DEs There is a class of problems for which numerical techniques can be unreliable. A thorough study of this class of problems is complex, so our purpose here is to give some feeling for the difficulties that can arise when approximating the solution of certain initial-value problems. Differential equations whose solution contains a term of the form e^{-kt}, where k is a fairly large positive constant, together with a term that remains relatively constant or decays much more slowly, occur frequently in applications. DEs of this sort are called **stiff**. When the standard numerical methods are applied to stiff DEs, very small step sizes are required to retain reasonable accuracy, but this in turn could cause problems with round-off error.

(a) Solve the initial-value problem $y' + 120y = 0$, $y(0) = 1$ by hand. Determine the value of $y(1)$. Use the fourth-order Runge-Kutta method to approximate the solution on the interval [0, 1]. Determine a step size that yields reasonable results.

(b) Use the fourth-order Runge-Kutta method to approximate the solution of $y' = -30(y - t) + 2$, $y(0) = 4$ on the interval [0, 3]. Determine a step size that yields reasonable results. How can you check that your results are, in fact, reasonable?

(c) Examine the effects of plotting the solution of the IVP in part (b) using the fourth-order Runge-Kutta method with step sizes ranging from $h = 0.1$ to $h = 0.01$. Compare the graphs with the results obtained in part (b).

31. An Immigration Model During the nineteenth century, many midwestern cities grew at phenomenal rates because of an influx of immigrants attracted by an abundance of land, innovations in construction, and improved means of travel. The autonomous differential equation

$$\frac{dP}{dt} = P(1 - P) + \alpha e^{-\beta P}, \quad \alpha > 0, \beta > 0$$

is a model of the population of a city that increases because of immigration. Here P is measured in millions and t in years.

(a) Suppose $\alpha = 0.2$ and $\beta = 0.5$. Use a CAS or a calculator with a root-finding function to estimate, to three decimal places, the critical points of the given equation.

(b) Use a phase line to determine whether these critical points are asymptotically stable or unstable. If possible, use this information to predict the population of the city over a longer period of time.

(c) Suppose $P(0) = 0.1$. Use an ODE solver that can implement the Runge-Kutta method to construct a table of population values over a 10-year period. Use $h = 0.1$, but only print out every tenth step, that is, at $t = 1, t = 2$, and so on. Using the logistic equation ($\alpha = 0$), repeat the calculation over the same period.

(d) Use an ODE solver to obtain the graphs of these two solutions in part (c) on the same set of coordinate axes.

(e) Use the graphs in part (d) to estimate, for each model, the time at which the rate of change of the population is the greatest.

(f) Using the numerical data and the graphs, do either of the times estimated in part (e) correspond to the time at which the difference in populations predicted by the two models is the greatest?

(g) At $t = 10$, compute the percentage increase in the population predicted by the immigration model as compared with that population predicted by the logistic model.

32. The Discrete Logistic Equation

(a) Show that Euler's method applied to the logistic differential equation $P' = P(a - bP)$ yields the **discrete logistic equation**

$$P_{n+1} = P_n(\alpha - \beta P_n), \quad n = 0, 1, 2, \ldots,$$

where $\alpha = 1 + ah$ and $\beta = bh$.

(b) The equation in part (a) is called a difference equation and defines a sequence recursively. For $n > 0$, the terms in the sequence P_1, P_2, P_3, ... are approximations to a population $P(t)$ predicted by the differential equation at times $t_1 = h$, $t_2 = 2h$, $t_3 = 3h$, and so on. Use a CAS, a programmable calculator, or write a short computer program for the iteration of the difference equation. Use $a = 0.2$, $b = 0.025$, $h = 1$, and $n = 20$ to construct a table of iterates corresponding to each of the following initial values: $P_0 = -0.2$, $P_0 = 1$, $P_0 = 5$, and $P_0 = 12$.

(c) Plot, using a CAS or by hand, the points (n, P_n) obtained in each table in part (b). Use the same set of coordinate axes and connect the points with straight lines forming four graphs. Compare these four graphs with the solution curves of the logistic differential equation $P' = P(a - bP)$ with $a = 0.2$, $b = 0.025$, subject to $P(0) = P_0$. Use the values of P_0 given in part (b).

33. Period Doubling and Chaos With a and b fixed, the behavior of the P_n defined by the difference equation in part (a) of Problem 32 is sensitive to slight variations of the parameter h. This time use $a = 1$, $b = 1$, $P_0 = 0.1$, and $n = 200$.

(a) Show that for $0 < h \leq 1.9$, the terms of the sequence appear to converge to a single value for the population. Is it the same number for each h? Experiment with values $1.9 < h \leq 2$ and report what you observe. Use values of h having two or three decimal places.

(b) Show that at $h = 2.1$, the terms of the sequence tend toward two numbers. What are these numbers (approximately)? Because the terms oscillate between these two numbers as $n \to \infty$, they are called a **population cycle of period 2**.

(c) Show that at $h = 2.5$ there is a population cycle of period 4, in other words, the terms of the sequence oscillate between four numbers as $n \to \infty$. What are these numbers (approximately)?

(d) Do the results in parts (a)–(c) depend on the initial value P_0? Experiment with values $0 < P_0 < 1$.

(e) Determine whether there are any population cycles of period greater than 4 when h satisfies $2.5 < h < 2.6$. You will need to experiment with values of h having three (or more) decimal places as well as substantially increasing the size of n (for example, $n = 1000$, 1200, and so on).

(f) Show that for $2.6 \leq h \leq 2.8$, the terms of the sequence exhibit no distinguishable population cycles.

The behavior illustrated in parts (b)–(e) is called **period doubling**. As the parameter h is increased, an instability sets in. If we think in terms of graphing P_n as a function of h, the sequence bifurcates between a single population as $n \to \infty$, to two populations, four populations, and so on, and eventually into the phenomenon in part (f), which is called **chaos**.

(g) Experiment with values $2.8 < h \leq 3.0$ and report what you observe.

Writing Assignments

34. Exact Equations Go to the library and find an older text on differential equations. Write a short report on **exact** first-order differential equations. Include a discussion on the concept of **integrating factors** and their relationship with exact equations. Illustrate these concepts by solving the following problem.

A uniform 10-foot-long rope is coiled loosely on the ground. One end of the rope is then pulled vertically by means of constant upward force of 5 lb. The rope weighs 1 lb/ft. Let x denote the height of the end of the rope in the air at time t, and $v = dx/dt$. Show that x and v are related by the differential equation $(v^2 + 32x - 160)\,dx + xv\,dv = 0$. Intuitively, how much rope can the force lift? Solve the DE and then find x as a function of time t. Use $x(t)$ to determine how much of the rope actually leaves the ground. Try to explain any differences between the last answer and your intuition.

Since the mass of rope in the air is variable, Newton's second law states that the rate of change of momentum equals the net force F, that is, $F = \frac{d}{dt}[mv]$. Also use $\frac{dv}{dt} = \frac{dv}{dx}\frac{dx}{dt} = v\frac{dv}{dx}$.

35. Ricatti's Equation Write a short report on the solution of **Ricatti's differential equation**. Show how Ricatti's equation is related to Bernoulli's equation and a linear first-order equation. Illustrate the solution of a specific Ricatti equation.

36. Diets Who of us has not at one or more times resolved to stay on a diet? Write a report on the article: "A Linear Diet Model," by Arthur C. Segal, *The College Mathematics Journal*, January (1987) 44–45. Elaborate on the author's explanation of "why so many dieters give up in frustration."

37. Art Forgeries Go to the library and find the text: *Differential Equations and Their Applications*, by M. Braun, Springer-Verlag, New York. Prepare a report, written or oral, on *The van Meegeren Art Forgeries*.

38. Snowplows The "snowplow problem" is a classic and appears in many differential equations texts.

One day it started snowing at a heavy and steady rate. A snowplow started out at noon, going 2 miles the first hour and 1 mile the second hour. What time did it start snowing?

It is a good first problem for you to study; it illustrates the part that differential equations play in the modeling of certain physical situations. The differential equation that arises is first-order and separable. Write a report on the content and solution of the "snowplow problem." Try to find one of the original texts that made the problem famous: *Differential Equations*, by Ralph Palmer Agnew, McGraw-Hill, New York, 1960. For those interested in a numerical solution, see "A Computational Solution to the Snowplow Problem," by J. L. Lewis, *ACM SIGCSE Bulletin* 18 (1986) 9–12.

39. More Snow The snowplow problem stated in Problem 38 was first posed by J. A. Benner, Problem E 275, *American Mathematical Monthly*, 44 (1937) 245. A variation of the problem, involving three snowplows, was posed by M. S. Klamkin, Problem E 963, *American Mathematical Monthly*, 58 (1951) 260. Write a report on the solution of the latter problem by first reading "The Meeting of the Plows: A Simulation" by Jerome L. Lewis, *The College Mathematics Journal*, 26 (1995) 395–400.

40. Time of Death Again In case you had difficulty starting Problem 14 in this section, you might start by reading a *very* short (four paragraphs) and amusing article on criminology: "Elementary (My Dear Watson) Differential Equation," by Brian J. Winkel, *Mathematics Magazine*, 52 (1979) 315. Write a conclusion to the article by filling in just what Sherlock Holmes tells Dr. Watson. See also: "The Homicide Problem Revisited," by David A. Smith, *The Two Year College Mathematics Journal*, 9 (1978) 141–145, and "An Application of Newton's Law of Cooling," by J. F. Hurley, *Mathematics Teacher*, 67 (1974) 141–142.

41. Brachistochrone Problem Write a short report on the history and solution of the brachistochrone problem introduced on page xiii. A good starting place is the text *Differential Equations with Applications and Historical Notes*, by George F. Simmons, McGraw-Hill, New York.

42. Multistep Method The Euler and Runge-Kutta methods discussed in Section 2.5 are examples of **single-step methods** in that each successive value y_{n+1} is computed based only on information about the immediately preceding value y_n. A **multistep method**, on the other hand, uses the values from several previously computed steps to obtain the value y_{n+1}. Write a report on the **Adams-Bashforth/Adams-Moulton method**. Be sure to include the advantages and disadvantages of multistep methods in your report.

43. Adaptive Method Write a report on an **adaptive numerical method** such as the **Runge-Kutta-Fehlberg method** for approximating the solution of a first-order initial-value problem. Discuss the formulas used and the method by which the step size is determined at each step. Contrast the Runge-Kutta-Fehlberg method with the Runge-Kutta methods of orders four and five in terms of accuracy and number of function evaluations required.

44. Period Doubling Again

(a) Do a little library research to find a biological interpretation of period doubling considered in Problem 33. You might start with texts on mathematical modeling in biology.

(b) Investigate part (g) of Problem 33. Start with an article by Li and Yorke, *American Mathematical Monthly*, 82 (1975) 985–992.

Chapter 3

Higher-Order Equations

3.1 Linear Equations

We turn now to the solution of DEs of order two or higher. In this and subsequent sections, we examine some of the basic theory and methods for solving *linear* equations. As we did in Section 2.3, we set forth conditions on the differential equation under which we can obtain its general solution, that is, a family of solutions that contains all the solutions of the equation on some interval.

3.1.1 Initial-Value and Boundary-Value Problems

Initial-Value Problem In Section 1.2, we defined an initial-value problem (IVP) for a general nth-order differential equation. For a linear equation, an nth-order initial-value problem is

$$\begin{aligned} &\textit{Solve:} && a_n(t)\frac{d^n y}{dt^n} + a_{n-1}(t)\frac{d^{n-1}y}{dt^{n-1}} + \cdots + a_1(t)\frac{dy}{dt} + a_0(t)y = g(t) \\ &\textit{Subject to:} && y(t_0) = y_0, \quad y'(t_0) = y_1, \quad \ldots, \quad y^{(n-1)}(t_0) = y_{n-1} \end{aligned} \tag{1}$$

For a problem such as this, we seek a function defined on some interval I containing t_0 that satisfies the differential equation and the n initial conditions that are specified at t_0.

Existence and Uniqueness In Section 1.2, we stated a theorem that gave conditions under which the existence and uniqueness of a solution of a first-order initial-value problem were guaranteed. The theorem that follows gives sufficient conditions for the existence of a unique solution of the problem in (1).

Theorem 3.1 Existence of a Unique Solution

Let $a_n(t), a_{n-1}(t), \ldots, a_1(t), a_0(t)$, and $g(t)$ be continuous on an interval I and let $a_n(t) \neq 0$ for every t in this interval. If $t = t_0$ is any point in this interval, then a solution of the initial-value problem (1) exists on the interval and is unique.

■ **EXAMPLE 1 Unique Solution of an IVP**

Observe that the initial-value problem

$$3y''' + 5y'' - y' + 7y = 0, \qquad y(1) = 0, \quad y'(1) = 0, \quad y''(1) = 0$$

possesses the trivial solution $y = 0$. Since the third-order equation is linear with constant coefficients, it follows that all the conditions of Theorem 3.1

are fulfilled. Hence $y = 0$ is the *only* solution on any interval containing $x = 1$. ■

■ EXAMPLE 2 **Unique Solution of an IVP**

You should verify that the function $y = 3e^{2t} + e^{-2t} - 3t$ is a solution of the initial-value problem

$$y'' - 4y = 12t, \qquad y(0) = 4, \quad y'(0) = 1$$

Now the differential equation is linear, the coefficients as well as $g(t) = 12t$ are continuous, and $a_2(t) = 1 \neq 0$ on any interval containing $t = 0$. We conclude from Theorem 3.1 that the given function is the unique solution. ■

The requirements in Theorem 3.1 that $a_i(t)$, $i = 0, 1, 2, \ldots, n$ be continuous and $a_n(t) \neq 0$ for every t in I are both important. Specifically, if $a_n(t) = 0$ for some t in the interval, then the solution of a linear initial-value problem may not be unique or even exist. For example, you should verify that the function $y = ct^2 + t + 3$ is a solution of the initial-value problem

$$t^2y'' - 2ty' + 2y = 6, \qquad y(0) = 3, \quad y'(0) = 1$$

on the interval $(-\infty, \infty)$ for any choice of the parameter c. In other words, there is no unique solution of the problem. Although most of the conditions of Theorem 3.1 are satisfied, the obvious difficulties are that $a_2(t) = t^2$ is zero at $t = 0$ and that the initial conditions are also imposed at $t = 0$.

Boundary-Value Problem Another type of problem consists of solving a differential equation of order two or greater in which the dependent variable y or its derivatives are specified at *different points.* A problem such as

$$\textit{Solve:} \quad a_2(t)\frac{d^2y}{dt^2} + a_1(t)\frac{dy}{dt} + a_0(t)y = g(t)$$

$$\textit{Subject to:} \quad y(a) = y_0, \quad y(b) = y_1$$

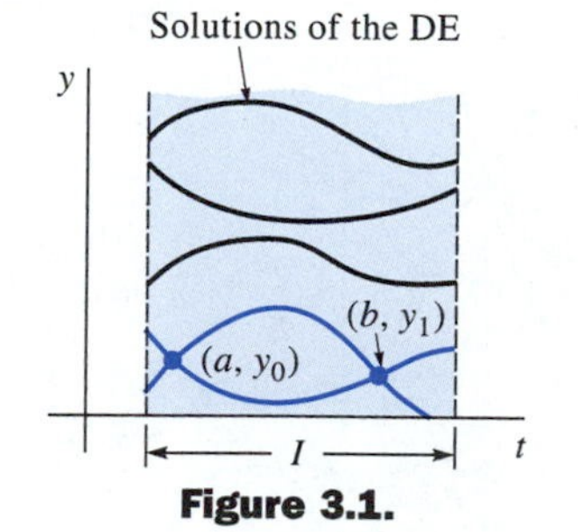

Figure 3.1.

is called a **boundary-value problem** (BVP). The prescribed values $y(a) = y_0$ and $y(b) = y_1$ are called **boundary conditions.** A solution of the foregoing problem is a function satisfying the differential equation on some interval I, containing a and b, whose graph passes through the two points (a, y_0) and (b, y_1). See Figure 3.1.

For a second-order differential equation, other pairs of boundary conditions could be

$$y'(a) = y_0, \qquad y(b) = y_1;$$

$$y(a) = y_0, \qquad y'(b) = y_1;$$

or

$$y'(a) = y_0, \qquad y'(b) = y_1$$

where y_0 and y_1 denote arbitrary constants. These three pairs of conditions are just special cases of the general boundary conditions

$$\alpha_1 y(a) + \beta_1 y'(a) = \gamma_1$$

$$\alpha_2 y(b) + \beta_2 y'(b) = \gamma_2$$

The next examples show that even when the conditions of Theorem 3.1 are fulfilled, a boundary-value problem may have (i) several solutions (as suggested in Figure 3.1), (ii) a unique solution, or (iii) no solution at all.

EXAMPLE 3 A BVP Can Have Many, One, or No Solutions

In Example 2 of Section 1.2 we saw that the two-parameter family of solutions of the differential equation $x'' + 16x = 0$ is

$$x = c_1 \cos 4t + c_2 \sin 4t \tag{2}$$

(a) Suppose we now wish to determine that solution of the equation that further satisfies the boundary conditions $x(0) = 0$, $x(\pi/2) = 0$. Observe that the first condition $0 = c_1 \cos 0 + c_2 \sin 0$ implies $c_1 = 0$, so that $x = c_2 \sin 4t$. But when $t = \pi/2$, $0 = c_2 \sin 2\pi$ is satisfied for any choice of c_2 since $\sin 2\pi = 0$. Hence the boundary-value problem

$$x'' + 16x = 0, \qquad x(0) = 0, \quad x\left(\frac{\pi}{2}\right) = 0 \tag{3}$$

has infinitely many solutions. Figure 3.2 shows the graphs of some of the members of the one-parameter family $x = c_2 \sin 4t$ that pass through the two points $(0, 0)$ and $(\pi/2, 0)$.

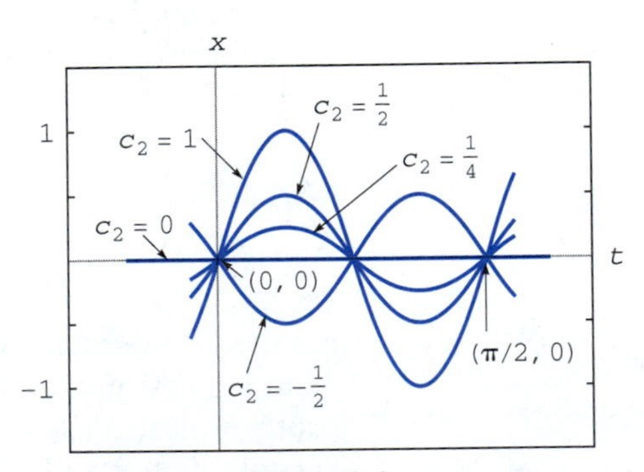

Figure 3.2.

(b) If the boundary-value problem in (3) is changed to

$$x'' + 16x = 0, \qquad x(0) = 0, \quad x\left(\frac{\pi}{8}\right) = 0 \tag{4}$$

then $x(0) = 0$ still requires $c_1 = 0$ in the solution (2). But applying $x(\pi/8) = 0$ to $x = c_2 \sin 4t$ demands that $0 = c_2 \sin(\pi/2) = c_2 \cdot 1$.

Hence $x = 0$ is a solution of this new boundary-value problem. Indeed, it can be proved that $x = 0$ is the *only* solution of (4).

(c) Finally, if we change the problem to

$$x'' + 16x = 0, \qquad x(0) = 0, \quad x\left(\frac{\pi}{2}\right) = 1 \tag{5}$$

we find again that $c_1 = 0$ from $x(0) = 0$, but that applying $x(\pi/2) = 1$ to $x = c_2 \sin 4t$ leads to the contradiction $1 = c_2 \sin 2\pi = c_2 \cdot 0 = 0$. Hence the boundary-value problem (5) has no solution. ■

3.1.2 Homogeneous Equations

A linear nth-order differential equation of the form

$$a_n(t)\frac{d^n y}{dt^n} + a_{n-1}(t)\frac{d^{n-1}y}{dt^{n-1}} + \cdots + a_1(t)\frac{dy}{dt} + a_0(t)\,y = 0 \tag{6}$$

is said to be **homogeneous**, whereas an equation of the form

$$a_n(t)\frac{d^n y}{dt^n} + a_{n-1}(t)\frac{d^{n-1}y}{dt^{n-1}} + \cdots + a_1(t)\frac{dy}{dt} + a_0(t)\,y = g(t) \tag{7}$$

with $g(t)$ not identically zero, is said to be **nonhomogeneous**. For example, $2y'' + 3y' - 5y = 0$ is a homogeneous linear second-order differential equation, whereas $t^3y''' + 6y' + 10y = e^t$ is a nonhomogeneous linear third-order differential equation. The function g in (7) is referred to as the **input** or **driving function** and a corresponding solution of the differential is called the **output** or **response**.

We shall see that in order to solve a nonhomogeneous equation (7), we must first be able to solve the associated homogeneous equation (6).

Note To avoid needless repetition throughout the remainder of this text we shall, as a matter of course, make the following very important assumptions when stating definitions and theorems about the linear equations (6) and (7). On some common interval I,

- the coefficients $a_i(t)$, $i = 0, 1, 2, \ldots, n$ are continuous,
- the right-hand member $g(t)$ is continuous, and
- $a_n(t) \neq 0$ for every t in the interval.

Superposition Principle In the next theorem we see that the sum, or **superposition**, of two or more solutions of a homogeneous linear differential equation is also a solution.

Theorem 3.2 Superposition Principle—Homogeneous Equations

Let $y_1, y_2, \ldots, y_k$ be solutions of the homogeneous linear nth-order differential equation (6) on an interval I. Then the linear combination

$$y = c_1 y_1(t) + c_2 y_2(t) + \cdots + c_k y_k(t)$$

where the c_i, $i = 1, 2, \ldots, k$ are arbitrary constants, is also a solution on the interval.

Proof We prove the case when $n = k = 2$. Let $y_1(t)$ and $y_2(t)$ be solutions of $a_2(t)y'' + a_1(t)y' + a_0(t)y = 0$. If we define $y = c_1 y_1(t) + c_2 y_2(t)$, then

$$\begin{aligned} a_2(t)[c_1 y_1'' + c_2 y_2''] &+ a_1(t)[c_1 y_1' + c_2 y_2'] + a_0(t)[c_1 y_1 + c_2 y_2] \\ &= c_1 \underbrace{[a_2(t)y_1'' + a_1(t)y_1' + a_0(t)y_1]}_{\text{zero}} + c_2 \underbrace{[a_2(t)y_2'' + a_1(t)y_2' + a_0(t)y_2]}_{\text{zero}} \\ &= c_1 \cdot 0 + c_2 \cdot 0 = 0 \end{aligned}$$

■

Corollaries

(A) A constant multiple $y = c_1 y_1(t)$ of a solution $y_1(t)$ of a homogeneous linear differential equation is also a solution.

(B) A homogeneous linear differential equation always possesses the trivial solution $y = 0$.

■ EXAMPLE 4 Superposition/Homogeneous DE

The functions $y_1 = t^2$ and $y_2 = t^2 \ln t$ are both solutions of the homogeneous linear equation $t^3 y''' - 2ty' + 4y = 0$ on the interval $(0, \infty)$. By the superposition principle, the linear combination $y = c_1 t^2 + c_2 t^2 \ln t$ is also a solution of the equation on the interval. ■

The function $y = e^{7t}$ is a solution of $y'' - 9y' + 14y = 0$. Since the differential equation is linear and homogeneous, the constant multiple $y = ce^{7t}$ is also a solution. For various values of c we see that $y = 9e^{7t}$, $y = 0$, $y = -\sqrt{5}\, e^{7t} \ldots$ are all solutions of the equation.

Linear Dependence and Linear Independence The two concepts defined next are a keystone in the study of linear differential equations.

> **Definition 3.1** **Linear Dependence/Independence**
>
> A set of functions $f_1(t), f_2(t), \ldots, f_n(t)$ is said to be **linearly dependent** on an interval I if there exist constants $c_1, c_2, \ldots, c_n$, not all zero, such that
>
> $$c_1 f_1(t) + c_2 f_2(t) + \cdots + c_n f_n(t) = 0$$
>
> for every t in the interval. If the set of functions is not linearly dependent on the interval, it is said to be **linearly independent**.

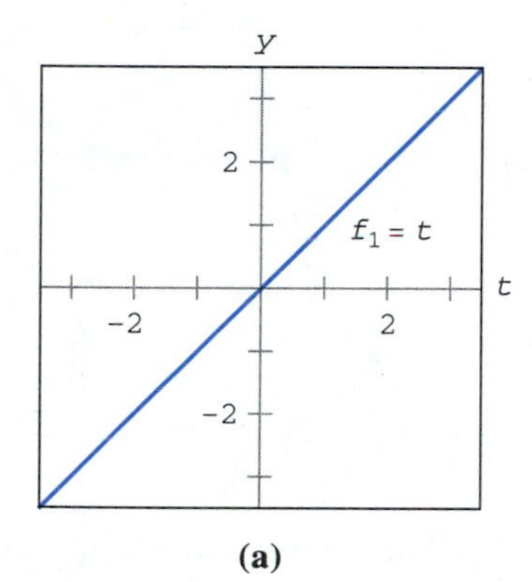

(a)

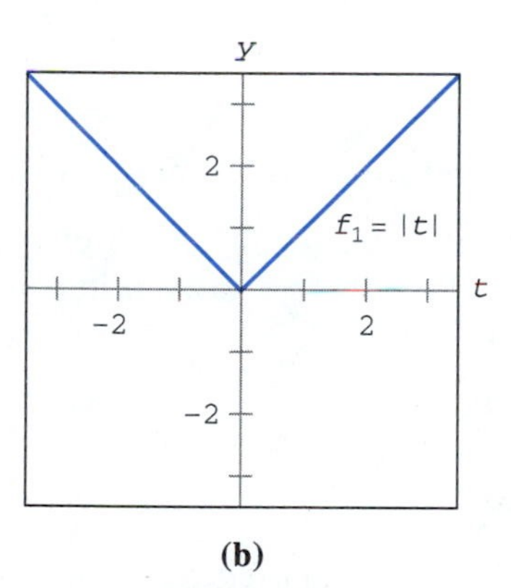

(b)

Figure 3.3.

In other words, a set of functions is linearly independent on an interval if the only constants for which

$$c_1 f_1(t) + c_2 f_2(t) + \cdots + c_n f_n(t) = 0$$

for every t in the interval, are $c_1 = c_2 = \cdots = c_n = 0$.

It is easy to understand these definitions in the case of two functions $f_1(t)$ and $f_2(t)$. If the functions are linearly dependent on an interval, then there exist constants c_1 and c_2 that are not both zero such that for every t in the interval, $c_1 f_1(t) + c_2 f_2(t) = 0$. Therefore, if we assume that $c_1 \neq 0$, it follows that $f_1(t) = -(c_2/c_1) f_2(t)$; that is, *if two functions are linearly dependent, then one is simply a constant multiple of the other*. Conversely, if $f_1(t) = c_2 f_2(t)$ for some constant c_2, then $(-1) \cdot f_1(t) + c_2 f_2(t) = 0$ for every t on some interval. Hence the functions are linearly dependent since at least one of the constants (namely, $c_1 = -1$) is not zero. We conclude that *two functions are linearly independent when neither is a constant multiple of the other* on an interval. For example, the functions $f_1(t) = \sin 2t$ and $f_2(t) = \sin t \cos t$ are linearly dependent on $(-\infty, \infty)$ because $f_1(t)$ is a constant multiple of $f_2(t)$. Recall from the double angle formula for the sine that $\sin 2t = 2 \sin t \cos t$. On the other hand, the functions $f_1(t) = t$ and $f_2(t) = |t|$ are linearly independent on $(-\infty, \infty)$. Inspection of Figure 3.3 should convince you that neither function is a constant multiple of the other on the interval.

It follows from the preceding discussion that the ratio $f_2(t)/f_1(t)$ is not a constant on an interval on which $f_1(t)$ and $f_2(t)$ are linearly independent. This little fact will be used in the next section.

A set of functions $f_1(t), f_2(t), \ldots, f_n(t)$ is linearly dependent on an interval if at least one function can be expressed as a linear combination of the remaining functions.

■ EXAMPLE 5 **Linearly Dependent Functions**

The functions $f_1(t) = \sqrt{t} + 5$, $f_2(t) = \sqrt{t} + 5t$, $f_3(t) = t - 1$, $f_4(t) = t^2$ are linearly dependent on the interval $(0, \infty)$ since f_2 can be written as a linear combination of f_1, f_3, and f_4. Observe that

$$f_2(t) = 1 \cdot f_1(t) + 5 \cdot f_3(t) + 0 \cdot f_4(t)$$

for every t in the interval $(0, \infty)$. ■

Solutions of Differential Equations We are primarily interested in linearly independent functions or, more to the point, linearly independent solutions of a linear differential equation. Although we could always appeal directly to Definition 3.1, it turns out that the question of whether n solutions $y_1, y_2, \ldots, y_n$ of a homogeneous linear nth-order differential equation (6) are linearly independent can be settled somewhat mechanically using a determinant.

Definition 3.2 Wronskian

Suppose each of the functions $f_1(t), f_2(t), \ldots, f_n(t)$ possess at least $n - 1$ derivatives. The determinant

$$W(f_1, f_2, \ldots, f_n) = \begin{vmatrix} f_1 & f_2 & \cdots & f_n \\ f_1' & f_2' & \cdots & f_n' \\ \vdots & \vdots & & \vdots \\ f_1^{(n-1)} & f_2^{(n-1)} & \cdots & f_n^{(n-1)} \end{vmatrix}$$

where the primes denote derivatives, is called the **Wronskian** of the functions.

Theorem 3.3 Criterion for Linearly Independent Solutions

Let $y_1, y_2, \ldots, y_n$ be n solutions of the homogeneous linear nth-order differential equation (6) on an interval I. Then the set of solutions is **linearly independent** on I if and only if $W(y_1, y_2, \ldots, y_n) \neq 0$ for every t in the interval.

It follows from Theorem 3.3 that when $y_1, y_2, \ldots, y_n$ are n solutions of (6) on an interval I, the Wronskian $W(y_1, y_2, \ldots, y_n)$ is either identically zero or never zero on the interval.

A set of n linearly independent solutions of a homogeneous linear nth-order differential equation is given a special name.

Definition 3.3 Fundamental Set of Solutions

Any set $y_1, y_2, \ldots, y_n$ of n linearly independent solutions of the homogeneous linear nth-order differential equation (6) on an interval I is said to be a **fundamental set of solutions** on the interval.

The question of whether a fundamental set of solutions exists for a linear equation is answered in Theorem 3.4.

Theorem 3.4 Existence of a Fundamental Set

There exists a fundamental set of solutions for the homogeneous linear nth-order differential equation (6) on an interval I.

Analogous to the fact that any vector in three dimensions can be expressed as a linear combination of the *linearly independent* vectors **i**, **j**, **k**, any solution of an nth-order homogeneous linear differential equation on an interval I can be expressed as a linear combination of n linearly independent solutions on I. In other words, n linearly independent solutions $y_1, y_2, \ldots, y_n$ are the building blocks for the general solution of the equation.

Theorem 3.5 General Solution—Homogeneous Equations

Let $y_1, y_2, \ldots, y_n$ be a fundamental set of solutions of the homogeneous linear nth-order differential equation (6) on an interval I. Then the **general solution** of the equation on the interval is

$$y = c_1y_1(t) + c_2y_2(t) + \cdots + c_ny_n(t)$$

where c_i, $i = 1, 2, \ldots, n$ are arbitrary constants.

Theorem 3.5 states that if $Y(t)$ is any solution of (6) on the interval I, then Y must be a member of the n-parameter family that represents the general solution of the equation on I; in other words, constants $C_1, C_2, \ldots, C_n$ can always be found so that

$$Y(t) = C_1y_1(t) + C_2y_2(t) + \cdots + C_ny_n(t)$$

We will prove the case when $n = 2$.

Proof Let Y be a solution and y_1 and y_2 be linearly independent solutions of $a_2y'' + a_1y' + a_0y = 0$ on an interval I. Suppose $t = x$ is a point in I for which $W(y_1(x), y_2(x)) \neq 0$. Suppose also that $Y_1(x) = k_1$ and $Y_1'(x) = k_2$. If we now examine the equations

$$C_1y_1(x) + C_2y_2(x) = k_1$$

$$C_1y_1'(x) + C_2y_2'(x) = k_2$$

it follows that we can determine C_1 and C_2 uniquely, provided that the determinant of the coefficients satisfies $\begin{vmatrix} y_1(x) & y_2(x) \\ y_1'(x) & y_2'(x) \end{vmatrix} \neq 0$. But this determinant is simply the Wronskian evaluated at $t = x$, and by assumption $W \neq 0$. If we define $G(t) = C_1y_1(t) + C_2y_2(t)$, we observe (i) $G(t)$ satisfies the differential equation since it is a superposition of two known solutions; (ii) $G(t)$ satisfies the initial conditions

$$G(x) = C_1y_1(x) + C_2y_2(x) = k_1 \quad \text{and} \quad G'(x) = C_1y_1'(x) + C_2y_2'(x) = k_2$$

(iii) $Y(t)$ satisfies the *same* linear equation and the *same* initial conditions. Since the solution of this second-order initial-value problem is unique (Theorem 3.1), we have $Y(t) = G(t)$ or $Y(t) = C_1y_1(t) + C_2y_2(t)$. ■

■ EXAMPLE 6 General Solution of a Homogeneous DE

The functions $y_1 = e^t$, $y_2 = e^{2t}$, and $y_3 = e^{3t}$ satisfy the homogeneous linear third-order equation $y''' - 6y'' + 11y' - 6y = 0$ on the interval $(-\infty, \infty)$. (Verify this.) Since the Wronskian

$$W(e^t, e^{2t}, e^{3t}) = \begin{vmatrix} e^t & e^{2t} & e^{3t} \\ e^t & 2e^{2t} & 3e^{3t} \\ e^t & 4e^{2t} & 9e^{3t} \end{vmatrix} = 2e^{6t} \neq 0$$

for every real value of t, we conclude that y_1, y_2, and y_3 form a fundamental set of solutions on $(-\infty, \infty)$ and that the three-parameter family

$$y = c_1e^t + c_2e^{2t} + c_3e^{3t}$$

is the general solution of the equation on the interval. ■

■ EXAMPLE 7 A Solution Obtained from a General Solution

The two-parameter family $y = c_1e^{3t} + c_2e^{-3t}$ is the general solution of the homogeneous linear second-order equation $y'' - 9y = 0$. Moreover, the function $y = 4\sinh 3t - 5e^{-3t}$ is a particular solution of the same differential equation. (Verify these two statements.) In view of Theorem 3.5, we must be able to obtain the particular solution from the general solution. Observe that if we choose $c_1 = 2$ and $c_2 = -7$, then $y = 2e^{3t} - 7e^{-3t}$ can be rewritten as

$$y = 2e^{3t} - 2e^{-3t} - 5e^{-3t} = 4\left(\frac{e^{3t} - e^{-3t}}{2}\right) - 5e^{-3t}$$

The last expression is recognized as $y = 4\sinh 3t - 5e^{-3t}$. ■

3.1.3 Nonhomogeneous Equations

Any function y_p, free of arbitrary parameters, that satisfies (7) is said to be a **particular solution** or **particular integral** of the equation. For example, it is a straightforward task to show that the constant function $y_p = 3$ is a particular solution of the nonhomogeneous equation $y'' + 9y = 27$.

Now if $y_1, y_2, \ldots, y_k$ are solutions of the **associated** homogeneous equation (6) on an interval I and y_p is any particular solution of (7) on I, then the linear combination

$$y = c_1y_1(t) + c_2y_2(t) + \cdots + c_ky_k(t) + y_p \tag{8}$$

is also a solution of the nonhomogeneous equation (7). If you think about it, this makes sense because the linear combination $c_1y_1(t) + c_2y_2(t) + \cdots + c_ky_k(t)$ is mapped into 0 by $a_n(t)y^{(n)} + a_{n-1}(t)y^{(n-1)} + \cdots + a_1(t)y' + a_0(t)y$, whereas y_p is mapped into $g(t)$. If we use $k = n$ linearly independent solutions of the nth-order equation (6), then the expression in (8) becomes the general solution of (7).

Theorem 3.6 General Solution—Nonhomogeneous Equations

Let y_p be any particular solution of the nonhomogeneous linear nth-order differential equation (7) on an interval I and let $y_1, y_2, \ldots, y_n$ be a fundamental set of solutions of the associated homogeneous differential equation (6) on I. Then the **general solution** of the equation on the interval is

$$y = c_1y_1(t) + c_2y_2(t) + \cdots + c_ny_n(t) + y_p$$

where the c_i, $i = 1, 2, \ldots, n$ are arbitrary constants.

The proof of this result follows from Theorem 3.5. See Problem 41 in Exercises 3.1.

Complementary Function We see in Theorem 3.6 that the general solution of a nonhomogeneous linear equation consists of a sum of two functions:

$$y = c_1y_1(t) + c_2y_2(t) + \cdots + c_ny_n(t) + y_p = y_c(t) + y_p(t).$$

The linear combination $y_c(t) = c_1y_1(t) + c_2y_2(t) + \cdots + c_ny_n(t)$, which is the general solution of (6), is called the **complementary function** for equation (7). In other words, to solve a nonhomogeneous linear differential equation, we first solve the associated homogeneous equation and then we must find any particular solution of the nonhomogeneous equation. The general solution of the nonhomogeneous equation is then

$$y = \textit{complementary function} + \textit{any particular solution}$$

EXAMPLE 8 General Solution of a Nonhomogeneous DE

By substitution, the function $y_p = -\frac{11}{12} - \frac{1}{2}t$ is readily shown to be a particular solution of the nonhomogeneous equation

$$\frac{d^3y}{dt^3} - 6\frac{d^2y}{dt^2} + 11\frac{dy}{dt} - 6y = 3t \tag{9}$$

In order to write the general solution of (9), we must also be able to solve the associated homogeneous equation

$$\frac{d^3y}{dt^3} - 6\frac{d^2y}{dt^2} + 11\frac{dy}{dt} - 6y = 0$$

But in Example 6, we saw that the general solution of this latter equation on the interval $(-\infty, \infty)$ was $y_c = c_1e^t + c_2e^{2t} + c_3e^{3t}$. Hence the general solution of (9) on the interval is

$$y = y_c + y_p = c_1e^t + c_2e^{2t} + c_3e^{3t} - \frac{11}{12} - \frac{1}{2}t$$

Another Superposition Principle Linear differential equations have the remarkable property that the response to a superposition of inputs is a superposition of outputs. Stated in formal terms in the next theorem, this property will prove useful in Section 3.4 when we find particular solutions of nonhomogeneous equations.

Theorem 3.7 Superposition Principle—Nonhomogeneous Equations

Let $y_{p_1}, y_{p_2}, \ldots, y_{p_k}$ be k particular solutions of the nonhomogeneous linear nth-order differential equation (7) on an interval I corresponding, in turn, to k distinct functions $g_1, g_2, \ldots, g_k$. That is, suppose y_{p_i} denotes a particular solution of the corresponding differential equation

$$a_n(t)y^{(n)} + a_{n-1}(t)y^{(n-1)} + \cdots + a_1(t)y' + a_0(t)y = g_i(t) \tag{10}$$

where $i = 1, 2, \ldots, k$. Then

$$y_p = y_{p_1}(t) + y_{p_2}(t) + \cdots + y_{p_k}(t) \tag{11}$$

is a particular solution of

$$\begin{aligned} &a_n(t)y^{(n)} + a_{n-1}(t)y^{(n-1)} + \cdots + a_1(t)y' + a_0(t)y \\ &= g_1(t) + g_2(t) + \cdots + g_k(t) \end{aligned} \tag{12}$$

We leave the proof of this result when $k = 2$ as an exercise. See Problem 42 in Exercises 3.1.

EXAMPLE 9 Superposition/Nonhomogeneous DE

You should verify that

$y_{p_1} = -4t^2$ is a particular solution of $y'' - 3y' + 4y = -16t^2 + 24t - 8$,

$y_{p_2} = e^{2t}$ is a particular solution of $y'' - 3y' + 4y = 2e^{2t}$, and

$y_{p_3} = te^t$ is a particular solution of $y'' - 3y' + 4y = 2te^t - e^t$

It follows from Theorem 3.7 that the superposition of y_{p_1}, y_{p_2}, and y_{p_3},

$$y = y_{p_1} + y_{p_2} + y_{p_3} = -4t^2 + e^{2t} + te^t$$

is a solution of

$$y'' - 3y' + 4y = \underbrace{-16t^2 + 24t - 8}_{g_1(t)} + \underbrace{2e^{2t}}_{g_2(t)} + \underbrace{2te^t - e^t}_{g_3(t)}$$

■

Note If the y_{p_i} are particular solutions of (10) for $i = 1, 2, \ldots, k$, then it also follows that the linear combination

$$y_p = c_1 y_{p_1} + c_2 y_{p_2} + \cdots + c_k y_{p_k}$$

where the c_i are constants, is a particular solution of (12) when the right-hand member of the equation is the linear combination

$$c_1 g_1(t) + c_2 g_2(t) + \cdots + c_k g_k(t)$$

EXERCISES 3.1

Answers to odd-numbered problems begin on page 441.

3.1.1

1. Given that $y = c_1 e^t + c_2 e^{-t}$ is a two-parameter family of solutions of $y'' - y = 0$ on the interval $(-\infty, \infty)$, find a member of the family satisfying the initial conditions $y(0 = 0)$, $y'(0) = 1$.
2. Find a solution of the differential equation in Problem 1 satisfying the boundary conditions $y(0) = 0$, $y(1) = 1$.
3. Given that $y = c_1 e^{4t} + c_2 e^{-t}$ is a two-parameter family of solutions of $y'' - 3y' - 4y = 0$ on the interval $(-\infty, \infty)$, find a member of the family satisfying the initial conditions $y(0) = 1$, $y'(0) = 2$.
4. Given that $y = c_1 + c_2 \cos t + c_3 \sin t$ is a three-parameter family of solutions of $y''' + y' = 0$ on the interval $(-\infty, \infty)$, find a member of the family satisfying the initial conditions $y(\pi) = 0$, $y'(\pi) = 2$, $y''(\pi) = -1$.
5. Given that $y = c_1 t + c_2 t \ln t$ is a two-parameter family of solutions of $t^2 y'' - ty' + y = 0$ on the interval $(0, \infty)$, find a member of the family satisfying the initial conditions $y(1) = 3$, $y'(1) = -1$.
6. Given that $y = c_1 + c_2 t^2$ is a two-parameter family of solutions of $ty'' - y' = 0$ on the interval $(-\infty, \infty)$, show that constants c_1 and c_2 cannot be found so that a member of the family satisfies the initial conditions $y(0) = 0$, $y'(0) = 1$. Explain why this does not violate Theorem 3.1.
7. Find two members of the family of solutions of $ty'' - y' = 0$ given in Problem 6 satisfying the initial conditions $y(0) = 0$, $y'(0) = 0$.
8. Find a member of the family of solutions of $ty'' - y' = 0$ given in Problem 6 satisfying the boundary conditions

$y(0) = 1$, $y'(1) = 6$. Does Theorem 3.1 guarantee that this solution is unique?

9. Given that $y = c_1e^t \cos t + c_2e^t \sin t$ is a two-parameter family of solutions of $y'' - 2y' + 2y = 0$ on the interval $(-\infty, \infty)$, determine whether a member of the family can be found that satisfies the boundary conditions
(a) $y(0) = 1$, $y'(0) = 0$ **(b)** $y(0) = 1$, $y(\pi) = -1$
(c) $y(0) = 1$, $y(\pi/2) = 1$ **(d)** $y(0) = 0$, $y(\pi) = 0$

10. Given that $y = c_1t^2 + c_2t^4 + 3$ is a two-parameter family of solutions of $t^2y'' - 5ty' + 8y = 24$ on the interval $(-\infty, \infty)$, determine whether a member of the family can be found that satisfies the boundary conditions
(a) $y(-1) = 0$, $y(1) = 4$ **(b)** $y(0) = 1$, $y(1) = 2$
(c) $y(0) = 3$, $y(1) = 0$ **(d)** $y(1) = 3$, $y(2) = 15$

In Problems 11 and 12, find an interval around $t = 0$ for which the given initial-value problem has a unique solution.

11. $(t - 2)y'' + 3y = t$, $y(0) = 0$, $y'(0) = 1$

12. $y'' + (\tan t)y = e^t$, $y(0) = 1$, $y'(0) = 0$

3.1.2

In Problems 13–20, determine whether the given functions are linearly independent or dependent on $(-\infty, \infty)$.

13. $f_1(t) = t$, $f_2(t) = t^2$, $f_3(t) = 4t - 3t^2$

14. $f_1(t) = 1$, $f_2(t) = t$, $f_3(t) = e^t$

15. $f_1(t) = 5$, $f_2(t) = \cos^2 t$, $f_3(t) = \sin^2 t$

16. $f_1(t) = \cos 2t$, $f_2(t) = 1$, $f_3(t) = \cos^2 t$

17. $f_1(t) = t$, $f_2(t) = t - 1$, $f_3(t) = t + 3$

18. $f_1(t) = 2 + t$, $f_2(t) = 2 + |t|$

19. $f_1(t) = 1 + t$, $f_2(t) = t$, $f_3(t) = t^2$

20. $f_1(t) = e^t$, $f_2(t) = e^{-t}$, $f_3(t) = \sinh t$

In Problems 21–28, verify that the given functions form a fundamental set of solutions of the differential equation on the indicated interval. Form the general solution.

21. $y'' - y' - 12y = 0$; $e^{-3t}, e^{4t}, (-\infty, \infty)$

22. $y'' - 4y = 0$; $\cosh 2t, \sinh 2t, (-\infty, \infty)$

23. $y'' - 2y' + 5y = 0$; $e^t \cos 2t, e^t \sin 2t, (-\infty, \infty)$

24. $4y'' - 4y' + y = 0$; $e^{t/2}, te^{t/2}, (-\infty, \infty)$

25. $t^2y'' - 6ty' + 12y = 0$; $t^3, t^4, (0, \infty)$

26. $t^2y'' + ty' + y = 0$; $\cos(\ln t), \sin(\ln t), (0, \infty)$

27. $t^3y''' + 6t^2y'' + 4ty' - 4y = 0$; $t, t^{-2}, t^{-2}\ln t, (0, \infty)$

28. $y^{(4)} + y'' = 0$; $1, t, \cos t, \sin t, (-\infty, \infty)$

3.1.3

In Problems 29–34, verify that the given two-parameter family of functions is the general solution of the nonhomogeneous differential equation on the indicated interval.

29. $y'' - 7y' + 10y = 24e^t$
$y = c_1e^{2t} + c_2e^{5t} + 6e^t, (-\infty, \infty)$

30. $y'' + y = \sec t$
$y = c_1 \cos t + c_2 \sin t + t \sin t + (\cos t)\ln(\cos t)$, $(-\pi/2, \pi/2)$

31. $y'' - 4y' + 4y = 2e^{2t} + 4t - 12$
$y = c_1e^{2t} + c_2te^{2t} + t^2e^{2t} + t - 2, (-\infty, \infty)$

32. $2t^2y'' + 5ty' + y = t^2 - t$

$$y = c_1t^{-1/2} + c_2t^{-1} + \frac{1}{15}t^2 - \frac{1}{6}t, \ (0, \infty)$$

33. Given that $y_{p_1} = 3e^{2t}$ and $y_{p_2} = t^2 + 3t$ are particular solutions of

$$y'' - 6y' + 5y = -9e^{2t}$$

and

$$y'' - 6y' + 5y = 5t^2 + 3t - 16$$

respectively, find particular solutions of

$$y'' - 6y' + 5y = 5t^2 + 3t - 16 - 9e^{2t}$$

and

$$y'' - 6y' + 5y = -10t^2 - 6t + 32 + e^{2t}$$

34. **(a)** By inspection, determine a particular solution of $y'' + 2y = 10$.
(b) By inspection, determine a particular solution of $y'' + 2y = -4t$.

(c) Find a particular solution of $y'' + 2y = -4t + 10$.
(d) Find a particular solution of $y'' + 2y = 8t + 5$.

Discussion Problems

35. Suppose $y_1(t)$ and $y_2(t)$ are solutions of the differential equation $y'' + ty' + (t + 5)y = 0$ on the interval $(-\infty, \infty)$ and have the properties $y_1(0) = 1, y_1'(0) = -1$, and $y_2(0) = -2, y_2'(0) = 2$. Discuss: Are the solutions y_1 and y_2 linearly independent on $(-\infty, \infty)$?

36. (a) Verify that $y_1 = t^3$ and $y_2 = |t|^3$ are linearly independent solutions of the differential equation $t^2y'' - 4ty' + 6y = 0$ on the interval $(-\infty, \infty)$.
(b) Show that $W(y_1, y_2) = 0$ for every real number t. Discuss: Does this result violate Theorem 3.3?

37. Suppose f is a differentiable function on an interval I and $f(t) \neq 0$ for every t in the interval. Discuss: Are $f(t)$ and $tf(t)$ linearly independent on I?

38. The functions $y_1 = t, y_2 = e^{-t}, y_3 = 0$ are three solutions of the third-order equation $y''' + y'' = 0$ on the interval $(-\infty, \infty)$. Discuss whether the following statement is true or false: y_1, y_2, and y_3 form a fundamental set of solutions on the interval.

39. The functions $y_1 = e^t$ and $y_2 = e^{-t}$ form a fundamental set of solutions of the second-order equation $y'' - y = 0$ on the interval $(-\infty, \infty)$. Explain: $y_1 = \cosh t$ and $y_2 = \sinh t$ form a fundamental set of solutions of the same equation.

40. A linear second-order differential equation with constant coefficients can be solved using the procedure discussed in Section 2.3 whenever the y term is missing. Illustrate by solving $y'' + y' = e^t$. Point out y_c and y_p. [*Hint*: $(d/dt)y' = y''$]

41. In this problem, you are asked to prove Theorem 3.6 in the case when $n = 2$. To get started, suppose that Y and y_p are any two solutions of $a_2(t)y'' + a_1(t)y' + a_0(t)y = g(t)$. See what happens when $u(t) = Y(t) - y_p(t)$ is substituted into this differential equation. Now discuss the relevance of Theorem 3.5 in completing your proof.

42. Prove Theorem 3.7 in the case when $k = 2$.

3.2 Homogeneous Linear Equations with Constant Coefficients

We have seen that the linear first-order DE $y' + ay = 0$, where a is a constant, possesses the exponential solution $y = c_1e^{-at}$ on the interval $(-\infty, \infty)$. Therefore, it is natural to ask whether exponential solutions exist on $(-\infty, \infty)$ for homogeneous linear higher-order DEs with constant coefficients. The surprising fact is that *all* solutions of these higher-order equations are exponential functions or are constructed out of exponential functions.

Auxiliary Equation We begin by considering the special case of a second-order equation

$$ay'' + by' + cy = 0, \quad a \neq 0 \tag{1}$$

where the coefficients a, b, and c are real constants. If we try a solution of the form $y = e^{mt}$, then after substituting $y' = me^{mt}$ and $y'' = m^2e^{mt}$, equation (1) becomes

$$am^2e^{mt} + bme^{mt} + ce^{mt} = 0 \quad \text{or} \quad e^{mt}(am^2 + bm + c) = 0$$

Since e^{mt} is never zero for real values of t, it is apparent that the only way that this exponential function can satisfy the differential equation (1) is to choose m as a root of the quadratic equation

$$am^2 + bm + c = 0 \tag{2}$$

This last equation is called the **auxiliary equation** of the differential equation (1). Since the two roots of (2) are $m_1 = (-b + \sqrt{b^2 - 4ac})/2a$, and $m_2 = (-b - \sqrt{b^2 - 4ac})/2a$, there will be three forms of the general solution of (1) corresponding to the three cases:

- m_1 and m_2 real and distinct ($b^2 - 4ac > 0$),
- m_1 and m_2 real and equal ($b^2 - 4ac = 0$), and
- m_1 and m_2 are conjugate complex numbers ($b^2 - 4ac < 0$).

We discuss each of these cases in turn.

CASE I **Distinct Real Roots** Under the assumption that the auxiliary equation (2) has two unequal real roots m_1 and m_2, we find two different solutions $y_1 = e^{m_1 t}$ and $y_2 = e^{m_2 t}$. We see that these two functions are linearly independent on $(-\infty, \infty)$ and hence form a fundamental set. It follows that the general solution of (1) on this interval is

$$y = c_1 e^{m_1 t} + c_2 e^{m_2 t} \tag{3}$$

CASE II **Repeated Real Roots** When $m_1 = m_2$, the single exponential solution is $y_1 = e^{m_1 t}$. In this case, we must have $b^2 - 4ac = 0$ and so $m_1 = -b/2a$. Now we seek a second solution y_2 of (1) so that y_1 and y_2 are linearly independent on $(-\infty, \infty)$. Recall that if y_1 and y_2 are linearly independent, then their ratio y_2/y_1 is *nonconstant* on the interval; that is, $y_2/y_1 = u(t)$ or $y_2(t) = u(t)y_1(t) = u(t)e^{m_1 t}$. We can find $u(t)$ in a straightforward manner by differentiating $y = u(t)e^{m_1 t}$, substituting

$$y' = u(t)e^{m_1 t}m_1 + u'(t)e^{m_1 t}, \quad y'' = u(t)e^{m_1 t}m_1^2 + 2u'(t)e^{m_1 t}m_1 + u''(t)e^{m_1 t}$$

into (1), dividing out $e^{m_1 t}$, and simplifying:

$$au'' + \underbrace{(2am_1 + b)}_{\substack{\text{zero since} \\ m_1 = -b/2a}}u' + \underbrace{(am_1^2 + bm_1 + c)}_{\substack{\text{zero since } m_1 \text{ is a root} \\ \text{of the auxiliary equation}}}u = 0$$

From the last equation, we see that $u'' = 0$ and so by integrating twice we find $u = C_1 + C_2 t$. Thus $y = u(t)y_1 = (C_1 + C_2 t)e^{m_1 t}$. But since $C_1 e^{m_1 t}$ is already a solution, we obtain the desired second solution of (1) by choosing $C_1 = 0$ and $C_2 = 1$; in other words, $y_2 = te^{m_1 t}$. The general solution is then

$$y = c_1 e^{m_1 t} + c_2 t e^{m_1 t} \tag{4}$$

CASE III **Conjugate Complex Roots** If m_1 and m_2 are complex, then we can write $m_1 = \alpha + i\beta$ and $m_2 = \alpha - i\beta$, where α and $\beta > 0$ are real and $i^2 = -1$. Formally, there is no difference between this case and

Case I, and hence

$$y = C_1 e^{(\alpha + i\beta)t} + C_2 e^{(\alpha - i\beta)t}$$

However, in practice we prefer to work with real functions instead of complex exponentials. To this end we use Euler's formula:

$$e^{i\theta} = \cos\theta + i\sin\theta$$

where θ is any real number. It follows from this formula that

$$e^{i\beta t} = \cos\beta t + i\sin\beta t \quad \text{and} \quad e^{-i\beta t} = \cos\beta t - i\sin\beta t \tag{5}$$

where we have used $\cos(-\beta t) = \cos\beta t$ and $\sin(-\beta t) = -\sin\beta t$. Note that by first adding and then subtracting the two equations in (5), we obtain, respectively,

$$e^{i\beta t} + e^{-i\beta t} = 2\cos\beta t \quad \text{and} \quad e^{i\beta t} - e^{-i\beta t} = 2i\sin\beta t$$

Since $y = C_1 e^{(\alpha + i\beta)t} + C_2 e^{(\alpha - i\beta)t}$ is a solution of (1) for any choice of the constants C_1 and C_2, the choices $C_1 = C_2 = 1$ and $C_1 = 1$, $C_2 = -1$ give, in turn, two solutions:

$$y_1 = e^{(\alpha + i\beta)t} + e^{(\alpha - i\beta)t} \quad \text{and} \quad y_2 = e^{(\alpha + i\beta)t} - e^{(\alpha - i\beta)t}$$

But

$$y_1 = e^{\alpha t}(e^{i\beta t} + e^{-i\beta t}) = 2e^{\alpha t}\cos\beta t$$

and

$$y_2 = e^{\alpha t}(e^{i\beta t} - e^{-i\beta t}) = 2ie^{\alpha t}\sin\beta t$$

Hence, from Corollary (A) of Theorem 3.2, the last two results show that the *real* functions $e^{\alpha t}\cos\beta t$ and $e^{\alpha t}\sin\beta t$ are solutions of (1). Moreover, these solutions form a fundamental set on $(-\infty, \infty)$. Consequently, the general solution is

$$\begin{aligned} y &= c_1 e^{\alpha t}\cos\beta t + c_2 e^{\alpha t}\sin\beta t \\ &= e^{\alpha t}(c_1\cos\beta t + c_2\sin\beta t) \end{aligned} \tag{6}$$

EXAMPLE 1 Second-Order DEs

Solve the following differential equations:

(a) $2y'' - 5y' - 3y = 0$ **(b)** $y'' - 10y' + 25y = 0$ **(c)** $y'' + y' + y = 0$

Solution We give the auxiliary equations, roots, and the corresponding general solutions.

(a) $2m^2 - 5m - 3 = (2m + 1)(m - 3) = 0$, $m_1 = -\frac{1}{2}$, $m_2 = 3$,
$y = c_1 e^{-t/2} + c_2 e^{3t}$

(b) $m^2 - 10m + 25 = (m - 5)^2 = 0$, $m_1 = m_2 = 5$,
$y = c_1 e^{5t} + c_2 t e^{5t}$

(c) $m^2 + m + 1 = 0$, $m_1 = -\frac{1}{2} + \frac{\sqrt{3}}{2}i$, $m_2 = -\frac{1}{2} - \frac{\sqrt{3}}{2}i$,
$y = e^{-t/2}\left(c_1\cos\frac{\sqrt{3}}{2}t + c_2\sin\frac{\sqrt{3}}{2}t\right)$ ■

EXAMPLE 2 An Initial-Value Problem

Solve the initial-value problem $y'' - 4y' + 13y = 0$, $y(0) = -1$, $y'(0) = 2$.

Solution The roots of the auxiliary equation $m^2 - 4m + 13 = 0$ are $m_1 = 2 + 3i$ and $m_2 = 2 - 3i$ so that

$$y = e^{2t}(c_1 \cos 3t + c_2 \sin 3t)$$

Applying the condition $y(0) = -1$ we see from $-1 = e^0(c_1 \cos 0 + c_2 \sin 0)$ that $c_1 = -1$. Differentiating $y = e^{2t}(-\cos 3t + c_2 \sin 3t)$ and then using $y'(0) = 2$ give $2 = 3c_2 - 2$ or $c_2 = \frac{4}{3}$. Hence the solution is

$$y = e^{2t}\left(-\cos 3t + \frac{4}{3}\sin 3t\right)$$

Two Equations Worth Knowing The two differential equations

$$y'' + k^2y = 0 \quad \text{and} \quad y'' - k^2y = 0$$

k real, are important in applied mathematics. For the first equation, the auxiliary equation $m^2 + k^2 = 0$ has imaginary roots $m_1 = ki$ and $m_2 = -ki$. From (6) with $\alpha = 0$ and $\beta = k$, we obtain the general solution

$$y = c_1 \cos kt + c_2 \sin kt \tag{7}$$

The auxiliary equation of the second equation $m^2 - k^2 = 0$ has distinct real roots $m_1 = k$ and $m_2 = -k$. Hence its general solution is

$$y = c_1e^{kt} + c_2e^{-kt} \tag{8}$$

Notice that if we choose $c_1 = c_2 = \frac{1}{2}$ and then $c_1 = \frac{1}{2}$, $c_2 = -\frac{1}{2}$ in (8) we get the particular solutions $y = (e^{kt} + e^{-kt})/2 = \cosh kt$ and $y = (e^{kt} - e^{-kt})/2 = \sinh kt$. Because $\cosh kt$ and $\sinh kt$ are linearly independent on any interval of the t-axis, an alternative form for the general solution of $y'' - k^2y = 0$ is

$$y = c_1 \cosh kt + c_2 \sinh kt$$

Higher-Order Equations In general, to solve a linear nth-order differential equation

$$a_ny^{(n)} + a_{n-1}y^{(n-1)} + \cdots + a_2y'' + a_1y' + a_0y = 0 \tag{9}$$

where the a_i, $i = 0, 1, \ldots, n$, are real constants, we must solve an nth-degree polynomial equation

$$a_nm^n + a_{n-1}m^{n-1} + \cdots + a_2m^2 + a_1m + a_0 = 0 \tag{10}$$

If all the roots of (10) are real and distinct, then the general solution of (9) is

$$y = c_1e^{m_1t} + c_2e^{m_2t} + \cdots + c_ne^{m_nt}.$$

It is somewhat harder to summarize the analogues of Cases II and III because the roots of an auxiliary equation of degree greater than two can occur in many combinations. For example, a fifth-degree equation could have five distinct real roots, or three distinct real and two complex roots, or one real and four complex roots, or five real but equal roots, or five real roots but two of them equal, and so on. When m_1 is a root of multiplicity k of an nth-degree auxiliary equation (that is, k roots are equal to m_1), it can be shown that the linearly independent solutions are

$$e^{m_1 t}, \quad te^{m_1 t}, \quad t^2 e^{m_1 t}, \ldots, \quad t^{k-1} e^{m_1 t}$$

and the general solution must contain the linear combination

$$c_1 e^{m_1 t} + c_2 t e^{m_1 t} + c_3 t^2 e^{m_1 t} + \cdots + c_k t^{k-1} e^{m_1 t}$$

Lastly, it should be remembered that when the coefficients are real, complex roots of an auxiliary equation always appear in conjugate pairs. Thus, for example, a cubic polynomial equation can have at most two complex roots.

EXAMPLE 3 **Third-Order DE**

Solve $y''' + 3y'' - 4y = 0$.

Solution It should be apparent from inspection of $m^3 + 3m^2 - 4 = 0$ that one root is $m_1 = 1$. Now if we divide $m^3 + 3m^2 - 4$ by $m - 1$, we find

$$m^3 + 3m^2 - 4 = (m - 1)(m^2 + 4m + 4) = (m - 1)(m + 2)^2$$

and so the other roots are $m_2 = m_3 = -2$. Thus the general solution is

$$y = c_1 e^t + c_2 e^{-2t} + c_3 t e^{-2t}$$ ■

EXAMPLE 4 **Fourth-Order DE**

Solve $\dfrac{d^4 y}{dt^4} + 2\dfrac{d^2 y}{dt^2} + y = 0$.

Solution The auxiliary equation $m^4 + 2m^2 + 1 = (m^2 + 1)^2 = 0$ has roots $m_1 = m_3 = i$ and $m_2 = m_4 = -i$. Thus from Case II the solution is

$$y = C_1 e^{it} + C_2 e^{-it} + C_3 t e^{it} + C_4 t e^{-it}$$

By Euler's formula, the grouping $C_1 e^{it} + C_2 e^{-it}$ can be rewritten as

$$c_1 \cos t + c_2 \sin t$$

after a relabeling of constants. Similarly, $t(C_3 e^{it} + C_4 e^{-it})$ can be expressed as $t(c_3 \cos t + c_4 \sin t)$. Hence the general solution is

$$y = c_1 \cos t + c_2 \sin t + c_3 t \cos t + c_4 t \sin t$$ ■

Example 4 illustrates a special case when the auxiliary equation has repeated complex roots. In general, if $m_1 = \alpha + i\beta$, $\beta > 0$, is a complex root of multiplicity k of an auxiliary equation with real coefficients, then its conjugate, $m_2 = \alpha - i\beta$ is also a root of multiplicity k. For the $2k$ complex-valued solutions

$$e^{(\alpha + i\beta)t}, \; te^{(\alpha + i\beta)t}, \; t^2 e^{(\alpha + i\beta)t}, \ldots, \; t^{k-1} e^{(\alpha + i\beta)t}$$

$$e^{(\alpha - i\beta)t}, \; te^{(\alpha - i\beta)t}, \; t^2 e^{(\alpha - i\beta)t}, \ldots, \; t^{k-1} e^{(\alpha - i\beta)t}$$

we conclude, with the aid of Euler's formula, that the general solution of the corresponding differential equation must then contain a linear combination of the $2k$ real linearly independent solutions

$$e^{\alpha t}\cos\beta t,\quad te^{\alpha t}\cos\beta t,\quad t^2e^{\alpha t}\cos\beta t,\ldots,\quad t^{k-1}e^{\alpha t}\cos\beta t$$
$$e^{\alpha t}\sin\beta t,\quad te^{\alpha t}\sin\beta t,\quad t^2e^{\alpha t}\sin\beta t,\ldots,\quad t^{k-1}e^{\alpha t}\sin\beta t$$

In Example 4, we identify $k = 2$, $\alpha = 0$, and $\beta = 1$.

Of course the most difficult aspect of solving constant-coefficient differential equations is finding roots of auxiliary equations of degree greater than two. For example, to solve $3y''' + 5y'' + 10y' - 4y = 0$ we must solve $3m^3 + 5m^2 + 10m - 4 = 0$. Something we can try is to test the auxiliary equation for rational roots. Recall, if $m_1 = p/q$ is a rational root (expressed in lowest terms) of an auxiliary equation $a_nm^n + \cdots + a_1m + a_0 = 0$ with integer coefficients, then p is a factor of a_0 and q is a factor of a_n. For our specific cubic auxiliary equation, all the factors of $a_0 = -4$ and $a_n = 3$ are p: $\pm1, \pm2, \pm4$ and q: $\pm1, \pm3$, so the possible rational roots are p/q: $\pm1, \pm2, \pm4, \pm\frac{1}{3}, \pm\frac{2}{3}, \pm\frac{4}{3}$. Each of these numbers can then be tested, say, by synthetic division. In this way we discover both the root $m_1 = \frac{1}{3}$ and the factorization

$$3m^3 + 5m^2 + 10m - 4 = \left(m - \frac{1}{3}\right)(3m^2 + 6m + 12)$$

The quadratic formula then yields the remaining roots $m_2 = -1 + \sqrt{3}\,i$ and $m_3 = -1 - \sqrt{3}\,i$. Therefore, the general solution of $3y''' + 5y'' + 10y' - 4y = 0$ is

$$y = c_1e^{t/3} + e^{-t}(c_2\cos\sqrt{3}t + c_3\sin\sqrt{3}t)$$

Use of Computers Finding roots, or approximations of roots, of equations is a routine problem with computer software or with an appropriate graphic calculator. Moreover, some computer algebra systems are capable of giving explicit solutions of homogeneous linear constant-coefficient differential equations. For example, using *Mathematica*, the application

DSolve [y″[t] + 2y′[t] + 2y[t] = = 0, y[t], t]

yields

$$\text{y[t]} \rightarrow \frac{\text{C[2] Cos [t]} - \text{C[1] Sin [t]}}{\text{E}^{\text{t}}} \tag{11}$$

Translated, this means that $y = c_2e^{-t}\cos t + c_1e^{-t}\sin t$ is a solution of $y'' + 2y' + 2y = 0$. See Problem 2 in *Projects and Computer Lab Experiments for Chapter 3*.

Incidentally, notice that when we rewrote (11), −C[1] was changed to $+c_1$. Why can we do that?

EXERCISES 3.2

Answers to odd-numbered problems begin on page 441.

In Problems 1–32, find the general solution of the given differential equation.

1. $4y'' + y' = 0$

2. $y'' - 36y = 0$

3. $y'' + 9y = 0$

4. $3y'' + y = 0$

5. $y'' - y' - 6y = 0$

6. $y'' - 3y' + 2y = 0$

7. $\dfrac{d^2y}{dt^2} + 8\dfrac{dy}{dt} + 16y = 0$

8. $\dfrac{d^2y}{dt^2} - 10\dfrac{dy}{dt} + 25y = 0$

9. $y'' + 3y' - 5y = 0$

10. $y'' + 4y' - y = 0$

11. $12y'' - 5y' - 2y = 0$

12. $8y'' + 2y' - y = 0$

13. $y'' - 4y' + 5y = 0$

14. $2y'' - 3y' + 4y = 0$

15. $y''' - 4y'' - 5y' = 0$

16. $4y''' + 4y'' + y' = 0$

17. $y''' - y = 0$

18. $y''' + 5y'' = 0$

19. $y''' - 5y'' + 3y' + 9y = 0$

20. $y''' + 3y'' - 4y' - 12y = 0$

21. $y''' + y'' - 2y = 0$

22. $y''' - y'' - 4y = 0$

23. $y''' + 3y'' + 3y' + y = 0$

24. $y''' - 6y'' + 12y' - 8y = 0$

25. $\dfrac{d^4y}{dt^4} + \dfrac{d^3y}{dt^3} + \dfrac{d^2y}{dt^2} = 0$

26. $\dfrac{d^4y}{dt^4} - 2\dfrac{d^2y}{dt^2} + y = 0$

27. $16\dfrac{d^4y}{dx^4} + 24\dfrac{d^2y}{dx^2} + 9y = 0$

28. $\dfrac{d^4y}{dx^4} - 7\dfrac{d^2y}{dx^2} - 18y = 0$

29. $\dfrac{d^5y}{dx^5} - 16\dfrac{dy}{dx} = 0$

30. $\dfrac{d^5y}{dx^5} - 2\dfrac{d^4y}{dx^4} + 17\dfrac{d^3y}{dx^3} = 0$

31. $\dfrac{d^5y}{dr^5} + 5\dfrac{d^4y}{dr^4} - 2\dfrac{d^3y}{dr^3} - 10\dfrac{d^2y}{dr^2} + \dfrac{dy}{dr} + 5y = 0$

32. $2\dfrac{d^5y}{ds^5} - 7\dfrac{d^4y}{ds^4} + 12\dfrac{d^3y}{ds^3} + 8\dfrac{d^2y}{ds^2} = 0$

In Problems 33–42, solve the given differential equation subject to the indicated initial conditions.

33. $2y'' - 2y' + y = 0,\quad y(0) = -1,\quad y'(0) = 0$

34. $y'' - 2y' + y = 0,\quad y(0) = 5,\quad y'(0) = 10$

35. $y'' + y' + 2y = 0,\quad y(0) = y'(0) = 0$

36. $4y'' - 4y' - 3y = 0,\quad y(0) = 1,\quad y'(0) = 5$

37. $y'' - 3y' + 2y = 0,\quad y(1) = 0,\quad y'(1) = 1$

38. $y'' + y = 0,\quad y(\pi/3) = 0,\quad y'(\pi/3) = 2$

39. $y''' + 12y'' + 36y' = 0,\quad y(0) = 0,\quad y'(0) = 1,$
$y''(0) = -7$

40. $y''' + 2y'' - 5y' - 6y = 0,\quad y(0) = y'(0) = 0,\quad y''(0) = 1$

41. $\dfrac{d^4y}{dt^4} - 3\dfrac{d^3y}{dt^3} + 3\dfrac{d^2y}{dt^2} - \dfrac{dy}{dt} = 0,\quad y(0) = y'(0) = 0,$
$y''(0) = y'''(0) = 1$

42. $\dfrac{d^4y}{dt^4} - y = 0,\quad y(0) = y'(0) = y''(0) = 0,\quad y'''(0) = 1$

In Problems 43–46, solve the given differential equation subject to the indicated boundary conditions.

43. $y'' - 10y' + 25y = 0,\quad y(0) = 1,\quad y(1) = 0$

44. $y'' + 4y = 0,\quad y(0) = 0,\quad y(\pi) = 0$

45. $y'' + y = 0,\quad y'(0) = 0,\quad y'(\pi/2) = 2$

46. $y'' - y = 0,\quad y(0) = 1,\quad y'(1) = 0$

Discussion Problems

47. **(a)** The roots of an auxiliary equation are $m_1 = 4$, $m_2 = m_3 = -5$. What is the corresponding homogeneous differential equation? Discuss: Is your answer unique?

(b) The roots of an auxiliary equation are $m_1 = -\frac{1}{2}$, $m_2 = 3 + i$, $m_3 = 3 - i$. What is the corresponding homogeneous differential equation?

48. Find the general solution of $y''' + 6y'' + y' - 34y = 0$ if it is known that $y_1 = e^{-4t}\cos t$ is one solution.

49. Consider the second-order equation with constant coefficients $y'' + by' + cy = 0$.

(a) If $y(t)$ is a solution of the equation, discuss what conditions should be put on b and c so that $\lim_{t\to\infty} y(t) = 0$.

(b) Discuss what conditions should be put on b and c so that the equation possesses a nontrivial solution satisfying the boundary-conditions $y(0) = 0$, $y(1) = 0$.

50. In order to solve $y^{(4)} + y = 0$ we must solve $m^4 + 1 = 0$. This is a trivial problem using a CAS, but can also be done by hand, working with complex numbers. Observe that $m^4 + 1 = (m^2 + 1)^2 - 2m^2$. How does this help? Solve the differential equation.

51. Write a differential equation of the form $ay'' + by' + cy = 0$, where a, b, and c are nonzero real constants, for which a solution curve has the shape consistent with the figure. There may be no unique answer.

(a)

Figure 3.4.

(b)

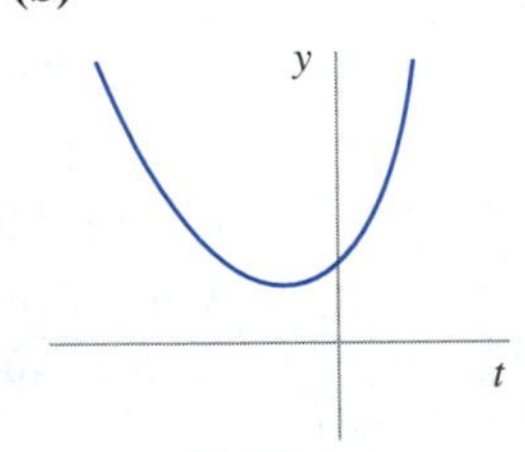

Figure 3.5.

(c)

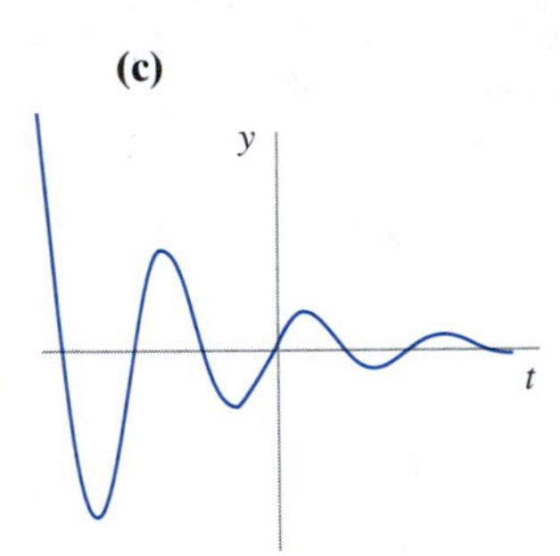

Figure 3.6.

(d)

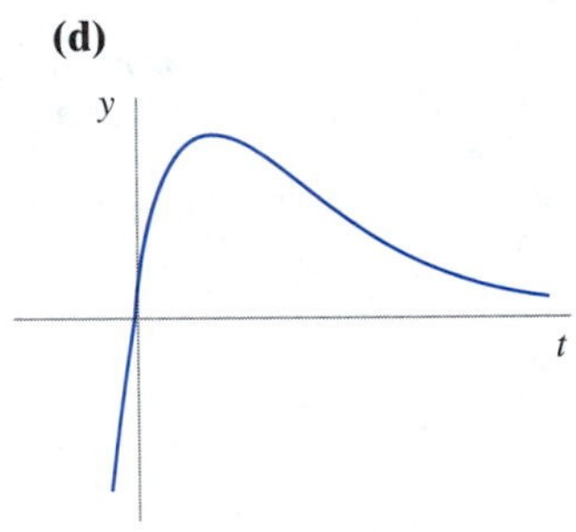

Figure 3.7.

(e)

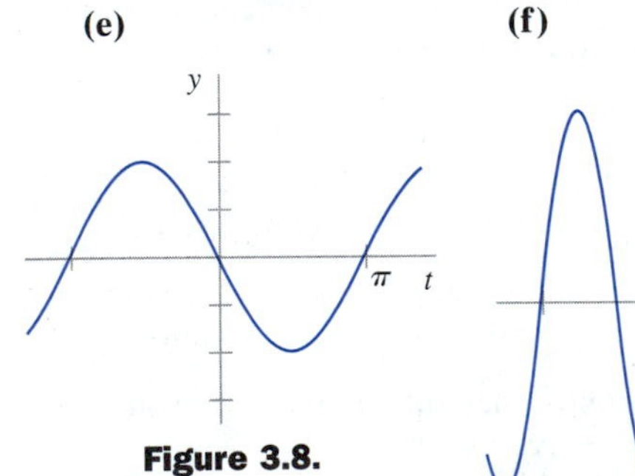

Figure 3.8.

(f)

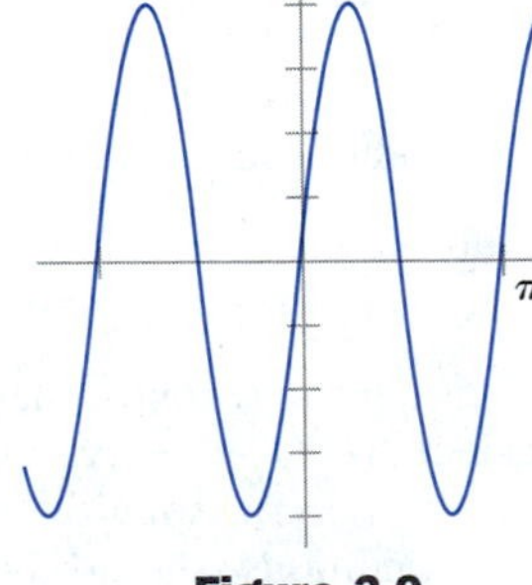

Figure 3.9.

52. (a) Suppose the linear second-order differential equation

$$a_2(t)y'' + a_1(t)y' + a_0(t)y = 0$$

is put into the standard form $y'' + P(t)y' + Q(t)y = 0$ by dividing through by $a_2(t)$. Assume that P and Q are continuous on some interval I. If $y_1(t)$ is a known solution of the last equation, use the substitution $y = u(t)y_1(t)$ to derive the second solution

$$y_2(t) = y_1(t)\int \frac{e^{-\int P(t)\,dt}}{y_1^2(t)}\,dt$$

(b) Verify that $y_1(t)$ and $y_2(t)$ are linearly independent on any interval I for which $y_1(t)$ is not zero.

(c) Find the general solution of the equation $t^2y'' - 3ty' + 4y = 0$ on the interval $(0, \infty)$, if it is known that $y_1 = t^2$ is a solution.

In Problems 53–55, the symbols y_0, y_1, L, and λ represent real numbers.

53. Consider the boundary-value problem

$$y'' + 16y = 0, \quad y(0) = y_0, \quad y\left(\frac{\pi}{2}\right) = y_1$$

Discuss: Is it possible to determine values of y_0 and y_1 so that the problem possesses

(a) precisely one nontrivial solution,
(b) more than one solution,
(c) no solution,
(d) the trivial solution?

54. Consider the boundary-value problem

$$y'' + 16y = 0, \quad y(0) = 1, \quad y(L) = 1$$

Discuss: Is it possible to determine values of $L > 0$ so that the problem possesses

(a) precisely one nontrivial solution,
(b) more than one solution,
(c) no solution,
(d) the trivial solution?

55. Consider the boundary-value problem

$$y'' + \lambda y = 0, \quad y(0) = 0, \quad y\left(\frac{\pi}{2}\right) = 0$$

Discuss: Is it possible to determine values of λ so that the problem possesses

(a) trivial solutions, **(b)** nontrivial solutions?

56. In the study of techniques of integration in calculus, certain indefinite integrals of the form $\int e^{ax}f(x)\,dx$ could be evaluated by applying integration by parts twice, recovering the original integral on the right side, solving for the original integral, and obtaining a constant multiple $k\int e^{ax}f(x)\,dx$ on the left side. The value of the integral is found then by dividing by k. Discuss: For what kind of functions f does the described procedure work? Your solution should lead to a differential equation. Carefully analyze this equation and solve for f.

3.3 Phase Portraits and Stability

In Section 2.1, we saw that we could discern certain qualitative properties of solutions of an autonomous first-order DE without even solving the equation. We turn now to autonomous second-order equations, although at this point we are going to consider only autonomous *linear* second-order equations that are homogeneous with constant coefficients. Since there is no mystery about the qualitative properties of solutions of such equations—we can always write down a general solution—our goal in this section is simply to introduce certain procedures and terminology that will be useful in Section 4.5. We are setting the stage for that future discussion.

Autonomous DEs Like their first-order relatives discussed in Section 2.1, we say that a second-order differential equation is **autonomous** if the independent variable does not appear explicitly. If x denotes the dependent variable, an autonomous second-order equation can be represented symbolically, either as $F(x, x', x'') = 0$, or in normal form as

$$\frac{d^2x}{dt^2} = g(x, x') \tag{1}$$

For example, $x'' + 2x' + x = 0$ (or $x'' = -2x' - x$) is an autonomous equation, but both $x'' = t^2x$ and $x'' + 2x' + x = \sin t$ are nonautonomous. We shall assume that the function g in (1) is continuously differentiable in some region in the xx'-plane.

In our first example, we consider a linear autonomous equation with constant coefficients. Using the methods of the last section we can easily obtain solutions of this equation and their graphs—but the focus will not be on the integral curves of the equation, but rather on curves that are determined by the variables x and dx/dt.

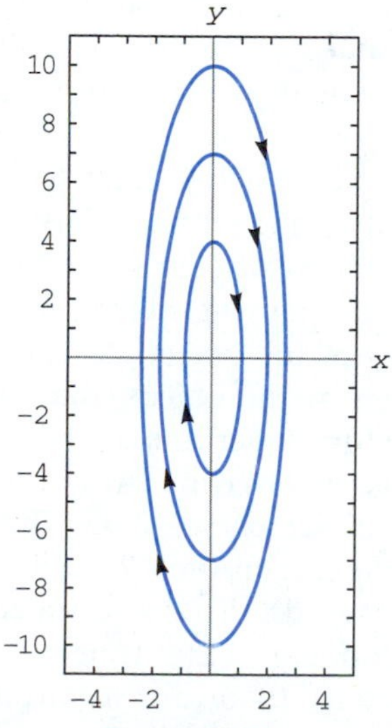

Figure 3.10.

■ EXAMPLE 1 Autonomous Second-Order DE

The function

$$x = c_1 \cos 4t + c_2 \sin 4t \tag{2}$$

is a two-parameter family of solutions of the autonomous differential equation $x'' + 16x = 0$. Let us denote the derivative dx/dt of (2) by the symbol y:

$$y = -4c_1 \sin 4t + 4c_2 \cos 4t \tag{3}$$

For each choice of c_1 and c_2, equations (2) and (3) constitute a set of parametric equations of a family of curves in the xy-plane. With the aid of a graphing utility, we get the curves shown in Figure 3.10. The apparent fact that these curves are ellipses is confirmed by eliminating the parameter

t from (2) and (3). Squaring these expressions, adding, and simplifying give

$$x^2 + \frac{y^2}{4^2} = c_1^2 + c_2^2 \tag{4}$$

The arrowheads were added to indicate the orientation induced on the curves by taking increasing values of t from the interval $-\infty < t < \infty$. ■

In general, the plane of the variables x and y, that is, x and dx/dt, is called the **phase plane** of the autonomous second-order differential equation $F(x, x', x'') = 0$. If $x = x(t)$ represents a real solution of $F(x, x', x'') = 0$ on some interval, and $y = dx/dt = y(t)$, then curves in the phase plane defined parametrically by the equations $x = x(t)$, $y = y(t)$, are called **trajectories**. The direction corresponding to increasing values of the parameter t is said to be the **positive direction** on the trajectory. The graphs of the family of trajectories are variously called a **two-dimensional phase portrait**, **phase portrait**, or **phase-plane diagram** of the differential equation $F(x, x', x'') = 0$.

You should not think that there is a one-to-one correspondence between solutions of a differential equation and trajectories in the phase plane. In Example 1, we saw that $x = c_2 \cos 4t + c_2 \sin 4t$ is a solution of the equation $x'' + 16x = 0$ for any choice of c_1 and c_2. For the choices $c_1 = 1, c_2 = 0$, and $c_1 = 0, c_2 = 1$ we obtain, in turn, the particular solutions $x = \cos 4t$ and $x = \sin 4t$. But (4) shows that these different choices for c_1 and c_2 yield the same ellipse $x^2 + y^2/4^2 = 1$. In other words, many different solutions of the original differential equation $F(x, x', x'') = 0$ could have the same trajectory in the phase plane.

Although we cannot do justice to the importance of the phase plane at this point in the course, nevertheless it is possible to give some meaning to these concepts. Let's once again consider the differential equation $x'' + 16x = 0$ in Example 1, but this time let us ignore the fact that we possess an explicit two-parameter family of solutions and suppose that we have on hand only the phase portrait of the differential equation. What information does Figure 3.10 give?

To interpret the phase portrait in Figure 3.10 it is helpful to think, as we did in the discussion in Section 2.1, in dynamic terms by letting the variable t represent time. A trajectory, then, is a set of points (x, y) where the first coordinate $x(t)$ gives the *position* of a moving object, and the second coordinate $y(t) = dx/dt$ is the *velocity* of the object. A trajectory then is not a representation of the path taken by the object (that would be a graph of x versus t), but rather a trajectory describes a *state* of the object, in other words, where it is and how fast it is moving. Also, the fact that the trajectories in Figure 3.10 are simple closed curves shows that the moving object returns to a given point after a certain time T, and keeps returning to that point subsequently. In other words, the closed trajectory indicates that the unknown solutions corresponding to that trajectory are *periodic*. Note, too, that the common center (0, 0) of all the orbits or trajectories corresponds to a constant solution ($x = 0$) of the equation

$x'' + 16x = 0$. Moreover, by substituting $y = dx/dt$ into the original differential equation, we find $dy/dt = -16x$, which is an equation that gives the *acceleration* of the object. Thus, in our dynamic interpretation, the center $(0, 0)$ could be called an **equilibrium** or **stationary point** since the object is at rest (velocity $y = 0$) and is free of forces (acceleration $dy/dt = 0$).

Writing a Second-Order DE as a System of DEs A phase portrait of an autonomous second-order differential equation can be obtained directly from the differential equation, in other words, we do not need an explicit solution. To do this, we need to express the second-order differential equation (1) as an equivalent system of two first-order equations. If we substitute $y = dx/dt$ and use $d^2x/dt^2 = dy/dt = g(x, y)$, then the autonomous equation (1) becomes

$$\begin{aligned} \frac{dx}{dt} &= y \\ \frac{dy}{dt} &= g(x, y) \end{aligned} \tag{5}$$

For example, to write $2x'' - 3x' + 5x = 0$ as a system, we solve the equation for the highest derivative: $x'' = -\frac{5}{2}x + \frac{3}{2}x'$ and then let $x' = y$. Since $x'' = y'$, the equivalent system of first-order equations is

$$\begin{aligned} \frac{dx}{dt} &= y \\ \frac{dy}{dt} &= -\frac{5}{2}x + \frac{3}{2}y \end{aligned}$$

Use of Computers When we enter system (5) along with initial conditions $x(t_0) = x_0$, $y(t_0) = y_0$, computer software—an ODE solver adapted to systems—enables us to plot a solution of the system (that is, graphs of x versus t and y versus t), or a trajectory (that is, a graph of y versus x) that passes through the specified point (x_0, y_0). Note that the graph of x versus t is the solution curve corresponding to the initial-value problem: $d^2x/dt^2 = g(x, x')$, $x(t_0) = x_0$, $x'(t_0) = y_0$.

■ EXAMPLE 2 Phase Portrait

Find the phase portrait of the differential equation $x'' + 2x' + x = 0$.

Solution If $y = dx/dt$, then after solving the given equation for $x'' = dy/dt$, we get $dy/dt = -x - 2y$. The phase portrait of the equation, or equivalently, the phase portrait of the system

$$\begin{aligned} \frac{dx}{dt} &= y \\ \frac{dy}{dt} &= -x - 2y \end{aligned}$$

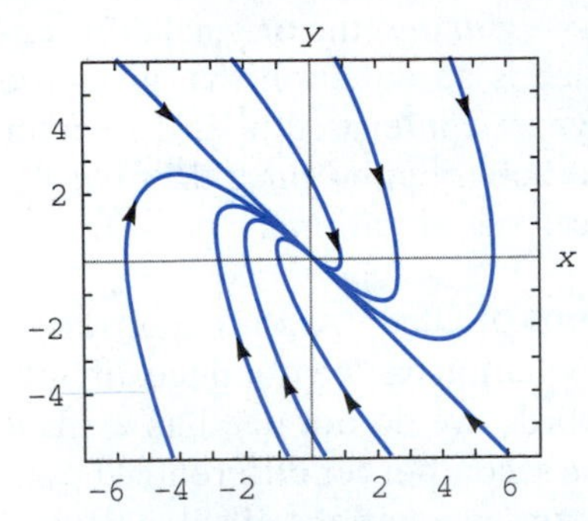

Figure 3.11.

in Figure 3.11 was obtained with the aid of computer software by choosing to plot y versus x for $-10 \le t \le 10$. The figure shows nine trajectories chosen to pass through the points $(5, 4)$, $(2, 3)$, $(1, 0)$, $(-4, 4)$, $(-5, 2)$, $(-2, -2)$, $(-1, -3)$, $(1, -4)$, and $(5, -5)$. The arrowheads mark these points and also indicate the positive direction on each trajectory. Note that since the trajectories are not closed curves, the corresponding solutions of the original differential equation are not periodic. If $x = x(t)$ is a solution of the equation corresponding to any of these trajectories, observe that $x \to 0$ as $t \to \infty$. ■

Critical Points Since equation (1) is autonomous, the equivalent system (5) is also said to be autonomous. The system (5) is just a special case of a more general autonomous system of two first-order equations

$$\begin{aligned} \frac{dx}{dt} &= f(x, y) \\ \frac{dy}{dt} &= g(x, y) \end{aligned} \tag{6}$$

that will be studied in greater detail in Chapter 4. Points where both dx/dt and dy/dt are zero in an autonomous system of two equations are said to be **critical points of the system**. In other words, (x_1, y_1) is a critical point of the system (6) if $f(x_1, y_1) = 0$ and $g(x_1, y_1) = 0$. As in the discussion of autonomous first-order equations, a critical point corresponds to a *constant solution*, in this case, the constant-valued functions $x(t) = x_1$, $y(t) = y_1$ is a solution of the system of equations (6). A constant-valued solution of the system (6) is called an **equilibrium solution**. In general,

A critical point of an autonomous second-order differential equation $F(x, x', x'') = 0$ is a critical point of its equivalent autonomous system.

Unless stated to the contrary, we shall assume hereafter that f and g in (6) are continuous and possess continuous first partial derivatives in a neighborhood of a critical point.

Linear Equations Because we have just studied them, our examples have employed homogeneous linear second-order differential equations with constant coefficients. It should be obvious that every equation of this kind,

$$ax'' + bx' + cx = 0, \quad a \neq 0 \tag{7}$$

a, b, and c real constants, has the form $F(x, x', x'') = 0$ and so is autonomous. In normal form $x'' = g(x, x')$, equation (7) is $x'' = -\frac{c}{a}x - \frac{b}{a}x'$. With the aid of this last form and the substitution $y = x'$, we see that (7) is equivalent to the following system of two first-order equations

$$\begin{aligned} \frac{dx}{dt} &= y \\ \frac{dy}{dt} &= -\frac{c}{a}x - \frac{b}{a}y \end{aligned} \tag{8}$$

When we set the right sides in (8) equal to zero and solve the system of algebraic equations $y = 0$, $(c/a)x + (b/a)y = 0$, we find a solution to be $x = 0$, $y = 0$. Indeed, by making the added assumption that $a \neq 0$ *and* $c \neq 0$, then the only critical point of the system (8) is the origin (0, 0), and so (0, 0) is the only critical point of the homogeneous linear equation (7). The restriction $c \neq 0$ guarantees that the critical point (0, 0) of (7) is **isolated**. This simply means that there exists some circular neighborhood around the critical point that is free of all other critical points. See Problem 19 in Exercises 3.3.

A Gallery of Phase Portraits The phase portrait of a homogeneous linear differential equation (7) depends on the nature of the roots of its auxiliary equation. Within the three cases for the general solution of (7) we can distinguish eight subcases. For example, in Case I, where the roots m_1 and m_2 are real and unequal, we have three possibilities: The roots can be both negative, both positive, or can have opposite algebraic signs. (Remember, $c \neq 0$ in (7) guarantees that $m = 0$ is not a root of its auxiliary equation.) All of the subcases are summarized in the table that follows.

CASE I Distinct Real Roots $m_1 \neq m_2$	*CASE II* Repeated Real Roots $m_1 = m_2$	*CASE III* Conjugate Complex Roots $m_1 = \alpha + i\beta$, $m_2 = \alpha - i\beta$
(i) Both negative: $m_1 < m_2 < 0$	(i) Negative: $m_1 < 0$	(i) Complex: $\alpha < 0$, $\beta \neq 0$
(ii) Both positive: $m_1 > m_2 > 0$	(ii) Positive: $m_1 > 0$	(ii) Complex: $\alpha > 0$, $\beta \neq 0$
(iii) Opposite signs: $m_1 < 0 < m_2$	—	(iii) Pure imaginary: $\alpha = 0$, $\beta \neq 0$

The next three figures show the phase portrait of the differential equation (y versus x) along with a typical solution curve (x versus t) corresponding to the different subcases listed. To understand properly the nature of a phase portrait, it is recommended that you verify that the graph of $x(t)$ is consistent with the pattern of the trajectories. For example, in Figure 3.12(b), observe that a solution $x(t) \to 0$ and its slope $x'(t) \to 0$ as $t \to \infty$; in Figure 3.12(a), if $(x(t), y(t))$ denotes a point on a trajectory, then the direction indicated by the arrowheads shows $x(t) \to 0$ and $y(t) = x'(t) \to 0$ as $t \to \infty$.

As was done in the caption of each phase portrait, it is standard practice to give the critical point (0, 0) a descriptive name depending on

CASE I General solution $x(t) = c_1 e^{m_1 t} + c_2 e^{m_2 t}$

(i) $m_1 < m_2 < 0$

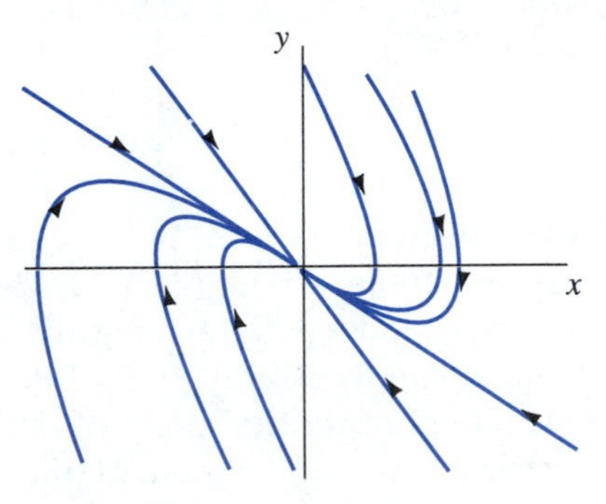

(a) The point (0, 0) is called a **node.**

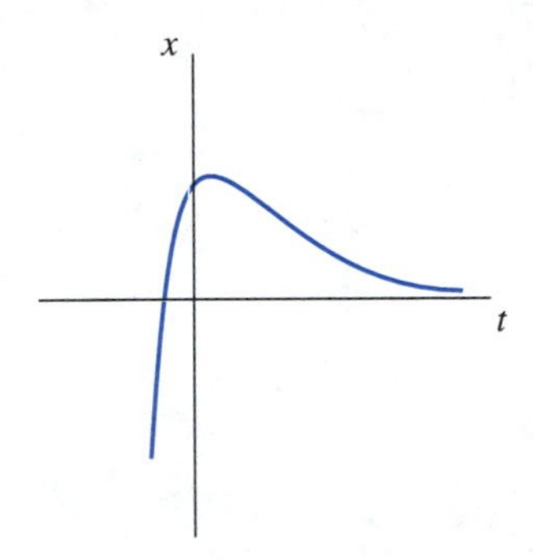

(b) Typical graph of $x(t)$.

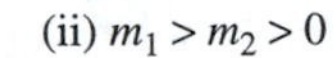

(ii) $m_1 > m_2 > 0$

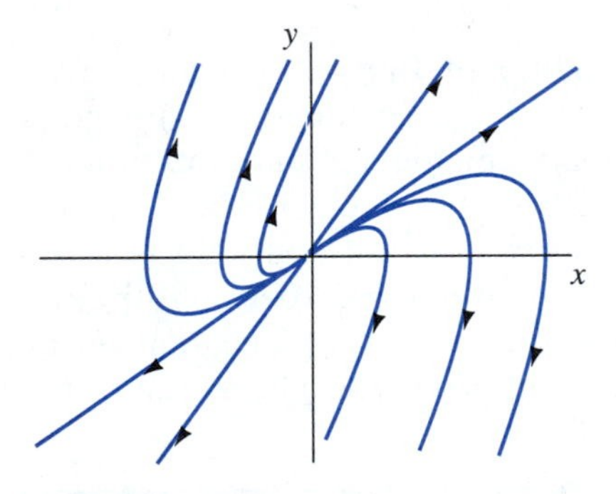

(c) The point (0, 0) is called a **node.**

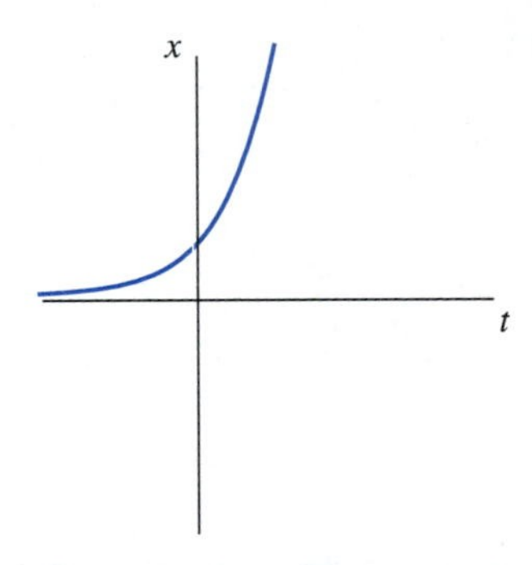

(d) Typical graph of $x(t)$.

(iii) $m_1 < 0 < m_2$

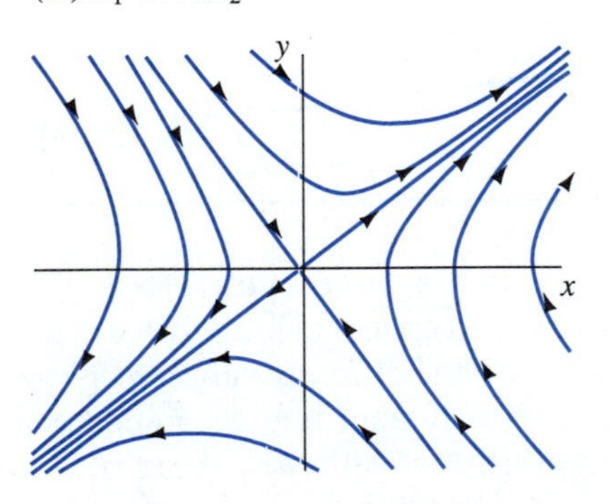

(e) The point (0, 0) is called a **saddle point.**

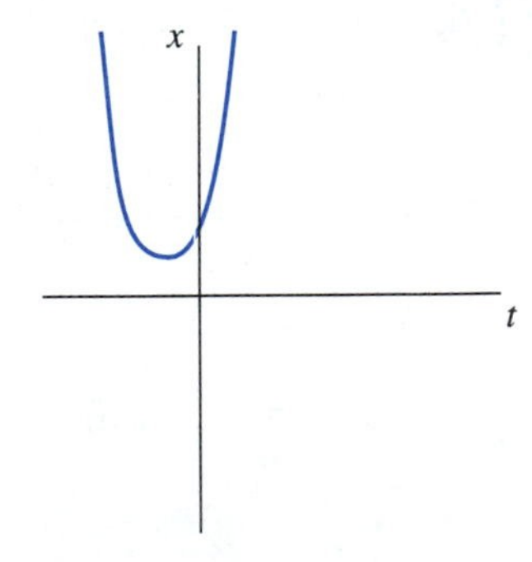

(f) Typical graph of $x(t)$.

Figure 3.12.

the behavior of the trajectories near this point. The *saddle point* derives its name from the resemblance of the phase portrait with the level curves of a hyperbolic paraboloid. (Recall the graph of that surface has a saddle point.) We will say more about these names in Section 4.5.

CASE II General solution $x(t) = c_1 e^{m_1 t} + c_2 t e^{m_1 t}$

(i) $m_1 < 0$

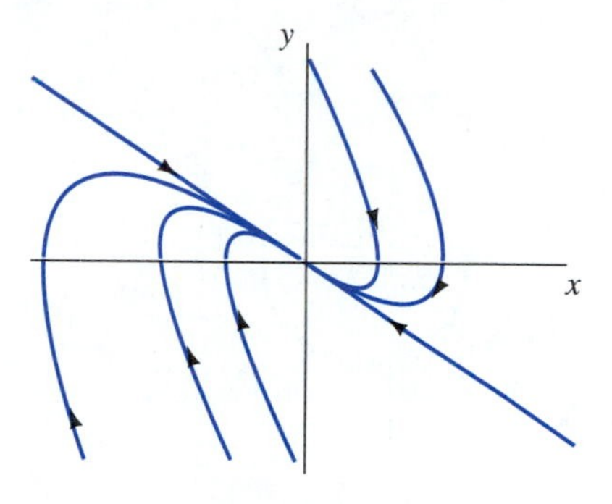

(a) The point (0, 0) is called a **node.**

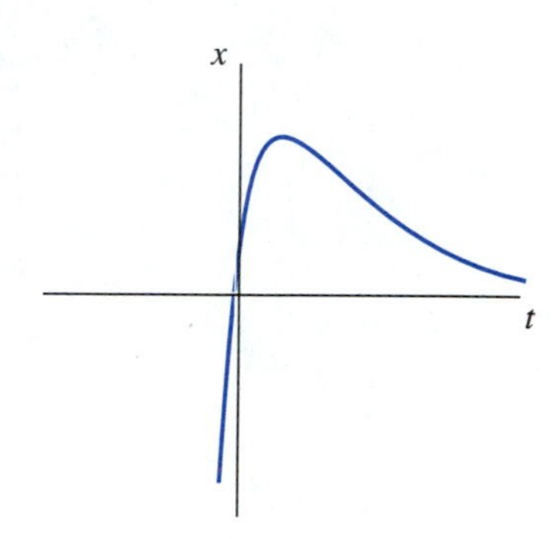

(b) Typical graph of $x(t)$.

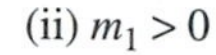

(ii) $m_1 > 0$

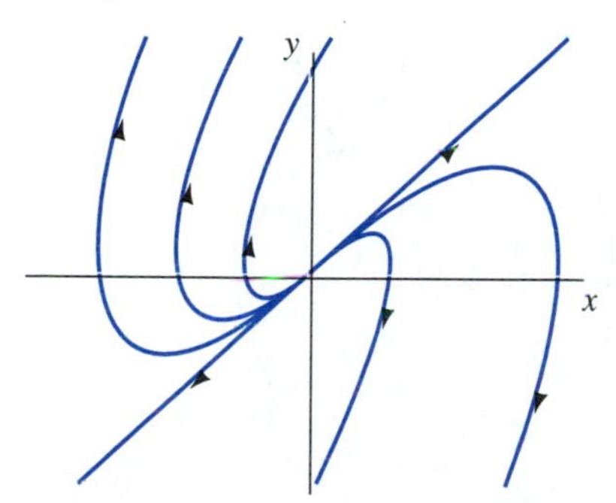

(c) The point (0, 0) is called a **node.**

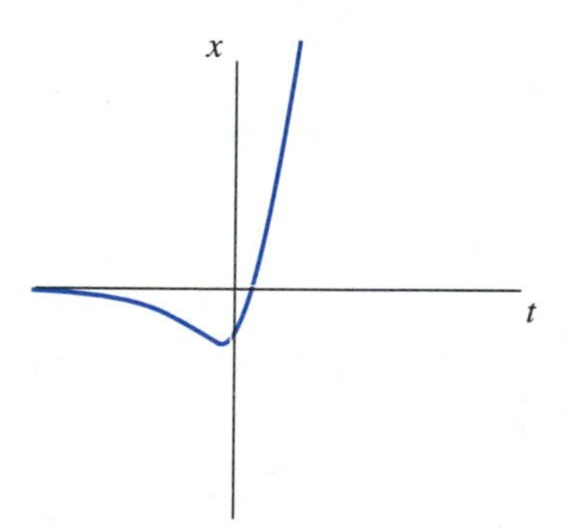

(d) Typical graph of $x(t)$.

Figure 3.13.

Stability In Section 2.1, we classified a critical point x_1 of an autonomous first-order differential equation as either asymptotically stable or unstable. The critical point (0, 0) of an autonomous *linear* second-order equation (7) can be classified in one of three different ways. The arrowheads on the trajectories in Figures 3.12(a), 3.13(a), and 3.14(a) indicate that *all* trajectories tend to the origin, that is, $(x(t), y(t)) \to (0, 0)$, as $t \to \infty$. In each of these cases, we say that the origin (0, 0) is an **asymptotically stable** critical point, or more specifically, (0, 0) is an asymptotically stable node in Figures 3.12(a) and 3.13(a) and an asymptotically stable spiral point in Figure 3.14(a). In Figures 3.12(c), 3.13(c), and 3.14(c), the arrowheads on the trajectories indicate that the points on all trajectories move away, or recede, from (0, 0) as $t \to \infty$. Correspondingly notice that a typical solution $x(t)$, graphed in Figures 3.12(d), 3.13(d), and 3.14(d), becomes unbounded as $t \to \infty$. In each of these cases, we say that the origin (0, 0) is an **unstable** critical point; the origin is an unstable node in Figures 3.12(c) and 3.13(c) and an unstable spiral point in Figure 3.14(c). In Figure 3.12(e), the origin is a saddle point; a saddle point is always unstable, even though it is apparent that the points on *two* trajectories—namely the straight-line trajectories in

CASE III General solution $x(t) = c_1e^{\alpha t}\cos\beta t + c_2e^{\alpha t}\sin\beta t$

(i) $\alpha < 0$, $\beta \neq 0$

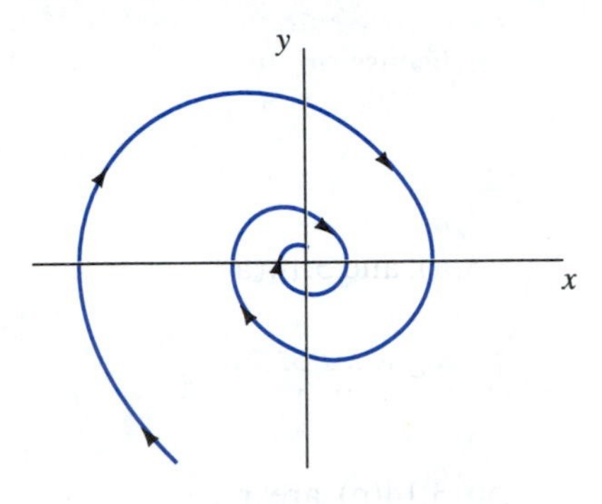

(a) The point (0, 0) is called a **spiral point**

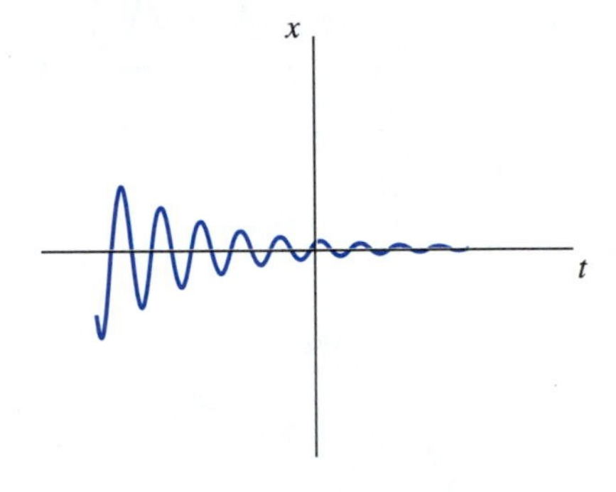

(b) Typical graph of $x(t)$.

(ii) $\alpha > 0$, $\beta \neq 0$

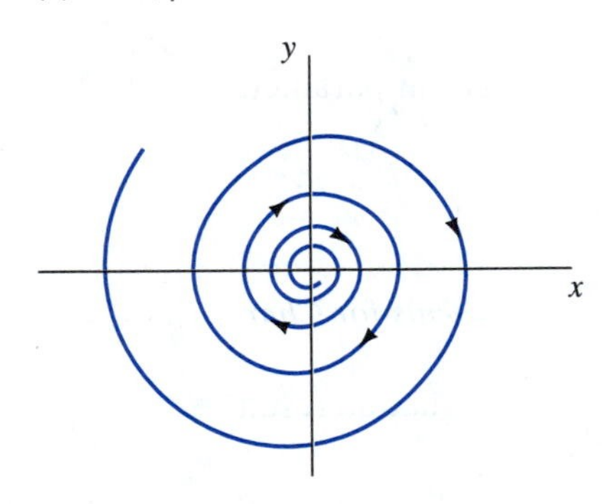

(c) The point (0, 0) is called a **spiral point.**

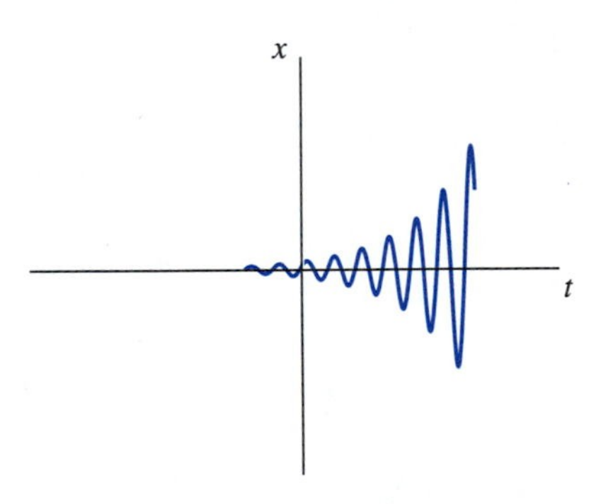

(d) Typical graph of $x(t)$.

(iii) $\alpha = 0$, $\beta \neq 0$

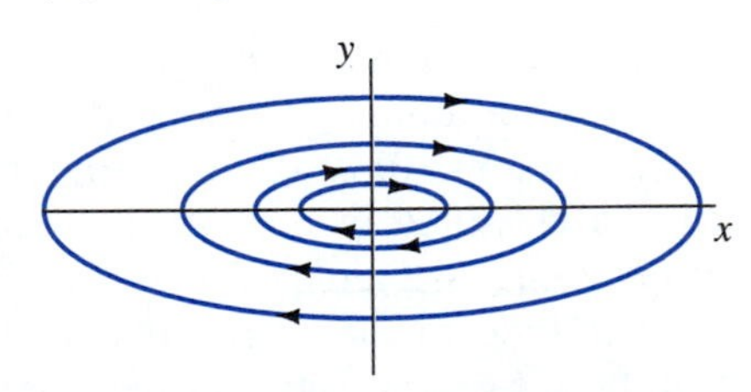

(e) The point (0, 0) is called a **center.**

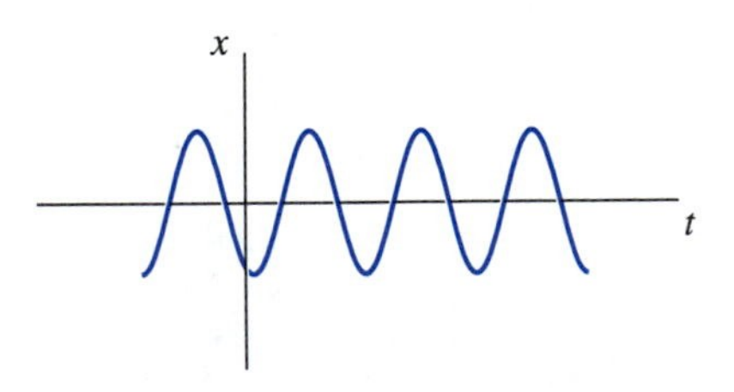

(f) Typical graph of $x(t)$.

Figure 3.14.

the second and fourth quadrants—approach or "enter" (0, 0) as $t \to \infty$. Finally, we are left with the case where (0, 0) is a center as illustrated in Figure 3.14(e). The closed trajectories reflect the fact that solutions of (8) and the corresponding solutions of (7) are periodic—that is, there exists a positive number T such that $x(t + T) = x(t)$, $y(t + T) = y(t)$. The trajectories neither converge to the origin nor diverge away from it, and so (0, 0) is said to be a **stable** critical point. The center in Figure 3.14(e) is not, however, asymptotically stable because, regardless of how close an ellipse in that family of closed trajectories may be to the origin, a point $(x(t), y(t))$ must move around that curve in a periodic manner, in other words, $(x(t), y(t))$ remains nearby but cannot approach the critical point $(0, 0)$ as $t \to \infty$. In light of this discussion, we see that $(0, 0)$ in Example 1 is a (stable) center, whereas in Example 2 $(0, 0)$ is an asymptotically stable node.

Attractors and Repellers The direction of the arrowheads in Figures 3.12(a), 3.13(a), and 3.14(a) indicate that points $(x(t), y(t))$ on all trajectories are drawn into the origin as $t \to \infty$, in a manner reminiscent of iron filings pulled to a magnet. For this reason, asymptotically stable nodes and spiral points are also called **attractors**. Analogous to particles with the same electric charge, points $(x(t), y(t))$ on all trajectories in Figures 3.12(c), 3.13(c), and 3.14(c) are repulsed from the origin. Accordingly, unstable nodes and unstable spiral points are referred to as **repellers**.

Bifurcation Solutions of a differential equation are often sensitive to small changes in parameters (coefficients) in the equation; a slight change in the value of a parameter may result in a **bifurcation**. The word *bifurcation* means to split into two parts, and so the original meaning of the word referred to the phenomenon of a single critical point splitting into two or more critical points. See Problems 24, 25, and 33 in *Projects and Computer Lab Experiments for Chapter 2*. Nowadays the word bifurcation is also used to refer to an *abrupt change in the qualitative nature of the solutions* of a differential equation resulting from a small change in the value of a parameter. Consider the simple second-order equation $y'' + 2\lambda y' + y = 0$, where λ represents any real number. The auxiliary roots of the equation are $m_1 = -\lambda + \sqrt{\lambda^2 - 1}$ and $m_2 = -\lambda - \sqrt{\lambda^2 - 1}$. The value $\lambda = 0$ is an important transition number called a **bifurcation value**; *at* this number, the origin $(0, 0)$ is a stable center; changing λ slightly in either direction abruptly changes $(0, 0)$ to an unstable critical point ($\lambda < 0$) or to an asymptotically stable critical point ($\lambda > 0$). Specifically, for $\lambda = 0$, all the solutions $y(t)$ of the equation are bounded as $t \to \infty$, but if we change λ to $\lambda = -0.001$, for example, then all solutions $y(t)$ are unbounded as $t \to \infty$; changing λ a little in the opposite direction to, say, $\lambda = 0.001$, not only are all solutions bounded but $y(t) \to 0$ as $t \to \infty$. Also see the Remarks at the end of Section 3.7.

Remarks

(*i*) In Example 1, we obtained a Cartesian equation for the family of trajectories by using a known solution $x(t)$ to find $y(t) = dx/dt$ and then eliminating the parameter t from a set of parametric equations. It is sometimes possible to obtain such a Cartesian equation without the knowledge of an explicit solution. We can divide the second equation in (5) by the first and solve, if possible, the resulting first-order differential equation $dy/dx = g(x, y)/y$. The one-parameter family of solution curves of that equation is a one-parameter Cartesian equation of the family of trajectories for (1). For example, the autonomous system equivalent to $x'' + 16x = 0$ in Example 1 is

$$\frac{dx}{dt} = y$$

$$\frac{dy}{dt} = -16x$$

By dividing the second equation of this system by the first equation and using the fact that

$$\frac{dy}{dx} = \frac{dy/dt}{dx/dt}$$

we obtain

$$\frac{dy}{dx} = -\frac{16x}{y} \tag{9}$$

A solution of the first-order equation (9) is a trajectory of the second-order equation $x'' + 16x = 0$. Equation (9) is separable and so integration of $16x\,dx + y\,dy = 0$ yields $16x^2 + y^2 = c^2$ or $x^2 + y^2/4^2 = c_1^2$. The last equation represents the same family of ellipses with center (0, 0) defined by the parametric equations (2) and (3) except that it does not assign any direction to a trajectory.

(*ii*) If you explore the topic of stability, you will find that there is a vast amount of terminology, and texts will often differ on what things are called. For example, a critical point is variously called an **equilibrium point**, a **singular point**, a **stationary point**, or a **rest point**. A spiral point is called a **focus**, **focal point**, or **vortex point**. Trajectories are called **paths** or **orbits**. Some texts use the word "orbit" only in reference to the trajectories around a center. Asymptotically stable nodes and spiral points are also called **sinks** and their unstable counterparts are **sources**.

EXERCISES 3.3

Answers to odd-numbered problems begin on page 441.

In Problems 1 and 2, solve the given autonomous differential equation. Proceed as in Example 1 and obtain parametric equations of the trajectories. Use the parametric equations and a graphing utility to obtain a phase portrait of the differential equation. Use arrows to indicate the positive direction on the trajectories.

1. $x'' + x' - 2x = 0$

2. $x'' - 4x' + 4x = 0$

3. Find two different particular solutions of the differential equation in Problem 1 that correspond to the same trajectory.

4. Find two different particular solutions of the differential equation in Problem 2 that correspond to the same trajectory.

In Problems 5 and 6, do not solve the given autonomous differential equation. Proceed as in Example 2 and use computer software to obtain a phase portrait of the differential equation. Use arrows to indicate the positive direction on the trajectories.

5. $x'' + x = 0$

6. $x'' - 2x' + 10x = 0$

In Problems 7–14, use the gallery of phase portraits in this section to classify the critical point (0, 0) as a node, saddle

point, spiral point, or center. Classify (0, 0) as stable (but not asymptotically stable), asymptotically stable, or unstable.

7. $x'' - 9x = 0$ **8.** $x'' + 5x = 0$

9. $x'' - 7x' + 10x = 0$ **10.** $x'' + 8x' + 7x = 0$

11. $x'' - 2x' + 37x = 0$ **12.** $x'' - 6x' + 9x = 0$

13. $x'' + 10x' + 25x = 0$ **14.** $3x'' + 2x' + 3x = 0$

In Problems 15 and 16, do not solve the given differential equation, but use computer software to plot the trajectory of the equation corresponding to the given initial point. Plot the corresponding solution of the initial-value problem either by hand or by using an ODE solver as instructed. Indicate the initial point on each graph.

15. $x'' - x = 0$
 (a) $x(0) = 0, x'(0) = 1$ **(b)** $x(0) = 1, x'(0) = 0$
 (c) $x(0) = 1, x'(0) = -1$

16. $x'' + 2x' + x = 0$
 (a) $x(0) = 2, x'(0) = 1$ **(b)** $x(0) = -1, x'(0) = 1$
 (c) $x(0) = -4, x'(0) = -2$

Discussion Problems

17. Without graphing, discuss the phase portrait of the autonomous equation $x'' = 0$.

18. Four of the trajectories in the phase portrait of $x'' - 9x' + 18x = 0$ are straight line segments.
 (a) Find Cartesian equations of these lines.
 (b) Find four initial-value problems whose solutions correspond to these straight-line trajectories.

19. Consider the autonomous equation $ax'' + bx' = 0$, where a and b are nonzero real constants. Discuss: Why isn't (0, 0) an isolated critical point of the equation?

20. Consider equation (7) in the form $x'' + bx' + cx = 0$, where b and c are real constants and $c \neq 0$. Discuss what conditions should be put on b and c such that (0, 0) is
 (a) a node, which is either unstable or asymptotically stable,
 (b) a saddle point,
 (c) a spiral point, which is either unstable or asymptotically stable, or
 (d) a center.

21. **(a)** Let $a \neq 0$, b, $c \neq 0$, and $d \neq 0$ be constants. Show that the critical point of $ax'' + bx' + cx = d$ is $(d/c, 0)$.
 (b) Show that the change of variables $u(t) = x(t) - d/c$ gives $au'' + bu' + cu = 0$. Briefly discuss the significance of this result.

22. Find the critical points of the autonomous nonlinear equation $x'' + x + x^2 = 0$.

In Problems 23 and 24, the given curve is a trajectory in the phase plane for an autonomous second-order differential equation $F(x, x', x'') = 0$, where $dx/dt = y$. Assume $t_i > 0$, $i = 0, 1, 2, 3, 4$. Use the data from the figure to graph a corresponding solution curve.

23. **24.**

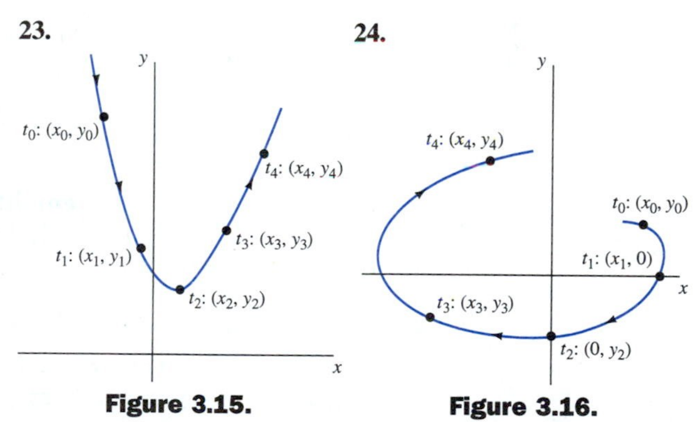

Figure 3.15. **Figure 3.16.**

3.4 Nonhomogeneous Linear Equations

To solve a nonhomogeneous linear DE, we must do two things: Find the complementary function y_c, which is the general solution of the associated homogeneous equation, and then find any particular solution y_p of the nonhomogeneous equation. The general solution of the nonhomogeneous DE is then $y = y_c + y_p$. In the last section we saw how to obtain y_c when the coefficients of the equation were constants. In this section we are going to examine two different methods for finding a particular solution y_p.

As we did in Section 3.2, we begin the discussion by considering the special case of a linear second-order differential equation

$$a_2(t)y'' + a_1(t)y' + a_0(t)y = g(t) \tag{1}$$

3.4.1 Undetermined Coefficients

Method of Solution The first of two ways we shall consider for obtaining a particular solution y_p is called the **method of undetermined coefficients.** The underlying idea in this method is a conjecture or educated guess about the form of y_p motivated by the kinds of functions that comprise $g(t)$. Basically straightforward, the method is, however, limited to nonhomogeneous linear equations such as (1) where

- the coefficients a_i of the equation are constants, and
- the input $g(t)$ is a polynomial function, an exponential function $e^{\alpha t}$, a sine or cosine function $\sin \beta t$, $\cos \beta t$, or a finite sum or product of these functions.

Note A constant function $g(t) = k$ is a polynomial function of degree 0.

The following functions are some examples of the types of inputs $g(t)$ that are appropriate for this discussion:

$$g(t) = 10, \quad g(t) = t^2 - 5t, \quad g(t) = 15t - 6 + 8e^{-t}$$

$$g(t) = \sin 3t - 5t \cos 2t, \quad g(t) = te^t \sin t + (3t^2 - 1)e^{-4t}$$

That is, $g(t)$ is a linear combination of functions of the type

$$P(t) = a_n t^n + a_{n-1}t^{n-1} + \cdots + a_1 t + a_0, \quad P(t)e^{\alpha t}, \quad P(t)e^{\alpha t} \sin \beta t, \quad \text{and} \quad P(t)e^{\alpha t} \cos \beta t$$

where n is a nonnegative integer and α and β are real numbers.

The set of functions that consists of polynomials, exponentials $e^{\alpha t}$, sines, and cosines has the remarkable property that derivatives of their sums and products are again sums and products of polynomials, exponentials $e^{\alpha t}$, sines, and cosines. Since a linear differential equation is simply a linear combination of derivatives, and since $a_2 y_p'' + a_1 y_p' + a_0 y_p$ must be identical to $g(t)$, it seems reasonable to assume then that y_p *has the same form as* $g(t)$. As a consequence of this and the fact that the coefficients are constants, each general solution $y = y_c + y_p$ will be defined on the interval $(-\infty, \infty)$.

The next two examples illustrate the idea.

EXAMPLE 1 General Solution Using Undetermined Coefficients

Solve

$$y'' + 4y' - 2y = 2t^2 - 3t + 6 \tag{2}$$

Solution **Step 1.** The first step is to solve the homogeneous equation $y'' + 4y' - 2y = 0$. From the quadratic formula, we find that the roots of the auxiliary equation $m^2 + 4m - 2 = 0$ are $m_1 = -2 + \sqrt{6}$ and $m_2 = -2 - \sqrt{6}$. Hence the complementary function is

$$y_c = c_1 e^{(-2 + \sqrt{6})t} + c_2 e^{(-2 - \sqrt{6})t}$$

Step 2. Since the function $g(t)$ is a quadratic polynomial, let us assume a particular that is also in the form of a quadratic polynomial:

$$y_p = At^2 + Bt + C$$

We seek to determine *specific* coefficients A, B, and C for which y_p is a solution of (2). Substituting y_p and the derivatives $y_p' = 2At + B$ and $y_p'' = 2A$ into the given differential equation (2), we get

$$y_p'' + 4y_p' - 2y_p = 2A + 8At + 4B - 2At^2 - 2Bt - 2C = 2t^2 - 3t + 6$$

Since the last equation is supposed to be an identity, the coefficients of like powers of t must be equal:

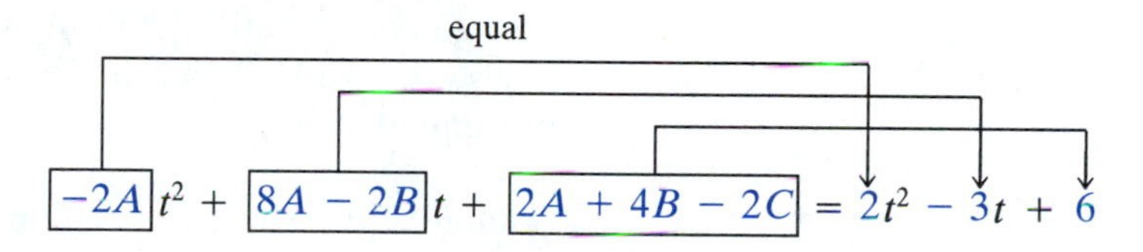

That is, $\quad -2A = 2, \quad 8A - 2B = -3, \quad 2A + 4B - 2C = 6$

Solving this system of equations leads to the values $A = -1$, $B = -\frac{5}{2}$, and $C = -9$. Thus a particular solution is

$$y_p = -t^2 - \frac{5}{2}t - 9$$

Step 3. The general solution of the given equation is

$$y = y_c + y_p = c_1 e^{(-2 + \sqrt{6})t} + c_2 e^{-(2 + \sqrt{6})t} - t^2 - \frac{5}{2}t - 9$$ ■

■ EXAMPLE 2 **Particular Solution Using Undetermined Coefficients**

Find a particular solution of $\quad y'' - y' + y = 2 \sin 3t$.

Solution A natural first guess for a particular solution would be $A \sin 3t$. But since successive differentiations of $\sin 3t$ produce $\sin 3t$ *and* $\cos 3t$, we are prompted instead to assume a particular solution that includes both of these terms:

$$y_p = A \cos 3t + B \sin 3t$$

Differentiating y_p and substituting the results into the differential equation give, after regrouping,

$$y_p'' - y_p' + y_p = (-8A - 3B) \cos 3t + (3A - 8B) \sin 3t = 2 \sin 3t$$

or

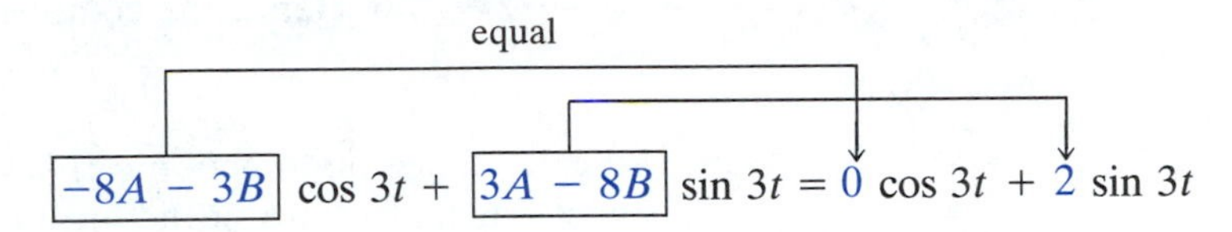

From the resulting system of equations,

$$-8A - 3B = 0, \quad 3A - 8B = 2$$

we get $A = \frac{6}{73}$ and $B = -\frac{16}{73}$. A particular solution of the equation is

$$y_p = \frac{6}{73}\cos 3t - \frac{16}{73}\sin 3t$$ ■

As we mentioned, the form that we assume for the particular solution y_p is an educated guess; it is not a blind guess. This educated guess must take into consideration not only the types of functions that make up $g(t)$, but also, as we shall see in Example 4, the functions that make up the complementary function y_c.

■ EXAMPLE 3 Forming y_p by Superposition

Solve

$$y'' - 2y' - 3y = 4t - 5 + 6te^{2t} \tag{3}$$

Solution **Step 1.** First, the solution of the associated homogeneous equation $y'' - 2y' - 3y = 0$ is found to be $y_c = c_1e^{-t} + c_2e^{3t}$.

Step 2. Next, the presence of $4t - 5$ in $g(t)$ suggests that the particular solution includes a linear polynomial. Furthermore, since the derivative of the product te^{2t} produces $2te^{2t}$ and e^{2t}, we also assume that the particular solution includes both te^{2t} and e^{2t}. In other words, g is the sum of two basic kinds of functions:

$$g(t) = g_1(t) + g_2(t) = \textit{polynomial} + \textit{exponentials}$$

Correspondingly, the superposition principle for nonhomogeneous equations (Theorem 3.7) suggests that we seek a particular solution

$$y_p = y_{p_1} + y_{p_2}$$

where $y_{p_1} = At + B$ and $y_{p_2} = Cte^{2t} + De^{2t}$. Substituting

$$y_p = At + B + Cte^{2t} + De^{2t}$$

into the given equation (3) and grouping like terms give

$$y_p'' - 2y_p' - 3y_p = -3At - 2A - 3B - 3Cte^{2t} + (2C - 3D)e^{2t} = 4t - 5 + 6te^{2t} \tag{4}$$

From this identity, we obtain the four equations

$$-3A = 4, \quad -2A - 3B = -5, \quad -3C = 6, \quad 2C - 3D = 0$$

The last equation in this system results from the interpretation that the coefficient of e^{2t} in the right member of (4) is zero. Solving, we find $A = -\frac{4}{3}$, $B = \frac{23}{9}$, $C = -2$, and $D = -\frac{4}{3}$. Consequently,

$$y_p = -\frac{4}{3}t + \frac{23}{9} - 2te^{2t} - \frac{4}{3}e^{2t}$$

Step 3. The general solution of the equation is

$$y = c_1e^{-t} + c_2e^{3t} - \frac{4}{3}t + \frac{23}{9} - \left(2t + \frac{4}{3}\right)e^{2t}$$

■

In light of the superposition principle (Theorem 3.7), we can also approach Example 3 from the viewpoint of solving two simpler problems. You should verify that substituting

$$y_{p_1} = At + B \quad \text{into} \quad y'' - 2y' - 3y = 4t - 5$$

and

$$y_{p_2} = Cte^{2t} + De^{2t} \quad \text{into} \quad y'' - 2y' - 3y = 6te^{2t}$$

yields in turn $y_{p_1} = -\frac{4}{3}t + \frac{23}{9}$ and $y_{p_2} = -(2t + \frac{4}{3})e^{2t}$. A particular solution of (3) is then $y_p = y_{p_1} + y_{p_2}$.

The next example illustrates that sometimes the "obvious" assumption for the form of y_p is not a correct assumption.

■ **EXAMPLE 4 A Glitch in the Method**

Find a particular solution of $y'' - 5y' + 4y = 8e^t$.

Solution Differentiation of e^t produces no new functions. Thus, proceeding as we did in the earlier examples, we can reasonably assume a particular solution of the form $y_p = Ae^t$. But substitution of this expression into the differential equation yields the contradictory statement $0 = 8e^t$, and so we have clearly made the wrong guess for y_p.

The difficulty here is apparent upon examining the complementary function $y_c = c_1e^t + c_2e^{4t}$. Observe that our assumption Ae^t is already present in y_c. This means that e^t is a solution of the associated homogeneous differential equation, and a constant multiple Ae^t, when substituted into the differential equation, necessarily produces zero.

What then should be the form of y_p? Inspired by Case II of Section 3.2, let's see whether we can find a particular solution of the form

$$y_p = Ate^t$$

Substituting $y_p' = Ate^t + Ae^t$ and $y_p'' = Ate^t + 2Ae^t$ into the differential equation and simplifying give

$$y_p'' - 5y_p' + 4y_p = -3Ae^t = 8e^t$$

From the last equality we see that the value of A is now determined as $A = -\frac{8}{3}$. Therefore a particular solution of the given equation is

$$y_p = -\frac{8}{3}te^t \qquad \blacksquare$$

The difference in the procedures used in Examples 1–3 and in Example 4 suggests that we consider two cases. The first case reflects the situation in Examples 1–3.

CASE I No function in the assumed particular solution is a solution of the associated homogeneous differential equation.

In Table 3.1, we illustrate some specific examples of $g(t)$ in (1) along with the corresponding form of the particular solution. We are, of course, taking for granted that no function in the assumed particular solution y_p is duplicated by a function in the complementary function y_c.

Table 3.1 Trial Particular Solutions

$g(t)$	Form of y_p
1. 1 (any constant)	A
2. $5t + 7$	$At + B$
3. $3t^2 - 2$	$At^2 + Bt + C$
4. $t^3 - t + 1$	$At^3 + Bt^2 + Ct + D$
5. $\sin 4t$	$A \cos 4t + B \sin 4t$
6. $\cos 4t$	$A \cos 4t + B \sin 4t$
7. e^{5t}	Ae^{5t}
8. $(9t - 2)e^{5t}$	$(At + B)e^{5t}$
9. t^2e^{5t}	$(At^2 + Bt + C)e^{5t}$
10. $e^{3t} \sin 4t$	$Ae^{3t} \cos 4t + Be^{3t} \sin 4t$
11. $5t^2 \sin 4t$	$(At^2 + Bt + C) \cos 4t + (Dt^2 + Et + F) \sin 4t$
12. $te^{3t} \cos 4t$	$(At + B)e^{3t} \cos 4t + (Ct + D)e^{3t} \sin 4t$

■ EXAMPLE 5 Forms of Particular Solutions—Case I

Determine the form of a particular solution of

$$\textbf{(a)}\ y'' - 8y' + 25y = 5t^3e^{-t} - 7e^{-t} \quad \text{and} \quad \textbf{(b)}\ y'' + 4y = t \cos t$$

Solution

(a) We can write $g(t) = (5t^3 - 7)e^{-t}$. Using entry 9 in Table 3.1 as a model (with a cubic polynomial replacing the quadratic polynomial), we assume a particular solution of the form

$$y_p = (At^3 + Bt^2 + Ct + D)e^{-t}$$

Note that there is no duplication between the terms in y_p and the terms in the complementary function $y_c = e^{4t}(c_1 \cos 3t + c_2 \sin 3t)$.

(b) The function $g(t) = t \cos t$ is similar to entry 11 in Table 3.1, except of course we use a linear rather than a quadratic polynomial and $\cos t$ and $\sin t$ instead of $\cos 4t$ and $\sin 4t$ in the form of y_p:

$$y_p = (At + B) \cos t + (Ct + D) \sin t$$

Again, observe that there is no duplication of terms between y_p and $y_c = c_1 \cos 2t + c_2 \sin 2t$. ■

If $g(t)$ consists of a sum of, say, m terms of the kind listed in Table 3.1, then (as in Example 3) the assumption for a particular solution y_p consists of the sum of the trial forms $y_{p_1}, y_{p_2}, \ldots, y_{p_m}$ corresponding to these terms:

$$y_p = y_{p_1} + y_{p_2} + \cdots + y_{p_m}$$

The foregoing sentence can be put another way.

Form Rule for Case I *The form of y_p is a linear combination of all linearly independent functions that are generated by repeated differentiations of $g(t)$.*

■ EXAMPLE 6 Forming y_p by Superposition—Case I

Determine the form of a particular solution of

$$y'' - 9y' + 14y = 3t^2 - 5 \sin 2t + 7te^{6t}$$

Solution Corresponding to $3t^2$, we assume: $y_{p_1} = At^2 + Bt + C$.

Corresponding to $-5 \sin 2t$, we assume: $y_{p_2} = D \cos 2t + E \sin 2t$.

Corresponding to $7te^{6t}$, we assume: $y_{p_3} = (Ft + G)e^{6t}$.

The assumption for the particular solution is then

$$y_p = y_{p_1} + y_{p_2} + y_{p_3} = At^2 + Bt + C + D \cos 2t + E \sin 2t + (Ft + G)e^{6t}$$

No term in this assumption duplicates a term in $y_c = c_1e^{2t} + c_2e^{7t}$. ■

CASE II A function in the assumed particular solution is also a solution of the associated homogeneous differential equation.

The next example is similar to Example 4.

■ EXAMPLE 7 Particular Solution—Case II

Find a particular solution of $y'' - 2y' + y = e^t$.

Solution The complementary function is $y_c = c_1e^t + c_2te^t$. As in Example 4, the assumption $y_p = Ae^t$ will fail since it is apparent from y_c that e^t is a solution of the associated homogeneous equation $y'' - 2y' + y = 0$. Moreover, we will not be able to find a particular solution of the form $y_p = Ate^t$ since the term te^t is also duplicated in y_c. We next try

$$y_p = At^2e^t$$

Substituting into the given differential equation yields

$$2Ae^t = e^t \quad \text{and so} \quad A = \frac{1}{2}$$

Thus a particular solution is $y_p = \frac{1}{2}t^2e^t$. ■

Suppose again that $g(t)$ consists of m terms of the kind given in Table 3.1 and suppose further that the usual assumption for a particular solution is

$$y_p = y_{p_1} + y_{p_2} + \cdots + y_{p_m}$$

where the y_{p_i}, $i = 1, 2, \ldots, m$, are the trial particular solution forms corresponding to these terms. Under the circumstances described in Case II, we can make up the following general rule.

Multiplication Rule for Case II *If any y_{p_i} contains terms that duplicate terms in y_c, then that y_{p_i} must be multiplied by t^n, where n is the smallest positive integer that eliminates that duplication.*

■ EXAMPLE 8 An Initial-Value Problem

Solve the initial-value problem

$$y'' + y = 4t + 10 \sin t, \qquad y(\pi) = 0, \quad y'(\pi) = 2$$

Solution The solution of the associated homogeneous equation $y'' + y = 0$ is $y_c = c_1 \cos t + c_2 \sin t$. Since $g(t) = 4t + 10 \sin t$ is the sum of a linear polynomial and a sine function, our normal assumption for y_p, from entries 2 and 5 of Table 3.1, would be the sum of $y_{p_1} = At + B$ and $y_{p_2} = C \cos t + D \sin t$:

$$y_p = At + B + C \cos t + D \sin t \tag{5}$$

But there is an obvious duplication of the terms $\cos t$ and $\sin t$ in this assumed form and two terms in the complementary function. This duplication can be eliminated by simply multiplying y_{p_2} by t. Instead of (5) we now use

$$y_p = At + B + Ct \cos t + Dt \sin t \tag{6}$$

Differentiating this expression and substituting the results into the differential equation give

$$y_p'' + y_p = At + B - 2C \sin t + 2D \cos t = 4t + 10 \sin t$$

and so $A = 4$, $B = 0$, $-2C = 10$, $2D = 0$. The solutions of the system are immediate: $A = 4$, $B = 0$, $C = -5$, and $D = 0$. Therefore from (6) we obtain $y_p = 4t - 5t \cos t$. The general solution of the given equation is

$$y = y_c + y_p = c_1 \cos t + c_2 \sin t + 4t - 5t \cos t$$

We now apply the prescribed initial conditions to the general solution of the equation. First, $y(\pi) = c_1 \cos \pi + c_2 \sin \pi + 4\pi - 5\pi \cos \pi = 0$ yields $c_1 = 9\pi$, since $\cos \pi = -1$ and $\sin \pi = 0$. Next, from the derivative

$$y' = -9\pi \sin t + c_2 \cos t + 4 + 5t \sin t - 5 \cos t$$

and $\quad y'(\pi) = -9\pi \sin \pi + c_2 \cos \pi + 4 + 5\pi \sin \pi - 5 \cos \pi = 2$

we find $c_2 = 7$. The solution of the initial value is then

$$y = 9\pi \cos t + 7 \sin t + 4t - 5t \cos t$$ ■

■ **EXAMPLE 9 Using the Multiplication Rule**

Solve $\quad y'' - 6y' + 9y = 6t^2 + 2 - 12e^{3t}$.

Solution The complementary function is $y_c = c_1 e^{3t} + c_2 t e^{3t}$. And so, based on entries 3 and 7 of Table 3.1, the usual assumption for a particular solution would be

$$y_p = \underbrace{At^2 + Bt + C}_{y_{p_1}} + \underbrace{De^{3t}}_{y_{p_2}}$$

Inspection of these functions shows that the one term in y_{p_2} is duplicated in y_c. If we multiply y_{p_2} by t, we note that the term te^{3t} is still part of y_c. But multiplying y_{p_2} by t^2 eliminates all duplications. Thus the operative form of a particular solution is

$$y_p = At^2 + Bt + C + Dt^2 e^{3t}$$

Differentiating this last form, substituting into the differential equation, and collecting like terms give

$$y_p'' - 6y_p' + 9y_p = 9At^2 + (-12A + 9B)t + 2A - 6B + 9C + 2De^{3t} = 6t^2 + 2 - 12e^{3t}$$

It follows from this identity that $A = \frac{2}{3}$, $B = \frac{8}{9}$, $C = \frac{2}{3}$, and $D = -6$. Hence the general solution $y = y_c + y_p$ is

$$y = c_1 e^{3t} + c_2 t e^{3t} + \frac{2}{3}t^2 + \frac{8}{9}t + \frac{2}{3} - 6t^2 e^{3t}$$ ■

Higher-Order Equations Up to now we have applied the method of undetermined coefficients to second-order equations. The next two examples illustrate that the method extends to higher-order equations without modification.

■ **EXAMPLE 10 Third-Order DE—Case I**

Solve $y''' + y'' = e^t \cos t$.

Solution From the characteristic equation $m^3 + m^2 = 0$, we find $m_1 = m_2 = 0$ and $m_3 = -1$. Hence the complementary solution of the equation is $y_c = c_1 + c_2 t + c_3 e^{-t}$. With $g(t) = e^t \cos t$ we see from entry 10 of Table 3.1 that we should assume $y_p = Ae^t \cos t + Be^t \sin t$. Since there are no functions in y_p that duplicate functions in the complementary solution, we proceed in the usual manner. From

$$y_p''' + y_p'' = (-2A + 4B)e^t \cos t + (-4A - 2B)e^t \sin t = e^t \cos t$$

we get $-2A + 4B = 1$, $-4A - 2B = 0$. This system gives $A = -\frac{1}{10}$ and $B = \frac{1}{5}$, so that a particular solution is

$$y_p = -\frac{1}{10}e^t \cos t + \frac{1}{5}e^t \sin t$$

The general solution of the equation is

$$y = y_c + y_p = c_1 + c_2 t + c_3 e^{-t} - \frac{1}{10}e^t \cos t + \frac{1}{5}e^t \sin t$$ ■

■ **EXAMPLE 11 Fourth-Order DE—Case II**

Determine the form of a particular solution of $y^{(4)} + y''' = 1 - t^2 e^{-t}$.

Solution Comparing $y_c = c_1 + c_2 t + c_3 t^2 + c_4 e^{-t}$ with our normal assumption for a particular solution

$$y_p = \underbrace{A}_{y_{p_1}} + \underbrace{Bt^2 e^{-t} + Cte^{-t} + De^{-t}}_{y_{p_2}}$$

we see that the duplications between y_c and y_p are eliminated when y_{p_1} is multiplied by t^3 and y_{p_2} is multiplied by t. The correct assumption of a particular solution is

$$y_p = At^3 + Bt^3 e^{-t} + Ct^2 e^{-t} + Dte^{-t}$$ ■

As we pointed out earlier, the method of undetermined coefficients has its shortcomings: it cannot be used when the coefficients of the differential equation are variable, nor can it be used when the input function in (1) is a function such as

$$g(t) = \ln t, \quad g(t) = \frac{1}{t}, \quad g(t) = \tan t, \quad g(t) = \sin^{-1} t$$

and so on. The method that we consider next does not have these limitations; it has the distinct advantage that it will *always* yield a particular solution y_p provided the associated homogeneous equation can be solved.

3.4.2 Variation of Parameters

Standard Form The procedure that we used in Section 2.3 to find a particular solution y_p of a linear first-order differential equation is applicable to linear higher-order equations as well. To adapt the method of **variation of parameters** to the linear second-order differential equation (1) we begin, as we did in that section, by putting the differential equation into **standard form**

$$y'' + P(t)y' + Q(t)y = f(t) \tag{7}$$

by dividing through by the lead coefficient $a_2(t)$. Here we assume that $P(t)$, $Q(t)$, and $f(t)$ are continuous on some common interval I.

The Assumptions In Section 2.3, we used the assumption $y_p = u_1(t)y_1(t)$ to find a particular solution of the linear first-order equation $y' + P(t)y = f(t)$ when y_1 was a known solution of $y' + P(t)y = 0$. Analogously, for the nonhomogeneous linear second-order equation (7), we seek a particular solution

$$y_p = u_1(t)y_1(t) + u_2(t)y_2(t) \tag{8}$$

where y_1 and y_2 are known to be linearly independent of solutions of the homogeneous equation $y'' + P(t)y' + Q(t)y = 0$ on the interval I. Substituting

$$y_p' = u_1y_1' + y_1u_1' + u_2y_2' + y_2u_2'$$

$$y_p'' = u_1y_1'' + y_1'u_1' + y_1u_1'' + u_1'y_1' + u_2y_2'' + y_2'u_2' + y_2u_2'' + u_2'y_2'$$

and (8) into (7), and then grouping terms, yields

$$y_p'' + P(t)y_p' + Q(t)y_p = u_1\underbrace{[y_1'' + Py_1' + Qy_1]}_{\text{zero}} + u_2\underbrace{[y_2'' + Py_2' + Qy_2]}_{\text{zero}}$$

$$+ y_1u_1'' + u_1'y_1' + y_2u_2'' + u_2'y_2' + P[y_1u_1' + y_2u_2'] + y_1'u_1' + y_2'u_2'$$

$$= \frac{d}{dt}[y_1u_1'] + \frac{d}{dt}[y_2u_2'] + P[y_1u_1' + y_2u_2'] + y_1'u_1' + y_2'u_2'$$

$$= \frac{d}{dt}[y_1u_1' + y_2u_2'] + P[y_1u_1' + y_2u_2'] + y_1'u_1' + y_2'u_2' = f(t) \tag{9}$$

Because we seek to determine two unknown functions u_1 and u_2, reason dictates that we need two equations to accomplish this. We can obtain these equations by making the further assumption that the functions u_1 and u_2 satisfy $y_1u_1' + y_2u_2' = 0$. This assumption does not come out of the blue, but is prompted by the first two terms in (9), since, if we demand that $y_1u_1' + y_2u_2' = 0$, then (9) reduces to $y_1'u_1' + y_2'u_2' = f(t)$. We now have our desired two equations, albeit two equations for determining the derivatives u_1' and u_2'. By Cramer's rule, the solution of the system

$$y_1u_1' + y_2u_2' = 0$$
$$y_1'u_1' + y_2'u_2' = f(t)$$

can be expressed in terms of determinants:

$$u_1' = \frac{W_1}{W} = -\frac{y_2 f(t)}{W} \quad \text{and} \quad u_2' = \frac{W_2}{W} = \frac{y_1 f(t)}{W} \tag{10}$$

where
$$W = \begin{vmatrix} y_1 & y_2 \\ y_1' & y_2' \end{vmatrix}, \quad W_1 = \begin{vmatrix} 0 & y_2 \\ f(t) & y_2' \end{vmatrix}, \quad W_2 = \begin{vmatrix} y_1 & 0 \\ y_1' & f(t) \end{vmatrix} \tag{11}$$

The functions u_1 and u_2 are found by integrating the results in (10). The determinant W is recognized as the Wronskian of y_1 and y_2. By linear independence of y_1 and y_2 on I, we know that $W(y_1(t), y_2(t)) \neq 0$ for every t in the interval I.

Method of Solution Usually it is not a good idea to memorize formulas in lieu of understanding a procedure. However, the foregoing procedure is too long and complicated to use each time we wish to solve a differential equation. In this case, it is more efficient simply to use the formulas in (10). Thus to solve $a_2y'' + a_1y' + a_0y = g(t)$, first find the complementary function $y_c = c_1y_1 + c_2y_2$ and then compute the Wronskian $W(y_1, y_2)$. By dividing by a_2, we put the equation into the standard form $y'' + Py' + Qy = f(t)$ in order to identify $f(t)$. We find u_1 and u_2 by integrating $u_1' = W_1/W$ and $u_2' = W_2/W$ where W_1 and W_2 are defined in (11). A particular solution is $y_p = u_1y_1 + u_2y_2$ and the general solution is $y = y_c + y_p$.

EXAMPLE 12 General Solution Using Variation of Parameters

Solve $y'' - 4y' + 4y = (t + 1)e^{2t}$.

Solution From the auxiliary equation $m^2 - 4m + 4 = (m - 2)^2 = 0$, we have $y_c = c_1e^{2t} + c_2te^{2t}$. With the identifications $y_1 = e^{2t}$ and $y_2 = te^{2t}$, we next compute the Wronskian:

$$W(e^{2t}, te^{2t}) = \begin{vmatrix} e^{2t} & te^{2t} \\ 2e^{2t} & 2te^{2t} + e^{2t} \end{vmatrix} = e^{4t}$$

Since the given differential equation is already in form (7) (that is, the coefficient of y'' is 1), we identify $f(t) = (t + 1)e^{2t}$. From (11) we obtain

$$W_1 = \begin{vmatrix} 0 & te^{2t} \\ (t+1)e^{2t} & 2te^{2t} + e^{2t} \end{vmatrix} = -(t+1)te^{4t}, \quad W_2 = \begin{vmatrix} e^{2t} & 0 \\ 2e^{2t} & (t+1)e^{2t} \end{vmatrix} = (t+1)e^{4t}$$

and so from (10),

$$u_1' = -\frac{(t+1)te^{4t}}{e^{4t}} = -t^2 - t, \quad u_2' = \frac{(t+1)e^{4t}}{e^{4t}} = t + 1$$

It follows that $$u_1 = -\frac{t^3}{3} - \frac{t^2}{2} \quad \text{and} \quad u_2 = \frac{t^2}{2} + t$$

Hence $$y_p = \left(-\frac{t^3}{3} - \frac{t^2}{2}\right)e^{2t} + \left(\frac{t^2}{2} + t\right)te^{2t} = \left(\frac{t^3}{6} + \frac{t^2}{2}\right)e^{2t}$$

and $$y = y_c + y_p = c_1e^{2t} + c_2te^{2t} + \left(\frac{t^3}{6} + \frac{t^2}{2}\right)e^{2t}$$ ■

When computing the indefinite integrals of u_1' and u_2', we do not need to introduce constants of integration. See Problem 57 in Exercises 3.4.

■ **EXAMPLE 13 General Solution Using Variation of Parameters**

Solve $4y'' + 36y = \csc 3t$.

Solution We first put the equation into standard form by dividing by 4:

$$y'' + 9y = \frac{1}{4}\csc 3t$$

Since the roots of the auxiliary equation $m^2 + 9 = 0$ are $m_1 = 3i$ and $m_2 = -3i$, the complementary function is $y_c = c_1 \cos 3t + c_2 \sin 3t$. Using $y_1 = \cos 3t$, $y_2 = \sin 3t$, and $f(t) = \frac{1}{4}\csc t$, we obtain

$$W(\cos 3t, \sin 3t) = \begin{vmatrix} \cos 3t & \sin 3t \\ -3\sin 3t & 3\cos 3t \end{vmatrix} = 3$$

$$W_1 = \begin{vmatrix} 0 & \sin 3t \\ \frac{1}{4}\csc 3t & 3\cos 3t \end{vmatrix} = -\frac{1}{4}, \quad W_2 = \begin{vmatrix} \cos 3t & 0 \\ -3\sin 3t & \frac{1}{4}\csc 3t \end{vmatrix} = \frac{1}{4}\frac{\cos 3t}{\sin 3t}$$

Integrating $$u_1' = \frac{W_1}{W} = -\frac{1}{12} \quad \text{and} \quad u_2' = \frac{W_2}{W} = \frac{1}{12}\frac{\cos 3t}{\sin 3t}$$

gives $$u_1 = -\frac{1}{12}t \quad \text{and} \quad u_2 = \frac{1}{36}\ln|\sin 3t|$$

Thus a particular solution is

$$y_p = -\frac{1}{12}t\cos 3t + \frac{1}{36}(\sin 3t)\ln|\sin 3t|$$

The general solution of the equation is

$$y = y_c + y_p = c_1\cos 3t + c_2\sin 3t - \frac{1}{12}t\cos 3t + \frac{1}{36}(\sin 3t)\ln|\sin 3t| \quad \textbf{(12)}$$ ■

The expression in (12) represents the general solution of the differential equation on, say, the interval $(0, \pi/6)$.

EXAMPLE 14 General Solution Using Variation of Parameters

Solve $y'' - y = \dfrac{1}{t}$.

Solution The auxiliary equation $m^2 - 1 = 0$ yields $m_1 = -1$ and $m_2 = 1$. Therefore $y_c = c_1e^t + c_2e^{-t}$. Now $W(e^t, e^{-t}) = -2$ and

$$u_1' = -\frac{e^{-t}(1/t)}{-2}, \qquad u_1 = \frac{1}{2}\int_{t_0}^{t} \frac{e^{-v}}{v}\,dv$$

$$u_2' = \frac{e^{t}(1/t)}{-2}, \qquad u_2 = -\frac{1}{2}\int_{t_0}^{t} \frac{e^{v}}{v}\,dv$$

It is well known that the integrals defining u_1 and u_2 cannot be expressed in terms of elementary functions. Hence we write

$$y_p = \frac{1}{2}e^t\int_{t_0}^{t} \frac{e^{-v}}{v}\,dv - \frac{1}{2}e^{-t}\int_{t_0}^{t} \frac{e^{v}}{v}\,dv$$

and so $$y = y_c + y_p = c_1e^t + c_2e^{-t} + \frac{1}{2}e^t\int_{t_0}^{t} \frac{e^{-v}}{v}\,dv - \frac{1}{2}e^{-t}\int_{t_0}^{t} \frac{e^{v}}{v}\,dv$$ ■

In Example 14 we can integrate on any interval $t_0 \leq v \leq t$ not containing the origin.

Higher-Order Equations The method we have just examined for nonhomogeneous second-order differential equations can be generalized to linear nth-order equations that have been put into the form

$$y^{(n)} + P_{n-1}(t)y^{(n-1)} + \cdots + P_1(t)y' + P_0(t)y = f(t) \tag{13}$$

If $y = c_1y_1 + c_2y_2 + \cdots + c_ny_n$ is the complementary function for (13), then a particular solution is

$$y_p = u_1(t)y_1(t) + u_2(t)y_2(t) + \cdots + u_n(t)y_n(t)$$

where the u_k', $k = 1, 2, \ldots, n$, are determined by the n equations

$$\begin{aligned}
y_1u_1' + \quad y_2u_2' + \cdots + \quad y_nu_n' &= 0\\
y_1'u_1' + \quad y_2'u_2' + \cdots + \quad y_n'u_n' &= 0\\
&\vdots\\
y_1^{(n-1)}u_1' + y_2^{(n-1)}u_2' + \cdots + y_n^{(n-1)}u_n' &= f(t)
\end{aligned}$$

The first $n - 1$ equations in this system, like $y_1u_1' + y_2u_2' = 0$ in (9), are assumptions made to simplify the resulting equation after $y_p =$

$u_1(t)y_1(t) + \cdots + u_n(t)y_n(t)$ is substituted in (17). In this case, Cramer's rule gives

$$u'_k = \frac{W_k}{W}, \qquad k = 1, 2, \ldots, n$$

where W is the Wronskian of $y_1, y_2, \ldots, y_n$, and W_k is the determinant obtained by replacing the kth column of the Wronskian by the column $(0, 0, \ldots, f(t))$.

Remarks

(*i*) In the problems that follow, do not hesitate to simplify the form of y_p. Depending on how the antiderivatives of u'_1 and u'_2 are found, you may not obtain the same y_p as given in the answer section. For example, both $y_p = \frac{1}{2}\sin t - \frac{1}{2}t\cos t$ and $y_p = \frac{1}{4}\sin t - \frac{1}{2}t\cos t$ are particular solutions of $y'' + y = \sin t$. In either case, the general solution $y = y_c + y_p$ simplifies to $y = c_1\cos t + c_2\sin t - \frac{1}{2}t\cos t$. Why?

(*ii*) When solving an initial-value problem, be sure to apply the initial conditions to the general solution $y = y_c + y_p$. Students often make the mistake of applying the initial conditions only to the complementary function y_c since it is that part of the solution that contains the constants.

EXERCISES 3.4

Answers to odd-numbered problems begin on page 442.

3.4.1

In Problems 1–24, solve the given differential equation by undetermined coefficients.

1. $y'' + 3y' + 2y = 6$

2. $4y'' + 9y = 15$

3. $y'' - 10y' + 25y = 30t + 3$

4. $y'' + y' - 6y = 2t$

5. $\frac{1}{4}y'' + y' + y = t^2 - 2t$

6. $y'' - 8y' + 20y = 100t^2 - 26te^t$

7. $y'' + 3y = -48t^2e^{3t}$

8. $4y'' - 4y' - 3y = \cos 2t$

9. $y'' - y' = -3$

10. $y'' + 2y' = 2t + 5 - e^{-2t}$

11. $y'' - y' + \frac{1}{4}y = 3 + e^{t/2}$

12. $y'' - 16y = 2e^{4t}$

13. $y'' + 4y = 3\sin 2t$

14. $y'' + 4y = (t^2 - 3)\sin 2t$

15. $y'' + y = 2t\sin t$

16. $y'' - 5y' = 2t^3 - 4t^2 - t + 6$

17. $y'' - 2y' + 5y = e^t\cos 2t$

18. $y'' - 2y' + 2y = e^{2t}(\cos t - 3\sin t)$

19. $y'' + 2y' + y = \sin t + 3\cos 2t$

20. $y'' + 2y' - 24y = 16 - (t + 2)e^{4t}$

21. $y''' - 6y'' = 3 - \cos t$

22. $y''' - 2y'' - 4y' + 8y = 6te^{2t}$

23. $y''' - 3y'' + 3y' - y = t - 4e^t$

24. $y''' - y'' - 4y' + 4y = 5 - e^t + e^{2t}$

In Problems 25–28, solve the given differential equation subject to the indicated initial conditions.

25. $y'' + 4y = -2, \quad y(\pi/8) = \frac{1}{2}, \quad y'(\pi/8) = 2$

26. $2y'' + 3y' - 2y = 14t^2 - 4t - 11,$
$y(0) = 0, \quad y'(0) = 0$

27. $5y'' + y' = -6t, \quad y(0) = 0, \quad y'(0) = -10$

28. $y'' + 4y' + 4y = (3 + t)e^{-2t}, \quad y(0) = 2, \quad y'(0) = 5$

3.4.2

In Problems 29–52, solve the given differential equation by variation of parameters. State an interval on which the general solution is defined.

29. $y'' + y = \sec t$

30. $y'' + y = \tan t$

31. $y'' + y = \sin t$

32. $y'' + y = \sec t \tan t$

33. $y'' + y = \cos^2 t$

34. $y'' + y = \sec^2 t$

35. $y'' - y = \cosh t$

36. $y'' - y = \sinh 2t$

37. $y'' - 4y = e^{2t}/t$

38. $y'' - 9y = 9t/e^{3t}$

39. $y'' + 3y' + 2y = 1/(1 + e^t)$

40. $y'' - 3y' + 2y = e^{3t}/(1 + e^t)$

41. $y'' + 3y' + 2y = \sin e^t$

42. $y'' - 2y' + y = e^t \arctan t$

43. $y'' - 2y' + y = e^t/(1 + t^2)$

44. $y'' - 2y' + 2y = e^t \sec t$

45. $y'' + 2y' + y = e^{-t} \ln t$

46. $y'' + 10y' + 25y = e^{-5t}/t^2$

47. $3y'' - 6y' + 30y = e^t \tan 3t$

48. $4y'' - 4y' + y = e^{t/2}\sqrt{1 - t^2}$

49. $y''' + y' = \tan t$

50. $y''' + 4y' = \sec 2t$

51. $y''' - 2y'' - y' + 2y = e^{3t}$

52. $2y''' - 6y'' = t^2$

In Problems 53–56, solve each differential equation by variation of parameters subject to the initial conditions $y(0) = 1$, $y'(0) = 0$.

53. $4y'' - y = te^{t/2}$

54. $2y'' + y' - y = t + 1$

55. $y'' + 2y' - 8y = 2e^{-2t} - e^{-t}$

56. $y'' - 4y' + 4y = (12t^2 - 6t)e^{2t}$

Discussion Problems

57. Discuss: Why it is unnecessary to use constants of integration when evaluating the indefinite integrals of u_1' and u_2'?

58. Discuss: How can the methods of undetermined coefficients and variation of parameters be combined to solve

$$y'' + 2y' + y = 4t^2 - 3 + \frac{e^{-t}}{t}$$

Solve the differential equation.

59. Given that $y_1 = \cos(\ln t)$ and $y_2 = \sin(\ln t)$ are known linearly independent solutions of $t^2y'' + ty' + y = 0$ on the interval $(0, \infty)$, find the general solution of

$$t^2y'' + ty' + y = \sec(\ln t)$$

and state an interval over which this solution is defined.

60. **(a)** Find an exponential solution $y_1 = e^{mt}$ for

$$ty'' - (5t + 1)y' + 5y = 0$$

(b) Find the general solution of the equation in part (a). [*Hint*: To find y_2, use the method illustrated in Case II in Section 3.2.]

(c) Find the general solution of

$$ty'' - (5t + 1)y' + 5y = t^2, \quad t > 0$$

61. Without solving, match the solution curve of $y'' + y = f(t)$ shown in each figure with one of the functions

(*i*) $f(t) = 1$, (*ii*) $f(t) = e^{-t}$, (*iii*) $f(t) = e^t$
(*iv*) $f(t) = \sin 2t$, (*v*) $f(t) = e^t \sin t$, (*vi*) $f(t) = \sin t$

Briefly discuss your reasoning.

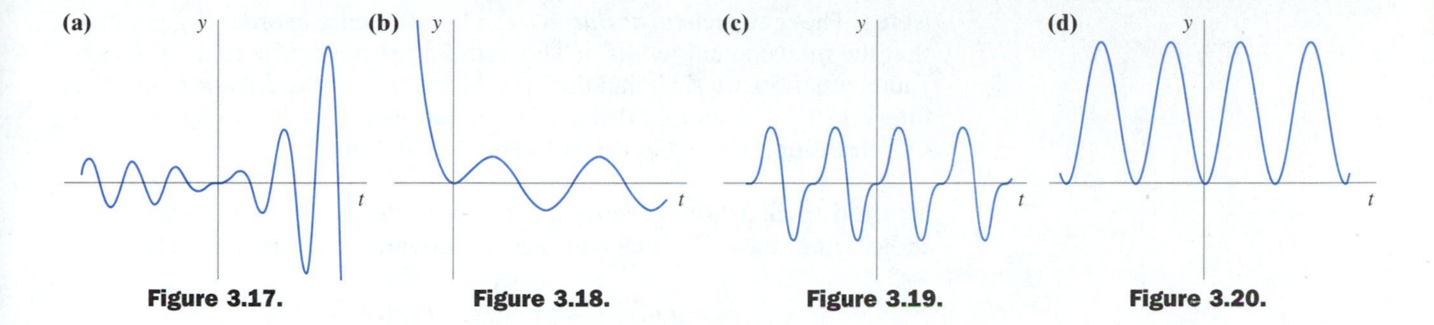

Figure 3.17. **Figure 3.18.** **Figure 3.19.** **Figure 3.20.**

3.5 CAUCHY-EULER EQUATION

The relative ease with which we solved linear higher-order DEs with constant coefficients does not carry over to linear equations with variable coefficients. Generally, we cannot expect to be able to solve even a simple linear equation such as $y'' + ty = 0$ in terms of elementary functions. However, there is one type of variable-coefficient differential equation, called a *Cauchy-Euler equation*, that is an exception to this rule.

Monomial Coefficients Any linear differential equation of the form

$$a_n t^n \frac{d^n y}{dt^n} + a_{n-1} t^{n-1} \frac{d^{n-1} y}{dt^{n-1}} + \cdots + a_1 t \frac{dy}{dt} + a_0 y = g(t)$$

where $a_n, a_{n-1}, \ldots, a_0$ are constants, is known as a **Cauchy-Euler equation**, a **Euler-Cauchy equation**, a **Euler equation**, or an **equidimensional equation**. The observable characteristic of this type of equation is that the degree k of the monomial coefficients t^k matches the order k of differentiation $d^k y/dt^k$:

$$\overset{\text{same}}{\overbrace{a_n t^n \frac{d^n y}{dt^n}}} + \overset{\text{same}}{\overbrace{a_{n-1} t^{n-1} \frac{d^{n-1} y}{dt^{n-1}}}} + \cdots$$

As we did in Section 3.2, we start the discussion with a homogeneous second-order equation

$$at^2 \frac{d^2 y}{dt^2} + bt \frac{dy}{dt} + cy = 0$$

The solution of higher-order equations follows analogously. Also, we can solve the nonhomogeneous equation $at^2 y'' + bty' + cy = g(t)$ by variation of parameters, once we have determined the complementary function y_c.

Note The coefficient of d^2y/dt^2 is zero at $t = 0$. Hence in order to guarantee that the fundamental results of Theorem 3.1 are applicable to the Cauchy-Euler equation, we shall find the general solution of the equation over the interval $(0, \infty)$. Solutions defined on the interval $(-\infty, 0)$ can be obtained by substituting $x = -t$ in the differential equation.

Method of Solution We try a solution of the form $y = t^m$, where m is to be determined. The first and second derivatives are, respectively,

$$\frac{dy}{dt} = mt^{m-1} \quad \text{and} \quad \frac{d^2y}{dt^2} = m(m-1)t^{m-2}$$

Consequently

$$at^2\frac{d^2y}{dt^2} + bt\frac{dy}{dt} + cy = at^2 \cdot m(m-1)t^{m-2} + bt \cdot mt^{m-1} + ct^m$$

$$= am(m-1)t^m + bmt^m + ct^m = t^m(am(m-1) + bm + c)$$

Thus $y = t^m$ is a solution of the differential equation whenever m is a solution of the **auxiliary equation**

$$am(m-1) + bm + c = 0 \quad \text{or} \quad am^2 + (b-a)m + c = 0 \tag{1}$$

There are three different cases to be considered, depending on whether the roots of this quadratic equation are real and distinct, real and equal, or complex. In the last case, the roots appear as a conjugate pair.

CASE I **Distinct Real Roots** Let m_1 and m_2 denote the real roots of (1) such that $m_1 \neq m_2$. Then $y_1 = t^{m_1}$ and $y_2 = t^{m_2}$ form a fundamental set of solutions. Hence the general solution is

$$y = c_1t^{m_1} + c_2t^{m_2} \tag{2}$$

EXAMPLE 1 Cauchy-Euler Equation: Distinct Roots

Solve $t^2\dfrac{d^2y}{dt^2} - 2t\dfrac{dy}{dt} - 4y = 0$.

Solution Rather than just memorizing equation (1), it is preferable to assume $y = t^m$ as the solution a few times in order to understand the origin and the difference between this new form of the auxiliary equation and that obtained in Section 3.2. Differentiate twice

$$\frac{dy}{dt} = mt^{m-1}, \quad \frac{d^2y}{dt^2} = m(m-1)t^{m-2}$$

and substitute back into the differential equation:

$$t^2\frac{d^2y}{dt^2} - 2t\frac{dy}{dt} - 4y = t^2 \cdot m(m-1)t^{m-2} - 2t \cdot mt^{m-1} - 4t^m$$

$$= t^m(m(m-1) - 2m - 4) = t^m(m^2 - 3m - 4) = 0$$

if $m^2 - 3m - 4 = 0$. Now $(m + 1)(m - 4) = 0$ implies $m_1 = -1$, $m_2 = 4$ so that

$$y = c_1 t^{-1} + c_2 t^4$$ ■

CASE II **Repeated Real Roots** If the roots of (1) are repeated, that is, $m_1 = m_2$, then the single solution is $y_1 = t^{m_1}$. By the quadratic formula, it follows that $m_1 = -(b - a)/2a$. We can now employ the same procedure used in Case II on page 132 to construct a second solution y. If we substitute $y = u(t)t^{m_1}$ into $at^2y'' + bty + cy = 0$ and group terms, we obtain

$$at^{m_1+2}u'' + \underbrace{(2am_1 + b)}_{\substack{a \text{ since} \\ 2am_1 = -b + a}}t^{m_1+1}u' + \underbrace{(am_1^2 + (b - a)m_1 + c)}_{\substack{\text{zero since } m_1 \text{ is a root} \\ \text{of the auxiliary equation}}}t^{m_1}u$$

or $tu'' + u' = 0$. At this point, it does not appear that we have gained much since the last equation still has a variable coefficient. However, if we let $v = u'$, the equation reduces to the first-order equation $tv' + v = 0$. Separating variables and integrating give $v = C_1/t$. Since $u' = v$, we integrate again to find $u(t) = C_1 \ln t + C_2$. Finally, by choosing $C_1 = 1$ and $C_2 = 0$ in $y = (C_1 \ln t + C_2)t^{m_1}$, we find that $y_2 = t^{m_1} \ln t$ is our second solution. The general solution of the differential equation is then

$$y = c_1 t^{m_1} + c_2 t^{m_1} \ln t \tag{3}$$

■ **EXAMPLE 2** **Cauchy-Euler Equation: Repeated Roots**

Solve $4t^2 \dfrac{d^2y}{dt^2} + 8t\dfrac{dy}{dt} + y = 0.$

Solution Using the substitution $y = t^m$ we find that

$$4t^2 \frac{d^2y}{dt^2} + 8t\frac{dy}{dt} + y = t^m(4m(m - 1) + 8m + 1) = t^m(4m^2 + 4m + 1) = 0$$

when $4m^2 + 4m + 1 = (2m + 1)^2 = 0$. Therefore, $m_1 = -\frac{1}{2}$ is a repeated root, so from (3) we see that the general solution of the equation is

$$y = c_1 t^{-1/2} + c_2 t^{-1/2} \ln t$$ ■

For higher-order equations, if m_1 is a root of multiplicity k, then it can be shown that

$$t^{m_1}, \quad t^{m_1} \ln t, \quad t^{m_1}(\ln t)^2, \quad \ldots, \quad t^{m_1}(\ln t)^{k-1}$$

are k linearly independent solutions. Correspondingly, the general solution of the differential equation must then contain a linear combination of these k solutions.

CASE III **Conjugate Complex Roots** If the roots of (1) are the conjugate pair $m_1 = \alpha + i\beta$, $m_2 = \alpha - i\beta$, where α and $\beta > 0$ are real, then a solution is

$$y = C_1 t^{\alpha + i\beta} + C_2 t^{\alpha - i\beta}$$

But, as in the case of equations with constant coefficients, when the roots of the auxiliary equation are complex, we wish to write the solution in terms of real functions only. We note the identity

$$t^{i\beta} = (e^{\ln t})^{i\beta} = e^{i\beta \ln t}$$

which, by Euler's formula, is the same as

$$t^{i\beta} = \cos(\beta \ln t) + i \sin(\beta \ln t)$$

Similarly, $$t^{-i\beta} = \cos(\beta \ln t) - i \sin(\beta \ln t)$$

Adding and subtracting the last two results yield

$$t^{i\beta} + t^{-i\beta} = 2\cos(\beta \ln t) \quad \text{and} \quad t^{i\beta} - t^{-i\beta} = 2i \sin(\beta \ln t)$$

respectively. From the fact that $y = C_1 t^{\alpha + i\beta} + C_2 t^{\alpha - i\beta}$ is a solution for any values of the constants, we see, in turn, for $C_1 = C_2 = 1$ and $C_1 = 1$, $C_2 = -1$ that

$$y_1 = t^\alpha(t^{i\beta} + t^{-i\beta}) \quad \text{and} \quad y_2 = t^\alpha(t^{i\beta} - t^{-i\beta})$$

or $$y_1 = 2t^\alpha \cos(\beta \ln t) \quad \text{and} \quad y_2 = 2it^\alpha \sin(\beta \ln t)$$

are also solutions. Since $W(t^\alpha\cos(\beta \ln t), t^\alpha\sin(\beta \ln t)) = \beta t^{2\alpha - 1} \neq 0$, $\beta > 0$, on the interval $(0, \infty)$, we conclude that

$$y_1 = t^\alpha \cos(\beta \ln t) \quad \text{and} \quad y_2 = t^\alpha \sin(\beta \ln t)$$

constitute a fundamental set of real solutions of the differential equation. Hence, the general solution is

$$y = t^\alpha[c_1 \cos(\beta \ln t) + c_2 \sin(\beta \ln t)] \tag{4}$$

EXAMPLE 3 An Initial-Value Problem

Solve the initial-value problem

$$t^2 \frac{d^2y}{dt^2} + 3t\frac{dy}{dt} + 3y = 0, \qquad y(1) = 1, \quad y'(1) = -5$$

Solution We have

$$t^2 \frac{d^2y}{dt^2} + 3t\frac{dy}{dt} + 3y = t^m(m(m-1) + 3m + 3) = t^m(m^2 + 2m + 3) = 0$$

when $m^2 + 2m + 3 = 0$. From the quadratic formula we find $m_1 = -1 + \sqrt{2}i$ and $m_2 = -1 - \sqrt{2}i$. If we make the identifications $\alpha = -1$ and $\beta = \sqrt{2}$, we see from (4) that the general solution of the differential equation is

$$y = t^{-1}[c_1 \cos(\sqrt{2} \ln t) + c_2 \sin(\sqrt{2} \ln t)]$$

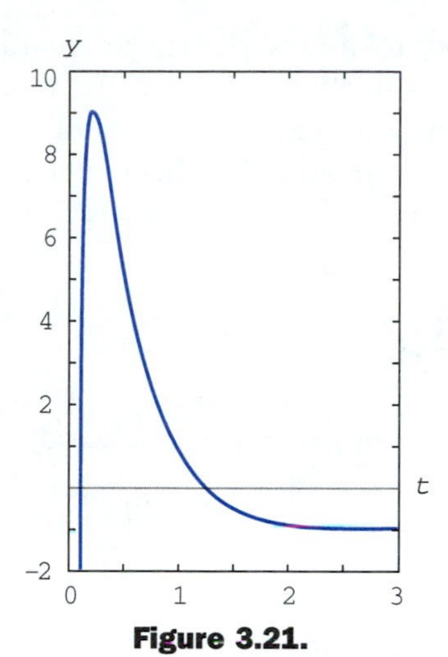

Figure 3.21.

By applying the conditions $y(1) = 1$, $y'(1) = -5$ to the foregoing solution, we find, in turn, $c_1 = 1$ and $c_2 = -2\sqrt{2}$. Thus the solution to the initial-value problem is

$$y = t^{-1}[\cos(\sqrt{2}\ln t) - 2\sqrt{2}\sin(\sqrt{2}\ln t)]$$

The graph of this solution, obtained with the aid of computer software, is given in Figure 3.21. ■

The next example illustrates the solution of a third-order Cauchy-Euler equation.

■ EXAMPLE 4 Third-Order Cauchy-Euler Equation

Solve $t^3\dfrac{d^3y}{dt^3} + 5t^2\dfrac{d^2y}{dt^2} + 7t\dfrac{dy}{dt} + 8y = 0.$

Solution The first three derivatives of $y = t^m$ are

$$\frac{dy}{dt} = mt^{m-1}, \quad \frac{d^2y}{dt^2} = m(m-1)t^{m-2}, \quad \frac{d^3y}{dt^3} = m(m-1)(m-2)t^{m-3}$$

so that the given differential equation becomes

$$\begin{aligned} t^3\frac{d^3y}{dt^3} + 5t^2\frac{d^2y}{dt^2} + 7t\frac{dy}{dt} + 8y &= t^3m(m-1)(m-2)t^{m-3} + 5t^2m(m-1)t^{m-2} + 7tmt^{m-1} + 8t^m \\ &= t^m(m(m-1)(m-2) + 5m(m-1) + 7m + 8) \\ &= t^m(m^3 + 2m^2 + 4m + 8) = t^m(m+2)(m^2+4) = 0 \end{aligned}$$

In this case, we see that $y = t^m$ will be a solution of the differential equation for $m_1 = -2$, $m_2 = 2i$, and $m_3 = -2i$. Hence the general solution is

$$y = c_1t^{-2} + c_2\cos(2\ln t) + c_3\sin(2\ln t)$$ ■

Because the method of undetermined coefficients is applicable only to differential equations with constant coefficients, it cannot be applied *directly* to a nonhomogeneous Cauchy-Euler equation (see the Remarks at the end of the section). In our last example, the method of variation of parameters is employed.

■ EXAMPLE 5 Variation of Parameters

Solve $t^2y'' - 3ty' + 3y = 2t^4e^t$.

Solution The substitution $y = t^m$ leads to the auxiliary equation

$$m(m-1) - 3m + 3 = 0 \quad \text{or} \quad (m-1)(m-3) = 0$$

Thus $$y_c = c_1t + c_2t^3$$

Before using variation of parameters to find a particular solution $y_p = u_1y_1 + u_2y_2$, recall that the formulas $u_1' = W_1/W$ and $u_2' = W_2/W$, where W_1, W_2, and W are the determinants defined on page 160, were derived under the assumption that the differential equation has been put into standard form $y'' + P(t)y' + Q(t)y = f(t)$. Therefore we divide the given equation by t^2, and from

$$y'' - \frac{3}{t}y' + \frac{3}{t^2}y = 2t^2e^t$$

we make the identification $f(t) = 2t^2e^t$. Now with $y_1 = t$, $y_2 = t^3$, and

$$W = \begin{vmatrix} t & t^3 \\ 1 & 3t^2 \end{vmatrix} = 2t^3, \quad W_1 = \begin{vmatrix} 0 & t^3 \\ 2t^2e^t & 3t^2 \end{vmatrix} = -2t^5e^t, \quad W_2 = \begin{vmatrix} t & 0 \\ 1 & 2t^2e^t \end{vmatrix} = 2t^3e^t$$

we find $\qquad u_1' = -\dfrac{2t^5e^t}{2t^3} = -t^2e^t \quad \text{and} \quad u_2' = \dfrac{2t^3e^t}{2t^3} = e^t$

The integral of the latter function is immediate, but in the case of u_1' we integrate by parts twice. The results are $u_1 = -t^2e^t + 2te^t - 2e^t$ and $u_2 = e^t$. Hence

$$y_p = u_1y_1 + u_2y_2 = (-t^2e^t + 2te^t - 2e^t)t + e^tt^3 = 2t^2e^t - 2te^t$$

Finally we have $y = y_c + y_p = c_1t + c_2t^3 + 2t^2e^t - 2te^t$. ■

Remarks

The similarities between the forms of solutions of Cauchy-Euler equations and solutions of linear equations with constant coefficients is not just a coincidence. For example, when the roots of the auxiliary equations for $ay'' + by' + cy = 0$ and $at^2y'' + bty' + cy = 0$ are distinct and real, the respective general solutions are

$$y = c_1e^{m_1t} + c_2e^{m_2t} \quad \text{and} \quad y = c_1t^{m_1} + c_2t^{m_2},\ t > 0 \tag{5}$$

In view of the identity $e^{\ln t} = t$, $t > 0$, the second solution in (5) can be expressed in the same form as the first solution:

$$y = c_1e^{m_1\ln t} + c_2e^{m_2\ln t} = c_1e^{m_1x} + c_2e^{m_2x}$$

where $x = \ln t$. This last result illustrates the fact that any Cauchy-Euler equation can *always* be rewritten as a linear differential equation with constant coefficients by means of the substitution $t = e^x$. The idea is to solve the new differential equation in terms of the variable x using the methods of the previous sections and, once the general solution is obtained, then resubstitute $x = \ln t$. This procedure provides a good review of the Chain Rule of differentiation so you are urged to work Problems 35–40 in Exercises 3.5.

EXERCISES 3.5

Answers to odd-numbered problems begin on page 442.

In Problems 1–22, solve the given differential equation.

1. $t^2y'' - 2y = 0$
2. $4t^2y'' + y = 0$
3. $ty'' + y' = 0$
4. $ty'' - y' = 0$
5. $t^2y'' + ty' + 4y = 0$
6. $t^2y'' + 5ty' + 3y = 0$
7. $t^2y'' - 3ty' - 2y = 0$
8. $t^2y'' + 3ty' - 4y = 0$
9. $25t^2y'' + 25ty' + y = 0$
10. $4t^2y'' + 4ty' - y = 0$
11. $t^2y'' + 5ty' + 4y = 0$
12. $t^2y'' + 8ty' + 6y = 0$
13. $t^2y'' - ty' + 2y = 0$
14. $t^2y'' - 7ty' + 41y = 0$
15. $3t^2y'' + 6ty' + y = 0$
16. $2t^2y'' + ty' + y = 0$
17. $t^3y''' - 6y = 0$
18. $t^3y''' + ty' - y = 0$
19. $t^3\dfrac{d^3y}{dt^3} - 2t^2\dfrac{d^2y}{dt^2} - 2t\dfrac{dy}{dt} + 8y = 0$
20. $t^3\dfrac{d^3y}{dt^3} - 2t^2\dfrac{d^2y}{dt^2} + 4t\dfrac{dy}{dt} - 4y = 0$
21. $t\dfrac{d^4y}{dt^4} + 6\dfrac{d^3y}{dt^3} = 0$
22. $t^4\dfrac{d^4y}{dt^4} + 6t^3\dfrac{d^3y}{dt^3} + 9t^2\dfrac{d^2y}{dt^2} + 3t\dfrac{dy}{dt} + y = 0$

In Problems 23–26, solve the given differential equation subject to the indicated initial conditions.

23. $t^2y'' + 3ty' = 0, \quad y(1) = 0, \quad y'(1) = 4$
24. $t^2y'' - 5ty' + 8y = 0, \quad y(2) = 32, \quad y'(2) = 0$
25. $t^2y'' + ty' + y = 0, \quad y(1) = 1, \quad y'(1) = 2$
26. $t^2y'' - 3ty' + 4y = 0, \quad y(1) = 5, \quad y'(1) = 3$

In Problems 27 and 28, solve the given differential equation subject to the indicated initial conditions. [*Hint*: Let $x = -t$.]

27. $4t^2y'' + y = 0, \quad y(-1) = 2, \quad y'(-1) = 4$
28. $t^2y'' - 4ty' + 6y = 0, \quad y(-2) = 8, \quad y'(-2) = 0$

In Problems 29–34, solve the given differential equation by variation of parameters.

29. $ty'' + y' = t$
30. $ty'' - 4y' = t^4$
31. $2t^2y'' + 5ty' + y = t^2 - t$
32. $t^2y'' - 2ty' + 2y = t^4e^t$
33. $t^2y'' - ty' + y = 2t$
34. $t^2y'' - 2ty' + 2y = t^3 \ln t$

Any Cauchy-Euler differential equation can be reduced to an equation with constant coefficients by means of the substitution $t = e^x$. In Problems 35–40, solve the given differential equation by means of this substitution and undetermined coefficients.

35. $t^2\dfrac{d^2y}{dt^2} + 10t\dfrac{dy}{dt} + 8y = t^2$
36. $t^2y'' - 4ty' + 6y = \ln t^2$
37. $t^2y'' - 3ty' + 13y = 4 + 3t$
38. $2t^2y'' - 3ty' - 3y = 1 + 2t + t^2$
39. $t^2y'' + 9ty' - 20y = 5/t^3$
40. $t^3\dfrac{d^3y}{dt^3} - 3t^2\dfrac{d^2y}{dt^2} + 6t\dfrac{dy}{dt} - 6y = 3 + \ln t^3$

Discussion Problems

41. Appropriately modify the procedure used in this section to solve

$$(t+2)^2\frac{d^2y}{dt^2} + (t+2)\frac{dy}{dt} + y = 0$$

State an interval over which the general solution is defined.

42. Solve the initial-value problem

$$t^2y'' - 4ty' + 4y = 0, \quad y(0) = 0, \; y'(0) = 0$$

Discuss why your solution does not violate Theorem 3.1.

43. Show that the linear combination $y = c_1t^3 + c_2t^4$ is the general solution of $t^2y'' - 6ty' + 12y = 0$ for $t > 0$. Show that the same linear combination can serve as the general solution of the differential equation for $t < 0$. Verify that $y_1 = t^3$ and $y_2 = t^4$ are linearly independent solutions of the differential equation for $-\infty < t < \infty$. Discuss: Does $y = c_1t^3 + c_2t^4$ represent the general solution of the equation on $(-\infty, \infty)$?

3.6 Mathematical Models: Initial-Value Problems

A branch of mathematics currently in vogue is the study of *dynamical systems*—roughly, systems that change or evolve with time. In more precise terms, a dynamical system consists of a set of time-dependent variables, called *state variables*, together with a rule that enables us to determine the state of the system (this may be a past, present, or future state) in terms of a state prescribed at some time t_0.

In this section we concentrate on a mathematical model of one such dynamical system—a spring/mass system—whose state at any future time $t > 0$ (the position and the velocity of the mass) depends on initial conditions $x(0) = x_0$ and $x'(0) = x_1$, which represent the state of the mass at $t = 0$.

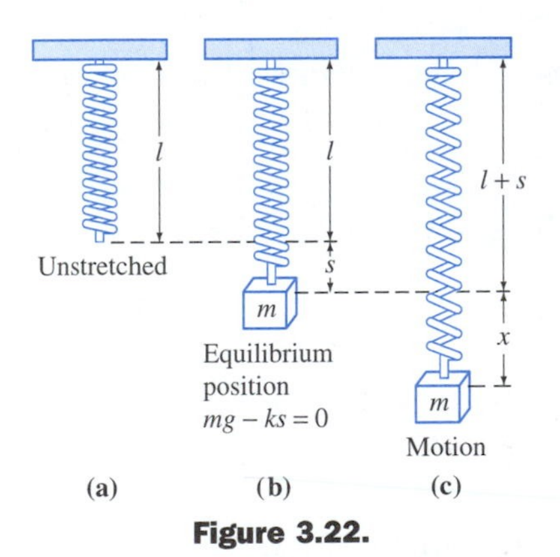

Figure 3.22.

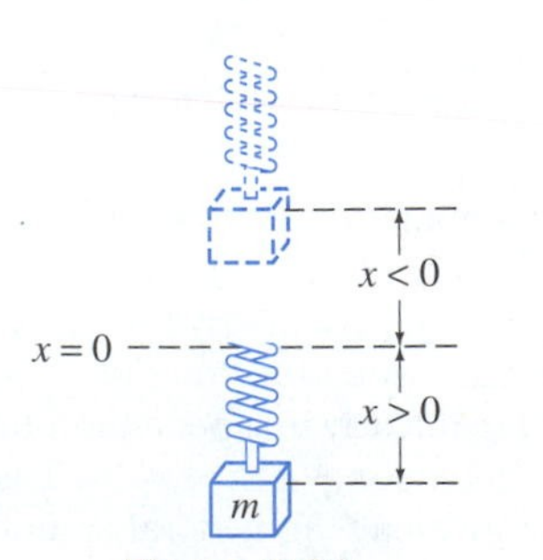

Figure 3.23.

Hooke's Law Suppose a flexible spring is suspended vertically from a rigid support and then a mass m is attached to its free end. The amount of stretch, or elongation, of the spring will of course depend on the mass; masses with different weights stretch the spring by differing amounts. By Hooke's law, the spring itself exerts a restoring force F opposite to the direction of elongation and proportional to the amount of elongation s. Simply stated, $F = ks$, where k is a constant of proportionality called the **spring constant**. The spring is essentially characterized by the number k. For example, if a mass weighing 10 lb stretches a spring $\frac{1}{2}$ ft, then $10 = k(\frac{1}{2})$ implies $k = 20$ lb/ft. Necessarily then, a mass weighing, say, 8 lb stretches the same spring only $\frac{2}{5}$ ft.

Newton's Second Law After a mass m is attached to a spring, it stretches the spring by an amount s and attains a position of equilibrium at which its weight W is balanced by the restoring force ks. Recall that weight is defined by $W = mg$, where m is measured in slugs, kilograms, or grams, and g is the acceleration due to gravity (32 ft/s^2, 9.8 m/s^2, or 980 cm/s^2). As indicated in Figure 3.22, the condition of equilibrium is $mg = ks$ or $mg - ks = 0$. Now suppose the mass on the spring is set in motion by giving it an initial displacement and velocity. Let us assume that the motion takes place in a vertical line, and that the displacements $x(t)$ of the mass are measured along this line such that $x = 0$ corresponds to the equilibrium position, and that displacements measured *below* the equilibrium position are *positive*. See Figure 3.23. To construct a mathematical model that describes this dynamic case, we employ Newton's second law of motion. We saw in Section 2.4 that the net or resultant force acting on a moving body of mass m is given by $\Sigma F_k = ma$, where a is its acceleration. Now if the elongation of the spring at any time t is $x + s$, and the velocity and acceleration of the mass are dx/dt and d^2x/dt^2, respectively, then

$ma = \Sigma_{k=1}^{4} F_k$ is

$$m\frac{d^2x}{dt^2} = -k(s + x) + mg - \beta\frac{dx}{dt} + f(t) \tag{1}$$

The first term $F_1 = -k(s + x)$, on the right side of the equation in (1), is the restoring force of the spring; the negative sign indicates that this force acts opposite to the direction of motion. The second term $F_2 = mg$ is the weight W of the mass, which always acts in the downward or positive direction. In the discussion of the falling body in Section 2.4, we stated that resistance to motion is taken to be proportional to a power of the instantaneous velocity of the body. Thus in (1) we have assumed that any retarding force acting on the mass is given by the linear law $F_3 = -\beta\, dx/dt$, where β is a positive *damping constant* and the negative sign is a consequence of the fact that the retarding force acts in a direction opposite to the motion. The last expression, $F_4 = f(t)$, represents all other time-dependent external forces that act on the mass. Note in this formulation that we have ignored the mass of the spring itself; we assume that the mass of the spring is negligible when compared to the mass m attached to it. Finally, by using $mg - ks = 0$, observe that (1) simplifies to

$$m\frac{d^2x}{dt^2} + \beta\frac{dx}{dt} + kx = f(t) \tag{2}$$

Two obvious initial conditions associated with (2) are $x(0) = x_0$—the amount of initial displacement—and $x'(0) = x_1$—the initial velocity of the mass. For example, if $x_0 > 0$, $x_1 < 0$, the mass starts from a point *below* the equilibrium position with an imparted *upward* velocity. When $x_1 = 0$, the mass is said to be released from *rest*. For example, if $x_0 < 0$, $x_1 = 0$, the mass is released from rest from a point $|x_0|$ units *above* the equilibrium position.

Free Motion If $f(t) = 0$, equation (2) is homogeneous and the motion of the mass is said to be **free**. For $\beta = 0$, a model for **free undamped motion** is obtained by dividing (2) by m and identifying $\omega^2 = k/m$:

$$\frac{d^2x}{dt^2} + \omega^2 x = 0 \tag{3}$$

Similarly, for $\beta \neq 0$, a model for **free damped motion** is

$$\frac{d^2x}{dt^2} + 2\lambda\frac{dx}{dt} + \omega^2 x = 0 \tag{4}$$

where $2\lambda = \beta/m$ and $\omega^2 = k/m$.

Solution of (3) To solve (3), we note that the solutions of the auxiliary equation $m^2 + \omega^2 = 0$ are the complex numbers $m_1 = \omega i$, $m_2 = -\omega i$. Hence the general solution of the differential equation is

$$x(t) = c_1 \cos \omega t + c_2 \sin \omega t \tag{5}$$

The obvious and immediate conclusion we can draw is that for any set of initial conditions $x(0) = x_0$, $x'(0) = x_1$, with x_0 and x_1 not both zero, (5) determines a particular solution—the **equation of motion**—that is sinusoidal. Thus in the absence of any damping after the system is set in motion, it stays in motion with the mass bouncing back and forth a fixed number A of units on either side of the equilibrium position. Such oscillatory motion is called **simple harmonic motion** with **period** $T = 2\pi/\omega$ and **frequency** $f = 1/T = \omega/2\pi$. The number ω is referred to as the **circular** or **angular frequency**.

The period $2\pi/\omega$ is the time that it takes the mass to execute one complete cycle, that is, the time between two successive maxima or two successive minima; or put another way, the graph of $x(t)$ repeats every $2\pi/\omega$ units of time. Because we have chosen the downward direction as the positive direction, it is important to keep in mind that a maximum of $x(t)$ is a displacement corresponding to the mass attaining a maximum distance *below* the equilibrium position, whereas a minimum of $x(t)$ is a displacement corresponding to the mass attaining a maximum height *above* the equilibrium position. We shall refer to a maximum or minimum displacement as an **extreme displacement** of the mass.

Alternative Form of the Solution When $c_1 \neq 0$ and $c_2 \neq 0$, the actual amplitude A of oscillations is not obvious from (5). Hence it is convenient to convert a solution of the form given in (5) to the simpler form:

$$x(t) = A\sin(\omega t + \phi) \tag{6}$$

where $A = \sqrt{c_1^2 + c_2^2}$ and ϕ is a **phase angle** defined by

$$\left.\begin{aligned} \sin\phi &= \frac{c_1}{A} \\ \cos\phi &= \frac{c_2}{A} \end{aligned}\right\} \tan\phi = \frac{c_1}{c_2} \tag{7}$$

To verify this, we expand (6) by the addition formula for the sine function and use $A\sin\phi = c_1$ and $A\cos\phi = c_2$ from (7):

$$\begin{aligned} A\sin(\omega t + \phi) &= A\sin\omega t\cos\phi + A\cos\omega t\sin\phi \\ &= (A\cos\phi)\sin\omega t + (A\sin\phi)\cos\omega t \\ &= c_2\sin\omega t + c_1\cos\omega t = x(t) \end{aligned}$$

EXAMPLE 1 Free Undamped Motion

A mass weighing 2 lb stretches a spring 6 in. At $t = 0$, the mass is released from a point 8 in. below the equilibrium position with an upward velocity of $\frac{4}{3}$ ft/s. Determine the equation of free undamped motion. Express this solution in form (6) and graph.

Solution Because we are using the engineering system of units, the measurements given in terms of inches must be converted into feet: 6 in. $= \frac{1}{2}$ ft, 8 in. $= \frac{2}{3}$ ft. In addition, we must convert the units of weight given in pounds into units of mass. From $m = W/g$ we have $m = \frac{2}{32} = \frac{1}{16}$ slug. Also, Hooke's law, $2 = k(\frac{1}{2})$, implies that the spring constant is $k = 4$ lb/ft. Hence (2) gives

$$\frac{1}{16}\frac{d^2x}{dt^2} + 4x = 0 \quad \text{or} \quad \frac{d^2x}{dt^2} + 64x = 0$$

The initial displacement and initial velocity are $x(0) = \frac{2}{3}$, $x'(0) = -\frac{4}{3}$, where the negative sign in the last condition is a consequence of the fact that the mass is given an initial velocity in the upward or negative direction.

Now $\omega^2 = 64$ or $\omega = 8$, so the general solution of the differential equation is $x(t) = c_1 \cos 8t + c_2 \sin 8t$. Applying the initial conditions to $x(t)$ and $x'(t)$ gives $c_1 = \frac{2}{3}$ and $c_2 = -\frac{1}{6}$. Thus the equation of motion is

$$x(t) = \frac{2}{3}\cos 8t - \frac{1}{6}\sin 8t \tag{8}$$

To express (8) in the form (6), we see that the amplitude of motion is

$$A = \sqrt{\left(\frac{2}{3}\right)^2 + \left(-\frac{1}{6}\right)^2} = \frac{\sqrt{17}}{6} \approx 0.69 \text{ ft}$$

Care should be exercised when computing the phase angle ϕ defined by (7). With $c_1 = \frac{2}{3}$ and $c_2 = -\frac{1}{6}$, we find $\tan \phi = -4$, and a calculator then gives $\tan^{-1}(-4) = -1.326$ radians. But since the range of the inverse tangent is the interval $(-\pi/2, \pi/2)$, the angle $\tan^{-1}(-4)$ is located in the fourth quadrant and therefore contradicts the fact that $\sin \phi > 0$ and $\cos \phi < 0$ (recall, $c_1 > 0$ and $c_2 < 0$). Hence we must take ϕ to be the second-quadrant angle $\phi = \pi + (-1.326) = 1.816$ radians. Thus an alternative form of the solution in (8) is

$$x(t) = \frac{\sqrt{17}}{6}\sin(8t + 1.816) \tag{9}$$

The period of this function is $T = 2\pi/8 = \pi/4$.

Figure 3.24(a) shows the mass going through approximately two complete cycles of motion. Reading left to right, the first five positions marked with black dots correspond to: the initial position of the mass below the equilibrium position ($x = \frac{2}{3}$), the mass passing through the equilibrium position for the first time heading upward ($x = 0$), the mass at its extreme displacement above the equilibrium position ($x = -\sqrt{17}/6$), the mass at the equilibrium position for the second time heading downward ($x = 0$), and the mass at its extreme displacement below the equilibrium position ($x = \sqrt{17}/6$). The dots on the graph of (9), given in Figure 3.24(b), also agree with the five positions just given. Note, however, that in Figure 3.24(b), the positive direction in the tx-plane is the usual upward direction and so is opposite to the positive direction indicated in Figure 3.24(a).

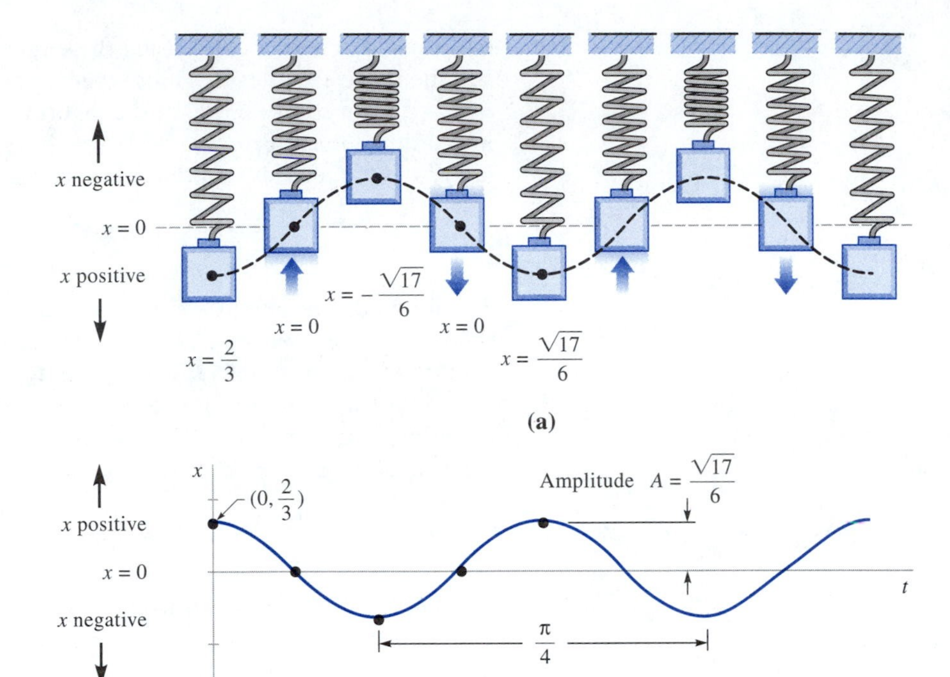

Figure 3.24.

Hence the solid colored graph representing the motion of the mass in Figure 3.24(b) is the mirror image through the t-axis of the black dashed curve in Figure 3.24(a). ■

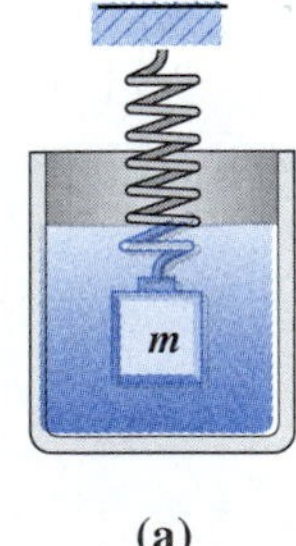

(a)

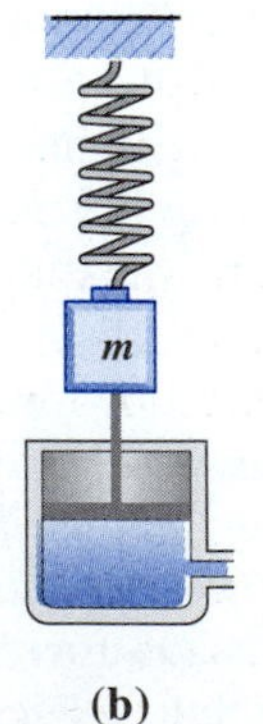

(b)

Figure 3.25.

The alternative form (6) is useful since it is easy to find values of time for which the mass passes through the equilibrium position $x = 0$, that is, when the graph of $x(t)$ crosses the t-axis. We observe that $\sin(\omega t + \phi) = 0$ when $\omega t + \phi = n\pi$, where n is a nonnegative integer.

The discussion of free harmonic motion is somewhat unrealistic; unless the mass is suspended in a perfect vacuum, there will be at least a resisting force due the surrounding medium. Or, as shown in Figure 3.25, the mass m could be suspended in a viscous medium or connected to a dashpot damping device.

Solution of (4) The symbol 2λ is used in (4) only for algebraic convenience since the roots of the auxiliary equation $m^2 + 2\lambda m + \omega^2 = 0$ are then

$$m_1 = -\lambda + \sqrt{\lambda^2 - \omega^2}, \quad m_2 = -\lambda - \sqrt{\lambda^2 - \omega^2} \tag{10}$$

As in the discussion in Section 3.2, we can now distinguish three forms of the general solution of (4) depending on the algebraic sign of $\lambda^2 - \omega^2$. But before writing anything, note that the roots in (10) indicate that each form of the general solution must necessarily contain $e^{-\lambda t}$, $\lambda > 0$. This exponential term acts as a *damping factor* and, consequently, free damped motion is characterized by the fact that the mass must return to the equilibrium position over time, that is, $x(t) \to 0$ as $t \to \infty$.

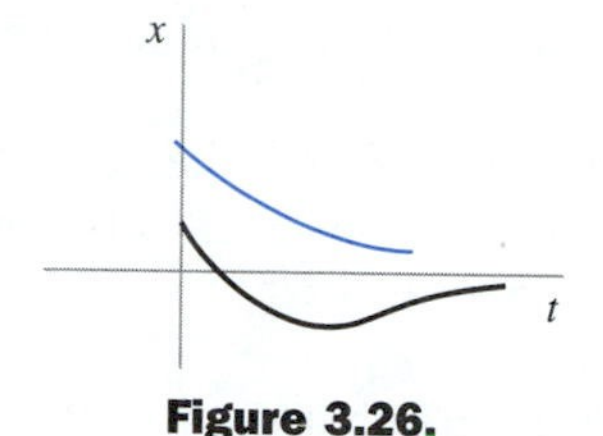

Figure 3.26.

CASE I $\lambda^2 - \omega^2 > 0$ In this situation, the spring/mass system is said to be **overdamped** since the damping coefficient β is large when compared to the spring constant k. The roots m_1 and m_2 are real and distinct, so the corresponding general solution of (4) is $x(t) = c_1 e^{m_1 t} + c_2 e^{m_2 t}$ or

$$x(t) = e^{-\lambda t}\left(c_1 e^{\sqrt{\lambda^2 - \omega^2}\,t} + c_2 e^{-\sqrt{\lambda^2 - \omega^2}\,t}\right) \tag{11}$$

This equation represents a smooth and nonoscillatory motion. Figure 3.26 shows two possible graphs of $x(t)$.

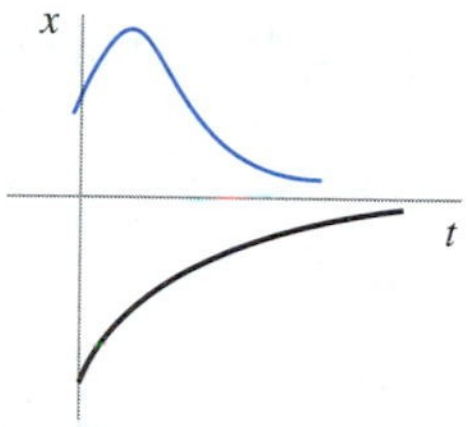

Figure 3.27.

CASE II $\lambda^2 - \omega^2 = 0$ The system is said to be **critically damped** because any slight decrease in the damping force would result in oscillatory motion. Using $m_1 = m_2 = -\lambda$, the general solution of (4) is $x(t) = c_1 e^{m_1 t} + c_2 t e^{m_1 t}$ or

$$x(t) = e^{-\lambda t}(c_1 + c_2 t) \tag{12}$$

Two graphs illustrating typical motion are given in Figure 3.27. The motion is similar to that of the overdamped case: the mass returns smoothly to the equilibrium position as $t \to \infty$ and the mass either does not pass through the equilibrium position or does so one time at most.

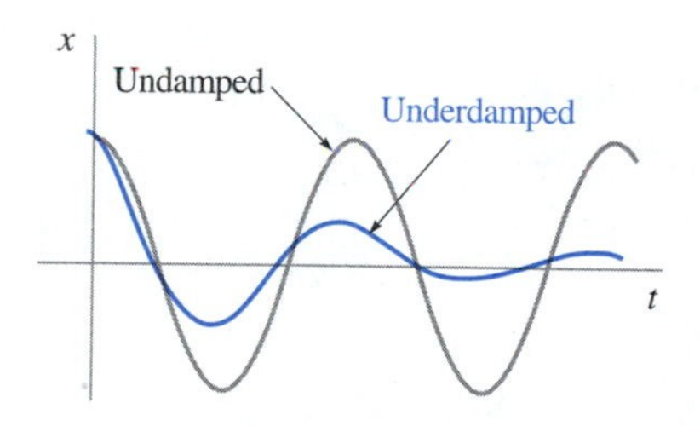

Figure 3.28.

CASE III $\lambda^2 - \omega^2 < 0$ In this case the system is said to be **underdamped**, since the damping coefficient is small compared to the spring constant. The roots m_1 and m_2 are now complex:

$$m_1 = -\lambda + \sqrt{\omega^2 - \lambda^2}\,i, \quad m_2 = -\lambda - \sqrt{\omega^2 - \lambda^2}\,i$$

and so the general solution of equation (4) is

$$x(t) = e^{-\lambda t}\left(c_1 \cos \sqrt{\omega^2 - \lambda^2}\,t + c_2 \sin \sqrt{\omega^2 - \lambda^2}\,t\right) \tag{13}$$

As illustrated in Figure 3.28, the motion described by (13) is oscillatory; but because of the coefficient $e^{-\lambda t}$, the amplitudes of vibration $\to 0$ as $t \to \infty$.

■ EXAMPLE 2 **Overdamped Motion**

It is readily verified that the solution of the initial-value problem

$$\frac{d^2x}{dt^2} + 5\frac{dx}{dt} + 4x = 0, \quad x(0) = 1, \quad x'(0) = 1$$

is

$$x(t) = \frac{5}{3}e^{-t} - \frac{2}{3}e^{-4t} \tag{14}$$

The problem can be interpreted as representing the overdamped motion of a mass on a spring. The mass starts from a position 1 unit *below* the equilibrium position with a *downward* velocity of 1 ft/s.

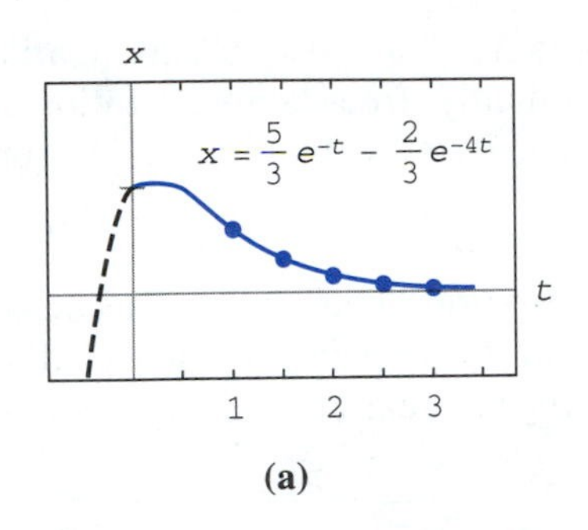

(a)

t	$x(t)$
1.0	0.601
1.5	0.370
2.0	0.225
2.5	0.137
3.0	0.083

(b)

Figure 3.29.

To graph $x(t)$ we find the value of t for which the function has an extremum, that is, the value of time for which the first derivative (velocity) is zero. Differentiating (14) gives $x'(t) = -\frac{5}{3}e^{-t} + \frac{8}{3}e^{-4t}$ so that $x'(t) = 0$ implies $e^{3t} = \frac{8}{5}$ or $t = \frac{1}{3}\ln\frac{8}{5} = 0.157$. It follows from the first derivative test, as well as our physical intuition, that $x(0.157) = 1.069$ ft is actually a maximum. In other words, the mass attains an extreme displacement of 1.069 ft below the equilibrium position.

We should also check to see whether the graph crosses the t-axis, that is, whether the mass passes through the equilibrium position. This cannot happen in this instance since the equation $x(t) = 0$, or $e^{3t} = \frac{2}{5}$, has the physically irrelevant solution $t = \frac{1}{3}\ln\frac{2}{5} = -0.305$.

The graph of $x(t)$, along with some other pertinent data, is given in Figure 3.29. ■

■ EXAMPLE 3 Critically Damped Motion

An 8-lb weight stretches a spring 2 ft. Assuming that a damping force numerically equal to two times the instantaneous velocity acts on the system, determine the equation of motion if the weight is released from the equilibrium position with an upward velocity of 3 ft/s.

Solution From Hooke's law, we see that $8 = k(2)$ gives $k = 4$ lb/ft and that $W = mg$ gives $m = \frac{8}{32} = \frac{1}{4}$ slug. The differential equation of motion is then

$$\frac{1}{4}\frac{d^2x}{dt^2} + 2\frac{dx}{dt} + 4x = 0 \quad \text{or} \quad \frac{d^2x}{dt^2} + 8\frac{dx}{dt} + 16x = 0 \tag{15}$$

The auxiliary equation for (15) is $m^2 + 8m + 16 = (m + 4)^2 = 0$ so that $m_1 = m_2 = -4$. Hence the system is critically damped and

$$x(t) = c_1e^{-4t} + c_2te^{-4t} \tag{16}$$

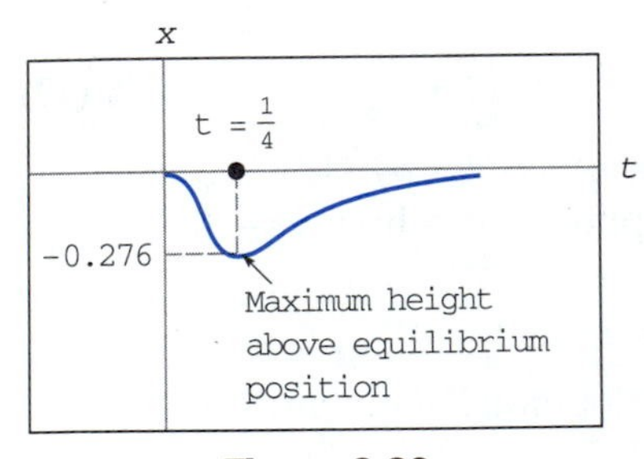

Figure 3.30.

Applying the initial conditions $x(0) = 0$ and $x'(0) = -3$, we find, in turn, that $c_1 = 0$ and $c_2 = -3$. Thus the equation of motion is

$$x(t) = -3te^{-4t} \tag{17}$$

To graph $x(t)$, we proceed as in Example 2. From $x'(t) = -3e^{-4t}(1 - 4t)$ we see that $x'(t) = 0$ when $t = \frac{1}{4}$. The extreme displacement is then $x(\frac{1}{4}) = -3(\frac{1}{4})e^{-1} = -0.276$ ft. As shown in Figure 3.30, we interpret this value to mean that the weight reaches a maximum height of 0.276 ft above the equilibrium position. ■

■ EXAMPLE 4 Underdamped Motion

A 16-lb weight is attached to a 5-ft-long spring. At equilibrium, the spring measures 8.2 ft. If the weight is pushed up and released from rest at a point 2 ft above the equilibrium position, find the displacements $x(t)$ if it is further known that the surrounding medium offers a resistance numerically equal to the instantaneous velocity.

Solution The elongation of the spring after the weight is attached is $8.2 - 5 = 3.2$ ft, so it follows from Hooke's law that $16 = k(3.2)$ or $k = 5$ lb/ft. In addition, $m = \frac{16}{32} = \frac{1}{2}$ slug, so that the differential equation is given by

$$\frac{1}{2}\frac{d^2x}{dt^2} + \frac{dx}{dt} + 5x = 0 \quad \text{or} \quad \frac{d^2x}{dt^2} + 2\frac{dx}{dt} + 10x = 0 \tag{18}$$

Proceeding, we find that the roots of $m^2 + 2m + 10 = 0$ are $m_1 = -1 + 3i$ and $m_2 = -1 - 3i$, which then implies the system is underdamped and

$$x(t) = e^{-t}(c_1 \cos 3t + c_2 \sin 3t) \tag{19}$$

Finally, the initial conditions $x(0) = -2$ and $x'(0) = 0$ yield $c_1 = -2$ and $c_2 = -\frac{2}{3}$, so the equation of motion is

$$x(t) = e^{-t}\left(-2 \cos 3t - \frac{2}{3} \sin 3t\right) \tag{20}$$ ■

Alternative Form of the Solution In a manner identical to the procedure used on page 174, we can write any solution

$$x(t) = e^{-\lambda t}(c_1 \cos \sqrt{\omega^2 - \lambda^2}\,t + c_2 \sin \sqrt{\omega^2 - \lambda^2}\,t)$$

in the alternative form

$$x(t) = Ae^{-\lambda t} \sin(\sqrt{\omega^2 - \lambda^2}\,t + \phi) \tag{21}$$

where $A = \sqrt{c_1^2 + c_2^2}$ and the phase angle ϕ is determined from the equations

$$\sin \phi = \frac{c_1}{A}, \quad \cos \phi = \frac{c_2}{A}, \quad \tan \phi = \frac{c_1}{c_2}$$

The coefficient $Ae^{-\lambda t}$ is sometimes called the **damped amplitude** of vibrations. Because (21) is not a periodic function, the number $2\pi/\sqrt{\omega^2 - \lambda^2}$ is called the **quasi period** and $\sqrt{\omega^2 - \lambda^2}/2\pi$ is the **quasi frequency**. The quasi period is the time interval between two successive maxima of $x(t)$. You should verify, for the equation of motion in Example 4, that $A = 2\sqrt{10}/3$ and $\phi = 4.931$. Therefore an equivalent form of (20) is

$$x(t) = \frac{2\sqrt{10}}{3} e^{-t} \sin(3t + 4.931)$$

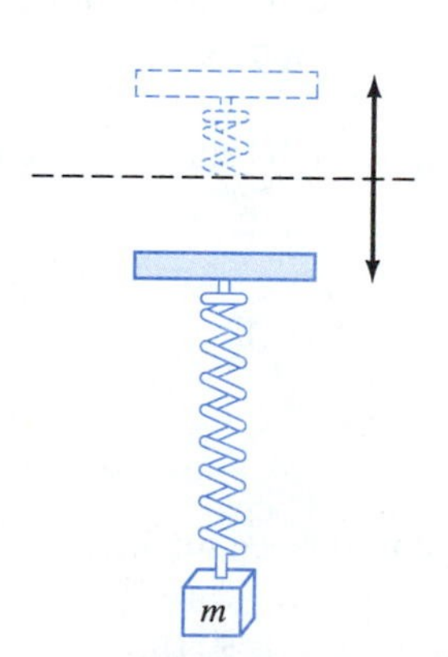

Figure 3.31.

Driven Motion Suppose we now take into consideration an external force $f(t)$ acting on a vibrating spring/mass system. As illustrated in Figure 3.31, the function $f(t)$ could represent a driving force causing an oscillatory vertical motion of the support of the spring. Dividing (2) by m gives

$$\frac{d^2x}{dt^2} + 2\lambda\frac{dx}{dt} + \omega^2 x = F(t) \tag{22}$$

where $F(t) = f(t)/m$ and, as in the preceding section, $2\lambda = \beta/m$, $\omega^2 = k/m$. To solve the latter nonhomogeneous equation, we can use either the method of undetermined coefficients or variation of parameters.

EXAMPLE 5 Interpretation of an Initial-Value Problem

Interpret and solve the initial-value problem

$$\frac{1}{5}\frac{d^2x}{dt^2} + 1.2\frac{dx}{dt} + 2x = 5\cos 4t, \qquad x(0) = \frac{1}{2}, \quad x'(0) = 0 \tag{23}$$

Solution We can interpret the problem to represent a vibrational system consisting of a mass ($m = \frac{1}{5}$ slug or kg) attached to a spring ($k = 2$ lb/ft or N/m). The mass is released from rest $\frac{1}{2}$ unit (foot or meter) below the equilibrium position. The motion is damped ($\beta = 1.2$) and is being driven by an external periodic ($T = \pi/2$ s) force beginning at $t = 0$. Intuitively we would expect that even with damping the system would remain in motion until such time as the forcing function is "turned off," in which case the amplitudes diminish. However, as the problem is given, $f(t) = 5\cos 4t$ will remain "on" forever.

We first multiply the differential equation in (23) by 5 and solve

$$\frac{dx^2}{dt^2} + 6\frac{dx}{dt} + 10x = 0$$

by the usual methods. Since $m_1 = -3 + i$, $m_2 = -3 - i$, it follows that

$$x_c(t) = e^{-3t}(c_1 \cos t + c_2 \sin t)$$

Using the method of undetermined coefficients, we assume a particular solution of the form $x_p(t) = A\cos 4t + B\sin 4t$. Now from

$$x_p'' + 6x_p' + 10x_p = (-6A + 24B)\cos 4t + (-24A - 6B)\sin 4t = 25\cos 4t$$

the resulting system of equations

$$-6A + 24B = 25, \quad -24A - 6B = 0$$

yields $A = -\frac{25}{102}$ and $B = \frac{50}{51}$. It follows that

$$x(t) = e^{-3t}(c_1 \cos t + c_2 \sin t) - \frac{25}{102}\cos 4t + \frac{50}{51}\sin 4t \tag{24}$$

When we set $t = 0$ in the above equation, we obtain $c_1 = \frac{38}{51}$. By differentiating the expression and then setting $t = 0$, we also find that $c_2 = -\frac{86}{51}$. Therefore the equation of motion is

$$x(t) = e^{-3t}\left(\frac{38}{51}\cos t - \frac{86}{51}\sin t\right) - \frac{25}{102}\cos 4t + \frac{50}{51}\sin 4t \tag{25}$$ ■

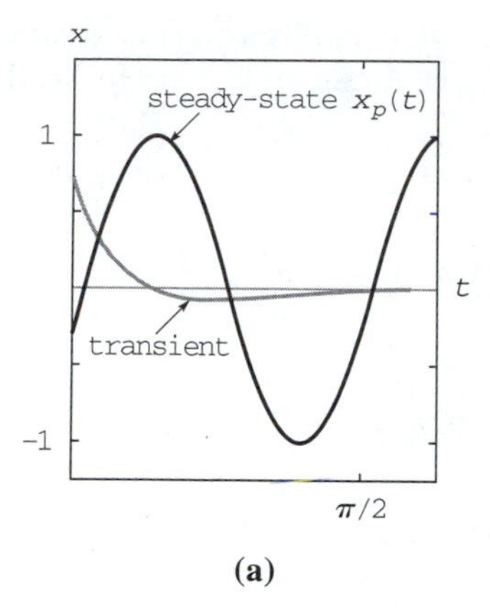

(a)

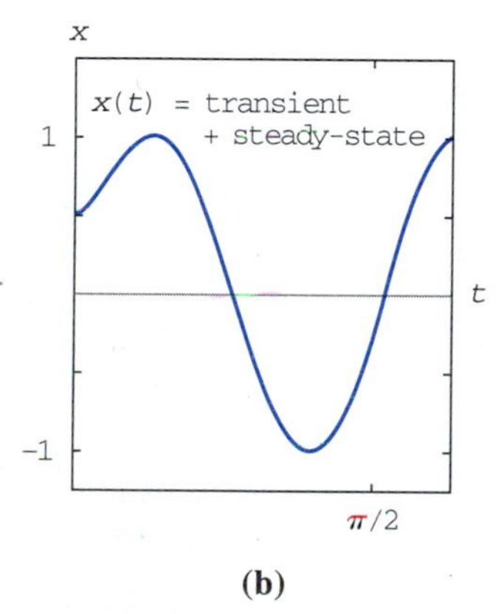

(b)

Figure 3.32.

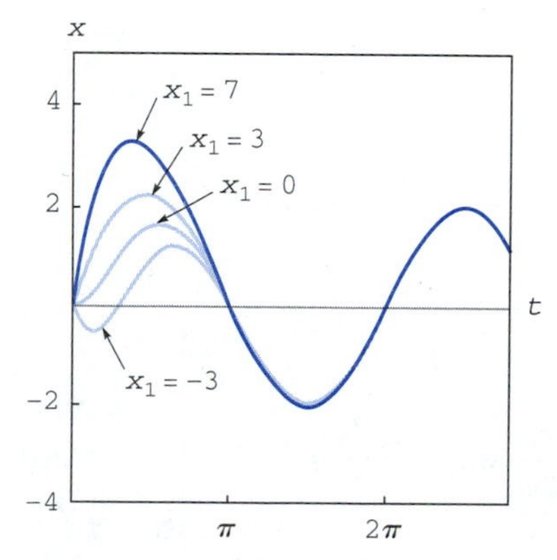

Figure 3.33.

Transient and Steady-State Terms When F is a periodic function, such as $F(t) = F_0 \sin \gamma t$ or $F(t) = F_0 \cos \gamma t$, the general solution of (22) for $\lambda > 0$ is the sum of a nonperiodic function $x_c(t)$ and a periodic function $x_p(t)$. Moreover, $x_c(t)$ dies off as time increases, that is, $\lim_{t\to\infty} x_c(t) = 0$. Thus for large time, the displacements of the mass are closely approximated by the particular solution $x_p(t)$. The complementary function $x_c(t)$ is said to be a **transient term** or **transient solution**, and the function $x_p(t)$, the part of the solution that remains after an interval of time, is called a **steady-state term** or **steady-state solution**. Note therefore that the effect of the initial conditions on a spring/mass system driven by F is transient. In the particular solution (25), $e^{-3t}(\frac{38}{51}\cos t - \frac{86}{51}\sin t)$ is a transient term and $x_p(t) = -\frac{25}{102}\cos 4t + \frac{50}{51}\sin 4t$ is a steady-state term. The graphs of these two terms and the solution (25) are given Figures 3.32(a) and 3.32(b), respectively.

EXAMPLE 6 Transient/Steady State Solutions

The solution of the initial-value problem

$$\frac{d^2x}{dt^2} + 2\frac{dx}{dt} + 2x = 4\cos t + 2\sin t, \qquad x(0) = 0, \quad x'(0) = x_1$$

where x_1 is constant, is given by

$$x(t) = \underbrace{(x_1 - 2)e^{-t}\sin t}_{\text{transient}} + \underbrace{2\sin t}_{\text{steady-state}}$$

For selected values of the initial velocity x_1, we obtain with the aid of a graphing utility, the solution curves shown in Figure 3.33. The graphs show that the influence of the transient term is negligible for about $t > 3\pi/2$. ■

With a periodic impressed force and no damping, there is no transient term in the solution of the problem. Also, we shall see that severe problems may result when a periodic force impressed on an oscillatory mechanical system has a frequency near, or perhaps equal to, the frequency of free undamped vibrations of the system.

EXAMPLE 7 Undamped Forced Motion

Solve the initial-value problem

$$\frac{d^2x}{dt^2} + \omega^2 x = F_0 \sin \gamma t, \qquad x(0) = 0, \quad x'(0) = 0 \tag{26}$$

where F_0 is a constant and $\gamma \neq \omega$.

Solution The complementary function is $x_c(t) = c_1 \cos \omega t + c_2 \sin \omega t$. To obtain a particular solution, we assume $x_p(t) = A \cos \gamma t + B \sin \gamma t$, so that

$$x_p'' + \omega^2 x_p = A(\omega^2 - \gamma^2) \cos \gamma t + B(\omega^2 - \gamma^2) \sin \gamma t = F_0 \sin \gamma t$$

Equating coefficients immediately gives $A = 0$ and $B = F_0/(\omega^2 - \gamma^2)$. Therefore

$$x_p(t) = \frac{F_0}{\omega^2 - \gamma^2} \sin \gamma t$$

Applying the given initial conditions to the general solution

$$x(t) = c_1 \cos \omega t + c_2 \sin \omega t + \frac{F_0}{\omega^2 - \gamma^2} \sin \gamma t$$

yields $c_1 = 0$ and $c_2 = -\gamma F_0/\omega(\omega^2 - \gamma^2)$. Thus the solution is

$$x(t) = \frac{F_0}{\omega(\omega^2 - \gamma^2)} (-\gamma \sin \omega t + \omega \sin \gamma t), \quad \gamma \neq \omega \qquad \textbf{(27)}$$ ■

Pure Resonance Although equation (27) is not defined for $\gamma = \omega$, its limiting value as $\gamma \to \omega$ can be obtained by applying L'Hôpital's rule. This limiting process is analogous to "tuning in" the frequency of the driving force ($\gamma/2\pi$) to the frequency of free vibrations ($\omega/2\pi$). Intuitively we expect that over a length of time we should be able to increase substantially the amplitudes of vibration. For $\gamma = \omega$ we define the solution to be

$$\begin{aligned} x(t) = \lim_{\gamma \to \omega} F_0 \frac{-\gamma \sin \omega t + \omega \sin \gamma t}{\omega(\omega^2 - \gamma^2)} &= F_0 \lim_{\gamma \to \omega} \frac{\dfrac{d}{d\gamma}(-\gamma \sin \omega t + \omega \sin \gamma t)}{\dfrac{d}{d\gamma}(\omega^3 - \omega\gamma^2)} \\ &= F_0 \lim_{\gamma \to \omega} \frac{-\sin \omega t + \omega t \cos \gamma t}{-2\omega\gamma} \\ &= F_0 \frac{-\sin \omega t + \omega t \cos \omega t}{-2\omega^2} \\ &= \frac{F_0}{2\omega^2} \sin \omega t - \frac{F_0}{2\omega} t \cos \omega t \end{aligned} \qquad \textbf{(28)}$$

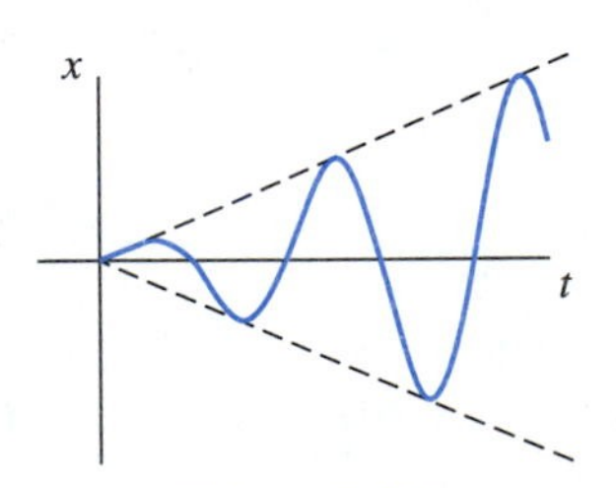

Figure 3.34.

As suspected, when $t \to \infty$, the displacements become large; in fact, $|x(t_n)| \to \infty$ when $t_n = n\pi/\omega$, $n = 1, 2, \ldots$. The phenomenon we have just described is known as **pure resonance**. The graph given in Figure 3.34 shows typical motion in this case.

In conclusion, it should be noted that there is no actual need to use a limiting process on (27) to obtain the solution for $\gamma = \omega$. Alternatively, equation (28) follows by solving the initial-value problem

$$\frac{d^2x}{dt^2} + \omega^2 x = F_0 \sin \omega t, \qquad x(0) = 0, \quad x'(0) = 0$$

directly by conventional methods.

If the displacements of a spring/mass system were actually described by a function such as (28), the system would necessarily fail. Large oscillations of the mass would eventually force the spring beyond its elastic limit. One might argue, too, that the resonating model presented in Figure 3.34 is completely unrealistic, because it ignores the retarding effects of ever-present damping forces. Although it is true that pure resonance cannot occur when the smallest amount of damping is taken into consideration, large and equally destructive amplitudes of vibration (although bounded as $t \to \infty$) can occur.

An Aging Spring In the spring/mass model just discussed, we assumed an ideal world—a world in which the physical characteristics of the system do not change over time. In the nonideal world, however, it seems reasonable to expect that when such a spring/mass system is in motion for a long period, the spring would weaken, in other words, the spring "constant" would decay with time. In one model for an **aging spring**, the spring constant k is replaced by the function $K(t) = ke^{-\alpha t}$, $k > 0$, $\alpha > 0$. The differential equation of free motion, $mx'' + \beta x' + ke^{-\alpha t}x = 0$, cannot be solved in terms of elementary functions. Nevertheless, we can obtain two linearly independent solutions of the model $mx'' + ke^{-\alpha t}x = 0$ for a free undamped aging spring using the methods of Chapter 6. See Example 3 in Section 6.3.

When a spring/mass system is subjected to an environment in which temperature is rapidly decreasing, it might make sense to replace the constant k with $K(t) = kt$, $k > 0$, a function that increases with time. The resulting model, $mx'' + ktx = 0$, is a form of **Airy's differential equation**. Like the equation for an aging spring, Airy's equation can be solved by the methods of Chapter 6. See Example 3 in Section 6.1 and Problems 25–27 in Exercises 6.3.

An Analogous System As mentioned in the introduction to this chapter, many different systems can be described by a linear second-order differential equation similar to (2), which is the equation of forced motion with damping. If $i(t)$ denotes current in an ***L-R-C* series electrical circuit**, shown in Figure 3.35(a), then the voltage drops across the inductor, resistor, and capacitor are as shown in Figure 3.35(b). By Kirchhoff's second law, the sum of these voltages equals the voltage $E(t)$ impressed on the circuit; that is,

$$L\frac{di}{dt} + Ri + \frac{1}{C}q = E(t) \tag{29}$$

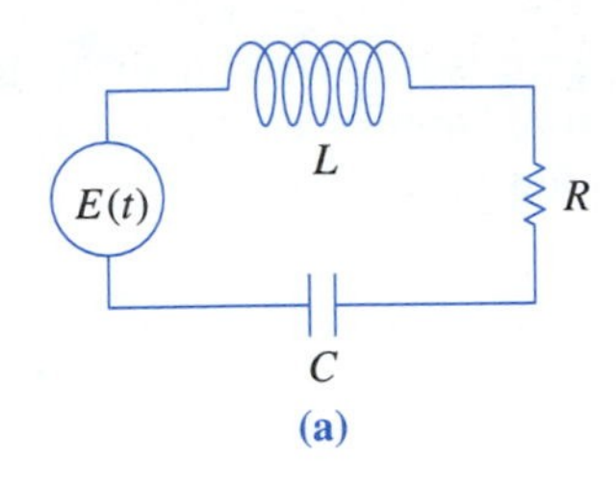

Inductor
Inductance L: henrys (h)
voltage drop across: $L\dfrac{di}{dt}$

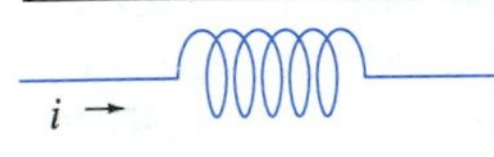

Resistor
Resistance R: ohms (Ω)
voltage drop across: iR

Capacitor
Capacitance C: farads (f)
voltage drop across: $\dfrac{1}{C}q$

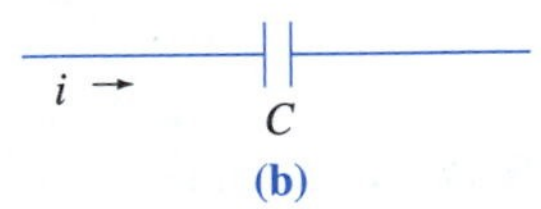

Figure 3.35.

But the charge $q(t)$ on the capacitor is related to the current $i(t)$ by $i = dq/dt$, so (29) becomes the linear second-order differential equation

$$L\frac{d^2q}{dt^2} + R\frac{dq}{dt} + \frac{1}{C}q = E(t) \tag{30}$$

Except for the physical interpretation of the symbols in (2) and (30), there is absolutely no difference between the mathematics of vibrating springs and simple series circuits. Even the nomenclature used in circuit analysis is similar to that used to describe spring/mass systems.

If $E(t) = 0$, the **electrical vibrations** of the circuit are said to be **free**. Since the auxiliary equation for (4) is $Lm^2 + Rm + 1/C = 0$, there will be three forms of the solution when $R \neq 0$, depending on the value of the discriminant $R^2 - 4L/C$. We say that the circuit is

	overdamped if	$R^2 - 4L/C > 0$,
	critically damped if	$R^2 - 4L/C = 0$,
and	**underdamped** if	$R^2 - 4L/C < 0$

In each of these three cases, the general solution of (30) contains the factor $e^{-Rt/2L}$ and so $q(t) \to 0$ as $t \to \infty$. In the underdamped case when $q(0) = q_0$, the charge on the capacitor oscillates as it decays; in other words, the capacitor is charging and discharging as $t \to \infty$. When $E(t) = 0$ and $R = 0$, the circuit is said to be undamped and the electrical vibrations do not approach zero as t increases without bound; the response of the circuit is **simple harmonic**.

When there is an impressed voltage $E(t)$ on the circuit, the electrical vibrations are said to be **forced**. In the case when $R \neq 0$, the complementary function $q_c(t)$ of (30) is called a **transient solution**. If $E(t)$ is periodic or a constant, then the particular solution $q_p(t)$ of (30) is a **steady-state solution**.

For example, we leave it to you to show that an L-R-C circuit governed by

$$L\frac{d^2q}{dt^2} + R\frac{dq}{dt} + \frac{1}{C}q = E_0 \sin \gamma t \tag{31}$$

has the steady-state solution or steady-state charge

$$q_p(t) = -\frac{E_0X}{\gamma Z^2}\sin \gamma t - \frac{E_0R}{\gamma Z^2}\cos \gamma t$$

From this solution and $i_p(t) = q_p'(t)$, we obtain the **steady-state current**:

$$i_p(t) = \frac{E_0}{Z}\left(\frac{R}{Z}\sin \gamma t - \frac{X}{Z}\cos \gamma t\right) \tag{32}$$

The quantities $X = L\gamma - 1/C\gamma$ and $Z = \sqrt{X^2 + R^2}$ are called the **reactance** and **impedance** of the circuit, respectively. Both the reactance and impedance are measured in ohms. See Problem 36 in Exercises 3.6.

Remarks

(*i*) The model in (3) has the form $md^2x/dt^2 + F(x) = 0$, where F represents the restoring force. Naturally, for a restoring force given by the linear function $F(x) = kx$, the spring is referred to as a **linear spring**. But springs are seldom perfectly linear and so a better model for the restoring force F may be a nonlinear function such as $F(x) = kx + k_1x^3$. **Nonlinear springs**—springs acting under nonlinear restoring forces—will be discussed in Section 3.10.

(*ii*) In general, any spring/mass system whose mathematical model is $md^2x/dt^2 + F(x) = 0$ is said to be a **conservative system**. For example, the trajectories of the phase portrait of (3) are defined by the one-parameter family of solutions of

$$\frac{dy}{dx} = -\frac{kx}{my}$$

where $y = dx/dt$ is the velocity of the mass. Separating variables in this first-order equation and integrating give $\frac{1}{2}my^2 + \frac{1}{2}kx^2 = E$, where E is a constant. This last equation expresses the fact that the law of conservation of energy holds in a free *undamped* spring/mass system since the sum of the kinetic energy of the mass $E_k = \frac{1}{2}my^2$ and the potential energy of the spring $E_p = \frac{1}{2}kx^2$ is a constant E. Thus the trajectories defined by this equation represent various energy states of the system and so they are also called **energy curves**. Because all the trajectories are closed curves (ellipses), (0, 0) is a stable center of (3) and all solutions are periodic. A free but *damped* spring/mass system is a **nonconservative system** since energy is dissipated. In the case of the linear model (4), when $\beta > 0$, none of the trajectories is closed and so there exist no periodic solutions. For $\beta > 0$, (0, 0) is *always* an asymptotically stable critical point—an attractor—of the linear equation (4). In the cases of overdamped and critically damped spring/mass systems, (0, 0) is an asymptotically stable node. For an underdamped system, (0, 0) is an asymptotically stable spiral point.

EXERCISES 3.6

Answers to odd-numbered problems begin on page 442.

Problems 1–14 are on free undamped motion.

1. A 4-lb weight is attached to a spring whose spring constant is 16 lb/ft. What is the period of simple harmonic motion?

2. A 20-kg mass is attached to a spring. If the frequency of simple harmonic motion is $2/\pi$ vibrations/second, what is the spring constant k? What is the frequency of simple harmonic motion if the original mass is replaced with an 80-kg mass?

3. A 24-lb weight, attached to the end of a spring, stretches it 4 in. Find the equation of motion if the weight is released from rest from a point 3 in. above the equilibrium position.

4. Determine the equation of motion if the weight in Problem 3 is released from the equilibrium position with an initial downward velocity of 2 ft/s.

5. A 20-lb weight stretches a spring 6 in. The weight is released from rest 6 in. below the equilibrium position.
   (a) Find the position of the weight at $t = \pi/12$, $\pi/8$, $\pi/6$, $\pi/4$, $9\pi/32$ s.
   (b) What is the velocity of the weight when $t = 3\pi/16$ s? In which direction is the weight heading at this instant?
   (c) At what times does the weight pass through the equilibrium position?

6. A force of 400 N stretches a spring 2 m. A mass of 50 kg is attached to the end of the spring and released from the equilibrium position with an upward velocity of 10 m/s. Find the equation of motion.

7. Another spring whose constant is 20 N/m is suspended from the same rigid support but parallel to the spring/mass system in Problem 6. A mass of 20 kg is attached to the second spring and both masses are released from the equilibrium position with an upward velocity of 10 m/s.
   (a) Which mass exhibits the greater amplitude of motion?
   (b) Which mass is moving faster at $t = \pi/4$ s? at $\pi/2$ s?
   (c) At what times are the two masses in the same position? Where are the masses at these times? In which directions are they moving?

8. A 32-lb weight stretches a spring 2 ft. Determine the amplitude and period of motion if the weight is released 1 ft above the equilibrium position with an initial upward velocity of 2 ft/s. How many complete vibrations will the weight have completed at the end of 4π s?

9. An 8-lb weight attached to a spring exhibits simple harmonic motion. Determine the equation of motion if the spring constant is 1 lb/ft and if the weight is released 6 in. below the equilibrium position with a downward velocity of $\frac{3}{2}$ ft/s. Express the solution in form (6).

10. A mass weighing 10 lb stretches a spring $\frac{1}{4}$ ft. This mass is removed and replaced with a mass of 1.6 slugs, which is released $\frac{1}{3}$ ft above the equilibrium position with a downward velocity of $\frac{5}{4}$ ft/s. Express the solution in form (6). At what times does the mass attain a displacement below the equilibrium position numerically equal to one-half the amplitude?

11. A 64-lb weight attached to the end of a spring stretches it 0.32 ft. From a position 8 in. above the equilibrium position, the weight is given a downward velocity of 5 ft/s.
    (a) Find the equation of motion.
    (b) What are the amplitude and period of motion?
    (c) How many complete vibrations will the weight have completed at the end of 3π s?
    (d) At what time does the weight pass through the equilibrium position heading downward for the second time?
    (e) At what time does the weight attain its extreme displacement on either side of the equilibrium position?
    (f) What is the position of the weight at $t = 3$ s?
    (g) What is the instantaneous velocity at $t = 3$ s?
    (h) What is the acceleration at $t = 3$ s?
    (i) What is the instantaneous velocity at the times when the weight passes through the equilibrium position?
    (j) At what times is the weight 5 in. below the equilibrium position?
    (k) At what times is the weight 5 in. below the equilibrium position heading in the upward direction?

12. A mass of 1 slug is suspended from a spring whose characteristic spring constant is 9 lb/ft. Initially the mass starts from a point 1 ft above the equilibrium position with an upward velocity of $\sqrt{3}$ ft/s. Find the times for which the mass is heading downward at a velocity of 3 ft/s.

13. Under some circumstances, when two parallel springs, with constants k_1 and k_2, support a single weight W, the **effective spring constant** of the system is given by $k = 4k_1k_2/(k_1 + k_2)$. A 20-lb weight stretches one spring 6 in. and another spring 2 in. The springs are attached to a common rigid support and then to a metal plate. As shown in Figure 3.36, the 20-lb weight is attached to the center of the plate in the double spring arrangement. Determine the effective spring constant of this system. Find the equation of motion if the weight is released from the equilibrium position with a downward velocity of 2 ft/s.

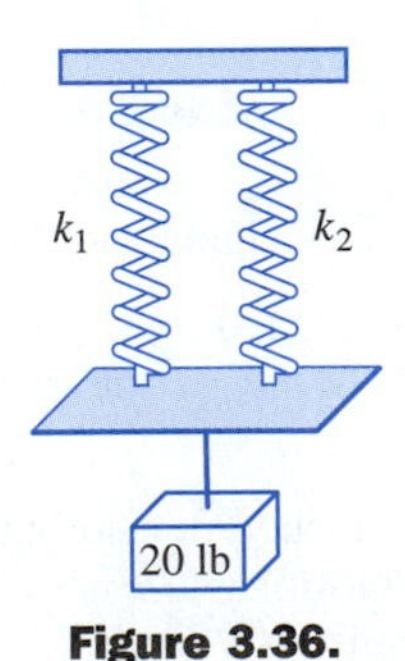

Figure 3.36.

14. A certain weight stretches one spring $\frac{1}{3}$ ft and another spring $\frac{1}{2}$ ft. The two springs are attached to a common rigid support in a manner indicated in Problem 13 and Figure 3.36. The first weight is set aside, and an 8-lb weight is attached to the double spring arrangement and the system is set in motion. If the period of motion is $\pi/15$ s, determine the numerical value of the first weight.

Problems 15–22 are on free damped motion.

15. A 4-lb weight is attached to a spring whose constant is 2 lb/ft. The medium offers a resistance to the motion of the weight numerically equal to the instantaneous velocity. If the weight is released from a point 1 ft above the equilibrium position with a downward velocity of 8 ft/s, determine the time that the weight passes through the equilibrium position. Find the time for which the weight attains its extreme displacement from the equilibrium position. What is the position of the weight at this instant?

16. A 4-ft spring measures 8 ft long after an 8-lb weight is attached to it. The medium through which the weight moves offers a resistance numerically equal to $\sqrt{2}$ times the instantaneous velocity. Find the equation of motion if the weight is released from the equilibrium position with a downward velocity of 5 ft/s. Find the time for which the weight attains its extreme displacement from the equilibrium position. What is the position of the weight at this instant?

17. A 1-kg mass is attached to a spring whose constant is 16 N/m and the entire system is then submerged in a liquid that imparts a damping force numerically equal to 10 times the instantaneous velocity. Determine the equations of motion if

(a) the weight is released from rest 1 m below the equilibrium position; and

(b) the weight is released 1 m below the equilibrium position with an upward velocity of 12 m/s.

18. In parts (a) and (b) of Problem 17, determine whether the weight passes through the equilibrium position. In each case, find the time at which the weight attains its extreme displacement from the equilibrium position. What is the position of the weight at this instant?

19. A force of 2 lb stretches a spring 1 ft. A 3.2-lb weight is attached to the spring and the system is then immersed in a medium that imparts a damping force numerically equal to 0.4 times the instantaneous velocity.

(a) Find the equation of motion if the weight is released from rest 1 ft above the equilibrium position.

(b) Express the equation of motion in the form given in (21).

(c) Find the first time for which the weight passes through the equilibrium position heading upward.

20. After a 10-lb weight is attached to a 5-ft spring, the spring measures 7 ft long. The 10-lb weight is removed and replaced with an 8-lb weight and the entire system is placed in a medium offering a resistance numerically equal to the instantaneous velocity.

(a) Find the equation of motion if the weight is released $\frac{1}{2}$ ft below the equilibrium position with a downward velocity of 1 ft/s.

(b) Express the equation of motion in the form given in (21).

(c) Find the times for which the weight passes through the equilibrium position heading downward.

(d) Graph the equation of motion.

21. A 10-lb weight attached to a spring stretches it 2 ft. The weight is attached to a dashpot damping device that offers a resistance numerically equal to $\beta\,(\beta > 0)$ times the instantaneous velocity. Determine the values of the damping constant β so that the subsequent motion is **(a)** overdamped, **(b)** critically damped, and **(c)** underdamped.

22. A 24-lb weight stretches a spring 4 ft. The subsequent motion takes place in a medium offering a resistance numerically equal to $\beta\,(\beta > 0)$ times the instantaneous velocity. If the weight starts from the equilibrium position with an upward velocity of 2 ft/s, show that if $\beta > 3\sqrt{2}$, the equation of motion is

$$x(t) = \frac{-3}{\sqrt{\beta^2 - 18}}\, e^{-2\beta t/3} \sinh \frac{2}{3}\sqrt{\beta^2 - 18}\, t$$

Problems 23–30 are on forced motion.

23. A 16-lb weight stretches a spring $\frac{8}{3}$ ft. Initially the weight starts from rest 2 ft below the equilibrium position and the subsequent motion takes place in a medium that offers a damping force numerically equal to $\frac{1}{2}$ the instantaneous velocity. Find the equation of motion if the weight is driven by an external force equal to $f(t) = 10 \cos 3t$.

24. A mass of 1 slug is attached to a spring whose constant is 5 lb/ft. Initially the mass is released 1 ft below the equilibrium position with a downward velocity of 5 ft/s, and the subsequent motion takes place in a medium that offers a damping force numerically equal to 2 times the instantaneous velocity.

(a) Find the equation of motion if the mass is driven by an external force equal to $f(t) = 12 \cos 2t + 3 \sin 2t$.
(b) Graph the transient and steady-state solutions on the same coordinate axes.
(c) Graph the equation of motion.

25. A mass of 1 slug, when attached to a spring, stretches it 2 ft and then comes to rest in the equilibrium position. Starting at $t = 0$, an external force equal to $f(t) = 8 \sin 4t$ is applied to the system. Find the equation of motion if the surrounding medium offers a damping force numerically equal to 8 times the instantaneous velocity.

26. In Problem 25, determine the equation of motion if the external force is $f(t) = e^{-t} \sin 4t$. Analyze the displacements for $t \to \infty$.

27. When a mass of 2 kg is attached to a spring whose constant is 32 N/m, it comes to rest in the equilibrium position. Starting at $t = 0$, a force equal to $f(t) = 68e^{-2t} \cos 4t$ is applied to the system. Find the equation of motion in the absence of damping.

28. In Problem 27, write the equation of motion in the form $x(t) = A \sin(\omega t + \phi) + Be^{-2t} \sin(4t + \theta)$. What is the amplitude of vibrations after a very long time?

29. A mass m is attached to the end of a spring whose constant is k. After the mass reaches equilibrium, its support begins to oscillate vertically about a horizontal line L according to a formula $h(t)$. The value of h represents the distance in feet measured from L. See Figure 3.37.

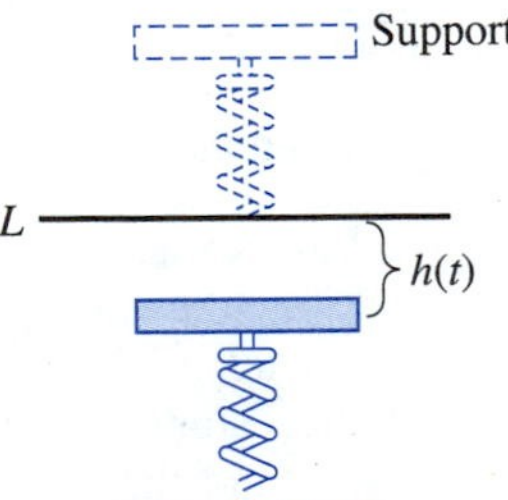

Figure 3.37.

(a) Determine the differential equation of motion if the entire system moves through a medium offering a damping force numerically equal to $\beta(dx/dt)$.
(b) Solve the differential equation in part (a) if the spring is stretched 4 ft by a weight of 16 lb, and $\beta = 2$, $h(t) = 5 \cos t$, $x(0) = x'(0) = 0$.

30. A mass of 100 g is attached to a spring whose constant is 1600 dynes/cm. After the mass reaches equilibrium, its support oscillates according to the formula $h(t) = \sin 8t$, where h represents displacement from its original position. See Problem 29 and Figure 3.37.
(a) In the absence of damping, determine the equation of motion if the mass starts from rest from the equilibrium position.
(b) At what times does the mass pass through the equilibrium position?
(c) At what times does the mass attain its extreme displacements?
(d) What are the maximum and minimum displacements?
(e) Graph the equation of motion.

Problems 31–38 are on series electrical circuits.

31. Find the charge on the capacitor in an *L-R-C* series circuit at $t = 0.01$ s when $L = 0.05$ henry, $R = 2$ ohms, $C = 0.01$ farad, $E(t) = 0$ volts, $q(0) = 5$ coulombs, and $i(0) = 0$ amperes. Determine the first time at which the charge on the capacitor is equal to zero.

32. Find the charge on the capacitor in an *L-R-C* series circuit when $L = \frac{1}{4}$ henry, $R = 20$ ohms, $C = \frac{1}{300}$ farad, $E(t) = 0$ volts, $q(0) = 4$ coulombs, and $i(0) = 0$ amperes. Is the charge on the capacitor ever equal to zero?

In Problems 33 and 34, find the charge on the capacitor and the current in the given *L-R-C* series circuit. Find the maximum charge on the capacitor.

33. $L = \frac{5}{3}$ henry, $R = 10$ ohms, $C = \frac{1}{30}$ farad, $E(t) = 300$ volts, $q(0) = 0$ coulombs, $i(0) = 0$ amperes.

34. $L = 1$ henry, $R = 100$ ohms, $C = 0.0004$ farad, $E(t) = 30$ volts, $q(0) = 0$ coulombs, $i(0) = 2$ amperes.

35. Find the steady-state charge and the steady-state current in an *L-R-C* series circuit when $L = 1$ henry, $R = 2$ ohms, $C = 0.25$ farad, and $E(t) = 50 \cos t$ volts.

36. **(a)** Use undetermined coefficients to find the steady-state charge $q_p(t)$ for the *L-R-C* series circuit governed by equation (31).
(b) Show that the amplitude of the steady-state current $i_p(t)$ in (32) is given by E_0/Z, where Z is the impedance of the circuit defined on page 184.

37. Show that the steady-state current in an *L-R-C* series circuit when $L = \frac{1}{2}$ henry, $R = 20$ ohms, $C = 0.001$ farad, and $E(t) = 100 \sin 60t$ is given by $i_p(t) = (4.160)\sin(60t - 0.588)$. [*Hint*: Use Problem 36.]

38. **(a)** Show that if L, R, C, and E_0 are constant, then the amplitude of the steady-state current in (32) is a maximum when $\gamma = 1/\sqrt{LC}$. What is the maximum amplitude?

(b) Show that if L, R, E_0, and γ are constant, then the amplitude of the steady-state current in (32) is maximum when $C = 1/L\gamma^2$.

Discussion Problems

In Problems 39–42, the given figure represents the graph of an equation of motion for a mass on a spring. The spring/mass system is damped. Discuss: Is the initial displacement of the mass above or below the equilibrium position? Is the mass initially released from rest, heading downward, or heading upward?

39.

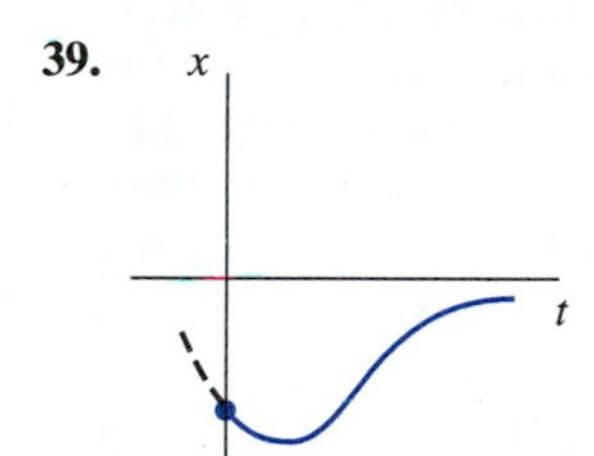

Figure 3.38.

40.

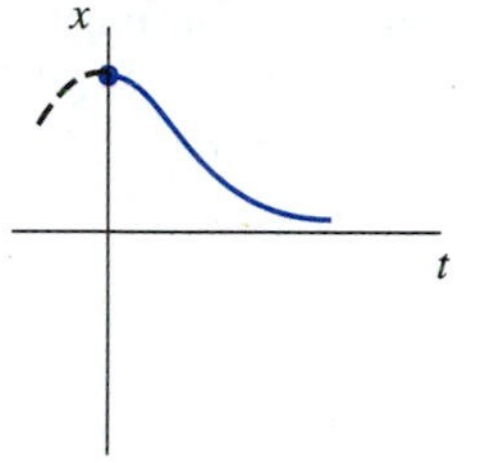

Figure 3.39.

41.

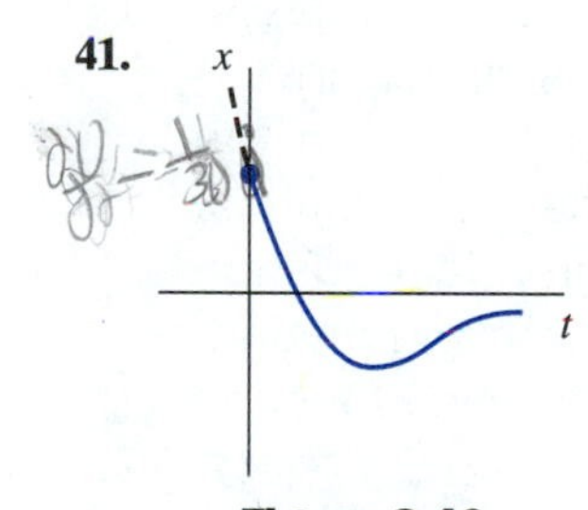

Figure 3.40.

42.

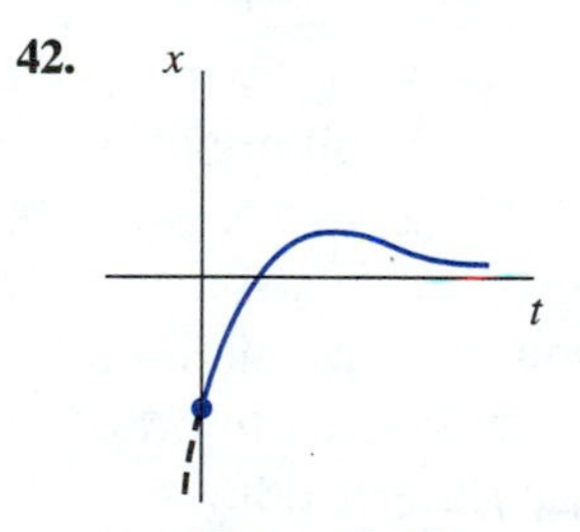

Figure 3.41.

43. A spring/mass system is described by the given mathematical model. By inspection of the differential equation only, discuss the motion of the mass over a long period of time.

(a) $4x'' + e^{-0.1t}x = 0$ **(b)** $4x'' + tx = 0$

44. Discuss possible alternative mathematical models for an aging spring. Back your argument by using an ODE solver to illustrate solution curves.

45. Two springs are attached in series as shown in Figure 3.42. If the mass of each spring is ignored, then the effective spring constant k of the system is given by $1/k = 1/k_1 + 1/k_2$. Discuss: How can the preceding statement be proved?

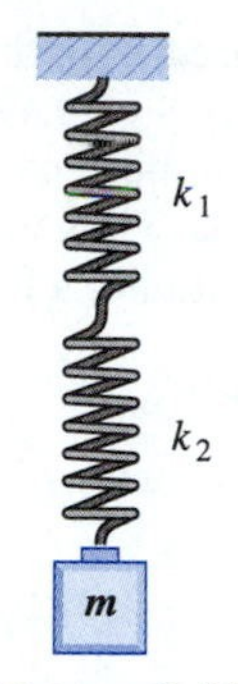

Figure 3.42.

46. In ballistics it is often important to be able to determine the muzzle velocity of gun, that is, the speed of a bullet as it leaves the barrel. This can be determined indirectly with the aid of a **ballistic pendulum** (invented in 1742), which is simply a plane pendulum consisting of a rod of negligible mass to which a block of wood of mass m_w is attached. The system is set in motion by the impact of a bullet which is moving horizontally at the unknown velocity v_b; at the time of impact, which we shall take to be $t = 0$, the combined mass is $m_w + m_b$, where m_b is the mass of the bullet embedded in the wood. In Section 3.10, we shall see that in the case of small oscillations, the angular displacement $\theta(t)$ of a plane pendulum shown in Figure 3.43 is given by the same basic equation as in (3): $d^2\theta/dt^2 + (g/l)\theta = 0$, where $\theta > 0$ corresponds to motion to the right of vertical. The velocity v_b can be found by measuring the height h of the mass $m_w + m_b$ at the maximum displacement angle θ_{max} shown in Figure 3.43. Intuitively, the horizontal velocity V of the combined mass (wood plus bullet) after impact is only a fractional amount of the velocity v_b of the bullet, that is, $V = \dfrac{m_b}{m_w + m_b} v_b$. Recalling that a distance s traveled

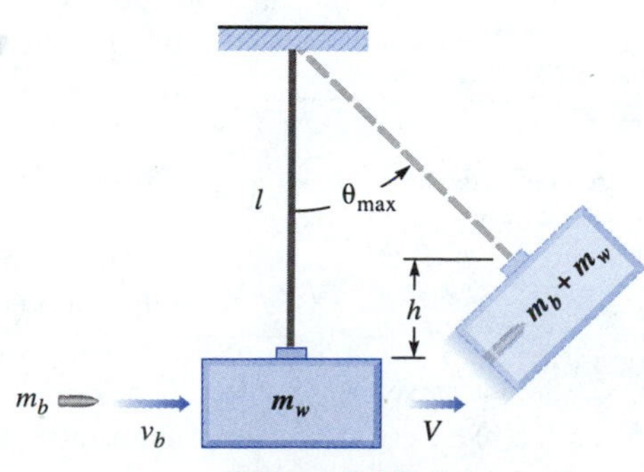

Figure 3.43.

by a point along a circular path is related to the radius l and central angle θ by $s = l\theta$, it follows that the angular velocity ω of the combined mass and its linear velocity v are related by $v = l\omega$. Thus the initial angular velocity ω_0 at the time of the bullet's impact is related to V by $V = l\omega_0$, in other words, $\omega_0 = \dfrac{m_b}{m_w + m_b}\dfrac{v_b}{l}$.

(a) Solve the initial-value problem

$$\frac{d^2\theta}{dt^2} + \frac{g}{l}\theta = 0, \ \ \theta(0) = 0, \theta'(0) = \omega_0$$

(b) Use the result from part (a) to show that

$$v_b = \frac{m_w + m_b}{m_b}\sqrt{lg}\,\theta_{\max}$$

(c) Now use Figure 3.43 to express $\cos\theta_{\max}$ in terms of l and h. Then use the first two terms of the Maclaurin expansion for $\cos\theta_{\max}$ to express $\theta_{\max}$ in terms of l and h. Finally, show that v_b is given (approximately) by

$$v_b = \frac{m_w + m_b}{m_b}\sqrt{2gh}$$

(d) Find v_b if $m_b = 5$ g, $m_w = 1$ kg, and $h = 6$ cm.

47. (a) In the Remarks at the end of this section, we saw that the total energy $E = \frac{1}{2}my^2 + \frac{1}{2}kx^2$ for a free undamped spring/mass system is a constant. Without solving, discuss the value of E for the problem consisting of equation (3) and the initial conditions $x(0) = x_0, x'(0) = x_1$.

(b) Solve the initial-value problem consisting of (3) with $m = k = 1$ and the initial conditions $x(0) = 1$, $x'(0) = \frac{1}{2}$. Use a graphing utility to obtain the graphs of $x(t)$, $y(t) = dx/dt$, and $\frac{1}{2}my^2 + \frac{1}{2}kx^2$ on the same set of coordinate axes. What is E in this case?

(c) Use your physical intuition and appropriate graphs to discuss, in general, how the kinetic energy E_k and the potential energy E_p for a free undamped spring/mass system vary over time.

3.7 Mathematical Models: Boundary-Value Problems

In the preceding section, we examined some second-order mathematical models involving prescribed initial conditions, that is, side conditions that are specified on the unknown function and its first derivative at a single point. But often the mathematical description of a mathematical system demands that we solve a differential equation subject to boundary conditions, that is, conditions specified on the unknown function, or on one of its derivatives, or even on a linear combination of the unknown function and one of its derivatives, at two (or more) different points.

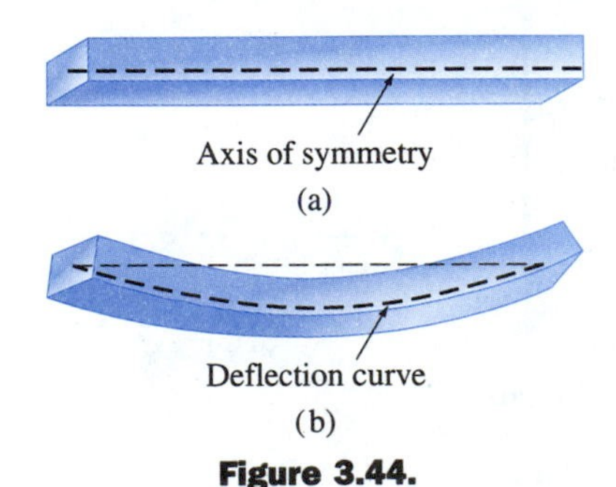

Figure 3.44.

Deflection of a Beam Many structures are constructed using girders or beams and these beams deflect or distort under their own weight or under the influence of some external force. As we shall now see, this deflection $y(x)$ can be modeled by a relatively simple linear fourth-order differential equation.

To begin, let us assume that a beam of length L is homogeneous and has uniform cross sections along its length. In the absence of any load on the beam (including its weight), a curve joining the centroids of all its cross sections is a straight line called the **axis of symmetry**. See Figure 3.44(a). If a load is applied to the beam in a vertical plane containing the axis of symmetry, then, as shown in Figure 3.44(b), the beam undergoes a distortion and the curve connecting the centroids of all cross sections is called the

deflection curve or **elastic curve**. The deflection curve approximates the shape of the beam. Now suppose the x-axis coincides with the axis of symmetry, and that the deflection $y(x)$, measured from this axis, is positive if downward. In the theory of elasticity, it is argued that a good model for the bending moment $M(x)$ at a point x along the beam is related to the load per unit length $w(x)$ by the equation

$$\frac{d^2M}{dx^2} = w(x) \tag{1}$$

In addition, the bending moment $M(x)$ is proportional to the curvature κ of the elastic curve

$$M(x) = EI\kappa \tag{2}$$

where E and I are constants; E is Young's modulus of elasticity of the material of the beam and I is the moment of inertia of a cross-section of the beam (about an axis known as the neutral axis). The product EI is called the **flexural rigidity** of the beam.

Now, from calculus, curvature is given by $\kappa = y''/[1 + (y')^2]^{3/2}$. When the deflection $y(x)$ is small, the slope $y' \approx 0$ and so $[1 + (y')^2]^{3/2} \approx 1$. If we let $\kappa = y''$, equation (2) becomes $M = EIy''$. The second derivative of this last expression is

$$\frac{d^2M}{dx^2} = EI\frac{d^2}{dx^2}y'' = EI\frac{d^4y}{dx^4} \tag{3}$$

Using the given result in (1) to replace d^2M/dx^2 in (3), we see that the deflection $y(x)$ satisfies the fourth-order differential equation

$$EI\frac{d^4y}{dx^4} = w(x) \tag{4}$$

Boundary conditions associated with equation (4) depend on how the ends of the beam are supported. A cantilever beam is **embedded** or **clamped** at one end and **free** at the other. A diving board, an outstretched arm, an airplane wing, and a balcony are common examples of such beams; but even trees, flagpoles, skyscrapers, and the George Washington monument can act as cantilever beams because they are embedded at one end and are subject to the bending force of the wind. For a cantilever beam, the deflection $y(x)$ must satisfy the following two conditions at the embedded end $x = 0$:

- $y(0) = 0$ since there is no deflection, and
- $y'(0) = 0$ since the deflection curve is tangent to the x-axis (in other words, the slope of the deflection curve is zero at this point).

At $x = L$, the free end conditions are

- $y''(L) = 0$ since the bending moment is zero, and
- $y'''(L) = 0$ since the shear force is zero.

(a) Embedded at both ends.

(b) Cantilever beam: embedded at the left end, free at the right end.

(c) Simply supported at both ends.

Figure 3.45.

The function $F(x) = dM/dx = EI\, d^3y/dx^3$ is called the shear force. If an end of a beam is **simply supported** (also called **pin supported**, **fulcrum supported**, and **hinged**) then we must have $y = 0$ and $y'' = 0$ at that end. The following table summarizes the boundary conditions that are associated with (4). See Figure 3.45.

Ends of the Beam	Boundary Conditions
embedded	$y = 0,\ y' = 0$
free	$y'' = 0,\ y''' = 0$
simply supported	$y = 0,\ y'' = 0$

EXAMPLE 1 Embedded Beam

A beam of length L is embedded at both ends. Find the deflection of the beam if a constant load w_0 is uniformly distributed along its length, that is, $w(x) = w_0,\ 0 < x < L$.

Solution From the preceding discussion, we see that the deflection $y(x)$ satisfies

$$EI\frac{d^4y}{dx^4} = w_0$$

Because the beam is embedded at both its left ($x = 0$) and its right end ($x = L$) there is no vertical deflection and the line of deflection is horizontal at these points, and so the boundary conditions are

$$y(0) = 0,\ y'(0) = 0, \qquad y(L) = 0,\ y'(L) = 0$$

We can solve the nonhomogeneous differential equation in the usual manner (find y_c by observing that $m = 0$ is a root of multiplicity four of the auxiliary equation $m^4 = 0$ and then find a particular solution y_p by undetermined coefficients), or we can simply integrate the equation $d^4y/dx^4 = w_0/EI$ four times in succession. Either way, we find the general solution of the equation to be

$$y(x) = c_1 + c_2x + c_3x^2 + c_4x^3 + \frac{w_0}{24EI}x^4$$

Now the conditions $y(0) = 0$ and $y'(0) = 0$ give in turn $c_1 = 0$ and $c_2 = 0$. The remaining conditions $y(L) = 0$ and $y'(L) = 0$ applied to $y(x) = c_3x^2 + c_4x^3 + \dfrac{w_0}{24EI}x^4$ yield the two equations

$$c_3L^2 + c_4L^3 + \frac{w_0}{24EI}L^4 = 0$$

$$2c_3L + 3c_4L^2 + \frac{w_0}{6EI}L^3 = 0$$

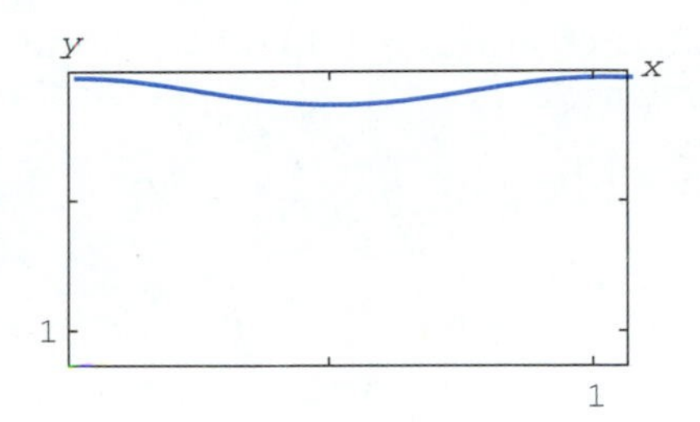

Figure 3.46.

Solving this system gives $c_3 = w_0L^2/24EI$ and $c_4 = -w_0L/12EI$. Thus the deflection is

$$y(x) = \frac{w_0L^2}{24EI}x^2 - \frac{w_0L}{12EI}x^3 + \frac{w_0}{24EI}x^4 = \frac{w_0}{24EI}x^2(x-L)^2$$

The deflection curve, corresponding to the choice $w_0 = 24EI$ and $L = 1$, is shown in Figure 3.46. ■

See Problem 14 in *Projects and Computer Lab Experiments for Chapter 3* for an interesting application of the beam equation (4).

Eigenvalues and Eigenfunctions Many applied problems demand that we solve a two-point boundary-value problem involving a linear differential equation that contains a parameter λ. Note that in the following example, the trivial solution $y = 0$ is a solution of the given boundary-value problem. We are not interested in this solution; but rather we want to determine whether there are values of λ for which the problem has nontrivial solutions.

■ EXAMPLE 2 Nontrivial Solutions of a BVP

Solve the boundary-value problem

$$y'' + \lambda y = 0, \qquad y(0) = 0, \quad y(L) = 0$$

Solution We consider three cases: $\lambda = 0$, $\lambda < 0$, and $\lambda > 0$.

(*i*) For $\lambda = 0$, the solution of $y'' = 0$ is $y = c_1x + c_2$. The conditions $y(0) = 0$ and $y(L) = 0$ imply, in turn, $c_2 = 0$ and $c_1 = 0$. Hence for $\lambda = 0$, the only solution of the boundary-value problem is the trivial solution $y = 0$.

(*ii*) For $\lambda < 0$, we have $y = c_1 \cosh \sqrt{-\lambda}x + c_2 \sinh \sqrt{-\lambda}x$.* Again, $y(0) = 0$ gives $c_1 = 0$, and so $y = c_2 \sinh \sqrt{-\lambda}x$. The second condition $y(L) = 0$ dictates that $c_2 \sinh \sqrt{-\lambda}L = 0$. Since $\sinh \sqrt{-\lambda}L \neq 0$, we must have $c_2 = 0$. Thus $y = 0$.

(*iii*) For $\lambda > 0$, the general solution of $y'' + \lambda y = 0$ is $y = c_1 \cos \sqrt{\lambda}x + c_2 \sin \sqrt{\lambda}x$. As before, $y(0) = 0$ yields $c_1 = 0$, but $y(L) = 0$ implies

$$c_2 \sin \sqrt{\lambda}L = 0$$

If $c_2 = 0$, then necessarily $y = 0$. However, if $c_2 \neq 0$, then $\sin \sqrt{\lambda}L = 0$. The last condition implies that the argument of the sine function must be an integer multiple of π:

$$\sqrt{\lambda}L = n\pi \quad \text{or} \quad \lambda = \frac{n^2\pi^2}{L^2}, \qquad n = 1, 2, 3, \ldots$$

Therefore, for any real nonzero c_2, $y = c_2 \sin(n\pi x/L)$ is a solution of the problem for each n. Since the differential equation is

*$\sqrt{-\lambda}$ looks a little strange, but bear in mind that $\lambda < 0$ is equivalent to $-\lambda > 0$.

homogeneous, we may, if desired, choose $c_2 = 1$. In other words, for a given number in the sequence

$$\frac{\pi^2}{L^2}, \quad \frac{4\pi^2}{L^2}, \quad \frac{9\pi^2}{L^2}, \ldots$$

the *corresponding* function in the sequence

$$\sin\frac{\pi}{L}x, \quad \sin\frac{2\pi}{L}x, \quad \sin\frac{3\pi}{L}x, \ldots$$

is a nontrivial solution of the original problem. ■

The numbers $\lambda_n = n^2\pi^2/L^2$, $n = 1, 2, 3, \ldots$, for which the boundary-value problem in Example 2 has a nontrivial solution, are known as **characteristic values** or, more commonly, **eigenvalues**. The nontrivial solutions that depend on these values of λ_n, namely $y_n = c_2 \sin(n\pi x/L)$, or simply $y_n = \sin(n\pi x/L)$, are called **characteristic functions**, or **eigenfunctions**.

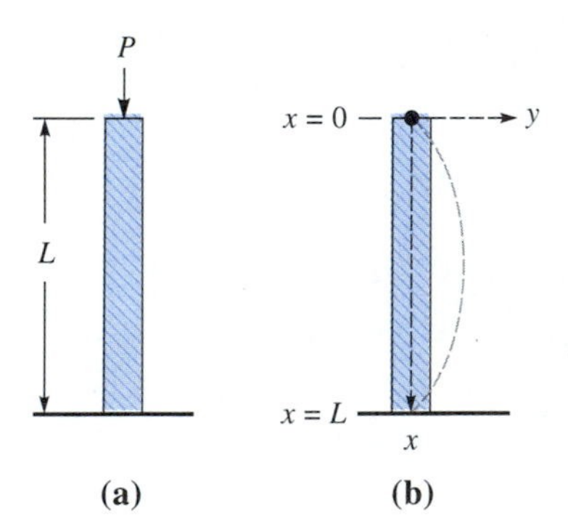

Figure 3.47.

Buckling of a Thin Vertical Column In the eighteenth century, Leonhard Euler was one of the first mathematicians to study an eigenvalue problem in analyzing how a thin elastic column buckles under a compressive axial force.

Consider a long slender vertical column of uniform cross section and length L. Let $y(x)$ denote the deflection of the column when a constant vertical compressive force, or load, P is applied to its top, as shown in Figure 3.47. By comparing bending moments at any point along the column we obtain

$$EI\frac{d^2y}{dx^2} = -Py \quad \text{or} \quad EI\frac{d^2y}{dx^2} + Py = 0 \tag{5}$$

where again E is Young's modulus of elasticity and I is the moment of inertia of a cross section about a vertical line through its centroid.

■ EXAMPLE 3 The Euler Load

Find the deflection of a thin vertical homogeneous column of length L subjected to a constant axial load P if the column is hinged at both ends.

Solution The boundary-value problem to be solved is

$$EI\frac{d^2y}{dx^2} + Py = 0, \qquad y(0) = 0, \quad y(L) = 0 \tag{6}$$

First note that $y = 0$ is a perfectly good solution of this problem. This solution has a simple intuitive interpretation: if the load P is not great enough, there is no deflection. The question then is this: For what values of P will the column bend? In mathematical terms: For what values of P does the given boundary-value problem possess *nontrivial* solutions? By

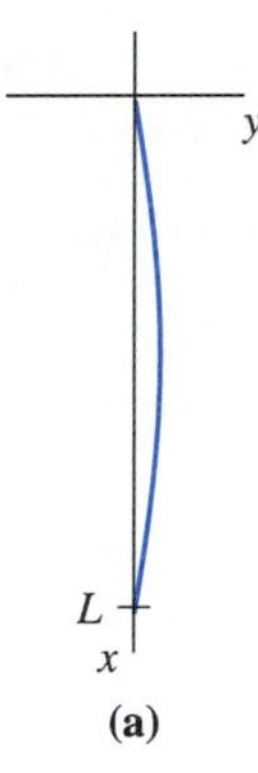

(a)

writing $\lambda = P/EI$ we see that (6) is the same as

$$y'' + \lambda y = 0, \qquad y(0) = 0, \quad y(L) = 0$$

which, in turn, is recognized as the problem in Example 2. From part (*iii*) of that discussion, we see that the deflections or **buckling modes** corresponding to the eigenvalues $\lambda_n = P_n/EI = n^2\pi^2/L^2$ are $y_n(x) = c_2 \sin(n\pi x/L)$, $n = 1, 2, 3, \ldots$. Physically, this means the column will buckle or deflect only when the compressive force is one of the values $P_n = n^2\pi^2 EI/L^2$, $n = 1, 2, 3, \ldots$. These different forces are called **critical loads**. The deflection curve corresponding to the smallest critical load $P_1 = \pi^2 EI/L^2$, called the **Euler load**, is $y_1(x) = c_2 \sin(\pi x/L)$ and is known as the **first buckling mode**. ■

(b)

The deflection curves in Example 3, corresponding to $n = 1$, $n = 2$, and $n = 3$, are shown in Figure 3.48. Note that if the original column has some sort of physical restraint put on it at $x = L/2$, then the smallest critical load would be $P_2 = 4\pi^2 EI/L^2$ and the deflection curve would be as in Figure 3.48(b). If restraints are put on the column at $x = L/3$ and at $x = 2L/3$, then the column will not buckle until the critical load $P_3 = 9\pi^2 EI/L^2$ is applied and the deflection curve would be as in Figure 3.48(c).

For the thin column discussed in Example 3, we assumed that the moment of inertia I of a cross section of the column was constant. For two applications where the moment of inertia of a cross section of the column is a function $I(x)$, see Problems 31 and 32 in Exercises 6.2 and Problems 29 and 30 in Exercises 6.3.*

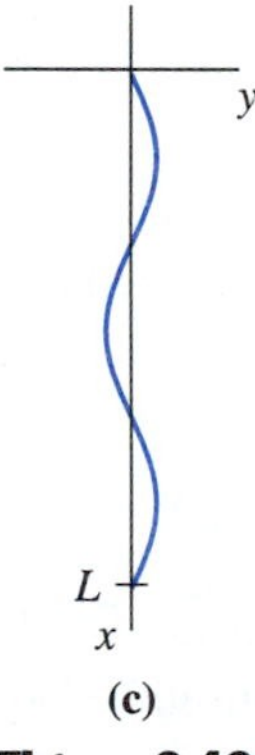

(c)

Figure 3.48.

Rotating String The simple linear second-order differential equation $y'' + \lambda y = 0$ occurs again and again as a mathematical model. Earlier in this section we have seen this equation in the forms $d^2x/dt^2 + (k/m)x = 0$ and $d^2q/dt^2 + (1/LC)q = 0$ as models for, respectively, simple harmonic motion of a spring/mass system and the simple harmonic response of a series circuit; we have just seen it in the form $d^2y/dx^2 + (P/EI)y = 0$ in the preceding example. We encounter the equation $y'' + \lambda y = 0$ one more time in this section, this time as a model that defines the deflection curve or the shape $y(x)$ assumed by a rotating string. The physical situation is analogous to two persons holding a jump rope and then twirling it in a synchronous manner. See Figure 3.49(a) and Figure 3.49(b).

Suppose a string of length L with constant linear density ρ (mass per unit length) is stretched along the x-axis and fixed at $x = 0$ and $x = L$. Suppose the string is then rotated about that axis at a constant angular speed ω. Consider a portion of the string on the interval $[x, x + \Delta x]$, where Δx is small. If the magnitude T of the tension $\mathbf{T}$, acting tangential to the string, is constant along the string, then the desired differential equation

*See *Advanced Calculus for Applications* by Francis Hildebrand, Prentice-Hall, Inc. for a more complete discussion of this theory.

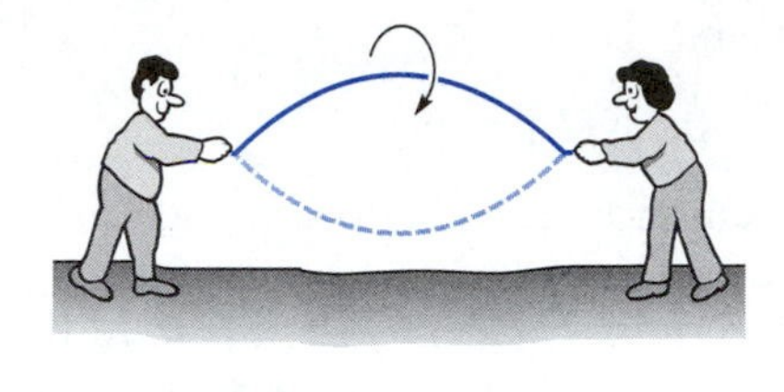

(a)

ω

$y(x)$

$x = 0$ $x = L$

(b)

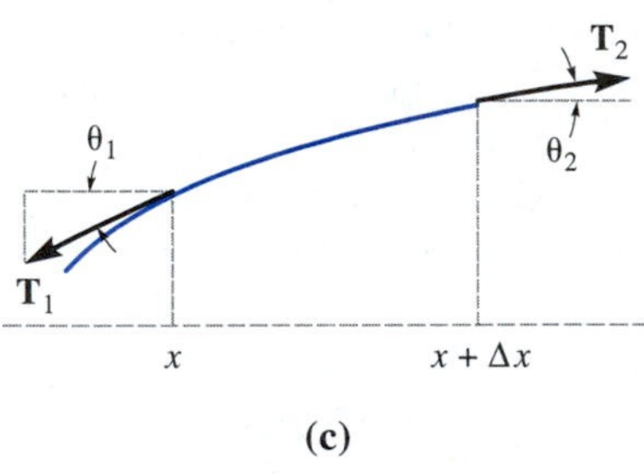

(c)

Figure 3.49.

can be obtained by equating two different formulations of the net force acting on the string on the interval $[x, x + \Delta x]$. First, we see from Figure 3.49(c) that the net vertical force is

$$F = T \sin \theta_2 - T \sin \theta_1 \tag{7}$$

When the angles θ_1 and θ_2 (measured in radians) are small, we have $\sin \theta_2 \approx \tan \theta_2$ and $\sin \theta_1 \approx \tan \theta_1$. Moreover, since $\tan \theta_2$ and $\tan \theta_1$ are, in turn, slopes of the lines containing the vectors $\mathbf{T}_2$ and $\mathbf{T}_1$ we can also write

$$\tan \theta_2 = y'(x + \Delta x) \quad \text{and} \quad \tan \theta_1 = y'(x)$$

Thus (7) becomes

$$F \approx T[y'(x + \Delta x) - y'(x)] \tag{8}$$

Second, we can obtain a different form of this same net force using Newton's second law $F = ma$. Here the mass of the string on the interval is $m = \rho\, \Delta x$; the centripetal acceleration of a body rotating with angular speed ω in a circle of radius r is $a = r\omega^2$. With Δx small, we take $r = y$. Thus the net vertical force is also approximated by

$$F \approx -(\rho\, \Delta x) y\omega^2 \tag{9}$$

where the minus sign comes from the fact that the acceleration points in the direction opposite to the positive y-direction. Now by equating (8) and (9),

$$T[y'(x + \Delta x) - y'(x)] \approx -(\rho\, \Delta x) y\omega^2 \quad \text{or} \quad T\frac{y'(x + \Delta x) - y'(x)}{\Delta x} \approx -\rho\omega^2 y \tag{10}$$

For Δx close to zero, the difference quotient $[y'(x + \Delta x) - y'(x)]/\Delta x$ in (10) is approximated by the second derivative d^2y/dx^2. Finally, we arrive at the model

$$T\frac{d^2y}{dx^2} = -\rho\omega^2 y \quad \text{or} \quad T\frac{d^2y}{dx^2} + \rho\omega^2 y = 0 \tag{11}$$

Since the string is anchored at its ends $x = 0$ and $x = L$ we expect that the solution $y(x)$ of the last equation in (11) should also satisfy the boundary conditions $y(0) = 0$ and $y(L) = 0$.

Remarks

In Example 3, the basic model is a differential equation containing a parameter λ. Let us assume for the sake of discussion that the length of the column is taken to be $L = 1$. From the perspective of piling different loads onto the column, the column does not buckle, that is, the solution of the boundary-value problem $y'' + \lambda y = 0$, $y(0) = 0$, $y(1) = 0$, is the trivial solution $y = 0$ for $\lambda = 0.5, 1, 4.6, 8.3, 9.5, 9.7, 9.8, \ldots$, and so on, *except* when we get to the load that corresponds to $\lambda = \pi^2 = 9.86960\ldots$ the solution of the problem—the deflection curve of the column—changes abruptly from $y = 0$ to a constant multiple of $y = \sin \pi x$. As pointed out in Section 3.3, such a drastic change in the predicted qualitative behavior of a physical system for a small change of a parameter in its mathematical model is called a **bifurcation**.

EXERCISES 3.7

Answers to odd-numbered problems begin on page 443.

In Problems 1–4, solve (4) subject to the appropriate boundary conditions. The beam is of length L and w_0 is a constant.

1. **(a)** The beam is embedded at its left end and is free at its right end and $w(x) = w_0$, $0 < x < L$.
   (b) Use a graphing utility to obtain the graph of the deflection curve of the beam when $w_0 = 24EI$ and $L = 1$.

2. **(a)** The beam is simply supported at both ends and $w(x) = w_0$, $0 < x < L$.
   (b) Use a graphing utility to obtain the graph of the deflection curve of the beam when $w_0 = 24EI$ and $L = 1$.

3. **(a)** The beam is embedded at its left end and is simply supported at its right end, and $w(x) = w_0$, $0 < x < L$.
   (b) Use a graphing utility to obtain the graph of the deflection curve of the beam when $w_0 = 48EI$ and $L = 1$.

4. **(a)** The beam is embedded at its left end and is simply supported at its right end, and $w(x) = w_0 \sin(\pi x/L)$, $0 < x < L$.
   (b) Use a graphing utility to obtain the graph of the deflection curve of the beam when $w_0 = 2\pi^3 EI$ and $L = 1$.

5. **(a)** Find the maximum deflection of the cantilever beam in Problem 1.
   (b) How does the maximum deflection of a beam that is half as long compare with the value in part (a)?

6. **(a)** Find the maximum deflection of the simply supported beam in Problem 2.
   (b) How does the maximum deflection of the simply supported beam compare with the value of maximum deflection of the embedded beam in Example 1?

7. A cantilever beam of length L is embedded at its right end and a horizontal tensile force of P pounds is applied to its free left end. When the origin is taken at its free end as shown in Figure 3.50, the deflection $y(x)$ of the beam can be shown to satisfy the differential equation

$$EI\, y'' = Py - w(x)\frac{x}{2}$$

Find the deflection of the cantilever beam if $w(x) = w_0 x$, $0 < x < L$, and $y(0) = 0$, $y'(L) = 0$.

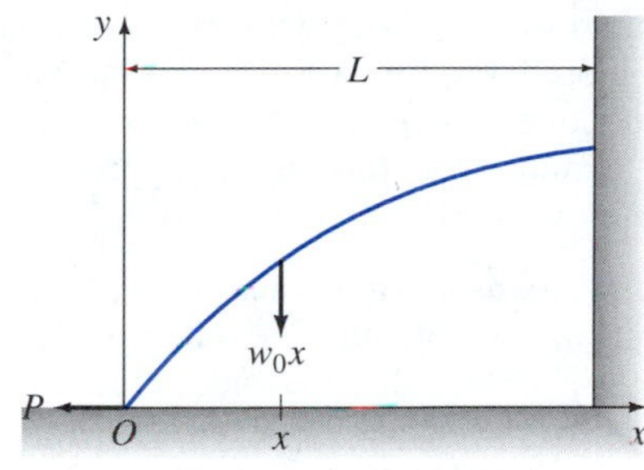

Figure 3.50.

8. When a compressive instead of a tensile force is applied at the free end of the beam in Problem 7, the differential equation of the deflection is

$$EI\, y'' = -Py - w(x)\frac{x}{2}$$

Solve this equation if $w(x) = w_0 x$, $0 < x < L$, and $y(0) = 0$, $y'(L) = 0$.

In Problems 9–22, find the eigenvalues and eigenfunctions for the given boundary-value problem.

9. $y'' + \lambda y = 0$, $y(0) = 0$, $y(\pi) = 0$
10. $y'' + \lambda y = 0$, $y(0) = 0$, $y(\pi/4) = 0$
11. $y'' + \lambda y = 0$, $y'(0) = 0$, $y(L) = 0$
12. $y'' + \lambda y = 0$, $y(0) = 0$, $y'(\pi/2) = 0$
13. $y'' + \lambda y = 0$, $y'(0) = 0$, $y'(\pi) = 0$
14. $y'' + \lambda y = 0$, $y(-\pi) = 0$, $y(\pi) = 0$
15. $y'' + 2y' + (\lambda + 1)y = 0$, $y(0) = 0$, $y(5) = 0$
16. $y'' + (\lambda + 1)y = 0$, $y'(0) = 0$, $y'(1) = 0$

17. $y'' + \lambda^2 y = 0,\ y(0) = 0,\ y(L) = 0$

18. $y'' + \lambda^2 y = 0,\ y(0) = 0,\ y'(3\pi) = 0$

19. $x^2y'' + xy' + \lambda y = 0,\ y(1) = 0,\ y(e^{\pi}) = 0$

20. $x^2y'' + xy' + \lambda y = 0,\ y'(e^{-1}) = 0,\ y(1) = 0$

21. $x^2y'' + xy' + \lambda y = 0,\ y'(1) = 0,\ y'(e^2) = 0$

22. $x^2y'' + 2xy' + \lambda y = 0,\ y(1) = 0,\ y(e^2) = 0$

23. **(a)** Show that the eigenfunctions of the boundary-value problem $y'' + \lambda y = 0,\ y(0) = 0,\ y(1) + y'(1) = 0$ are $y = \sin\sqrt{\lambda_n}\,x$, where the eigenvalues λ_n of the problem are $\lambda_n = x_n^2$, where x_n, $n = 1, 2, 3, \ldots$, are the consecutive *positive* roots of the equation $\tan\sqrt{\lambda} = -\sqrt{\lambda}$.

(b) Use a graphing utility to convince yourself that the equation $\tan x = -x$ has an infinite number of roots. Explain why the negative roots of the equation can be ignored. Explain why $\lambda = 0$ is not an eigenvalue even though $x = 0$ is an obvious root of the equation.

(c) Use a numerical procedure to approximate the first four eigenvalues λ_1, λ_2, λ_3, and λ_4.

24. The critical loads of thin columns depend on the end conditions of the column. The value of the Euler load P_1 in Example 3 was derived under the assumption that the column was hinged at both ends.

Suppose a thin vertical homogeneous column is embedded at its base ($x = 0$) and is free at its top ($x = L$) and that an axial load P is applied to its free end. Either this load causes a small deflection δ, as shown in Figure 3.51, or it does not cause such a deflection. In either case, the differential equation for the deflection $y(x)$ is

$$EI\frac{d^2y}{dx^2} + Py = P\delta$$

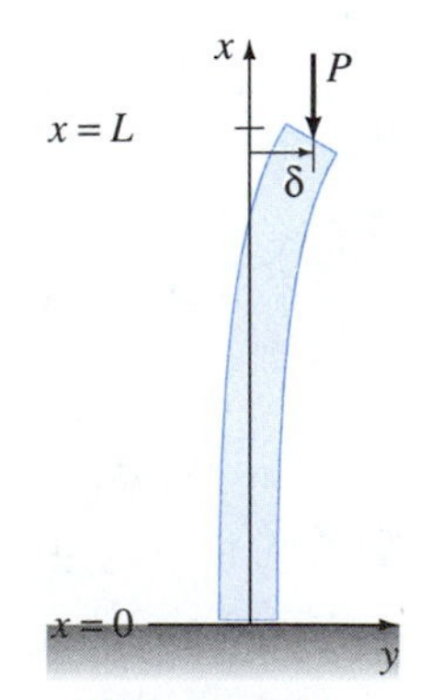

Figure 3.51.

(a) What is the predicted deflection when $\delta = 0$?

(b) When $\delta \neq 0$, show that the Euler load for this column is one-fourth of the Euler load for the hinged column in Example 3.

25. Consider the boundary-value problem introduced in the discussion of the rotating string:

$$T\frac{d^2y}{dx^2} + \rho\omega^2 y = 0, \quad y(0) = 0, y(L) = 0$$

Find the critical speeds ω_n and corresponding deflection curves of the problem.

26. The differential equation for the displacement $y(x)$ of a rotating string in which the tension T is not constant is given by

$$\frac{d}{dx}\left[T(x)\frac{dy}{dx}\right] + \rho\omega^2 y = 0$$

Suppose $1 < x < e$ and that the tension is given by $T(x) = x^2$.

(a) If $y(1) = 0$, $y(e) = 0$, and $\rho\omega^2 > 0.25$, show that the critical speeds of angular rotation are $\omega_n = \frac{1}{2}\sqrt{(4n^2\pi^2 + 1)/\rho}$ and the corresponding deflection curves are $y_n(x) = c_2x^{-1/2}\sin(n\pi\ln x)$, $n = 1, 2, 3, \ldots$.

(b) Use a graphing utility to obtain the graphs of the deflection curves on the interval $[1, e]$ for $n = 1, 2, 3$. Choose $c_2 = 1$.

27. Consider two concentric spheres of radius $r = a$ and $r = b$, $a < b$, as shown in Figure 3.52. The temperature $u(r)$ in the region between the spheres is determined from the boundary-value problem

$$r\frac{d^2u}{dr^2} + 2\frac{du}{dr} = 0, \qquad u(a) = u_0, \quad u(b) = u_1$$

where u_0 and u_1 are constants. Solve for $u(r)$.

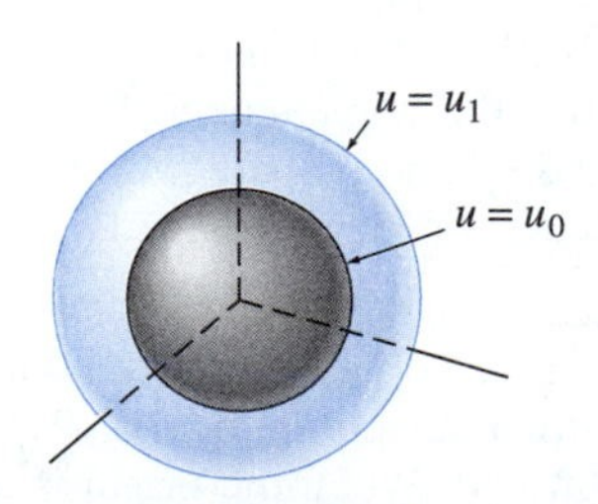

Figure 3.52.

28. The temperature $u(r)$ in the circular ring shown in Figure 3.53 is determined from the boundary-value problem

$$r\frac{d^2u}{dr^2}+\frac{du}{dr}=0, \qquad u(a)=u_0, \quad u(b)=u_1$$

where u_0 and u_1 are constants. Show that

$$u(r)=\frac{u_0\ln(r/b)-u_1\ln(r/a)}{\ln(a/b)}$$

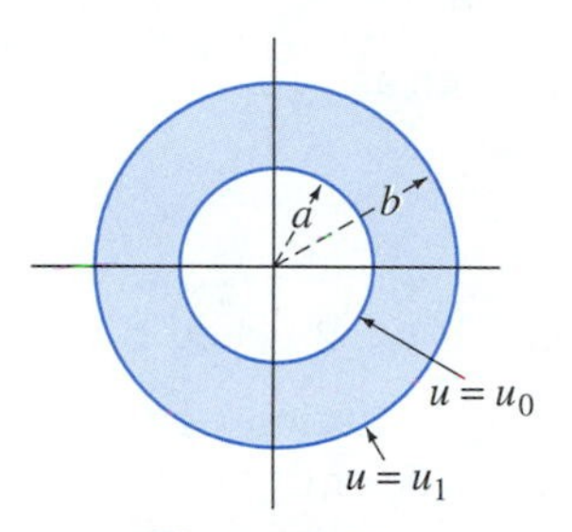

Figure 3.53.

Discussion Problems

29. Consider Figure 3.48. Where should physical restraints be placed on the column if we want the critical load to be P_4? Give a sketch of the deflection curve corresponding to this load.

30. Consider the boundary-value problem

$$y''+\lambda^2y=0, \quad y(-\pi)=y(\pi), \quad y'(-\pi)=y'(\pi)$$

(a) Discuss a geometric interpretation of the boundary conditions.

(b) Find the eigenvalues and eigenfunctions of the problem.

(c) Use a graphing utility to graph some of the eigenfunctions. Verify the geometric interpretation of the boundary conditions.

3.8 Systems of Differential Equations—Elimination Method

Simultaneous ordinary differential equations involve two or more equations that contain derivatives of two or more dependent variables—the unknown functions—with respect to a single independent variable. The method of systematic elimination for solving systems of linear equations with constant coefficients is based on the algebraic principle of elimination of variables. We shall see that the analog of *multiplying* an algebraic equation by a constant is *operating* on a DE with some combination of derivatives. Since this method simply uncouples the system into distinct linear DEs in each independent variable, this section gives you an opportunity to practice what you learned earlier in the chapter.

Differential Operators In calculus, the symbol D^n is often used to denoted the nth derivative of a function, that is, $D^ny = d^ny/dt^n$. Hence a linear nth-order differential equation

$$a_ny^{(n)}+a_{n-1}y^{(n-1)}+\cdots+a_1y'+a_0y=g(t)$$

where the a_i, $i=0, 1, \ldots, n$ are constants, can be rewritten as

$$(a_nD^n+a_{n-1}D^{n-1}+\cdots+a_1D+a_0)y=g(t)$$

An expression of the form

$$a_nD^n+a_{n-1}D^{n-1}+\cdots+a_1D+a_0$$

is called a **linear *n*th-order differential operator** and will be abbreviated as $P(D)$. Since $P(D)$ is a polynomial in the symbol D, we may be able to factor it into differential operators of lower order. Moreover, the factors of $P(D)$ commute. For example, the linear second-order differential operator $D^2 - 9$ is the same as $(D + 3)(D - 3)$ or $(D - 3)(D + 3)$.

Before we solve it in Example 2, observe that the system of differential equations

$$\begin{aligned} \frac{dx}{dt} + \frac{dy}{dt} + 2y &= 0 \\ \frac{dx}{dt} - 3x + \frac{dy}{dt} &= 0 \end{aligned} \quad \text{can be written as} \quad \begin{aligned} Dx + (D + 2)y &= 0 \\ (D - 3)x + Dy &= 0 \end{aligned}$$

Solution of a System In Section 1.1, we touched briefly on the notion of a solution of a system. Recall that a solution of a system of ordinary differential equations is a set of sufficiently differentiable functions $x = \phi_1(t)$, $y = \phi_2(t)$, $z = \phi_3(t)$, and so on, that satisfies each equation of the system on some common interval I.

Elimination Method Consider the simple system of two linear first-order differential equations

$$\begin{aligned} Dy &= 2x \\ Dx &= 3y \end{aligned} \tag{1}$$

or equivalently,

$$\begin{aligned} 2x - Dy &= 0 \\ Dx - 3y &= 0 \end{aligned} \tag{2}$$

Operating on the first equation in (2) by D, while multiplying the second by 2 and then subtracting, eliminates x from the system. It follows that

$$-D^2y + 6y = 0 \quad \text{or} \quad D^2y - 6y = 0$$

Since the roots of the auxiliary equation are $m_1 = \sqrt{6}$ and $m_2 = -\sqrt{6}$, we obtain

$$y(t) = c_1e^{\sqrt{6}t} + c_2e^{-\sqrt{6}t} \tag{3}$$

Multiplying the first equation by -3, while operating on the second by D and then adding gives, the differential equation for x, $D^2x - 6x = 0$. It follows immediately that

$$x(t) = c_3e^{\sqrt{6}t} + c_4e^{-\sqrt{6}t} \tag{4}$$

Now (3) and (4) do not satisfy the system (1) for every choice of c_1, c_2, c_3, and c_4. Substituting $x(t)$ and $y(t)$ into the first equation of the original system (1) gives, after we simplify,

$$(\sqrt{6}c_1 - 2c_3)e^{\sqrt{6}t} + (-\sqrt{6}c_2 - 2c_4)e^{-\sqrt{6}t} = 0$$

Since the latter expression is to be zero for all values of t, we must have

$$\sqrt{6}c_1 - 2c_3 = 0 \quad \text{and} \quad -\sqrt{6}c_2 - 2c_4 = 0$$

or

$$c_3 = \frac{\sqrt{6}}{2}c_1, \quad c_4 = -\frac{\sqrt{6}}{2}c_2 \tag{5}$$

Hence we conclude that a solution of the system must be

$$x(t) = \frac{\sqrt{6}}{2}c_1e^{\sqrt{6}t} - \frac{\sqrt{6}}{2}c_2e^{-\sqrt{6}t}, \qquad y(t) = c_1e^{\sqrt{6}t} + c_2e^{-\sqrt{6}t}$$

You are urged to substitute (3) and (4) into the second equation of (1) and verify that the same relationship (5) holds between the constants.

■ **EXAMPLE 1 Solution by Elimination**

Solve

$$\begin{aligned} Dx + (D+2)y &= 0 \\ (D-3)x - \quad 2y &= 0 \end{aligned} \tag{6}$$

Solution Operating on the first equation by $D - 3$ and on the second by D and subtracting, eliminate x from the system. It follows that the differential equation for y is

$$[(D-3)(D+2) + 2D]y = 0 \quad \text{or} \quad (D^2 + D - 6)y = 0$$

Since the characteristic equation of this last differential equation is $m^2 + m - 6 = (m-2)(m+3) = 0$, we obtain the solution

$$y(t) = c_1e^{2t} + c_2e^{-3t} \tag{7}$$

Eliminating y in a similar manner yields $(D^2 + D - 6)x = 0$, from which we find

$$x(t) = c_3e^{2t} + c_4e^{-3t} \tag{8}$$

As we noted in the foregoing discussion, a solution of (6) does not contain four independent constants since the system itself puts a constraint on the actual number that can be chosen arbitrarily. Substituting (7) and (8) into the first equation of (6) gives

$$(4c_1 + 2c_3)e^{2t} + (-c_2 - 3c_4)e^{-3t} = 0$$

From $4c_1 + 2c_3 = 0$ and $-c_2 - 3c_4 = 0$, we get $c_3 = -2c_1$ and $c_4 = -\frac{1}{3}c_2$. Accordingly, a solution of the system is

$$x(t) = -2c_1e^{2t} - \frac{1}{3}c_2e^{-3t}, \qquad y(t) = c_1e^{2t} + c_2e^{-3t}$$ ■

Since we could just as easily solve for c_3 and c_4 in terms of c_1 and c_2, the solution in Example 1 can be written in the alternative form

$$x(t) = c_3e^{2t} + c_4e^{-3t}, \qquad y(t) = -\frac{1}{2}c_3e^{2t} - 3c_4e^{-3t}$$

Also, it sometimes pays to keep one's eyes open when solving systems. Had we solved for x first, then y could be found, along with the relationship between the constants by using the last equation in (6). You should verify that substituting $x(t)$ in $y = \frac{1}{2}(Dx - 3x)$ yields $y = -\frac{1}{2}c_3e^{2t} - 3c_4e^{-3t}$.

EXAMPLE 2 Solution by Elimination

Solve

$$\begin{aligned} x' - 4x + y'' &= t^2 \\ x' + x + y' &= 0 \end{aligned} \tag{9}$$

Solution First we rewrite the system in differential operator notation:

$$\begin{aligned} (D - 4)x + D^2y &= t^2 \\ (D + 1)x + Dy &= 0 \end{aligned} \tag{10}$$

Then, by eliminating x, we obtain

$$[(D + 1)D^2 - (D - 4)D]y = (D + 1)t^2 - (D - 4)0$$

or

$$(D^3 + 4D)y = t^2 + 2t$$

Since the roots of the auxiliary equation $m(m^2 + 4) = 0$ are $m_1 = 0$, $m_2 = 2i$, and $m_3 = -2i$, the complementary function is

$$y_c = c_1 + c_2 \cos 2t + c_3 \sin 2t$$

To determine the particular solution y_p, we use undetermined coefficients by assuming $y_p = At^3 + Bt^2 + Ct$. Therefore

$$y_p''' + 4y_p' = 12At^2 + 8Bt + 6A + 4C = t^2 + 2t$$

The last equality implies

$$12A = 1, \quad 8B = 2, \quad 6A + 4C = 0,$$

and hence $A = \frac{1}{12}$, $B = \frac{1}{4}$, $C = -\frac{1}{8}$. Thus

$$y = y_c + y_p = c_1 + c_2 \cos 2t + c_3 \sin 2t + \frac{1}{12}t^3 + \frac{1}{4}t^2 - \frac{1}{8}t \tag{11}$$

Eliminating y from the system (10) leads to

$$[(D - 4) - D(D + 1)]x = t^2 \quad \text{or} \quad (D^2 + 4)x = -t^2$$

It should be obvious that

$$x_c = c_4 \cos 2t + c_5 \sin 2t$$

and that undetermined coefficients can be applied to obtain a particular solution of the form $x_p = At^2 + Bt + C$. In this case the usual differentiations and algebra yield $x_p = -\frac{1}{4}t^2 + \frac{1}{8}$ and so

$$x = x_c + x_p = c_4 \cos 2t + c_5 \sin 2t - \frac{1}{4}t^2 + \frac{1}{8} \tag{12}$$

Now c_4 and c_5 can be expressed in terms of c_2 and c_3 by substituting (11) and (12) into either equation of (9). By using the second equation, we find, after combining terms,

$$(c_5 - 2c_4 - 2c_2)\sin 2t + (2c_5 + c_4 + 2c_3)\cos 2t = 0$$

so that $\quad c_5 - 2c_4 - 2c_2 = 0 \quad \text{and} \quad 2c_5 + c_4 + 2c_3 = 0$

Solving for c_4 and c_5 in terms of c_2 and c_3 gives

$$c_4 = -\frac{1}{5}(4c_2 + 2c_3) \quad \text{and} \quad c_5 = \frac{1}{5}(2c_2 - 4c_3)$$

Finally, a solution of (9) is found to be

$$x(t) = -\frac{1}{5}(4c_2 + 2c_3)\cos 2t + \frac{1}{5}(2c_2 - 4c_3)\sin 2t - \frac{1}{4}t^2 + \frac{1}{8}$$

$$y(t) = c_1 + c_2\cos 2t + c_3\sin 2t + \frac{1}{12}t^3 + \frac{1}{4}t^2 - \frac{1}{8}t$$

■

We will examine two other methods of solving systems of linear differential equations, along with systems as mathematical models, in Chapters 4 and 5.

EXERCISES 3.8

Answers to odd-numbered problems begin on page 444.

In Problems 1–20, solve the given system of differential equations by systematic elimination.

1. $\dfrac{dx}{dt} = 2x - y$, $\quad \dfrac{dy}{dt} = x$

2. $\dfrac{dx}{dt} = 4x + 7y$, $\quad \dfrac{dy}{dt} = x - 2y$

3. $\dfrac{dx}{dt} = -y + t$, $\quad \dfrac{dy}{dt} = x - t$

4. $\dfrac{dx}{dt} - 4y = 1$, $\quad \dfrac{dy}{dt} + x = 2$

5. $(D^2 + 5)x - 2y = 0$
$-2x + (D^2 + 2)y = 0$

6. $(D + 1)x + (D - 1)y = 2$
$3x + (D + 2)y = -1$

7. $\dfrac{d^2x}{dt^2} = 4y + e^t$, $\quad \dfrac{d^2y}{dt^2} = 4x - e^t$

8. $\dfrac{d^2x}{dt^2} + \dfrac{dy}{dt} = -5x$, $\quad \dfrac{dx}{dt} + \dfrac{dy}{dt} = -x + 4y$

9. $Dx + D^2y = e^{3t}$
$(D + 1)x + (D - 1)y = 4e^{3t}$

10. $D^2x - Dy = t$
$(D + 3)x + (D + 3)y = 2$

11. $(D^2 - 1)x - y = 0$
$(D - 1)x + Dy = 0$

12. $(2D^2 - D - 1)x - (2D + 1)y = 1$
$(D - 1)x + Dy = -1$

13. $2\dfrac{dx}{dt} - 5x + \dfrac{dy}{dt} = e^t$, $\quad \dfrac{dx}{dt} - x + \dfrac{dy}{dt} = 5e^t$

14. $\dfrac{dx}{dt} + \dfrac{dy}{dt} = e^t$, $\quad -\dfrac{d^2x}{dt^2} + \dfrac{dx}{dt} + x + y = 0$

15. $(D - 1)x + (D^2 + 1)y = 1$
$(D^2 - 1)x + (D + 1)y = 2$

16. $D^2x - 2(D^2 + D)y = \sin t$
$x + Dy = 0$

17. $Dx = y$
$Dy = z$
$Dz = x$

18.
$$\begin{aligned} Dx + z &= e^t \\ (D-1)x + Dy + Dz &= 0 \\ x + 2y + Dz &= e^t \end{aligned}$$

19.
$$\frac{dx}{dt} = 6y, \quad \frac{dy}{dt} = x + z, \quad \frac{dz}{dt} = x + y$$

20.
$$\frac{dx}{dt} = -x + z, \quad \frac{dy}{dt} = -y + z, \quad \frac{dz}{dt} = -x + y$$

In Problems 21 and 22, solve the given system of differential equations subject to the indicated initial conditions.

21.
$$\frac{dx}{dt} = -5x - y, \quad \frac{dy}{dt} = 4x - y, \quad x(1) = 0, y(1) = 1$$

22.
$$\frac{dx}{dt} = y - 1, \quad \frac{dy}{dt} = -3x + 2y, \quad x(0) = 0, y(0) = 0$$

Discussion Problems

23. Discuss the following system:

$$\begin{aligned} Dx - \quad 2Dy &= t^2 \\ (D+1)x - 2(D+1)y &= 1 \end{aligned}$$

3.9 Numerical Solutions

We saw in Section 3.3 that in order to analyze a second-order DE with an ODE solver, we had to express the equation as a system of two first-order equations. The reason for this is the fact that we solve systems of differential equations numerically by adapting an approximation technique to the system. We do this by simply applying a particular method, such as the Runge-Kutta method, to each first-order equation in the system.

Second-Order Initial-Value Problems In Section 2.5, we examined two numerical procedures that could be applied directly to a first-order initial-value problem $y' = f(t, y)$, $y(t_0) = y_0$. To approximate the solution of a second-order initial-value problem

$$y'' = f(t, y, y'), \qquad y(t_0) = y_0, \quad y'(t_0) = y_1 \tag{1}$$

we reduce the equation to a system of two first-order differential equations by means of the substitution $u = y'$. Observe then that $y'' = u'$ and $y'(t_0) = u(t_0)$ so that initial-value problem (1) is equivalent to

$$\textit{Solve:} \quad \begin{cases} y' = u \\ u' = f(t, y, u) \end{cases} \tag{2}$$
$$\textit{Subject to:} \quad y(t_0) = y_0, u(t_0) = y_1$$

For example, **Euler's method**, applied to the system in (2), would be

$$\begin{aligned} y_{n+1} &= y_n + hu_n \\ u_{n+1} &= u_n + hf(t_n, y_n, u_n) \end{aligned} \tag{3}$$

EXAMPLE 1 Euler's Method

Use Euler's method to obtain an approximate value of $y(0.2)$, where $y(t)$ is the solution of the initial-value problem

$$y'' + ty' + y = 0, \qquad y(0) = 1, \quad y'(0) = 2$$

Solution In terms of the substitution $u = y'$, the equation is equivalent to the system

$$\begin{aligned} y' &= u \\ u' &= -tu - y \end{aligned}$$

Thus from (3) we obtain

$$\begin{aligned} y_{n+1} &= y_n + hu_n \\ u_{n+1} &= u_n + h[-t_n u_n - y_n] \end{aligned}$$

Using the step size $h = 0.1$ and initial conditions $y_0 = 1$, $u_0 = 2$, we find

$$\begin{aligned} y_1 &= y_0 + (0.1)u_0 = 1 + (0.1)2 = 1.2 \\ u_1 &= u_0 + (0.1)[-t_0 u_0 - y_0] = 2 + (0.1)[-(0)(2) - 1] = 1.9 \\ y_2 &= y_1 + (0.1)u_1 = 1.2 + (0.1)(1.9) = 1.39 \\ u_2 &= u_1 + (0.1)[-t_1 u_1 - y_1] = 1.9 + (0.1)[-(0.1)(1.9) - 1.2] = 1.761 \end{aligned}$$

In other words, $y(0.2) \approx 1.39$ and $y'(0.2) \approx 1.761$. ■

In general, we approximate a solution of an nth-order differential equation $y^{(n)} = f(t, y, y', \ldots, y^{(n-1)})$ by reducing it to a system of n first-order equations using the substitutions $y = u_1$, $y' = u_2$, $y'' = u_3, \ldots, y^{(n-1)} = u_n$.

Numerical Solution of a System Analogous to what we have just illustrated for a single higher-order equation, a solution of a system of first-order differential equations

$$\begin{aligned} \frac{dx_1}{dt} &= g_1(t, x_1, x_2, \ldots, x_n) \\ \frac{dx_2}{dt} &= g_2(t, x_1, x_2, \ldots, x_n) \\ &\vdots \\ \frac{dx_n}{dt} &= g_n(t, x_1, x_2, \ldots, x_n) \end{aligned}$$

can be approximated by a version of the Euler or Runge-Kutta methods adapted to the system. For example, the **fourth-order Runge-Kutta method** applied to the system

$$x' = f(t, x, y)$$
$$y' = g(t, x, y) \tag{4}$$
$$x(t_0) = x_0, \quad y(t_0) = y_0$$

looks like this:

$$x_{n+1} = x_n + \frac{1}{6}(m_1 + 2m_2 + 2m_3 + m_4)$$
$$y_{n+1} = y_n + \frac{1}{6}(k_1 + 2k_2 + 2k_3 + k_4) \tag{5}$$

where

$$\begin{aligned} m_1 &= hf(t_n, x_n, y_n) & k_1 &= hg(t_n, x_n, y_n) \\ m_2 &= hf(t_n + \tfrac{1}{2}h, x_n + \tfrac{1}{2}m_1, y_n + \tfrac{1}{2}k_1) & k_2 &= hg(t_n + \tfrac{1}{2}h, x_n + \tfrac{1}{2}m_1, y_n + \tfrac{1}{2}k_1) \\ m_3 &= hf(t_n + \tfrac{1}{2}h, x_n + \tfrac{1}{2}m_2, y_n + \tfrac{1}{2}k_2) & k_3 &= hg(t_n + \tfrac{1}{2}h, x_n + \tfrac{1}{2}m_2, y_n + \tfrac{1}{2}k_2) \\ m_4 &= hf(t_n + h, x_n + m_3, y_n + k_3) & k_4 &= hg(t_n + h, x_n + m_3, y_n + k_3) \end{aligned} \tag{6}$$

EXAMPLE 2 **Runge-Kutta Method**

Consider the initial-value problem

$$x' = 2x + 4y$$
$$y' = -x + 6y$$
$$x(0) = -1, \quad y(0) = 6$$

Use the fourth-order Runge-Kutta method to approximate $x(0.6)$ and $y(0.6)$. Compare the results for $h = 0.2$ and $h = 0.1$.

Solution We illustrate the computations of x_1 and y_1 with the step size $h = 0.2$. With the identifications $f(t, x, y) = 2x + 4y$, $g(t, x, y) = -x + 6y$, $t_0 = 0$, $x_0 = -1$, and $y_0 = 6$, we see from (6) that

$$\begin{aligned} m_1 &= hf(t_0, x_0, y_0) = 0.2f(0, -1, 6) = 0.2[2(-1) + 4(6)] = 4.4000 \\ k_1 &= hg(t_0, x_0, y_0) = 0.2g(0, -1, 6) = 0.2[-1(-1) + 6(6)] = 7.4000 \\ m_2 &= hf(t_0 + \tfrac{1}{2}h, x_0 + \tfrac{1}{2}m_1, y_0 + \tfrac{1}{2}k_1) = 0.2f(0.1, 1.2, 9.7) = 8.2400 \\ k_2 &= hg(t_0 + \tfrac{1}{2}h, x_0 + \tfrac{1}{2}m_1, y_0 + \tfrac{1}{2}k_1) = 0.2g(0.1, 1.2, 9.7) = 11.4000 \\ m_3 &= hf(t_0 + \tfrac{1}{2}h, x_0 + \tfrac{1}{2}m_2, y_0 + \tfrac{1}{2}k_2) = 0.2f(0.1, 3.12, 11.7) = 10.6080 \\ k_3 &= hg(t_0 + \tfrac{1}{2}h, x_0 + \tfrac{1}{2}m_2, y_0 + \tfrac{1}{2}k_2) = 0.2g(0.1, 3.12, 11.7) = 13.4160 \\ m_4 &= hf(t_0 + h, x_0 + m_3, y_0 + k_3) = 0.2f(0.2, 8, 20.216) = 19.3760 \\ k_4 &= hg(t_0 + h, x_0 + m_3, y_0 + k_3) = 0.2g(0.2, 8, 20.216) = 21.3776 \end{aligned}$$

Therefore, from (5) we get

$$x_1 = x_0 + \frac{1}{6}(m_1 + 2m_2 + 2m_3 + m_4)$$

$$= -1 + \frac{1}{6}(4.4 + 2(8.24) + 2(10.608) + 19.3760) = 9.2453$$

$$y_1 = y_0 + \frac{1}{6}(k_1 + 2k_2 + 2k_3 + k_4)$$

$$= 6 + \frac{1}{6}(7.4 + 2(11.4) + 2(13.416) + 21.3776) = 19.0683$$

where, as usual, the computed values are rounded to four decimal places. These numbers give us the approximations $x_1 \approx x(0.2)$ and $y_1 \approx y(0.2)$. The subsequent values, obtained with the aid of a computer, are summarized in Tables 3.2 and 3.3.

Table 3.2 Runge-Kutta Method with $h = 0.2$

m_1	m_2	m_3	m_4	k_1	k_2	k_3	k_4	t_n	x_n	y_n
								0.00	−1.0000	6.0000
4.4000	8.2400	10.6080	19.3760	7.4000	11.4000	13.4160	21.3776	0.20	9.2453	19.0683
18.9527	31.1564	37.8870	63.6848	21.0329	31.7573	36.9716	57.8214	0.40	46.0327	55.1203
62.5093	97.7863	116.0063	187.3669	56.9378	84.8495	98.0688	151.4191	0.60	158.9430	150.8192

Table 3.3 Runge-Kutta Method with $h = 0.1$

m_1	m_2	m_3	m_4	k_1	k_2	k_3	k_4	t_n	x_n	y_n
								0.00	−1.0000	6.0000
2.2000	3.1600	3.4560	4.8720	3.7000	4.7000	4.9520	6.3256	0.10	2.3840	10.8883
4.8321	6.5742	7.0778	9.5870	6.2946	7.9413	8.3482	10.5957	0.20	9.3379	19.1332
9.5208	12.5821	13.4258	17.7609	10.5461	13.2339	13.8872	17.5358	0.30	22.5541	32.8539
17.6524	22.9090	24.3055	31.6554	17.4569	21.8114	22.8549	28.7393	0.40	46.5103	55.4420
31.4788	40.3496	42.6387	54.9202	28.6141	35.6245	37.2840	46.7207	0.50	88.5729	93.3006
54.6348	69.4029	73.1247	93.4107	46.5231	57.7482	60.3774	75.4370	0.60	160.7563	152.0025

■

You should verify that the solution of the initial-value problem in Example 2 is given by $x(t) = (26t - 1)e^{4t}$, $y(t) = (13t + 6)e^{4t}$. From these equations, we see that the exact values are $x(0.6) = 160.9384$ and $y(0.6) = 152.1198$.

In conclusion, we state **Euler's method** for the general system (4) as

$$x_{n+1} = x_n + hf(t_n, x_n, y_n)$$

$$y_{n+1} = y_n + hg(t_n, x_n, y_n)$$

EXERCISES 3.9

Answers to odd-numbered problems begin on page 444.

1. Use the Euler method to approximate $y(0.2)$, where $y(t)$ is the solution of the initial-value problem

$$y'' - 4y' + 4y = 0, \qquad y(0) = -2, \quad y'(0) = 1$$

Use $h = 0.1$. Find the exact solution of the problem and compare the exact value of $y(0.2)$ with y_2.

2. Use the Euler method to approximate $y(1.2)$, where $y(t)$ is the solution of the initial-value problem

$$t^2y'' - 2ty' + 2y = 0, \qquad y(1) = 4, \quad y'(1) = 9$$

where $t > 0$. Use $h = 0.1$. Find the exact solution of the problem and compare the exact value of $y(1.2)$ with y_2.

3. Repeat Problem 1 using the Runge-Kutta method with $h = 0.2$ and $h = 0.1$.

4. Repeat Problem 2 using the Runge-Kutta method with $h = 0.2$ and $h = 0.1$.

5. Use the Runge-Kutta method to obtain the approximate value of $y(0.2)$, where $y(t)$ is a solution of the initial-value problem

$$y'' - 2y' + 2y = e^t \cos t, \qquad y(0) = 1, \quad y'(0) = 2$$

Use $h = 0.2$ and $h = 0.1$.

6. When $E = 100$ volts, $R = 10$ ohms, and $L = 1$ henry, the system of differential equations for the currents $i_1(t)$ and $i_3(t)$ in the electrical network given in Figure 3.54 is

$$\frac{di_1}{dt} = -20i_1 + 10i_3 + 100$$

$$\frac{di_3}{dt} = 10i_1 - 20i_3$$

where $i_1(0) = 0$ and $i_3(0) = 0$. Use the Runge-Kutta method to approximate $i_1(t)$ and $i_3(t)$ at $t = 0.1, 0.2, 0.3, 0.4$, and 0.5. Use $h = 0.1$.

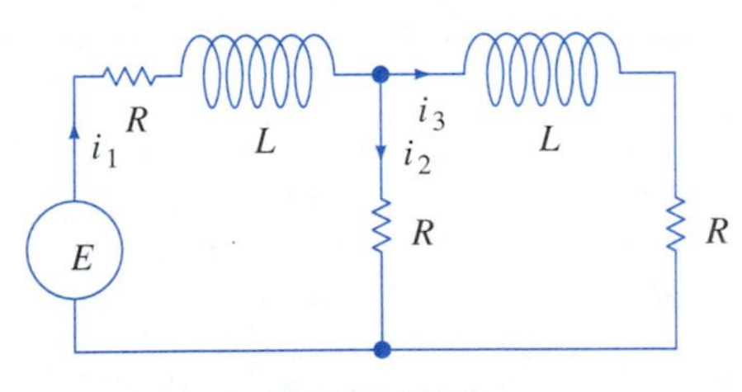

Figure 3.54.

In Problems 7–12, use the Runge-Kutta method to approximate $x(0.2)$ and $y(0.2)$. Compare the results for $h = 0.2$ and $h = 0.1$.

7. $x' = 2x - y$
$y' = x$
$x(0) = 6, y(0) = 2$

8. $x' = x + 2y$
$y' = 4x + 3y$
$x(0) = 1, y(0) = 1$

9. $x' = -y + t$
$y' = x - t$
$x(0) = -3, y(0) = 5$

10. $x' = 6x + y + 6t$
$y' = 4x + 3y - 10t + 4$
$x(0) = 0.5, y(0) = 0.2$

11. $x' + 4x - y' = 7t$
$x' + y' - 2y = 3t$
$x(0) = 1, y(0) = -2$

12. $x' + y' = 4t$
$-x' + y' + y = 6t^2 + 10$
$x(0) = 3, y(0) = -1$

Discussion Problems

13. Accept as a given that numerical methods for approximating an initial-value problem are applied to first-order equations. Discuss: How would you adapt the Euler and Runge-Kutta methods to approximate a solution of the system that follows?

$$x'' - x' + 5x + 2y'' = e^t$$

$$-2x + y'' + y = 3t^2$$

3.10 Nonlinear Equations

We have seen that there is quite a bit we can do when faced with the problem of solving a linear DE, especially if the equation has constant coefficients. In contrast, there is little we can do when faced with the similar problem of solving a higher-order nonlinear DE. This does not mean a nonlinear higher-order DE has no solution, but rather there are no general methods whereby either an explicit or implicit solution can be found.

In the first part of the discussion that follows, we briefly examine some of the differences between linear and nonlinear equations; the section concludes with nonlinear second-order mathematical models.

Some Distinctions There are several significant differences between linear and nonlinear differential equations. We saw in Section 3.1 that homogeneous linear equations of order two or higher have the property that a linear combination of solutions is also a solution (Theorem 3.2). Nonlinear equations do not possess this superposition property. For example, on the interval $(-\infty, \infty)$, $y_1 = e^t$, $y_2 = e^{-t}$, $y_3 = \cos t$, and $y_4 = \sin t$ are four linearly independent solutions of the nonlinear second-order differential equation $(y'')^2 - y^2 = 0$. But linear combinations, such as $y = c_1e^t + c_3 \cos t$, $y = c_2e^{-t} + c_4 \sin t$, and $y = c_1e^t + c_2e^{-t} + c_3 \cos t + c_4 \sin t$, are not solutions of the equation for arbitrary nonzero constants c_i.

In Chapter 2, we were able to solve some nonlinear first-order differential equations by separation of variables. Even though the solutions of these equations were in the form of a one-parameter family, this family did not, as a rule, represent the general solution of the differential equation. On the other hand, by paying attention to certain continuity conditions, we obtained general solutions of linear first-order differential equations. Stated another way, nonlinear first-order differential equations can possess singular solutions whereas linear equations can not. But the major difference between linear and nonlinear equations of order two or higher lies in the realm of solvability. Given a linear equation, there is a chance that we can find some form of a solution that we can look at—an explicit solution or perhaps a solution in the form of an infinite series (see Chapter 6). On the other hand, nonlinear higher-order differential equations virtually defy solution. Although this sounds disheartening, there are still things that can be done.

Let us make it clear at the outset that nonlinear higher-order differential equations are important, dare we say even more important than linear equations, because as we fine-tune a mathematical model of, say, a physical system, we also increase the likelihood that this higher-resolution model will be nonlinear. Because we can apply a numerical technique to, for example, a nonlinear second-order equation in the same manner as we did to a linear second-order equation, that is, by writing the equation as a system of two first-order differential equations, we can see graphical representations of solutions by utilizing an ODE solver. Answers to questions such as "Are there periodic solutions?" can be attacked by delving deeper into the qualitative analysis of autonomous differential equations that we began in Sections 2.1 and 3.3.

Nonlinear Springs We have already remarked on the fact that the autonomous second-order differential equation

$$m\frac{d^2x}{dt^2} + F(x) = 0 \qquad (1)$$

represents a general model for the free undamped motion of a mass m attached to a spring. Here x denotes the displacement of the mass from its equilibrium position and F represents the force exerted by the spring that tends to restore the mass to its equilibrium position. A spring acting under a linear restoring force $F(x) = kx$—in other words, Hooke's law—is called a **linear spring**. But, depending on how it is constructed and the material used, springs are seldom perfectly linear; a spring can range from "mushy" or soft to "stiff" or hard, so that its restorative force may vary from something below or above that given by the linear law. If we assume that a nonaging spring possesses some nonlinear characteristics, then it might be reasonable to assume that the restorative force $F(x)$ of a spring is proportional to, say, the cube of the displacement x of the mass beyond its equilibrium position, or that $F(x)$ is a linear combination of powers of the displacement such as that given by the nonlinear function $F(x) = kx + k_1x^3$. A spring whose mathematical model (1) incorporates a nonlinear restorative force, such as

$$m\frac{d^2x}{dt^2} + kx^3 = 0 \quad \text{or} \quad m\frac{d^2x}{dt^2} + kx + k_1x^3 = 0 \qquad (2)$$

is called a **nonlinear spring**. In Section 3.6, we also examined mathematical models in which damping imparted to the motion was proportional to the instantaneous velocity dx/dt—that is, linear damping. But this was simply an assumption; in more realistic situations, damping could be proportional to some power of the instantaneous velocity dx/dt. The nonlinear differential equation

$$m\frac{d^2x}{dt^2} + \beta\left|\frac{dx}{dt}\right|\frac{dx}{dt} + kx = 0 \qquad (3)$$

is one model of a free spring/mass system with a linear restoring force, but with damping proportional to the square of the velocity. One can then envision other kinds of models: linear damping and nonlinear restoring force, nonlinear damping and nonlinear restoring force, and so on. The point is, nonlinear characteristics of a physical system lead to a mathematical model that is nonlinear.

Notice in (2) that both $F(x) = kx^3$ and $F(x) = kx + k_1x^3$ are odd functions of x. To see why a polynomial function containing only odd powers of x provides a reasonable model for the restoring force, let us express F as power series centered at the equilibrium position $x = 0$:

$$F(x) = c_0 + c_1x + c_2x^2 + c_3x^3 + \ldots$$

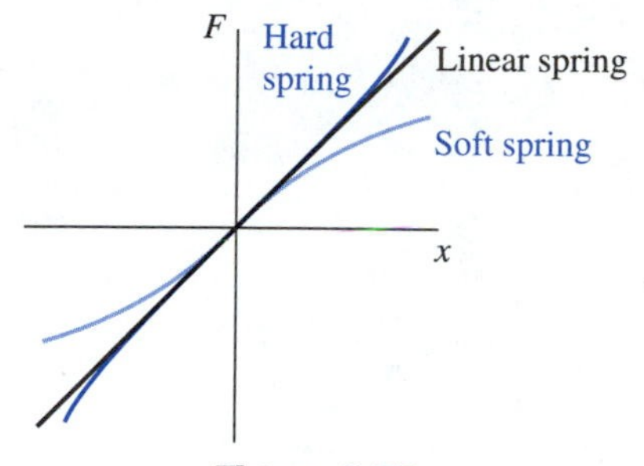

Figure 3.55.

When the displacements x are small, the values of x^n are negligible for n sufficiently large. If we truncate the series with, say, the fourth term, then

$$F(x) = c_0 + c_1x + c_2x^2 + c_3x^3$$

In order that the force at $x > 0$ ($F(x) = c_0 + c_1x + c_2x^2 + c_3x^3$) and the force at $-x$, ($F(-x) = c_0 - c_1x + c_2x^2 - c_3x^3$) have the same magnitude but act in opposite directions, we must have $F(-x) = -F(x)$. Since this means F is an odd function, we must have $c_0 = 0$ and $c_2 = 0$, and so $F(x) = c_1x + c_3x^3$. Had we used only the first two terms in the series, the same argument yields $F(x) = c_1x$. For discussion purposes, we shall write $c_1 = k$ and $c_2 = k_1$. A restoring force with mixed powers, such as $F(x) = kx + k_1x^2$, and the corresponding vibrations, are said to be unsymmetrical.

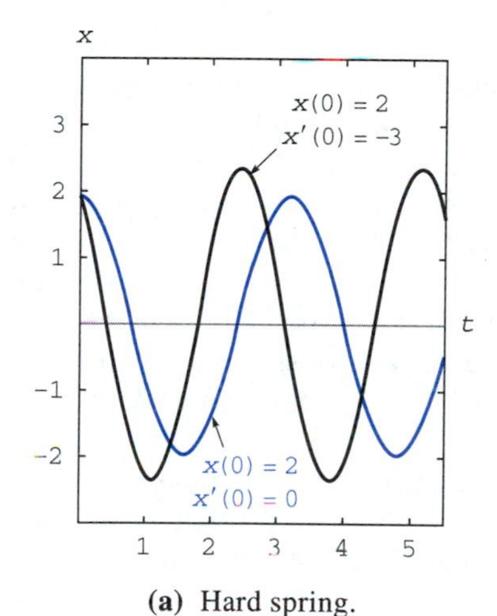

(a) Hard spring.

(b) Soft spring.

Figure 3.56.

Hard and Soft Springs Let us take a closer look at the equation in (1) in the case where the restoring force is given by $F(x) = kx + k_1x^3$, $k > 0$. The spring is said to be **hard** if $k_1 > 0$ and **soft** if $k_1 < 0$. Graphs of three types of restoring forces are illustrated in Figure 3.55. The next example illustrates these two special cases of the differential equation $md^2x/dt^2 + kx + k_1x^3 = 0$, $m > 0$, $k > 0$.

■ EXAMPLE 1 Comparison of Hard and Soft Springs

The autonomous differential equations

$$\frac{d^2x}{dt^2} + x + x^3 = 0 \tag{4}$$

$$\frac{d^2x}{dt^2} + x - x^3 = 0 \tag{5}$$

are models of a hard spring and a soft spring, respectively. Figure 3.56(a) shows two solutions of (4) and Figure 3.56(b) shows two solutions of (5), obtained from an ODE solver. The curves shown in black are solutions satisfying the initial conditions $x(0) = 2$, $x'(0) = -3$; the two curves in color are solutions satisfying $x(0) = 2$, $x'(0) = 0$. These solution curves certainly suggest that the motion of a mass on the hard spring is oscillatory, whereas the motion of a mass on the soft spring is not oscillatory. But we must be careful about drawing conclusions based on a couple of solution curves. A more complete picture of the nature of the solutions of both of these equations can be obtained from their phase portraits. ■

Phase Portraits We have seen in Section 3.3 that a phase portrait of an autonomous second-order differential equation can be obtained by expressing the equation as an autonomous system of two first-order equations and then using computer software to plot trajectories through specified initial points (x_0, y_0). If $dx/dt = y$, then the autonomous systems corresponding

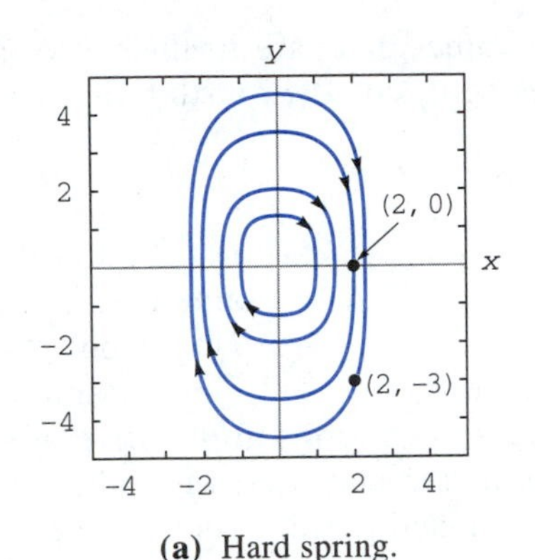

(a) Hard spring.

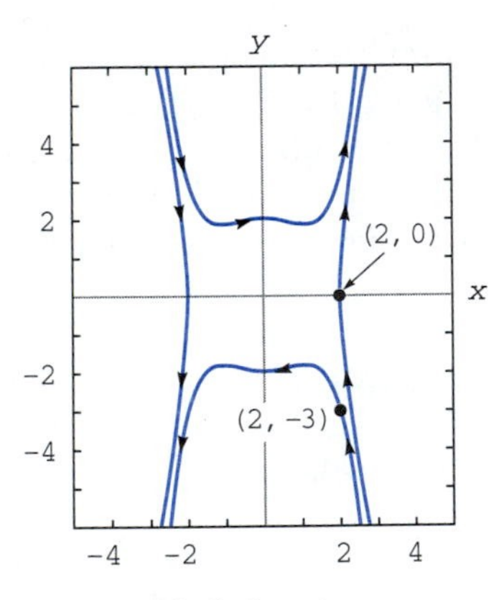

(b) Soft spring.

Figure 3.57.

to (4) and (5) are, respectively,

$$\begin{aligned} \frac{dx}{dt} &= y \\ \frac{dy}{dt} &= -x - x^3 \end{aligned} \tag{6}$$

and

$$\begin{aligned} \frac{dx}{dt} &= y \\ \frac{dy}{dt} &= -x + x^3 \end{aligned} \tag{7}$$

Recall from Section 3.3 that a critical point of an autonomous second-order differential equation is a critical point of its equivalent autonomous system. A critical point of an autonomous system is a point where both dx/dt and dy/dt are zero. In (6), for example, $dx/dt = 0$ at $y = 0$ and $dy/dt = 0$ at the solutions of $-x - x^3 = 0$ or $x(x^2 + 1) = 0$. Because 0 is the only real solution of the last equation, we conclude that $x = 0$, $y = 0$ is a constant or equilibrium solution of this system. In other words, (0, 0) is the only critical point of (6) and consequently the only critical point of equation (4). On the other hand, examination of (7) shows that (0, 0), (1, 0), and $(-1, 0)$ are three critical points of (5).

Using the systems (6) and (7) we obtain, in turn, the phase portraits in Figure 3.57(a) and Figure 3.57(b). We saw in Section 3.3 that trajectories corresponding to periodic solutions were closed curves; conversely, it can be proved that if a trajectory defined by a solution $x = x(t)$, $y = y(t)$ of an autonomous system is closed, then $x(t)$ and $y(t)$ are periodic, that is, there exists a real number $T > 0$ such that $x(t + T) = x(t)$ and $y(t + T) = y(t)$ for all t. Thus the closed trajectories in Figure 3.57(a) indicate what we had intuitively expected, namely, that each corresponding solution $x(t)$ is periodic. The solution curves for the hard spring given in Figure 3.56(a) correspond to the two trajectories in Figure 3.57(a) that pass through the indicated points (2, 0) and $(2, -3)$. The solution curves for the soft spring in Figure 3.56(b) correspond to the two trajectories passing through (2, 0) and $(2, -3)$ in Figure 3.57(b) and, of course, verify the unbounded motion of the spring/mass system for the stipulated initial conditions. However, Figure 3.57(b) is not an informative phase portrait of (12) since *it is important to examine trajectories close to the critical points.* Figure 3.58(a) shows trajectories of (7) obtained by choosing initial points close to the three critical points. Observe that we have discovered that there do indeed exist periodic solutions of (5) provided that the magnitudes of the initial displacement and initial velocity are not too large. Figure 3.58(b) illustrates the periodic solution of (5) subject to the initial conditions $x(0) = \frac{1}{2}$, $x'(0) = \frac{1}{2}$, and corresponds to the closed trajectory shown in black in Figure 3.58(a).

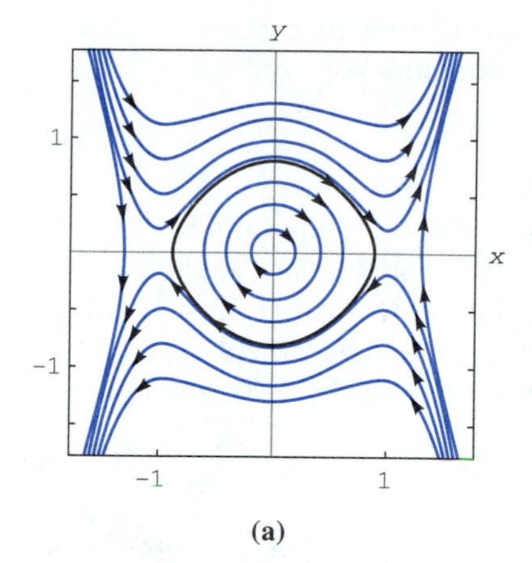

(a)

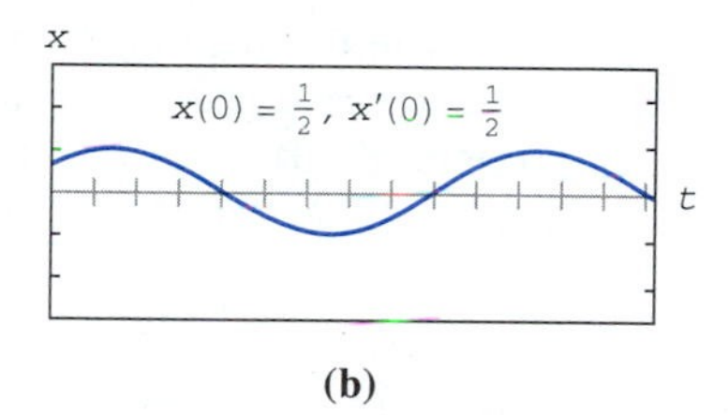

(b)

Figure 3.58.

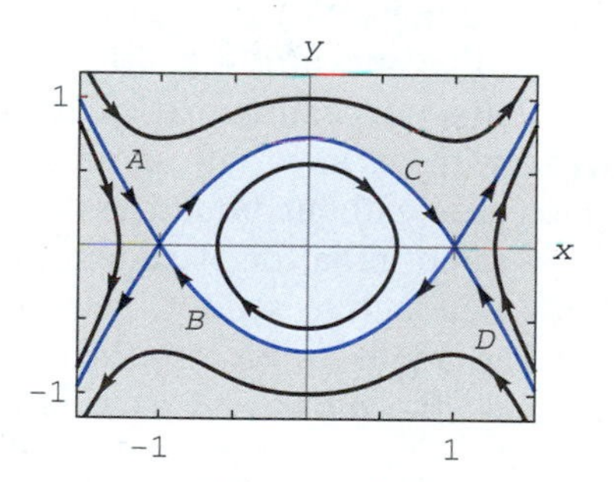

Figure 3.59.

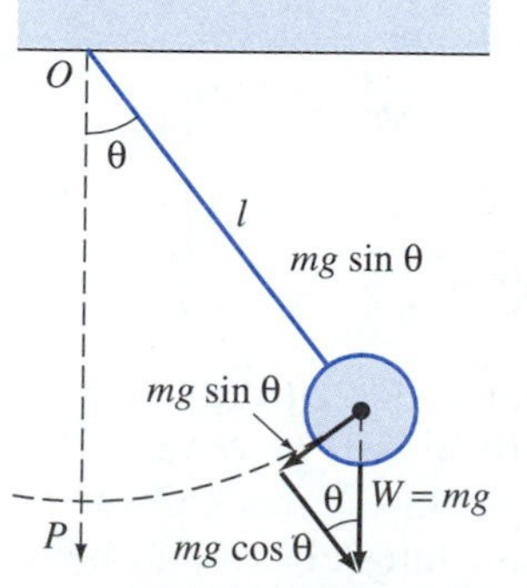

Figure 3.60.

Separatrices Several trajectories of importance are not shown in the phase portrait of the soft spring given Figure 3.58(a). Notice in Figure 3.59 that at each of the two critical points $(-1, 0)$ and $(1, 0)$ there exist two pairs of trajectories, such that on one pair of trajectories, points $(x(t), y(t))$ approach or "enter" the critical point as $t \to \infty$. These trajectories, with the arrows pointing toward the critical point, are labeled by the letters A and B at $(-1, 0)$ and by C and D at $(1, 0)$. Points $(x(t), y(t))$ on the remaining pair of trajectories approach or "enter" the critical point as $t \to -\infty$. Such trajectories are called **separatrices** (singular: separatrix). Forgetting about arrows and the like, the separatrices when viewed together are simply the two colored curves in the figure that appear to intersect at the critical points.

As their name connotes, separatrices are *dividing curves*—in the example of the soft spring, they divide the open and closed trajectories. In other words, they separate the phase plane into regions corresponding to the different behaviors—in Figure 3.59, the regions corresponding to periodic and nonperiodic motions of the spring/mass system are shaded in light color and light gray, respectively. Also, see Problems 20 and 21 in *Projects and Computer Lab Experiments for Chapter 3.*

Nonlinear Pendulum Any object that swings back and forth is called a **physical pendulum**. The **simple pendulum** is a special case of the physical pendulum and consists of a rod of length l to which a mass m is attached at one end.

In describing the motion of a simple pendulum in a vertical plane, we make the simplifying assumptions that the mass of the rod is negligible and that no external damping or driving forces act on the system. The displacement angle θ of the pendulum, measured from the vertical, as shown in Figure 3.60, is considered positive when measured to the right of OP and negative when to the left of OP. Now recall that the arc s of a circle of radius l is related to the central angle θ by the formula $s = l\theta$. Hence angular acceleration is

$$a = \frac{d^2s}{dt^2} = l\frac{d^2\theta}{dt^2}$$

From Newton's second law we then have

$$F = ma = ml\frac{d^2\theta}{dt^2}$$

From Figure 3.60, we see that the tangential component of the force due to the weight W is $mg \sin \theta$. We equate the two different formulations of the tangential force to obtain $ml\, d^2\theta/dt^2 = -mg \sin \theta$ or

$$\frac{d^2\theta}{dt^2} + \frac{g}{l}\sin \theta = 0 \tag{8}$$

Linearization Because of the presence of $\sin \theta$, equation (8) is nonlinear. In the study of nonlinear higher-order differential equations, we often try to simplify a problem by replacing nonlinear terms by certain approximations. For example, the Maclaurin series for $\sin\theta$ is given by

$$\sin\theta = \theta - \frac{\theta^3}{3!} + \frac{\theta^5}{5!} - \cdots$$

and so if we use the approximation $\sin\theta \approx \theta - \theta^3/6$, equation (8) becomes $d^2\theta/dt^2 + (g/l)\theta - (g/6l)\theta^3 = 0$. Observe that this last equation is the same as the second nonlinear equation in (2), with $m = 1$, $k = g/l$, and $k_1 = -g/6l$. However, if we assume that the displacements θ are small enough to justify using the replacement $\sin\theta \approx \theta$, then (8) becomes

$$\frac{d^2\theta}{dt^2} + \frac{g}{l}\theta = 0 \tag{9}$$

If we set $\omega^2 = g/l$, (9) has the same structure as the differential equation (3) of Section 3.6, governing the free undamped vibrations of a linear spring/mass system. Equation (9) is called a **linearization** of equation (8). Since the general solution of (9) is $\theta(t) = c_1 \cos\omega t + c_2 \sin\omega t$, this linearization suggests that, for initial conditions amenable to small oscillations, the motion of the pendulum described by (8) will be periodic.

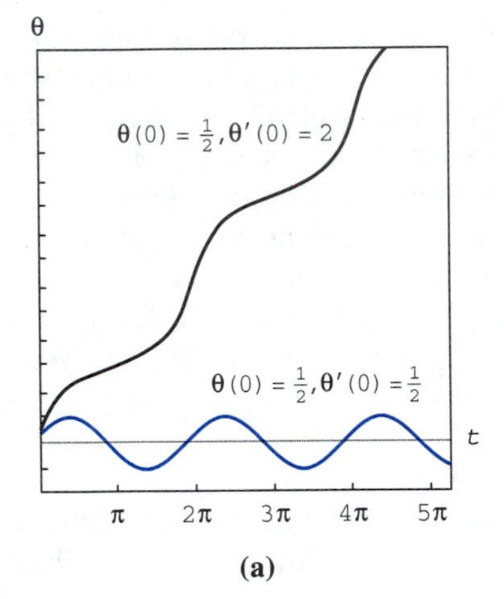

(a)

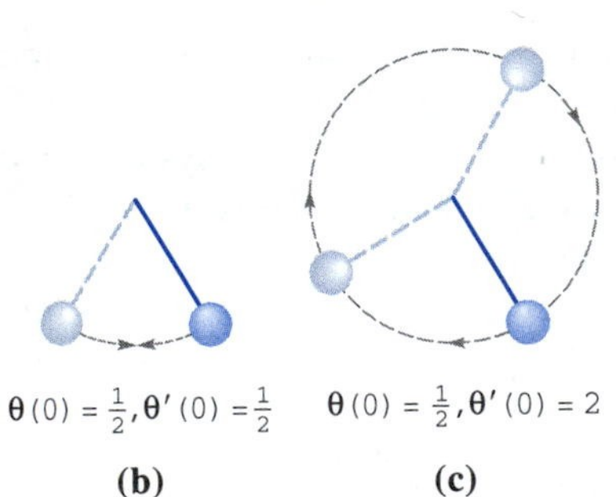

(b) (c)

Figure 3.61.

EXAMPLE 2 Nonlinear Pendulum

The graphs in Figure 3.61(a) were obtained with the aid of an ODE solver and represent solution curves of (8) when $\omega^2 = 1$. The colored curve depicts the solution of (8) that satisfies the initial conditions $\theta(0) = \frac{1}{2}$, $\theta'(0) = \frac{1}{2}$, whereas the black curve is the solution of (8) that satisfies $\theta(0) = \frac{1}{2}$, $\theta'(0) = 2$. The colored curve represents a periodic solution—the pendulum oscillating back and forth, as shown in Figure 3.61(b), with an apparent amplitude $A < 1$. The black curve shows that θ increases without bound as time increases—the pendulum, starting from the same initial displacement, is given an initial velocity of a magnitude great enough to send it over the top; in other words, the pendulum is whirling about its pivot, as shown in Figure 3.61(c). In the absence of damping, the motion in each case continues indefinitely. ■

Phase Portrait If we denote $x = \theta(t)$ and $\omega^2 = g/l$, then (8) is equivalent to the autonomous system of first-order equations

$$\begin{aligned}\frac{dx}{dt} &= y\\ \frac{dy}{dt} &= -\omega^2 \sin x\end{aligned} \tag{10}$$

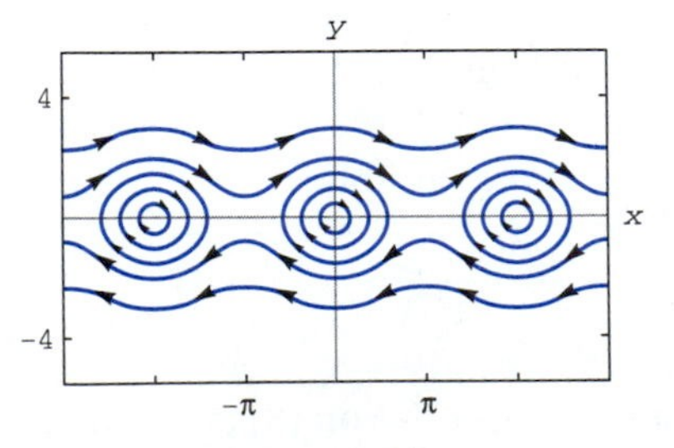

Figure 3.62.

Solving $y = 0$ and $\sin x = 0$ gives the critical points $(0, 0)$, $(\pi, 0)$, $(-\pi, 0)$, ..., or $(n\pi, 0)$, $n = 0, \pm 1, \pm 2, \ldots$. Using (10) with $\omega^2 = 1$, we obtain the phase portrait in Figure 3.62. The closed trajectories in Figure 3.62 correspond to periodic solutions of (8), whereas the wavy lines above and below the closed curves correspond to solutions that represent the whirling pendulum.

Stability As in the discussion of linear equations in Section 3.3, critical points of autonomous *nonlinear* second-order differential equations are classified as stable, asymptotically stable, or unstable. From the phase portrait, Figure 3.57(a), for the hard spring defined by (4), it should be apparent that any solution starting at an initial point (x_0, y_0) near the critical point $(0, 0)$ always stays close to the critical point. Thus $(0, 0)$ is stable. Moreover, because the behavior of the simple closed curves representing periodic solutions is similar to the ellipses in Figure 3.14(e), we also say that $(0, 0)$ is a center. From the phase portrait, Figure 3.58(a), for the soft spring defined by (5), we conclude that $(0, 0)$ is a center but that $(-1, 0)$ and $(1, 0)$ are unstable critical points. Because the behavior of the trajectories near these latter two points in Figure 3.58(a) *resembles* Figure 3.12(e), we further say that $(-1, 0)$ and $(1, 0)$ are saddle points. In other words, a critical point (x_1, y_1) of an autonomous nonlinear second-order differential equation is assigned one of the names: node, saddle point, spiral point, or center, if the behavior of the trajectories near (x_1, y_1) resembles the behavior of the trajectories near $(0, 0)$ in Figures 3.12, 3.13, and 3.14. Keep in mind that a center is always a stable critical point and that a saddle point is always an unstable critical point. We note in passing that, unlike Figure 3.14(e), the closed curves surrounding a center of a nonlinear differential equation, as in Figures 3.57(a) and 3.58(a), need not be ellipses.

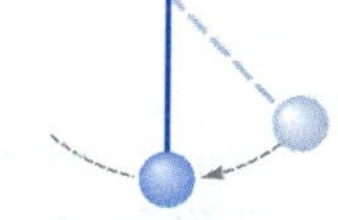

(a) $\theta = 0, \theta' = 0$

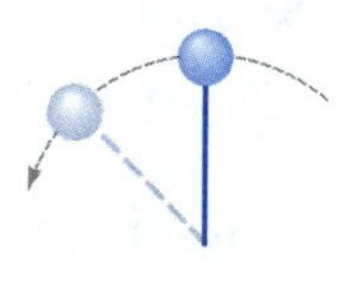

(b) $\theta = \pi, \theta' = 0$

Figure 3.63.

In the case of the nonlinear pendulum, the equilibrium solutions $\theta(t) = 0$ and $\theta(t) = \pi$ correspond, respectively, to the pendulum in its lowest position hanging vertically downward, as shown in Figure 3.63(a), and the pendulum in its highest position pointing vertically upward, as in Figure 3.63(b). Our physical intuition tells us an undamped pendulum that starts *near* $(0, 0)$, in other words that starts with an initial displacement and velocity of reasonably small magnitude, will exhibit periodic oscillations about $\theta = 0$, of reasonably small amplitude. This is, of course, the nature of a stable critical point. Similarly, we can argue that $(\pi, 0)$ is an unstable critical point, because when the pendulum is at rest at its highest position, the slightest displacement or the slightest push will result in it falling rapidly away from that position. These conjectures are verified by the portrait in Figure 3.62, $(0, 0)$ along with $(2k\pi, 0)$, $k = \pm1, \pm2, \ldots$, are centers; $(\pi, 0)$ along with $((2k + 1)\pi, 0)$, $k = \pm1, \pm2, \ldots$, are saddle points.

See Problems 22–25 in *Projects and Computer Lab Experiments for Chapter 3.*

Remark

Earlier statements notwithstanding, *some* nonlinear second-order equations can be solved. Nonlinear equations $F(t, y', y'') = 0$, where the dependent variable y is missing, and $F(y, y', y'') = 0$, where the independent variable t is missing, can be reduced to first-order differential equations by means of the substitution $u = y'$. If we can solve this first-order equation for u, we can find y by integrating $y' = u$. See Problems 13–18 in Exercises 3.10.

EXERCISES 3.10

Answers to odd-numbered problems begin on page 445.

In Problems 1–4, the given differential equation is a model of an undamped spring/mass system in which the restoring force $F(x)$ in (1) is nonlinear.

(a) For each equation, use an ODE solver to obtain the solution curves satisfying the given initial conditions. If the solution is periodic, use the solution curve to estimate the period T of oscillations.

(b) Find the critical points of each equation.

(c) Obtain a phase portrait of each equation and identify the trajectories corresponding to the solution curves obtained in part (a). Use the phase portrait to classify the critical points as stable or unstable. If possible, determine whether the descriptive names: node, saddle point, spiral point, or center, are applicable to the critical points.

1. $\dfrac{d^2x}{dt^2} + x^3 = 0$

$x(0) = 1,\ x'(0) = 1;$

$x(0) = \frac{1}{2},\ x'(0) = -1$

2. $\dfrac{d^2x}{dt^2} + 16x - 4x^3 = 0$

$x(0) = 1,\ x'(0) = 1;$

$x(0) = -2,\ x'(0) = 2$

3. $\dfrac{d^2x}{dt^2} + 2x - x^2 = 0$

$x(0) = 1,\ x'(0) = 1;$

$x(0) = \frac{3}{2},\ x'(0) = -1$

4. $\dfrac{d^2x}{dt^2} + xe^{0.01x} = 0$

$x(0) = 1,\ x'(0) = 1;$

$x(0) = 3,\ x'(0) = -1$

5. (a) In Problem 3, suppose that the mass is released from the initial position $x(0) = 1$ with an initial velocity $x'(0) = x_1$. Use a phase portrait of the equation to estimate the smallest value of $|x_1|$ at which the motion of the mass is nonperiodic.

(b) In Problem 3, suppose that the mass is released from the initial position $x(0) = x_0$ with an initial velocity $x'(0) = 1$. Use a phase portrait of the equation to estimate an interval $a \le x_0 \le b$ for which the motion is oscillatory.

6. In Problem 3, suppose that the mass is released from rest from the initial position $x(0) = 1$. Since the restoring force $F(x) = 2x - x^2$ is asymmetric, it makes sense to assume that the mass oscillates between $x = 1$ and some value $x = a < 0$, where $a \neq -1$.

(a) Use an ODE solver to find the solution curve and use this curve to estimate the extreme displacements of the mass from the line $x = 0$.

(b) Find a Cartesian equation of the trajectory corresponding to these initial conditions and use this trajectory to obtain the exact values of the extreme displacements of the mass from the line $x = 0$.

7. Find a linearization of the differential equation in Problem 4.

8. Consider the model of an undamped nonlinear spring/mass system given by

$$\frac{d^2x}{dt^2} + 8x - 6x^3 + x^5 = 0$$

Discuss in detail the nature of the oscillations of the system by examining critical points, a phase portrait, and solution curves obtained from an ODE solver. Use the phase portrait to classify the critical points as stable or unstable. If possible, determine whether the descriptive names: node, saddle point, spiral point, or center, are applicable to the critical points.

In Problems 9 and 10, the given differential equation is a model of a damped nonlinear spring/mass system.

(a) Predict the behavior of each system as $t \to \infty$.

(b) For each equation, use an ODE solver to obtain the solution curves satisfying the given initial conditions.

(c) Find the critical points of each equation.

(d) Obtain a phase portrait of each equation and identify the trajectories corresponding to the solution curves obtained in part (b). Use the phase portrait to classify the critical points as stable or unstable. If possible, determine whether the descriptive names: node, saddle point, spiral point, or center, are applicable to the critical points.

9. $\dfrac{d^2x}{dt^2} + \dfrac{dx}{dt} + x + x^3 = 0$

$x(0) = -3,\ x'(0) = 4;$

$x(0) = 0,\ x'(0) = -8$

10. $\dfrac{d^2x}{dt^2} + \dfrac{dx}{dt} + x - x^3 = 0$

$x(0) = 0,\ x'(0) = \frac{3}{2};$

$x(0) = -1,\ x'(0) = 1$

11. In Section 3.6, we saw that an undriven spring/mass system with linear damping was described by the differential equation $x'' + 2\lambda x' + \omega^2 x = 0$, $\lambda > 0$. Such a system is overdamped if $\lambda^2 - \omega^2 > 0$ and underdamped if $\lambda^2 - \omega^2 < 0$. A model of an undriven spring/mass system in which damping is nonlinear is given by (3), which can be written as

$$x'' + 2\lambda x'|x'| + \omega^2 x = 0$$

Use a computer graphing program to investigate this nonlinear model by means of phase-plane analysis for selected values of λ and ω to determine whether there exist overdamped and underdamped systems.

12. The model $mx'' + kx + k_1x^3 = F_0 \cos \omega t$ of an undamped, periodically driven spring/mass system is called **Duffing's**

differential equation. (Note that the equation is not autonomous.)

(a) Consider the initial-value problem

$$x'' + x + k_1x^3 = 5\cos t, \qquad x(0) = 1, \quad x'(0) = 0$$

Use an ODE solver to investigate the behavior of the system for values of $k_1 > 0$ ranging from $k_1 = 0.01$ to $k_1 = 100$.

(b) Find values of $k_1 < 0$ for which the system in part (a) is oscillatory.

(c) Consider the initial-value problem

$$x'' + x + k_1x^3 = \cos\frac{3}{2}t, \qquad x(0) = 0, \quad x'(0) = 0$$

Find the values of $k_1 < 0$ for which the system is oscillatory.

Miscellaneous Problems

In Problems 13 and 14, the given nonlinear second-order differential equation is of the form $F(t, y', y'') = 0$. Use the substitution $u = y'$ to find a two-parameter family of solutions of the equation.

13. $y'' = 2t(y')^2$

14. $t^2y'' + (y')^2 = 0$

In Problems 15 and 16, the given nonlinear second-order differential equation is of the form $F(y, y', y'') = 0$. Use the substitution $u = y'$ to find a two-parameter family of solutions of the equation. You will also need to use the Chain Rule to compute the second derivative:

$$y'' = \frac{du}{dt} = \frac{du}{dy}\frac{dy}{dt} = u\frac{du}{dy}$$

15. $yy'' = (y')^2$

16. $y'' + 2y(y')^3 = 0$

17. Suppose a suspended wire hangs under its own weight. As Figure 3.64 shows, a physical model for this could be a long telephone wire strung between two posts. A mathematical model that describes the shape that the hanging wire assumes is given by

$$\frac{d^2y}{dx^2} = \frac{w}{T}\sqrt{1 + \left(\frac{dy}{dx}\right)^2}$$

where T and w are positive constants. Solve this differential equation subject to $y(0) = T/w$, $y'(0) = 0$.

18. A mathematical model for the position $x(t)$ of a body moving rectilinearly on the x-axis in an inverse-square force field is given by

$$\frac{d^2x}{dt^2} = -\frac{k^2}{x^2}$$

Suppose the body starts from rest at $t = 0$ from the position $x = x_0$, $x_0 > 0$. Show that the velocity v of the body at time t is given by $v^2/2 = k^2(1/x - 1/x_0)$. Use the last equation and a CAS to carry out the integration to express time t in terms of x.

Discussion Problems

19. We saw that $\sin t$, $\cos t$, e^t, and e^{-t} are four solutions of the nonlinear equation $(y'')^2 - y^2 = 0$. Discuss: How can these explicit solutions be found using knowledge about linear equations? Without attempting to verify, explain why the two special linear combinations $y = c_1e^t + c_2e^{-t}$ and $y = c_3\cos t + c_4\sin t$ must satisfy the nonlinear differential equation.

20. Discuss: Why is the damping term in equation (3) written as

$$\beta\left|\frac{dx}{dt}\right|\frac{dx}{dt} \quad \text{instead of} \quad \beta\left(\frac{dx}{dt}\right)^2$$

21. A person skis down an incline as shown in Figure 3.65. What two obvious resistive forces are acting on the skier? Devise a *nonlinear* mathematical model for the distance s traversed down the slope in time t.

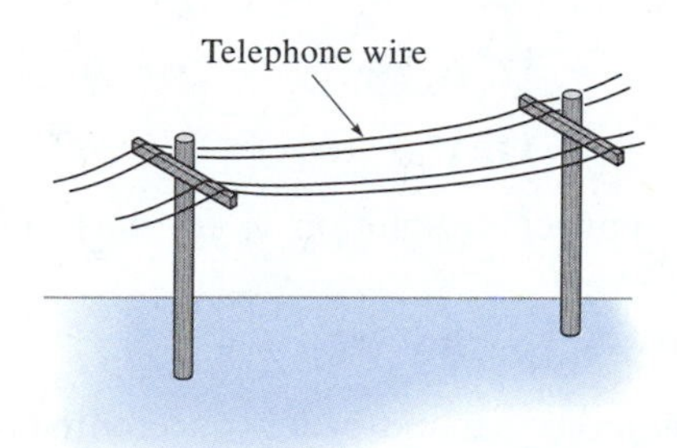

Figure 3.64.

Figure 3.65.

Chapter 3 in Review

Answers to this review are posted at our Web site diffeq.brookscole.com/books.html#zill 98

In Problems 1–4, answer true or false.

1. If y_1 is a solution of $y'' - y = 1$ and y_2 is a solution of $y'' - y = t - 1$, then $y_1 + y_2$ is a solution of $y'' - y = t$.

2. Two functions $f_1(t)$ and $f_2(t)$ are linearly independent on an interval I if one is not a constant multiple of the other.

3. A constant multiple of a solution of a linear differential equation is also a solution.

4. The boundary-value problem $y'' + y = 0$, $y(0) = 0$, $y(\pi/2) = 0$ has no solution.

In Problems 5–8, solve the given differential equation.

5. $\dfrac{d^3y}{dt^3} + 2\dfrac{dy}{dt} - 12y = 0$

6. $\dfrac{d^2r}{d\theta^2} - 2\dfrac{dr}{d\theta} + 26r = 1$

7. $ty'' - 2y' = t$

8. $2u^3\dfrac{d^3y}{du^3} + u^2\dfrac{d^2y}{du^2} + 7u\dfrac{dy}{du} - 4y = 0$

9. The auxiliary equation of a homogeneous linear ninth-order DE with constant coefficients has the roots 3, 3, 3, 0, 0, $1 \pm i$, $1 \pm i$. Write down the general solution of the differential equation.

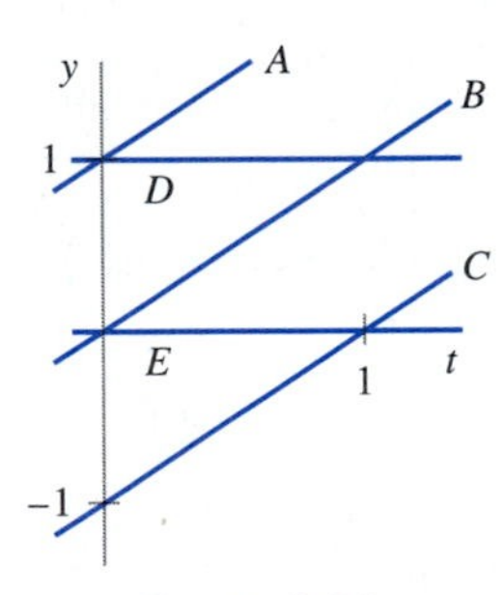

Figure 3.66.

10. Match a solution of the boundary-value problem consisting of the differential equation $y'' = 0$ and the given boundary conditions with one of the curves in Figure 3.66.

(a) $y(0) = 0$, $y(1) = 0$

(b) $y'(0) = 0$, $y'(1) = 0$

(c) $y(0) = 0$, $y'(1) = 1$

(d) $y'(0) = 1$, $y'(1) = 1$

(e) $y(0) = 1$, $y'(1) = 1$

(f) $y(0) = 1$, $y(1) = 1$

(g) $y'(0) = 1$, $y(1) = 0$

(h) $y'(0) = 0$, $y(1) = 0$

In Problems 11–14, match each of the given differential equations with the appropriate words: center, node, saddle point, spiral point, asymptotically stable, stable, unstable.

11. $y'' + 4y' + 20y = 0$

12. $6y'' + 7y' + 2y = 0$

13. $3y'' - 5y = 0$

14. $y'' + 100y = 0$

15. Write down the form of a particular solution y_p for each differential equation. Do not solve.

(a) $y'' + 9y = 10\cos 3t + 5t^2$

(b) $y'' + y = \sin^4 t$

16. Construct a linear autonomous second-order differential equation $F(x, x', x'') = 0$ with a critical point at $(2, 0)$ such that the critical point is a center.

17. Consider a free undamped spring/mass system whose mathematical model is $x'' + \omega^2 x = 0$. The amplitude A of the motion of the mass depends on its initial displacement and initial velocity. Determine A in the following cases:

(a) $x(0) = x_0,\ x'(0) = 0$ **(b)** $x(0) = 0,\ x'(0) = x_1$
(c) $x(0) = x_0,\ x'(0) = x_1$

18. Which of the three graphs in Figure 3.67 does *not* portray a critically damped spring/mass system? Briefly explain your reasoning.

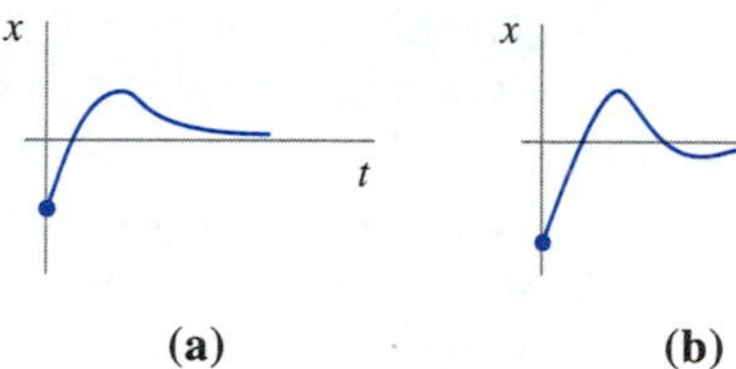

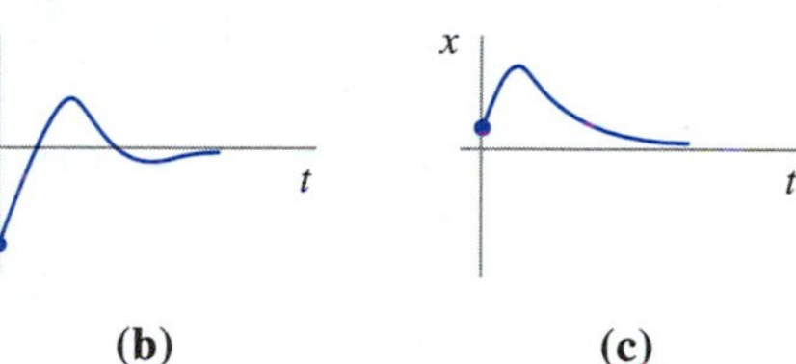

(a) (b) (c)

Figure 3.67.

19. A model for a free damped spring/mass system is given by the differential equation $x'' + 2\lambda x' + 2x = 0$, $\lambda > 0$. Find the values of λ for which the system is overdamped, critically damped, and underdamped.

20. A cylindrical barrel s ft in diameter of weight w lb is floating in water. After an initial depression, the barrel exhibits an up-and-down bobbing motion along a vertical line. Using Figure 3.68, construct a mathematical model for the vertical displacements $y(t)$ if the origin is taken to be on the vertical axis at the surface of the water when the barrel is at rest. Use Archimedes' principle that the buoyancy, or upward force of the water on the barrel, is equal to the weight of the water displaced, and the fact that the density of water is 62.4 lb/ft^3. Assume that the downward direction is positive and that the resistance of the water can be ignored.

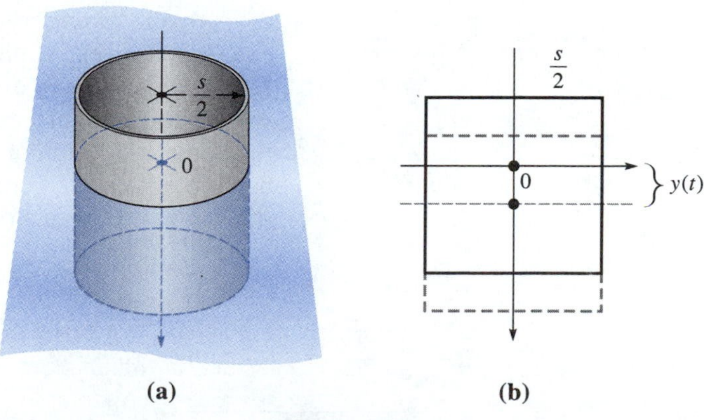

(a) (b)

Figure 3.68.

Projects and Computer Lab Experiments for Chapter 3

Section 3.2

1. **Euler's Formula** Here are two ways of motivating Euler's formula given on page 133.
   (a) Substitute $x = i\theta$ in the Maclaurin series

$$e^x = \sum_{k=0}^{\infty} \frac{x^k}{k!}$$

   and then use $i^2 = -1$, $i^3 = -i$, and so on. Examine what happens when you separate the series into real and imaginary parts.
   (b) Since the roots of the auxiliary equation $m^2 + 1 = 0$ are $\pm i$, suppose the general solution of $y'' + y = 0$ is defined by $y = c_1 e^{it} + c_2 e^{-it}$. Now verify that $\sin t$ is a particular solution of $y'' + y = 0$, $y(0) = 0$, $y'(0) = 1$, and that $\cos t$ is a particular solution of $y'' + y = 0$, $y(0) = 1$, $y'(0) = 0$. Apply the general solution to these two initial-value problems. Finally, use the basic principle that any particular solution must be obtainable from the general solution.

2. **Now It Is Reasonable**
   (a) In the classic text *Differential Equations*, by Ralph Palmer Agnew,* the following statement is made:

   It is not reasonable to expect students in this course to have computing skill and equipment necessary for efficient solving of equations such as

$$4.317\frac{d^4y}{dx^4} + 2.179\frac{d^3y}{dx^3} + 1.416\frac{d^2y}{dx^2} + 1.295\frac{dy}{dx} + 3.169y = 0$$

   Although it is debatable whether computing skills have improved in the intervening years since that text was written, it is a certainty that technology has. If one has access to a computer algebra system, then solving the given DE can now be considered a reasonable task. Use a CAS and an application such as **DSolve** in *Mathematica* or **dsolve** in *Maple* to find the general solution of the differential equation above. Simplify the output.
   (b) At your instructor's direction, use a computer either as an aid in solving the auxiliary equation or as a means of directly obtaining the general solution of the given differential equation. If a CAS is used to obtain the general solution, simplify the output and write the solution in terms of real functions.

$$y''' - 6y'' + 2y' + y = 0$$
$$6.11y''' + 8.59y'' + 7.93y' + 0.778y = 0$$
$$3.15y^{(4)} - 5.34y'' + 6.33y' - 2.03y = 0$$
$$y^{(4)} + 2y'' - y' + 2y = 0$$

*See Problem 38 in *Projects and Computer Lab Experiments for Chapter 2.*

(c) Use a CAS or graphic calculator to solve the auxiliary equation of the given differential equation. By hand, write down the general solution of the DE.

$$y^{(4)} - 2y''' - 6y'' + 16y' - 8y = 0$$
$$2y^{(5)} - y^{(4)} - 4y''' + 3y'' - 8y' - 12y = 0$$

Section 3.4

3. A Race If the method of variation of parameters always "works" whenever the associated homogeneous differential equation can be solved, then why bother with the method of undetermined coefficients? The point of this problem is to conduct a race between three teams to see which team is fastest in finding the general solution of the nonhomogeneous differential equation

$$2y'' + 2y' + 3y = t^2$$

(a) Team A is to solve the equation by hand using the method of undetermined coefficients.

(b) Team B is to use a computer algebra system and the method of variation of parameters. The team is to use the CAS to find roots of the auxiliary equation, but must form the complementary function y_c by hand. Next, the CAS must be used for differentiation, computation of the Wronskian, the integration of u_1' and u_2', simplification, and finally, to form the sum $y_p = u_1y_1 + u_2y_2$.

(c) Team C is also to use a CAS, but this time the idea is to try to solve the equation directly using functions such as **DSolve** and **dsolve** in *Mathematica* and *Maple*, respectively. Team C must also use the CAS to do all simplifications and must use the necessary commands or procedures to produce a *real solution output*.

4. Green's Function

(a) Suppose y_1 and y_2 form a fundamental set of solutions of $y'' + P(t)y' + Q(t)y = 0$. Show that a particular solution of $y'' + P(t)y' + Q(t)y = f(t)$ on the interval $[t_0, t]$ is

$$y_p(t) = \int_{t_0}^{t} G(t, u)f(u)\,du, \qquad \text{where} \qquad G(t, u) = \frac{\begin{vmatrix} y_1(u) & y_2(u) \\ y_1(t) & y_2(t) \end{vmatrix}}{W(u)}$$

The function $G(t, u)$ is called a **Green's function** for the nonhomogeneous equation. In this case, $W(u)$ denotes the Wronskian of $y_1(u)$ and $y_2(u)$.

(b) Verify that $y_p(t)$, defined in part (a), satisfies $y_p(t_0) = 0$ and $y_p'(t_0) = 0$.

(c) Use a Green's function to solve the initial-value problem

$$y'' + \omega^2 y = \sin \omega t, \quad y(0) = 0, \; y'(0) = 0$$

(d) Show how the particular solution obtained in part (c) can be used to solve the initial-value problem

$$y'' + \omega^2 y = \sin \omega t, \quad y(0) = 1,\, y'(0) = -1$$

[*Hint*: First consider the problem $y'' + \omega^2 y = 0$, $y(0) = 1$, $y'(0) = -1$, then think about what "superposition" means.]

Section 3.5

5. **Singular Point** The value of the lead coefficient $a_n t^n$ of any Cauchy-Euler equation is zero at $t = 0$. We say that 0 is a **singular point** of the differential equation. As remarked at the end of Section 2.3, a singular point of a linear differential equation may be troublesome in that the solutions of the equation *could* be unbounded or exhibit other peculiar behavior near the point. Discuss the nature of the pairs of roots m_1 and m_2 of the auxiliary equation of (1) in each of the three cases: distinct real (for example, m_1 and m_2 can both be positive), repeated real, and conjugate complex. Make up corresponding solutions and use a graphic calculator or graphing software to graph these solutions. Discuss the behavior of these solutions as $t \to 0^+$.

Section 3.6

6. **Overdamped Motion** The initial-value problem

$$x'' + 3x' + 2x = 0, \quad x(0) = x_0,\, x'(0) = x_1$$

is a model for an overdamped spring/mass system.

(a) Let $x_0 = 3$. Use an ODE solver to determine experimentally a range of values for the initial velocity x_1 such that the mass passes through the equilibrium position.

(b) Repeat part (a) for $x_0 = 2$ and $x_0 = 1$.

(c) For any $x_0 > 0$, use parts (a) and (b) to conjecture a range of values for x_1 such that the mass passes through the equilibrium position. Then prove your assertion analytically.

7. **Beats** When the frequency of a periodic impressed force is exactly the same as the frequency of free undamped vibration, then a spring/mass system is in a state of pure resonance. In this state, the displacements of the mass grow without bound as $t \to \infty$. But when the frequency $\gamma/2\pi$ of a periodic driving function is close to the frequency $\omega/2\pi$ of free vibrations, the mass undergoes complicated but bounded oscillations known as **beats**.

(a) To see this phenomenon, consider the initial-value problem

$$\frac{d^2x}{dt^2} + \omega^2 x = F_0 \cos \gamma t, \quad x(0) = 0, x'(0) = 0$$

Suppose $\omega = 2$ and $F_0 = \frac{1}{2}$. Use an ODE solver to graph, on separate coordinate axes, the solution curves corresponding to $\gamma = 1$,

$\gamma = 1.5$, $\gamma = 1.75$ and $\gamma = 1.9$. Use the intervals $0 \leq t < 60$, $-3 \leq x \leq 3$.

(b) Use undetermined coefficients to show that the solution of the general initial-value problem in part (a) is

$$x(t) = \frac{F_0}{\omega^2 - \gamma^2}(\cos \gamma t - \cos \omega t), \quad \gamma \neq \omega$$

(c) Use a trigonometric identity to show that the solution in part (b) can be written as the product

$$x(t) = \frac{2F_0}{\gamma^2 - \omega^2} \sin \tfrac{1}{2}(\gamma - \omega)t \ \sin \tfrac{1}{2}(\gamma + \omega)t$$

(d) If $\epsilon = \frac{1}{2}(\gamma - \omega)$, show that when ϵ is small, the solution in part (c) is approximately

$$x(t) = \frac{F_0}{2\epsilon\gamma} \sin \epsilon t \sin \gamma t$$

Use a graphing utility to graph this function with $\omega = 2$, $F_0 = \frac{1}{2}$, and $\gamma = 1.75$. Then graph $\pm(F_0/2\epsilon\gamma) \sin \epsilon t$ on the same set of coordinate axes. Use color or a different shading so that the curves can be distinguished. The graphs of $\pm(F_0/2\epsilon\gamma) \sin \epsilon t$ are called an *envelope* for the graph of $x(t)$; the function $(F_0/2\epsilon\gamma) \sin \epsilon t$ is called a *time-varying amplitude* for $\sin \gamma t$. Compare the graph of $x(t)$ obtained in this manner with the third graph in part (a).

(e) Write the solution in part (b) as a superposition of two functions: $x(t) = x_1(t) + x_2(t)$. Graph $x_1(t)$ and $x_2(t)$ with $\omega = 2$, $F_0 = \frac{1}{2}$, and $\gamma = 1.75$, on the same set of coordinate axes for $0 \leq t < 60$. Use color or a different shading so that the curves can be distinguished. Mark the times when the graphs of $x_1(t)$ and $x_2(t)$ coincide to the least and to the greatest extent; these times are called *nodes* and *antinodes*, respectively. Compare these graphs with the third graph obtained in part (a); on the latter graph mark the nodes, antinodes, and one beat, that is, the oscillations between two nodes.

(f) Consult a physics text and describe the phenomenon of beats in the context of the theory of sound. If possible, demonstrate beats using a tuning fork.

8. Beats Again Can there be beats oscillations when a damping force is added to the model in part (a) of Problem 7? Defend your position with graphs obtained either from the explicit solution of the problem

$$\frac{d^2x}{dt^2} + 2\lambda\frac{dx}{dt} + \omega^2 x = F_0 \cos \gamma t, \qquad x(0) = 0, \quad x'(0) = 0$$

or from solution curves obtained using an ODE solver.

9. Piecewise-Defined Forcing Function The only solution of the initial-value problem

$$\frac{d^2x}{dt^2} + \omega^2 x = F(t), \qquad x(0) = 0, \quad x'(0) = 0$$

when $F(t) = 0$, $t > 0$, is $x(t) = 0$. This makes physical sense since the mass has no external force driving it, and is initially at rest in the equilibrium position. We now wish to solve the same problem when the forcing function is a constant, acting over a short interval of time, say,

$$F(t) = \begin{cases} F_0/\epsilon, & 0 \le t \le \epsilon \\ 0, & t > \epsilon \end{cases}$$

and see whether we get $x(t) = 0$ as $\epsilon \to 0$.

(a) Use the concept of a Green's function given in Problem 4 to solve the initial-value problem with this piecewise-defined forcing function. Before integrating, use the trigonometric identity for $\sin \omega(t - u)$. Denote this solution as $x(t, \epsilon)$.

(b) Suppose $\omega = 1$ and $F_0 = 1$. Use a graphing utility to obtain the graph of the solution $x(t, \epsilon)$ obtained in part (a) for $\epsilon = 0.5$, $\epsilon = 0.25$, $\epsilon = 0.1$, and $\epsilon = 0.01$. Use a different set of coordinate axes for each value of ϵ so that the curves can be distinguished. Conjecture the value of $x(t) = \lim_{\epsilon \to 0} x(t, \epsilon)$.

(c) Finally, compute $x(t) = \lim_{\epsilon \to 0} x(t, \epsilon)$ by L'Hôpital's rule. Try to explain your answer in physical terms.

10. Pure Resonance Consider a driven undamped spring/mass system described by the initial-value problem

$$\frac{d^2x}{dt^2} + \omega^2 x = F_0 \sin^n \gamma t, \qquad x(0) = 0, \quad x'(0) = 0$$

(a) Demonstrate for $n = 2$ that there is a single frequency $\gamma_1/2\pi$ at which the system is in pure resonance.

(b) Demonstrate for $n = 3$ that there are two frequencies $\gamma_1/2\pi$ and $\gamma_2/2\pi$ at which the system is in pure resonance.

(c) Suppose $\omega = 1$ and $F_0 = 1$. Use an ODE solver to obtain the graph of the solution of the initial-value problem for $n = 2$ and $\gamma = \gamma_1$ in part (a). Obtain the graph of the solution of the initial-value problem for $n = 3$ corresponding, in turn, to $\gamma = \gamma_1$ and $\gamma = \gamma_2$ in part (b).

11. Resonance Curve

(a) Show that the general solution of

$$\frac{d^2x}{dt^2} + 2\lambda\frac{dx}{dt} + \omega^2 x = F_0 \sin \gamma t$$

is $x(t) = Ae^{-\lambda t}\sin(\sqrt{\omega^2 - \lambda^2}\,t + \phi) + \dfrac{F_0}{\sqrt{(\omega^2 - \gamma^2)^2 + 4\lambda^2\gamma^2}}\sin(\gamma t + \theta)$

where $A = \sqrt{c_1^2 + c_2^2}$ and the phase angles ϕ and θ are, respectively, defined by $\sin\phi = c_1/A$, $\cos\phi = c_2/A$ and

$$\sin\theta = \frac{-2\lambda\gamma}{\sqrt{(\omega^2-\gamma^2)^2 + 4\lambda^2\gamma^2}}, \qquad \cos\theta = \frac{\omega^2-\gamma^2}{\sqrt{(\omega^2-\gamma^2)^2 + 4\lambda^2\gamma^2}}$$

(b) The solution in part (a) has the form $x(t) = x_c(t) + x_p(t)$. Inspection shows that $x_c(t)$ is transient and hence, for large values of time, the solution is approximated by $x_p(t) = g(\gamma)\sin(\gamma t + \theta)$, where

$$g(\gamma) = \frac{F_0}{\sqrt{(\omega^2-\gamma^2)^2 + 4\lambda^2\gamma^2}}$$

Although the amplitude $g(\gamma)$ of $x_p(t)$ is bounded as $t \to \infty$, show that the maximum oscillations will occur at the value $\gamma_1 = \sqrt{\omega^2 - 2\lambda^2}$. What is the maximum value of g? The number $\sqrt{\omega^2 - 2\gamma^2}/2\pi$ is said to be the **resonance frequency** of the system.

(c) When $F_0 = 2$, $m = 1$, and $k = 4$, the function g becomes

$$g(\gamma) = \frac{2}{\sqrt{(4-\gamma^2)^2 + \beta^2\gamma^2}}$$

Construct a table of the values of γ_1 and $g(\gamma_1)$ corresponding to the damping coefficients $\beta = 2$, $\beta = 1$, $\beta = \frac{3}{4}$, $\beta = \frac{1}{2}$, and $\beta = \frac{1}{4}$. Use a computer graphing program to obtain the graphs of g corresponding to these damping coefficients. Use the same set of coordinate axes. This family of graphs is called the **resonance curve** or **frequency response curve** of the system. What value is γ_1 approaching as $\beta \to 0$? What is happening to the resonance curve as $\beta \to 0$?

12. Bungee Jumping In bungee jumping, it is obviously important to know in advance how far an elastic cord of unstretched length L will stretch for a given weight attached to its end. In one model for this elongation, the cord is simply considered to be a weak spring with constant k and the person tied to the end of the cord is a point mass m. Air resistance and any pendulum motion of the cord/mass system are ignored. Suppose a person jumps from a platform, suspended over water as shown in Figure 3.69, and falls freely until the entire slack cord is extended to length L; denote this point $x = 0$. After the person passes $x = 0$, the cord is stretched an amount $x(t)$.

(a) Devise a mathematical model for $x(t)$ defined only on the first interval of time $0 \le t \le T$ for which $x \ge 0$. Solve for $x(t)$ and then use (6) of Section 3.6 to determine the maximum elongation x_{max}.

(b) Suppose that $L = 30$ ft, the person weighs 150 lb, and that $k = 12$ lb/ft. How far above the level of the water should the platform be raised so that the person will clear the water before rebounding upward for the first time? Ignore the height of the person.

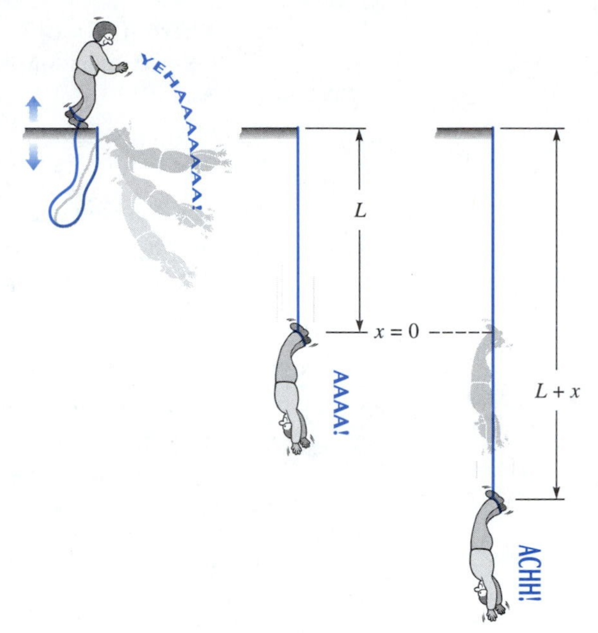

Figure 3.69.

(c) If T is the time indicated in part (a), then $x(T) = x_{\max}$. Use the data in part (b) to find T and the total time it takes the jumper to attain $L + x_{\max}$.

(d) Write a paragraph discussing the motion during the first rebound. Specifically address the role that the bungee cord plays during a rebound. (See also Problem 31.)

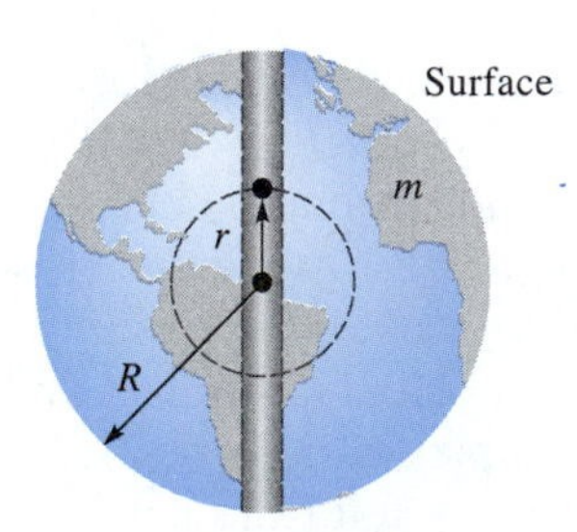

Figure 3.70.

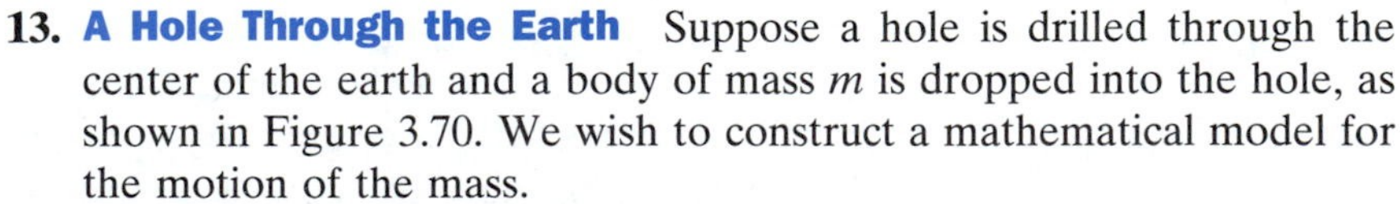

13. A Hole Through the Earth Suppose a hole is drilled through the center of the earth and a body of mass m is dropped into the hole, as shown in Figure 3.70. We wish to construct a mathematical model for the motion of the mass.

(a) Let the distance from the center of the earth to the mass at any time t be denoted by r. Let M denote the mass of the earth and M_r denote the mass of that portion of the earth indicated in Figure 3.70 within a sphere of radius r. The gravitational force of m is $F = -kM_r m/r^2$, where the minus sign indicates that the force is one of attraction. Use this fact to show that $F = -k\dfrac{mM}{R^3}r$.

[*Hint*: Assume that the earth is homogeneous, that is, has constant density δ. Use mass = density × volume.]

(b) Use Newton's second law together with the result in part (a) to arrive at a mathematical model for the motion of the mass m. Discuss the motion.

14. Blowing in the Wind In September 1989, Hurricane Hugo hammered the coast of South Carolina, at times with winds estimated to be as high as 60.4 m/s (135 mi/hr). Of the billions of dollars in damage, approximately $420 million of this was due to the market value of loblolly pine (*Pinus tacda*) lumber in the Francis Marion National Forest. One image from that storm remains hauntingly bizarre: all through the forest and surrounding region, thousands upon thousands of pine trees lay pointing exactly in the same direction, and all the trees were broken 5–8 m from their base. In September 1996, Hurricane Fran destroyed over 8.2 million acres of timber forest in eastern North Carolina. As had happened seven years earlier, the planted loblolly trees all broke at approximately the same height. This seems to be a reproducible phenomenon, brought on by the fact that the trees in these planted forests are approximately the same age and size.

In this problem, we are going to examine a mathematical model for the bending of loblolly pines in strong winds, and then use the model to predict the height at which a tree will break in hurricane-force winds.*

Wind hitting the branches of a tree transmits a force to the trunk of the tree. The trunk is approximately a big cylindrical beam of length L, and so we will model the deflection $y(x)$ of the tree with the static beam equation $EIy^{(4)} = w(x)$ (see (4) of Section 3.7), where x is distance measured in meters from ground level. Since the tree is rooted in the earth, the accompanying boundary conditions are those of a cantilevered beam: $y(0) = 0$, $y'(0) = 0$ at the rooted end, and $y''(L) = 0$, $y'''(L) = 0$ at the free end, which is the top of the tree.

(a) Loblolly pines in the forest have the majority of their crown (that is, branches and needles) in the upper 50% of their length, so let's ignore the force of the wind on the lower portion of the tree. Furthermore, let's assume that the wind hitting the tree's crown results in a uniform load per unit length w_0. In other words, the load on the tree is modeled by

$$w(x) = \begin{cases} 0, & 0 \le x < L/2 \\ w_0, & L/2 \le x \le L \end{cases}$$

We can determine $y(x)$ by integrating both sides of $EIy^{(4)} = w(x)$. Integrate $w(x)$ on $0 \le x < L/2$ and then on $L/2 \le x \le L$ to find an expression for $EIy'''(x)$ on each of these intervals. Let c_1 be the constant of integration on $[0, L/2)$ and c_2 be the constant of integration on $[L/2, L]$. Apply the boundary condition $y'''(L) = 0$ and solve for c_2. Then find the value of c_1 that ensures continuity of the third derivative y''' at the point $x = L/2$.

*For further information about the bending of trees in high winds, see the articles by W. Kubinec, *Phys. Teacher*, March, 1990, or F. Mergen, *J. Forest.* 52(2), 1954.

(b) Following the same procedure as in part (a), show that

$$EIy'(x) = \begin{cases} \frac{w_0}{8}(-2Lx^2 + 3L^2x), & 0 \leq x < L/2 \\ \frac{w_0}{48}(8x^3 - 24Lx^2 + 24L^2x - L^3), & L/2 \leq x \leq L \end{cases}$$

Integrate EIy' to obtain the deflection $y(x)$.

(c) Note that in our model, $y(L)$ describes the maximum amount by which the loblolly will bend. Compute this quantity in terms of the problem's parameters.

(d) By making some assumptions about the density of the crown's foliage, the total force F on the tree can be calculated using a formula from physics: $F = \rho A v^2/6$, where $\rho \approx 1.225$ kg/m^3 is the density of air, v is the wind speed in m/s, and A is the cross-sectional area of the tree's crown. If we assume that the crown is roughly cylindrical, then its cross section is a rectangle of area $A = (2R)(L/2) = RL$, where R is the average radius of the cylinder. The total force *per unit length* is then $w_0 = F/(L/2) = 0.408Rv^2$.

The cross-sectional moment of inertia for a uniform cylindrical beam is $I = \frac{1}{4}\pi r^4$, where r is the radius of the cylinder (tree trunk).

- By your answer to part (c), and the explanations above, the amount that a loblolly will bend depends on each of the parameters in the following table.

Symbol	Description	Typical Values
r	Radius of trunk	0.15–0.25 m
R	Radius of crown	3–4 m
L	Height of pine	15–20 m
E	Modulus of elasticity	11–14 kg/m^2
v	Hugo wind speeds	40–60 m/s (90–135 mi/hr)

  Mathematically show how each parameter affects the bending, and explain in physical terms why this makes sense. (For example, a large value of E results in less bending since E appears in the denominator of $y(L)$. This means that hard wood such as oak bends less than soft wood such as palm.)
- Graph $y(x)$ for 40 m/s winds for an "average" tree by choosing average values of each parameter (for example, $r = 0.2$, $R = 3.5$, and so on).
- Graph $y(x)$ for 60 m/s winds for a "tall" (but otherwise average) pine.
- Recall, in the derivation of the beam equation (see page 191) it was assumed that the deflection of the beam was small. What is

the *largest* possible value of $y(L)$ that is predicted by the model if all parameters are chosen from the given table? Is this prediction realistic, or is the mathematical model no longer valid for parameters in this range?

(e) The beam equation always predicts that a beam will bend, even if the load and flexural rigidity reflect an elephant standing on a toothpick! Different methods are used by engineers to predict when and where a beam will break. In particular, a beam subject to a load will break at the location where the stress function $y''(x)/I(x)$ reaches a maximum.

- Differentiate the function in part (b) to obtain EIy'', and use this to obtain the stress $y''(x)/I(x)$.
- Real pine tree trunks are not uniformly wide, but taper as they approach the top of the tree. Substitute $r(x) = 0.2 - x/(15L)$ into the equation for I and then use a graphing utility to graph the resulting stress as a function of height for an average loblolly. Where does the maximum stress occur? Does this location depend on the speed of the wind? On the radius of the crown? On the height of the pine? Compare the model to observed data from Hurricane Hugo.
- A mathematical model is *sensitive* to an assumption if small changes in the assumption lead to widely different predictions for the model. Repeat the previous question using $r(x) = 0.2 - x/(20L)$ and $r(x) = 0.2 - x/(10L)$ as formulas that describe the radius of a pine tree trunk that tapers. Is our model sensitive to our choice for these formulas?

Section 3.7

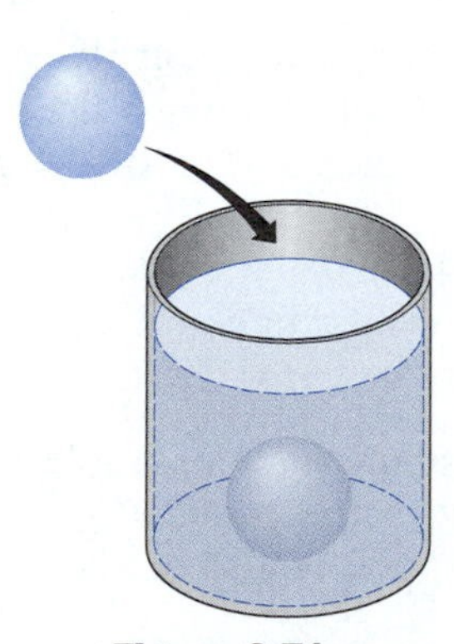

Figure 3.71.

15. Temperature in an Immersed Sphere A uniform solid sphere of radius $r = 1$ at a constant initial temperature throughout is dropped into a large container of fluid that is also kept at a constant, but higher, temperature for all time. See Figure 3.71. To find the temperature $u(r, t)$ inside the sphere for time $t > 0$, it is necessary to

$$\textit{Solve:} \quad r\frac{d^2R}{dr^2} + 2\frac{dR}{dr} + \lambda^2 rR = 0, \quad 0 < r < 1,$$

$$\textit{Subject to:} \quad R'(1) = -hR(1), \quad h > 0$$

This BVP determines the radial component of the temperature $u(r, t)$.

(a) Solve the differential equation by means of the substitution $R(r) = y(r)/r$.

(b) Do some extra reading to supply a physical interpretation for the given boundary condition.

(c) Use a calculator or CAS to find approximations for the first three positive eigenvalues where $h = 0.2$. But first supply a physically plausible but implicit condition that can be used to simplify the general solution of the differential equation.

Section 3.9

16. Finite Difference Equations There are ways of approximating the solution of a boundary-value problem. In this and the next two problems, we consider two such methods.

(a) Do some outside reading and find the definitions of: **forward difference**, **backward difference**, and **central differences**. Show how the first derivative $y'(t)$ can be approximated by **difference quotients** involving forward, backward, and central differences. Show how the second derivative $y''(t)$ can be approximated by a difference quotient involving a central difference.

(b) Consider the boundary-value problem

$$y'' + P(t)y' + Q(t)y = f(t), \quad y(a) = \alpha, \; y(b) = \beta \qquad \textbf{(1)}$$

and suppose $a = t_0 < t_1 < t_2 < \ldots < t_{n-1} < t_n = b$ is a regular partition of the interval $[a, b]$. For $h = (b - a)/n$, the points $t_i = a + ih$, $i = 1, 2, \ldots, n - 1$, are called interior mesh points of the interval. Let $y_i = y(t_i)$, $P_i = P(t_i)$, $Q_i = Q(t_i)$, $f_i = f(t_i)$, $y_0 = y(a) = \alpha$, and $y_n = y(b) = \beta$. Use the information in part (a) to show how the differential equation in (1) can be replaced by the **finite difference equation**

$$\left(1 + \frac{h}{2}P_i\right)y_{i+1} + (-2 + h^2Q_i)y_i + \left(1 - \frac{h}{2}P_i\right)y_{i-1} = h^2 f_i \qquad \textbf{(2)}$$

17. Using a Finite Difference Equation The difference equation (2) enables us to approximate the value of a solution $y(t)$ of (1) at selected points $t_1, t_2, \ldots, t_{n-1}$ within the interval $[a, b]$. The method results in $n - 1$ algebraic equations in $n - 1$ unknowns $y_1, y_2, \ldots, y_{n-1}$, where $y_1 \approx y(t_1)$, and so on. For reasonable values of n, say, 4, 6, or 10, systems of three, five, and nine equations are readily solved using a CAS. Use the difference equation (2) to approximate the solution of the given boundary-value problem.

(a) $y'' - y = t^2$, $y(0) = 0$, $y(1) = 0$; $n = 4$
(b) $y'' - 4y' + 4y = (t + 1)e^{2t}$, $y(0) = 3$, $y(1) = 0$; $n = 6$
(c) $y'' + (1 - t)y' + ty = t$, $y(0) = 0$, $y(1) = 2$; $n = 10$

18. The Shooting Method Another way of approximating a solution of a boundary-value problem $y'' = f(t, y, y')$, $y(a) = \alpha$, $y(b) = \beta$ is called the **shooting method**. The starting point in this method is the replacement of the BVP by an IVP

$$y'' = f(t, y, y'), \quad y(a) = \alpha, \; y'(a) = m_1 \qquad \textbf{(3)}$$

The number m_1 in (3) is simply a *guess* for the unknown slope of the solution curve at the known point $(a, y(a))$. We then apply one of the step-by-step numerical techniques such as the improved Euler method or the Runge-Kutta method to the second-order equation in (3) to find an approximation β_1 for the value of $y(b)$. If β_1 agrees with the given value $y(b) = \beta$ to some preassigned tolerance, we stop; otherwise the calculations are repeated starting with a different guess $y'(a) = m_2$ to obtain a second approximation β_2 for $y(b)$. This method can be continued in a trial-and-error manner, or the subsequent slopes $m_3, m_4, \ldots$, can be adjusted in some systematic way (linear interpolation is particularly successful when the differential equation in (3) is linear). The procedure is analogous to shooting (the "aim" is the choice of the initial slope) at a target until the bull's eye $y(b)$ is hit.

Consider the BVP: $y'' = y + e^y$, $y(0) = 0$, $y(1) = \frac{1}{2}$. Use an ODE solver and make guesses for the initial slope $y'(0)$ until you obtain a solution curve that appears to pass through the required point $(1, \frac{1}{2})$. Use that slope and a numerical method (chosen by your instructor) with step size $h = 0.2$ to estimate the value $y(1)$. Display your data from your two best guesses in the table that follows.

x	y	$u = y'$	x	y	$u = y'$
0.0000	0.0000		0.0000	0.0000	

19. Orbiting Moons A classic model in astrophysics is called the two-body problem. In this problem, we will investigate two versions of the model and the orbits for each.

(a) Suppose a planet has a single moon whose position at time t is (x, y), where the pair of functions $x(t)$ and $y(t)$ is a solution of the nonlinear system

$$x'' + \frac{x}{r^3} = 0$$

$$y'' + \frac{y}{r^3} = 0$$

Here $r = \sqrt{x^2 + y^2}$ is the distance between the moon and the center of the planet whose coordinates are taken to be $(0, 0)$. Solve the system of second-order DEs in the case when the distance between the moon and the planet is a constant; that is, take $r = 1$. Solve the resulting linear system with initial conditions $x(0) = 1$, $x'(0) = 0$, $y(0) = 0$, $y'(0) = 0.75$. Use a graphing utility to plot the curve defined by the parametric equations $x = x(t)$, $y = y(t)$ in the xy-plane. Describe the curve found. Where is the planet relative to the orbit?

(b) In the case when r is nonconstant, we must solve the system numerically. To do this we rewrite the system as a system of four first-order DEs. Show that the substitutions $u_1 = x$, $u_2 = x'$, $u_3 = y$, $u_4 = y'$ lead to the system

$$u_1' = u_2$$

$$u_2' = -\frac{u_1}{(u_1^2 + u_3^2)^{3/2}}$$

$$u_3' = u_4$$

$$u_4' = -\frac{u_3}{(u_1^2 + u_3^2)^{3/2}}$$

Solve the system numerically using the fourth-order Runge-Kutta method with $h = 0.1$ and the initial conditions given in part (a). Plot $x = x(t)$, $y = y(t)$ in the xy-plane. Describe the curve found and compare the results with part (a). Where is the planet relative to the orbit?

20. A Spring/Mass Pendulum Consider a spring/mass system, but instead of pulling the mass down and releasing it, pull it down and slightly to one side before releasing it. The resultant motion would seem to be almost random. But, as we shall see in this problem, a simple resonance condition leads to a remarkable phenomenon. That is, for a correct choice of a parameter in its mathematical model, the mass will continually switch between the spring/mass mode and the pendulum mode.

(a) A model for the spring/mass pendulum system is given by

$$(x + 1)\theta'' + 2x'\theta' + \sin\theta = 0$$

$$x'' - (x + 1)(\theta')^2 + kx - \cos\theta = 0$$

where $x(t)$ is the displacement of the mass from its equilibrium position, $\theta(t)$ is the angle of deflection from the vertical, and k is a parameter. Take $\theta = 0$, $k = 1$, and $x(t) = u(t) + 1$ and express the model in terms of $u(t)$. Use the fourth-order Runge-Kutta method with $h = 0.025$ to solve the resulting differential equation with initial conditions $u(0) = 1$, $u'(0) = 0$, on the interval [0, 100]. Compare with the exact solution.

(b) Use the substitutions $u_1 = \theta$, $u_2 = \theta'$, $u_3 = x$, $u_4 = x'$ to rewrite the model as a system of four first-order DEs. (See part (b) of Problem 19.) Use the fourth-order Runge-Kutta method to approximate a solution of the resulting system using the initial conditions $x(0) = \frac{1}{3}$, $x'(0) = 0$, $\theta(0) = \frac{1}{20}$, and $\theta'(0) = 0$ on the interval [0, 100] with $k = 2$. Plot your data.

(c) Repeat part (b) with $k = 3$. This time notice the resonance effect. The mass switches between the vertically bouncing spring/mass mode and the pendulum mode.

(d) What happens when $k = 3.1$? Is the system still in resonance? (See Problems 10 and 11.)

Section 3.10

21. Separatrices

(a) The spring/mass system described by the nonlinear differential equation $md^2x/dt^2 + kx + k_1x^3 = 0$ is a conservative system (see page 185). Show that a Cartesian equation of the phase-plane trajectories, or energy curves, of this DE is given by $\frac{1}{2}my^2 + \frac{1}{2}kx^2 + \frac{1}{4}k_1x^4 = E$, where E represents the constant energy of the system.

(b) In the case of the soft spring, where $m = 1$, $k = 1$, and $k_1 = -1$, show that

$$y = \pm\sqrt{2E - x^2 + \tfrac{1}{2}x^4}$$

(c) Use the equations in part (b) and a graphing utility to obtain the phase portrait of the differential equation. First, plot the trajectories by choosing values of E satisfying $0 < E < \frac{1}{4}$ and then values satisfying $E > \frac{1}{4}$. Then plot the trajectories corresponding to the energy $E = \frac{1}{4}$. The curves defined by this last value are called **separatrices** (see page 213). Finally, plot trajectories for $E < 0$.

(d) Explain where the four energy states $0 < E < \frac{1}{4}$, $E = \frac{1}{4}$, $E > \frac{1}{4}$, and $E < 0$, come from.

(e) For which of the energy states in part (d) does the mass *not* pass through the equilibrium position? Give an example of a set of initial conditions that corresponds to this state.

(f) Obtain the phase portrait of the DE in part (a) when $m = 1$, $k = 4$, and $k_1 = -4$ by considering four energy states analogous to those in part (d). Give the Cartesian equation that defines the separatrices.

22. Nonlinear Pendulum

(a) Proceed as in part (a) of Problem 21 to obtain a Cartesian equation of the trajectories for the model of the *undamped* nonlinear pendulum

$$\frac{d^2\theta}{dt^2} + \omega^2 \sin\theta = 0$$

where $\omega^2 = 1$. As in (10) of Section 3.10, let $x = \theta(t)$, $y = dx/dt$. Give the Cartesian equation that defines the separatrices. Use a graphing utility to plot a few representative trajectories and the separatrices.

(b) Consider the model of the *damped* nonlinear pendulum

$$\frac{d^2\theta}{dt^2} + 2\lambda\frac{d\theta}{dt} + \omega^2 \sin\theta = 0$$

Find the critical points of this differential equation.

(c) As in (10) of Section 3.10, let $x = \theta(t)$, $y = dx/dt$, and write the DE in part (b) as a system of first-order equations.

(d) Use computer software and the system in part (c) to obtain a phase portrait of the differential equation in part (b) in the underdamped case where $\lambda^2 - \omega^2 < 0$ by selecting $\lambda = \frac{1}{2}$ and $\omega^2 = 1$. Use the phase portrait to classify the critical points as stable or unstable. Use the phase portrait to determine which, if any, of the critical points can be described as a node, saddle point, spiral point, or center.

(e) Define a separatrix in the case of the underdamped nonlinear pendulum in part (d) as a pair of trajectories in the phase plane on which points approach or "enter" a *saddle point* as $t \to \infty$. Use computer software as an aid in locating the approximate graphs of the separatrix trajectories that enter the first saddle point P with a positive x-coordinate. Repeat for the first saddle point Q with a negative x-coordinate. Use a colored pencil to highlight the approximate separatrices and then shade the region R in the plane between these two separatrices. Then explain why the region R (or more generally the region between any two consecutive separatrices) is called a **region of attraction**.

(f) Describe the motion of the pendulum for a typical trajectory inside the region of attraction R in part (e). Describe the motion of the pendulum for a typical trajectory outside the region of attraction R. Describe the motion of the pendulum under the *assumption* that it is possible to prescribe initial conditions that yield a separatrix trajectory entering P.

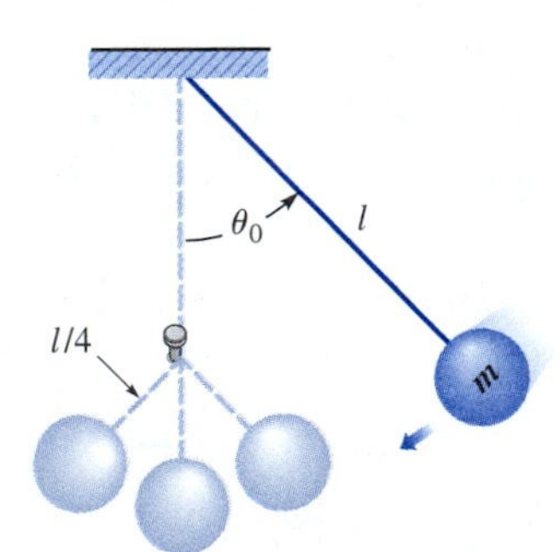

Figure 3.72.

23. Another Pendulum Suppose the massless rod in the discussion of the nonlinear pendulum is actually a string of length l. A mass m is attached to the end of the string and the pendulum is released from rest at a small displacement angle $\theta_0 > 0$. When the pendulum reaches the equilibrium position OP in Figure 3.60, the string hits a nail and gets caught at this point $l/4$ above the mass. The mass oscillates from this new pivot point as shown in Figure 3.72.

(a) Construct and solve a linear initial-value problem that gives the displacement angle, denote it $\theta_1(t)$, for $0 \le t < t_n$, where t_n represents the time when the string first hits the nail.

(b) Find the time t_n in part (a).

(c) Construct and solve a linear initial-value problem that gives the displacement angle, denote it $\theta_2(t)$, for $t \ge t_n$, where t_n is the time in part (a). Compare the amplitude and period of oscillations in this case with that predicted by the initial-value problem in part (a).

(d) Use a graphing utility to obtain the graphs of $\theta_1(t)$ and $\theta_2(t)$ on the same set of coordinate axes. Take $\theta_0 = 0.2$ radians and $l = 2$ ft.

(e) Use an ODE solver to obtain the graphs of the solutions $\theta_1(t)$ and $\theta_2(t)$ of the corresponding nonlinear initial-value problems and compare with part (d). Take $\theta_0 = 0.2$ radians and $l = 2$ ft.

(f) Experiment with values of θ_0 until you discern a noticeable difference between the solutions of the linear and nonlinear initial-value problems.

24. Period of a Nonlinear Pendulum From (9) of Section 3.10, it is seen that the period of small oscillations of a linear simple pendulum is given by $T = 2\pi\sqrt{l/g}$. Note that this period does *not* depend on the initial angular displacement θ_0. In contrast, the period of oscillations described by the initial-value problem for the nonlinear pendulum

$$\frac{d^2\theta}{dt^2} + \frac{g}{l}\sin\theta = 0, \quad \theta(0) = \theta_0, \quad \theta'(0) = 0$$

does depend on θ_0.

(a) While it is not possible to find a close-form solution of the foregoing equation, it is nevertheless possible to find a formula for the period of oscillations. First, show that by multiplying the foregoing differential equation by $2d\theta/dt$ we obtain the first-order equation

$$\left(\frac{d\theta}{dt}\right)^2 = \frac{2g}{l}(\cos\theta - \cos\theta_0)$$

(b) Now use the differential equation in part (a) to show that the period of large oscillations depends on the initial angular displacement θ_0 and is given by

$$T_{\theta_0} = 2\sqrt{\frac{2l}{g}}\int_0^{\theta_0}\frac{d\theta}{\sqrt{\cos\theta - \cos\theta_0}}$$

(c) Use the trigonometric identities

$$\cos\theta = 1 - 2\sin^2\frac{\theta}{2} \quad \text{and} \quad \cos\theta_0 = 1 - 2\sin^2\frac{\theta_0}{2}$$

to show that the integral in part (b) can be written as

$$T_{\theta_0} = 2\sqrt{\frac{2l}{g}}\int_0^{\theta_0}\frac{d\theta}{\sqrt{\sin^2(\theta_0/2) - \sin^2(\theta/2)}}$$

(d) Define $k = \sin(\theta_0/2)$ and $\sin(\theta/2) = k\sin\phi$ to show that

$$T_{\theta_0} = \frac{2T}{\pi}K(k),$$

where $K(k) = \int_0^{\pi/2}\frac{d\phi}{\sqrt{1 - k^2\sin\phi}}$ is called the **complete elliptic integral of the first kind** and T is the period of the linear simple pendulum.

(e) Use the result in part (d) to find k for $\theta_0 = 60°$ and $\theta_0 = 2°$. Use tables or a CAS to evaluate the elliptic integral for $\theta_0 = 60°$ and $\theta_0 = 2°$. Compare the periods of oscillation $T_{60°}$ and $T_{2°}$ of the nonlinear pendulum with the period T of the linear simple pendulum.

(f) Use the difference $(T_{2°} - T)/T$ to compare the time disparity between the nonlinear and linear pendulums over a time interval of one year (365 days).

25. A Pendulum on the Moon

(a) Suppose a nonlinear pendulum, whose oscillations are defined by (8) of Section 3.10, is located on the moon. Use an ODE solver as an aid in determining whether a pendulum of length l will oscillate faster on the earth or on the moon. Use the same initial conditions but choose these initial conditions so that the pendulum oscillates back and forth.

(b) Which pendulum in part (a) has the greater amplitude?

(c) Are the conclusions in parts (a) and (b) the same when the linear model (9) of Section 3.10 is used?

26. Coulomb Friction Consider a spring/mass system in which the mass m slides over a dry horizontal surface whose coefficient of kinetic friction is μ. If the force of kinetic friction has the constant magnitude $f_k = \mu mg$ and acts opposite to the direction of motion, then it is known as **coulomb friction** and the motion is described by the piecewise-defined differential equation

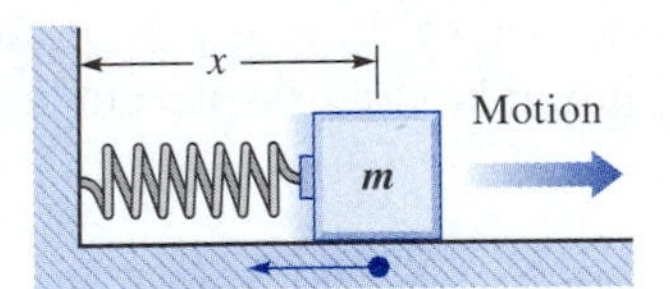

Figure 3.73.

$$mx'' + f_k \operatorname{sgn}(x') + kx = 0 \quad \text{where} \quad \operatorname{sgn}(x') = \begin{cases} 1, & x' > 0 \\ -1, & x' < 0 \end{cases}$$

See Figure 3.73. Although the differential equation consists of two linear equations, the problem is considered nonlinear since the motion cannot be represented by a single equation.

(a) Find Cartesian equations for the phase-plane trajectories in the special case $x'' + \operatorname{sgn}(x') + x = 0$. Find the phase portrait of the differential equation.

(b) Using the results in part (a), trace the motion of the mass using the appropriate trajectories in the phase plane if it is released from rest from the initial position $x(0) = 9$. Indicate each point where the mass comes to rest. Give the position of the mass at these points. Find the position where the motion of mass stops completely.

27. Pursuit Curve In a naval exercise, ship S_1 is pursued by submarine S_2, as shown in Figure 3.74. Ship S_1 departs point $(0, 0)$ at $t = 0$ and proceeds along a straight line course (the y-axis) at a constant speed v_1. The submarine S_2 keeps ship S_1 in visual contact, indicated by the straight dashed line L in the figure, while traveling at a constant speed v_2 along a curve C. Assume that S_2 starts at the point $(a, 0)$, $a > 0$, at $t = 0$ and that L is tangent to C. Determine the nonlinear second-order differential equation that describes the curve C. Find an explicit solution of the differential equation. For convenience, define $r = v_1/v_2$. Determine whether the paths of S_1 and S_2 will ever intersect by considering the cases $r > 1$, $r < 1$, and $r = 1$. [*Hint*: $\dfrac{dt}{dx} = \dfrac{dt}{ds}\dfrac{ds}{dx}$, where s is arc length measured along C.]

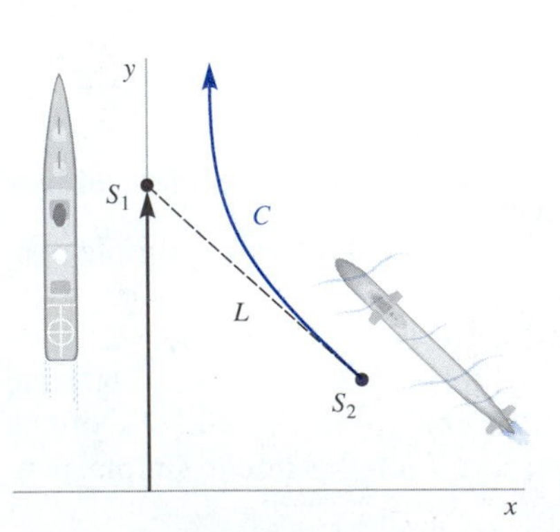

Figure 3.74.

28. An Airplane Delivery As shown in Figure 3.75, a plane flying horizontally at a constant speed v_0 drops a relief food supply pack to persons

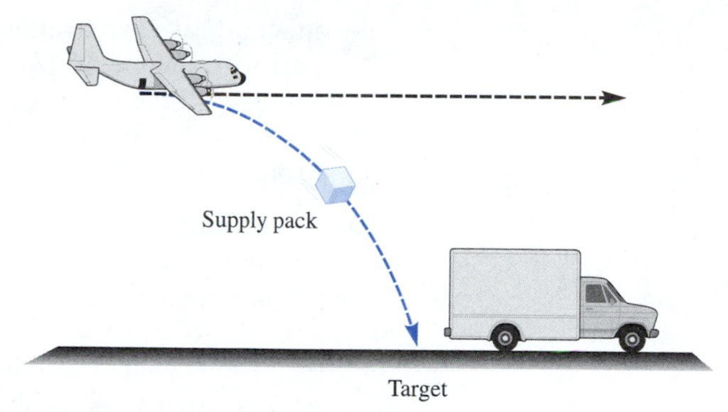

Figure 3.75.

on the ground. Assume that the origin is the point where the supply pack is released and that the positive y-axis points downward.

(a) Determine the system of differential equations and initial conditions that define the trajectory of the supply pack under the assumption that the horizontal and vertical components of the air resistance are proportional to $(dx/dt)^2$ and $(dy/dt)^2$, respectively. [*Hint*: Use vector methods.]

(b) Solve the differential equations in part (a).

(c) Suppose the plane flies at an altitude of 1000 ft and that its constant speed is 300 mi/hr. Assume that the constant of proportionality for air resistance is $k = 0.0053$ and that the supply pack weighs 256 lb. Determine the horizontal distance the pack travels, measured from its point of release to the point where it hits the ground.

Writing Assignments

29. Tacoma Narrows Bridge After half a century the collapse of the Tacoma Narrows bridge continues to be of interest to mathematicians. See Figure 3.76. Elaborate on the early theories that the collapse of the bridge in 1940 was due to a resonance phenomenon. Conclude your report with a discussion of some modern theories (*c.* 1990) that provide

Figure 3.76. (AP/Wide World Photos.)

an alternative explanation for the bridge's structural failure. You might start with the essay, *The Collapse of the Tacoma Narrows Suspension Bridge* by Gilbert N. Lewis in *Differential Equations with Modeling Applications,* Sixth Edition, by Dennis G. Zill (Pacific Grove, CA, Brooks/Cole, 1997).

30. Resonance and Airplane Crashes In late 1959 and early 1960, two commercial airplane crashes occurred that involved the relatively new propjet called the *Electra*. Write a report on how it was discovered that a resonance phenomenon was the probable cause for the crashes. You might start your research with the books *Loud and Clear* by Robert J. Serling (New York; Dell, 1970) and *Blind Trust* by John J. Nance (New York, William Morrow, 1986).

31. Bungee Jumping—Another Model In Problem 12, we assumed that there was on hand one bungee cord and that the dive platform could be raised to accommodate persons of different weights. A variation of this problem is that each person jumps from a fixed height, such as a bridge above a stream as shown in Figure 3.77, and that there are several bungee cords, each of the same length, but each with a different stiffness characterized by the number k. For your particular weight, which bungee cord should you choose so that you do not plunge into the stream? Write or give an oral presentation of this and other related problems found (at this writing) at the NSF sponsored Web-site of the Department of Pure and Applied Mathematics at Washington State University: IDEA (**I**nternet **D**ifferential **E**quations **A**ctivities; http://www.sci.wsu.edu/idea/). Be sure to explain the nonlinear model with a discontinuous restoring force

$$m\frac{d^2x}{dt^2} = mg - \beta\frac{dx}{dt} + F(x) \quad \text{where} \quad F(x) = \begin{cases} 0, & x \le 0 \\ -kx, & x > 0 \end{cases}$$

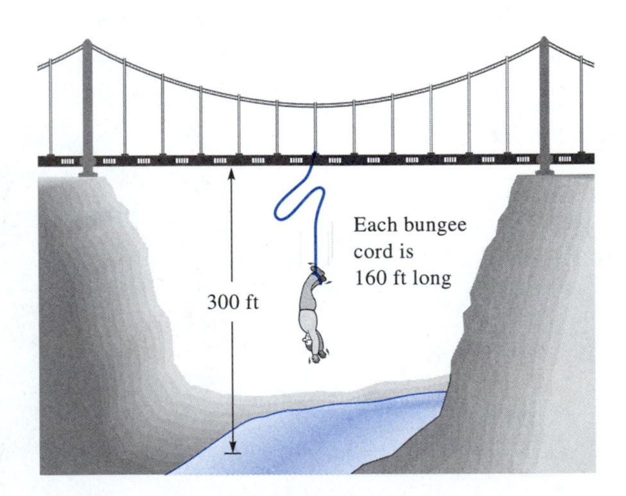

Figure 3.77.

Chapter 4

Systems of First-Order Equations

4.1 Linear Systems

We have already encountered systems of differential equations several times in Chapter 3. In Section 3.8, we solved some systems using differential operators and applying the algebraic principle of systematic elimination, and in Sections 3.3, 3.9, and 3.10 we saw that in order to analyze a higher-order DE qualitatively or numerically we had to express the differential equation as a system of first-order DEs.

Systems of first-order DEs are extremely important. Thus it is appropriate that we devote an entire chapter to the examination of such systems.

We begin with systems of linear first-order DEs. Since matrix notation and properties are used throughout this discussion, you should review Appendix I if you are unfamiliar with these concepts.

Systems of DEs Recall that in Section 3.8 we illustrated how to solve systems of n linear differential equations in n unknowns of the form

$$\begin{aligned} P_{11}(D)x_1 + P_{12}(D)x_2 + \cdots + P_{1n}(D)x_n &= b_1(t) \\ P_{21}(D)x_1 + P_{22}(D)x_2 + \cdots + P_{2n}(D)x_n &= b_2(t) \\ &\vdots \\ P_{n1}(D)x_1 + P_{n2}(D)x_2 + \cdots + P_{nn}(D)x_n &= b_n(t) \end{aligned} \tag{1}$$

where the P_{ij} were polynomials of various degrees in the differential operator D. In Sections 3.3, 3.9, and 3.10 we worked with systems of differential equations that were special cases of the form

$$\begin{aligned} \frac{dx_1}{dt} &= g_1(t, x_1, x_2, \ldots, x_n) \\ \frac{dx_2}{dt} &= g_2(t, x_1, x_2, \ldots, x_n) \\ &\vdots \\ \frac{dx_n}{dt} &= g_n(t, x_1, x_2, \ldots, x_n) \end{aligned} \tag{2}$$

A system such as (2) of n first-order equations is called an ***n*th-order system**.

Linear Systems When each of the functions $g_1, g_2, \ldots, g_n$ in (2) is linear in the dependent variables $x_1, x_2, \ldots, x_n$, we get the **normal form** of an nth-order system of linear equations:

$$\begin{aligned}
\frac{dx_1}{dt} &= a_{11}(t)x_1 + a_{12}(t)x_2 + \cdots + a_{1n}(t)x_n + f_1(t) \\
\frac{dx_2}{dt} &= a_{21}(t)x_1 + a_{22}(t)x_2 + \cdots + a_{2n}(t)x_n + f_2(t) \\
&\vdots \\
\frac{dx_n}{dt} &= a_{n1}(t)x_1 + a_{n2}(t)x_2 + \cdots + a_{nn}(t)x_n + f_n(t)
\end{aligned} \tag{3}$$

We refer to a system of the form (3) as an ***n*th-order linear system** or simply as a **linear system**. We assume that the coefficients a_{ij} and the functions f_i are continuous on a common interval I. When $f_i(t) = 0$, $i = 1, 2, \ldots, n$, the linear system is said to be **homogeneous**; otherwise it is **nonhomogeneous**.

Matrix Form of a Linear System If $\mathbf{X}$, $\mathbf{A}(t)$, and $\mathbf{F}(t)$ denote the respective matrices

$$\mathbf{X} = \begin{pmatrix} x_1(t) \\ x_2(t) \\ \vdots \\ x_n(t) \end{pmatrix}, \quad \mathbf{A}(t) = \begin{pmatrix} a_{11}(t) & a_{12}(t) & \cdots & a_{1n}(t) \\ a_{21}(t) & a_{22}(t) & \cdots & a_{2n}(t) \\ \vdots & & & \vdots \\ a_{n1}(t) & a_{n2}(t) & \cdots & a_{nn}(t) \end{pmatrix}, \quad \mathbf{F}(t) = \begin{pmatrix} f_1(t) \\ f_2(t) \\ \vdots \\ f_n(t) \end{pmatrix}$$

then the normal form of an nth-order system of linear equations (3) can be written as

$$\frac{d}{dt}\begin{pmatrix} x_1 \\ x_2 \\ \vdots \\ x_n \end{pmatrix} = \begin{pmatrix} a_{11}(t) & a_{12}(t) & \cdots & a_{1n}(t) \\ a_{21}(t) & a_{22}(t) & \cdots & a_{2n}(t) \\ \vdots & & & \vdots \\ a_{n1}(t) & a_{n2}(t) & \cdots & a_{nn}(t) \end{pmatrix}\begin{pmatrix} x_1 \\ x_2 \\ \vdots \\ x_n \end{pmatrix} + \begin{pmatrix} f_1(t) \\ f_2(t) \\ \vdots \\ f_n(t) \end{pmatrix}$$

or simply

$$\mathbf{X}' = \mathbf{A}\mathbf{X} + \mathbf{F} \tag{4}$$

If the system is homogeneous, its matrix form is then

$$\mathbf{X}' = \mathbf{A}\mathbf{X} \tag{5}$$

■ EXAMPLE 1 Systems Written in Matrix Notation

(a) If $\mathbf{X} = \begin{pmatrix} x \\ y \end{pmatrix}$, then the matrix form of the homogeneous system

$$\begin{aligned} \frac{dx}{dt} &= 3x + 4y \\ \frac{dy}{dt} &= 5x - 7y \end{aligned} \qquad \text{is} \quad \mathbf{X}' = \begin{pmatrix} 3 & 4 \\ 5 & -7 \end{pmatrix}\mathbf{X}$$

(b) If $\mathbf{X} = \begin{pmatrix} x \\ y \\ z \end{pmatrix}$, then the matrix form of the nonhomogeneous system

$$\begin{aligned} \frac{dx}{dt} &= 6x + y + z + t \\ \frac{dy}{dt} &= 8x + 7y - z + 10t \\ \frac{dz}{dt} &= 2x + 9y - z + 6t \end{aligned} \quad \text{is} \quad \mathbf{X}' = \begin{pmatrix} 6 & 1 & 1 \\ 8 & 7 & -1 \\ 2 & 9 & -1 \end{pmatrix}\mathbf{X} + \begin{pmatrix} t \\ 10t \\ 6t \end{pmatrix}$$

■

Definition 4.1 Solution Vector

A **solution vector** on an interval I is any column matrix

$$\mathbf{X} = \begin{pmatrix} x_1(t) \\ x_2(t) \\ \vdots \\ x_n(t) \end{pmatrix}$$

whose entries are differentiable functions satisfying the system (4) on the interval.

■ EXAMPLE 2 Verification of Solutions

Verify that on the interval $(-\infty, \infty)$

$$\mathbf{X}_1 = \begin{pmatrix} 1 \\ -1 \end{pmatrix} e^{-2t} = \begin{pmatrix} e^{-2t} \\ -e^{-2t} \end{pmatrix} \quad \text{and} \quad \mathbf{X}_2 = \begin{pmatrix} 3 \\ 5 \end{pmatrix} e^{6t} = \begin{pmatrix} 3e^{6t} \\ 5e^{6t} \end{pmatrix}$$

are solutions of

$$\mathbf{X}' = \begin{pmatrix} 1 & 3 \\ 5 & 3 \end{pmatrix}\mathbf{X} \tag{6}$$

Solution From $\mathbf{X}_1' = \begin{pmatrix} -2e^{-2t} \\ 2e^{-2t} \end{pmatrix}$ and $\mathbf{X}_2' = \begin{pmatrix} 18e^{6t} \\ 30e^{6t} \end{pmatrix}$, we see

$$\mathbf{AX}_1 = \begin{pmatrix} 1 & 3 \\ 5 & 3 \end{pmatrix}\begin{pmatrix} e^{-2t} \\ -e^{-2t} \end{pmatrix} = \begin{pmatrix} e^{-2t} - 3e^{-2t} \\ 5e^{-2t} - 3e^{-2t} \end{pmatrix} = \begin{pmatrix} -2e^{-2t} \\ 2e^{-2t} \end{pmatrix} = \mathbf{X}_1'$$

and

$$\mathbf{AX}_2 = \begin{pmatrix} 1 & 3 \\ 5 & 3 \end{pmatrix}\begin{pmatrix} 3e^{6t} \\ 5e^{6t} \end{pmatrix} = \begin{pmatrix} 3e^{6t} + 15e^{6t} \\ 15e^{6t} + 15e^{6t} \end{pmatrix} = \begin{pmatrix} 18e^{6t} \\ 30e^{6t} \end{pmatrix} = \mathbf{X}_2'$$

■

Much of the theory of systems of n linear first-order differential equations is similar to that of linear nth-order differential equations.

Initial-Value Problem Let t_0 denote a point on an interval I and

$$\mathbf{X}(t_0) = \begin{pmatrix} x_1(t_0) \\ x_2(t_0) \\ \vdots \\ x_n(t_0) \end{pmatrix} \quad \text{and} \quad \mathbf{X}_0 = \begin{pmatrix} \gamma_1 \\ \gamma_2 \\ \vdots \\ \gamma_n \end{pmatrix}$$

where the γ_i, $i = 1, 2, \ldots, n$, are given constants. Then the problem

$$\begin{aligned} \textit{Solve:} \quad & \mathbf{X}' = \mathbf{A}(t)\mathbf{X} + \mathbf{F}(t) \\ \textit{Subject to:} \quad & \mathbf{X}(t_0) = \mathbf{X}_0 \end{aligned} \qquad \textbf{(7)}$$

is an **initial-value problem** on the interval.

Theorem 4.1 Existence of a Unique Solution

Let the entries of the matrices $\mathbf{A}(t)$ and $\mathbf{F}(t)$ be functions continuous on a common interval I that contains the point t_0. Then there exists a unique solution of the initial-value problem (7) on the interval.

Homogeneous Systems In the next several definitions and theorems, we are concerned only with homogeneous systems. Without stating it, we shall always assume that the a_{ij} and the f_i are continuous functions of t on some common interval I.

Superposition Principle The following result is a **superposition principle** for solutions of linear systems.

Theorem 4.2 Superposition Principle

Let $\mathbf{X}_1, \mathbf{X}_2, \ldots, \mathbf{X}_k$ be a set of solution vectors of the homogeneous system (5) on an interval I. Then the linear combination

$$\mathbf{X} = c_1\mathbf{X}_1 + c_2\mathbf{X}_2 + \cdots + c_k\mathbf{X}_k$$

where the c_i, $i = 1, 2, \ldots, k$, are arbitrary constants, is also a solution on the interval.

It follows from Theorem 4.2 that a constant multiple of any solution vector of a homogeneous system of linear first-order differential equations is also a solution.

EXAMPLE 3 Using the Superposition Principle

You should practice by verifying that the two vectors

$$\mathbf{X}_1 = \begin{pmatrix} \cos t \\ -\frac{1}{2}\cos t + \frac{1}{2}\sin t \\ -\cos t - \sin t \end{pmatrix} \quad \text{and} \quad \mathbf{X}_2 = \begin{pmatrix} 0 \\ e^t \\ 0 \end{pmatrix}$$

are solutions of the system

$$\mathbf{X}' = \begin{pmatrix} 1 & 0 & 1 \\ 1 & 1 & 0 \\ -2 & 0 & -1 \end{pmatrix} \mathbf{X} \tag{8}$$

By the superposition principle, the linear combination

$$\mathbf{X} = c_1\mathbf{X}_1 + c_2\mathbf{X}_2 = c_1 \begin{pmatrix} \cos t \\ -\frac{1}{2}\cos t + \frac{1}{2}\sin t \\ -\cos t - \sin t \end{pmatrix} + c_2 \begin{pmatrix} 0 \\ e^t \\ 0 \end{pmatrix}$$

is yet another solution of the system. ■

Linear Dependence and Linear Independence We are primarily interested in linearly independent solutions of the homogeneous system (5).

Definition 4.2 Linear Dependence/Independence

Let $\mathbf{X}_1, \mathbf{X}_2, \ldots, \mathbf{X}_k$ be a set of solution vectors of the homogeneous system (5) on an interval I. We say that the set is **linearly dependent** on the interval if there exist constants $c_1, c_2, \ldots, c_k$, not all zero, such that

$$c_1\mathbf{X}_1 + c_2\mathbf{X}_2 + \cdots + c_k\mathbf{X}_k = \mathbf{0}$$

for every t in the interval. If the set of vectors is not linearly dependent on the interval, it is said to be **linearly independent**.

The case when $k = 2$ should be clear; two solution vectors $\mathbf{X}_1$ and $\mathbf{X}_2$ are linearly dependent if one is a constant multiple of the other, and conversely. For $k > 2$, a set of solution vectors is linearly dependent if we can express at least one solution vector as a linear combination of the remaining vectors.

Wronskian As in our earlier consideration of the theory of a single ordinary differential equation, we can introduce the concept of the **Wronskian** determinant as a test for linear independence. We state the following theorem without proof.

Theorem 4.3 Criterion for Linearly Independent Solutions

Let
$$\mathbf{X}_1 = \begin{pmatrix} x_{11} \\ x_{21} \\ \vdots \\ x_{n1} \end{pmatrix}, \quad \mathbf{X}_2 = \begin{pmatrix} x_{12} \\ x_{22} \\ \vdots \\ x_{n2} \end{pmatrix}, \ldots, \quad \mathbf{X}_n = \begin{pmatrix} x_{1n} \\ x_{2n} \\ \vdots \\ x_{nn} \end{pmatrix}$$

be n solution vectors of the homogeneous system (5) on an interval I. Then the set of solution vectors is linearly independent on I if and only if the **Wronskian**

$$W(\mathbf{X}_1, \mathbf{X}_2, \ldots, \mathbf{X}_n) = \begin{vmatrix} x_{11} & x_{12} & \cdots & x_{1n} \\ x_{21} & x_{22} & \cdots & x_{2n} \\ \vdots & \vdots & & \vdots \\ x_{n1} & x_{n2} & \cdots & x_{nn} \end{vmatrix} \neq 0 \tag{9}$$

for every t in the interval.

In fact, it can be shown that if $\mathbf{X}_1, \mathbf{X}_2, \ldots, \mathbf{X}_n$ are solution vectors of (5), then for every t in I, either $W(\mathbf{X}_1, \mathbf{X}_2, \ldots, \mathbf{X}_n) \neq 0$ or $W(\mathbf{X}_1, \mathbf{X}_2, \ldots, \mathbf{X}_n) = 0$. Thus if we can show that $W \neq 0$ for some t_0 in I, then $W \neq 0$ for every t, and hence the solutions are linearly independent on the interval.

Notice that, unlike our previous definition of the Wronskian, the determinant (9) does not involve differentiation.

■ EXAMPLE 4 **Linearly Independent Solutions**

In Example 2 we saw that $\mathbf{X}_1 = \begin{pmatrix} 1 \\ -1 \end{pmatrix} e^{-2t}$ and $\mathbf{X}_2 = \begin{pmatrix} 3 \\ 5 \end{pmatrix} e^{6t}$ are solutions of system (6). Clearly, $\mathbf{X}_1$ and $\mathbf{X}_2$ are linearly independent on the interval $(-\infty, \infty)$ since neither vector is a constant multiple of the other. In addition, we have

$$W(\mathbf{X}_1, \mathbf{X}_2) = \begin{vmatrix} e^{-2t} & 3e^{6t} \\ -e^{-2t} & 5e^{6t} \end{vmatrix} = 8e^{4t} \neq 0$$

for all real values of t. ■

Definition 4.3 Fundamental Set of Solutions

Any set $\mathbf{X}_1, \mathbf{X}_2, \ldots, \mathbf{X}_n$ of n linearly independent solution vectors of the homogeneous system (5) on an interval I is said to be a **fundamental set of solutions** on the interval.

Theorem 4.4 Existence of a Fundamental Set

There exists a fundamental set of solutions for the homogeneous system (5) on an interval I.

Theorem 4.5 General Solution—Homogeneous Systems

Let $\mathbf{X}_1, \mathbf{X}_2, \ldots, \mathbf{X}_n$ be a fundamental set of solutions of the homogeneous system (5) on an interval I. Then the **general solution** of the system on the interval is

$$\mathbf{X} = c_1\mathbf{X}_1 + c_2\mathbf{X}_2 + \cdots + c_n\mathbf{X}_n$$

where the c_i, $i = 1, 2, \ldots, n$, are arbitrary constants.

EXAMPLE 5 General Solution of System (6)

From Example 2, we know that $\mathbf{X}_1 = \begin{pmatrix} 1 \\ -1 \end{pmatrix} e^{-2t}$ and $\mathbf{X}_2 = \begin{pmatrix} 3 \\ 5 \end{pmatrix} e^{6t}$ are linearly independent solutions of (6) on $(-\infty, \infty)$. Hence $\mathbf{X}_1$ and $\mathbf{X}_2$ form a fundamental set of solutions on the interval. The general solution of the system on the interval is then

$$\mathbf{X} = c_1\mathbf{X}_1 + c_2\mathbf{X}_2 = c_1\begin{pmatrix} 1 \\ -1 \end{pmatrix} e^{-2t} + c_2\begin{pmatrix} 3 \\ 5 \end{pmatrix} e^{6t} \tag{10}$$

■

EXAMPLE 6 General Solution of System (8)

The vectors

$$\mathbf{X}_1 = \begin{pmatrix} \cos t \\ -\frac{1}{2}\cos t + \frac{1}{2}\sin t \\ -\cos t - \sin t \end{pmatrix}, \quad \mathbf{X}_2 = \begin{pmatrix} 0 \\ 1 \\ 0 \end{pmatrix} e^t, \quad \mathbf{X}_3 = \begin{pmatrix} \sin t \\ -\frac{1}{2}\sin t - \frac{1}{2}\cos t \\ -\sin t + \cos t \end{pmatrix}$$

are solutions of the system (8) in Example 3 (see Problem 16 in Exercises 4.1). Now

$$W(\mathbf{X}_1, \mathbf{X}_2, \mathbf{X}_3) = \begin{vmatrix} \cos t & 0 & \sin t \\ -\frac{1}{2}\cos t + \frac{1}{2}\sin t & e^t & -\frac{1}{2}\sin t - \frac{1}{2}\cos t \\ -\cos t - \sin t & 0 & -\sin t + \cos t \end{vmatrix} = e^t \neq 0$$

for all real values of t. We conclude that $\mathbf{X}_1, \mathbf{X}_2$, and $\mathbf{X}_3$ form a fundamental set of solutions on $(-\infty, \infty)$. Thus the general solution of the system on the interval is the linear combination $\mathbf{X} = c_1\mathbf{X}_1 + c_2\mathbf{X}_2 + c_3\mathbf{X}_3$, that is,

$$\mathbf{X} = c_1 \begin{pmatrix} \cos t \\ -\frac{1}{2}\cos t + \frac{1}{2}\sin t \\ -\cos t - \sin t \end{pmatrix} + c_2 \begin{pmatrix} 0 \\ 1 \\ 0 \end{pmatrix} e^t + c_3 \begin{pmatrix} \sin t \\ -\frac{1}{2}\sin t - \frac{1}{2}\cos t \\ -\sin t + \cos t \end{pmatrix}$$ ■

Nonhomogeneous Systems For nonhomogeneous systems a **particular solution $\mathbf{X}_p$** on an interval I is any vector, free of arbitrary parameters, whose entries are functions that satisfy the system (4).

Theorem 4.6 General Solution—Nonhomogeneous Systems

Let $\mathbf{X}_p$ be a given solution of the nonhomogeneous system (4) on an interval I, and let $\mathbf{X}_c = c_1\mathbf{X}_1 + c_2\mathbf{X}_2 + \cdots + c_n\mathbf{X}_n$ denote the general solution on the same interval of the associated homogeneous system (5). Then the **general solution** of the nonhomogeneous system on the interval is

$$\mathbf{X} = \mathbf{X}_c + \mathbf{X}_p$$

The general solution $\mathbf{X}_c$ of the homogeneous system (5) is called the **complementary function** of the nonhomogeneous system (4).

■ EXAMPLE 7 General Solution—Nonhomogeneous System

The vector $\mathbf{X}_p = \begin{pmatrix} 3t - 4 \\ -5t + 6 \end{pmatrix}$ is a particular solution of the nonhomogeneous system

$$\mathbf{X}' = \begin{pmatrix} 1 & 3 \\ 5 & 3 \end{pmatrix}\mathbf{X} + \begin{pmatrix} 12t - 11 \\ -3 \end{pmatrix} \tag{11}$$

on the interval $(-\infty, \infty)$. (Verify this.) The complementary function of (11) on the same interval, or the general solution of $\mathbf{X}' = \begin{pmatrix} 1 & 3 \\ 5 & 3 \end{pmatrix}\mathbf{X}$, was seen in (10) of Example 5 to be $\mathbf{X}_c = c_1\begin{pmatrix} 1 \\ -1 \end{pmatrix}e^{-2t} + c_2\begin{pmatrix} 3 \\ 5 \end{pmatrix}e^{6t}$. Hence by Theorem 4.6,

$$\mathbf{X} = \mathbf{X}_c + \mathbf{X}_p = c_1\begin{pmatrix} 1 \\ -1 \end{pmatrix}e^{-2t} + c_2\begin{pmatrix} 3 \\ 5 \end{pmatrix}e^{6t} + \begin{pmatrix} 3t - 4 \\ -5t + 6 \end{pmatrix}$$

is the general solution of (11) on $(-\infty, \infty)$. ■

EXERCISES 4.1 Answers to odd-numbered problems begin on page 446.

In Problems 1–6, write the given system in matrix form.

1. $\dfrac{dx}{dt} = 3x - 5y$

$\dfrac{dy}{dt} = 4x + 8y$

2. $\dfrac{dx}{dt} = 4x - 7y$

$\dfrac{dy}{dt} = 5x$

3. $\dfrac{dx}{dt} = -3x + 4y - 9z$

$\dfrac{dy}{dt} = 6x - y$

$\dfrac{dz}{dt} = 10x + 4y + 3z$

4. $\dfrac{dx}{dt} = x - y$

$\dfrac{dy}{dt} = x + 2z$

$\dfrac{dz}{dt} = -x + z$

5. $\dfrac{dx}{dt} = x - y + z + t - 1$

$\dfrac{dy}{dt} = 2x + y - z - 3t^2$

$\dfrac{dz}{dt} = x + y + z + t^2 - t + 2$

6. $\dfrac{dx}{dt} = -3x + 4y + e^{-t}\sin 2t$

$\dfrac{dy}{dt} = 5x + 9y + 4e^{-t}\cos 2t$

In Problems 7–10, write the given system without the use of matrices.

7. $\mathbf{X}' = \begin{pmatrix} 4 & 2 \\ -1 & 3 \end{pmatrix}\mathbf{X} + \begin{pmatrix} 1 \\ -1 \end{pmatrix} e^t$

8. $\mathbf{X}' = \begin{pmatrix} 7 & 5 & -9 \\ 4 & 1 & 1 \\ 0 & -2 & 3 \end{pmatrix}\mathbf{X} + \begin{pmatrix} 0 \\ 2 \\ 1 \end{pmatrix} e^{5t} - \begin{pmatrix} 8 \\ 0 \\ 3 \end{pmatrix} e^{-2t}$

9. $\dfrac{d}{dt}\begin{pmatrix} x \\ y \\ z \end{pmatrix} = \begin{pmatrix} 1 & -1 & 2 \\ 3 & -4 & 1 \\ -2 & 5 & 6 \end{pmatrix}\begin{pmatrix} x \\ y \\ z \end{pmatrix} + \begin{pmatrix} 1 \\ 2 \\ 2 \end{pmatrix} e^{-t} - \begin{pmatrix} 3 \\ -1 \\ 1 \end{pmatrix} t$

10. $\dfrac{d}{dt}\begin{pmatrix} x \\ y \end{pmatrix} = \begin{pmatrix} 3 & -7 \\ 1 & 1 \end{pmatrix}\begin{pmatrix} x \\ y \end{pmatrix} + \begin{pmatrix} 4 \\ 8 \end{pmatrix}\sin t + \begin{pmatrix} t-4 \\ 2t+1 \end{pmatrix} e^{4t}$

In Problems 11–16, verify that the vector X is a solution of the given system.

11. $\dfrac{dx}{dt} = 3x - 4y$

$\dfrac{dy}{dt} = 4x - 7y; \quad \mathbf{X} = \begin{pmatrix} 1 \\ 2 \end{pmatrix} e^{-5t}$

12. $\dfrac{dx}{dt} = -2x + 5y$

$\dfrac{dy}{dt} = -2x + 4y; \quad \mathbf{X} = \begin{pmatrix} 5\cos t \\ 3\cos t - \sin t \end{pmatrix} e^t$

13. $\mathbf{X}' = \begin{pmatrix} -1 & \frac{1}{4} \\ 1 & -1 \end{pmatrix}\mathbf{X}; \quad \mathbf{X} = \begin{pmatrix} -1 \\ 2 \end{pmatrix} e^{-3t/2}$

14. $\mathbf{X}' = \begin{pmatrix} 2 & 1 \\ -1 & 0 \end{pmatrix}\mathbf{X}; \quad \mathbf{X} = \begin{pmatrix} 1 \\ 3 \end{pmatrix} e^t + \begin{pmatrix} 4 \\ -4 \end{pmatrix} te^t$

15. $\mathbf{X}' = \begin{pmatrix} 1 & 2 & 1 \\ 6 & -1 & 0 \\ -1 & -2 & -1 \end{pmatrix}\mathbf{X}; \quad \mathbf{X} = \begin{pmatrix} 1 \\ 6 \\ -13 \end{pmatrix}$

16. $\mathbf{X}' = \begin{pmatrix} 1 & 0 & 1 \\ 1 & 1 & 0 \\ -2 & 0 & -1 \end{pmatrix}\mathbf{X}; \quad \mathbf{X} = \begin{pmatrix} \sin t \\ -\frac{1}{2}\sin t - \frac{1}{2}\cos t \\ -\sin t + \cos t \end{pmatrix}$

In Problems 17–20, the given vectors are solutions of a system $\mathbf{X}' = \mathbf{AX}$. Determine whether the vectors form a fundamental set on $(-\infty, \infty)$.

17. $\mathbf{X}_1 = \begin{pmatrix} 1 \\ 1 \end{pmatrix} e^{-2t}, \mathbf{X}_2 = \begin{pmatrix} 1 \\ -1 \end{pmatrix} e^{-6t}$

18. $\mathbf{X}_1 = \begin{pmatrix} 1 \\ -1 \end{pmatrix} e^t, \mathbf{X}_2 = \begin{pmatrix} 2 \\ 6 \end{pmatrix} e^t + \begin{pmatrix} 8 \\ -8 \end{pmatrix} te^t$

19. $\mathbf{X}_1 = \begin{pmatrix} 1 \\ -2 \\ 4 \end{pmatrix} + t\begin{pmatrix} 1 \\ 2 \\ 2 \end{pmatrix}, \mathbf{X}_2 = \begin{pmatrix} 1 \\ -2 \\ 4 \end{pmatrix},$

$\mathbf{X}_3 = \begin{pmatrix} 3 \\ -6 \\ 12 \end{pmatrix} + t\begin{pmatrix} 2 \\ 4 \\ 4 \end{pmatrix}$

20. $\mathbf{X}_1 = \begin{pmatrix} 1 \\ 6 \\ -13 \end{pmatrix}, \mathbf{X}_2 = \begin{pmatrix} 1 \\ -2 \\ -1 \end{pmatrix} e^{-4t}, \mathbf{X}_3 = \begin{pmatrix} 2 \\ 3 \\ -2 \end{pmatrix} e^{3t}$

In Problems 21–24, verify that the vector $\mathbf{X}_p$ is a particular solution of the given system.

21. $\dfrac{dx}{dt} = x + 4y + 2t - 7$

$\dfrac{dy}{dt} = 3x + 2y - 4t - 18;\quad \mathbf{X}_p = \begin{pmatrix} 2 \\ -1 \end{pmatrix} t + \begin{pmatrix} 5 \\ 1 \end{pmatrix}$

22. $\mathbf{X}' = \begin{pmatrix} 2 & 1 \\ 1 & -1 \end{pmatrix}\mathbf{X} + \begin{pmatrix} -5 \\ 2 \end{pmatrix};\quad \mathbf{X}_p = \begin{pmatrix} 1 \\ 3 \end{pmatrix}$

23. $\mathbf{X}' = \begin{pmatrix} 2 & 1 \\ 3 & 4 \end{pmatrix}\mathbf{X} - \begin{pmatrix} 1 \\ 7 \end{pmatrix} e^t;\quad \mathbf{X}_p = \begin{pmatrix} 1 \\ 1 \end{pmatrix} e^t + \begin{pmatrix} 1 \\ -1 \end{pmatrix} te^t$

24. $\mathbf{X}' = \begin{pmatrix} 1 & 2 & 3 \\ -4 & 2 & 0 \\ -6 & 1 & 0 \end{pmatrix}\mathbf{X} + \begin{pmatrix} -1 \\ 4 \\ 3 \end{pmatrix} \sin 3t;$

$\mathbf{X}_p = \begin{pmatrix} \sin 3t \\ 0 \\ \cos 3t \end{pmatrix}$

25. Prove that the general solution of

$$\mathbf{X}' = \begin{pmatrix} 0 & 6 & 0 \\ 1 & 0 & 1 \\ 1 & 1 & 0 \end{pmatrix}\mathbf{X}$$

on the interval $(-\infty, \infty)$ is

$$\mathbf{X} = c_1 \begin{pmatrix} 6 \\ -1 \\ -5 \end{pmatrix} e^{-t} + c_2 \begin{pmatrix} -3 \\ 1 \\ 1 \end{pmatrix} e^{-2t} + c_3 \begin{pmatrix} 2 \\ 1 \\ 1 \end{pmatrix} e^{3t}$$

26. Prove that the general solution of

$$\mathbf{X}' = \begin{pmatrix} -1 & -1 \\ -1 & 1 \end{pmatrix}\mathbf{X} + \begin{pmatrix} 1 \\ 1 \end{pmatrix} t^2 + \begin{pmatrix} 4 \\ -6 \end{pmatrix} t + \begin{pmatrix} -1 \\ 5 \end{pmatrix}$$

on the interval $(-\infty, \infty)$ is

$$\mathbf{X} = c_1 \begin{pmatrix} 1 \\ -1 - \sqrt{2} \end{pmatrix} e^{\sqrt{2}t} + c_2 \begin{pmatrix} 1 \\ -1 + \sqrt{2} \end{pmatrix} e^{-\sqrt{2}t} + \begin{pmatrix} 1 \\ 0 \end{pmatrix} t^2 + \begin{pmatrix} -2 \\ 4 \end{pmatrix} t + \begin{pmatrix} 1 \\ 0 \end{pmatrix}$$

Discussion Problems

27. The system of linear first-order differential equations

$$\frac{dx}{dt} - \frac{dy}{dt} - x + 3y = 2t$$

$$\frac{dx}{dt} + \frac{dy}{dt} + 2x + y = 10$$

is not in the normal form $\mathbf{X}' = \mathbf{AX} + \mathbf{F}$. Discuss: Can the system be put into normal form? Can anything be done with the following observation?

$$\begin{pmatrix} 1 & -1 \\ 1 & 1 \end{pmatrix}\begin{pmatrix} x' \\ y' \end{pmatrix} + \begin{pmatrix} -1 & 3 \\ 2 & 1 \end{pmatrix}\begin{pmatrix} x \\ y \end{pmatrix} = \begin{pmatrix} 2t \\ 10 \end{pmatrix}$$

28. Given that $\mathbf{X}_1 = \begin{pmatrix} 3 \\ 2 \end{pmatrix} e^t$ and $\mathbf{X}_2 = \begin{pmatrix} 1 \\ 1 \end{pmatrix} e^{2t}$ form a fundamental set of solutions of a linear system $\mathbf{X}' = \mathbf{AX}$ on $(-\infty, \infty)$. Discuss: Can the system be determined? Carry out your ideas.

4.2 Homogeneous Linear Systems with Constant Coefficients

The systematic solution of a homogeneous linear system $\mathbf{X}' = \mathbf{AX}$ with constant coefficients depends on our ability to find the *eigenvalues* and *eigenvectors* of the coefficient matrix $\mathbf{A}$. To find the eigenvalues we must solve a polynomial equation. (See Appendix I for a review of this theory.) Analogous to the case of a single homogeneous linear nth-order DE with constant coefficients, the form that the general solution of the system $\mathbf{X}' = \mathbf{AX}$ takes on depends on whether the eigenvalues of $\mathbf{A}$ are distinct and real, are repeated and real, or appear in conjugate complex pairs.

Eigenvalues and Eigenvectors We saw in Example 5 of Section 4.1 that the general solution of the homogeneous system $\mathbf{X}' = \begin{pmatrix} 1 & 3 \\ 5 & 3 \end{pmatrix}\mathbf{X}$ is $\mathbf{X} = c_1\mathbf{X}_1 + c_2\mathbf{X}_2 = c_1\begin{pmatrix} 1 \\ -1 \end{pmatrix}e^{-2t} + c_2\begin{pmatrix} 3 \\ 5 \end{pmatrix}e^{6t}$. Since both solution vectors have the form $\mathbf{X}_i = \begin{pmatrix} k_1 \\ k_2 \end{pmatrix}e^{\lambda_i t}$, $i = 1, 2$, where k_1 and k_2 are constants, we are prompted to ask whether we can always find a solution of the form

$$\mathbf{X} = \begin{pmatrix} k_1 \\ k_2 \\ \vdots \\ k_n \end{pmatrix} e^{\lambda t} = \mathbf{K}e^{\lambda t} \tag{1}$$

for the general homogeneous linear first-order system

$$\mathbf{X}' = \mathbf{AX} \tag{2}$$

where $\mathbf{A}$ is an $n \times n$ matrix of constants.

If (1) is to be a solution vector of the system, then $\mathbf{X}' = \mathbf{K}\lambda e^{\lambda t}$, so that (2) becomes $\mathbf{K}\lambda e^{\lambda t} = \mathbf{AK}e^{\lambda t}$. After dividing out $e^{\lambda t}$ and rearranging, we obtain $\mathbf{AK} = \lambda\mathbf{K}$ or $\mathbf{AK} - \lambda\mathbf{K} = \mathbf{0}$. Since $\mathbf{K} = \mathbf{IK}$, the last equation is the same as

$$(\mathbf{A} - \lambda\mathbf{I})\mathbf{K} = \mathbf{0} \tag{3}$$

Equation (3) is equivalent to the simultaneous algebraic equations

$$\begin{aligned} (a_{11} - \lambda)k_1 + \quad a_{12}k_2 + \cdots + \quad a_{1n}k_n &= 0 \\ a_{21}k_1 + (a_{22} - \lambda)k_2 + \cdots + \quad a_{2n}k_n &= 0 \\ \vdots \qquad\qquad\qquad\qquad & \ \ \vdots \\ a_{n1}k_1 + \quad a_{n2}k_2 + \cdots + (a_{nn} - \lambda)k_n &= 0 \end{aligned}$$

Thus to find a nontrivial solution $\mathbf{X}$ of (2), we must first find a nontrivial solution of the foregoing system; in other words, we must find a nontrivial vector $\mathbf{K}$ that satisfies (3). But in order for (3) to have solutions other than the obvious solution $k_1 = k_2 = \ldots = k_n = 0$ we must have

$$\det(\mathbf{A} - \lambda\mathbf{I}) = 0$$

This polynomial equation in λ is called the **characteristic equation** of the matrix $\mathbf{A}$; its solutions are the **eigenvalues** of $\mathbf{A}$. A solution $\mathbf{K} \neq \mathbf{0}$ of (3), corresponding to an eigenvalue λ, is called an **eigenvector** of $\mathbf{A}$. A solution of the homogeneous system (2) is then $\mathbf{X} = \mathbf{K}e^{\lambda t}$.

4.2.1 Distinct Real Eigenvalues

When the $n \times n$ matrix $\mathbf{A}$ possesses n distinct real eigenvalues $\lambda_1, \lambda_2, \ldots, \lambda_n$, then a set of n linearly independent eigenvectors $\mathbf{K}_1, \mathbf{K}_2, \ldots, \mathbf{K}_n$ can always be found and

$$\mathbf{X}_1 = \mathbf{K}_1 e^{\lambda_1 t}, \quad \mathbf{X}_2 = \mathbf{K}_2 e^{\lambda_2 t}, \quad \ldots, \quad \mathbf{X}_n = \mathbf{K}_n e^{\lambda_n t}$$

is a fundamental set of solutions of (2) on $(-\infty, \infty)$.

Theorem 4.7 General Solution—Homogeneous Systems

Let $\lambda_1, \lambda_2, \ldots, \lambda_n$ be n distinct real eigenvalues of the coefficient matrix $\mathbf{A}$ of the homogeneous system (2), and let $\mathbf{K}_1, \mathbf{K}_2, \ldots, \mathbf{K}_n$ be the corresponding eigenvectors. Then the **general solution** of (2) on the interval $(-\infty, \infty)$ is given by

$$\mathbf{X} = c_1 \mathbf{K}_1 e^{\lambda_1 t} + c_2 \mathbf{K}_2 e^{\lambda_2 t} + \cdots + c_n \mathbf{K}_n e^{\lambda_n t}$$

■ EXAMPLE 1 **Distinct Eigenvalues**

Solve

$$\begin{aligned}\frac{dx}{dt} &= 2x + 3y \\ \frac{dy}{dt} &= 2x + y\end{aligned} \tag{4}$$

Solution We first find the eigenvalues and eigenvectors of the matrix of coefficients.

From the characteristic equation

$$\det(\mathbf{A} - \lambda\mathbf{I}) = \begin{vmatrix} 2-\lambda & 3 \\ 2 & 1-\lambda \end{vmatrix} = \lambda^2 - 3\lambda - 4 = (\lambda + 1)(\lambda - 4) = 0$$

we see that the eigenvalues are $\lambda_1 = -1$ and $\lambda_2 = 4$.

Now for $\lambda_1 = -1$, (3) is equivalent to

$$\begin{aligned}3k_1 + 3k_2 &= 0 \\ 2k_1 + 2k_2 &= 0\end{aligned}$$

Thus $k_1 = -k_2$. When $k_2 = -1$, the related eigenvector is

$$\mathbf{K}_1 = \begin{pmatrix} 1 \\ -1 \end{pmatrix}$$

For $\lambda_2 = 4$ we have

$$\begin{aligned}-2k_1 + 3k_2 &= 0 \\ 2k_1 - 3k_2 &= 0\end{aligned}$$

so that $k_1 = 3k_2/2$, and therefore with $k_2 = 2$, the corresponding eigenvector is

$$\mathbf{K}_2 = \begin{pmatrix} 3 \\ 2 \end{pmatrix}$$

Since the matrix of coefficients $\mathbf{A}$ is a 2×2 matrix, and since we have found two linearly independent solutions of (4), namely,

$$\mathbf{X}_1 = \begin{pmatrix} 1 \\ -1 \end{pmatrix} e^{-t} \quad \text{and} \quad \mathbf{X}_2 = \begin{pmatrix} 3 \\ 2 \end{pmatrix} e^{4t}$$

we conclude that the general solution of the system is

$$\mathbf{X} = c_1\mathbf{X}_1 + c_2\mathbf{X}_2 = c_1 \begin{pmatrix} 1 \\ -1 \end{pmatrix} e^{-t} + c_2 \begin{pmatrix} 3 \\ 2 \end{pmatrix} e^{4t} \tag{5}$$ ■

For the sake of review, you should keep firmly in mind that a solution of a system of first-order differential equations, when written in terms of matrices, is simply an alternative to the method that we employed in Section 3.8, namely, listing the individual functions and the relationships between the constants. If we add the vectors on the right side in (5) and equate the entries with the corresponding entries in the vector on the left, we obtain the more familiar statement

$$x(t) = c_1e^{-t} + 3c_2e^{4t}, \quad y(t) = -c_1e^{-t} + 2c_2e^{4t}$$

■ EXAMPLE 2 Distinct Eigenvalues

Solve

$$\begin{aligned} \frac{dx}{dt} &= -4x + y + z \\ \frac{dy}{dt} &= x + 5y - z \\ \frac{dz}{dt} &= y - 3z \end{aligned} \tag{6}$$

Solution Using the cofactors of the third row, we find

$$\det(\mathbf{A} - \lambda\mathbf{I}) = \begin{vmatrix} -4-\lambda & 1 & 1 \\ 1 & 5-\lambda & -1 \\ 0 & 1 & -3-\lambda \end{vmatrix} = -(\lambda + 3)(\lambda + 4)(\lambda - 5) = 0$$

and so the eigenvalues are $\lambda_1 = -3$, $\lambda_2 = -4$, $\lambda_3 = 5$.

For $\lambda_1 = -3$, Gauss-Jordan elimination gives

$$(\mathbf{A} + 3\mathbf{I}|\mathbf{0}) = \left(\begin{array}{ccc|c} -1 & 1 & 1 & 0 \\ 1 & 8 & -1 & 0 \\ 0 & 1 & 0 & 0 \end{array}\right) \xrightarrow{\text{row operations}} \left(\begin{array}{ccc|c} 1 & 0 & -1 & 0 \\ 0 & 1 & 0 & 0 \\ 0 & 0 & 0 & 0 \end{array}\right)$$

Therefore $k_1 = k_3$ and $k_2 = 0$. The choice $k_3 = 1$ gives an eigenvector and corresponding solution vector

$$\mathbf{K}_1 = \begin{pmatrix} 1 \\ 0 \\ 1 \end{pmatrix}, \quad \mathbf{X}_1 = \begin{pmatrix} 1 \\ 0 \\ 1 \end{pmatrix} e^{-3t} \tag{7}$$

Similarly, for $\lambda_2 = -4$,

$$(\mathbf{A} + 4\mathbf{I}|\mathbf{0}) = \left(\begin{array}{ccc|c} 0 & 1 & 1 & 0 \\ 1 & 9 & -1 & 0 \\ 0 & 1 & 1 & 0 \end{array}\right) \xrightarrow{\text{row operations}} \left(\begin{array}{ccc|c} 1 & 0 & -10 & 0 \\ 0 & 1 & 1 & 0 \\ 0 & 0 & 0 & 0 \end{array}\right)$$

implies $k_1 = 10k_3$ and $k_2 = -k_3$. Choosing $k_3 = 1$, we get a second eigenvector and solution vector

$$\mathbf{K}_2 = \begin{pmatrix} 10 \\ -1 \\ 1 \end{pmatrix}, \quad \mathbf{X}_2 = \begin{pmatrix} 10 \\ -1 \\ 1 \end{pmatrix} e^{-4t} \tag{8}$$

Finally, when $\lambda_3 = 5$, the augmented matrices

$$(\mathbf{A} - 5\mathbf{I}|\mathbf{0}) = \left(\begin{array}{ccc|c} -9 & 1 & 1 & 0 \\ 1 & 0 & -1 & 0 \\ 0 & 1 & -8 & 0 \end{array}\right) \xrightarrow{\text{row operations}} \left(\begin{array}{ccc|c} 1 & 0 & -1 & 0 \\ 0 & 1 & -8 & 0 \\ 0 & 0 & 0 & 0 \end{array}\right)$$

yield

$$\mathbf{K}_3 = \begin{pmatrix} 1 \\ 8 \\ 1 \end{pmatrix}, \quad \mathbf{X}_3 = \begin{pmatrix} 1 \\ 8 \\ 1 \end{pmatrix} e^{5t} \tag{9}$$

The general solution of (6) is a linear combination of the solution vectors in (7), (8), and (9):

$$\mathbf{X} = c_1 \begin{pmatrix} 1 \\ 0 \\ 1 \end{pmatrix} e^{-3t} + c_2 \begin{pmatrix} 10 \\ -1 \\ 1 \end{pmatrix} e^{-4t} + c_3 \begin{pmatrix} 1 \\ 8 \\ 1 \end{pmatrix} e^{5t}$$ ■

Use of Computers Software packages such as MATLAB, *MacMath*, *Mathematica*, *Maple*, and DERIVE can be real time-savers when finding eigenvalues and eigenvectors of a matrix. For example, to find the eigenvalues and eigenvectors of the matrix of coefficients in (6) using *Mathematica*, we first enter the definition of the matrix by rows

m = {{−4, 1, 1}, {1, 5, −1}, {0, 1, −3}}

The commands: **Eigenvalues[m]** and **Eigenvectors[m]**, entered in sequence, yield

$$\{-4, -3, 5\} \quad \text{and} \quad \{\{10, -1, 1\}, \{1, 0, 1\}, \{1, 8, 1\}\}$$

respectively. Alternatively, we can obtain both the eigenvalues and the corresponding eigenvectors by using **Eigensystem[m]**.

4.2.2 Repeated Eigenvalues

Of course, not all of the n eigenvalues $\lambda_1, \lambda_2, \ldots, \lambda_n$ of an $n \times n$ matrix **A** may be distinct; that is, some of the eigenvalues may be repeated. For example, the characteristic equation of the coefficient matrix in the system

$$\mathbf{X}' = \begin{pmatrix} 3 & -18 \\ 2 & -9 \end{pmatrix} \mathbf{X} \tag{10}$$

is readily shown to be $(\lambda + 3)^2 = 0$, and therefore $\lambda_1 = \lambda_2 = -3$ is a root of *multiplicity two*. For this value, we find the single eigenvector

$$\mathbf{K}_1 = \begin{pmatrix} 3 \\ 1 \end{pmatrix}, \quad \text{so} \quad \mathbf{X}_1 = \begin{pmatrix} 3 \\ 1 \end{pmatrix} e^{-3t} \tag{11}$$

is one solution of (10). But since we are obviously interested in forming the general solution of the system, we need to pursue the question of finding a second solution.

In general, if m is a positive integer and $(\lambda - \lambda_1)^m$ is a factor of the characteristic equation, while $(\lambda - \lambda_1)^{m+1}$ is not a factor, then λ_1 is said to be an **eigenvalue of multiplicity m**. The next three examples illustrate the cases:

(i) For some $n \times n$ matrices **A**, it may be possible to find m linearly independent eigenvectors $\mathbf{K}_1, \mathbf{K}_2, \ldots, \mathbf{K}_m$ corresponding to an eigenvalue λ_1 of multiplicity $m \leq n$. In this case, the general solution of the system contains the linear combination

$$c_1\mathbf{K}_1 e^{\lambda_1 t} + c_2\mathbf{K}_2 e^{\lambda_1 t} + \cdots + c_m\mathbf{K}_m e^{\lambda_1 t}$$

(ii) If there is only one eigenvector corresponding to the eigenvalue λ_1 of multiplicity m, then m linearly independent solutions, of the form

$$\mathbf{X}_1 = \mathbf{K}_{11} e^{\lambda_1 t}$$

$$\mathbf{X}_2 = \mathbf{K}_{21} t e^{\lambda_1 t} + \mathbf{K}_{22} e^{\lambda_1 t}$$

$$\vdots$$

$$\mathbf{X}_m = \mathbf{K}_{m1} \frac{t^{m-1}}{(m-1)!} e^{\lambda_1 t} + \mathbf{K}_{m2} \frac{t^{m-2}}{(m-2)!} e^{\lambda_1 t} + \cdots + \mathbf{K}_{mm} e^{\lambda_1 t}$$

where $\mathbf{K}_{ij}$ are column vectors, can always be found.

Eigenvalue of Multiplicity Two We begin by considering eigenvalues of multiplicity two. In the first example, we illustrate a matrix for which we can find two distinct eigenvectors corresponding to a double eigenvalue.

EXAMPLE 3 Repeated Eigenvalues

Solve $\mathbf{X}' = \begin{pmatrix} 1 & -2 & 2 \\ -2 & 1 & -2 \\ 2 & -2 & 1 \end{pmatrix}\mathbf{X}.$

Solution Expanding the determinant in the characteristic equation

$$\det(\mathbf{A} - \lambda\mathbf{I}) = \begin{vmatrix} 1-\lambda & -2 & 2 \\ -2 & 1-\lambda & -2 \\ 2 & -2 & 1-\lambda \end{vmatrix} = 0$$

yields $-(\lambda + 1)^2(\lambda - 5) = 0$. We see that $\lambda_1 = \lambda_2 = -1$ and $\lambda_3 = 5$.

For $\lambda_1 = -1$, Gauss-Jordan elimination gives immediately

$$(\mathbf{A} + \mathbf{I}|\mathbf{0}) = \left(\begin{array}{ccc|c} 2 & -2 & 2 & 0 \\ -2 & 2 & -2 & 0 \\ 2 & -2 & 2 & 0 \end{array}\right) \xrightarrow{\text{row operations}} \left(\begin{array}{ccc|c} 1 & -1 & 1 & 0 \\ 0 & 0 & 0 & 0 \\ 0 & 0 & 0 & 0 \end{array}\right)$$

The first row of the last matrix means $k_1 - k_2 + k_3 = 0$ or $k_1 = k_2 - k_3$. The choices $k_2 = 1$, $k_3 = 0$ and $k_2 = 1$, $k_3 = 1$, yield, in turn, $k_1 = 1$ and $k_1 = 0$. Thus two eigenvectors corresponding to $\lambda_1 = -1$ are

$$\mathbf{K}_1 = \begin{pmatrix} 1 \\ 1 \\ 0 \end{pmatrix} \quad \text{and} \quad \mathbf{K}_2 = \begin{pmatrix} 0 \\ 1 \\ 1 \end{pmatrix}$$

Since neither eigenvector is a constant multiple of the other, we have found, corresponding to the same eigenvalue, two linearly independent solutions

$$\mathbf{X}_1 = \begin{pmatrix} 1 \\ 1 \\ 0 \end{pmatrix} e^{-t} \quad \text{and} \quad \mathbf{X}_2 = \begin{pmatrix} 0 \\ 1 \\ 1 \end{pmatrix} e^{-t}$$

Last, for $\lambda_3 = 5$, the reduction

$$(\mathbf{A} - 5\mathbf{I}|\mathbf{0}) = \left(\begin{array}{ccc|c} -4 & -2 & 2 & 0 \\ -2 & -4 & -2 & 0 \\ 2 & -2 & -4 & 0 \end{array}\right) \xrightarrow{\text{row operations}} \left(\begin{array}{ccc|c} 1 & 0 & -1 & 0 \\ 0 & 1 & 1 & 0 \\ 0 & 0 & 0 & 0 \end{array}\right)$$

implies $k_1 = k_3$ and $k_2 = -k_3$. Picking $k_3 = 1$ gives $k_1 = 1$, $k_2 = -1$, and thus a third eigenvector and corresponding solution are

$$\mathbf{K}_3 = \begin{pmatrix} 1 \\ -1 \\ 1 \end{pmatrix}, \quad \mathbf{X}_3 = \begin{pmatrix} 1 \\ -1 \\ 1 \end{pmatrix} e^{5t}$$

We conclude that the general solution of the system is

$$\mathbf{X} = c_1\begin{pmatrix}1\\1\\0\end{pmatrix}e^{-t} + c_2\begin{pmatrix}0\\1\\1\end{pmatrix}e^{-t} + c_3\begin{pmatrix}1\\-1\\1\end{pmatrix}e^{5t}$$ ■

The matrix of coefficients $\mathbf{A}$ in Example 3 is a special kind of matrix known as a symmetric matrix. An $n \times n$ matrix $\mathbf{A}$ is said to be **symmetric** if its transpose $\mathbf{A}^T$ (where the rows and columns are interchanged) is the same as $\mathbf{A}$, that is, $\mathbf{A}^T = \mathbf{A}$. It can be proved that if the matrix $\mathbf{A}$ in the system $\mathbf{X}' = \mathbf{AX}$ is symmetric and has real entries, then we can always find n linearly independent eigenvectors $\mathbf{K}_1, \mathbf{K}_2, \ldots, \mathbf{K}_n$, and so the general solution of such a system is the linear combination given in Theorem 4.7. As illustrated in Example 3, this result holds even when some of the eigenvalues are repeated.

Second Solution Now suppose that λ_1 is an eigenvalue of multiplicity two and that there is only one eigenvector associated with this value. A second solution can be found of the form

$$\mathbf{X}_2 = \mathbf{K}te^{\lambda_1 t} + \mathbf{P}e^{\lambda_1 t} \tag{12}$$

where

$$\mathbf{K} = \begin{pmatrix}k_1\\k_2\\\vdots\\k_n\end{pmatrix} \quad \text{and} \quad \mathbf{P} = \begin{pmatrix}p_1\\p_2\\\vdots\\p_n\end{pmatrix}$$

To see this, we substitute (12) into the system $\mathbf{X}' = \mathbf{AX}$ and simplify:

$$(\mathbf{AK} - \lambda_1\mathbf{K})te^{\lambda_1 t} + (\mathbf{AP} - \lambda_1\mathbf{P} - \mathbf{K})e^{\lambda_1 t} = \mathbf{0}$$

Since this last equation is to hold for all values of t, we must have

$$(\mathbf{A} - \lambda_1\mathbf{I})\mathbf{K} = \mathbf{0} \tag{13}$$

and

$$(\mathbf{A} - \lambda_1\mathbf{I})\mathbf{P} = \mathbf{K} \tag{14}$$

Equation (13) simply states that $\mathbf{K}$ must be an eigenvector of $\mathbf{A}$ associated with λ_1. By solving (13), we find one solution $\mathbf{X}_1 = \mathbf{K}e^{\lambda_1 t}$. To find the second solution $\mathbf{X}_2$ we need only solve the additional system (14) for the vector $\mathbf{P}$.

■ EXAMPLE 4 **Repeated Eigenvalues**

Find the general solution of the system given in (10).

Solution From (11), we know that $\lambda_1 = -3$ and that one solution is $\mathbf{X}_1 = \begin{pmatrix}3\\1\end{pmatrix}e^{-3t}$. Identifying $\mathbf{K} = \begin{pmatrix}3\\1\end{pmatrix}$ and $\mathbf{P} = \begin{pmatrix}p_1\\p_2\end{pmatrix}$, we find from (14) that we must now solve

$$(\mathbf{A} + 3\mathbf{I})\mathbf{P} = \mathbf{K} \quad \text{or} \quad \begin{aligned} 6p_1 - 18p_2 &= 3 \\ 2p_1 - 6p_2 &= 1 \end{aligned}$$

Since this system is obviously equivalent to one equation, we have an infinite number of choices for p_1 and p_2. For example, by choosing $p_1 = 1$, we find $p_2 = \frac{1}{6}$. However, for simplicity, we shall choose $p_1 = \frac{1}{2}$ so that $p_2 = 0$. Hence $\mathbf{P} = \begin{pmatrix} \frac{1}{2} \\ 0 \end{pmatrix}$. Thus from (12) we find

$$\mathbf{X}_2 = \begin{pmatrix} 3 \\ 1 \end{pmatrix} te^{-3t} + \begin{pmatrix} \frac{1}{2} \\ 0 \end{pmatrix} e^{-3t}$$

The general solution of (10) is then

$$\mathbf{X} = c_1 \begin{pmatrix} 3 \\ 1 \end{pmatrix} e^{-3t} + c_2 \left[\begin{pmatrix} 3 \\ 1 \end{pmatrix} te^{-3t} + \begin{pmatrix} \frac{1}{2} \\ 0 \end{pmatrix} e^{-3t} \right]$$ ■

Eigenvalues of Multiplicity Three When a matrix **A** has only one eigenvector associated with an eigenvalue λ_1 of multiplicity three, we can find a second solution of the form (12) and a third solution of the form

$$\mathbf{X}_3 = \mathbf{K}\frac{t^2}{2}e^{\lambda_1 t} + \mathbf{P}te^{\lambda_1 t} + \mathbf{Q}e^{\lambda_1 t} \tag{15}$$

where $$\mathbf{K} = \begin{pmatrix} k_1 \\ k_2 \\ \vdots \\ k_n \end{pmatrix}, \quad \mathbf{P} = \begin{pmatrix} p_1 \\ p_2 \\ \vdots \\ p_n \end{pmatrix}, \quad \text{and} \quad \mathbf{Q} = \begin{pmatrix} q_1 \\ q_2 \\ \vdots \\ q_n \end{pmatrix}$$

By substituting (15) into the system $\mathbf{X}' = \mathbf{AX}$, we find the column vectors **K**, **P**, and **Q** must satisfy

$$(\mathbf{A} - \lambda_1\mathbf{I})\mathbf{K} = \mathbf{0} \tag{16}$$

$$(\mathbf{A} - \lambda_1\mathbf{I})\mathbf{P} = \mathbf{K} \tag{17}$$

and $$(\mathbf{A} - \lambda_1\mathbf{I})\mathbf{Q} = \mathbf{P} \tag{18}$$

Of course the solutions of (16) and (17) can be used in forming the solutions $\mathbf{X}_1$ and $\mathbf{X}_2$.

■ **EXAMPLE 5 Repeated Eigenvalues**

Solve $\mathbf{X}' = \begin{pmatrix} 2 & 1 & 6 \\ 0 & 2 & 5 \\ 0 & 0 & 2 \end{pmatrix}\mathbf{X}$.

Solution The characteristic equation $(\lambda - 2)^3 = 0$ shows that $\lambda_1 = 2$ is an eigenvalue of multiplicity three. But solving $(\mathbf{A} - 2\mathbf{I})\mathbf{K} = \mathbf{0}$, we find the single eigenvector and the corresponding solution:

$$\mathbf{K} = \begin{pmatrix} 1 \\ 0 \\ 0 \end{pmatrix}, \quad \mathbf{X}_1 = \begin{pmatrix} 1 \\ 0 \\ 0 \end{pmatrix} e^{2t}$$

We next solve the systems $(\mathbf{A} - 2\mathbf{I})\mathbf{P} = \mathbf{K}$ and $(\mathbf{A} - 2\mathbf{I})\mathbf{Q} = \mathbf{P}$ in succession and find that

$$\mathbf{P} = \begin{pmatrix} 0 \\ 1 \\ 0 \end{pmatrix} \quad \text{and} \quad \mathbf{Q} = \begin{pmatrix} 0 \\ -\frac{6}{5} \\ \frac{1}{5} \end{pmatrix}$$

Thus from (12) and (15) we find the additional solutions

$$\mathbf{X}_2 = \begin{pmatrix} 1 \\ 0 \\ 0 \end{pmatrix} te^{2t} + \begin{pmatrix} 0 \\ 1 \\ 0 \end{pmatrix} e^{2t} \quad \text{and} \quad \mathbf{X}_3 = \begin{pmatrix} 1 \\ 0 \\ 0 \end{pmatrix} \frac{t^2}{2} e^{2t} + \begin{pmatrix} 0 \\ 1 \\ 0 \end{pmatrix} te^{2t} + \begin{pmatrix} 0 \\ -\frac{6}{5} \\ \frac{1}{5} \end{pmatrix} e^{2t}$$

The general solution of the system is then $\mathbf{X} = c_1\mathbf{X}_1 + c_2\mathbf{X}_2 + c_3\mathbf{X}_3$. ■

Remarks

When an eigenvalue λ_1 has multiplicity m, either we can find m linearly independent eigenvectors or the number of corresponding eigenvectors is less than m. Hence the two cases listed on pages 256–257 are not all the possibilities under which a repeated eigenvalue can occur. It could happen, say, that a 5×5 matrix has an eigenvalue of multiplicity 5 and there exist three linearly independent eigenvectors. See Problems 29 and 30 in Exercises 4.2.

4.2.3 Complex Eigenvalues

If $\lambda_1 = \alpha + i\beta$ and $\lambda_2 = \alpha - i\beta$, $i^2 = -1$, are complex eigenvalues of the coefficient matrix $\mathbf{A}$, we can then certainly expect their corresponding eigenvectors to also have complex entries.*

For example, the characteristic equation of the system

$$\begin{aligned} \frac{dx}{dt} &= 6x - y \\ \frac{dy}{dt} &= 5x + 4y \end{aligned} \tag{19}$$

*When the characteristic equation has real coefficients, complex eigenvalues always appear in conjugate pairs.

is
$$\det(\mathbf{A} - \lambda\mathbf{I}) = \begin{vmatrix} 6-\lambda & -1 \\ 5 & 4-\lambda \end{vmatrix} = \lambda^2 - 10\lambda + 29 = 0$$

From the quadratic formula, we find $\lambda_1 = 5 + 2i$, $\lambda_2 = 5 - 2i$.

Now for $\lambda_1 = 5 + 2i$, we must solve

$$\begin{aligned} (1-2i)k_1 - \qquad\qquad k_2 &= 0 \\ 5k_1 - (1+2i)k_2 &= 0 \end{aligned}$$

Since $k_2 = (1 - 2i)k_1$,* the choice $k_1 = 1$ gives the following eigenvector and a solution vector

$$\mathbf{K}_1 = \begin{pmatrix} 1 \\ 1-2i \end{pmatrix}, \quad \mathbf{X}_1 = \begin{pmatrix} 1 \\ 1-2i \end{pmatrix} e^{(5+2i)t}$$

In like manner, for $\lambda_2 = 5 - 2i$ we find

$$\mathbf{K}_2 = \begin{pmatrix} 1 \\ 1+2i \end{pmatrix}, \quad \mathbf{X}_2 = \begin{pmatrix} 1 \\ 1+2i \end{pmatrix} e^{(5-2i)t}$$

We can verify by means of the Wronskian that these solution vectors are linearly independent and so the general solution of (19) is

$$\mathbf{X} = c_1 \begin{pmatrix} 1 \\ 1-2i \end{pmatrix} e^{(5+2i)t} + c_2 \begin{pmatrix} 1 \\ 1+2i \end{pmatrix} e^{(5-2i)t} \tag{20}$$

Note that the entries in $\mathbf{K}_2$ corresponding to λ_2 are the conjugates of the entries in $\mathbf{K}_1$ corresponding to λ_1. The conjugate of λ_1 is, of course, λ_2. We write this as $\lambda_2 = \overline{\lambda}_1$ and $\mathbf{K}_2 = \overline{\mathbf{K}}_1$. We have illustrated the following general result.

Theorem 4.8 Solutions Corresponding to a Complex Eigenvalue

Let $\mathbf{A}$ be the coefficient matrix having real entries of the homogeneous system (2), and let $\mathbf{K}_1$ be an eigenvector corresponding to the complex eigenvalue $\lambda_1 = \alpha + i\beta$, α and β real. Then

$$\mathbf{K}_1 e^{\lambda_1 t} \quad \text{and} \quad \overline{\mathbf{K}}_1 e^{\overline{\lambda}_1 t}$$

are solutions of (2).

It is desirable and relatively easy to rewrite a solution such as (20) in terms of real functions. To this end we first use Euler's formula to write

$$\begin{aligned} e^{(5+2i)t} &= e^{5t}e^{2ti} = e^{5t}(\cos 2t + i \sin 2t) \\ e^{(5-2i)t} &= e^{5t}e^{-2ti} = e^{5t}(\cos 2t - i \sin 2t) \end{aligned}$$

*Note that the second equation is simply $(1 + 2i)$ times the first.

Then, after we multiply complex numbers, collect terms, and replace $c_1 + c_2$ by C_1 and $(c_1 - c_2)i$ by C_2, (20) becomes

$$\mathbf{X} = C_1\mathbf{X}_1 + C_2\mathbf{X}_2 \tag{21}$$

where
$$\mathbf{X}_1 = \left[\begin{pmatrix}1\\1\end{pmatrix}\cos 2t - \begin{pmatrix}0\\-2\end{pmatrix}\sin 2t\right]e^{5t}$$

$$\mathbf{X}_2 = \left[\begin{pmatrix}0\\-2\end{pmatrix}\cos 2t + \begin{pmatrix}1\\1\end{pmatrix}\sin 2t\right]e^{5t}$$

It is now important to realize that the two vectors $\mathbf{X}_1$ and $\mathbf{X}_2$ in (21) are themselves linearly independent *real* solutions of the original system. Consequently, we are justified in ignoring the relationship between C_1, C_2 and c_1, c_2 and we regard C_1 and C_2 as completely arbitrary and real. In other words, the linear combination (21) is an alternative general solution of (19).

The foregoing process can be generalized. Let $\mathbf{K}_1$ be an eigenvector of the coefficient matrix $\mathbf{A}$ (with real entries) corresponding to the complex eigenvalue $\lambda_1 = \alpha + i\beta$. Then the two solution vectors in Theorem 4.8 can be written as

$$\mathbf{K}_1 e^{\lambda_1 t} = \mathbf{K}_1 e^{\alpha t} e^{i\beta t} = \mathbf{K}_1 e^{\alpha t}(\cos \beta t + i \sin \beta t)$$

$$\overline{\mathbf{K}}_1 e^{\overline{\lambda}_1 t} = \overline{\mathbf{K}}_1 e^{\alpha t} e^{-i\beta t} = \overline{\mathbf{K}}_1 e^{\alpha t}(\cos \beta t - i \sin \beta t)$$

By the superposition principle, Theorem 4.2, the following vectors are also solutions

$$\mathbf{X}_1 = \frac{1}{2}(\mathbf{K}_1 e^{\lambda_1 t} + \overline{\mathbf{K}}_1 e^{\overline{\lambda}_1 t}) = \frac{1}{2}(\mathbf{K}_1 + \overline{\mathbf{K}}_1)e^{\alpha t}\cos \beta t - \frac{i}{2}(-\mathbf{K}_1 + \overline{\mathbf{K}}_1)e^{\alpha t}\sin \beta t$$

$$\mathbf{X}_2 = \frac{i}{2}(-\mathbf{K}_1 e^{\lambda_1 t} + \overline{\mathbf{K}}_1 e^{\overline{\lambda}_1 t}) = \frac{i}{2}(-\mathbf{K}_1 + \overline{\mathbf{K}}_1)e^{\alpha t}\cos \beta t + \frac{1}{2}(\mathbf{K}_1 + \overline{\mathbf{K}}_1)e^{\alpha t}\sin \beta t$$

For *any* complex number $z = a + ib$, both $\frac{1}{2}(z + \overline{z}) = a$ and $\frac{i}{2}(-z + \overline{z}) = b$ are *real* numbers. Therefore, the entries in the column vectors $\frac{1}{2}(\mathbf{K}_1 + \overline{\mathbf{K}}_1)$ and $\frac{i}{2}(-\mathbf{K}_1 + \overline{\mathbf{K}}_1)$ are real numbers. By defining

$$\mathbf{B}_1 = \frac{1}{2}(\mathbf{K}_1 + \overline{\mathbf{K}}_1) \quad \text{and} \quad \mathbf{B}_2 = \frac{i}{2}(-\mathbf{K}_1 + \overline{\mathbf{K}}_1) \tag{22}$$

we are led to the following theorem.

Theorem 4.9 Real Solutions Corresponding to a Complex Eigenvalue

Let $\lambda_1 = \alpha + i\beta$ be a complex eigenvalue of the coefficient matrix $\mathbf{A}$ in the homogeneous system (2), and let $\mathbf{B}_1$ and $\mathbf{B}_2$ denote the column vectors defined in (22). Then

$$\begin{aligned} \mathbf{X}_1 &= [\mathbf{B}_1 \cos \beta t - \mathbf{B}_2 \sin \beta t]e^{\alpha t} \\ \mathbf{X}_2 &= [\mathbf{B}_2 \cos \beta t + \mathbf{B}_1 \sin \beta t]e^{\alpha t} \end{aligned} \tag{23}$$

are linearly independent solutions of (2) on $(-\infty, \infty)$.

The matrices $\mathbf{B}_1$ and $\mathbf{B}_2$ in (22) are often denoted by

$$\mathbf{B}_1 = \text{Re}(\mathbf{K}_1) \quad \text{and} \quad \mathbf{B}_2 = \text{Im}(\mathbf{K}_1) \tag{24}$$

since these vectors are, in turn, the *real* and *imaginary* parts of the eigenvector $\mathbf{K}_1$. For example, (21) follows from (23) with

$$\mathbf{K}_1 = \begin{pmatrix} 1 \\ 1 - 2i \end{pmatrix} = \begin{pmatrix} 1 \\ 1 \end{pmatrix} + i \begin{pmatrix} 0 \\ -2 \end{pmatrix}$$

$$\mathbf{B}_1 = \text{Re}(\mathbf{K}_1) = \begin{pmatrix} 1 \\ 1 \end{pmatrix} \quad \text{and} \quad \mathbf{B}_2 = \text{Im}(\mathbf{K}_1) = \begin{pmatrix} 0 \\ -2 \end{pmatrix}$$

EXAMPLE 6 Complex Eigenvalues

Solve $\mathbf{X}' = \begin{pmatrix} 2 & 8 \\ -1 & -2 \end{pmatrix} \mathbf{X}$.

Solution First we obtain the eigenvalues from

$$\det(\mathbf{A} - \lambda \mathbf{I}) = \begin{vmatrix} 2 - \lambda & 8 \\ -1 & -2 - \lambda \end{vmatrix} = \lambda^2 + 4 = 0$$

Thus the eigenvalues are $\lambda_1 = 2i$ and $\lambda_2 = \overline{\lambda_1} = -2i$. For λ_1, the system

$$\begin{aligned} (2 - 2i)k_1 + \qquad 8k_2 &= 0 \\ -k_1 + (-2 - 2i)k_2 &= 0 \end{aligned}$$

gives $k_1 = -(2 + 2i)k_2$. By choosing $k_2 = -1$, we get

$$\mathbf{K}_1 = \begin{pmatrix} 2 + 2i \\ -1 \end{pmatrix} = \begin{pmatrix} 2 \\ -1 \end{pmatrix} + i \begin{pmatrix} 2 \\ 0 \end{pmatrix}$$

Now from (24) we form

$$\mathbf{B}_1 = \text{Re}(\mathbf{K}_1) = \begin{pmatrix} 2 \\ -1 \end{pmatrix} \quad \text{and} \quad \mathbf{B}_2 = \text{Im}(\mathbf{K}_1) = \begin{pmatrix} 2 \\ 0 \end{pmatrix}$$

Since $\alpha = 0$, it follows from (23) that the general solution of the system is

$$\begin{aligned}\mathbf{X} &= c_1\left[\begin{pmatrix} 2 \\ -1 \end{pmatrix}\cos 2t - \begin{pmatrix} 2 \\ 0 \end{pmatrix}\sin 2t\right] + c_2\left[\begin{pmatrix} 2 \\ 0 \end{pmatrix}\cos 2t + \begin{pmatrix} 2 \\ -1 \end{pmatrix}\sin 2t\right] \\ &= c_1\begin{pmatrix} 2\cos 2t - 2\sin 2t \\ -\cos 2t \end{pmatrix} + c_2\begin{pmatrix} 2\cos 2t + 2\sin 2t \\ -\sin 2t \end{pmatrix}\end{aligned}$$ ■

Remarks

Some of the systems considered in Section 3.8 can be solved by the eigenvalue method developed in this section. For example, examination of Problems 1 and 2 in Exercises 3.8 shows that each system is homogeneous and can be written in the normal form $\mathbf{X}' = \mathbf{AX}$. A system such as in Problem 5 of Exercises 3.8, two equations each involving a second derivative, can be rewritten as a homogeneous fourth-order linear system, that is, a system of the form $\mathbf{X}' = \mathbf{AX}$ involving four linear first-order differential equations. See Problems 3 and 10 in *Projects and Computer Lab Experiments for Chapter 4.*

EXERCISES 4.2

Answers to odd-numbered problems begin on page 446.

4.2.1

In Problems 1–12, find the general solution of the given system.

1. $\dfrac{dx}{dt} = x + 2y$, $\quad \dfrac{dy}{dt} = 4x + 3y$

2. $\dfrac{dx}{dt} = 2y$, $\quad \dfrac{dy}{dt} = 8x$

3. $\dfrac{dx}{dt} = -4x + 2y$, $\quad \dfrac{dy}{dt} = -\dfrac{5}{2}x + 2y$

4. $\dfrac{dx}{dt} = \dfrac{1}{2}x + 9y$, $\quad \dfrac{dy}{dt} = \dfrac{1}{2}x + 2y$

5. $\mathbf{X}' = \begin{pmatrix} 10 & -5 \\ 8 & -12 \end{pmatrix}\mathbf{X}$

6. $\mathbf{X}' = \begin{pmatrix} -6 & 2 \\ -3 & 1 \end{pmatrix}\mathbf{X}$

7. $\dfrac{dx}{dt} = x + y - z$, $\quad \dfrac{dy}{dt} = 2y$, $\quad \dfrac{dz}{dt} = y - z$

8. $\dfrac{dx}{dt} = 2x - 7y$, $\quad \dfrac{dy}{dt} = 5x + 10y + 4z$, $\quad \dfrac{dz}{dt} = 5y + 2z$

9. $\mathbf{X}' = \begin{pmatrix} -1 & 1 & 0 \\ 1 & 2 & 1 \\ 0 & 3 & -1 \end{pmatrix}\mathbf{X}$

10. $\mathbf{X}' = \begin{pmatrix} 1 & 0 & 1 \\ 0 & 1 & 0 \\ 1 & 0 & 1 \end{pmatrix}\mathbf{X}$

11. $\mathbf{X}' = \begin{pmatrix} -1 & -1 & 0 \\ \frac{3}{4} & -\frac{3}{2} & 3 \\ \frac{1}{8} & \frac{1}{4} & -\frac{1}{2} \end{pmatrix}\mathbf{X}$

12. $\mathbf{X}' = \begin{pmatrix} -1 & 4 & 2 \\ 4 & -1 & -2 \\ 0 & 0 & 6 \end{pmatrix} \mathbf{X}$

In Problems 13 and 14, solve the given system subject to the indicated initial condition.

13. $\mathbf{X}' = \begin{pmatrix} \frac{1}{2} & 0 \\ 1 & -\frac{1}{2} \end{pmatrix} \mathbf{X}, \quad \mathbf{X}(0) = \begin{pmatrix} 3 \\ 5 \end{pmatrix}$

14. $\mathbf{X}' = \begin{pmatrix} 1 & 1 & 4 \\ 0 & 2 & 0 \\ 1 & 1 & 1 \end{pmatrix} \mathbf{X}, \quad \mathbf{X}(0) = \begin{pmatrix} 1 \\ 3 \\ 0 \end{pmatrix}$

In Problems 15 and 16, use a CAS or linear algebra software as an aid in finding the general solution of the given system.

15. $\mathbf{X}' = \begin{pmatrix} 0.9 & 2.1 & 3.2 \\ 0.7 & 6.5 & 4.2 \\ 1.1 & 1.7 & 3.4 \end{pmatrix} \mathbf{X}$

16. $\mathbf{X}' = \begin{pmatrix} 1 & 0 & 2 & -1.8 & 0 \\ 0 & 5.1 & 0 & -1 & 3 \\ 1 & 2 & -3 & 0 & 0 \\ 0 & 1 & -3.1 & 4 & 0 \\ -2.8 & 0 & 0 & 1.5 & 1 \end{pmatrix} \mathbf{X}$

4.2.2

In Problems 17–26, find the general solution of the given system.

17. $\dfrac{dx}{dt} = 3x - y$

$\dfrac{dy}{dt} = 9x - 3y$

18. $\dfrac{dx}{dt} = -6x + 5y$

$\dfrac{dy}{dt} = -5x + 4y$

19. $\dfrac{dx}{dt} = -x + 3y$

$\dfrac{dy}{dt} = -3x + 5y$

20. $\dfrac{dx}{dt} = 12x - 9y$

$\dfrac{dy}{dt} = 4x$

21. $\dfrac{dx}{dt} = 3x - y - z$

$\dfrac{dy}{dt} = x + y - z$

$\dfrac{dz}{dt} = x - y + z$

22. $\dfrac{dx}{dt} = 3x + 2y + 4z$

$\dfrac{dy}{dt} = 2x + 2z$

$\dfrac{dz}{dt} = 4x + 2y + 3z$

23. $\mathbf{X}' = \begin{pmatrix} 5 & -4 & 0 \\ 1 & 0 & 2 \\ 0 & 2 & 5 \end{pmatrix} \mathbf{X}$

24. $\mathbf{X}' = \begin{pmatrix} 1 & 0 & 0 \\ 0 & 3 & 1 \\ 0 & -1 & 1 \end{pmatrix} \mathbf{X}$

25. $\mathbf{X}' = \begin{pmatrix} 1 & 0 & 0 \\ 2 & 2 & -1 \\ 0 & 1 & 0 \end{pmatrix} \mathbf{X}$

26. $\mathbf{X}' = \begin{pmatrix} 4 & 1 & 0 \\ 0 & 4 & 1 \\ 0 & 0 & 4 \end{pmatrix} \mathbf{X}$

In Problems 27 and 28, solve the given system subject to the indicated initial condition.

27. $\mathbf{X}' = \begin{pmatrix} 2 & 4 \\ -1 & 6 \end{pmatrix} \mathbf{X}, \quad \mathbf{X}(0) = \begin{pmatrix} -1 \\ 6 \end{pmatrix}$

28. $\mathbf{X}' = \begin{pmatrix} 0 & 0 & 1 \\ 0 & 1 & 0 \\ 1 & 0 & 0 \end{pmatrix} \mathbf{X}, \quad \mathbf{X}(0) = \begin{pmatrix} 1 \\ 2 \\ 5 \end{pmatrix}$

29. Show that the 5×5 matrix

$$\mathbf{A} = \begin{pmatrix} 2 & 1 & 0 & 0 & 0 \\ 0 & 2 & 0 & 0 & 0 \\ 0 & 0 & 2 & 0 & 0 \\ 0 & 0 & 0 & 2 & 1 \\ 0 & 0 & 0 & 0 & 2 \end{pmatrix}$$

has an eigenvalue λ_1 of multiplicity five. Show that three linearly independent eigenvectors corresponding to λ_1 can be found.

30. Consider the 5×5 matrix given in Problem 29. Solve the system $\mathbf{X}' = \mathbf{AX}$ without the aid of matrix methods, but write the general solution using the matrix notation.

4.2.3

In Problems 31–42, find the general solution of the given system.

31. $\dfrac{dx}{dt} = 6x - y$

$\dfrac{dy}{dt} = 5x + 2y$

32. $\dfrac{dx}{dt} = x + y$

$\dfrac{dy}{dt} = -2x - y$

33. $\dfrac{dx}{dt} = 5x + y$

$\dfrac{dy}{dt} = -2x + 3y$

34. $\dfrac{dx}{dt} = 4x + 5y$

$\dfrac{dy}{dt} = -2x + 6y$

35. $\mathbf{X}' = \begin{pmatrix} 4 & -5 \\ 5 & -4 \end{pmatrix} \mathbf{X}$

36. $\mathbf{X}' = \begin{pmatrix} 1 & -8 \\ 1 & -3 \end{pmatrix} \mathbf{X}$

37. $\dfrac{dx}{dt} = z$

$\dfrac{dy}{dt} = -z$

$\dfrac{dz}{dt} = y$

38. $\dfrac{dx}{dt} = 2x + y + 2z$

$\dfrac{dy}{dt} = 3x + 6z$

$\dfrac{dz}{dt} = -4x - 3z$

39. $\mathbf{X}' = \begin{pmatrix} 1 & -1 & 2 \\ -1 & 1 & 0 \\ -1 & 0 & 1 \end{pmatrix}\mathbf{X}$

40. $\mathbf{X}' = \begin{pmatrix} 4 & 0 & 1 \\ 0 & 6 & 0 \\ -4 & 0 & 4 \end{pmatrix}\mathbf{X}$

41. $\mathbf{X}' = \begin{pmatrix} 2 & 5 & 1 \\ -5 & -6 & 4 \\ 0 & 0 & 2 \end{pmatrix}\mathbf{X}$

42. $\mathbf{X}' = \begin{pmatrix} 2 & 4 & 4 \\ -1 & -2 & 0 \\ -1 & 0 & -2 \end{pmatrix}\mathbf{X}$

In Problems 43 and 44, solve the given system subject to the indicated initial condition.

43. $\mathbf{X}' = \begin{pmatrix} 1 & -12 & -14 \\ 1 & 2 & -3 \\ 1 & 1 & -2 \end{pmatrix}\mathbf{X}, \quad \mathbf{X}(0) = \begin{pmatrix} 4 \\ 6 \\ -7 \end{pmatrix}$

44. $\mathbf{X}' = \begin{pmatrix} 6 & -1 \\ 5 & 4 \end{pmatrix}\mathbf{X}, \quad \mathbf{X}(0) = \begin{pmatrix} -2 \\ 8 \end{pmatrix}$

Discussion Problems

45. Use the general solution in Problem 30 as a basis for a discussion on how the linear system in Problem 29 can be solved using the matrix methods of this section. Carry out your ideas.

46. A linear system such as $t\mathbf{X}' = \begin{pmatrix} 1 & 3 \\ -1 & 5 \end{pmatrix}\mathbf{X}$ is analogous to a Cauchy-Euler equation. Find a solution of the system having the form $\mathbf{X} = t^m\mathbf{K}$, where $t > 0$ and $\mathbf{K}$ is a matrix of constants.

47. **(a)** Suppose λ_1 is an eigenvalue of multiplicity two and that there is only one eigenvector associated with that value. Show that the vector $\mathbf{P}$ defined in (14) also satisfies $(\mathbf{A} - \lambda_1\mathbf{I})^2\mathbf{P} = \mathbf{0}$ and that the solutions corresponding to λ_1 can be rewritten in terms of $\mathbf{P}$ as

$$\mathbf{X}_1 = (\mathbf{A} - \lambda_1\mathbf{I})\mathbf{P}e^{\lambda_1 t}$$

and

$$\mathbf{X}_2 = [\mathbf{P} + t(\mathbf{A} - \lambda_1\mathbf{I})\mathbf{P}]e^{\lambda_1 t}.$$

(b) Use $\lambda_1 = -3$ in Example 4, and solve the algebraic system $(\mathbf{A} + 3\mathbf{I})^2\mathbf{P} = \mathbf{0}$ where $\mathbf{A}$ is defined in (10). Form $\mathbf{X}_1$ and $\mathbf{X}_2$ as given in part (a). Discuss several points of significance in your result.

48. If $\mathbf{A}$ is a nonsingular matrix, that is, $\mathbf{A}^{-1}$ exists, then it can be shown that none of its eigenvalues are zero. Now multiplying $\mathbf{AK} = \lambda\mathbf{K}$ by $\mathbf{A}^{-1}$ and dividing by λ gives

$$\mathbf{A}^{-1}\mathbf{K} = \frac{1}{\lambda}\mathbf{K}$$

Discuss what this means in terms of the eigenvalues and eigenvectors of $\mathbf{A}^{-1}$.

4.3 Nonhomogeneous Linear Systems

Nonhomogeneous linear systems $\mathbf{X}' = \mathbf{AX} + \mathbf{F}(t)$ with constant coefficients can be solved by methods that are generalizations of the methods that we discussed in Section 3.4 for solving single nonhomogeneous linear equations.

In this section, we will develop a matrix version of the method of *variation of parameters*. But before doing this we need to examine a special matrix that is formed out of the solution vectors of the corresponding homogeneous linear system $\mathbf{X}' = \mathbf{AX}$.

A Fundamental Matrix If $\mathbf{X}_1, \mathbf{X}_2, \ldots, \mathbf{X}_n$ is a fundamental set of solutions of the homogeneous system $\mathbf{X}' = \mathbf{AX}$ on an interval I, then its general solution on the interval is $\mathbf{X} = c_1\mathbf{X}_1 + c_2\mathbf{X}_2 + \cdots + c_n\mathbf{X}_n$, or

$$\mathbf{X} = c_1\begin{pmatrix} x_{11} \\ x_{21} \\ \vdots \\ x_{n1} \end{pmatrix} + c_2\begin{pmatrix} x_{12} \\ x_{22} \\ \vdots \\ x_{n2} \end{pmatrix} + \cdots + c_n\begin{pmatrix} x_{1n} \\ x_{2n} \\ \vdots \\ x_{nn} \end{pmatrix} = \begin{pmatrix} c_1x_{11} + c_2x_{12} + \cdots + c_nx_{1n} \\ c_1x_{21} + c_2x_{22} + \cdots + c_nx_{2n} \\ \vdots \\ c_1x_{n1} + c_2x_{n2} + \cdots + c_nx_{nn} \end{pmatrix} \quad \mathbf{(1)}$$

The matrix in (1) is recognized as the product of an $n \times n$ matrix with an $n \times 1$ matrix. In other words, the general solution (1) can be written as

$$\mathbf{X} = \mathbf{\Phi}(t)\mathbf{C} \quad \mathbf{(2)}$$

where $\mathbf{C}$ is an $n \times 1$ column vector of arbitrary constants $c_1, c_2, \ldots, c_n$, and the $n \times n$ matrix, whose columns consist of the entries of the solution vectors of system $\mathbf{X}' = \mathbf{AX}$,

$$\mathbf{\Phi}(t) = \begin{pmatrix} x_{11} & x_{12} & \cdots & x_{1n} \\ x_{21} & x_{22} & \cdots & x_{2n} \\ \vdots & \vdots & & \vdots \\ x_{n1} & x_{n2} & \cdots & x_{nn} \end{pmatrix}$$

is called a **fundamental matrix** of the system on the interval.

In the discussion that follows, we need to use two properties of a fundamental matrix:

- A fundamental matrix $\mathbf{\Phi}(t)$ is nonsingular.
- If $\mathbf{\Phi}(t)$ is a fundamental matrix of the system $\mathbf{X}' = \mathbf{AX}$, then

$$\mathbf{\Phi}'(t) = \mathbf{A\Phi}(t) \quad \mathbf{(3)}$$

A reexamination of (9) of Theorem 4.3 shows that det $\mathbf{\Phi}(t)$ is the same as the Wronskian $W(\mathbf{X}_1, \mathbf{X}_2, \ldots, \mathbf{X}_n)$. Hence the linear independence of the columns of $\mathbf{\Phi}(t)$ on the interval I guarantees that det $\mathbf{\Phi}(t) \neq 0$ for every t in the interval. Since $\mathbf{\Phi}(t)$ is nonsingular, the multiplicative inverse $\mathbf{\Phi}^{-1}(t)$ exists for every t in the interval. The result given in (3) follows immediately from the fact that every column of $\mathbf{\Phi}(t)$ is a solution vector of $\mathbf{X}' = \mathbf{AX}$.

Variation of Parameters The method of **variation of parameters** for nonhomogeneous linear systems is analogous to the procedure that we employed in Sections 2.3 and 3.4 to find a particular solution of a single differential equation. But instead of replacing a constant by a "variable parameter," we ask here whether it is possible to replace the matrix of constants $\mathbf{C}$ in (2) by a column matrix of functions

$$\mathbf{U}(t) = \begin{pmatrix} u_1(t) \\ u_2(t) \\ \vdots \\ u_n(t) \end{pmatrix} \quad \text{so that} \quad \mathbf{X}_p = \mathbf{\Phi}(t)\mathbf{U}(t) \quad \mathbf{(4)}$$

is a particular solution of the nonhomogeneous system

$$\mathbf{X}' = \mathbf{AX} + \mathbf{F}(t) \tag{5}$$

By the product rule, the derivative of the last expression in (4) is

$$\mathbf{X}_p' = \mathbf{\Phi}(t)\mathbf{U}'(t) + \mathbf{\Phi}'(t)\mathbf{U}(t) \tag{6}$$

Note that the order of the products in (6) is important. Since $\mathbf{U}(t)$ is a column matrix, the products $\mathbf{U}'(t)\mathbf{\Phi}(t)$ and $\mathbf{U}(t)\mathbf{\Phi}'(t)$ are not defined. Substituting (4) and (6) into (5) gives

$$\mathbf{\Phi}(t)\mathbf{U}'(t) + \mathbf{\Phi}'(t)\mathbf{U}(t) = \mathbf{A\Phi}(t)\mathbf{U}(t) + \mathbf{F}(t) \tag{7}$$

Now using (3) to replace $\mathbf{\Phi}'(t)$, (7) becomes

$$\mathbf{\Phi}(t)\mathbf{U}'(t) + \mathbf{A\Phi}(t)\mathbf{U}(t) = \mathbf{A\Phi}(t)\mathbf{U}(t) + \mathbf{F}(t)$$

or

$$\mathbf{\Phi}(t)\mathbf{U}'(t) = \mathbf{F}(t) \tag{8}$$

Multiplying both sides of equation (8) by $\mathbf{\Phi}^{-1}(t)$ gives

$$\mathbf{U}'(t) = \mathbf{\Phi}^{-1}(t)\,\mathbf{F}(t) \quad \text{and so} \quad \mathbf{U}(t) = \int \mathbf{\Phi}^{-1}(t)\,\mathbf{F}(t)\,dt$$

Since $\mathbf{X}_p = \mathbf{\Phi}(t)\mathbf{U}(t)$, we conclude that a particular solution of (5) is

$$\mathbf{X}_p = \mathbf{\Phi}(t) \int \mathbf{\Phi}^{-1}(t)\mathbf{F}(t)\,dt \tag{9}$$

To calculate the indefinite integral of the column matrix $\mathbf{\Phi}^{-1}(t)\mathbf{F}(t)$ in (9), we integrate each entry. Thus the general solution of the system (5) is $\mathbf{X} = \mathbf{X}_c + \mathbf{X}_p$ or

$$\mathbf{X} = \mathbf{\Phi}(t)\mathbf{C} + \mathbf{\Phi}(t) \int \mathbf{\Phi}^{-1}(t)\mathbf{F}(t)\,dt \tag{10}$$

■ EXAMPLE 1 Variation of Parameters

Find the general solution of the nonhomogeneous system

$$\mathbf{X}' = \begin{pmatrix} -3 & 1 \\ 2 & -4 \end{pmatrix}\mathbf{X} + \begin{pmatrix} 3t \\ e^{-t} \end{pmatrix} \tag{11}$$

on the interval $(-\infty, \infty)$.

Solution We first solve the homogeneous system

$$\mathbf{X}' = \begin{pmatrix} -3 & 1 \\ 2 & -4 \end{pmatrix}\mathbf{X} \tag{12}$$

The characteristic equation of the coefficient matrix is

$$\det(\mathbf{A} - \lambda\mathbf{I}) = \begin{vmatrix} -3-\lambda & 1 \\ 2 & -4-\lambda \end{vmatrix} = (\lambda + 2)(\lambda + 5) = 0$$

so the eigenvalues are $\lambda_1 = -2$ and $\lambda_2 = -5$. By the usual method, we find that the eigenvectors corresponding to λ_1 and λ_2 are, respectively,

$$\begin{pmatrix}1\\1\end{pmatrix} \quad \text{and} \quad \begin{pmatrix}1\\-2\end{pmatrix}$$

The solution vectors of the system (12) are then

$$\mathbf{X}_1 = \begin{pmatrix}1\\1\end{pmatrix}e^{-2t} = \begin{pmatrix}e^{-2t}\\e^{-2t}\end{pmatrix} \quad \text{and} \quad \mathbf{X}_2 = \begin{pmatrix}1\\-2\end{pmatrix}e^{-5t} = \begin{pmatrix}e^{-5t}\\-2e^{-5t}\end{pmatrix}$$

The entries in $\mathbf{X}_1$ form the first column of $\mathbf{\Phi}(t)$, and the entries in $\mathbf{X}_2$ form the second column of $\mathbf{\Phi}(t)$. Hence

$$\mathbf{\Phi}(t) = \begin{pmatrix}e^{-2t} & e^{-5t}\\e^{-2t} & -2e^{-5t}\end{pmatrix} \quad \text{and} \quad \mathbf{\Phi}^{-1}(t) = \begin{pmatrix}\frac{2}{3}e^{2t} & \frac{1}{3}e^{2t}\\\frac{1}{3}e^{5t} & -\frac{1}{3}e^{5t}\end{pmatrix}$$

From (9) we obtain

$$\begin{aligned}
\mathbf{X}_p = \mathbf{\Phi}(t)\int \mathbf{\Phi}^{-1}(t)\mathbf{F}(t)\,dt &= \begin{pmatrix}e^{-2t} & e^{-5t}\\e^{-2t} & -2e^{-5t}\end{pmatrix}\int \begin{pmatrix}\frac{2}{3}e^{2t} & \frac{1}{3}e^{2t}\\\frac{1}{3}e^{5t} & -\frac{1}{3}e^{5t}\end{pmatrix}\begin{pmatrix}3t\\e^{-t}\end{pmatrix}dt\\
&= \begin{pmatrix}e^{-2t} & e^{-5t}\\e^{-2t} & -2e^{-5t}\end{pmatrix}\int \begin{pmatrix}2te^{2t} + \frac{1}{3}e^{t}\\te^{5t} - \frac{1}{3}e^{4t}\end{pmatrix}dt\\
&= \begin{pmatrix}e^{-2t} & e^{-5t}\\e^{-2t} & -2e^{-5t}\end{pmatrix}\begin{pmatrix}te^{2t} - \frac{1}{2}e^{2t} + \frac{1}{3}e^{t}\\\frac{1}{5}te^{5t} - \frac{1}{25}e^{5t} - \frac{1}{12}e^{4t}\end{pmatrix}\\
&= \begin{pmatrix}\frac{6}{5}t - \frac{27}{50} + \frac{1}{4}e^{-t}\\\frac{3}{5}t - \frac{21}{50} + \frac{1}{2}e^{-t}\end{pmatrix}
\end{aligned}$$

Hence from (10) the general solution of (11) on the interval is

$$\begin{aligned}
\mathbf{X} &= \begin{pmatrix}e^{-2t} & e^{-5t}\\e^{-2t} & -2e^{-5t}\end{pmatrix}\begin{pmatrix}c_1\\c_2\end{pmatrix} + \begin{pmatrix}\frac{6}{5}t - \frac{27}{50} + \frac{1}{4}e^{-t}\\\frac{3}{5}t - \frac{21}{50} + \frac{1}{2}e^{-t}\end{pmatrix}\\
&= c_1\begin{pmatrix}1\\1\end{pmatrix}e^{-2t} + c_2\begin{pmatrix}1\\-2\end{pmatrix}e^{-5t} + \begin{pmatrix}\frac{6}{5}\\\frac{3}{5}\end{pmatrix}t - \begin{pmatrix}\frac{27}{50}\\\frac{21}{50}\end{pmatrix} + \begin{pmatrix}\frac{1}{4}\\\frac{1}{2}\end{pmatrix}e^{-t}
\end{aligned}$$

■

Initial-Value Problem The general solution of the nonhomogeneous system (5) on an interval can be written in an alternative manner

$$\mathbf{X} = \mathbf{\Phi}(t)\mathbf{C} + \mathbf{\Phi}(t)\int_{t_0}^{t} \mathbf{\Phi}^{-1}(s)\mathbf{F}(s)\,ds \tag{13}$$

where t and t_0 are points in the interval. The last form is useful in solving (5) subject to an initial condition $\mathbf{X}(t_0) = \mathbf{X}_0$, because the limits of integration are chosen so that the particular solution vanishes at $t = t_0$. Substituting $t = t_0$ in (13) yields $\mathbf{X}_0 = \mathbf{\Phi}(t_0)\mathbf{C}$, from which we get $\mathbf{C} = \mathbf{\Phi}^{-1}(t_0)\mathbf{X}_0$. Substituting this last result in (13) gives the following solution of the initial-value problem:

$$\mathbf{X} = \mathbf{\Phi}(t)\mathbf{\Phi}^{-1}(t_0)\mathbf{X}_0 + \mathbf{\Phi}(t)\int_{t_0}^{t} \mathbf{\Phi}^{-1}(s)\mathbf{F}(s)\,ds \tag{14}$$

EXERCISES 4.3

Answers to odd-numbered problems begin on page 447.

In Problems 1–20, use variation of parameters to solve the given system.

1. $\dfrac{dx}{dt} = 3x - 3y + 4$

   $\dfrac{dy}{dt} = 2x - 2y - 1$

2. $\dfrac{dx}{dt} = 2x - y$

   $\dfrac{dy}{dt} = 3x - 2y + 4t$

3. $\mathbf{X}' = \begin{pmatrix} 3 & -5 \\ \frac{3}{4} & -1 \end{pmatrix}\mathbf{X} + \begin{pmatrix} 1 \\ -1 \end{pmatrix} e^{t/2}$

4. $\mathbf{X}' = \begin{pmatrix} 2 & -1 \\ 4 & 2 \end{pmatrix}\mathbf{X} + \begin{pmatrix} \sin 2t \\ 2\cos 2t \end{pmatrix} e^{2t}$

5. $\mathbf{X}' = \begin{pmatrix} 0 & 2 \\ -1 & 3 \end{pmatrix}\mathbf{X} + \begin{pmatrix} 1 \\ -1 \end{pmatrix} e^{t}$

6. $\mathbf{X}' = \begin{pmatrix} 0 & 2 \\ -1 & 3 \end{pmatrix}\mathbf{X} + \begin{pmatrix} 2 \\ e^{-3t} \end{pmatrix}$

7. $\mathbf{X}' = \begin{pmatrix} 1 & 8 \\ 1 & -1 \end{pmatrix}\mathbf{X} + \begin{pmatrix} 12 \\ 12 \end{pmatrix} t$

8. $\mathbf{X}' = \begin{pmatrix} 1 & 8 \\ 1 & -1 \end{pmatrix}\mathbf{X} + \begin{pmatrix} e^{-t} \\ te^{t} \end{pmatrix}$

9. $\mathbf{X}' = \begin{pmatrix} 3 & 2 \\ -2 & -1 \end{pmatrix}\mathbf{X} + \begin{pmatrix} 2e^{-t} \\ e^{-t} \end{pmatrix}$

10. $\mathbf{X}' = \begin{pmatrix} 3 & 2 \\ -2 & -1 \end{pmatrix}\mathbf{X} + \begin{pmatrix} 1 \\ 1 \end{pmatrix}$

11. $\mathbf{X}' = \begin{pmatrix} 0 & -1 \\ 1 & 0 \end{pmatrix}\mathbf{X} + \begin{pmatrix} \sec t \\ 0 \end{pmatrix}$

12. $\mathbf{X}' = \begin{pmatrix} 1 & -1 \\ 1 & 1 \end{pmatrix}\mathbf{X} + \begin{pmatrix} 3 \\ 3 \end{pmatrix} e^{t}$

13. $\mathbf{X}' = \begin{pmatrix} 1 & -1 \\ 1 & 1 \end{pmatrix}\mathbf{X} + \begin{pmatrix} \cos t \\ \sin t \end{pmatrix} e^{t}$

14. $\mathbf{X}' = \begin{pmatrix} 2 & -2 \\ 8 & -6 \end{pmatrix}\mathbf{X} + \begin{pmatrix} 1 \\ 3 \end{pmatrix} \dfrac{e^{-2t}}{t}$

15. $\mathbf{X}' = \begin{pmatrix} 0 & 1 \\ -1 & 0 \end{pmatrix}\mathbf{X} + \begin{pmatrix} 0 \\ \sec t \tan t \end{pmatrix}$

16. $\mathbf{X}' = \begin{pmatrix} 0 & 1 \\ -1 & 0 \end{pmatrix}\mathbf{X} + \begin{pmatrix} 1 \\ \cot t \end{pmatrix}$

17. $\mathbf{X}' = \begin{pmatrix} 1 & 2 \\ -\frac{1}{2} & 1 \end{pmatrix}\mathbf{X} + \begin{pmatrix} \csc t \\ \sec t \end{pmatrix} e^{t}$

18. $\mathbf{X}' = \begin{pmatrix} 1 & -2 \\ 1 & -1 \end{pmatrix}\mathbf{X} + \begin{pmatrix} \tan t \\ 1 \end{pmatrix}$

19. $\mathbf{X}' = \begin{pmatrix} 1 & 1 & 0 \\ 1 & 1 & 0 \\ 0 & 0 & 3 \end{pmatrix}\mathbf{X} + \begin{pmatrix} e^{t} \\ e^{2t} \\ te^{3t} \end{pmatrix}$

20. $\mathbf{X}' = \begin{pmatrix} 3 & -1 & -1 \\ 1 & 1 & -1 \\ 1 & -1 & 1 \end{pmatrix}\mathbf{X} + \begin{pmatrix} 0 \\ t \\ 2e^{t} \end{pmatrix}$

In Problems 21 and 22, use (14) to solve the given system subject to the indicated initial condition.

21. $\mathbf{X}' = \begin{pmatrix} 3 & -1 \\ -1 & 3 \end{pmatrix}\mathbf{X} + \begin{pmatrix} 4e^{2t} \\ 4e^{4t} \end{pmatrix}, \quad \mathbf{X}(0) = \begin{pmatrix} 1 \\ 1 \end{pmatrix}$

22. $\mathbf{X}' = \begin{pmatrix} 1 & -1 \\ 1 & -1 \end{pmatrix}\mathbf{X} + \begin{pmatrix} 1/t \\ 1/t \end{pmatrix}, \quad \mathbf{X}(1) = \begin{pmatrix} 2 \\ -1 \end{pmatrix}$

4.4 Mathematical Models

All the mathematical models that we have considered up to now have been single differential equations. In this section, we are going to reexamine some of the topics already discussed in Section 2.4, namely, radioactive decay, mixtures, and population growth. But this time our assumptions will lead to mathematical models that are *systems* of differential equations, or more accurately, systems of *first-order* DEs. Analogous to Section 2.4, where the models for radioactive decay and mixtures were linear and the important logistic model for population growth was nonlinear, our second look at these topics yields linear and nonlinear systems.

The linear systems of this section can be solved using the eigenvalue/eigenvector method of this chapter, but like nonlinear equations, nonlinear systems generally defy solution.

A single differential equation could describe a single-species population in an environment; but if there are, say, two species living, interacting, and perhaps competing in the same environment (for example, rabbits and foxes), then a model for their populations $x(t)$ and $y(t)$ *might* be a system of two first-order differential equations such as

$$\frac{dx}{dt} = g_1(t, x, y)$$
$$\frac{dy}{dt} = g_2(t, x, y) \tag{1}$$

When g_1 and g_2 are linear in the variables x and y, that is, $g_1(x, y) = c_1x + c_2y + f_1(t)$ and $g_2(x, y) = c_3x + c_4y + f_2(t)$, then (1) is said to be a **linear system**. A system of differential equations that is not linear is said to be **nonlinear**.

Radioactive Series In the discussion of radioactive decay in Section 2.4, we assumed that the rate of decay was proportional to the number $A(t)$ of nuclei of the substance present at time t. When a substance decays by radioactivity, it usually doesn't just transmute into one stable substance and the process stops; rather, the first substance decays into another radioactive substance, this second substance in turn decays into a third substance, and so on. This process, called a **radioactive decay series**, continues until a stable element is reached. For example, the uranium decay series is U-238 → Th-234 → · · · → Pb-206, where Pb-206 is a stable isotope of lead. The half-lives of the various elements in a radioactive series can range from billions of years (4.5×10^9 years for U-238) to a fraction of a second. Suppose a radioactive series is described schematically by $X \xrightarrow{-\lambda_1} Y \xrightarrow{-\lambda_2} Z$, where $k_1 = -\lambda_1 < 0$ and $k_2 = -\lambda_2 < 0$ are the decay constants for substances X and Y, respectively, and Z is a stable element. Suppose too, that $x(t)$, $y(t)$, and $z(t)$ denote the amounts of the substances X, Y, and Z, respectively, remaining at time t. The decay of element X is described by

$$\frac{dx}{dt} = -\lambda_1 x$$

whereas the rate at which the second element Y decays is the net rate,

$$\frac{dy}{dt} = \lambda_1 x - \lambda_2 y$$

since it is *gaining* atoms from the decay of X and at the same time *losing* atoms because of its own decay. Since Z is a stable element, it is simply gaining atoms from the decay of element Y:

$$\frac{dz}{dt} = \lambda_2 y$$

In other words, a model of the radioactive decay series for three elements is the linear system of three first-order differential equations

$$\begin{aligned}\frac{dx}{dt} &= -\lambda_1 x \\ \frac{dy}{dt} &= \lambda_1 x - \lambda_2 y \\ \frac{dz}{dt} &= \lambda_2 y\end{aligned} \tag{2}$$

Mixtures Consider the two tanks shown in Figure 4.1. Let us suppose, for the sake of discussion, that tank A contains 50 gal of water in which 25 lb of salt are dissolved. Suppose tank B contains 50 gal of pure water. Liquid is pumped in and out of the tanks as indicated in the figure; the mixture exchanged between the two tanks and the liquid pumped out of tank B are assumed to be well stirred. We wish to construct a mathematical model that describes at time t the number of pounds $x_1(t)$ of salt in tank A and the number of pounds $x_2(t)$ of salt in tank B.

By an analysis similar to that in Example 5 of Section 2.4, we see for tank A that the net rate of change of $x_1(t)$ is given by

$$\begin{aligned}\frac{dx_1}{dt} &= \overbrace{(3\text{ gal/min})\cdot(0\text{ lb/gal}) + (1\text{ gal/min})\cdot\left(\frac{x_2}{50}\text{ lb/gal}\right)}^{\text{Input rate of salt}} - \overbrace{(4\text{ gal/min})\cdot\left(\frac{x_1}{50}\text{ lb/gal}\right)}^{\text{Output rate of salt}} \\ &= -\frac{2}{25}x_1 + \frac{1}{50}x_2\end{aligned}$$

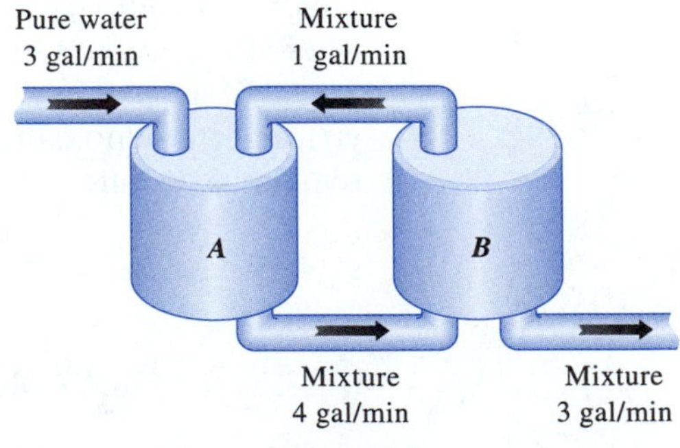

Figure 4.1.

Similarly, for tank B the net rate of change of $x_2(t)$ is

$$\frac{dx_2}{dt} = 4 \cdot \frac{x_1}{50} - 3 \cdot \frac{x_2}{50} - 1 \cdot \frac{x_2}{50} = \frac{2}{25}x_1 - \frac{2}{25}x_2$$

Thus we obtain the linear system

$$\begin{aligned} \frac{dx_1}{dt} &= -\frac{2}{25}x_1 + \frac{1}{50}x_2 \\ \frac{dx_2}{dt} &= \frac{2}{25}x_1 - \frac{2}{25}x_2 \end{aligned} \tag{3}$$

Observe that the foregoing system is accompanied by the initial conditions $x_1(0) = 25$, $x_2(0) = 0$.

Population Dynamics In Chapter 2, we considered models for the population of a single species: the malthusian model $dP/dt = kP$ that describes exponential growth $(k > 0)$ or decline $(k < 0)$, and the logistic model $dP/dt = P(a - bP)$, $a > 0$, $b > 0$, that can describe bounded growth. Underlying these two models was the assumption that the species of animals occupied an environment in which there was no interaction whatsoever with any other species of animals. In the discussion that follows, we consider some of the classic models that describe the populations of two species occupying the same environment: predator–prey interactions and two-species competition.

A Predator–Prey Model Suppose that two different species of animals interact within the same habitat and suppose further that one species is a predator and the other is its prey. For example, wolves hunt grass-eating caribou, sharks devour smaller fish, and the snowy owl pursues an arctic rodent called the lemming. For our purposes, let us imagine that the predators are foxes and the prey are rabbits. The rabbits eat only vegetation and the foxes eat only the rabbits.

Let $x(t)$ and $y(t)$ denote, respectively, the fox and rabbit populations at time t. If there were no rabbits, then one might expect that the foxes, lacking an adequate food supply, would decline in number according to

$$\frac{dx}{dt} = -ax, \quad a > 0 \tag{4}$$

However, when rabbits are present in the environment it seems reasonable that the number of encounters or interactions between these two species per unit time is jointly proportional to their populations x and y, that is, proportional to the product xy. Thus when rabbits are present, there is a supply of food and so foxes are added to the system at a rate bxy, $b > 0$. Adding this last rate to (4) gives a model for the fox population

$$\frac{dx}{dt} = -ax + bxy \tag{5}$$

On the other hand, were there no foxes, then the rabbits would, with an added assumption of unlimited food supply, grow at a rate that is proportional to the number of rabbits present at time t:

$$\frac{dy}{dt} = dy, \quad d > 0 \tag{6}$$

But when foxes are present, a model for the rabbit population is (6) decreased by cxy, $c > 0$, that is, decreased by the rate at which the rabbits are eaten during their encounters with the foxes:

$$\frac{dy}{dt} = dy - cxy \tag{7}$$

Equations (5) and (7) constitute a system of nonlinear first-order differential equations

$$\begin{aligned} \frac{dx}{dt} &= -ax + bxy = x(-a + by) \\ \frac{dy}{dt} &= dy - cxy = y(d - cx) \end{aligned} \tag{8}$$

where a, b, c, and d are positive constants. This famous system of equations is known as the **Lotka-Volterra predator-prey model**.

Except for two constant solutions, $x(t) = 0$, $y(t) = 0$ and $x(t) = d/c$, $y(t) = a/b$, the nonlinear system (8) cannot be solved in terms of elementary functions. However, we can analyze systems such as these numerically and qualitatively.

■ EXAMPLE 1 Predator–Prey Model

Suppose
$$\frac{dx}{dt} = -0.16x + 0.08xy$$
$$\frac{dy}{dt} = 4.5y - 0.9xy$$

represents a predator–prey model. Since we are dealing with populations, we have $x(t) \geq 0$, $y(t) \geq 0$. Figure 4.2, obtained with the aid of an ODE solver, shows typical population curves of the predators and prey for this model superimposed on the same coordinate axes. The initial conditions used were $x(0) = 4$, $y(0) = 4$. The curve in black represents the population $x(t)$ of the predator (foxes) and the colored curve is the population $y(t)$ of the prey (rabbits). Observe that the model seems to predict that both populations $x(t)$ and $y(t)$ are periodic in time. This would make intuitive sense, since as the number of prey decreases, the predator population eventually decreases because of a diminished food supply. But attendant to a decrease in the number of predators, the number of prey increase, and this in turn gives rise to an increased number of predators, which ultimately brings about another decrease in the number of prey. ■

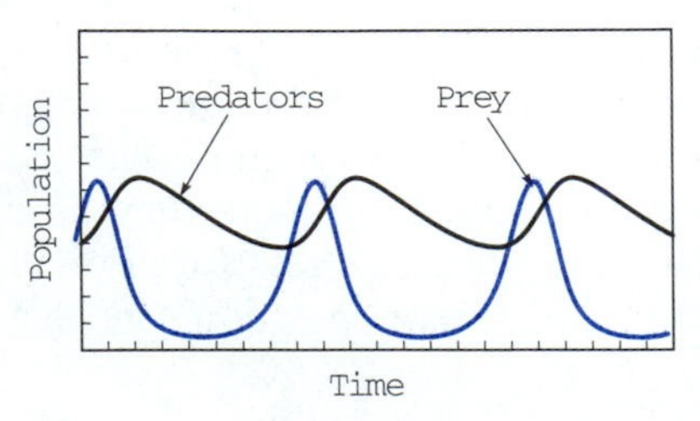

Figure 4.2.

We will reexamine the model in Example 1 from a qualitative viewpoint in the next section.

Competition Model Now suppose two different species of animals occupy the same ecosystem, not as predator and prey, but as competitors for the same resources (such as food, habitat, territory) in the system. In the absence of the other, let us assume that the rate at which each population grows is given by

$$\frac{dx}{dt} = ax \quad \text{and} \quad \frac{dy}{dt} = cy \tag{9}$$

respectively. Since the two species compete, another assumption might be that each of these rates is diminished by the influence, or put in simpler terms, the *existence,* of the other population. Thus a model for the two populations is given by the linear system

$$\begin{aligned} \frac{dx}{dt} &= ax - by \\ \frac{dy}{dt} &= cy - dx \end{aligned} \tag{10}$$

where a, b, c, and d are positive constants. Other the other hand, we might assume as we did in (5), that each growth rate in (9) should be reduced by a rate proportional to the number of interactions between the two species:

$$\begin{aligned} \frac{dx}{dt} &= ax - bxy \\ \frac{dy}{dt} &= cy - dxy \end{aligned} \tag{11}$$

Inspection shows that this nonlinear system is similar to the Lotka-Volterra predator–prey model. Lastly, it might be more realistic to replace the rates in (9) which indicate that the population of each species, in isolation, grows exponentially with rates that indicate each population grows logistically (that is, over a long time the population is bounded),

$$\frac{dx}{dt} = a_1x - b_1x^2 \quad \text{and} \quad \frac{dy}{dt} = a_2y - b_2y^2 \tag{12}$$

When each of these new rates is decreased by rates proportional to the number of interactions, we obtain another nonlinear model

$$\begin{aligned} \frac{dx}{dt} &= a_1x - b_1x^2 - c_1xy = x(a_1 - b_1x - c_1y) \\ \frac{dy}{dt} &= a_2y - b_2y^2 - c_2xy = y(a_2 - b_2y - c_2x) \end{aligned} \tag{13}$$

where all coefficients are positive. The linear system (10) and the nonlinear systems (11) and (13) are called **competition models**.

EXERCISES 4.4 Answers to odd-numbered problems begin on page 448.

1. Use the method of Section 4.2 to solve the system of equations (2) subject to the initial conditions $x(0) = x_0$, $y(0) = 0$, $z(0) = 0$.

2. Construct a mathematical model for a radioactive series of four elements W, X, Y, and Z, where Z is a stable element.

3. Use the method of Section 4.2 to solve the system of equations (3) subject to the initial conditions $x_1(0) = 25$, $x_2(0) = 0$.

4. Consider the two tanks A and B with liquid being pumped in and out at the same rates described by the system of equations (3). What is the system of differential equations if, instead of pure water, a brine solution containing 2 lb of salt per gallon is pumped into tank A?

5. Use the method of Section 4.2 to solve the system of equations in Problem 4 subject to the initial conditions $x_1(0) = 25$, $x_2(0) = 0$. [*Hint:* Instead of using (10) of Section 4.3, consider using undetermined coefficients by assuming a particular solution $\mathbf{X}_p = \begin{pmatrix} a \\ b \end{pmatrix}$, where a and b are constants.]

6. Use the information given in Figure 4.3 to construct a mathematical model for the number of pounds of salt $x_1(t)$, $x_2(t)$, and $x_3(t)$ at time t in tanks A, B, and C, respectively.

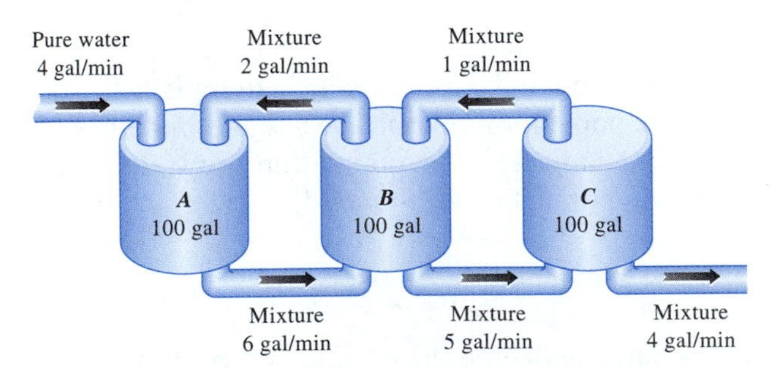

Figure 4.3.

7. Two large tanks A and B each hold 100 gal of brine. Initially, 100 lb of salt is dissolved in the solution in tank A and 50 lb of salt is dissolved in the solution in tank B. The system is closed, in that the well-stirred liquid is pumped only between the tanks as shown in Figure 4.4.

   (a) Use the information given in the figure to construct a mathematical model for the number of pounds of salt $x_1(t)$ and $x_2(t)$ at time t in tanks A and B, respectively.

   (b) Find a relationship between the variables $x_1(t)$ and $x_2(t)$ that holds at time t. Explain why this relationship makes intuitive sense. Use this relationship to help find the amount of salt in tank B at $t = 30$ min.

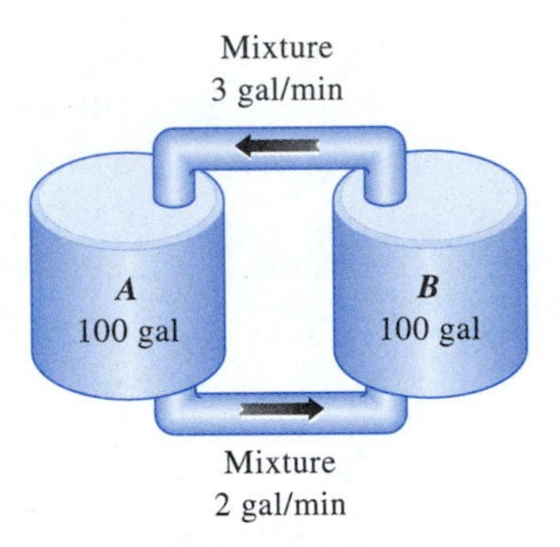

Figure 4.4.

8. Three large tanks contain brine as shown in Figure 4.5. Use the information in the figure to construct a mathematical model for the number of pounds of salt $x_1(t)$, $x_2(t)$, $x_3(t)$ at time t in tanks A, B, and C, respectively. Without solving the system, predict limiting values of $x_1(t)$, $x_2(t)$, and $x_3(t)$ as $t \to \infty$.

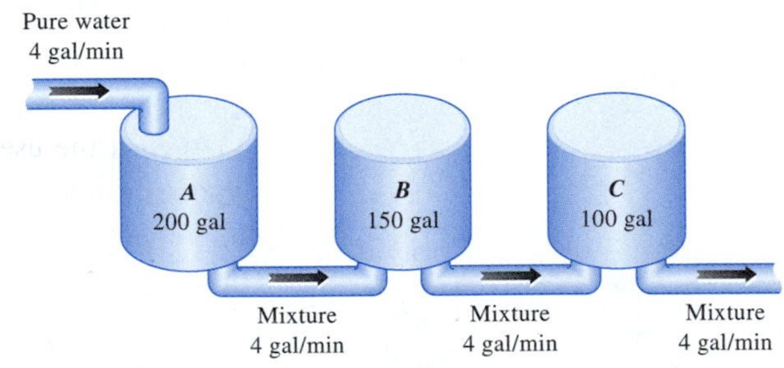

Figure 4.5.

9. Consider the Lotka-Volterra predator–prey model defined by

$$\frac{dx}{dt} = -0.1x + 0.02xy$$

$$\frac{dy}{dt} = 0.2y - 0.025xy$$

where $x(t)$ and $y(t)$ denote the populations of the predators and prey, respectively, at time $t > 0$. Use an ODE solver with $x(0) = 6$, $y(0) = 6$ to approximate the time $t > 0$ when the two populations are first equal. Use the graphs to approximate the period of each population.

10. Consider the competition model defined by

$$\frac{dx}{dt} = x(2 - 0.4x - 0.3y)$$

$$\frac{dy}{dt} = y(1 - 0.1y - 0.3x)$$

where the populations $x(t)$ and $y(t)$ are measured in thousands and time t is measured in years. Use an ODE solver to analyze the populations over a long period of time for each of the cases:

(a) $x(0) = 1.5$, $y(0) = 3.5$ **(b)** $x(0) = 1$, $y(0) = 1$
(c) $x(0) = 2$, $y(0) = 7$ **(d)** $x(0) = 4.5$, $y(0) = 0.5$

11. Consider the competition model defined by

$$\frac{dx}{dt} = x(1 - 0.1x - 0.05y)$$

$$\frac{dy}{dt} = y(1.7 - 0.1y - 0.15x)$$

where the populations $x(t)$ and $y(t)$ are measured in thousands and time t in years. Use an ODE solver to analyze the populations over a long period of time for each of the cases:

(a) $x(0) = 1$, $y(0) = 1$ **(b)** $x(0) = 4$, $y(0) = 10$
(c) $x(0) = 9$, $y(0) = 4$ **(d)** $x(0) = 5.5$, $y(0) = 3.5$

Discussion Problems

12. Discuss: The system (2) can be solved without the use of matrices or the use of elimination by differential operators. Show how this can be done.

13. A communicable disease is spread throughout a small community, with a fixed population of n people, by contact between infected persons and persons who are susceptible to the disease. Suppose initially that everyone is susceptible to the disease and that no one leaves the community while the epidemic is spreading. At time t, let $s(t)$, $i(t)$, and $r(t)$ denote, in turn, the number of people in the community (measured in hundreds) that are *susceptible* to the disease but not yet infected with it, the number of people who are *infected* with the disease, and the number of people who have *recovered* from the disease. Discuss: Why is the system of differential equations

$$\frac{ds}{dt} = -k_1 si$$

$$\frac{di}{dt} = -k_2 i + k_1 si$$

$$\frac{dr}{dt} = k_2 i$$

where $k_1 > 0$ (called the *infection rate*) and $k_2 > 0$ (called the *removal rate*) are constants, a reasonable mathematical model, commonly called a **SIR model**, for the spread of the epidemic throughout the community? Give plausible initial conditions associated with this system of equations.

14. Suppose compartments A and B shown in Figure 4.6 are filled with fluids and are separated by a permeable membrane. The figure is a compartmental representation of the exterior and interior of a cell. Suppose too, that a nutrient necessary for cell growth passes through the membrane. A model for the concentrations $x(t)$ and $y(t)$ of the nutrient in compartments A and B, respectively, at time t, is given by the linear system of differential equations

$$\frac{dx}{dt} = \frac{\kappa}{V_A}(y - x)$$

$$\frac{dy}{dt} = \frac{\kappa}{V_B}(x - y)$$

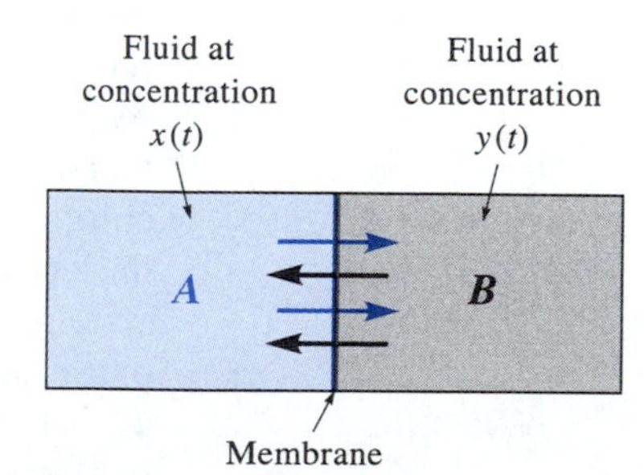

Figure 4.6.

where V_A and V_B are the volumes of the compartments, and $\kappa > 0$ is a permeability factor. Let $x(0) = x_0$ and $y(0) = y_0$ denote the initial concentrations of the nutrient. Based solely on the equations in the system and the assumption $x_0 > y_0 > 0$, sketch, on the same set of coordinate axes, possible solution curves of the system. Explain your reasoning. Discuss the behavior of the solutions over a long period of time.

15. **(a)** Discuss: The system in Problem 14, like the system in (2), can be solved with no advanced knowledge of matrices or elimination by differential operators. Explain.

(b) The system in Problem 14 can also be solved using the method of Section 4.2. Solve for $x(t)$ and $y(t)$ and compare their graphs with your conjecture in Problem 14. Determine the limiting values of $x(t)$ and $y(t)$ as $t \to \infty$. Discuss: Why do these limiting values make intuitive sense?

16. Let $x(t)$ denote the population of a predator and $y(t)$ and $z(t)$ the populations of two animals that are its prey. Discuss the form of a mathematical model consisting of three first-order differential equations for the populations if:

The predator population grows from eating the two prey and, lacking any enemies, can decline only due to natural deaths. The prey population would grow exponentially in the absence of the predator. The prey population would grow logistically in the absence of the predator.

4.5 Phase Portraits and Stability

We introduced the notion of stability of a critical point of autonomous first-order DEs in Section 2.1, of autonomous linear second-order DEs in Section 3.3, and autonomous nonlinear second-order DEs in Section 3.10. These earlier considerations of stability were purposely kept at a fairly intuitive level; it is now time to give the precise definition of this concept. We shall also see that the gallery of phase portraits in Section 3.3 is just a special case of phase portraits of autonomous systems of two linear first-order equations. The discussion concludes with the qualitative analysis of autonomous systems and models involving two nonlinear first-order equations.

Autonomous Systems The importance of systems of first-order differential equations cannot be overemphasized. We have seen that it is necessary at times to express a single differential equation as a system of differential equations. To solve a differential equation of order $n \geq 2$ numerically, we have to apply a numerical procedure to each equation in a system of n first-order equations. Likewise, to investigate the qualitative aspects of an autonomous second-order differential equation, we find the critical points of the equation by finding the critical points of an equivalent autonomous system of two first-order equations. We also use this system to find the phase portrait of the equation.

In the discussion that follows, we shall be concerned with the qualitative aspects of autonomous systems of two first-order equations

$$\frac{dx}{dt} = f(x, y) \qquad \frac{dy}{dt} = g(x, y) \tag{1}$$

The word **autonomous** indicates that the independent variable t does not appear explicitly in the functions f and g. Also, a **critical point** of (1) is a point (x_1, y_1) for which we have $f(x_1, y_1) = 0$ and $g(x_1, y_1) = 0$. As in Section 3.10, we shall suppose that a critical point (x_1, y_1) is isolated and that the functions f and g in the general system (1) are continuous and possess continuous first partial derivatives in a neighborhood of (x_1, y_1).

A system such as (1) is often referred to as a **plane autonomous system** since a solution $(x(t), y(t))$ of the system can be interpreted as a parametrized curve or **trajectory** in the xy-plane or **phase plane** We shall use the words *solution* and *trajectory* interchangeably.

4.5.1 Linear Systems

We have already seen in Section 3.3 that the autonomous linear second-order differential equation $ax'' + bx' + cx = 0$, $c \neq 0$, is equivalent to the linear system of two equations

$$\begin{aligned} \frac{dx}{dt} &= y \\ \frac{dy}{dt} &= -\frac{c}{a}x - \frac{b}{a}y \end{aligned} \tag{2}$$

The system (2) is a special case of the general autonomous linear system of two first-order equations

$$\begin{aligned} \frac{dx}{dt} &= ax + by \\ \frac{dy}{dt} &= cx + dy \end{aligned} \tag{3}$$

Of course (3) itself is just a special case of (1) when f and g are the linear functions $f(x, y) = ax + by$ and $g(x, y) = cx + dy$.

Since $x = 0$ and $y = 0$ are the only values that make the right-hand members of system (2) simultaneously zero, the only critical point of the equation $ax'' + bx' + cx = 0$, $c \neq 0$, is $(0, 0)$. Similarly, inspection of (3) reveals that $(0, 0)$ is the only critical point of the system provided that $ad - bc \neq 0$. We shall assume here that the coefficients a, b, c, and d in (3) are real numbers.

Classification of a Critical Point For a linear system (3), the origin $(0, 0)$ can be one of four **types** of critical points: node, saddle point, spiral point, or center. In addition, $(0, 0)$ can be assigned one of three **stability** classifications: asymptotically stable, stable (but not asymptotically stable), or unstable. Because (2) is a specific form of (3), the discussion on stability in Section 3.3 is subsumed in the current discussion. Consequently, a phase portrait of a linear system, except in one instance, is basically one of those in the gallery of phase portraits given in Figures 3.12, 3.13, and 3.14.

With the prior insight gained in Sections 2.1 and 3.3, the precise meaning of stability, as set forth in the following definition, should make sense. We refer to the general system (1) so that the definition applies to linear systems as well as nonlinear systems.

Definition 4.4 Stability of an Autonomous System

Let $(x(t), y(t))$ be a solution of the autonomous system (1) that satisfies $x(t_0) = x_0$, $y(t_0) = y_0$. Then a critical point (x_1, y_1) of the system is said to be

(i) **stable** if for every $\varepsilon > 0$ there exists a $\delta > 0$ such that

$$[x(t_0) - x_1]^2 + [y(t_0) - y_1]^2 < \delta^2 \text{ implies } [x(t) - x_1]^2 + [y(t) - y_1]^2 < \varepsilon^2$$

for all $t > t_0$;

(ii) **asymptotically stable** if it is stable and a $\delta_1 > 0$ can be chosen so that

$$[x(t_0) - x_1]^2 + [y(t_0) - y_1]^2 < \delta_1^2 \text{ implies } (x(t), y(t)) \to (x_1, y_1)$$

as $t \to \infty$; or

(iii) **unstable** if it is not stable.

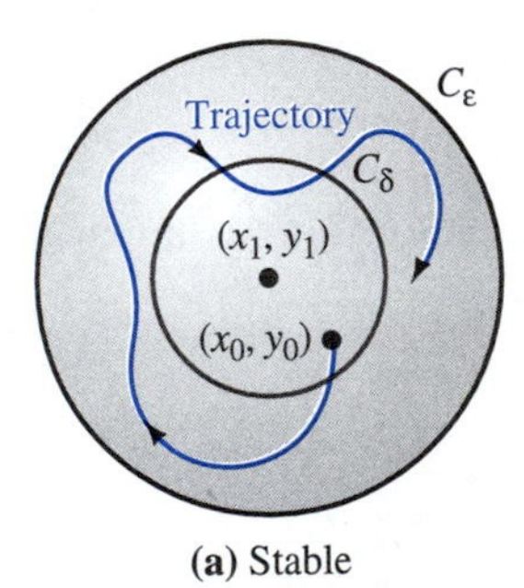

(a) Stable

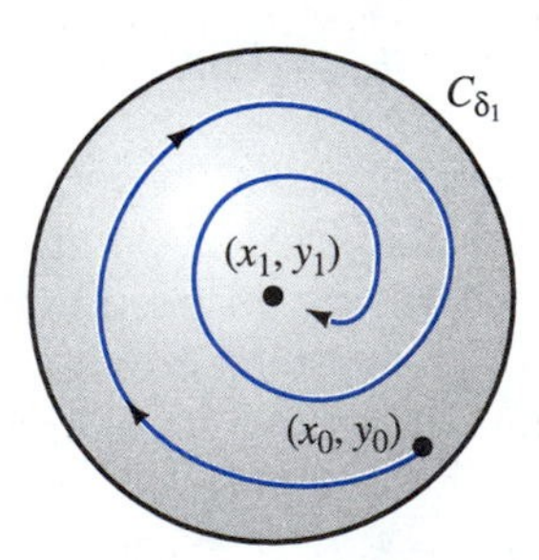

(b) Asymptotically stable

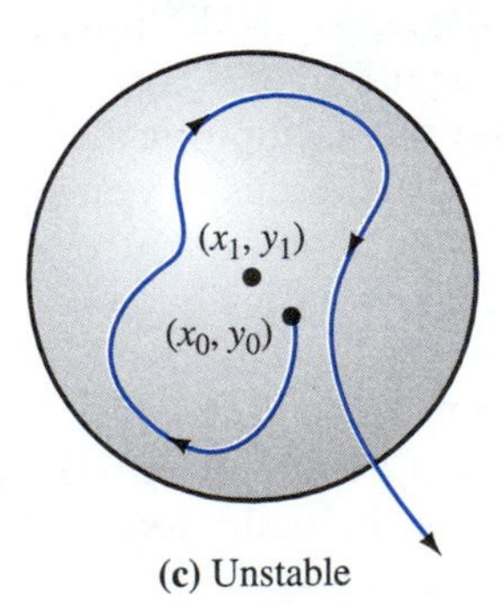

(c) Unstable

Figure 4.7.

An inequality such as $[x(t_0) - x_1]^2 + [y(t_0) - y_1]^2 < \delta^2$ indicates that the point $(x(t_0), y(t_0))$ is in the interior of a circle of radius δ centered at (x_1, y_1). Interpreted geometrically, a critical point (x_1, y_1) is stable provided that, for every circle C_ε we draw of radius ε centered at (x_1, y_1), we can find another circle C_δ of radius δ centered at (x_1, y_1) such that any trajectory starting at a point in C_δ never leaves C_ε. See Figure 4.7(a). If all trajectories that start sufficiently close, that is, within some circle C_{δ_1} centered at a stable critical point (x_1, y_1), also approach (x_1, y_1) as $t \to \infty$, then (x_1, y_1) is asymptotically stable. See Figure 4.7(b). A critical point is unstable if there exist trajectories that move away, or recede, from (x_1, y_1), regardless of how close we choose an initial point (x_0, y_0) to (x_1, y_1). See Figure 4.2(c).

Three Cases

When (3) is written as $\mathbf{X}' = \mathbf{AX}$, the characteristic equation of the coefficient matrix $\mathbf{A} = \begin{pmatrix} a & b \\ c & d \end{pmatrix}$ is the quadratic equation

$$\lambda^2 - (a + d)\lambda + ad - bc = 0 \tag{4}$$

where $\det \mathbf{A} = ad - bc \neq 0$. The roots λ_1 and λ_2 of (4) are the eigenvalues of $\mathbf{A}$. As in Section 4.2, we have the following three major cases.

CASE I Distinct real eigenvalues: $\lambda_1 \neq \lambda_2$.

CASE II Repeated real eigenvalues: $\lambda_1 = \lambda_2$.

CASE III Conjugate complex eigenvalues: $\lambda_1 = \alpha + i\beta$, $\lambda_2 = \alpha - i\beta$.

Under each of these cases, we can further distinguish subcases. These various subcases, except in one instance, are analogous to those listed in the table on page 143 in our discussion of the autonomous equation $ax'' + bx' + cx = 0$.

CASE I General solution $\quad \mathbf{X} = c_1\mathbf{K}_1e^{\lambda_1 t} + c_2\mathbf{K}_2e^{\lambda_2 t}$ **(5)**

(i) When both eigenvalues are negative, $\lambda_1 < \lambda_2 < 0$, then for any values of c_1 and c_2, $\mathbf{X} \to \mathbf{0}$ as $t \to \infty$. In other words, points (that is, solutions) $(x(t), y(t))$ on every trajectory approach, or are attracted to, the critical point $(0, 0)$:

$$\lim_{t\to\infty} x(t) = 0 \quad \text{and} \quad \lim_{t\to\infty} y(t) = 0$$

To understand this a little better, let us suppose for the sake of discussion that $\mathbf{K}_1 = \begin{pmatrix} k_{11} \\ k_{21} \end{pmatrix}$ and $\mathbf{K}_2 = \begin{pmatrix} k_{12} \\ k_{22} \end{pmatrix}$. When $c_2 = 0$, then $\mathbf{X} = c_1\mathbf{K}_1e^{\lambda_1 t}$ is the same as

$$x = c_1k_{11}e^{\lambda_1 t}, \quad y = c_1k_{21}e^{\lambda_1 t} \tag{6}$$

Eliminating the parameter t from these two equations shows that $\mathbf{X} \to \mathbf{0}$ along a half-line trajectory defined by $y = (k_{21}/k_{11})x$. The last equation defines only a half-line, since $e^{\lambda_1 t}$ is positive for all t and therefore x in (6) is either always positive or always negative. Moreover, there are *two* half-line trajectories with slope k_{21}/k_{11}, since we can choose $c_1 > 0$ or $c_1 < 0$ in (6). If we interpret the eigenvector $\mathbf{K}_1$ as a point (k_{11}, k_{21}), then we have shown that $\mathbf{X} \to \mathbf{0}$ along a *half-line determined by the eigenvector*, because the slope of a line through $(0, 0)$ and (k_{11}, k_{21}) is also k_{21}/k_{11}. The second half-line is determined by $-\mathbf{K}_1$. When $c_1 = 0$, a similar argument shows that $\mathbf{X} \to \mathbf{0}$ along the two half-line trajectories defined by $y = (k_{22}/k_{12})x$, that is, half-lines determined by the eigenvectors $\mathbf{K}_2$ and $-\mathbf{K}_2$. Finally, when $c_1 \neq 0$ and $c_2 \neq 0$, it follows from the quotient

$$\frac{y}{x} = \frac{c_1k_{21}e^{\lambda_1 t} + c_2k_{22}e^{\lambda_2 t}}{c_1k_{11}e^{\lambda_1 t} + c_2k_{12}e^{\lambda_2 t}} = \frac{c_1k_{21}e^{(\lambda_1 - \lambda_2)t} + c_2k_{22}}{c_1k_{11}e^{(\lambda_1 - \lambda_2)t} + c_2k_{12}}$$

and $\lambda_1 - \lambda_2 < 0$, that $y/x \to k_{22}/k_{12}$ as $t \to \infty$. We can interpret this result to mean that points on all other curved paths or trajectories approach $(0, 0)$ *tangentially* as $t \to \infty$, either to the half-line determined by the eigenvector $\mathbf{K}_2$ or to the half-line determined by $-\mathbf{K}_2$. The point $(0, 0)$ is **asymptotically stable** and is called a **node**.

(ii) When both eigenvalues are positive, $\lambda_1 > \lambda_2 > 0$, the situation described in (i) is now reversed; that is, points $(x(t), y(t))$, on every trajectory, move away or are repelled from $(0, 0)$ as $t \to \infty$. Every solution $\mathbf{X}$ of the system becomes unbounded as $t \to \infty$. By repeating the argument of choosing $c_2 = 0$, $c_1 \neq 0$, and then $c_1 = 0$, $c_2 \neq 0$, we see that there are again four half-line trajectories determined by the eigenvectors. The critical point $(0, 0)$ is now an **unstable node**.

(iii) Suppose the eigenvalues have opposite signs, $\lambda_1 < 0 < \lambda_2$. If $c_2 = 0$ then, since $\lambda_1 < 0$, $\mathbf{X} = c_1\mathbf{K}_1e^{\lambda_1 t} \to \mathbf{0}$ along the half-line trajectories determined by $\mathbf{K}_1$ and $-\mathbf{K}_1$. If $c_1 = 0$ then, since $\lambda_2 > 0$, $\mathbf{X} = c_2\mathbf{K}_2e^{\lambda_2 t}$ becomes unbounded on the half-line trajectories determined by $\mathbf{K}_2$ and $-\mathbf{K}_2$. If $c_1 \neq 0$ and $c_2 \neq 0$, then the trajectories are curved paths. Observe from (5) that as $t \to \infty$, $e^{\lambda_1 t} \to 0$, and so $\mathbf{X} \approx c_2\mathbf{K}_2e^{\lambda_2 t}$. This means $\mathbf{X}$ becomes unbounded but does so in a manner that is asymptotic to the half-line trajectories determined by $\mathbf{K}_2$ and $-\mathbf{K}_2$. On the other hand, if we let $t \to -\infty$, then $e^{\lambda_2 t} \to 0$ in (5) and so $\mathbf{X} \approx c_1\mathbf{K}_1e^{\lambda_1 t}$ shows $\mathbf{X}$ is asymptotic to the half-line trajectories determined by $\mathbf{K}_1$ and $-\mathbf{K}_1$. Roughly, this means that the curved trajectories must "start" (that is, as $t \to -\infty$) near one set of half-lines and then "finish" (that is, as $t \to \infty$) near the other set of half-lines. The critical point (0, 0) is a **saddle point**.

Recall from the discussions in Sections 3.3 and 3.10 that a node is characterized by the fact that points $(x(t), y(t))$ on *every* trajectory are either attracted to it or repelled from it as $t \to \infty$. We have also seen that a saddle point is always unstable; points $(x(t), y(t))$ on every trajectory, *except two*, recede from a saddle point as $t \to \infty$. A saddle point is characterized by the existence of two pairs of important trajectories. On these two pairs of trajectories points $(x(t), y(t))$ approach the saddle point from opposite directions as $t \to \infty$. As we have just seen in the case of linear systems, these trajectories are the half-lines determined by $\mathbf{K}_1$ and $-\mathbf{K}_1$. While not delving deeper into the precise meaning of the word, we will say that these two trajectories **enter** the saddle point. There are two additional trajectories that enter the saddle point from opposite directions but they do so as $t \to -\infty$; for linear systems, these trajectories are the half-lines determined by $\mathbf{K}_2$ and $-\mathbf{K}_2$.

Each of the three foregoing subcases is illustrated in the next example.

EXAMPLE 1 Distinct Eigenvalues/Phase Portraits

Classify the critical point (0, 0) and obtain the phase portrait of each system:

$$\textbf{(a)}\ \frac{dx}{dt} = -2x + 3y \qquad \textbf{(b)}\ \frac{dx}{dt} = -x + 2y \qquad \textbf{(c)}\ \frac{dx}{dt} = 2x + 3y$$

$$\frac{dy}{dt} = x - 4y \qquad \frac{dy}{dt} = -7x + 8y \qquad \frac{dy}{dt} = 2x + y$$

Solution **(a)** The eigenvalues of the matrix $\begin{pmatrix} -2 & 3 \\ 1 & -4 \end{pmatrix}$ and the corresponding eigenvectors are found to be

$$\lambda_1 = -5, \quad \mathbf{K}_1 = \begin{pmatrix} -1 \\ 1 \end{pmatrix}, \qquad \lambda_2 = -1, \quad \mathbf{K}_2 = \begin{pmatrix} 3 \\ 1 \end{pmatrix}$$

Since the eigenvalues are both negative, the critical point (0, 0) is an asymptotically stable node. With the aid of computer software we obtain the phase portrait of the system given in Figure 4.8(a).

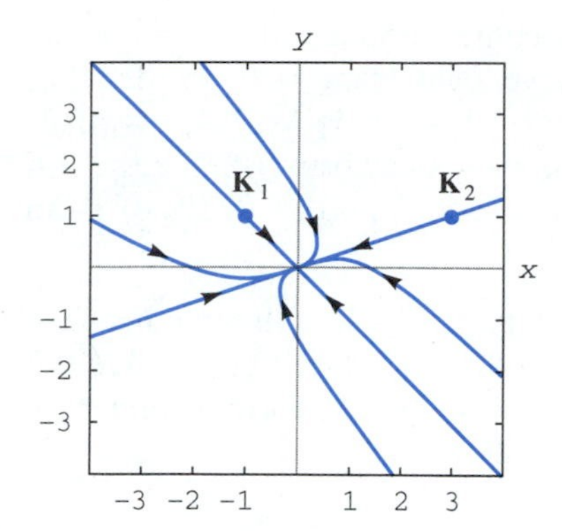

(a) (0, 0) is an asymptotically stable node.

(b) The eigenvalues of the matrix $\begin{pmatrix} -1 & 2 \\ -7 & 8 \end{pmatrix}$ and the corresponding eigenvectors are found to be

$$\lambda_1 = 6, \quad \mathbf{K}_1 = \begin{pmatrix} 2 \\ 7 \end{pmatrix}, \qquad \lambda_2 = 1, \quad \mathbf{K}_2 = \begin{pmatrix} 1 \\ 1 \end{pmatrix}$$

Since the eigenvalues are both positive, (0, 0) is an unstable node. A phase portrait of the system is given in Figure 4.8(b).

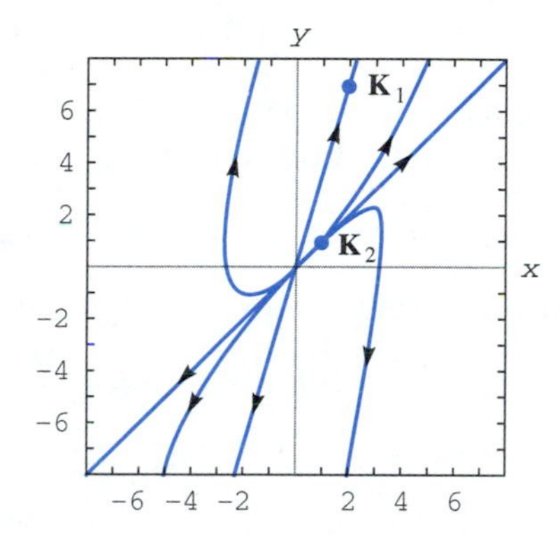

(b) (0, 0) is an unstable node.

(c) The eigenvalues of the matrix $\begin{pmatrix} 2 & 3 \\ 2 & 1 \end{pmatrix}$ and the corresponding eigenvectors are found to be

$$\lambda_1 = -1, \quad \mathbf{K}_1 = \begin{pmatrix} 1 \\ -1 \end{pmatrix}, \qquad \lambda_2 = 4, \quad \mathbf{K}_2 = \begin{pmatrix} 3 \\ 2 \end{pmatrix}$$

Because the eigenvalues have opposite signs, (0, 0) is a saddle point. A phase portrait of the system is given in Figure 4.8(c). ∎

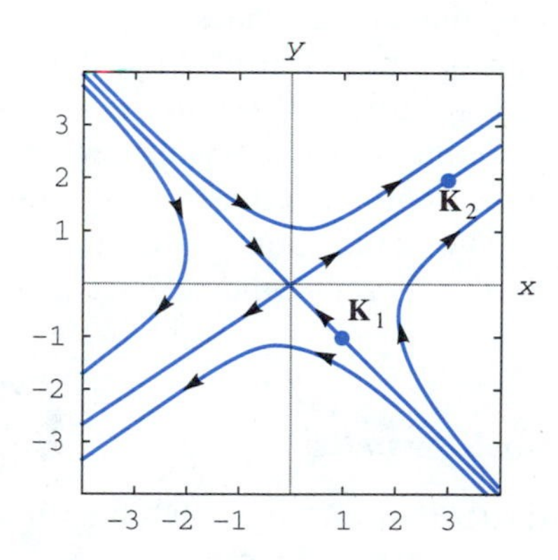

(c) (0, 0) is a saddle point.

Figure 4.8.

CASE II General solution $\quad \mathbf{X} = c_1\mathbf{K}_1 e^{\lambda_1 t} + c_2(\mathbf{K}_1 t e^{\lambda_1 t} + \mathbf{P}e^{\lambda_1 t}) \quad$ **(7)**

(i) For the moment, let us rule out the possibility that $a = d \neq 0$ and $b = c = 0$ in the linear system (3). For all other choices of a, b, c, and d in (3) that lead to a repeated eigenvalue $\lambda_1 = \lambda_2$, the form of the general solution of the system is as given in (7). Recall that $\mathbf{P}$ is a solution of $(\mathbf{A} - \lambda\mathbf{I})\mathbf{K}_1 = \mathbf{P}$. When $c_2 = 0$, $\mathbf{X} = c_1\mathbf{K}_1 e^{\lambda_1 t}$, and so, as in the discussion of Case I, the trajectories are two half-lines determined by $\mathbf{K}_1$ and $-\mathbf{K}_1$. When $c_2 \neq 0$, the trajectories are curved paths. If $\mathbf{K}_1 = \begin{pmatrix} k_1 \\ k_2 \end{pmatrix}$ and $\mathbf{P} = \begin{pmatrix} p_1 \\ p_2 \end{pmatrix}$, then we see from the quotient

$$\frac{y}{x} = \frac{c_1 k_2 e^{\lambda_1 t} + c_2(k_2 t e^{\lambda_1 t} + p_2 e^{\lambda_1 t})}{c_1 k_1 e^{\lambda_1 t} + c_2(k_1 t e^{\lambda_1 t} + p_1 e^{\lambda_1 t})} = \frac{c_1 k_2/t + c_2(k_2 + p_2/t)}{c_1 k_1/t + c_2(k_1 + p_1/t)} \qquad (8)$$

that $y/x \to k_2/k_1$ as $t \to \infty$. If $\lambda_1 < 0$, then $e^{\lambda_1 t} \to 0$ and $te^{\lambda_1 t} \to 0$ as $t \to \infty$ and so $\mathbf{X} \to \mathbf{0}$. In particular, the quotient shows that points on the curved trajectories approach (0, 0) tangentially to the half-lines determined by $\mathbf{K}_1$ and $-\mathbf{K}_1$. Thus for $\lambda_1 < 0$, the critical point (0, 0) is **asymptotically stable**. Because the points on the trajectories approach the origin either on or tangent to two half-lines, rather than four half-lines as in (i) and (ii) of Case I, (0, 0) is called a **degenerate node**.

(ii) If $\lambda_1 > 0$ then $e^{\lambda_1 t}$ and $te^{\lambda_1 t}$ become unbounded as $t \to \infty$, so points on all the trajectories move away from (0, 0). Notice that the quotient (8) also shows that $y/x \to k_2/k_1$ as $t \to -\infty$. This means that the points "start" moving away from (0, 0) either on or near the two half-lines. Therefore, for $\lambda_1 > 0$, (0, 0) is an **unstable degenerate node**.

(iii) This subcase has no analogue on page 143. It is the one instance in which a linear system (3) can have a repeated eigenvalue but the general solution of the system is *not* of the form given in (7). If we let $a = d \neq 0$ and $b = c = 0$, then (3) is

$$\begin{aligned}\frac{dx}{dt} &= ax \\ \frac{dy}{dt} &= ay\end{aligned} \tag{9}$$

Since the differential equations are independent of each other, the system is actually *uncoupled*. Nevertheless, if we write (9) in the matrix form $\mathbf{X}' = \begin{pmatrix} a & 0 \\ 0 & a \end{pmatrix}\mathbf{X}$, we see that the characteristic equation is $(\lambda - a)^2 = 0$ so $\lambda = a$ is an eigenvalue of multiplicity two. It is worth verifying that *any* vector $\begin{pmatrix} c_1 \\ c_2 \end{pmatrix}$ is an eigenvector corresponding to $\lambda = a$. Hence the general solution of (9) is given by $\mathbf{X} = \begin{pmatrix} c_1 \\ c_2 \end{pmatrix}e^{at} = c_1\begin{pmatrix} 1 \\ 0 \end{pmatrix}e^{at} + c_2\begin{pmatrix} 0 \\ 1 \end{pmatrix}e^{at}$. Notice that this solution is not of the form given in (7). By eliminating t from $x = c_1e^{at}$ and $y = c_2e^{at}$ we see that the trajectories of the system are half-lines defined by $c_1y = c_2x$. But since c_1 and c_2 can be chosen arbitrarily, there are an infinite number of half-line trajectories. If $\lambda = a < 0$, then $\mathbf{X} \to \mathbf{0}$ as $t \to \infty$ and the critical point (0, 0) is **asymptotically stable**. If $\lambda = a > 0$, then $\mathbf{X}$ becomes unbounded as $t \to \infty$ and (0, 0) is an **unstable** critical point. The critical point (0, 0) in this special case is a **degenerate node** that is sometimes referred to as a **star node**.

EXAMPLE 2 Repeated Eigenvalues/Phase Portraits

Classify the critical point (0, 0) and obtain the phase portrait of each system:

(a) $$\begin{aligned}\frac{dx}{dt} &= 3x - 18y \\ \frac{dy}{dt} &= 2x - 9y\end{aligned}$$

(b) $$\begin{aligned}\frac{dx}{dt} &= x \\ \frac{dy}{dt} &= y\end{aligned}$$

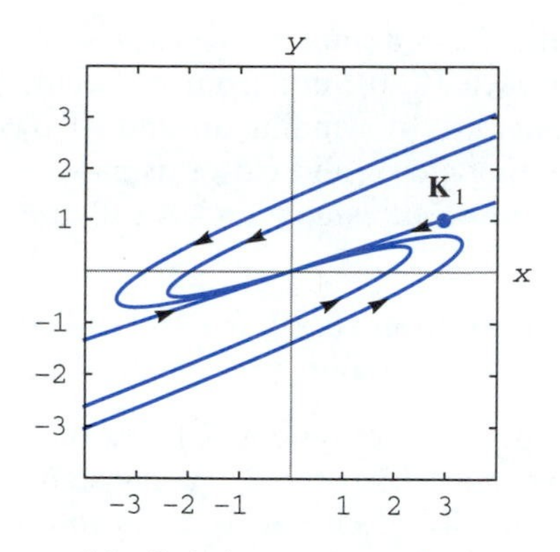

(a) (0, 0) is an asymptotically stable degenerate node.

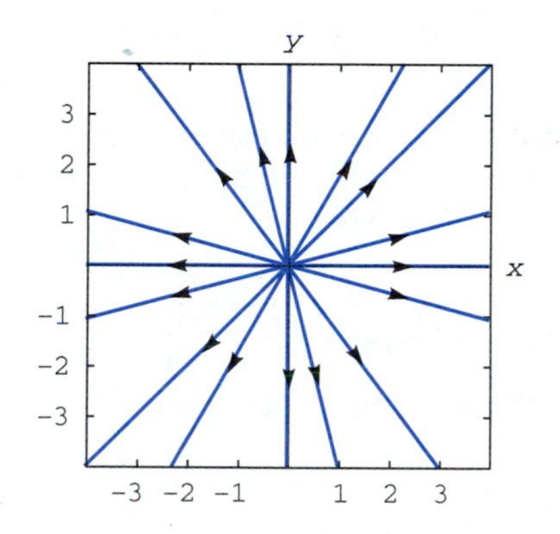

(b) (0, 0) is an unstable degenerate node.

Figure 4.9.

Solution **(a)** The repeated eigenvalue of this system is $\lambda_1 = \lambda_2 = -3 < 0$ and so, from the foregoing discussion, (0, 0) is an asymptotically stable degenerate node. The eigenvector indicated in the phase portrait in Figure 4.9(a) is $\mathbf{K}_1 = \begin{pmatrix} 3 \\ 1 \end{pmatrix}$.

(b) By treating these two independent equations as a system, we find $\lambda_1 = \lambda_2 = 1$. A computer is not necessary to obtain the phase portrait of the system; the phase portrait consists simply of half-lines with every possible slope. The arrows on the trajectories in Figure 4.9(b) point away from (0, 0); since the eigenvalue is positive, the origin is an unstable degenerate star node. ■

CASE III General solution

$$\mathbf{X} = c_1[\mathbf{B}_1 \cos \beta t - \mathbf{B}_2 \sin \beta t]e^{\alpha t} + c_2[\mathbf{B}_2 \cos \beta t + \mathbf{B}_1 \sin \beta t]e^{\alpha t} \quad \textbf{(10)}$$

where $\mathbf{B}_1 = \text{Re}(\mathbf{K}_1)$, $\mathbf{B}_2 = \text{Im}(\mathbf{K}_1)$, and $\mathbf{K}_1$, is an eigenvector corresponding to $\lambda_1 = \alpha + i\beta$.

Before considering the three subcases of Case III, let us examine the special linear system

$$\begin{aligned} \frac{dx}{dt} &= \alpha x - \beta y \\ \frac{dy}{dt} &= \beta x + \alpha y \end{aligned} \quad \textbf{(11)}$$

System (11) should be considered as the archetype of every linear system with complex eigenvalues; you should verify that its coefficient matrix has the eigenvalues $\lambda_1 = \alpha + i\beta$, $\lambda_2 = \alpha - i\beta$. It is now helpful to introduce polar coordinates. When we substitute $x = r \cos \theta$, $y = r \sin \theta$, where r and θ are functions of t, (11) becomes

$$\begin{aligned} (-r \sin \theta)\theta' + (\cos \theta)r' &= \alpha r \cos \theta - \beta r \sin \theta \\ (r \cos \theta)\theta' + (\sin \theta)r' &= \beta r \cos \theta + \alpha r \sin \theta \end{aligned}$$

Solving this last system (say, by Cramer's rule) for the symbols θ' and r', we find, after simplification, the first-order differential equations

$$r' = \alpha r \quad \text{and} \quad \theta' = \beta \quad \textbf{(12)}$$

Dividing the first equation in (12) by the second gives $dr/d\theta = (\alpha/\beta)r$. Solving yields

$$r = c_1 e^{(\alpha/\beta)\theta} \quad \textbf{(13)}$$

For $\alpha \neq 0$, $c_1 \neq 0$, (13) is a polar equation of a logarithmic spiral.

(i) If $\alpha < 0$, $\beta \neq 0$, we see in (10) that $\mathbf{X} \to \mathbf{0}$ since $e^{\alpha t} \to 0$ as $t \to \infty$. Points on each trajectory approach (0, 0), not from a specific direction as in the case of a node, but by winding around (0, 0) an infinite number of times, getting closer and closer as $t \to \infty$. The critical point (0, 0) is **asymptotically stable** and is called a **spiral point**

(ii) If $\alpha > 0$, $\beta \neq 0$, points spiral away from (0, 0) as $t \to \infty$. The critical point (0, 0) is an **unstable spiral point**

(iii) If $\alpha = 0$, $\beta \neq 0$, the eigenvalues of the system (2) are pure imaginary. Equation (13) shows that when $\alpha = 0$, r is constant, which is a polar equation of a family of circles. Bear in mind that the last remark only pertains to the special system (12). In general, when $\alpha = 0$ the solutions in (10) have the form

$$x = c_{11} \cos \beta t + c_{12} \sin \beta t, \quad y = c_{21} \cos \beta t + c_{22} \sin \beta t$$

which are parametric equations of a family of ellipses with center at the origin. The critical point (0, 0) is **stable** (but not asymptotically stable) and is naturally called a **center**

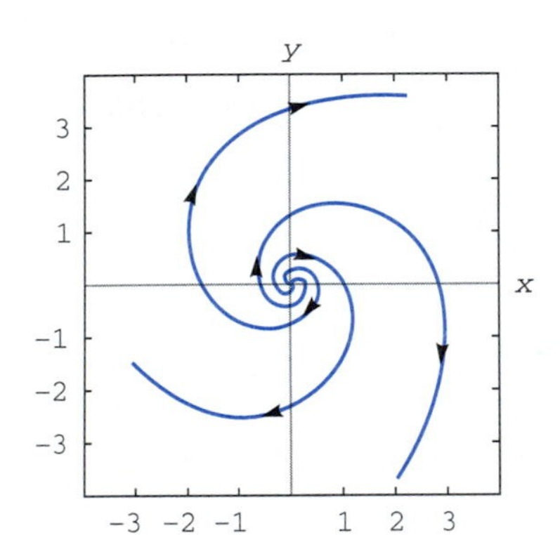

(a) (0, 0) is an unstable spiral point.

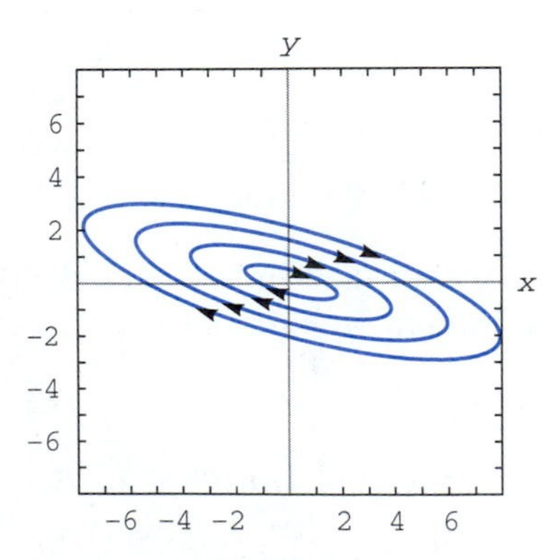

(b) (0, 0) is a center.

Figure 4.10.

EXAMPLE 3 Complex Eigenvalues/Phase Portraits

Classify the critical point (0, 0) and obtain the phase portrait of each system:

$$\text{(a)}\ \frac{dx}{dt} = x + 2y \qquad \text{(b)}\ \frac{dx}{dt} = 2x + 8y$$

$$\frac{dy}{dt} = -2x + y \qquad \frac{dy}{dt} = -x - 2y$$

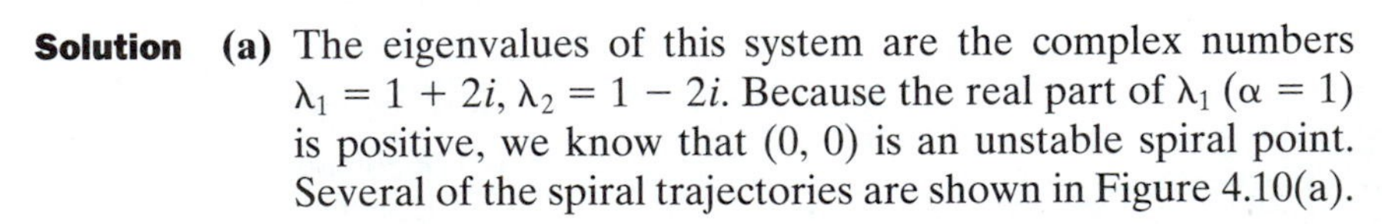

Solution **(a)** The eigenvalues of this system are the complex numbers $\lambda_1 = 1 + 2i$, $\lambda_2 = 1 - 2i$. Because the real part of λ_1 ($\alpha = 1$) is positive, we know that (0, 0) is an unstable spiral point. Several of the spiral trajectories are shown in Figure 4.10(a).

(b) The eigenvalues of the system are the pure imaginary numbers $\lambda_1 = 2i$, $\lambda_2 = -2i$. Hence from (iii) of Case III we know that (0, 0) is a center. The phase portrait of the system consists of the family of ellipses shown in Figure 4.10(b). ■

For convenience, we summarize the main results of the discussion in the form of a theorem.

Theorem 4.10 Stability of Linear Systems

Let λ_1 and λ_2 be the eigenvalues of the coefficient matrix **A** of system (3) with $ad - bc \neq 0$. Then the critical point (0, 0) of the system is:

(i) asymptotically stable if λ_1 and λ_2 are real and both negative, or if λ_1 and λ_2 are complex with negative real parts;

(ii) stable (but not asymptotically stable) if λ_1 and λ_2 are pure imaginary numbers; and

(iii) unstable if λ_1 and λ_2 are real and both positive, if λ_1 and λ_2 are real and have opposite signs, or if λ_1 and λ_2 are complex with positive real parts.

In case (i) of Theorem 4.10, (0, 0) is an **attractor**, and in case (ii), when the two eigenvalues are either both positive or are complex with positive real parts, (0, 0) is a **repeller**. In part (a) of Examples 1 and 2, (0, 0) is an attractor; in part (b) of Examples 1 and 2 and in part (a) of Example 3, (0, 0) is a repeller. In part (c) of Example 1 and part (b) of Example 3, (0, 0) is neither an attractor nor a repeller.

4.5.2 Nonlinear Systems

We shall be concerned only with nonlinear systems from this point on. For example

$$\frac{dx}{dt} = \sin y \qquad \frac{dx}{dt} = -0.16x + 0.08xy$$
$$\text{and} \qquad \textbf{(14)}$$
$$\frac{dy}{dt} = \cos x \qquad \frac{dy}{dt} = 4.5y - 0.9xy$$

are nonlinear systems.

We dealt briefly with special nonlinear systems in Section 3.10, where we expressed autonomous nonlinear second-order differential equations as equivalent autonomous nonlinear systems. It is worth remarking that one of the significant differences between linear differential equations and nonlinear differential equations also carries over to systems of differential equations: We are always able to find an explicit solution of a linear system of the form given in (3), but it is rarely possible to find, in the sense of displaying, any solution of a nonlinear system.

Critical Points In contrast with the linear systems that we have just studied, a nonlinear system can have more than one critical point, or no critical points at all. As usual, we shall consider systems that possess only isolated critical points. Note for example, that (0, 0) and (5, 2) are isolated critical points of the second system in (14). As in the linear case, a critical point of an autonomous nonlinear system is classified, *when it is possible to do so*, according to type (node, saddle point, spiral point, center) and stability (stable, asymptotically stable, unstable).

Global and Local Stability A critical point is stable if all solutions that start close to the critical point stay close to that point for all $t > t_0$. Stability for a linear system of the form (3) is a **global** stability. If the critical point (0, 0) of (3) is, say, asymptotically stable, closeness of an initial starting point (x_0, y_0) to the critical point really does not matter. All solutions or points $(x(t), y(t))$ must approach (0, 0) as $t \to \infty$. We see this global stability in Figures 4.8(a) and 4.10(b). Put another way, the entire phase portrait of a linear system is characterized by the behavior of a few trajectories near the critical point.

On the other hand, stability for nonlinear systems is often a **local** phenomenon. For example, in the case of the soft spring discussed in Section 3.10, we found that there exist periodic solutions of the nonlinear model $x'' + x - x^3 = 0$ by discovering, in Figure 3.58(a), that (0, 0) is a stable center only in a neighborhood of the origin. It might be worthwhile to go back and verify that points $(x(t), y(t))$ on the trajectory whose initial point is (0.5, 0.6) move away from the critical point but the trajectory whose initial point is (0.5, 0.5) is a closed path surrounding (0, 0).

Classifying a critical point of a nonlinear system is not as straightforward as it is for a linear system. But in this brief introduction to nonlinear systems, we will primarily consider those systems for which the behavior of the trajectories near a critical point is comparable to, or is *approximated* by, trajectories of associated linear systems near the point.

Linearization If f is a function of a single variable x and differentiable at x_1, then an equation of the tangent line to the graph at $(x_1, f(x_1))$ is $y = f(x_1) + f'(x_1)(x - x_1)$. The tangent line, which we write as $L(x) = f(x_1) + f'(x_1)(x - x_1)$, is said to be a **linearization** of f at x_1. When x is close to x_1 and the tangent line is close to the graph of f, we can approximate the function value $f(x)$ using the corresponding y-coordinate on the tangent line. We say $f(x) \approx L(x)$ is a **local linear approximation** of $f(x)$. Similarly, if f is a function of two variables x and y that has continuous first partial derivatives in a neighborhood of (x_1, y_1), then using an equation of the tangent plane at $(x_1, y_1, f(x_1, y_1))$, the linearization of f at (x_1, y_1) is

$$L(x, y) = f(x_1, y_1) + f_x(x_1, y_1)(x - x_1) + f_y(x_1, y_1)(y - y_1) \tag{15}$$

where f_x and f_y denote the first partial derivatives. Thus when (x, y) is close to (x_1, y_1),

$$f(x, y) \approx L(x, y) \tag{16}$$

is a local linear approximation of $f(x, y)$.

Now suppose that (x_1, y_1) is an isolated critical point of a nonlinear system (1). If we use (15) on both f and g then, since $f(x_1, y_1) = 0$ and $g(x_1, y_1) = 0$, (16) gives

$$\begin{aligned} f(x,y) &\approx f_x(x_1,y_1)(x-x_1) + f_y(x_1,y_1)(y-y_1) \\ g(x,y) &\approx g_x(x_1,y_1)(x-x_1) + g_y(x_1,y_1)(y-y_1) \end{aligned} \tag{17}$$

whenever (x, y) is close to (x_1, y_1). By replacing the right-hand members in (1) by the right-hand members in (17), we obtain

$$\begin{aligned} \frac{dx}{dt} &= f_x(x_1,y_1)(x-x_1) + f_y(x_1,y_1)(y-y_1) \\ \frac{dy}{dt} &= g_x(x_1,y_1)(x-x_1) + g_y(x_1,y_1)(y-y_1) \end{aligned} \tag{18}$$

If we define $\mathbf{X} = \begin{pmatrix} x - x_1 \\ y - y_1 \end{pmatrix}$ then, because x_1 and y_1 are constants, $\mathbf{X}' = \begin{pmatrix} x' \\ y' \end{pmatrix}$ and so the system in (18) has the familiar matrix linear form $\mathbf{X}' = \mathbf{AX}$. For convenience, we write the coefficient matrix $\mathbf{A}$ as

$$\mathbf{A}(x_1,y_1) = \begin{pmatrix} f_x(x_1,y_1) & f_y(x_1,y_1) \\ g_x(x_1,y_1) & g_y(x_1,y_1) \end{pmatrix} \tag{19}$$

We shall assume, as we did in (3), that the determinant of the coefficient matrix is not zero, that is, $\det \mathbf{A} \neq 0$. The system given in (18) is said to be a **linearization** of the system (1) at the critical point (x_1, y_1). The matrix $\mathbf{A}(x_1, y_1)$ in (19) is called the **linearization matrix** or **Jacobian matrix** of the nonlinear system at (x_1, y_1).

The next theorem asserts that in some instances we can determine the type and stability of a critical point of a nonlinear system from the type and stability of the linearization of the system at that point.

Theorem 4.11 Stability of Nonlinear Systems

A critical point (x_1, y_1) of the nonlinear system (1) is of the same general type (node, saddle point, spiral point) and stability (stable, asymptotically stable, unstable) as the critical point (x_1, y_1) of the linearized system (18). There are two possible exceptions:

(i) If the eigenvalues λ_1 and λ_2 of the linearization matrix (19) are equal, then the critical point (x_1, y_1) of the nonlinear system (1) may either be a node or a spiral point and is asymptotically stable if $\lambda_1 = \lambda_2 < 0$ and unstable if $\lambda_1 = \lambda_2 > 0$.

(ii) If the eigenvalues λ_1 and λ_2 of the linearization matrix (19) are pure imaginary, then the critical point (x_1, y_1) of the nonlinear system (1) may either be a center or a spiral point. The stability of the spiral point is not determined by the stability of (x_1, y_1) of the linearized system.

Note that in (i) of Theorem 4.11, we can determine the *stability* but not the *type* of critical point of the nonlinear system from the linearized system. Part (ii) of the theorem is completely inconclusive; we can neither determine the stability nor the type of critical point of the nonlinear system from the fact that the linearized system has a center.

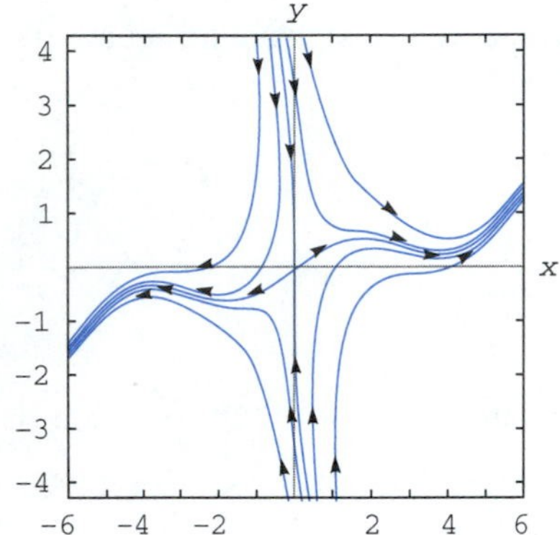

(a) Nonlinear system

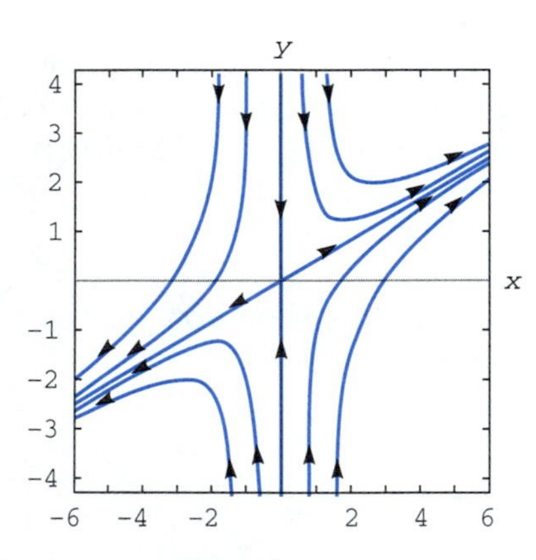

(b) Linearized system

Figure 4.11.

■ EXAMPLE 4 Linearization of a Nonlinear System

Observe that (0, 0) is the only critical point of the nonlinear system:

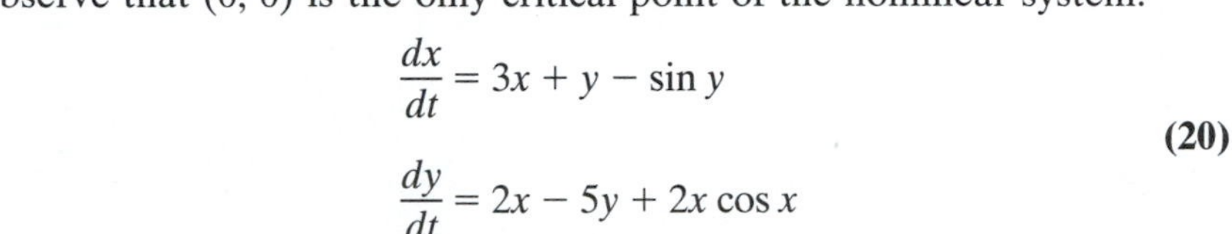

$$\frac{dx}{dt} = 3x + y - \sin y$$
$$\frac{dy}{dt} = 2x - 5y + 2x\cos x \tag{20}$$

With

$$f(x, y) = 3x + y - \sin y,\ g(x, y) = 2x - 5y + 2x\cos x$$
$$f_x(x, y) = 3,\ f_y(x, y) = 1 - \cos y,\ g_x(x, y) = 2 - 2x\sin x + 2\cos x,\ g_y(x, y) = -5$$

we have

$$\mathbf{A}(x, y) = \begin{pmatrix} 3 & 1 - \cos y \\ 2 - 2x\sin x + 2\cos x & -5 \end{pmatrix} \quad \text{and} \quad \mathbf{A}(0, 0) = \begin{pmatrix} 3 & 0 \\ 4 & -5 \end{pmatrix}$$

Hence from (18), the linearization of (20) at (0, 0) is

$$\frac{dx}{dt} = 3x$$
$$\frac{dy}{dt} = 4x - 5y \tag{21}$$

In the usual manner, we find that the eigenvalues of the linearization matrix $\mathbf{A}(0, 0)$ are $\lambda_1 = 3$ and $\lambda_2 = -5$. Since these eigenvalues are real and distinct, but with opposite signs, it follows from Case I (iii) and Theorem 4.10 that (0, 0) is an (unstable) saddle point. Continuing, it follows from Theorem 4.11 that (0, 0) is likewise a saddle point for the nonlinear system (20).

With computer software, we get the phase portraits in Figure 4.11. Notice that the trajectories near the origin behave in a similar manner. ■

■ EXAMPLE 5 Linearization of a Nonlinear System

Consider the nonlinear system

$$\frac{dx}{dt} = 2x + 8y$$
$$\frac{dy}{dt} = -x - 2y + \frac{1}{2}y^3$$

Again the only critical point of the system is (0, 0). With

$$f(x, y) = 2x + 8y,\quad g(x, y) = -x - 2y + \frac{1}{2}y^3,$$

$$f_x(x, y) = 2, \quad f_y(x, y) = 8, \quad g_x(x, y) = -1, \quad g_y(x, y) = -2 + \frac{3}{2}y^2,$$

$$\mathbf{A}(x, y) = \begin{pmatrix} 2 & 8 \\ -1 & -2 + \frac{3}{2}y^2 \end{pmatrix} \quad \text{and} \quad \mathbf{A}(0, 0) = \begin{pmatrix} 2 & 8 \\ -1 & -2 \end{pmatrix}$$

we see that a linearization of the given system at (0, 0) is

$$\frac{dx}{dt} = 2x + 8y$$

$$\frac{dy}{dt} = -x - 2y$$

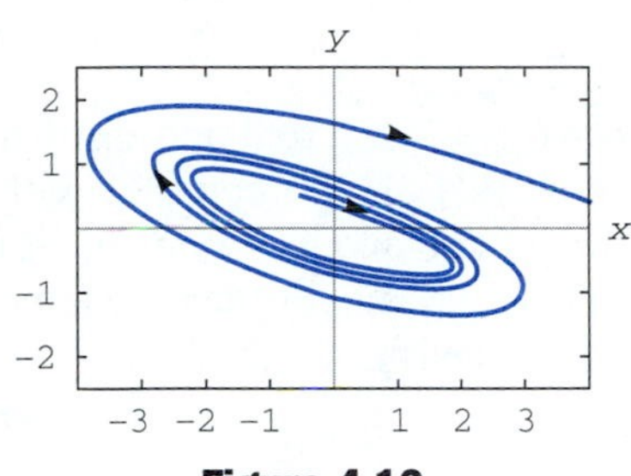

Figure 4.12.

We saw in part (b) of Example 3 that the linearization matrix $\mathbf{A}(0, 0)$ has imaginary eigenvalues $\lambda_1 = 2i$ and $\lambda_2 = -2i$, and so (0, 0) is a center. Although we can draw no conclusion from Theorem 4.11 regarding the nature of the critical point (0, 0) of the nonlinear system, it *appears* that (0, 0) is an unstable spiral point in Figure 4.12. ■

If a nonlinear system has more than one critical point, then the linearization must be carried out at each point. Also, it should be apparent that we need not actually write down the linearized system at a critical point (x_1, y_1) to apply Theorem 4.11. We need only examine the eigenvalues of the linearization matrix $\mathbf{A}(x_1, y_1)$.

■ EXAMPLE 6 Critical Points Not at Origin

Find and classify the critical points of the system:

$$\frac{dx}{dt} = 4x - x^2 - xy$$

$$\frac{dy}{dt} = 4x - 2y + 2xy - 4$$

Solution After factoring, inspection of the algebraic system

$$x(4 - x - y) = 0$$

$$2(y + 2)(x - 1) = 0$$

reveals that (0, −2), (6, −2), and (1, 3) are critical points of the given nonlinear system. The first partial derivatives are

$$f_x(x, y) = 4 - 2x - y, \quad f_y(x, y) = -x, \quad g_x(x, y) = 4 + 2y, \quad g_y(x, y) = -2 + 2x,$$

and so

$$\mathbf{A}(x, y) = \begin{pmatrix} 4 - 2x - y & -x \\ 4 + 2y & -2 + 2x \end{pmatrix} \tag{22}$$

We evaluate the matrix (22) at the three critical points and find the eigenvalues of each linearization matrix. The results are:

$$\mathbf{A}(0,-2)=\begin{pmatrix}6 & 0\\ 0 & -2\end{pmatrix};\quad \lambda_1=6, \lambda_2=-2$$

$$\mathbf{A}(6,-2)=\begin{pmatrix}-6 & -6\\ 0 & 10\end{pmatrix};\quad \lambda_1=10, \lambda_2=-6$$

$$\mathbf{A}(1,3)=\begin{pmatrix}-1 & -1\\ 10 & 0\end{pmatrix};\quad \lambda_1=-\frac{1}{2}+\frac{\sqrt{39}}{2}i, \lambda_2=-\frac{1}{2}-\frac{\sqrt{39}}{2}i$$

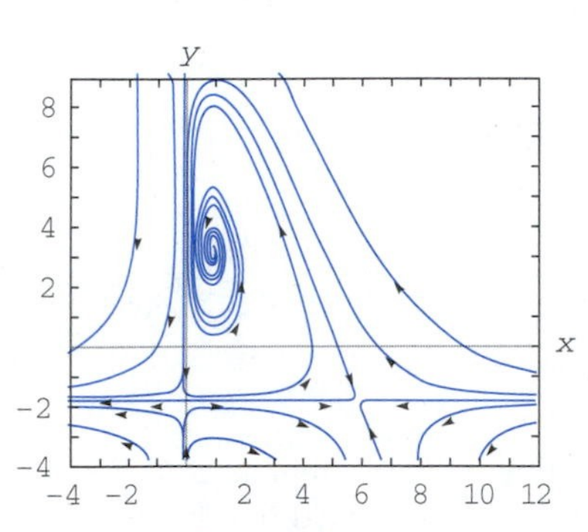

Figure 4.13.

As we see, the eigenvalues of $\mathbf{A}(0,-2)$ and $\mathbf{A}(6,-2)$ are real and distinct but with opposite signs, and the eigenvalues of $\mathbf{A}(1,3)$ are complex with negative real parts. We conclude from Theorems 4.10 and 4.11 that $(0,-2)$ and $(6,-2)$ are saddle points and that $(1,3)$ is an asymptotically stable spiral point. A phase portrait of the nonlinear system is given in Figure 4.13. ■

Critical Points that Cannot Be Classified by Type You should not get the impression from the foregoing discussion that linearization and phase portraits will tell us everything we want to know about a nonlinear system. Linearization has its shortcomings; as mentioned earlier, we can draw no conclusion about the nature of a critical point of a nonlinear system from Theorem 4.11 when $(0,0)$ is a center of the linearized system. Consider, too, the following example. The only critical point of the nonlinear system

$$\frac{dx}{dt}=x^2-y^2$$
$$\frac{dy}{dt}=2xy \tag{23}$$

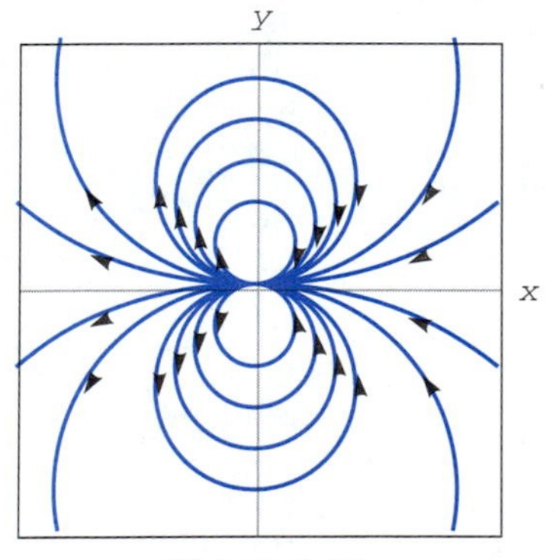

Figure 4.14.

is the origin $(0, 0)$. But since all first partials of $f(x, y) = x^2 - y^2$ and $g(x, y) = 2xy$ are zero at $(0, 0)$, the linearization of this system is $dx/dt = 0$, $dy/dt = 0$. We can extract no useful information from this trivial linear system. See Problem 46 in Exercises 4.5. With computer software we can, however, obtain the phase portrait of the nonlinear system that is shown in Figure 4.14. By dividing the second equation of (23) by the first and solving the homogeneous equation $dy/dx = 2xy/(x^2 - y^2)$, we get a Cartesian equation that appears to be a family of circles passing through $(0, 0)$. See Problem 39 in Exercises 4.5. The trajectories are circles but with one point missing. Points $(x(t), y(t))$ on the trajectories "start" arbitrarily close to the origin, loop around, and "finish" again arbitrarily close to the origin. The trajectories do not pass through the origin. For this system $(0, 0)$ is obviously not one of the four types: node, saddle point, spiral point, and center. This last example illustrates a fundamental fact of life in the study of nonlinear systems, namely, the local behavior of trajectories near the isolated critical points of a nonlinear system may differ widely from the behavior of trajectories of linear systems near the isolated critical point

(0, 0). We simply may not be able to assign to a critical point of a nonlinear system a classification according to type; *for general nonlinear systems there are an unlimited variety of types of critical points.*

Using the tools at hand, namely linearization and phase portraits, it also may not be possible to classify a critical point according to stability. A phase portrait can be so complicated or so subtle that the stability of a critical point is not obvious. For example, the phase portrait in Figure 4.14 clearly indicates that points $(x(t), y(t))$ on the trajectories approach the critical point (0, 0) for increasing values of t. Isn't this sufficient to say that (0, 0) is asymptotically stable? The answer is no. It must be remembered that to be asymptotically stable, a critical point (x_1, y_1) must first be stable and then $(x(t), y(t)) \to (x_1, y_1)$ as $t \to \infty$ on all trajectories. In this example, it is easy to show, using a parametrization, that the trajectories approach the origin for a finite value of t. Moreover—and this is worth thinking about—the critical point (0, 0) for (23) actually is unstable. See again Problem 39 in Exercises 4.5. We note in passing that there exist examples of nonlinear systems for which (x_1, y_1) is a critical point and $(x(t), y(t)) \to (x_1, y_1)$ as $t \to \infty$ on all trajectories and yet (x_1, y_1) is unstable.

We often have to appeal to an advanced technique to answer the question of stability.

A Predator–Prey Model We introduced the Lotka-Volterra predator–prey model in the preceding section:

$$\begin{aligned} \frac{dx}{dt} &= -ax + bxy = x(-a + by) \\ \frac{dy}{dt} &= dy - cxy = y(d - cx) \end{aligned} \tag{24}$$

where $a, b, c,$ and d are positive constants. From the equations $x(a - by) = 0$, $y(-c + dx) = 0$, we see that the critical points of the system are (0, 0) and $(d/c, a/b)$. Linearization of (24) at (0, 0) demonstrates that this critical point is a saddle point, whereas linearization at $(d/c, a/b)$ leads to a center which, unfortunately, is the inconclusive case (ii) in Theorem 4.11. Nonetheless, by dividing the second equation in (24) by the first, and solving the differential equation $dy/dx = (dy - cxy)/(-ax + bxy)$, it can be proved from the Cartesian equation of the trajectories that $(d/c, a/b)$ is indeed a center. See Problem 40 in Exercises 4.5.

■ EXAMPLE 7 **Predator–Prey Model**

In Example 1 of Section 4.4 we examined solution curves of the predator–prey model

$$\begin{aligned} \frac{dx}{dt} &= -0.16x + 0.08xy \\ \frac{dy}{dt} &= 4.5y - 0.9xy \end{aligned}$$

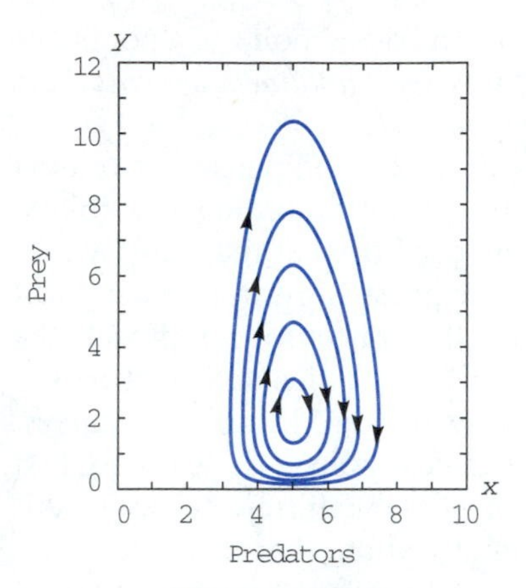

Figure 4.15.

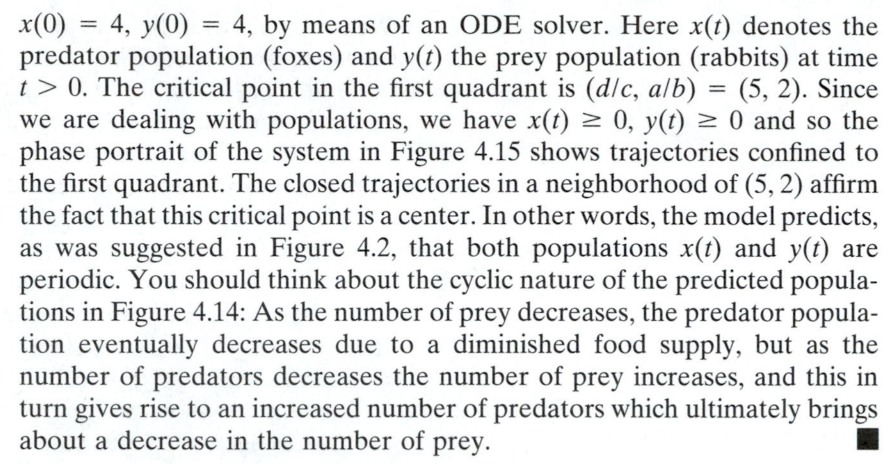

$x(0) = 4$, $y(0) = 4$, by means of an ODE solver. Here $x(t)$ denotes the predator population (foxes) and $y(t)$ the prey population (rabbits) at time $t > 0$. The critical point in the first quadrant is $(d/c, a/b) = (5, 2)$. Since we are dealing with populations, we have $x(t) \geq 0$, $y(t) \geq 0$ and so the phase portrait of the system in Figure 4.15 shows trajectories confined to the first quadrant. The closed trajectories in a neighborhood of (5, 2) affirm the fact that this critical point is a center. In other words, the model predicts, as was suggested in Figure 4.2, that both populations $x(t)$ and $y(t)$ are periodic. You should think about the cyclic nature of the predicted populations in Figure 4.14: As the number of prey decreases, the predator population eventually decreases due to a diminished food supply, but as the number of predators decreases the number of prey increases, and this in turn gives rise to an increased number of predators which ultimately brings about a decrease in the number of prey. ■

Vector Fields Matrix notation provides an invaluable way of both working with and discussing linear systems. Although the autonomous system (1) can also be expressed in terms of column matrices, it is common practice to use vector notation. If $\mathbf{X}(t) = (x(t), y(t))$ and $\mathbf{G}(\mathbf{X}) = (f(x, y), g(x, y))$ are two-dimensional vector functions, then $d\mathbf{X}/dt = (x'(t), y'(t))$ and so the system (1) is the same as

$$\frac{d\mathbf{X}}{dt} = \mathbf{G}(\mathbf{X})$$

Notice that this representation of an autonomous system retains the form of an autonomous first-order differential equation given in (1) of Section 2.1. The vector function $\mathbf{X}(t) = (x(t), y(t))$ represents a curve in 2-space and the derivative $d\mathbf{X}/dt = (x'(t), y'(t))$ represents a vector tangent to the curve. In dynamic terms, if $\mathbf{X}(t)$ denotes the position of a particle in the plane at time t, then $d\mathbf{X}/dt$ is its velocity vector. In a region of the xy-plane over which it is defined, the function $\mathbf{G}$ defines a collection of vectors varying in direction and length, in other words—a **vector field**. At a point (x, y), the components of the vector are given by $\mathbf{G}(\mathbf{X}) = \mathbf{G}(x, y) = (f(x, y), g(x, y))$. For example, for the linear system in (a) of Example 3, we have $\mathbf{G}(\mathbf{X}) = (x + 2y, -2x + y)$; at, say, (1, 1), $\mathbf{G}(1, 1) = (3, -1)$. To visualize this vector, start at (1, 1), move horizontally to the right three units and then down vertically one unit. The vector extends from the initial point (1, 1) to the terminal point (4, 0); its length is $\sqrt{10}$. Figure 4.16, obtained with the aid of a computer algebra system, shows the plot of the vector field on a rectangular grid of points defined by $-6 \leq x \leq 6$, $-6 \leq y \leq 6$. You should compare the flow of the vectors in Figure 4.16 with the phase portrait in Figure 4.10(a). Note, too, in Figure 4.16 that all the vectors have the same length; this was accomplished by adding a scaling function to the options in the plot command. In *Mathematica*, the plot command is **PlotVectorField**.

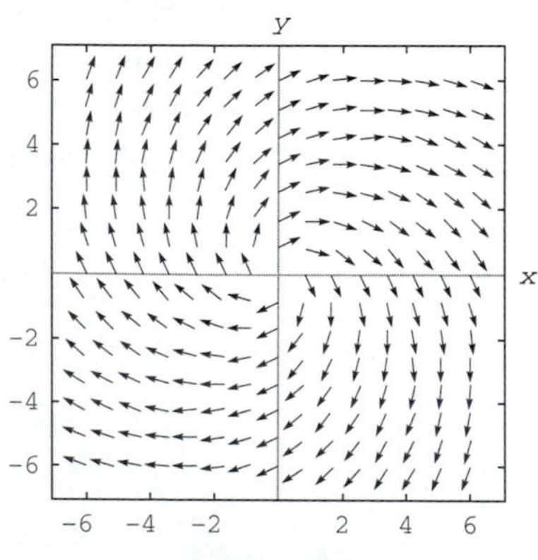

Figure 4.16.

If we divide the second equation in (1) by the first, we obtain the first-order differential equation

$$\frac{dy}{dx} = \frac{dy/dt}{dx/dt} = \frac{g(x, y)}{f(x, y)} \tag{25}$$

Thus we see that a vector field for an autonomous system of differential equations (1) is the same as a direction field of the first-order differential equation (25). See Problem 18 in *Projects and Computer Lab Experiments for Chapter 4.*

Remarks

As we pointed out in Section 3.3, the study of stability is larded with terminology. The definition of asymptotic stability just given is, in some texts, called **strict stability**. A stable critical point that is not strictly stable is said to be **neutrally stable**. Since asymptotic stability is often referred to as a **strong stability**, neutral stability is naturally also called **weak stability**

EXERCISES 4.5

Answers to odd-numbered problems begin on page 449.

4.5.1

In Problems 1–8, classify the critical point (0, 0) for the given autonomous linear system. Use a computer graphing program to obtain a phase portrait of each system.

1. $\dfrac{dx}{dt} = x + y$, $\dfrac{dy}{dt} = y$

2. $\dfrac{dx}{dt} = x - y$, $\dfrac{dy}{dt} = 9x - y$

3. $\dfrac{dx}{dt} = -4x + 2y$, $\dfrac{dy}{dt} = 2x - 4y$

4. $\dfrac{dx}{dt} = 2y$, $\dfrac{dy}{dt} = 5x + 7y$

5. $\dfrac{dx}{dt} = -3x + 6y$, $\dfrac{dy}{dt} = -2x + 3y$

6. $\dfrac{dx}{dt} = -x + y$, $\dfrac{dy}{dt} = -3x - 3y$

7. $\dfrac{dx}{dt} = -2x + 5y$, $\dfrac{dy}{dt} = -2x + 4y$

8. $\dfrac{dx}{dt} = 7x + \dfrac{3}{2}y$, $\dfrac{dy}{dt} = -4x - 2y$

In Problems 9–12, find the critical point (x_1, y_1) of the given autonomous linear system. Use the change of variables $x = X + x_1$, $y = Y + y_1$ as an aid in classifying the critical point. Use a computer graphing program to obtain a phase portrait of each system.

9. $\dfrac{dx}{dt} = 2x + 3y - 6$, $\dfrac{dy}{dt} = -x - 2y + 5$

10. $\dfrac{dx}{dt} = -5x + 9y + 13$, $\dfrac{dy}{dt} = -x - 11y - 23$

11. $\dfrac{dx}{dt} = 0.1x - 0.2y + 0.35$, $\dfrac{dy}{dt} = 0.1x + 0.1y - 0.25$

12. $\dfrac{dx}{dt} = 3x - 2y - 1$, $\dfrac{dy}{dt} = 5x - 3y - 2$

13. Determine whether it is possible to find values of c for the linear system

$$\frac{dx}{dt} = x + y$$

$$\frac{dy}{dt} = cx + y$$

so that (0, 0) is **(a)** a spiral point, **(b)** an unstable node, **(c)** an asymptotically stable node, **(d)** a degenerate node, and **(e)** a saddle point.

14. Determine values of c so that (0, 0) is a center for the linear system

$$\frac{dx}{dt} = 2x - cy$$

$$\frac{dy}{dt} = cx - 2y$$

15. Determine values of a so that (0, 0) is an asymptotically stable spiral point for the linear system

$$\frac{dx}{dt} = ax - 4y$$

$$\frac{dy}{dt} = x + y$$

16. Free damped motion of a mass on a spring is described by

$$\frac{d^2x}{dt^2} + 2\lambda\frac{dx}{dt} + \omega^2 x = 0, \quad \omega > 0, \lambda \geq 0$$

(a) Use the substitution $v = dx/dt$ to express this differential equation as an equivalent autonomous linear system.

(b) Interpret the following four cases physically and at the same time classify the critical point (0, 0) of the differential equation as to type and stability: $\lambda = 0$, $\lambda > \omega$, $\lambda = \omega$, $0 < \lambda < \omega$.

4.5.2

In Problems 17–24, (0, 0) is an isolated critical point of the given system. Use linearization at (0, 0) to classify the critical point, if possible, as to type and stability. If Theorem 4.11 is not applicable, use a computer graphing program to obtain a phase portrait of the system. Use the phase portrait as an aid in classifying the point.

17. $\dfrac{dx}{dt} = 2e^{2x} + \dfrac{1}{2}y - 2$, $\quad \dfrac{dy}{dt} = y^2 + 4y + \sin 2x$

18. $\dfrac{dx}{dt} = \sin(-x + y)$, $\quad \dfrac{dy}{dt} = e^{x-2y} + \sin 5y$

19. $\dfrac{dx}{dt} = 3x - 6y - xy$, $\quad \dfrac{dy}{dt} = x - y + x^2y$

20. $\dfrac{dx}{dt} = -y - x^3$, $\quad \dfrac{dy}{dt} = x - y^3$

21. $\dfrac{dx}{dt} = \sin x + 3\sin y$, $\quad \dfrac{dy}{dt} = 2x\cos x - 2y + y^2$

22. $\dfrac{dx}{dt} = -x - 4y + xy^3$, $\quad \dfrac{dy}{dt} = x - 6y - x^2 + y^2$

23. $\dfrac{dx}{dt} = x(2 + y)^2$, $\quad \dfrac{dy}{dt} = e^{x+4y} - 1$

24. $\dfrac{dx}{dt} = -x + 3y - \sin x$, $\quad \dfrac{dy}{dt} = -x + 2ye^y$

In Problems 25–28, find all critical points of the given system. Use linearization as illustrated in Example 6 to classify all the critical points as to type and stability. Obtain a phase portrait of the system to verify your results or as an aid in classifying a critical point if Theorem 4.11 is not applicable.

25. $\dfrac{dx}{dt} = y + x^2 - 2$, $\quad \dfrac{dy}{dt} = y - x^2$

26. $\dfrac{dx}{dt} = y - x^2$, $\quad \dfrac{dy}{dt} = xy - 2x$

27. $\dfrac{dx}{dt} = (x - 1)(3 - x - y)$, $\quad \dfrac{dy}{dt} = (y - 1)(2x - 3)$

28. $\dfrac{dx}{dt} = -5x - 4y + x^2 + 14$, $\quad \dfrac{dy}{dt} = 2x - 5y + 4$

In Problems 29–32, find the critical points of the given system. Use a computer graphing program to obtain a phase portrait of each system. Use the phase portrait to classify the critical points, if possible, as to type and stability.

29. $\dfrac{dx}{dt} = \sin y$

$\dfrac{dy}{dt} = \cos x$

30. $\dfrac{dx}{dt} = 10 - y^2 + x$

$\dfrac{dy}{dt} = 2x + 2$

31. $\dfrac{dx}{dt} = -2xy$

$\dfrac{dy}{dt} = x^2 - y^3$

32. $\dfrac{dx}{dt} = -2x + y + 2x^2 - xy$

$\dfrac{dy}{dt} = (y - 2)^2$

33. **(a)** Show that the two nonlinear systems

$$\frac{dx}{dt} = y \qquad \qquad \frac{dx}{dt} = y$$
$$\text{and}$$
$$\frac{dy}{dt} = -x - y^2 \qquad \qquad \frac{dy}{dt} = -x - y^3$$

possess the same linearized system.

(b) Classify the critical point (0, 0) of the linearized system found in part (a).

(c) Classify, if possible, the critical point (0, 0) of each of the nonlinear systems. Obtain a phase portrait of each system.

34. A model of the nonlinear pendulum with a constant driving function is $d^2\theta/dt^2 + \sin\theta = \frac{1}{2}$.

(a) Let $x = \theta(t)$ and express the differential equation as an equivalent autonomous system.

(b) Show that $(\pi/6, 0)$ is a critical point of the equation. Find all other critical points.

(c) Verify that $(\pi/6, 0)$ cannot be classified using linearization.

(d) Obtain a phase portrait of the differential equation and use it as an aid in classifying the critical points.

(e) Use the phase portrait to determine whether the differential equation possesses a periodic solution satisfying $\theta(0) = 1$, $\theta'(0) = 1$. Satisfying $\theta(0) = 1$, $\theta'(0) = 1.1$. Use an ODE solver to obtain the solution curve corresponding to each set of initial conditions.

35. Investigate the nonlinear differential equation $d^2\theta/dt^2 + \sin\theta = 1$ to determine whether it possesses nonconstant periodic solutions.

36. In 1927, the Dutch electrical engineer Balthasar van der Pol showed that certain properties of electrical circuits containing vacuum tubes could be deduced from the nonlinear differential equation

$$\frac{d^2x}{dt^2} + \mu(x^2 - 1)\frac{dx}{dt} + x = 0$$

where μ is a positive constant. Show that (0, 0) is an unstable critical point of van der Pol's equation for $\mu > 0$.

In Problems 37 and 38, (0, 0) is an isolated critical point of the given nonlinear system. Use polar coordinates, $x = r\cos\theta$, $y = r\sin\theta$, where r and θ are functions of t, in the system. Classify the critical point at the origin by finding a polar equation of the trajectories.

37. $\dfrac{dx}{dt} = -y + xy^2 + x^3$

$\dfrac{dy}{dt} = x + x^2y + y^3$

38. $\dfrac{dx}{dt} = -y - x\sqrt{x^2 + y^2}$

$\dfrac{dy}{dt} = x - y\sqrt{x^2 + y^2}$

39. **(a)** Show that an equation of the trajectory passing through an initial point (x_0, y_0) for the system given in (23) is $cy = x^2 + y^2$, where $c = (x_0^2 + y_0^2)/y_0$, $y_0 \neq 0$.

(b) Use the result in part (a) to prove that (0, 0) is an unstable critical point of the system.

40. **(a)** Show that a Cartesian equation for the trajectories of the Lotka-Volterra equations (24) is $(x^d e^{-cx})(y^a e^{-by}) = C$, where C is a constant.

(b) Use a CAS with a contour plot application to convince yourself that the Cartesian equation in part (a) defines a family of closed trajectories for $x > 0$, $y > 0$.

41. Consider the Lotka-Volterra predator–prey model defined by

$$\frac{dx}{dt} = -0.1x + 0.02xy$$

$$\frac{dy}{dt} = 0.2y - 0.025xy$$

where the populations $x(t)$ (predators) and $y(t)$ (prey) are measured in the thousands.

(a) Find the critical point (x_1, y_1) in the first quadrant.

(b) Use a computer graphing program to obtain a phase portrait of the system. Find the trajectory passing through (6, 6) at $t = 0$.

(c) Approximate the period of $x(t)$ and $y(t)$ by solving the linearization of the system at (x_1, y_1) subject to $x(0) = 6$, $y(0) = 6$.

(d) Use an ODE solver with $x(0) = 6$, $y(0) = 6$ to approximate the time $t > 0$ when the two populations are first equal. Use the graphs to approximate the period of $x(t)$ and $y(t)$ and compare with the result of part (c).

42. Each of the following systems is a competition model. Find all critical points. Analyze and interpret the model.

(a) $\dfrac{dx}{dt} = x(2 - 0.4x - 0.3y)$

$\dfrac{dy}{dt} = y(1 - 0.1y - 0.3x)$

(b) $\dfrac{dx}{dt} = x(1 - 0.1x - 0.05y)$

$\dfrac{dy}{dt} = y(1.7 - 0.1y - 0.15x)$

43. Each of the following systems is a predator–prey model; the populations $x(t)$ and $y(t)$ are predators and prey, respectively. Find all critical points. Analyze and interpret the model.

(a) $\dfrac{dx}{dt} = x(-2 - x + y)$

$\dfrac{dy}{dt} = y(6 - y - x)$

(b) $\dfrac{dx}{dt} = x(-2 - x + y)$

$\dfrac{dy}{dt} = y(2 - y - x)$

Discussion Problems

44. Suppose the linear system (3) is written as $\mathbf{X}' = \mathbf{AX}$.

(a) Discuss: Show that the coefficients in the characteristic equation (4) are related to the eigenvalues of $\mathbf{A}$ by $a + d = \lambda_1 + \lambda_2$ and $ad - bc = \lambda_1\lambda_2$.

(b) Discuss: Show that the critical point (0, 0) is a saddle point if and only if $\det \mathbf{A} < 0$. Can you draw any conclusion about (0, 0) if $\det \mathbf{A} > 0$?

45. In the discussion of linear systems (3), we assumed throughout that $ad - bc \neq 0$. We see from (4) that this restriction precludes $\lambda = 0$ from being an eigenvalue of the coefficient matrix $\mathbf{A}$. Construct linear systems of the form given in (3) for which $\lambda = 0$ *is* an eigenvalue. Consider three cases (i) $\lambda_1 = 0$, $\lambda_2 > 0$, (ii) $\lambda_1 = 0$, $\lambda_2 < 0$, and (iii) $\lambda_1 = \lambda_2 = 0$. Find the eigenvalues, eigenvectors, and the solution of each system. Determine the critical points and the phase portrait of each system. Now discuss: The assumption that $ad - bc \neq 0$ is made because . . . .

46. Discuss: Why is (0, 0) an isolated critical point of the nonlinear system (23) but (0, 0) is not an isolated critical point of the linearization of (23) at that point?

47. Suppose an autonomous system of first-order differential equations has several critical points. Visualize an *unscaled* vector field of the system and then describe the vectors in a neighborhood of the critical points.

Chapter 4 in Review

Answers to this review are posted at our Web site diffeq.brookscole.com/books.html#zill 98

In Problems 1–4, fill in the blanks or answer true or false.

1. The vector $\mathbf{X} = k\begin{pmatrix}4\\5\end{pmatrix}$ is a solution of $\mathbf{X}' = \begin{pmatrix}1 & 4\\2 & -1\end{pmatrix}\mathbf{X} - \begin{pmatrix}8\\1\end{pmatrix}$ for $k =$ ___.

2. The vector $\mathbf{X} = c_1\begin{pmatrix}-1\\1\end{pmatrix}e^{-9t} + c_2\begin{pmatrix}5\\3\end{pmatrix}e^{7t}$ is solution of the initial-value problem $\mathbf{X}' = \begin{pmatrix}1 & 10\\6 & -3\end{pmatrix}\mathbf{X}$, $\mathbf{X}(0) = \begin{pmatrix}2\\0\end{pmatrix}$ for $c_1 =$ ___ and $c_2 =$ ___.

3. The autonomous system

$$\frac{dx}{dt} = x^2 + 9$$

$$\frac{dy}{dt} = x^3 + 4x$$

has no critical points.

4. If the critical point (0, 0) is a saddle point of the linear system $\mathbf{X}' = \mathbf{A}\mathbf{X}$, and $\mathbf{X}(t)$ is a solution of the system, then $\lim_{t\to\infty} \mathbf{X}(t)$ does not exist.

5. Consider the linear system $\mathbf{X}' = \begin{pmatrix} 4 & 6 & 6 \\ 1 & 3 & 2 \\ -1 & -4 & -3 \end{pmatrix}\mathbf{X}$. Without attempting to solve the system, which one of the following vectors:

$$\mathbf{K}_1 = \begin{pmatrix} 0 \\ 1 \\ 1 \end{pmatrix}, \mathbf{K}_2 = \begin{pmatrix} 1 \\ 1 \\ -1 \end{pmatrix}, \mathbf{K}_3 = \begin{pmatrix} 3 \\ 1 \\ -1 \end{pmatrix}, \mathbf{K}_4 = \begin{pmatrix} 6 \\ 2 \\ -5 \end{pmatrix}$$

is an eigenvector of the coefficient matrix? What is the solution of the system corresponding to this eigenvector?

6. Consider the linear system $\mathbf{X}' = \mathbf{A}\mathbf{X}$ of two differential equations where $\mathbf{A}$ is a real coefficient matrix. What is the general solution of the system if it is known that $\lambda_1 = 1 + 2i$ is an eigenvalue and $\mathbf{K}_1 = \begin{pmatrix} 1 \\ i \end{pmatrix}$ is a corresponding eigenvector?

In Problems 7 and 8, solve the given linear system.

7. $\mathbf{X}' = \begin{pmatrix} 1 & 1 & 1 \\ 1 & 1 & 1 \\ 1 & 1 & 1 \end{pmatrix}\mathbf{X}$

8. $\mathbf{X}' = \begin{pmatrix} -1 & 1 \\ -2 & 1 \end{pmatrix}\mathbf{X} + \begin{pmatrix} 1 \\ \cot t \end{pmatrix}$

9. Consider the linear system $\mathbf{X}' = \mathbf{A}\mathbf{X}$ of three differential equations where

$$\mathbf{A} = \begin{pmatrix} 5 & 3 & 3 \\ 3 & 5 & 3 \\ -5 & -5 & -3 \end{pmatrix}$$

and $\lambda = 2$ is known to be an eigenvalue of multiplicity two. Find two different solutions of the system corresponding to this eigenvalue without using any special formula (such as (12) of Section 4.2).

10. Construct a mathematical model that describes the populations $x(t)$ and $y(t)$ of two species of animals that occupy the same habitat, but interact in a symbiotic manner within that habitat. In other words, the interaction of the animals is *not* one of predation or competition, but rather, is one of possible mutual benefit.

11. The nonlinear system of three first-order equations

$$\frac{ds}{dt} = -k_1 si + k_2 r$$

$$\frac{di}{dt} = -k_3 i + k_1 si$$

$$\frac{dr}{dt} = k_3 i - k_2 r$$

where the k_i, $i = 1, 2, 3$ are positive constants, represents a **SIR model** (see page 275) for the spread of an epidemic through a community with a fixed population of n persons. Here $s(t)$, $i(t)$, and $r(t)$ denote the number of people in the community that are *susceptible* to the disease but not yet infected with it, the number of people who are *infected* with the disease, and the number of people who have *recovered* from the disease, respectively.

(a) Give a relationship between $s(t)$, $i(t)$, and $r(t)$ that holds for all time $t > 0$.
(b) Find the critical points of this autonomous system and try to interpret them physically.
(c) From part (b), explain why this model predicts that the epidemic is "established" only when the total population n satisfies $n > k_3/k_1$.

12. Consider the linear system

$$\frac{dx}{dt} = \alpha x - \beta y$$

$$\frac{dy}{dt} = \beta x + \alpha y$$

Match each of the cases

(a) $\alpha > 0, \beta > 0$, **(b)** $\alpha < 0, \beta > 0$, **(c)** $\alpha = 0, \beta > 0$,

with one or more of the following words relating to the critical point (0, 0): node, spiral point, saddle point, center, stable, asymptotically stable, unstable, attractor, repeller.

In Problems 13 and 14, use linearization to classify, if possible, all critical points of the given nonlinear system as to type and stability.

13. $$\frac{dx}{dt} = 5x - 3x^2 - xy$$

$$\frac{dy}{dt} = y - y^2 - xy$$

14. $$\frac{dx}{dt} = -2x + y + 10$$

$$\frac{dy}{dt} = 2x - y - \frac{15y}{y + 5}$$

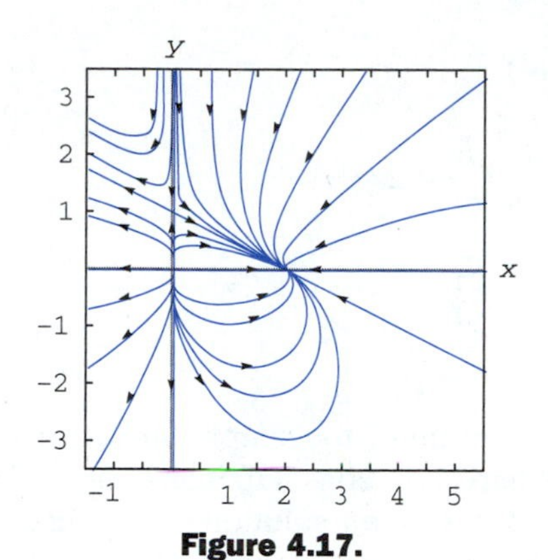

Figure 4.17.

15. **(a)** Find all critical points of the nonlinear system

$$\frac{dx}{dt} = 4x - 2x^2 - xy$$

$$\frac{dy}{dt} = y - y^2 - 2xy$$

(b) Use the phase portrait of the system in part (a) that is given in Figure 4.17 to classify, if possible, the critical points as to type and stability.

16. In Figure 4.18, the given curves are the graphs of a solution $x = x(t), y = y(t)$ of an autonomous system of two first-order differential equations. Use the data from the figure to sketch the corresponding portion of a trajectory in the phase plane.

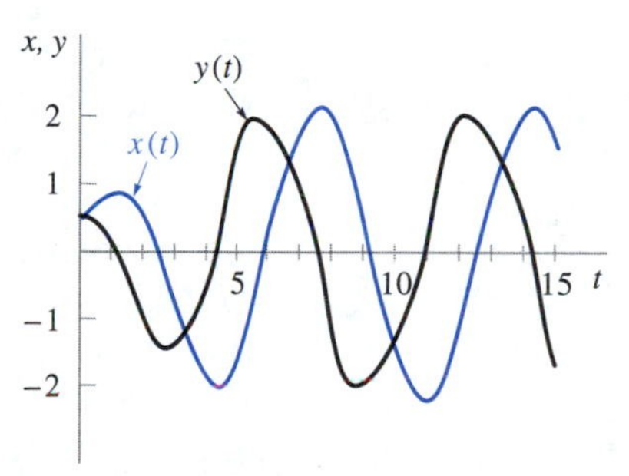

Figure 4.18.

Projects and Computer Lab Experiments for Chapter 4

Section 4.2

1. Using a CAS A solution of a linear system can sometimes be obtained directly from a computer algebra system through commands such as **DSolve** in *Mathematica* and **dsolve** in *Maple*. Use a computer algebra system to solve the following linear systems. Simplify each output and write a real general solution in one of the forms used in Section 4.2. [*Hint*: In part (d), it might help to write 2 = 2.0 and 3 = 3.0.]

(a) $\dfrac{dx}{dt} = 3x - 4y$

$\dfrac{dy}{dt} = 4x - 7y$

(b) $\dfrac{dx}{dt} = 3x + 2y$

$\dfrac{dy}{dt} = -5x + y$

(c) $\frac{dx}{dt} = x + 2y + z$

$\frac{dy}{dt} = 6x - y$

$\frac{dz}{dt} = -x - 2y - z$

(d) $\frac{dx}{dt} = x + 2y + z$

$\frac{dy}{dt} = 2x + y + z$

$\frac{dz}{dt} = 3x + y + 2z$

2. Matrix Exponential The general solution of the simple linear first-order differential equation $x' = ax$, where a is a constant, is an exponential function $x = ce^{at}$. Analogously, the general solution of the first-order linear system $\mathbf{X}' = \mathbf{AX}$ can be expressed in terms of a **matrix exponential** $e^{\mathbf{A}t}$. Although we shall not pursue the definition nor the various ways of computing a matrix exponential (see Problem 20 that follows), it suffices to say that for an $n \times n$ matrix $\mathbf{A}$ of constants, $e^{\mathbf{A}t}$ is an $n \times n$ matrix and a fundamental matrix of the system $\mathbf{X}' = \mathbf{AX}$. The general solution of this homogeneous system is $\mathbf{X} = e^{\mathbf{A}t}\mathbf{C}$, where $\mathbf{C}$ is an $n \times 1$ column matrix of arbitrary constants $c_1, c_2, c_3, \ldots, c_n$.

For those willing to momentarily trade understanding for speed of solution, $e^{\mathbf{A}t}$ can be computed in a mechanistic manner with the aid of computer software; for example, in *Mathematica*, the function **MatrixExp[A t]** computes the matrix exponential for a square matrix **At**; in *Maple*, the command is **exponential(A, t)**; in MATLAB, the function is `expm(At)`.

(a) Use a matrix exponential to find the general solution of $\mathbf{X}' = \begin{pmatrix} 4 & 2 \\ 3 & 3 \end{pmatrix}\mathbf{X}$. Then use the computer to find eigenvalues and eigenvectors of the coefficient matrix $\mathbf{A} = \begin{pmatrix} 4 & 2 \\ 3 & 3 \end{pmatrix}$ and form the general solution in the manner of Section 4.2. Finally, reconcile the two forms of the general solution of the system.

(b) Use a matrix exponential to find the general solution of $\mathbf{X}' = \begin{pmatrix} -3 & -1 \\ 2 & -1 \end{pmatrix}\mathbf{X}$. In the case of complex output, utilize the software to do the simplification; for example, in *Mathematica*, if **M** = **MatrixExp[A t]** has complex entries, then try the command **Simplify[ComplexExpand[M]]**.

(c) Use a matrix exponential to find the general solution of

$$\mathbf{X}' = \begin{pmatrix} -4 & 0 & 6 & 0 \\ 0 & -5 & 0 & -4 \\ -1 & 0 & 1 & 0 \\ 0 & 3 & 0 & 2 \end{pmatrix}\mathbf{X}.$$

3. Reducing a System to Normal Form Consider the system of differential equations

$$\frac{d^2x}{dt^2} + 2\frac{dx}{dt} - x - \frac{dy}{dt} - y = 0$$

$$\frac{d^2y}{dt^2} - \frac{dy}{dt} + y - 2\frac{dx}{dt} + x = 0$$

Use the substitutions $x = x_1$, $x' = x_2$, $y = x_3$, $y' = x_4$ to express the system as a linear system $\mathbf{X}' = \mathbf{AX}$ of four first-order equations, where $\mathbf{X} = \begin{pmatrix} x_1 \\ x_2 \\ x_3 \\ x_4 \end{pmatrix}$. Use a CAS or linear algebra software to find the eigenvalues and eigenvectors of the coefficient matrix $\mathbf{A}$. Form the general solution of $\mathbf{X}' = \mathbf{AX}$. From this result, find the solution of the original system.

Section 4.3

4. Eigenvalues/Eigenvectors Solving a nonhomogeneous linear system $\mathbf{X}' = \mathbf{AX} + \mathbf{F}(t)$ by variation of parameters when $\mathbf{A}$ is a 3×3 (or larger) matrix is almost an impossible task to do by hand. Consider the system

$$\mathbf{X}' = \begin{pmatrix} 2 & -2 & 2 & 1 \\ -1 & 3 & 0 & 3 \\ 0 & 0 & 4 & -2 \\ 0 & 0 & 2 & -1 \end{pmatrix} \mathbf{X} + \begin{pmatrix} te^t \\ e^{-t} \\ e^{2t} \\ 1 \end{pmatrix}$$

(a) Use a CAS or linear algebra software to find the eigenvalues and eigenvectors of the coefficient matrix.

(b) Form a fundamental matrix $\mathbf{\Phi}(t)$ and use the computer to find $\mathbf{\Phi}^{-1}(t)$.

(c) Use the computer to carry out the computations of $\mathbf{\Phi}^{-1}(t)\mathbf{F}(t)$, $\int \mathbf{\Phi}^{-1}(t)\, \mathbf{F}(t)\, dt$, $\mathbf{\Phi}(t) \int \mathbf{\Phi}^{-1}(t)\, \mathbf{F}(t)\, dt$, $\mathbf{\Phi}(t)\mathbf{C}$, and $\mathbf{\Phi}(t)\mathbf{C} + \mathbf{\Phi}(t) \int \mathbf{\Phi}^{-1}(t)\, \mathbf{F}(t)\, dt$, where $\mathbf{C}$ is a 4×1 column matrix of arbitrary constants c_1, c_2, c_3, c_4.

(d) Rewrite the computer output for the general solution of the system in the form $\mathbf{X} = \mathbf{X}_c + \mathbf{X}_p$, where $\mathbf{X}_c = c_1\mathbf{X}_1 + c_2\mathbf{X}_2 + c_3\mathbf{X}_3 + c_4\mathbf{X}_4$.

5. Matrix Exponential Again When a is constant, it follows from (4) of Section 2.3 that the general solution of the nonhomogeneous linear first-order differential equation $x' = ax + f(t)$ is $x = x_c + x_p = ce^{at} + e^{at} \int e^{-at} f(t)\, dt$, where c is a constant. Analogously, the general solution of the nonhomogeneous first-order linear system $\mathbf{X}' = \mathbf{AX} + \mathbf{F}(t)$ can be shown to be $\mathbf{X} = \mathbf{X}_c + \mathbf{X}_p = e^{\mathbf{A}t}\mathbf{C} + e^{\mathbf{A}t} \int e^{-\mathbf{A}t}\, \mathbf{F}(t)\, dt$, where $e^{\mathbf{A}t}$ is the $n \times n$ **matrix exponential** (see Problem 2), $\mathbf{C}$ is an $n \times 1$ column matrix of arbitrary constants, and $e^{-\mathbf{A}t}$ is the inverse of $e^{\mathbf{A}t}$. In parts (a) and (b), use the matrix exponential and computer software to carry out all computations and simplifications leading to the general solution of the given systems.

(a) $\mathbf{X}' = \begin{pmatrix} 3 & 1 \\ 4 & 3 \end{pmatrix}\mathbf{X} + \begin{pmatrix} 4e^t \\ te^{2t} \end{pmatrix}$ **(b)** $\mathbf{X}' = \begin{pmatrix} 2 & -1 \\ 4 & 6 \end{pmatrix}\mathbf{X} + \begin{pmatrix} te^{4t} \\ 0 \end{pmatrix}$

(c) Use a matrix exponential to find a particular solution of

$$\mathbf{X}' = \begin{pmatrix} 1 & -1 & -1 \\ 1 & -1 & 1 \\ 1 & -1 & -2 \end{pmatrix}\mathbf{X} + \begin{pmatrix} 5e^{2t} \\ 3t \\ 1 \end{pmatrix}$$

Section 4.4

6. Radioactive Series

(a) In the system (2) of Section 4.4, suppose time is measured in days, the decay constants are $k_1 = -0.138629$ and $k_2 = -0.004951$, and that $x_0 = 20$. Use a graphing utility to obtain the graphs of the solutions $x(t)$, $y(t)$, and $z(t)$ on the same set of coordinate axes. Use the graphs to approximate the half-lives of substances X and Y.

(b) Use the graphs in part (a) to approximate the times when the amounts $x(t)$ and $y(t)$ are the same, the times when the amounts $x(t)$ and $z(t)$ are the same, and the times when the amounts $y(t)$ and $z(t)$ are the same. Why does the time determined when the amounts $y(t)$ and $z(t)$ are the same make intuitive sense?

7. Mixtures

(a) Use a graphing utility to obtain the graphs of the solutions $x_1(t)$ and $x_2(t)$ of the system (3) of this section (Problem 3 in Exercises 4.4) on the same set of coordinate axes. Use these graphs to estimate the time after which the amount of salt in tank B is greater than in tank A. Explain why the behavior of $x_1(t)$ and $x_2(t)$ as $t \to \infty$ makes intuitive sense.

(b) Use a graphing utility to obtain the graphs $x_1(t)$ and $x_2(t)$ of the system in Problem 4 in Exercises 4.4. Use the same set of coordinate axes. Is there a time after which the amount of salt in tank B is greater than that in tank A? Explain why the behavior of $x_1(t)$ and $x_2(t)$ as $t \to \infty$ makes intuitive sense.

8. Mixtures Three large tanks A, B, and C hold 200, 150, and 100 gal of brine, respectively. The system is closed in that the well-stirred liquid is pumped only between the tanks, as shown in Figure 4.19.

(a) Use the information given in the figure to construct a mathematical model for the number of pounds of salt $x_1(t)$, $x_2(t)$, and $x_3(t)$ at time t in tanks A, B, and C, respectively.

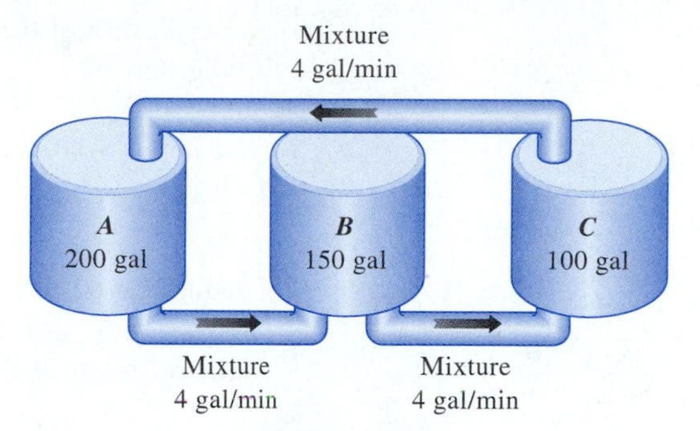

Figure 4.19.

(b) Write the system in part (a) in the form $\mathbf{X}' = \mathbf{AX}$. Use a CAS or linear algebra software to find the eigenvalues and eigenvectors of $\mathbf{A}$. Write down the general solution of the system.

(c) Use part (b) to find the particular solution of the system $\mathbf{X}' = \mathbf{AX}$ if initially: 50 lb of salt is dissolved in the solution in tank A, 25 lb of salt is dissolved in tank B, and tank C contains pure water.

(d) Use a graphing utility to obtain the graphs of $x_1(t)$, $x_2(t)$, $x_3(t)$ obtained in part (c). Use the same set of coordinate axes. What are the limiting values of $x_1(t)$, $x_2(t)$, and $x_3(t)$ as $t \to \infty$? Explain why these values make intuitive sense. Suppose t is measured in minutes. Is there a time at which the amount of salt in each tank is the same? Discuss.

(e) Verify that $x_1(t) + x_2(t) + x_3(t) = 75$ for any time t. Explain why this makes intuitive sense.

9. Spread of a Communicable Disease

(a) In the model of the spread of a communicable disease given in Problem 13 in Exercises 4.4, explain why it is sufficient to analyze only the two equations

$$\frac{ds}{dt} = -k_1 si$$

$$\frac{di}{dt} = -k_2 i + k_1 si$$

(b) Suppose $k_1 = 0.2$, $k_2 = 0.7$, and $n = 10$. Choose various values of $i(0) = i_0$, $0 < i_0 < 10$. Use an ODE solver to determine what the model predicts about the epidemic in the two cases $s_0 > k_2/k_1$ and $s_0 \leq k_2/k_1$. In the case of an epidemic, estimate the number of people that are eventually infected.

10. Coupled Springs

(a) Suppose two masses m_1 and m_2 are connected to two springs and the two springs are attached as shown in Figure 5.51 of Section

5.6. A model for the free undamped vertical motion of the masses is given by

$$m_1 x_1'' = -k_1 x_1 + k_2(x_2 - x_1)$$

$$m_2 x_2'' = -k_2(x_2 - x_1)$$

where $x_1(t)$ and $x_2(t)$ represent the displacements of the masses from their equilibrium positions. Use substitutions analogous to those given in Problem 3 to express the system as a linear system $\mathbf{Y}' = \mathbf{AY}$ of four first-order equations, where $\mathbf{Y} = \begin{pmatrix} y_1 \\ y_2 \\ y_3 \\ y_4 \end{pmatrix}$. Give the matrix $\mathbf{A}$ in terms of k_1, k_2, m_1, and m_2.

(b) Let $k_1 = 6$, $k_2 = 4$, and $m_1 = m_2 = 1$. Use a CAS or linear algebra software to find the eigenvalues of $\mathbf{A}$.

(c) Because the physical system is oscillatory and undamped, it stands to reason that the solution $x_1 = x_1(t)$, $x_2 = x_2(t)$ of the original system will involve sines and cosines of different frequencies. Without solving any system, use the results in part (b) to determine these frequencies.

Section 4.5

11. Fill in the Blanks If we let $p = a + d$ and $q = ad - bc$ in the characteristic equation (4) of Section 4.5, then by the quadratic formula the eigenvalues λ_1, λ_2 of the coefficient matrix $\mathbf{A}$ are given by $\dfrac{p \pm \sqrt{p^2 - 4q}}{2}$.

(a) Fill in each cell in the table with one of the symbols: > 0, < 0, $= 0$, —, to denote whether the given expression (that is, $p^2 - 4q$, p, or q) is positive, negative, equal to zero, or can be any value, in describing the critical point under the various cases.

CASE I $p^2 - 4q$ ☐			*CASE II* $p^2 - 4q$ ☐			*CASE III* $p^2 - 4q$ ☐		
Critical point (0, 0)	p	q	**Critical point** (0, 0)	p	q	**Critical point** (0, 0)	p	q
Stable node	☐	☐	Stable degenerate node	☐	☐	Stable spiral point	☐	☐
Unstable node	☐	☐	Unstable degenerate node	☐	☐	Unstable spiral point	☐	☐
Saddle point	☐	☐				Center	☐	☐

(b) Consider the pq-plane as shown in Figure 4.20. Use the table in part (a) to fill in the cells with a name such as *stable nodes*, *centers*, and so forth, for each of the eight indicated lines, curves, and regions.

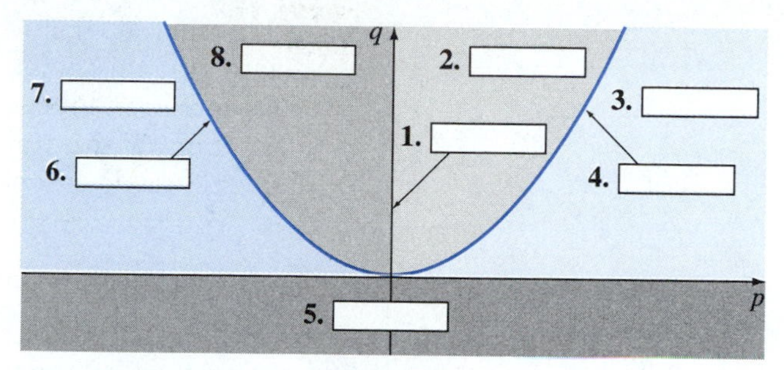

Figure 4.20.

12. Some Nonlinear Systems

(a) Use a computer graphing program to obtain a phase portrait of the nonlinear system

$$\frac{dx}{dt} = x^2 + 2x - y^2 - 5$$

$$\frac{dy}{dt} = x^2 + 3x + y^2 - 24$$

Based only on the phase portrait, approximate the coordinates of all critical points of the system. Based only on the phase portrait, classify the critical points as to stability. Finally, by hand, find the exact coordinates of each critical point.

(b) With the plot bounds set as $-6 \leq x \leq 6$, $-6 \leq y \leq 6$, use a computer graphing program to obtain a phase portrait of the nonlinear system

$$\frac{dx}{dt} = x + \sin(y - 2x) + 4$$

$$\frac{dy}{dt} = x + 2y - 5\sin(x + 2y) - 1$$

Determine the number and approximate location of all critical points within the given rectangular region. Then use a root-finding application in a CAS to approximate the coordinates of each critical point. Use a CAS to determine the linearization matrix $\mathbf{A}(x, y)$ (see (19) of Section 4.5) at each critical point. Then use a CAS to find the eigenvalues of each linearization matrix $\mathbf{A}(x, y)$. Then, if possible, apply Theorem 4.11 to classify the critical points as to type and stability.

(c) Show that (0, 0) is an isolated critical point of the nonlinear system

$$\frac{dx}{dt} = x^4 - 2xy^3$$

$$\frac{dy}{dt} = 2x^3y - y^4$$

Show that (0, 0), however, is not an isolated critical point of linearization at this point. Does the linearized system give any useful information? Proceed as on page 147 to show that a Cartesian equation of the trajectories is $x^3 + y^3 = 3cxy$. Verify that the parametric equations

$$x = \frac{3ct}{1 + t^3}, \quad y = \frac{3ct^2}{1 + t^3}$$

satisfy the Cartesian equation. These last equations define a classic curve known as the **folium of Descartes**. Use computer software and these parametric equations to obtain a phase portrait of the system in part (b). Based only on the phase portrait, can you classify the critical point as to type and stability?

13. **Limit Cycle** In the study of autonomous nonlinear differential equations or plane autonomous nonlinear systems, (1) of Section 4.5, the search for periodic solutions is equivalent to a search for closed trajectories in the phase plane. It is sometimes possible for a plane autonomous system to possess a single closed trajectory C with the property that nonclosed trajectories starting either from the interior of C or its exterior approach C in a *spiral* manner as $t \to \infty$. Such a curve C is called a **stable limit cycle** of the system.

(a) The nonlinear system

$$\frac{dx}{dt} = -y + x(1 - x^2 - y^2)$$

$$\frac{dy}{dt} = x + y(1 - x^2 - y^2)$$

has a single critical point at (0, 0). With the plot bounds set to $-4 \leq x \leq 4$, $-4 \leq y \leq 4$, and $-10 < t < 10$, use a computer graphing program to obtain the trajectories of the system whose initial points are (0.1, 0.1), (−0.1, 0.1), (−0.1, −0.1), (0.1, −0.1), (2, 2), (−2, 2), (−2, −2), (2, −2). Plot additional trajectories if desired.

(b) Use the phase portrait in part (a) to classify the critical point (0, 0) as to stability and type. Deduce the limit cycle C of the system.

(c) The system in part (a) can actually be solved by using polar coordinates. If $x = r\cos\theta$, $y = r\sin\theta$, show that

$$x\frac{dx}{dt} + y\frac{dy}{dt} = r\frac{dr}{dt}$$

$$x\frac{dy}{dt} - y\frac{dx}{dt} = r^2\frac{d\theta}{dt}$$

(d) Use the equations in part (c) to show that the original system becomes

$$\frac{dr}{dt} = r(1 - r^2)$$

$$\frac{d\theta}{dt} = 1$$

(e) Suppose $\theta = 0$ when $t = 0$. Use the equations in part (d) to show that a solution of the original system is

$$x = \frac{\cos t}{\sqrt{1 + ce^{-2t}}}, \quad y = \frac{\sin t}{\sqrt{1 + ce^{-2t}}}$$

(f) Use a graphing utility to obtain the representative graphs of the equations in part (e) for $c < 0$, $c > 0$, and $c = 0$. For $c > 0$, what happens to $(x(t), y(t))$ as $t \to \infty$? As $t \to -\infty$? For $c < 0$, what happens to $(x(t), y(t))$ as $t \to \infty$? As $t \to \ln \sqrt{|c|}$? Find an equation of the limit cycle C. Give an initial condition (x_0, y_0) so that the graph of the corresponding solution of the system is the limit cycle C.

14. Van der Pol's Equation

(a) By differentiating (29) of Section 3.6, it is seen that the current $i(t)$ in a free L-R-C series circuit is governed by

$$L\frac{d^2i}{dt^2} + R\frac{di}{dt} + \frac{1}{C}i = 0$$

If we replace i by the symbol x, replace R by the nonlinear term $x^2 - 1$, and change the time scale, the foregoing DE is transformed into **van der Pol's equation**

$$\frac{d^2x}{dt^2} + \mu(x^2 - 1)\frac{dx}{dt} + x = 0$$

where μ is a positive parameter. In Problem 36 in Exercises 4.5, you were asked to demonstrate that (0, 0) is an unstable critical point of this differential equation. What would a linearized version of van der Pol's DE predict for the behavior of the current x as $t \to \infty$? Specifically, are there any periodic solutions?

(b) Use a computer graphing program to obtain a phase portrait for van der Pol's equation in the cases: $\mu = 0.1$, $\mu = 1$, and $\mu = 6$.

(c) For each value of $\mu > 0$, a van der Pol equation possesses a periodic solution; this is revealed by the fact that each phase portrait has a single stable limit cycle. Unlike Problem 13, however, we cannot obtain an explicit formula for this cycle. Use a colored pencil to highlight the approximate limit cycle in the phase portrait of the DE in each of the cases: $\mu = 0.1$, $\mu = 1$, and $\mu = 6$. If necessary, redo the phase portraits with just a few carefully chosen trajectories. Adjust the plot bounds on x, y, and t as required.

(d) In each of the cases: $\mu = 0.1$, $\mu = 1$, and $\mu = 6$, use an ODE solver to graph x versus t for an initial point inside the limit cycle, outside the limit cycle, and (to a best estimate) on the limit cycle. In the last case, estimate the period of each of the three periodic solutions.

15. Predator–Prey Model In Problem 14, you were asked to estimate the period of a stable limit cycle by graphical means. In this problem, you are asked to estimate the period of a stable limit cycle by numerical means.

(a) The autonomous system of nonlinear equations

$$\frac{dx}{dt} = 1.5x\left(1 - \frac{x}{45}\right) - y\frac{5x}{18 + x}$$

$$\frac{dy}{dt} = -4y + 2y\frac{5x}{18 + x}$$

is a mathematical model for populations $y(t)$ of predators and $x(t)$ of prey, respectively. A model of this type is constructed with the expressed intention that it possess a single periodic solution or stable limit cycle. With the plot bounds set to $0 \le x \le 25$, $0 \le y \le 25$, use a computer graphing program to obtain a phase portrait of the system.

(b) Now use a computer program and a numerical routine, such as the fourth-order Runge-Kutta method, to construct a table of numerical values for the solution $x = x(t)$, $y = y(t)$. Experiment with different initial conditions $(x(0), y(0)) = (x_0, y_0)$ such as (9, 15), (10, 15), (11, 15), and so on, different time intervals, such as $0 \le t \le 20$, $0 \le t \le 30$, $0 \le t \le 40$, and so on, and with different step sizes, such as $h = 0.5$, $h = 0.25$, $h = 0.1$, and so on, and then select *one* of these solutions that you think is closest to being periodic. Use the numerical data to estimate the period T of the limit cycle.

(c) Use the graphing function of an ODE solver to obtain the graphs of x versus t and y versus t for the selected solution (and corresponding initial condition) in part (b). Do the graphs confirm your estimate of T from part (b)?

16. Competition Model

(a) The autonomous system of nonlinear equations

$$\frac{dx}{dt} = 0.001x(100 - x - 3y)$$

$$\frac{dy}{dt} = 0.004y\left(50 - y - \frac{3}{4}x\right)$$

represents a model for the populations of two competing species of animals (see (13) of Section 4.4). Find all critical points of the system. Use linearization to classify the critical points by stability and type.

(b) One of the critical points in part (a) is a saddle point. Define a separatrix as a pair of trajectories on which points approach or enter the saddle point as $t \to \infty$. See Problem 21, *Projects and Computer Lab Experiments for Chapter 3*. Use computer software to obtain a phase portrait of the system. Use a colored pencil to highlight the approximate separatrices.

(c) Analyze the populations $x(t)$ and $y(t)$ over a long period of time in the cases: the initial point (x_0, y_0) is in the region above the separatrix, (x_0, y_0) is in the region below the separatrix, (x_0, y_0) lies on the separatrix.

(d) Repeat the analysis above for the competition model

$$\frac{dx}{dt} = 0.001x\left(51 - \frac{3}{4}x - y\right)$$

$$\frac{dy}{dt} = 0.004y\left(72 - \frac{3}{2}y - x\right)$$

17. Rotating Bead A bead of mass m is free to slide on a wire circular hoop of radius l. If the hoop rotates with a constant angular velocity ω about a vertical diameter, then the angle θ from the vertical to a radius drawn to the bead, as shown in Figure 4.21, satisfies the nonlinear differential equation

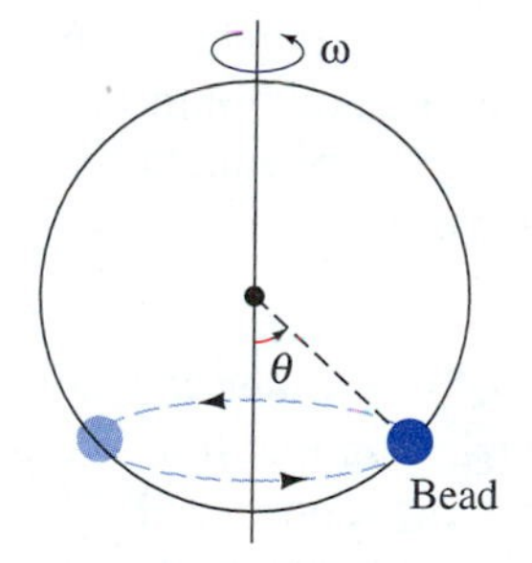

Figure 4.21.

$$\frac{d^2\theta}{dt^2} + \sin\theta - \omega^2 \sin\theta\cos\theta = 0, \quad \omega \geq 0$$

(a) Assume $-\pi \leq \theta \leq \pi$. Find all critical points of the differential equation. Consider the cases $0 < \omega < 1$, $\omega = 1$, $\omega > 1$.

(b) Use a computer graphing program to obtain a phase portrait for the differential equation in the cases: $\omega = \frac{1}{2}$, $\omega = 0.9$, $\omega = 1$, and $\omega = 1.1$.

(c) Using the results in part (b), elaborate on the meaning of the word **bifurcation** as it applies to this differential equation. See Problem 24, *Projects and Computer Lab Experiments for Chapter 2*.

18. Vector Fields

(a) Consider the vector field defined by the system of equations

$$\frac{dx}{dt} = y$$

$$\frac{dy}{dt} = -\frac{1}{2}x$$

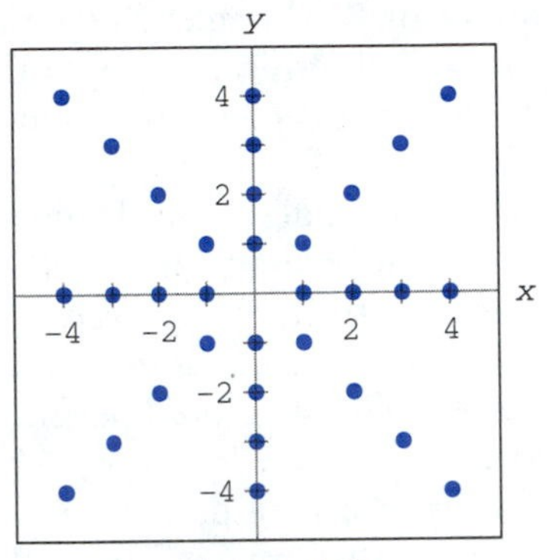

Figure 4.22.

Make two copies of the region given in Figure 4.22. On one copy sketch by hand the vectors at the points indicated by a colored dot. On the second copy, redraw the vectors scaled to unit vectors. Using only these two vector fields, conjecture the classification of the critical point (0, 0) as to type and stability.

(b) Use a CAS to plot the vector field for the system in part (a). Use scaled vectors.

(c) Use a CAS to plot the direction of the first-order differential equation defined by the system in part (a). Review (25) of Section 4.5.

(d) Use a CAS to plot the vector field for the system

$$\frac{dx}{dt} = xy - 3x - 2y + 6$$

$$\frac{dy}{dt} = -y + 4$$

Using only the vector field, determine the critical points of the system. Conjecture the classification of the critical points as to type and stability.

Writing Assignments

19. Undetermined Coefficients Write a short report on how the method of undetermined coefficients can be applied to a nonhomogeneous linear system $\mathbf{X}' = \mathbf{AX} + \mathbf{F}(t)$. Discuss the types of functions that can be entries in the matrix $\mathbf{F}(t)$ and any difficulties that one may encounter when using this method.

20. Computation of a Matrix Exponential In Problems 2 and 5, we used computer software to compute a matrix exponential. We did not actually *define* the $n \times n$ matrix $e^{\mathbf{A}t}$. Write a short report on the definition of a matrix exponential. Illustrate how to compute a matrix exponential using this definition. As directed by your instructor, discuss other matrix methods for computing a matrix exponential.

21. Unstable Critical Point Suppose (0, 0) is a critical point of a *nonlinear* system. It was stated on page 291 that even though on all trajectories of the system $(x(t), y(t)) \to (0, 0)$ as $t \to \infty$, the critical point (0, 0) may not be asymptotically stable. The point (0, 0) could still be unstable. Consult an advanced text and find an example of this and write up your findings in the form of a short report.

22. Limit Cycles Again In Problems 13 and 14 we introduced the notion of a *stable* limit cycle. There are, of course, different kinds of limit cycles. Do some additional reading on the subject, and then write a short report that elaborates on the various kinds of limit cycles that are not stable. If possible, illustrate each kind with an appropriate system of nonlinear differential equations and its phase portrait.

23. Solution by Diagonalization A homogeneous system $\mathbf{X}' = \mathbf{AX}$ with constant coefficients,

$$\begin{pmatrix} x_1' \\ x_2' \\ \vdots \\ x_n' \end{pmatrix} = \begin{pmatrix} a_{11} & a_{12} & \cdots & a_{1n} \\ a_{21} & a_{22} & \cdots & a_{2n} \\ \vdots & & & \vdots \\ a_{n1} & a_{n2} & \cdots & a_{nn} \end{pmatrix} \begin{pmatrix} x_1 \\ x_2 \\ \vdots \\ x_n \end{pmatrix}$$

in which each x_i' is expressed as a linear combination of $x_1, x_2, \ldots, x_n$, is said to be *coupled*. If the coefficient matrix $\mathbf{A}$ is **diagonalizable**, then the system can be *uncoupled* in that each x_i' can be expressed solely in terms of x_i. Write a short report that elaborates on how a linear system $\mathbf{X}' = \mathbf{AX}$ can be solved by **diagonalization**.

Chapter 5

The Laplace Transform

5.1 The Laplace Transform

In calculus, you learned that the operations of differentiation and integration are transforms, that is, these operations transform a function into another function. For example, $\frac{d}{dt} t^2 = 2t$ and $\int t^2\, dt = t^3/3 + c$. Moreover, these two transforms possess the linearity property that the transform of a linear combination of functions is a linear combination of the transforms:

$$\frac{d}{dt}[\alpha f(t) + \beta g(t)] = \alpha f'(t) + \beta g'(t) \qquad \int [\alpha f(t) + \beta g(t)]\, dt = \alpha \int f(t)\, dt + \beta \int g(t)\, dt$$

In this section, we will examine a special type of integral transform called the *Laplace transform*. In addition to possessing the linearity property, the Laplace transform has many other interesting properties that make it useful in solving linear initial-value problems.

5.1.1 Definition and Some Basic Properties

Definition If $f(t, y)$ is a function of two variables, then a definite integral of f with respect to one of the variables leads to a function of the other variable. For example, by holding y constant, we see that $\int_1^2 2ty^2\, dt = 3y^2$. Similarly, a definite integral such as $\int_a^b K(s, t) f(t)\, dt$ transforms a function f of the variable t into a function F of the variable s. We are particularly interested in an integral transform where the interval of integration is the unbounded interval $[0, \infty)$. Recall, an improper integral $\int_0^\infty K(s, t) f(t)\, dt$ is defined as a limit

$$\int_0^\infty K(s, t) f(t)\, dt = \lim_{b \to \infty} \int_0^b K(s, t) f(t)\, dt$$

If the limit exists, the integral exists or is convergent; if the limit does not exist, the integral does not exist and is said to be divergent. The foregoing limit will, in general, exist for only certain values of the variable s. The choice $K(s, t) = e^{-st}$ gives us an especially important integral transform.

Definition 5.1 Laplace Transform

Let f be a function defined for $t \geq 0$. The integral

$$\mathscr{L}\{f(t)\} = \int_0^\infty e^{-st} f(t)\, dt \tag{1}$$

is said to be the **Laplace transform** of f provided the integral converges.

When the defining integral (1) converges, the result is a function of s. As a rule, we shall use a lowercase letter to denote the function being transformed and the corresponding capital letter to denote its Laplace transform; for example,

$$\mathscr{L}\{f(t)\} = F(s), \quad \mathscr{L}\{g(t)\} = G(s), \quad \mathscr{L}\{y(t)\} = Y(s)$$

EXAMPLE 1 Applying Definition 5.1

Evaluate $\mathscr{L}\{1\}$.

Solution From (1),

$$\mathscr{L}\{1\} = \int_0^\infty e^{-st}(1)\,dt = \lim_{b\to\infty}\int_0^b e^{-st}\,dt$$

$$= \lim_{b\to\infty}\frac{-e^{-st}}{s}\bigg|_0^b = \lim_{b\to\infty}\frac{-e^{-sb}+1}{s} = \frac{1}{s}$$

provided $s > 0$. In other words, when $s > 0$, the exponent $-sb$ is negative and $e^{-sb} \to 0$ as $b \to \infty$. The integral diverges for $s < 0$. ■

The use of the limit sign becomes somewhat tedious, so we shall adopt the notation $|_0^\infty$ as a shorthand to writing $\lim_{b\to\infty}(\ )|_0^b$. For example,

$$\mathscr{L}\{1\} = \int_0^\infty e^{-st}(1)\,dt = \frac{-e^{-st}}{s}\bigg|_0^\infty = \frac{1}{s}, \quad s > 0$$

At the upper limit, it is understood we mean $e^{-st} \to 0$ as $t \to \infty$ for $s > 0$.

EXAMPLE 2 Applying Definition 5.1

Evaluate $\mathscr{L}\{t\}$.

Solution From Definition 5.1, we have $\mathscr{L}\{t\} = \int_0^\infty e^{-st}t\,dt$. Integrating by parts and using $\lim_{t\to\infty} te^{-st} = 0$, $s > 0$, along with the result from Example 1, we obtain

$$\mathscr{L}\{t\} = \frac{-te^{-st}}{s}\bigg|_0^\infty + \frac{1}{s}\int_0^\infty e^{-st}\,dt = \frac{1}{s}\mathscr{L}\{1\} = \frac{1}{s}\left(\frac{1}{s}\right) = \frac{1}{s^2}$$ ■

EXAMPLE 3 Applying Definition 5.1

Evaluate $\mathscr{L}\{e^{-3t}\}$.

Solution From Definition 5.1, we have

$$\mathscr{L}\{e^{-3t}\} = \int_0^\infty e^{-st}e^{-3t}\,dt = \int_0^\infty e^{-(s+3)t}\,dt$$

$$= \left.\frac{-e^{(s+3)t}}{s+3}\right|_0^\infty$$

$$= \frac{1}{s+3}, \quad s > -3$$

The result follows from the fact that $\lim_{t\to\infty} e^{-(s+3)t} = 0$ for $s + 3 > 0$ or $s > -3$. ■

■ **EXAMPLE 4 Applying Definition 5.1**

Evaluate $\mathscr{L}\{\sin 2t\}$.

Solution From Definition 5.1 and integration by parts, we have

$$\mathscr{L}\{\sin 2t\} = \int_0^\infty e^{-st}\sin 2t\,dt = \left.\frac{-e^{-st}\sin 2t}{s}\right|_0^\infty + \frac{2}{s}\int_0^\infty e^{-st}\cos 2t\,dt$$

$$= \frac{2}{s}\int_0^\infty e^{-st}\cos 2t\,dt, \quad s > 0$$

Laplace transform of $\sin 2t$ ↓

$$= \frac{2}{s}\left[\left.\frac{-e^{-st}\cos 2t}{s}\right|_0^\infty - \frac{2}{s}\int_0^\infty e^{-st}\sin 2t\,dt\right]$$

$$= \frac{2}{s^2} - \frac{4}{s^2}\mathscr{L}\{\sin 2t\}$$

At this point, we have an equation with $\mathscr{L}\{\sin 2t\}$ on both sides of the equality. Solving for that quantity yields the result

$$\mathscr{L}\{\sin 2t\} = \frac{2}{s^2+4}, \quad s > 0$$ ■

$\mathscr{L}$ Is a Linear Transform For a sum of functions, we can write

$$\int_0^\infty e^{-st}[\alpha f(t) + \beta g(t)]\,dt = \alpha\int_0^\infty e^{-st}f(t)\,dt + \beta\int_0^\infty e^{-st}g(t)\,dt$$

whenever both integrals converge for $s > c$. Hence it follows that

$$\mathscr{L}\{\alpha f(t) + \beta g(t)\} = \alpha\mathscr{L}\{f(t)\} + \beta\mathscr{L}\{g(t)\} = \alpha F(s) + \beta G(s) \tag{2}$$

Because of the property given in (2), $\mathscr{L}$ is said to be a **linear transform**. For example, from Examples 1 and 2,

$$\mathscr{L}\{1 + 5t\} = \mathscr{L}\{1\} + 5\mathscr{L}\{t\} = \frac{1}{s} + \frac{5}{s^2}$$

We state the generalizations of some of the preceding examples by means of the next theorem.

Theorem 5.1 Transforms of Some Basic Functions

	$f(t)$	$\mathscr{L}\{f(t)\}$
(a)	1	$\dfrac{1}{s}$
(b)	$t^n,\ n = 1, 2, 3, \ldots$	$\dfrac{n!}{s^{n+1}}$
(c)	e^{at}	$\dfrac{1}{s-a}$
(d)	$\sin kt$	$\dfrac{k}{s^2+k^2}$
(e)	$\cos kt$	$\dfrac{s}{s^2+k^2}$
(f)	$\sinh kt$	$\dfrac{k}{s^2-k^2}$
(g)	$\cosh kt$	$\dfrac{s}{s^2-k^2}$

Sufficient Conditions for Existence of $\mathscr{L}\{f(t)\}$ The integral in (1) that defines the Laplace transform does not have to converge. For example, neither $\mathscr{L}\{1/t\}$ nor $\mathscr{L}\{e^{t^2}\}$ exists. Sufficient conditions that guarantee the existence of $\mathscr{L}\{f(t)\}$ are that f be **piecewise continuous** on $[0, \infty)$ and that f be of exponential order for $t > T$. Recall that a function f is piecewise continuous on $[0, \infty)$ if, in any interval $0 \le a \le t \le b$, there are at most a finite number of points t_k, $k = 1, 2, \ldots, n$ $(t_{k-1} < t_k)$, at which f has finite discontinuities and is continuous on each open interval $t_{k-1} < t < t_k$. See Figure 5.1. The concept of **exponential order** is defined in the following manner.

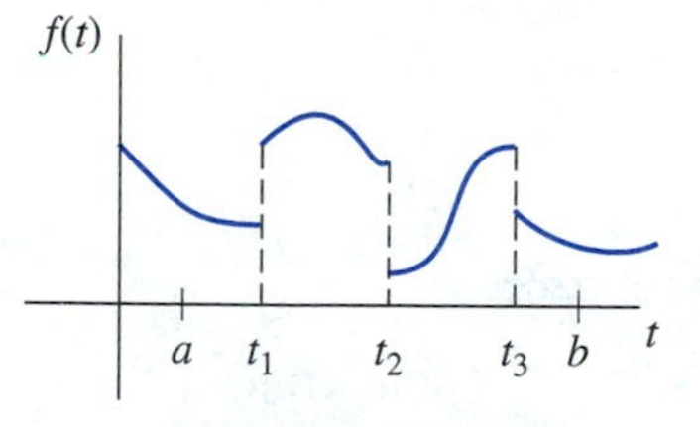

Figure 5.1.

Definition 5.2 Exponential Order

A function f is said to be of **exponential order** if there exist numbers c, $M > 0$, and $T > 0$ such that $|f(t)| \le Me^{ct}$ for $t > T$.

If, for example, f is an *increasing* function, then the condition $|f(t)| \le Me^{ct}$, $t > T$, simply states that the graph of f on the interval

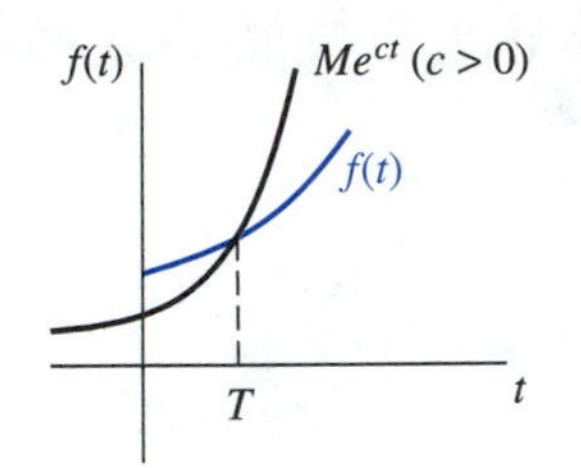

Figure 5.2.

(T, ∞) does not grow faster than the graph of the exponential function Me^{ct}, where c is a positive constant. See Figure 5.2. The functions $f(t) = t$, $f(t) = e^{-t}$, and $f(t) = 2\cos t$ are all of exponential order for $t > 0$ since we have, respectively,

$$|t| \le e^t, \quad |e^{-t}| \le e^t, \quad |2\cos t| \le 2e^t$$

A comparison of the graphs on the interval $[0, \infty)$ is given in Figure 5.3.

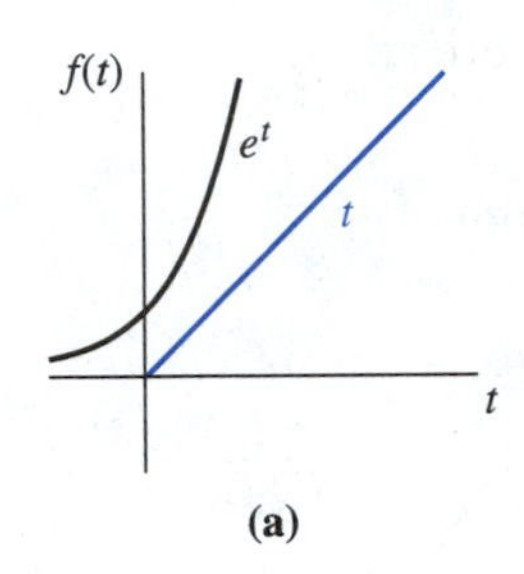

(a)

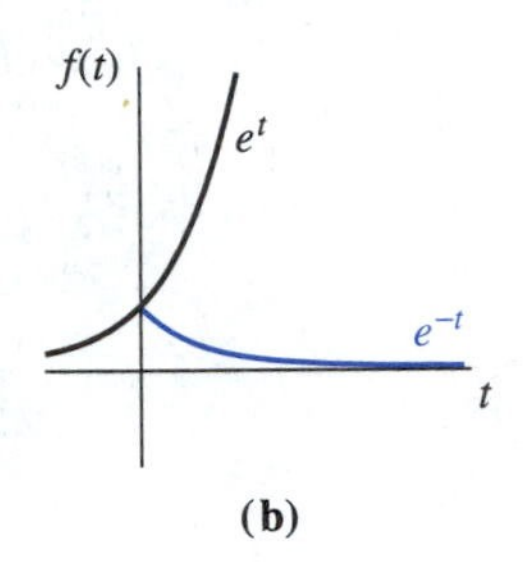

(b)

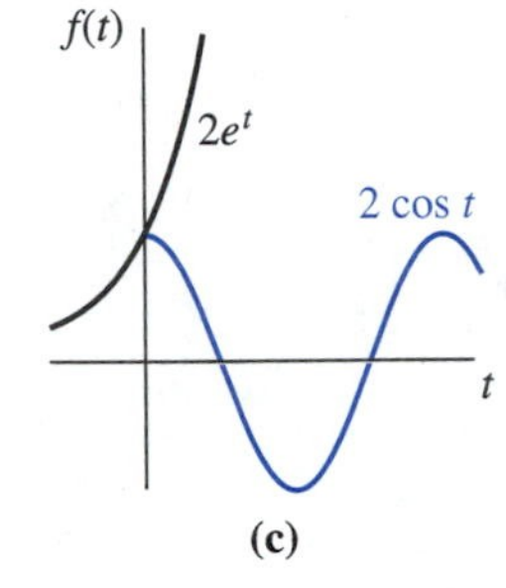

(c)

Figure 5.3.

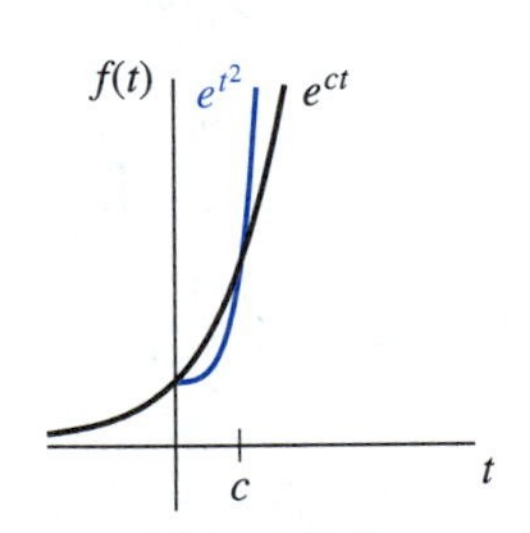

Figure 5.4.

A function, such as $f(t) = e^{t^2}$, is not of exponential order since, as shown in Figure 5.4, its graph grows faster than any positive linear power of e for $t > c > 0$.

A positive integral power of t is always of exponential order since for $c > 0$,

$$|t^n| \le Me^{ct} \quad \text{or} \quad \left|\frac{t^n}{e^{ct}}\right| \le M, \qquad \text{for } t > T$$

is equivalent to showing that $\lim_{t\to\infty} t^n/e^{ct}$ is finite for $n = 1, 2, 3, \ldots$. The result follows by n applications of L'Hôpital's rule.

Theorem 5.2 Sufficient Conditions for Existence

If f is piecewise continuous on $[0, \infty)$ and of exponential order for $t > T$, then $\mathscr{L}\{f(t)\}$ exists for $s > c$.

Proof $$\mathscr{L}\{f(t)\} = \int_0^T e^{-st}f(t)\,dt + \int_T^\infty e^{-st}f(t)\,dt = I_1 + I_2$$

The integral I_1 exists because it can be written as a sum of integrals over intervals for which $e^{-st}f(t)$ is continuous. Now

$$|I_2| \le \int_T^\infty |e^{-st}f(t)|\,dt \le M\int_T^\infty e^{-st}e^{ct}\,dt$$

$$= M\int_T^\infty e^{-(s-c)t}\,dt = -M\frac{e^{-(s-c)t}}{s-c}\Bigg|_T^\infty = M\frac{e^{-(s-c)T}}{s-c}$$

for $s > c$. Since $\int_T^\infty Me^{-(s-c)t}\,dt$ converges, the integral $\int_T^\infty |e^{-st}f(t)|\,dt$ converges by the comparison test for improper integrals. This, in turn, implies that I_2 exists for $s > c$. The existence of I_1 and I_2 implies that $\mathcal{L}\{f(t)\} = \int_0^\infty e^{-st}f(t)\,dt$ exists for $s > c$. □

■ EXAMPLE 5 **Transform of a Piecewise Continuous Function**

Evaluate $\mathcal{L}\{f(t)\}$ where $f(t) = \begin{cases} 0, & 0 \le t < 3 \\ 2, & t \ge 3 \end{cases}$

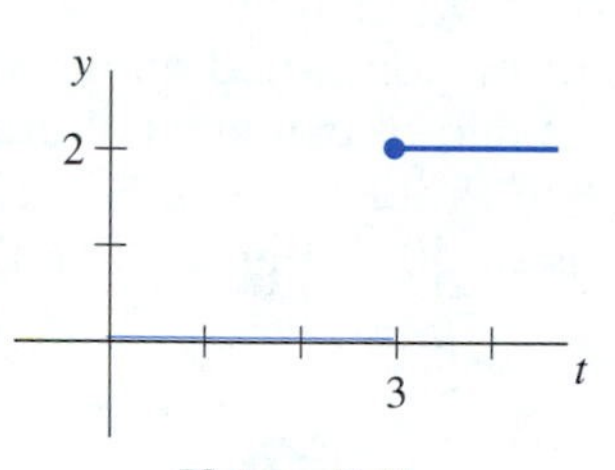

Figure 5.5.

Solution The function f, shown in Figure 5.5, is piecewise continuous, and of exponential order for $t > 0$. Since f is defined in two pieces, $\mathcal{L}\{f(t)\}$ is expressed as the sum of two integrals:

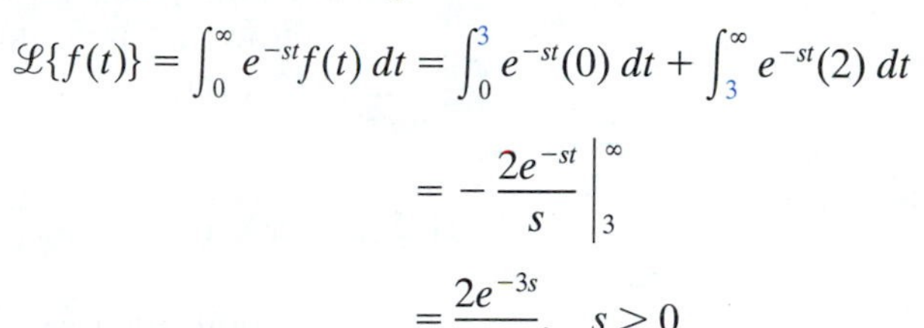

$$\mathcal{L}\{f(t)\} = \int_0^\infty e^{-st}f(t)\,dt = \int_0^3 e^{-st}(0)\,dt + \int_3^\infty e^{-st}(2)\,dt$$

$$= -\frac{2e^{-st}}{s}\Bigg|_3^\infty$$

$$= \frac{2e^{-3s}}{s}, \quad s > 0$$ ■

Throughout this chapter, we shall be concerned primarily with functions that are both piecewise continuous and of exponential order. We note, however, that these two conditions are sufficient but not necessary for the existence of a Laplace transform. The function $f(t) = t^{-1/2}$ is not piecewise continuous on the interval $[0, \infty)$, but its Laplace transform exists. See Problem 31 in Exercises 5.1.

As the next theorem indicates, not every arbitrary function of s is a Laplace transform of a piecewise continuous function of exponential order.

> **Theorem 5.3 Behavior of $F(s)$ as $s \to \infty$**
>
> If f is piecewise continuous on $[0, \infty)$ and of exponential order for $t > T$, then $\lim_{s\to\infty} \mathcal{L}\{f(t)\} = 0$.

Proof Since $f(t)$ is piecewise continuous on $0 \le t \le T$, it is necessarily bounded on the interval. That is, $|f(t)| \le M_1 = M_1 e^{0t}$. Also, $|f(t)| \le M_2 e^{\gamma t}$ for $t > T$. If M denotes the maximum of $\{M_1, M_2\}$ and c denotes the maximum of $\{0, \gamma\}$, then

$$|\mathcal{L}\{f(t)\}| \le \int_0^\infty e^{-st}|f(t)|\,dt \le M\int_0^\infty e^{-st}\cdot e^{ct}\,dt = -M\frac{e^{-(s-c)t}}{s-c}\Bigg|_0^\infty = \frac{M}{s-c}$$

for $s > c$. As $s \to \infty$, we have $|\mathcal{L}\{f(t)\}| \to 0$ and so $\to \mathcal{L}\{f(t)\} \to 0$. □

In view of Theorem 5.3, we can say that functions of s such as $F_1(s) = 1$ and $F_2(s) = s/(s+1)$ are not the Laplace transforms of piecewise continuous functions of exponential order since $F_1(s) \not\to 0$ and $F_2(s) \not\to 0$ as $s \to \infty$. You should not conclude from this, for example, that $F_1(s)$ and $F_2(s)$ are *not* Laplace transforms. There are other kinds of functions.

Remarks

In the discussion that follows, we want to find $f(t)$ given a Laplace transform $F(s)$. But the Laplace of a function $f(t)$ may not be unique, in other words, it is possible that $\mathscr{L}\{f_1(t)\} = \mathscr{L}\{f_2(t)\}$ and yet $f_1 \neq f_2$. For our purposes, this is not anything to be concerned about. If f_1 and f_2 are piecewise continuous on $[0, \infty)$ and of exponential order, then f_1 and f_2 are *essentially* the same. See Problem 33 in Exercises 5.1. However, if f_1 and f_2 are continuous on $[0, \infty)$ and $\mathscr{L}\{f_1(t)\} = \mathscr{L}\{f_2(t)\}$, then $f_1 = f_2$ on the interval.

5.1.2 The Inverse Transform and Solving DEs

Inverse Transform If $F(s)$ is the Laplace transform of a function $f(t)$, that is, $\mathscr{L}\{f(t)\} = F(s)$, we then say $f(t)$ is the **inverse Laplace transform** of $F(s)$ and write $\mathscr{L}^{-1}\{F(s)\} = f(t)$. For example, from Examples 1, 2, and 3, we have, respectively,

$$\mathscr{L}^{-1}\left\{\frac{1}{s}\right\} = 1, \quad \mathscr{L}^{-1}\left\{\frac{1}{s^2}\right\} = t, \quad \text{and} \quad \mathscr{L}^{-1}\left\{\frac{1}{s+3}\right\} = e^{-3t}$$

When evaluating inverse transforms, it often happens that a function of s under consideration does not match *exactly* the form of a Laplace transform $F(s)$ given in a table. It may be necessary to "fix up" the function of s by multiplying and dividing by an appropriate constant. Example 6 illustrates the method.

■ EXAMPLE 6 Applying Theorem 5.1

Evaluate $\mathscr{L}^{-1}\left\{\dfrac{1}{s^2+64}\right\}$.

Solution In order to match $1/(s^2 + 64)$ with the form in the right-hand column in part (d) of Theorem 5.1, we identify $k^2 = 64$. However, since the numerator lacks the required $k = 8$, we multiply and divide by 8:

↓ Now exactly $\dfrac{k}{s^2+k^2}$

$$\mathscr{L}^{-1}\left\{\frac{1}{s^2+64}\right\} = \frac{1}{8}\mathscr{L}^{-1}\left\{\frac{8}{s^2+64}\right\} = \frac{1}{8}\sin 8t$$

■

$\mathscr{L}^{-1}$ is a Linear Transform We assume that the inverse Laplace transform is itself a linear transform; that is, for constants α and β,

$$\mathscr{L}^{-1}\{\alpha F(s) + \beta G(s)\} = \alpha\mathscr{L}^{-1}\{F(s)\} + \beta\mathscr{L}^{-1}\{G(s)\}$$

where F and G are the transforms of some functions f and g.

Partial fractions play an important role in finding inverse Laplace transforms. You should consult either an older calculus text or a current precalculus text for a complete review of this theory.

EXAMPLE 7 Partial Fractions and Linearity

Evaluate $\mathscr{L}^{-1}\left\{\dfrac{1}{(s-1)(s+2)(s+4)}\right\}$.

Solution There exist unique constants A, B, and C so that

$$\frac{1}{(s-1)(s+2)(s+4)} = \frac{A}{s-1} + \frac{B}{s+2} + \frac{C}{s+4} = \frac{A(s+2)(s+4) + B(s-1)(s+4) + C(s-1)(s+2)}{(s-1)(s+2)(s+4)}$$

Since the denominators are identical, the numerators are identical:

$$1 = A(s+2)(s+4) + B(s-1)(s+4) + C(s-1)(s+2)$$

By comparing coefficients of powers of s on both sides of the equality, we know that the last equation is equivalent to a system of three equations in the three unknowns A, B, and C. However, you might recall the following shortcut for determining these unknowns. If we set $s = 1$, $s = -2$, and $s = -4$, the zeros of the common denominator $(s-1)(s+2)(s+4)$, we obtain, in turn,

$$1 = A(3)(5), \qquad 1 = B(-3)(2), \qquad 1 = C(-5)(-2)$$

and so $A = \frac{1}{15}$, $B = -\frac{1}{6}$, $C = \frac{1}{10}$. Hence we can write

$$\frac{1}{(s-1)(s+2)(s+4)} = \frac{1/15}{s-1} - \frac{1/6}{s+2} + \frac{1/10}{s+4}$$

and thus, from the linearity of $\mathscr{L}^{-1}$ and part (c) of Theorem 5.1,

$$\mathscr{L}^{-1}\left\{\frac{1}{(s-1)(s+2)(s+4)}\right\} = \frac{1}{15}\mathscr{L}^{-1}\left\{\frac{1}{s-1}\right\} - \frac{1}{6}\mathscr{L}^{-1}\left\{\frac{1}{s+2}\right\} + \frac{1}{10}\mathscr{L}^{-1}\left\{\frac{1}{s+4}\right\}$$

$$= \frac{1}{15}e^{t} - \frac{1}{6}e^{-2t} + \frac{1}{10}e^{-4t}$$

■

Transforming a Derivative Our goal is to use the Laplace transform to solve certain kinds of differential equations. To that end, we need to evaluate quantities such as $\mathscr{L}\{dy/dt\}$ and $\mathscr{L}\{d^2y/dt^2\}$. For example, if f' is continuous for $t \geq 0$, then integration by parts gives

$$\mathscr{L}\{f'(t)\} = \int_0^\infty e^{-st} f'(t)\, dt = e^{-st} f(t)\Big|_0^\infty + s\int_0^\infty e^{-st} f(t)\, dt$$

$$= -f(0) + s\mathscr{L}\{f(t)\}$$

or $$\mathscr{L}\{f'(t)\} = sF(s) - f(0) \quad \textbf{(3)}$$

Here we have assumed that $e^{-st}f(t) \to 0$ as $t \to \infty$. Similarly,

$$\mathscr{L}\{f''(t)\} = \int_0^\infty e^{-st} f''(t)\, dt = e^{-st} f'(t)\Big|_0^\infty + s\int_0^\infty e^{-st} f'(t)\, dt$$

$$= -f'(0) + s\mathscr{L}\{f'(t)\} \qquad \leftarrow \text{From (3)}$$

$$= s[sF(s) - f(0)] - f'(0)$$

or $$\mathscr{L}\{f''(t)\} = s^2F(s) - sf(0) - f'(0) \quad \textbf{(4)}$$

The results in (3) and (4) are special cases of the next theorem, which gives the Laplace transform of the nth derivative of f.

Theorem 5.4 Transform of a Derivative

If $f, f', \ldots, f^{(n-1)}$ are continuous on $[0, \infty)$ and are of exponential order, and if $f^{(n)}$ is piecewise continuous on $[0, \infty)$, then

$$\mathscr{L}\{f^{(n)}(t)\} = s^nF(s) - s^{n-1}f(0) - s^{n-2}f'(0) - \cdots - f^{(n-1)}(0)$$

where $F(s) = \mathscr{L}\{f(t)\}$

Solving DEs Because $\mathscr{L}\{d^ny/dt^n\}$ depends on $\mathscr{L}\{y\} = Y(s)$ and the $n - 1$ derivatives of y evaluated at $t = 0$, the Laplace transform is ideally suited to linear initial-value problems; the transform reduces a linear differential equation with constant coefficients to an algebraic equation. We simply solve the equation for the transformed function $Y(s)$ and then find the solution $y(t)$ of the initial-value problem by determining the inverse Laplace transform $\mathscr{L}^{-1}\{Y(s)\} = y(t)$. The next example illustrates the method.

EXAMPLE 8 Transforming a DE

Use the Laplace transform to solve

$$\frac{dy}{dt} - 3y = e^{2t}, \quad y(0) = 1$$

Solution We first take the transform of each member of the differential equation:

$$\mathcal{L}\left\{\frac{dy}{dt}\right\} - 3\mathcal{L}\{y\} = \mathcal{L}\{e^{2t}\} \tag{5}$$

But from (3), $\mathcal{L}\{dy/dt\} = sY(s) - y(0) = sY(s) - 1$, and from part (c) of Theorem 5.1, $\mathcal{L}\{e^{2t}\} = 1/(s-2)$, and so (5) is the same as

$$sY(s) - 1 - 3Y(s) = \frac{1}{s-2}$$

Solving the last equation for $Y(s)$ and then using partial fractions, we get

$$Y(s) = \frac{s-1}{(s-2)(s-3)} = \frac{-1}{s-2} + \frac{2}{s-3}$$

and so $$y(t) = -\mathcal{L}^{-1}\left\{\frac{1}{s-2}\right\} + \mathcal{L}^{-1}\left\{\frac{2}{s-3}\right\}$$

Again from part (c) of Theorem 5.1, it follows that the solution of the initial value problem is $y(t) = -e^{2t} + 2e^{3t}$. ∎

Example 8 illustrates the basic procedure for using the Laplace transform to solve an initial-value problem, but this example really doesn't demonstrate a method that is any better than the approach to such problems outlined in Section 2.3. Don't draw any negative conclusions from one simple problem, however; the Laplace transform method *is* ideally suited to *certain kinds* of initial-value problems: problems for which the traditional methods become somewhat unwieldy. We will illustrate problems of greater complexity in the sections that follow.

EXERCISES 5.1

Answers to odd-numbered problems begin on page 451.

5.1.1

In Problems 1–18, use Definition 5.1 to find $\mathcal{L}\{f(t)\}$.

1. $f(t) = \begin{cases} -1, & 0 \le t < 1 \\ 1, & t \ge 1 \end{cases}$

2. $f(t) = \begin{cases} 4, & 0 \le t < 2 \\ 0, & t \ge 2 \end{cases}$

3. $f(t) = \begin{cases} t, & 0 \le t < 1 \\ 1, & t \ge 1 \end{cases}$

4. $f(t) = \begin{cases} 2t+1, & 0 \le t < 1 \\ 0, & t \ge 1 \end{cases}$

5. $f(t) = \begin{cases} \sin t, & 0 \le t < \pi \\ 0, & t \ge \pi \end{cases}$

6. $f(t) = \begin{cases} 0, & 0 \le t < \pi/2 \\ \cos t, & t \ge \pi/2 \end{cases}$

7. 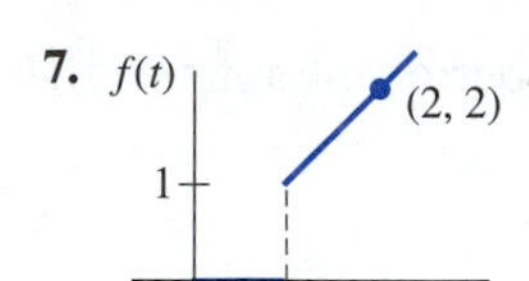

Figure 5.6.

8. 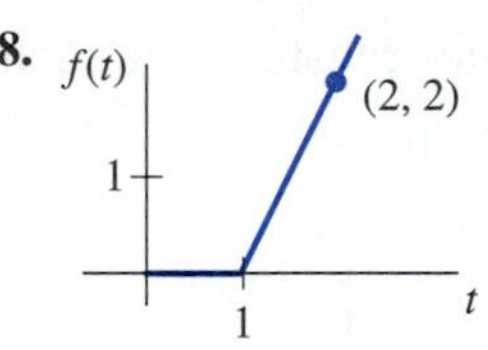

Figure 5.7.

9. 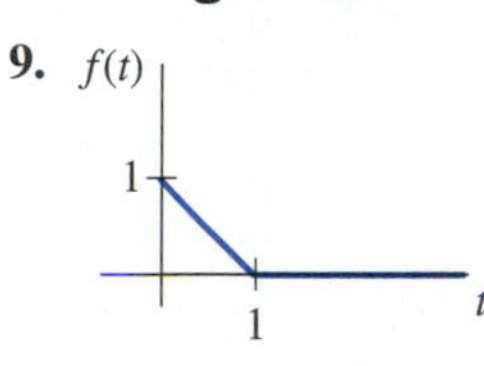

Figure 5.8.

10. 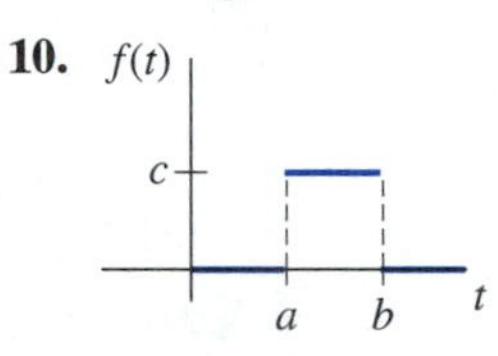

Figure 5.9.

11. $f(t) = e^{t+7}$ **12.** $f(t) = e^{-2t-5}$

13. $f(t) = te^{4t}$ **14.** $f(t) = t^2e^{3t}$

15. $f(t) = \sin kt$ **16.** $f(t) = \cos kt$

17. $f(t) = t \cos t$ **18.** $f(t) = t \sin t$

In Problems 19–26, use linearity and Theorem 5.1 to find $\mathcal{L}\{f(t)\}$.

19. $f(t) = 2t^4 - 10$ **20.** $f(t) = -t^5 + 3t + 9$

21. $f(t) = 1 + e^{4t}$ **22.** $f(t) = t^2 - e^{-9t} + 5$

23. $f(t) = (t + 1)^3$ **24.** $f(t) = (e^t - e^{-t})^2$

25. $f(t) = 4t^2 - 5 \sin 3t$ **26.** $f(t) = \cos 5t + \sin 2t$

In Problems 27–30, use linearity and parts (c) and (e) of Theorem 5.1 to find $\mathcal{L}\{f(t)\}$.

27. $f(t) = \sinh kt$ **28.** $f(t) = \cosh kt$

29. $f(t) = \cos^2 t$ **30.** $f(t) = \sin^2 4t$

31. Every student of mathematics should know something about the **gamma function**. One definition of this function is given by the improper integral

$$\Gamma(\alpha) = \int_0^\infty t^{\alpha-1}e^{-t}\,dt, \quad \alpha > 0$$

(a) Show that $\Gamma(\alpha + 1) = \alpha\,\Gamma(\alpha)$.

(b) Show that for $\alpha > -1$, $\mathcal{L}\{t^\alpha\} = \dfrac{\Gamma(\alpha+1)}{s^{\alpha+1}}$.

(c) Given that $\Gamma(\frac{1}{2}) = \sqrt{\pi}$, find the Laplace transforms of $f(t) = t^{-1/2}$, $f(t) = t^{1/2}$, and $f(t) = t^{3/2}$.

Discussion Problems

32. Make up a function $F(t)$ that is of exponential order, but $f(t) = F'(t)$ is not of exponential order. Make up a function f that is not of exponential order, but whose Laplace transform exists.

33. Make up two functions f and g that have the same Laplace transform. Do not think profound thoughts.

34. Find the Laplace transform of the staircase function f shown in Figure 5.10.

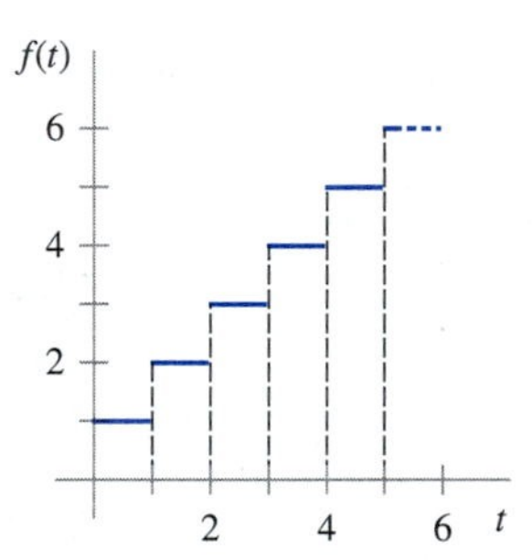

Figure 5.10.

35. Consider the gamma function in Problem 31. Show that $\Gamma(1) = 1$. Use this result and part (a) of Problem 31 as a basis for discussing why the gamma function is known as the "generalized factorial function."

36. Discuss: Suppose that $\mathcal{L}\{f_1(t)\} = F_1(s)$ for $s > c_1$ and $\mathcal{L}\{f_2(t)\} = F_2(s)$ for $s > c_2$. When does

$$\mathcal{L}\{f_1(t) + f_2(t)\} = F_1(s) + F_2(s)?$$

37. Discuss: Does the Laplace transform of $f(t) = e^{\sqrt{t}}$ exist?

38. Figure 5.4 suggests, but does not prove, that the function $f(t) = e^{t^2}$ is not of exponential order. Discuss: How does the observation that $t^2 > \ln M + ct$, for $M > 0$ and t sufficiently large, show that $e^{t^2} > Me^{ct}$ for any c?

5.1.2

In Problems 39–70, use Theorem 5.1 to find the given inverse transform.

39. $\mathcal{L}^{-1}\left\{\dfrac{1}{s^3}\right\}$ **40.** $\mathcal{L}^{-1}\left\{\dfrac{1}{s^4}\right\}$

41. $\mathcal{L}^{-1}\left\{\dfrac{1}{s^2} - \dfrac{48}{s^5}\right\}$ **42.** $\mathcal{L}^{-1}\left\{\left(\dfrac{2}{s} - \dfrac{1}{s^3}\right)^2\right\}$

43. $\mathcal{L}^{-1}\left\{\dfrac{(s+1)^3}{s^4}\right\}$ **44.** $\mathcal{L}^{-1}\left\{\dfrac{(s+2)^2}{s^3}\right\}$

45. $\mathscr{L}^{-1}\left\{\frac{1}{s^2}-\frac{1}{s}+\frac{1}{s-2}\right\}$

46. $\mathscr{L}^{-1}\left\{\frac{4}{s}+\frac{6}{s^5}-\frac{1}{s+8}\right\}$

47. $\mathscr{L}^{-1}\left\{\frac{1}{4s+1}\right\}$

48. $\mathscr{L}^{-1}\left\{\frac{1}{5s-2}\right\}$

49. $\mathscr{L}^{-1}\left\{\frac{5}{s^2+49}\right\}$

50. $\mathscr{L}^{-1}\left\{\frac{10s}{s^2+16}\right\}$

51. $\mathscr{L}^{-1}\left\{\frac{4s}{4s^2+1}\right\}$

52. $\mathscr{L}^{-1}\left\{\frac{1}{4s^2+1}\right\}$

53. $\mathscr{L}^{-1}\left\{\frac{1}{s^2-16}\right\}$

54. $\mathscr{L}^{-1}\left\{\frac{10s}{s^2-25}\right\}$

55. $\mathscr{L}^{-1}\left\{\frac{2s-6}{s^2+9}\right\}$

56. $\mathscr{L}^{-1}\left\{\frac{s-1}{s^2+2}\right\}$

57. $\mathscr{L}^{-1}\left\{\frac{1}{s^2+3s}\right\}$

58. $\mathscr{L}^{-1}\left\{\frac{s+1}{s^2-4s}\right\}$

59. $\mathscr{L}^{-1}\left\{\frac{s}{s^2+2s-3}\right\}$

60. $\mathscr{L}^{-1}\left\{\frac{1}{s^2+s-20}\right\}$

61. $\mathscr{L}^{-1}\left\{\frac{0.9s}{(s-0.1)(s+0.2)}\right\}$

62. $\mathscr{L}^{-1}\left\{\frac{s-3}{(s-\sqrt{3})(s+\sqrt{3})}\right\}$

63. $\mathscr{L}^{-1}\left\{\frac{s}{(s-2)(s-3)(s-6)}\right\}$

64. $\mathscr{L}^{-1}\left\{\frac{s^2+1}{s(s-1)(s+1)(s-2)}\right\}$

65. $\mathscr{L}^{-1}\left\{\frac{2s+4}{(s-2)(s^2+4s+3)}\right\}$

66. $\mathscr{L}^{-1}\left\{\frac{s+1}{(s^2-4s)(s+5)}\right\}$

67. $\mathscr{L}^{-1}\left\{\frac{1}{s^2(s^2+4)}\right\}$

68. $\mathscr{L}^{-1}\left\{\frac{s-1}{s^2(s^2+1)}\right\}$

69. $\mathscr{L}^{-1}\left\{\frac{s}{(s^2+4)(s+2)}\right\}$

70. $\mathscr{L}^{-1}\left\{\frac{1}{s^4-9}\right\}$

In Problems 71–80, use the Laplace transform to solve the given differential equation subject to the indicated initial conditions.

71. $\frac{dy}{dt}-y=1,\quad y(0)=0$

72. $\frac{dy}{dt}+2y=t,\quad y(0)=-1$

73. $y'+4y=e^{-4t},\quad y(0)=2$

74. $y'-y=\sin t,\quad y(0)=0$

75. $y''+5y'+4y=0,\quad y(0)=1,\ y'(0)=0$

76. $y''+9y=e^t,\quad y(0)=0,\quad y'(0)=0$

77. $2y'''+3y''-3y'-2y=e^{-t},\quad y(0)=0,\ y'(0)=0,\ y''(0)=1$

78. $y'''+2y''-y'-2y=\sin 3t,\quad y(0)=0,\ y'(0)=0,\ y''(0)=1$

79. $y^{(4)}-y=0,\quad y(0)=1,\quad y'(0)=0,\quad y''(0)=-1,\ y'''(0)=0$

80. $y^{(4)}-y=t,\quad y(0)=0,\quad y'(0)=0,\quad y''(0)=0,\ y'''(0)=0$

Discussion Problems

81. **(a)** Discuss how differentiation and (3) of this section can be used to obtain (a), (c), (d), (e), (f), and (g) in Theorem 5.1.
(b) Discuss how differentiation, (3), and induction can be used to obtain (b) in Theorem 5.1.
(c) Discuss how the result $\frac{d}{dt}\cos^2 t=-\sin 2t$, and (3) can be used to evaluate $\mathscr{L}\{\cos^2 t\}$.

5.2 Translation Theorems

One way of evaluating a Laplace transform such as $\mathscr{L}\{t^{10}e^{-t}\}$ is, of course, to use the basic integral definition and integrate by parts ten times. But there is an easier way. In this section, we present two labor saving theorems called the *translation theorems*. These theorems enable us to build up a more extensive list of transforms without the necessity of using the definition of the Laplace transform.

Consider This At this point, the application of the Laplace transform to initial-value problems is hampered due to the limited number of entries in Theorem 5.1. For example, the solution of the initial-value problem $y'' + 4y' + 6y = 1 + e^{-t}$, $y(0) = 0$, $y'(0) = 0$ requires the evaluation of

$$\mathscr{L}^{-1}\left\{\frac{-s/2 - 5/3}{s^2 + 4s + 6}\right\} \tag{1}$$

Although we have given a more extensive table of functions $f(t)$ and corresponding transforms $F(s)$ in Appendix II, none of the formulas in this table can be applied directly to the problem of finding the inverse transform (1). The evaluation of an inverse transform such as (1) demands both an expansion of our knowledge of the properties of the Laplace transform as well as the development of an algebraic dexterity for recasting a function of s into just the right form. We shall evaluate (1) in Example 4.

First Translation Theorem If we know $\mathscr{L}\{f(t)\} = F(s)$, we can compute the Laplace transform $\mathscr{L}\{e^{at}f(t)\}$ with no additional effort other than translating, or shifting, $F(s)$ to $F(s - a)$. The result is known as the **first translation theorem**.

Theorem 5.5 First Translation Theorem

If $\mathscr{L}\{f(t)\} = F(s)$ and a is a real number, then

$$\mathscr{L}\{e^{at}f(t)\} = F(s - a)$$

Proof The proof is immediate, since by Definition 5.1,

$$\mathscr{L}\{e^{at}f(t)\} = \int_0^\infty e^{-st}e^{at}f(t)\,dt = \int_0^\infty e^{-(s-a)t}f(t)\,dt = F(s-a) \qquad \square$$

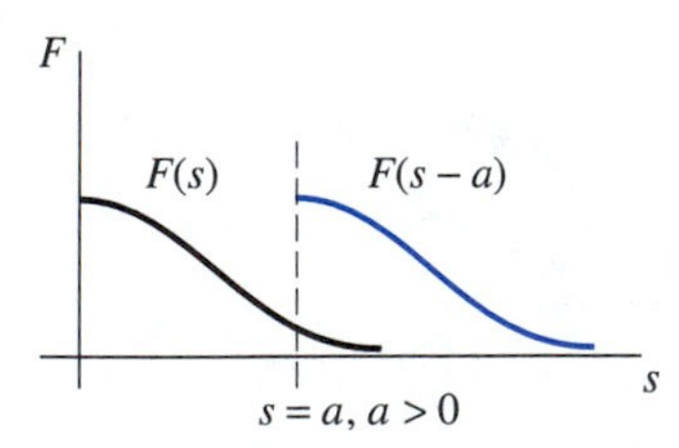

Figure 5.11. Shift on s-axis.

If we consider s a real variable, then the graph of $F(s - a)$ is the graph of $F(s)$ shifted on the s-axis by the amount $|a|$ units. For $a > 0$, the graph of $F(s)$ is shifted a units to the right, whereas for $a < 0$, the graph is shifted $|a|$ units to the left. See Figure 5.11.

For emphasis, it is sometimes useful to use the symbolism

$$\mathscr{L}\{e^{at}f(t)\} = \mathscr{L}\{f(t)\}_{s \to s-a}$$

where $s \to s - a$ means that we replace s in $F(s)$ by $s - a$.

EXAMPLE 1 First Translation Theorem

Evaluate **(a)** $\mathscr{L}\{e^{5t}t^3\}$ and **(b)** $\mathscr{L}\{e^{-2t}\cos 4t\}$.

Solution The results follow from Theorems 5.1 and 5.5.

(a) $\mathscr{L}\{e^{5t}t^3\} = \mathscr{L}\{t^3\}_{s \to s-5} = \left.\dfrac{3!}{s^4}\right|_{s \to s-5} = \dfrac{6}{(s-5)^4}$

(b) $\mathscr{L}\{e^{-2t}\cos 4t\} = \mathscr{L}\{\cos 4t\}_{s \to s+2}$ ← $a = -2$ so $s - a = s - (-2) = s + 2$

$$= \left.\frac{s}{s^2+16}\right|_{s \to s+2} = \frac{s+2}{(s+2)^2+16}$$

■

Inverse Form of the First Translation Theorem If $f(t) = \mathscr{L}^{-1}\{F(s)\}$, the inverse form of Theorem 5.5 is

$$\mathscr{L}^{-1}\{F(s-a)\} = \mathscr{L}^{-1}\{F(s)|_{s \to s-a}\} = e^{at}f(t) \tag{2}$$

■ EXAMPLE 2 Completing the Square to Find an Inverse

Evaluate $\mathscr{L}^{-1}\left\{\dfrac{s}{s^2+6s+11}\right\}$.

Solution If $s^2 + 6s + 11$ had real factors, we would use partial fractions. Since this quadratic term does not factor, we complete the square.

$$\mathscr{L}^{-1}\left\{\frac{s}{s^2+6s+11}\right\} = \mathscr{L}^{-1}\left\{\frac{s}{(s+3)^2+2}\right\} \quad \leftarrow \text{completion of square}$$

$$= \mathscr{L}^{-1}\left\{\frac{s+3-3}{(s+3)^2+2}\right\} \quad \leftarrow \text{adding zero in the numerator}$$

$$= \mathscr{L}^{-1}\left\{\frac{s+3}{(s+3)^2+2} - \frac{3}{(s+3)^2+2}\right\} \quad \leftarrow \text{termwise division}$$

$$= \mathscr{L}^{-1}\left\{\frac{s+3}{(s+3)^2+2}\right\} - 3\mathscr{L}^{-1}\left\{\frac{1}{(s+3)^2+2}\right\} \quad \leftarrow \text{linearity of } \mathscr{L}^{-1}$$

$$= \mathscr{L}^{-1}\left\{\left.\frac{s}{s^2+2}\right|_{s \to s+3}\right\} - \frac{3}{\sqrt{2}}\mathscr{L}^{-1}\left\{\left.\frac{\sqrt{2}}{s^2+2}\right|_{s \to s+3}\right\}$$

$$= e^{-3t}\cos\sqrt{2}t - \frac{3}{\sqrt{2}}e^{-3t}\sin\sqrt{2}t \quad \leftarrow \text{from (2) and Theorem 5.1}$$

■

■ EXAMPLE 3 An Initial-Value Problem

Solve $y'' - 6y' + 9y = t^2e^{3t}$, $y(0) = 2$, $y'(0) = 6$.

Solution $\mathscr{L}\{y''\} - 6\mathscr{L}\{y'\} + 9\mathscr{L}\{y\} = \mathscr{L}\{t^2e^{3t}\}$

$$\underbrace{s^2Y(s) - sy(0) - y'(0)}_{\mathscr{L}\{y''\}} - \underbrace{6[sY(s) - y(0)]}_{\mathscr{L}\{y'\}} + \underbrace{9Y(s)}_{\mathscr{L}\{y\}} = \underbrace{\frac{2}{(s-3)^3}}_{\mathscr{L}\{t^2e^{3t}\}}$$

Using the initial conditions and simplifying give

$$(s^2 - 6s + 9)Y(s) = 2s - 6 + \frac{2}{(s-3)^3}$$

$$(s-3)^2Y(s) = 2(s-3) + \frac{2}{(s-3)^3}$$

$$Y(s) = \frac{2}{s-3} + \frac{2}{(s-3)^5}$$

Thus $$y(t) = 2\mathscr{L}^{-1}\left\{\frac{1}{s-3}\right\} + \frac{2}{4!}\mathscr{L}^{-1}\left\{\frac{4!}{(s-3)^5}\right\}$$

From part (b) of Theorem 5.1 and the first translation theorem,

$$\mathscr{L}^{-1}\left\{\frac{4!}{s^5}\bigg|_{s \to s-3}\right\} = t^4e^{3t}$$

Hence we have $$y(t) = 2e^{3t} + \frac{1}{12}t^4e^{3t}$$ ■

■ EXAMPLE 4 **An Initial-Value Problem**

Solve $y'' + 4y' + 6y = 1 + e^{-t}$, $y(0) = 0$, $y'(0) = 0$.

Solution $\mathscr{L}\{y''\} + 4\mathscr{L}\{y'\} + 6\mathscr{L}\{y\} = \mathscr{L}\{1\} + \mathscr{L}\{e^{-t}\}$

$$s^2Y(s) - sy(0) - y'(0) + 4[sY(s) - y(0)] + 6Y(s) = \frac{1}{s} + \frac{1}{s+1}$$

$$(s^2 + 4s + 6)Y(s) = \frac{2s+1}{s(s+1)}$$

$$Y(s) = \frac{2s+1}{s(s+1)(s^2+4s+6)}$$

The partial fraction decomposition for $Y(s)$ is

$$Y(s) = \frac{1/6}{s} + \frac{1/3}{s+1} + \frac{-s/2 - 5/3}{s^2 + 4s + 6}$$

In preparation for taking the inverse transform, we fix up the numerator and denominator of the last term in $Y(s)$ in the following manner:

$$Y(s) = \frac{1/6}{s} + \frac{1/3}{s+1} + \frac{(-1/2)(s+2) - 2/3}{(s+2)^2 + 2}$$

$$= \frac{1/6}{s} + \frac{1/3}{s+1} - \frac{1}{2}\frac{s+2}{(s+2)^2+2} - \frac{2}{3}\frac{1}{(s+2)^2+2}$$

Finally, from parts (a), (c), (d) and (e) of Theorem 5.1 and the first translation theorem, we obtain

$$y(t) = \frac{1}{6}\mathscr{L}^{-1}\left\{\frac{1}{s}\right\} + \frac{1}{3}\mathscr{L}^{-1}\left\{\frac{1}{s+1}\right\} - \frac{1}{2}\mathscr{L}^{-1}\left\{\frac{s+2}{(s+2)^2+2}\right\} - \frac{2}{3\sqrt{2}}\mathscr{L}^{-1}\left\{\frac{\sqrt{2}}{(s+2)^2+2}\right\}$$

$$= \frac{1}{6} + \frac{1}{3}e^{-t} - \frac{1}{2}e^{-2t}\cos\sqrt{2}t - \frac{\sqrt{2}}{3}e^{-2t}\sin\sqrt{2}t$$ ■

Unit Step Function In engineering, one frequently encounters functions that can be either "on" or "off." For example, an external force acting on a mechanical system or a voltage impressed on a circuit can be turned off after a period of time. It is thus convenient to define a special function called the **unit step function**.

> **Definition 5.3 Unit Step Function**
>
> The function $\mathscr{U}(t-a)$ is defined to be
>
> $$\mathscr{U}(t-a) = \begin{cases} 0, & 0 \le t < a \\ 1, & t \ge a \end{cases}$$

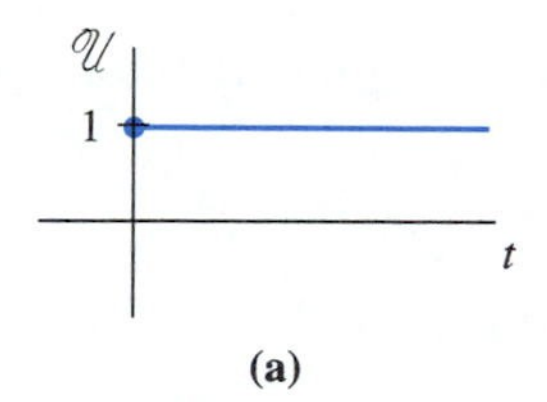

(a)

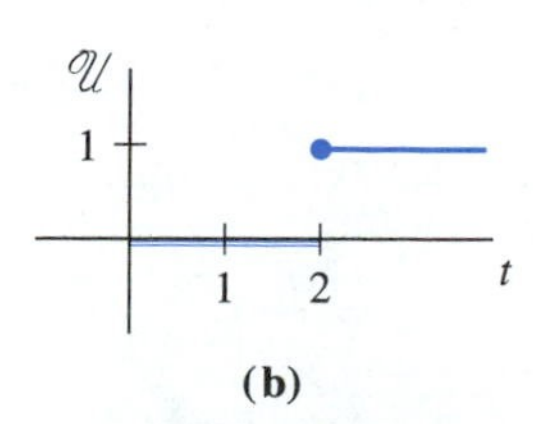

(b)

Figure 5.12.

Notice that we define $\mathscr{U}(t-a)$ only on the nonnegative t-axis since this is all that we are concerned with in the study of the Laplace transform. In a broader sense, $\mathscr{U}(t-a) = 0$ for $t < a$.

■ **EXAMPLE 5 Graphs of Unit Step Functions**

Graph **(a)** $\mathscr{U}(t)$ and **(b)** $\mathscr{U}(t-2)$.

Solution **(a)** $\mathscr{U}(t) = 1, \quad t \ge 0$ **(b)** $\mathscr{U}(t-2) = \begin{cases} 0, & 0 \le t < 2 \\ 1, & t \ge 2 \end{cases}$

The respective graphs are given in Figure 5.12. ■

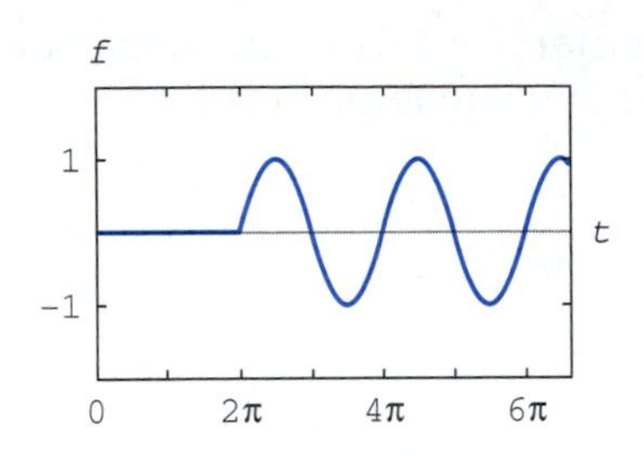

Figure 5.13.

When multiplied by another function defined for $t \geq 0$, the unit step function "turns off" a portion of the graph of the function. For example, Figure 5.13 illustrates the graph of $\sin t$, $t \geq 0$, when multiplied by $\mathcal{U}(t - 2\pi)$:

$$f(t) = \sin t\,\mathcal{U}(t - 2\pi) = \begin{cases} 0, & 0 \leq t < 2\pi \\ \sin t, & t \geq 2\pi \end{cases}$$

The unit step function can also be used to write piecewise-defined functions in a compact form. For instance, the piecewise-defined function

$$f(t) = \begin{cases} g(t), & 0 \leq t < a \\ h(t), & t \geq a \end{cases} \tag{3}$$

is the same as $\quad f(t) = g(t) - g(t)\mathcal{U}(t - a) + h(t)\mathcal{U}(t - a) \quad$ **(4)**

Similarly, a function of the type

$$f(t) = \begin{cases} 0, & 0 \leq t < a \\ g(t), & a \leq t < b \\ 0, & t \geq b \end{cases} \tag{5}$$

can be written $\quad f(t) = g(t)[\mathcal{U}(t - a) - \mathcal{U}(t - b)] \quad$ **(6)**

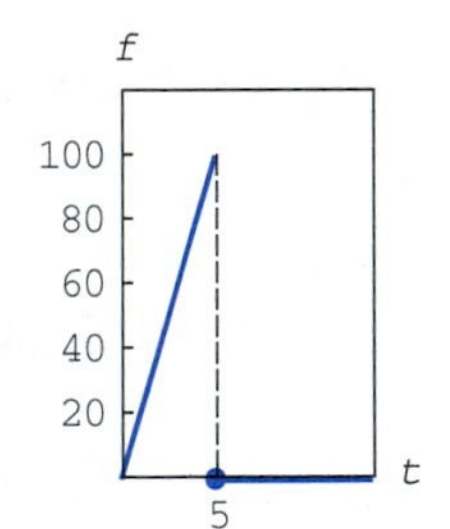

Figure 5.14.

■ EXAMPLE 6 Function Expressed in Terms of a Unit Step Function

Express $f(t) = \begin{cases} 20t, & 0 \leq t < 5 \\ 0, & t \geq 5 \end{cases}$ in terms of unit step functions. Graph.

Solution The graph of this piecewise-defined function is given in Figure 5.14. Now from (3) and (4) with $g(t) = 20t$ and $h(t) = 0$, we get

$$f(t) = 20t - 20t\mathcal{U}(t - 5)$$

■

■ EXAMPLE 7 Comparison of Functions

Consider the function $y = f(t)$ defined by $f(t) = t^3$. Compare the graphs of

(a) $f(t)$, $-\infty < t < \infty$, **(b)** $f(t)$, $t \geq 0$,
(c) $f(t - 2)$, $t \geq 0$, **(d)** $f(t - 2)\ \mathcal{U}(t - 2)$, $t \geq 0$

Solution The respective graphs are given in Figure 5.15.

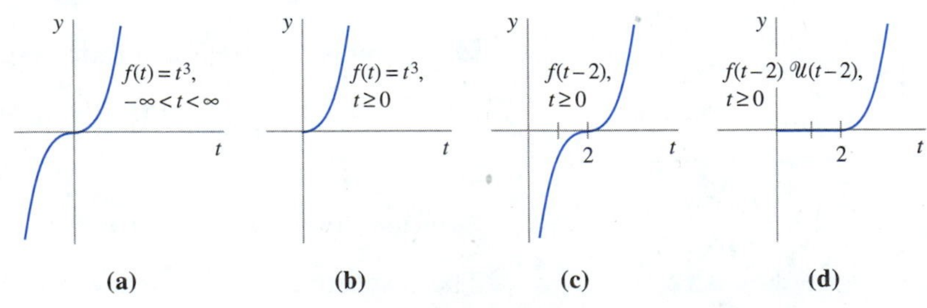

Figure 5.15.

■

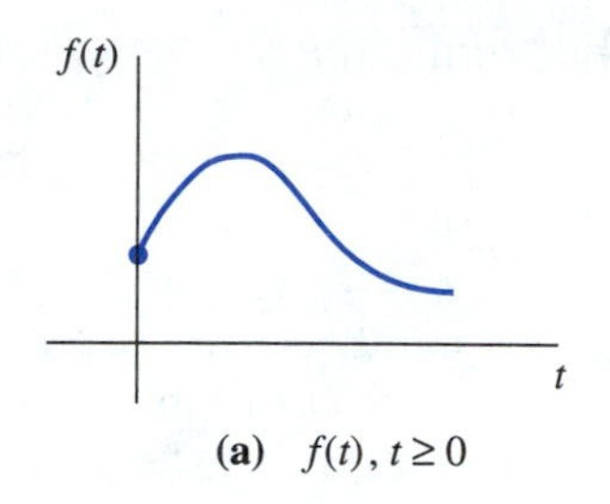

(a) $f(t), t \geq 0$

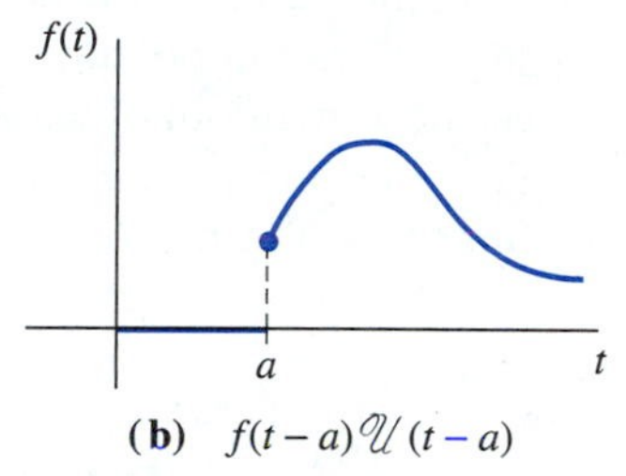

(b) $f(t-a)\mathcal{U}(t-a)$

Figure 5.16. Shift on t-axis.

In general, if $a > 0$, then the graph of $y = f(t - a)$ is the graph of $y = f(t)$, $t > 0$, shifted a units to the right on the t-axis. However, when $y = f(t - a)$ is multiplied by the unit step function $\mathcal{U}(t - a)$ in the manner illustrated in part (d) of Example 7, then the graph of the function $y = f(t - a)\mathcal{U}(t - a)$ coincides with the graph of $y = f(t - a)$ for $t \geq a$, but is identically zero for $0 \leq t < a$. See Figure 5.16.

Second Translation Theorem We saw in Theorem 5.5 that an exponential multiple of $f(t)$ results in a translation or shift of the transform $F(s)$ on the s-axis. In Theorem 5.6 that follows, we see that whenever $F(s)$ is multiplied by an appropriate exponential function, the inverse transform of this product is the shifted function $f(t - a)\mathcal{U}(t - a)$. The result is called the **second translation theorem**.

> **Theorem 5.6 Second Translation Theorem**
>
> If $F(s) = \mathcal{L}\{f(t)\}$ and $a > 0$, then
>
> $$\mathcal{L}\{f(t - a)\mathcal{U}(t - a)\} = e^{-as}F(s)$$

Proof We express $\int_0^\infty e^{-st}f(t - a)\mathcal{U}(t - a)\, dt$ as the sum of two integrals:

$$\mathcal{L}\{f(t-a)\mathcal{U}(t-a)\} = \int_0^a e^{-st}f(t-a)\underbrace{\mathcal{U}(t-a)}_{\substack{\text{zero for}\\ 0 \leq t < a}}\, dt + \int_a^\infty e^{-st}f(t-a)\underbrace{\mathcal{U}(t-a)}_{\substack{\text{one for}\\ t \geq a}}\, dt = \int_a^\infty e^{-st}f(t-a)\, dt$$

Now let $v = t - a$, $dv = dt$; then

$$\mathcal{L}\{f(t-a)\mathcal{U}(t-a)\} = \int_0^\infty e^{-s(v+a)}f(v)\, dv = e^{-as}\int_0^\infty e^{-sv}f(v)\, dv = e^{-as}\mathcal{L}\{f(t)\} \quad \square$$

■ **EXAMPLE 8 Second Translation Theorem**

Evaluate $\mathcal{L}\{(t - 2)^3\mathcal{U}(t - 2)\}$.

Solution With the identification $a = 2$, it follows from Theorem 5.6 that

$$\mathcal{L}\{(t-2)^3\mathcal{U}(t-2)\} = e^{-2s}\mathcal{L}\{t^3\} = e^{-2s}\frac{3!}{s^4} = \frac{6}{s^4}e^{-2s}$$ ■

We often wish to find the Laplace transform of just the unit step function. This can be found from either Definition 5.1 or Theorem 5.6. If we identify $f(t) = 1$ in Theorem 5.6, then $f(t - a) = 1$, $F(s) = \mathcal{L}\{1\} = 1/s$, and so

$$\mathcal{L}\{\mathcal{U}(t-a)\} = \frac{e^{-as}}{s} \tag{7}$$

Inverse Form of the Second Translation Theorem If $f(t) = \mathscr{L}^{-1}\{F(s)\}$, the inverse form of Theorem 5.6, $a > 0$, is

$$\mathscr{L}^{-1}\{e^{-as}F(s)\} = f(t-a)\mathscr{U}(t-a) \tag{8}$$

Alternative Form of Theorem 5.6 We are frequently confronted with the problem of finding the Laplace transform of a product of a function g and a unit step function $\mathscr{U}(t-a)$ where the function g lacks the precise shifted form $f(t-a)$ required in Theorem 5.6. To find the Laplace transform of $g(t)\,\mathscr{U}(t-a)$, it is possible to fix up $g(t)$ by algebraic manipulations to force it into the desired form $f(t-a)$. But since these manipulations are time consuming and often not obvious, it is simpler to devise an alternative version of Theorem 5.6. Using Definition 5.1, the definition of $\mathscr{U}(t-a)$, and the substitution $u = t - a$, we obtain

$$\mathscr{L}\{g(t)\,\mathscr{U}(t-a)\} = \int_a^\infty e^{-st}g(t)\,dt = \int_0^\infty e^{-s(u+a)}g(u+a)\,du$$

That is,

$$\mathscr{L}\{g(t)\,\mathscr{U}(t-a)\} = e^{-as}\mathscr{L}\{g(t+a)\} \tag{9}$$

■ **EXAMPLE 9** **Second Translation Theorem—Alternative Form**

Evaluate $\mathscr{L}\{\sin t\,\mathscr{U}(t-2\pi)\}$.

Solution With $g(t) = \sin t$, $a = 2\pi$, note that $g(t+2\pi) = \sin(t+2\pi) = \sin t$ because the sine function has period 2π. Now by (9),

$$\mathscr{L}\{\sin t\,\mathscr{U}(t-2\pi)\} = e^{-2\pi s}\mathscr{L}\{\sin t\} = \frac{e^{-2\pi s}}{s^2+1}$$ ■

■ **EXAMPLE 10** **Second Translation Theorem—Alternative Form**

Find the Laplace transform of the function shown in Figure 5.17.

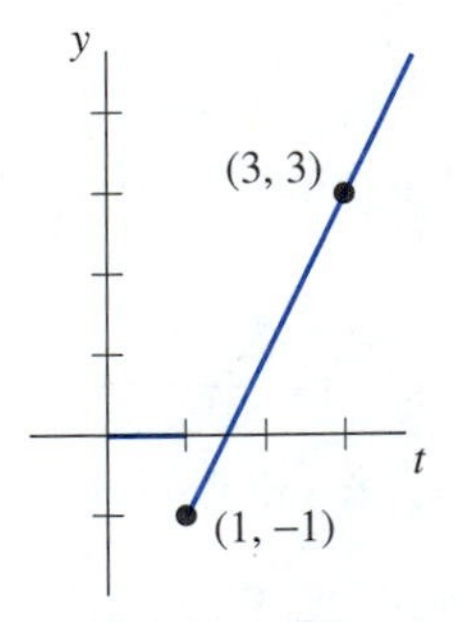

Figure 5.17.

Solution An equation of the straight line through the two points is found to be $y = 2t - 3$. To "turn off" the graph $y = 2t - 3$ on the interval $0 \le t < 1$, we employ the product $(2t-3)\mathscr{U}(t-1)$. In this case with $g(t) = 2t - 3$, $a = 1$, and $g(t+1) = 2(t+1) - 3 = 2t - 1$. It follows from (9) that

$$\mathscr{L}\{(2t-3)\mathscr{U}(t-1)\} = e^{-s}\mathscr{L}\{2t-1\} = e^{-s}\left(\frac{2}{s^2} - \frac{1}{s}\right)$$ ■

In the linear mathematical models for physical systems, such as a spring/mass system or a series electrical circuit, the right-hand member of the differential equations

$$m\frac{d^2x}{dt^2} + \beta\frac{dx}{dt} + kx = f(t) \quad \text{or} \quad L\frac{d^2q}{dt^2} + R\frac{dq}{dt} + \frac{1}{C}q = E(t)$$

is a driving function: $f(t)$ is an external force and $E(t)$ is an impressed voltage. In Section 3.6, we solved problems in which the functions f and E were continuous. However, discontinuous driving functions are not uncommon. As illustrated in the next example, the Laplace transform is an invaluable aid in solving problems in which f and E are piecewise continuous.

EXAMPLE 11 An Initial-Value Problem

Solve $x'' + 16x = f(t), x(0) = 0, x'(0) = 0$, where $f(t) = \begin{cases} 20t, & 0 \le t < 5 \\ 0, & t \ge 5. \end{cases}$

Solution The function $f(t)$ can be interpreted as an external force acting on a spring/mass system for a short period of time, which then is removed. In Example 6, we wrote f in terms of the unit step function as $f(t) = 20t - 20t\mathcal{U}(t-5)$, and so with $a = 5$, the alternative form of the second translation theorem given in (9) yields

$$\mathcal{L}\{f(t)\} = 20\mathcal{L}\{t\} - 20\mathcal{L}\{t\mathcal{U}(t-5)\} = 20\mathcal{L}\{t\} - 20e^{-5s}\mathcal{L}\{t+5\}$$

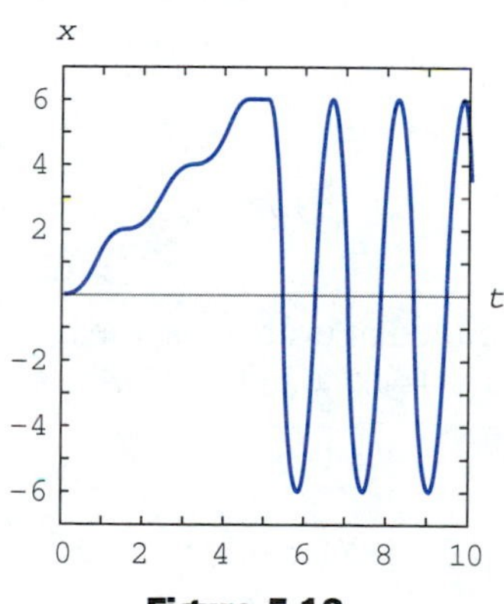

Figure 5.18.

Hence the Laplace transform of the differential equation is

$$\mathcal{L}\{x''\} + \mathcal{L}\{x\} = \mathcal{L}\{f(t)\}$$

$$s^2X(s) - sx(0) - x'(0) + 16X(s) = \frac{20}{s^2} - \left[\frac{20}{s^2} + \frac{100}{s}\right]e^{-5s}$$

$$(s^2+16)X(s) = \frac{20}{s^2} - \left[\frac{20}{s^2} + \frac{100}{s}\right]e^{-5s}$$

Solving for $X(s)$ and using partial fractions give us

$$X(s) = \frac{5}{4}\frac{1}{s^2} - \frac{5}{4}\frac{1}{s^2+16} - \frac{5}{4}\frac{1}{s^2}e^{-5s} + \frac{5}{4}\frac{1}{s^2+16}e^{-5s} - \frac{25}{4}\frac{1}{s}e^{-5s} + \frac{25}{4}\frac{s}{s^2+16}e^{-5s}$$

Therefore, with the help of (7) and (8),

$$x(t) = \frac{5}{4}\mathcal{L}^{-1}\left\{\frac{1}{s^2}\right\} - \frac{5}{16}\mathcal{L}^{-1}\left\{\frac{4}{s^2+16}\right\} - \frac{5}{4}\mathcal{L}^{-1}\left\{\frac{e^{-5s}}{s^2}\right\} + \frac{5}{16}\mathcal{L}^{-1}\left\{\frac{4}{s^2+16}e^{-5s}\right\} - \frac{25}{4}\mathcal{L}^{-1}\left\{\frac{e^{-5s}}{s}\right\} + \frac{25}{4}\mathcal{L}^{-1}\left\{\frac{s}{s^2+16}e^{-5s}\right\}$$

$$= \frac{5}{4}t - \frac{5}{16}\sin 4t - \frac{5}{4}(t-5)\mathcal{U}(t-5) + \frac{5}{16}\sin 4(t-5)\mathcal{U}(t-5) - \frac{25}{4}\mathcal{U}(t-5) + \frac{25}{4}\cos 4(t-5)\mathcal{U}(t-5)$$

or $$x(t) = \begin{cases} \frac{5}{4}t - \frac{5}{16}\sin 4t, & 0 \le t < 5 \\ -\frac{5}{16}\sin 4t + \frac{5}{16}\sin 4(t-5) + \frac{25}{4}\cos 4(t-5), & t \ge 5 \end{cases}$$

Using a graphing utility, we get the graph of $x(t)$ shown in Figure 5.18. Observe that the vibrations become steady and periodic as soon as the external force is turned off. ■

Beams In Section 3.7, we saw that the static deflection $y(x)$ of a uniform beam of length L carrying load $w(x)$ per unit length is found from the fourth-order differential equation

$$EI\frac{d^4y}{dx^4} = w(x) \tag{10}$$

where E is Young's modulus of elasticity and I is a moment of inertia of a cross section of the beam. The Laplace transform is particularly useful when $w(x)$ is piecewise defined, but in order to use the transform, we must tacitly assume that $y(x)$ and $w(x)$ are defined on $(0, \infty)$ rather than on $(0, L)$. Unlike the previous examples, where we solved initial-value problems, the next example illustrates how to apply the Laplace transform to a boundary-value problem.

EXAMPLE 12 A Boundary-Value Problem

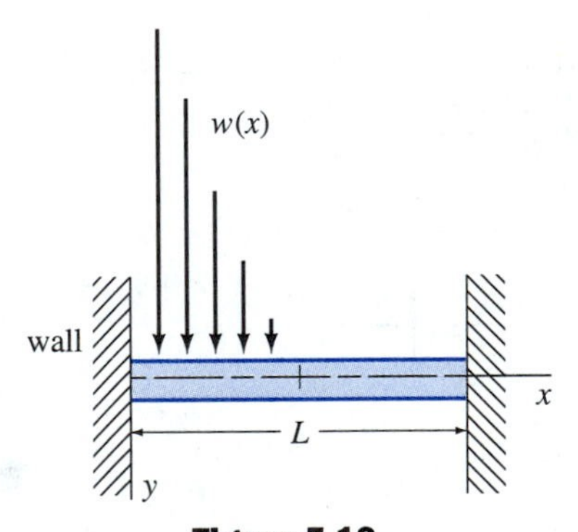

Figure 5.19.

A beam of length L is embedded at both ends as shown in Figure 5.19. Find the deflection of the beam when the load is given by

$$w(x) = \begin{cases} w_0\left(1 - \dfrac{2}{L}x\right), & 0 < x < L/2 \\ 0, & L/2 < x < L \end{cases}$$

Solution Recall that since the beam is embedded at both ends, the boundary conditions are $y(0) = 0$, $y'(0) = 0$, $y(L) = 0$, $y'(L) = 0$. Also, you should verify

$$w(x) = w_0\left(1 - \frac{2}{L}x\right) - w_0\left(1 - \frac{2}{L}x\right)\mathcal{U}\left(x - \frac{L}{2}\right)$$

$$= \frac{2w_0}{L}\left[\frac{L}{2} - x + \left(x - \frac{L}{2}\right)\mathcal{U}\left(x - \frac{L}{2}\right)\right]$$

Transforming (10) with respect to the variable x gives

$$EI(s^4Y(s) - s^3y(0) - s^2y'(0) - sy''(0) - y'''(0)) = \frac{2w_0}{EIL}\left[\frac{L/2}{s} - \frac{1}{s^2} + \frac{1}{s^2}e^{-Ls/2}\right]$$

or

$$s^4Y(s) - sy''(0) - y'''(0) = \frac{2w_0}{EIL}\left[\frac{L/2}{s} - \frac{1}{s^2} + \frac{1}{s^2}e^{-Ls/2}\right]$$

If we let $c_1 = y''(0)$ and $c_2 = y'''(0)$, then

$$Y(s) = \frac{c_1}{s^3} + \frac{c_2}{s^4} + \frac{2w_0}{EIL}\left[\frac{L/2}{s^5} - \frac{1}{s^6} + \frac{1}{s^6}e^{-Ls/2}\right]$$

and consequently

$$y(x) = \frac{c_1}{2!}\mathscr{L}^{-1}\left\{\frac{2!}{s^3}\right\} + \frac{c_2}{3!}\mathscr{L}^{-1}\left\{\frac{3!}{s^4}\right\} + \frac{2w_0}{EIL}\left[\frac{L/2}{4!}\mathscr{L}^{-1}\left\{\frac{4!}{s^5}\right\} - \frac{1}{5!}\mathscr{L}^{-1}\left\{\frac{5!}{s^6}\right\} + \frac{1}{5!}\mathscr{L}^{-1}\left\{\frac{5!}{s^6}e^{-Ls/2}\right\}\right]$$

$$= \frac{c_1}{2}x^2 + \frac{c_2}{6}x^3 + \frac{w_0}{60EIL}\left[\frac{5L}{2}x^4 - x^5 + \left(x - \frac{L}{2}\right)^5 \mathscr{U}\left(x - \frac{L}{2}\right)\right]$$

Applying the conditions $y(L) = 0$ and $y'(L) = 0$ to the last result yields a system of equations for c_1 and c_2:

$$c_1\frac{L^2}{2} + c_2\frac{L^3}{6} + \frac{49w_0L^4}{1920EI} = 0$$

$$c_1L + c_2\frac{L^2}{2} + \frac{85w_0L^3}{960EI} = 0$$

Solving, we find $c_1 = 23w_0L^2/960EI$ and $c_2 = -9w_0L/40EI$. Thus the deflection is given by

$$y(x) = \frac{23w_0L^2}{1920EI}x^2 - \frac{9w_0L}{240EI}x^3 + \frac{w_0}{60EIL}\left[\frac{5L}{2}x^4 - x^5 + \left(x - \frac{L}{2}\right)^5 \mathscr{U}\left(x - \frac{L}{2}\right)\right]$$ ■

EXERCISES 5.2

Answers to odd-numbered problems begin on page 452.

In Problems 1–36, find either $F(s)$ or $f(t)$ as indicated.

1. $\mathscr{L}\{te^{10t}\}$
2. $\mathscr{L}\{te^{-6t}\}$
3. $\mathscr{L}\{t^3e^{-2t}\}$
4. $\mathscr{L}\{t^{10}e^{-7t}\}$
5. $\mathscr{L}\{e^t \sin 3t\}$
6. $\mathscr{L}\{e^{-2t} \cos 4t\}$
7. $\mathscr{L}\{e^{5t} \sinh 3t\}$
8. $\mathscr{L}\left\{\frac{\cosh t}{e^t}\right\}$
9. $\mathscr{L}\{t(e^t + e^{2t})^2\}$
10. $\mathscr{L}\{e^{2t}(t - 1)^2\}$
11. $\mathscr{L}\{e^{-t} \sin^2 t\}$
12. $\mathscr{L}\{e^t \cos^2 3t\}$
13. $\mathscr{L}^{-1}\left\{\frac{1}{(s + 2)^3}\right\}$
14. $\mathscr{L}^{-1}\left\{\frac{1}{(s - 1)^4}\right\}$
15. $\mathscr{L}^{-1}\left\{\frac{1}{s^2 - 6s + 10}\right\}$
16. $\mathscr{L}^{-1}\left\{\frac{1}{s^2 + 2s + 5}\right\}$
17. $\mathscr{L}^{-1}\left\{\frac{s}{s^2 + 4s + 5}\right\}$
18. $\mathscr{L}^{-1}\left\{\frac{2s + 5}{s^2 + 6s + 34}\right\}$
19. $\mathscr{L}^{-1}\left\{\frac{s}{(s + 1)^2}\right\}$
20. $\mathscr{L}^{-1}\left\{\frac{5s}{(s - 2)^2}\right\}$
21. $\mathscr{L}^{-1}\left\{\frac{2s - 1}{s^2(s + 1)^3}\right\}$
22. $\mathscr{L}^{-1}\left\{\frac{(s + 1)^2}{(s + 2)^4}\right\}$
23. $\mathscr{L}\{(t - 1)\mathscr{U}(t - 1)\}$
24. $\mathscr{L}\{e^{2-t}\mathscr{U}(t - 2)\}$
25. $\mathscr{L}\{t\mathscr{U}(t - 2)\}$
26. $\mathscr{L}\{(3t + 1)\mathscr{U}(t - 3)\}$
27. $\mathscr{L}\{\cos 2t\,\mathscr{U}(t - \pi)\}$
28. $\mathscr{L}\left\{\sin t\,\mathscr{U}\left(t - \frac{\pi}{2}\right)\right\}$
29. $\mathscr{L}\{(t - 1)^3e^{t-1}\mathscr{U}(t - 1)\}$
30. $\mathscr{L}\{te^{t-5}\mathscr{U}(t - 5)\}$
31. $\mathscr{L}^{-1}\left\{\frac{e^{-2s}}{s^3}\right\}$
32. $\mathscr{L}^{-1}\left\{\frac{(1 + e^{-2s})^2}{s + 2}\right\}$
33. $\mathscr{L}^{-1}\left\{\frac{e^{-\pi s}}{s^2 + 1}\right\}$
34. $\mathscr{L}^{-1}\left\{\frac{se^{-\pi s/2}}{s^2 + 4}\right\}$
35. $\mathscr{L}^{-1}\left\{\frac{e^{-s}}{s(s + 1)}\right\}$
36. $\mathscr{L}^{-1}\left\{\frac{e^{-2s}}{s^2(s - 1)}\right\}$

In Problems 37–42, match the given graph with one of the functions in (a)–(f). The graph of $f(t)$ is given in Figure 5.20.

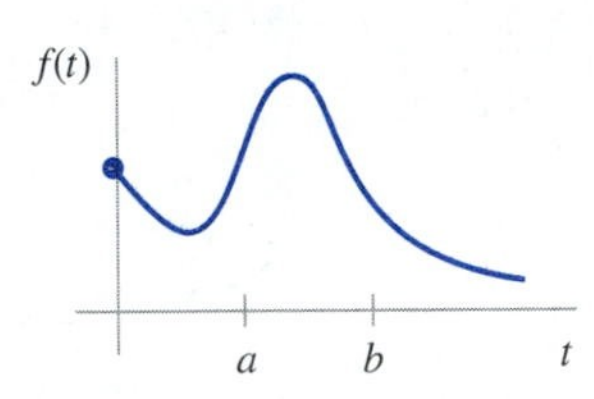

Figure 5.20.

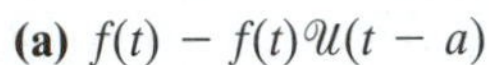

(a) $f(t) - f(t)\mathcal{U}(t - a)$
(b) $f(t - b)\mathcal{U}(t - b)$
(c) $f(t)\mathcal{U}(t - a)$
(d) $f(t) - f(t)\mathcal{U}(t - b)$
(e) $f(t)\mathcal{U}(t - a) - f(t)\mathcal{U}(t - b)$
(f) $f(t - a)\mathcal{U}(t - a) - f(t - a)\mathcal{U}(t - b)$

37.

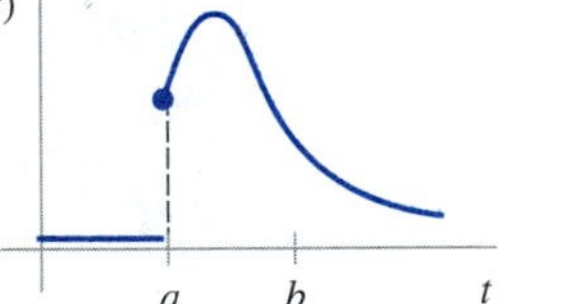

Figure 5.21.

38.

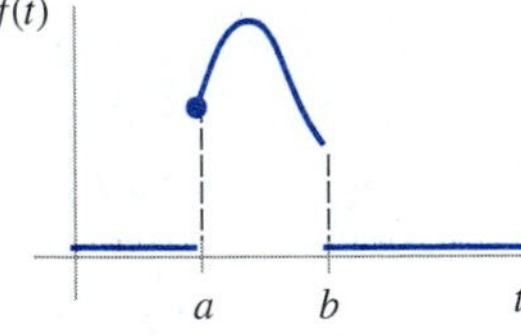

Figure 5.22.

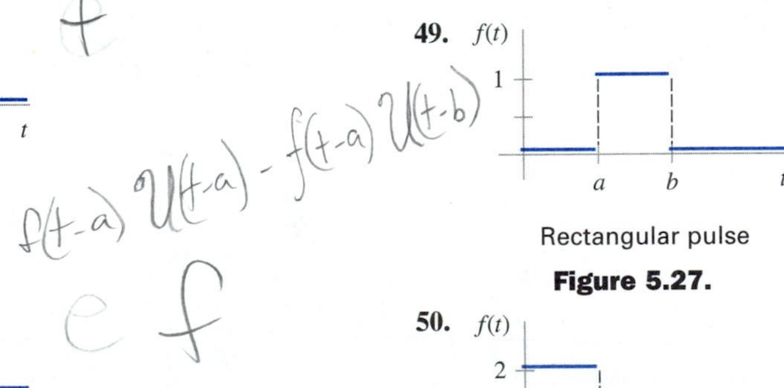

39.

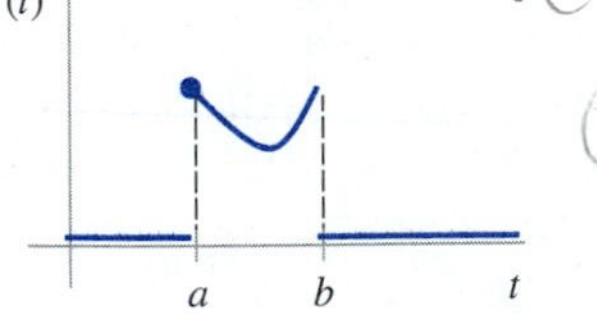

Figure 5.23.

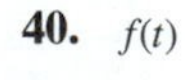

40.

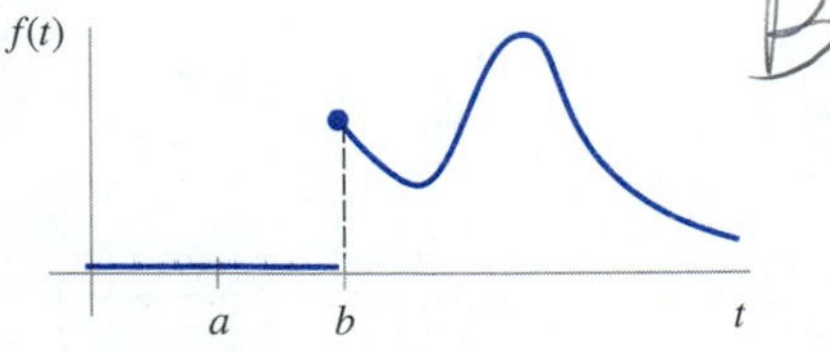

Figure 5.24.

41.

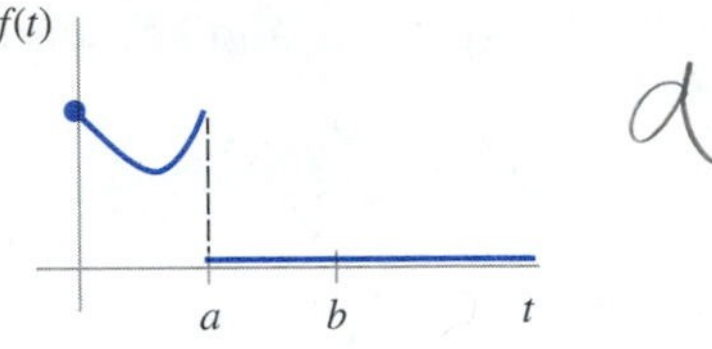

Figure 5.25.

42.

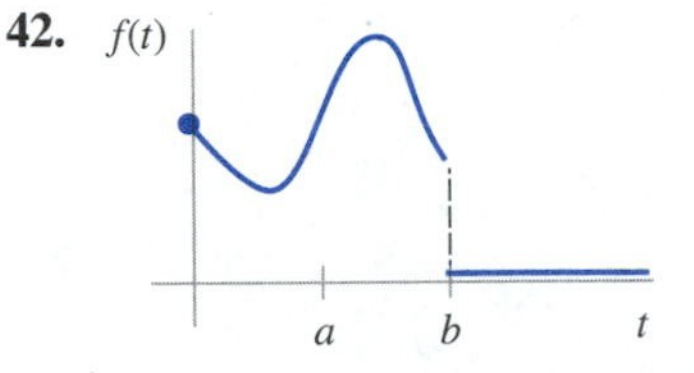

Figure 5.26.

In Problems 43–50, write each function in terms of unit step functions. Find the Laplace transform of the given function.

43. $f(t) = \begin{cases} 2, & 0 \le t < 3 \\ -2, & t \ge 3 \end{cases}$

44. $f(t) = \begin{cases} 1, & 0 \le t < 4 \\ 0, & 4 \le t < 5 \\ 1, & t \ge 5 \end{cases}$

45. $f(t) = \begin{cases} 0, & 0 \le t < 1 \\ t^2, & t \ge 1 \end{cases}$

46. $f(t) = \begin{cases} 0, & 0 \le t < \dfrac{3\pi}{2} \\ \sin t, & t \ge \dfrac{3\pi}{2} \end{cases}$

47. $f(t) = \begin{cases} t, & 0 \le t < 2 \\ 0, & t \ge 2 \end{cases}$

48. $f(t) = \begin{cases} \sin t, & 0 \le t < 2\pi \\ 0, & t \ge 2\pi \end{cases}$

49. $f(t)$

1, a, b, t

Rectangular pulse

Figure 5.27.

50.

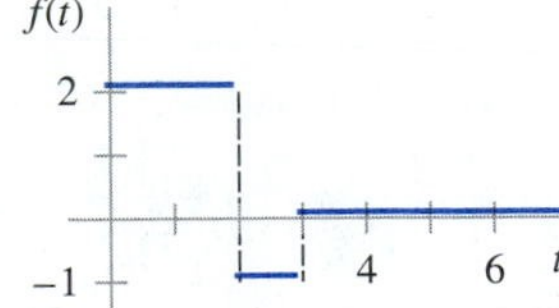

Figure 5.28.

In Problems 51–64, use the Laplace transform to solve the given differential equation subject to the indicated initial conditions. Where appropriate, write f in terms of unit step functions.

51. $y'' - 6y' + 13y = 0,\ y(0) = 0,\ y'(0) = -3$

52. $y'' - 6y' + 9y = t,\ y(0) = 0,\ y'(0) = 1$

53. $y'' - 4y' + 4y = t^3e^{2t}$, $y(0) = 0$, $y'(0) = 0$

54. $y'' - 2y' + 5y = 1 + t$, $y(0) = 0$, $y'(0) = 4$

55. $y'' - y' = e^t \cos t$, $y(0) = 0$, $y'(0) = 0$

56. $y'' - 2y' = e^t \sinh t$, $y(0) = 0$, $y'(0) = 0$

57. $y' + y = f(t)$, $y(0) = 0$ where $f(t) = \begin{cases} 0, & 0 \le t < 1 \\ 5, & t \ge 1 \end{cases}$

58. $y' + y = f(t)$, $y(0) = 0$ where $f(t) = \begin{cases} 1, & 0 \le t < 1 \\ -1, & t \ge 1 \end{cases}$

59. $y' + 2y = f(t)$, $y(0) = 0$ where $f(t) = \begin{cases} t, & 0 \le t < 1 \\ 0, & t \ge 1 \end{cases}$

60. $y'' + 4y = f(t)$, $y(0) = 0$, $y'(0) = -1$

where $f(t) = \begin{cases} 1, & 0 \le t < 1 \\ 0, & t \ge 1 \end{cases}$

61. $y'' + 4y = \sin t\,\mathcal{U}(t - 2\pi)$, $y(0) = 1$, $y'(0) = 0$

62. $y'' - 5y' + 6y = t\,\mathcal{U}(t - 1)$, $y(0) = 0$, $y'(0) = 1$

63. $y'' + y = f(t)$, $y(0) = 0$, $y'(0) = 1$

where $f(t) = \begin{cases} 0, & 0 \le t < \pi \\ 1, & \pi \le t < 2\pi \\ 0, & t \ge 2\pi \end{cases}$

64. $y'' + 4y' + 3y = 1 - \mathcal{U}(t - 2) - \mathcal{U}(t - 4) + \mathcal{U}(t - 6)$, $y(0) = 0$, $y'(0) = 0$

In Problems 65 and 66, use the Laplace Transform to solve the given differential equation subject to the indicated boundary conditions.

65. $y'' + 2y' + y = 0$, $y'(0) = 2$, $y(1) = 2$

66. $y'' - 9y' + 20y = 1$, $y(0) = 0$, $y'(1) = 0$

67. Recall that the differential equation for the charge $q(t)$ on the capacitor in an R-C series circuit is

$$R\frac{dq}{dt} + \frac{1}{C}q = E(t)$$

where $E(t)$ is the impressed voltage. See Section 2.4. Use the Laplace transform to determine the charge $q(t)$ when $q(0) = 0$ and $E(t) = E_0e^{-kt}$, $k > 0$. Consider two cases: $k \ne 1/RC$ and $k = 1/RC$.

68. Use the Laplace transform to determine the charge on the capacitor in an R-C series circuit if $q(0) = q_0$, $R = 10$ ohms, $C = 0.1$ farad, and $E(t)$ is as given in Figure 5.29.

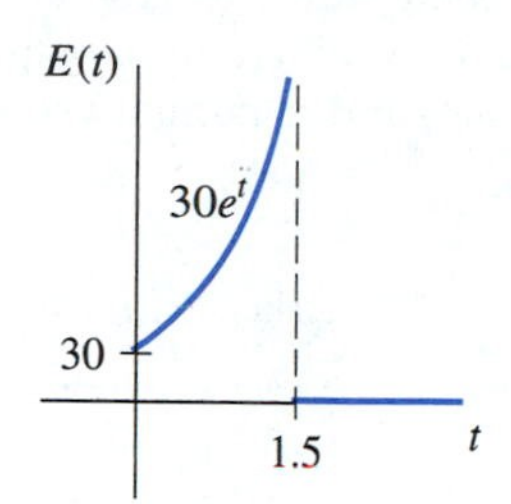

Figure 5.29.

69. Use the Laplace transform to determine the charge on the capacitor in an R-C series circuit if $q(0) = 0$, $R = 2.5$ ohms, $C = 0.08$ farad, and $E(t)$ is as given in Figure 5.30.

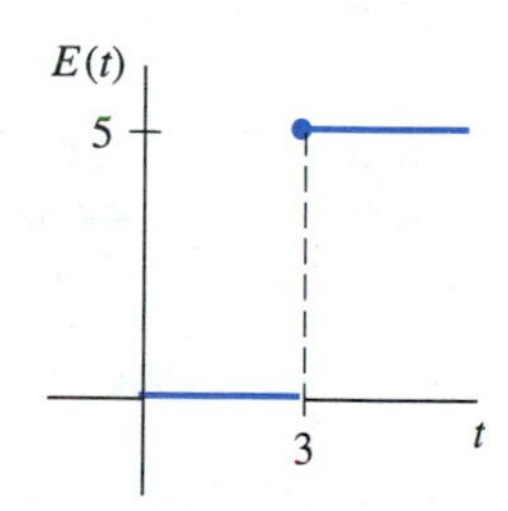

Figure 5.30.

70. **(a)** Use the Laplace transform to determine the charge $q(t)$ on the capacitor in an R-C series circuit when $q(0) = 0$, $R = 50$ ohms, $C = 0.01$ farad, and $E(t)$ is as given in Figure 5.31.

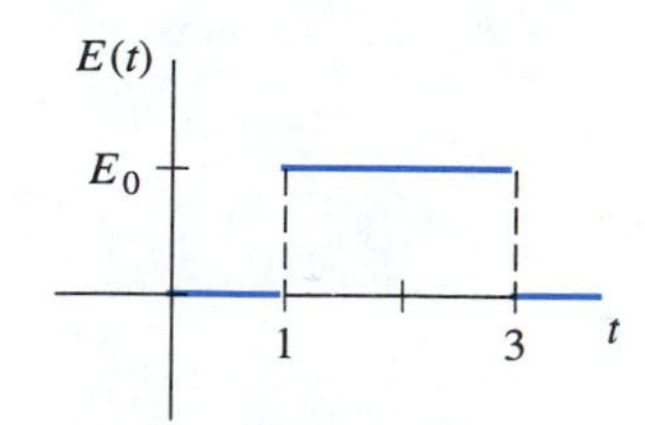

Figure 5.31.

(b) Assume $E_0 = 100$ volts. Use a graphing utility to graph $q(t)$ on the interval $0 \le t \le 6$. Use the graph to estimate q_{max}, the maximum value of the charge.

71. **(a)** Use the Laplace transform to determine the current $i(t)$ in a single loop L-R series circuit when $i(0) = 0$, $L = 1$ henry, $R = 10$ ohms, and $E(t)$ is as given in Figure 5.32.

(b) Use a graphing utility to graph $i(t)$ on the interval $0 \le t \le 6$. Use the graph to estimate i_{max} and i_{min}, the maximum and minimum values of the current.

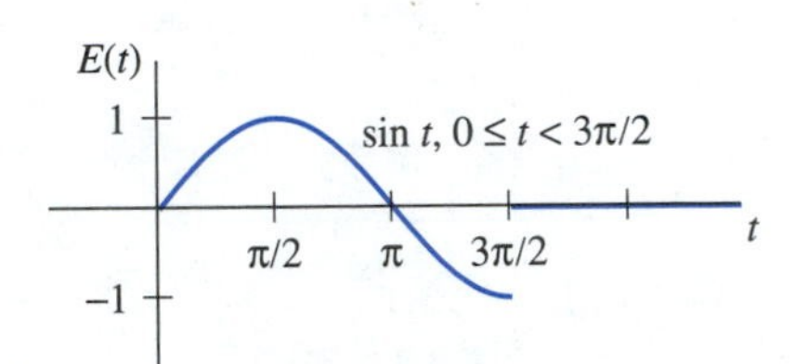

Figure 5.32.

72. Recall that the differential equation for the instantaneous charge $q(t)$ on the capacitor in an L-R-C series circuit is given by

$$L\frac{d^2q}{dt^2} + R\frac{dq}{dt} + \frac{1}{C}q = E(t) \qquad \textbf{(11)}$$

See Section 3.6. Use the Laplace transform to determine $q(t)$ when $L = 1$ henry, $R = 20$ ohms, $C = 0.005$ farad, $E(t) = 150$ volts, $t > 0$, $q(0) = 0$, and $i(0) = 0$. What is the current $i(t)$? What is the charge $q(t)$ if the same constant voltage is turned off for $t \ge 2$?

73. Determine the charge $q(t)$ and current $i(t)$ for a series circuit in which $L = 1$ henry, $R = 20$ ohms, $C = 0.01$ farad, $E(t) = 120 \sin 10t$ volts, $q(0) = 0$, and $i(0) = 0$. What is the steady-state current?

74. Consider the battery of constant voltage, E_0 that charges the capacitor shown in Figure 5.33. If we divide by L and define $\lambda = R/2L$ and $\omega^2 = 1/LC$, then (11) becomes

$$\frac{d^2q}{dt^2} + 2\lambda\frac{dq}{dt} + \omega^2 q = \frac{E_0}{L}$$

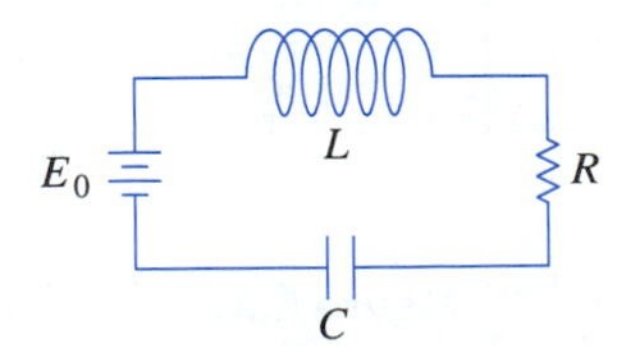

Figure 5.33.

Use the Laplace transform to show that the solution of this equation, subject to $q(0) = 0$ and $i(0) = 0$, is

$$q(t) = \begin{cases} E_0C[1 - e^{-\lambda t}(\cosh\sqrt{\lambda^2-\omega^2}\,t \\ \qquad + \dfrac{\lambda}{\sqrt{\lambda^2-\omega^2}}\sinh\sqrt{\lambda^2-\omega^2}\,t)], & \lambda > \omega \\ E_0C[1 - e^{-\lambda t}(1+\lambda t)], & \lambda = \omega \\ E_0C[1 - e^{-\lambda t}(\cos\sqrt{\omega^2-\lambda^2}\,t \\ \qquad + \dfrac{\lambda}{\sqrt{\omega^2-\lambda^2}}\sin\sqrt{\omega^2-\lambda^2}\,t)], & \lambda < \omega \end{cases}$$

75. A 4-lb weight stretches a spring 2 ft. The weight is released from rest 18 in. above the equilibrium position, and the resulting motion takes place in a medium that offers a damping force numerically equal to $\frac{7}{8}$ times the instantaneous velocity. Use the Laplace transform to determine the equation of motion.

76. Suppose a 32-lb weight stretches a spring 2 ft. If the weight is released from rest at the equilibrium position, determine the equation of motion if an impressed force $f(t) = \sin t$ acts on the system for $0 \le t < 2\pi$ and is then removed. Ignore any damping forces.

77. A cantilever beam is embedded at its left end and is free at its right end. Find the deflection $y(x)$ when the load is given by

$$w(x) = \begin{cases} 0, & 0 < x < L/3 \\ w_0, & L/3 < x < 2L/3 \\ 0, & 2L/3 < x < L \end{cases}$$

78. A beam is embedded at both ends. Find the deflection $y(x)$ when the load is given by

$$w(x) = \begin{cases} 2w_0\left(1 - \dfrac{1}{L}x\right), & 0 < x < L/2 \\ w_0, & L/2 < x < L \end{cases}$$

79. Find the deflection $y(x)$ of a cantilever beam embedded at its left end and free at its right end when the load is as given in Example 12.

80. A beam is embedded at its left end and is simply supported at its right end. Find the deflection $y(x)$ when the load is given by

$$w(x) = \begin{cases} w_0, & 0 < x < L/2 \\ 0, & L/2 < x < L \end{cases}$$

5.3 Derivative of a Transform, Transform of an Integral

So far we have examined only three of the many useful operational properties of the Laplace transform (Theorems 5.2, 5.4, 5.5). In this section, we consider two additional properties.

Consider This In Section 5.2, we extended the somewhat meager table of Laplace transforms given in Theorem 5.1 by showing how to find transforms of functions of the form $e^{at}f(t)$ and functions that we defined in a piecewise manner, that is, functions that could be expressed in terms of unit step functions. But as yet we have not seen the Laplace transforms of functions such as $t \sin t$ or $t^2 \cos 2t$. In Theorem 5.4, we saw how to transform the derivative $f'(t)$; we also need to know how to transform the integral $\int_0^t f(\tau)\, d\tau$.

We begin with functions of the form $t^n f(t)$.

Derivatives of a Transform If $F(s) = \mathscr{L}\{f(t)\}$, and if we assume that interchanging of differentiation and integration is possible, then

$$\frac{d}{ds}F(s) = \frac{d}{ds}\int_0^\infty e^{-st}f(t)\,dt = \int_0^\infty \frac{\partial}{\partial s}[e^{-st}f(t)]\,dt = -\int_0^\infty e^{-st}tf(t)\,dt = -\mathscr{L}\{tf(t)\}$$

that is,

$$\mathscr{L}\{tf(t)\} = -\frac{d}{ds}\mathscr{L}\{f(t)\}$$

Similarly

$$\mathscr{L}\{t^2f(t)\} = \mathscr{L}\{t \cdot tf(t)\} = -\frac{d}{ds}\mathscr{L}\{tf(t)\}$$
$$= -\frac{d}{ds}\left(-\frac{d}{ds}\mathscr{L}\{f(t)\}\right) = \frac{d^2}{ds^2}\mathscr{L}\{f(t)\}$$

The preceding two cases suggest the general result for $\mathscr{L}\{t^nf(t)\}$.

Theorem 5.7 Derivatives of Transforms

If $F(s) = \mathscr{L}\{f(t)\}$ and $n = 1, 2, 3, \ldots$, then

$$\mathscr{L}\{t^nf(t)\} = (-1)^n \frac{d^n}{ds^n}F(s)$$

■ **EXAMPLE 1 Applying Theorem 5.7**

Evaluate **(a)** $\mathscr{L}\{t \sin kt\}$, **(b)** $\mathscr{L}\{t^2 \sin kt\}$, and **(c)** $\mathscr{L}\{te^{-t} \cos t\}$.

Solution We make use of results (d), and (e) of Theorem 5.1.

(a) With $n = 1$ in Theorem 5.7,

$$\mathscr{L}\{t \sin kt\} = -\frac{d}{ds}\mathscr{L}\{\sin kt\} = -\frac{d}{ds}\left(\frac{k}{s^2 + k^2}\right) = \frac{2ks}{(s^2 + k^2)^2}$$

(b) With $n = 2$ in Theorem 5.7, this transform can be written as

$$\mathscr{L}\{t^2 \sin kt\} = \frac{d^2}{ds^2}\mathscr{L}\{\sin kt\}$$

and so by carrying out the two derivatives, we obtain the result. Alternatively, we can make use of the result already obtained in part (a). Since $t^2 \sin kt = t(t \sin kt)$, we have

$$\mathscr{L}\{t^2 \sin kt\} = -\frac{d}{ds}\mathscr{L}\{t \sin kt\} = -\frac{d}{ds}\left(\underbrace{\frac{2ks}{(s^2 + k^2)^2}}_{\text{from part (a)}}\right)$$

Differentiating and simplifying then give

$$\mathscr{L}\{t^2 \sin kt\} = \frac{6ks^2 - 2k^3}{(s^2 + k^2)^3}$$

(c)

$$\begin{aligned} \mathscr{L}\{te^{-t}\cos t\} &= -\frac{d}{ds}\mathscr{L}\{e^{-t}\cos t\} && \leftarrow \text{Theorem 5.7} \\ &= -\frac{d}{ds}\mathscr{L}\{\cos t\}_{s \to s+1} && \leftarrow \text{first translation theorem} \\ &= -\frac{d}{ds}\left(\frac{s+1}{(s+1)^2 + 1}\right) \\ &= \frac{(s+1)^2 - 1}{[(s+1)^2 + 1]^2} \end{aligned}$$

■

Note that to find transforms of functions $t^n e^{at}$, we can use either the first translation theorem or Theorem 5.7. For example,

$$\text{Theorem 5.5:} \quad \mathscr{L}\{te^{3t}\} = \mathscr{L}\{t\}_{s \to s-3} = \left.\frac{1}{s^2}\right|_{s \to s-3} = \frac{1}{(s-3)^2}$$

$$\text{Theorem 5.7:} \quad \mathscr{L}\{te^{3t}\} = -\frac{d}{ds}\mathscr{L}\{e^{3t}\} = -\frac{d}{ds}\frac{1}{s-3} = (s-3)^{-2} = \frac{1}{(s-3)^2}$$

■ **EXAMPLE 2** **An Initial-Value Problem**

Solve $x'' + 16x = \cos 4t, \quad x(0) = 0, \quad x'(0) = 1.$

Solution Recall that this initial-value problem could describe the forced, undamped, and resonant motion of a mass on a spring. The mass starts with an initial velocity of 1 ft/s in the downward direction from the equilibrium position.

Transforming the differential equation gives

$$(s^2+16)X(s) = 1 + \frac{s}{s^2+16} \quad \text{or} \quad X(s) = \frac{1}{s^2+16} + \frac{s}{(s^2+16)^2}$$

With the identification $k = 4$, part (d) of Theorem 5.1, and the inverse version of the formula in part (a) of Example 1, we obtain

$$\begin{aligned} x(t) &= \frac{1}{4}\mathcal{L}^{-1}\left\{\frac{4}{s^2+16}\right\} + \frac{1}{8}\mathcal{L}^{-1}\left\{\frac{8s}{(s^2+16)^2}\right\} \quad \leftarrow \text{(a) of Example 1} \\ &= \frac{1}{4}\sin 4t + \frac{1}{8}t\sin 4t \end{aligned}$$

■

■ **EXAMPLE 3 An Initial-Value Problem**

Solve $x'' + 16x = f(t), \quad x(0) = 0, \quad x'(0) = 1,$

where
$$f(t) = \begin{cases} \cos 4t, & 0 \le t < \pi \\ 0, & t \ge \pi \end{cases}$$

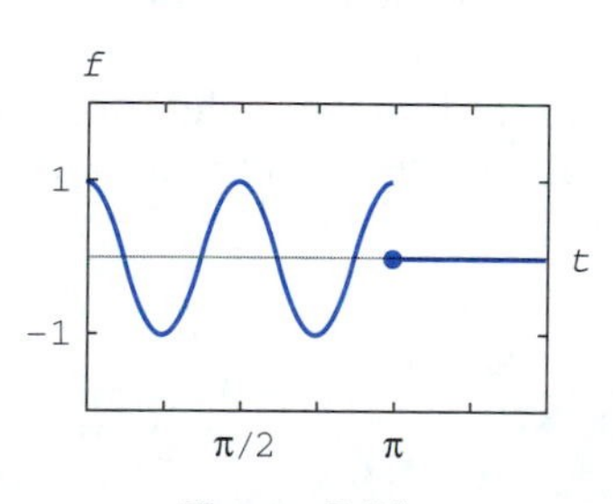

Figure 5.34.

Solution This is the same problem as in Example 2, except that the forcing function f is acting on the system only for a short time and then is removed. See Figure 5.34. In terms of a unit step function, f can be written as

$$f(t) = \cos 4t - \cos 4t\, \mathcal{U}(t-\pi)$$

Now with the aid of the periodicity of the cosine function, the alternative form of the second translation theorem, (9) of Section 5.2, then yields

$$\begin{aligned} \mathcal{L}\{x''\} + 16\mathcal{L}\{x\} &= \mathcal{L}\{f(t)\} \\ s^2X(s) - sx(0) - x'(0) + 16X(s) &= \frac{s}{s^2+16} - \frac{s}{s^2+16}e^{-\pi s} \\ (s^2+16)X(s) &= 1 + \frac{s}{s^2+16} - \frac{s}{s^2+16}e^{-\pi s} \\ X(s) &= \frac{1}{s^2+16} + \frac{s}{(s^2+16)^2} - \frac{s}{(s^2+16)^2}e^{-\pi s} \end{aligned}$$

Again, from part (a) of Example 1, we find

$$\begin{aligned} x(t) &= \frac{1}{4}\mathcal{L}^{-1}\left\{\frac{4}{s^2+16}\right\} + \frac{1}{8}\mathcal{L}^{-1}\left\{\frac{8s}{(s^2+16)^2}\right\} - \frac{1}{8}\mathcal{L}^{-1}\left\{\frac{8s}{(s^2+16)^2}e^{-\pi s}\right\} \\ &= \frac{1}{4}\sin 4t + \frac{1}{8}t\sin 4t - \frac{1}{8}(t-\pi)\sin 4(t-\pi)\mathcal{U}(t-\pi) \end{aligned}$$

■

Convolution If functions f and g are piecewise continuous on $[0, \infty)$, then the **convolution** of f and g, denoted by $f * g$, is defined by the integral

$$f * g = \int_0^t f(\tau)g(t-\tau)\,d\tau$$

For example, the convolution of $f(t) = e^t$ and $g(t) = \sin t$ is

$$e^t * \sin t = \int_0^t e^\tau \sin(t-\tau)\,d\tau = \frac{1}{2}(-\sin t - \cos t + e^t) \qquad \textbf{(1)}$$

It is left as an exercise to show that $\int_0^t f(\tau)g(t-\tau)\,d\tau = \int_0^t f(t-\tau)g(\tau)\,d\tau$; that is,

$$f * g = g * f$$

See Problem 33 in Exercises 5.3. This means that the convolution of two functions is commutative.

It is possible to find the Laplace transform of the convolution of two functions without actually evaluating the integral as we did in (1). The result that follows is known as the **convolution theorem.**

Theorem 5.8 Convolution Theorem

If $f(t)$ and $g(t)$ are piecewise continuous on $[0, \infty)$ and of exponential order, then

$$\mathscr{L}\{f * g\} = \mathscr{L}\{f(t)\}\mathscr{L}\{g(t)\} = F(s)G(s)$$

Proof Let $$F(s) = \mathscr{L}\{f(t)\} = \int_0^\infty e^{-s\tau}f(\tau)\,d\tau$$

and $$G(s) = \mathscr{L}\{g(t)\} = \int_0^\infty e^{-s\beta}g(\beta)\,d\beta$$

Proceeding formally, we have

$$\begin{aligned} F(s)G(s) &= \left(\int_0^\infty e^{-s\tau}f(\tau)\,d\tau\right)\left(\int_0^\infty e^{-s\beta}g(\beta)\,d\beta\right) \\ &= \int_0^\infty \int_0^\infty e^{-s(\tau+\beta)}f(\tau)g(\beta)\,d\tau\,d\beta \\ &= \int_0^\infty f(\tau)\,d\tau \int_0^\infty e^{-s(\tau+\beta)}g(\beta)\,d\beta \end{aligned}$$

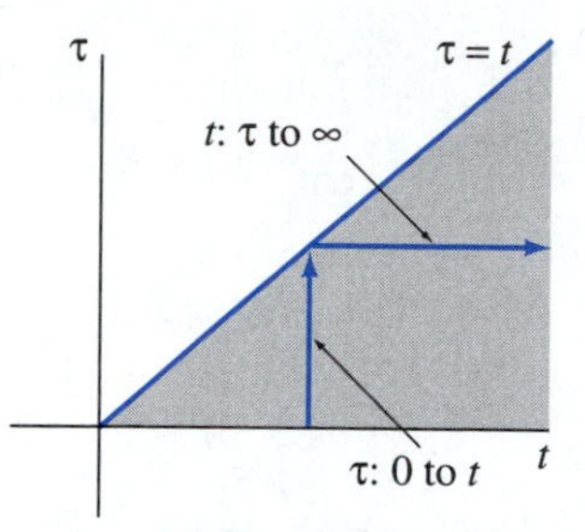

Figure 5.35.

Holding τ fixed, we let $t = \tau + \beta$, $dt = d\beta$ so that

$$F(s)G(s) = \int_0^\infty f(\tau)\, d\tau \int_\tau^\infty e^{-st} g(t-\tau)\, dt$$

In the $t\tau$-plane, we are integrating over the shaded region in Figure 5.35. Since f and g are piecewise continuous on $[0, \infty)$ and of exponential order, it is possible to interchange the order of integration:

$$F(s)G(s) = \int_0^\infty e^{-st}\, dt \int_0^t f(\tau)g(t-\tau)\, d\tau = \int_0^\infty e^{-st} \left\{ \int_0^t f(\tau)g(t-\tau)\, d\tau \right\} dt = \mathcal{L}\{f * g\} \quad \square$$

■ EXAMPLE 4 **Transform of a Convolution**

Evaluate $\mathcal{L}\left\{ \int_0^t e^\tau \sin(t-\tau)\, d\tau \right\}$.

Solution With $f(t) = e^t$ and $g(t) = \sin t$, the convolution theorem states that the Laplace transform of the convolution of f and g is the product of their Laplace transforms:

$$\mathcal{L}\left\{ \int_0^t e^\tau \sin(t-\tau)\, d\tau \right\} = \mathcal{L}\{e^t\} \cdot \mathcal{L}\{\sin t\} = \frac{1}{s-1} \cdot \frac{1}{s^2+1} = \frac{1}{(s-1)(s^2+1)}$$

■

Inverse Form of the Convolution Theorem The convolution theorem is sometimes useful in finding the inverse Laplace transform of a product of two Laplace transforms. From Theorem 5.8 we have

$$\mathcal{L}^{-1}\{F(s)G(s)\} = f * g \tag{2}$$

■ EXAMPLE 5 **Inverse Transform as a Convolution**

Evaluate $\mathcal{L}^{-1}\left\{ \frac{1}{(s-1)(s+4)} \right\}$.

Solution Partial fractions could be used, but if we identify

$$F(s) = \frac{1}{s-1} \quad \text{and} \quad G(s) = \frac{1}{s+4}$$

then $\quad \mathcal{L}^{-1}\{F(s)\} = f(t) = e^t \quad$ and $\quad \mathcal{L}^{-1}\{G(s)\} = g(t) = e^{-4t}$

Hence from (2) we obtain

$$\mathscr{L}^{-1}\left\{\frac{1}{(s-1)(s+4)}\right\} = \int_0^t f(\tau)g(t-\tau)\,d\tau = \int_0^t e^{\tau}e^{-4(t-\tau)}\,d\tau$$
$$= e^{-4t}\int_0^t e^{5\tau}\,d\tau$$
$$= e^{-4t}\frac{1}{5}e^{5\tau}\Big|_0^t$$
$$= \frac{1}{5}e^{t} - \frac{1}{5}e^{-4t} \qquad \blacksquare$$

EXAMPLE 6 Inverse Transform as a Convolution

Evaluate $\mathscr{L}^{-1}\left\{\dfrac{1}{(s^2+k^2)^2}\right\}$.

Solution Let

$$F(s) = G(s) = \frac{1}{s^2+k^2}$$

so that

$$f(t) = g(t) = \frac{1}{k}\mathscr{L}^{-1}\left\{\frac{k}{s^2+k^2}\right\} = \frac{1}{k}\sin kt$$

In this case, (2) gives

$$\mathscr{L}^{-1}\left\{\frac{1}{(s^2+k^2)^2}\right\} = \frac{1}{k^2}\int_0^t \sin k\tau \sin k(t-\tau)\,d\tau \tag{3}$$

Now recall from trigonometry that

$$\cos(A+B) = \cos A\cos B - \sin A\sin B$$

and

$$\cos(A-B) = \cos A\cos B + \sin A\sin B$$

Subtracting the first from the second gives the identity

$$\sin A\sin B = \frac{1}{2}[\cos(A-B) - \cos(A+B)]$$

If we set $A = k\tau$ and $B = k(t-\tau)$, we can carry out the integration in (3):

$$\mathscr{L}^{-1}\left\{\frac{1}{(s^2+k^2)^2}\right\} = \frac{1}{2k^2}\int_0^t [\cos k(2\tau - t) - \cos kt]\,d\tau$$
$$= \frac{1}{2k^2}\left[\frac{1}{2k}\sin k(2\tau - t) - \tau\cos kt\right]_0^t$$
$$= \frac{\sin kt - kt\cos kt}{2k^3} \qquad \blacksquare$$

Transform of an Integral When $g(t) = 1$ and $\mathscr{L}\{g(t)\} = G(s) = 1/s$, the convolution theorem implies that the Laplace transform of the integral of f is

$$\mathscr{L}\left\{\int_0^t f(\tau)\, d\tau\right\} = \frac{F(s)}{s} \tag{4}$$

The inverse form of (4),

$$\int_0^t f(\tau)\, d\tau = \mathscr{L}^{-1}\left\{\frac{F(s)}{s}\right\} \tag{5}$$

can be used at times in lieu of partial fractions when s^n is a factor of the denominator and $f(t) = \mathscr{L}^{-1}\{F(s)\}$ is easy to integrate. For example, we know for $f(t) = \sin t$ that $F(s) = 1/(s^2 + 1)$, and so by (5),

$$\mathscr{L}^{-1}\left\{\frac{1}{s(s^2+1)}\right\} = \int_0^t \sin\tau\, d\tau = 1 - \cos t$$

$$\mathscr{L}^{-1}\left\{\frac{1}{s^2(s^2+1)}\right\} = \int_0^t (1 - \cos\tau)\, d\tau = t - \sin t$$

$$\mathscr{L}^{-1}\left\{\frac{1}{s^3(s^2+1)}\right\} = \int_0^t (\tau - \sin\tau)\, d\tau = \frac{1}{2}t^2 - 1 + \cos t$$

and so on.

Volterra Integral Equation The convolution theorem is useful in solving other types of equations in which an unknown function appears under an integral sign. In the next example, we solve a **Volterra integral equation**

$$f(t) = g(t) + \int_0^t f(\tau)h(t - \tau)\, d\tau$$

for $f(t)$. The functions $g(t)$ and $h(t)$ are known.

EXAMPLE 7 An Integral Equation

Solve $f(t) = 3t^2 - e^{-t} - \int_0^t f(\tau)e^{t-\tau}\, d\tau$ for $f(t)$.

Solution It follows from Theorem 5.8 that

$$\mathscr{L}\{f(t)\} = 3\mathscr{L}\{t^2\} - \mathscr{L}\{e^{-t}\} - \mathscr{L}\{f(t)\}\mathscr{L}\{e^t\}$$

$$F(s) = 3 \cdot \frac{2}{s^3} - \frac{1}{s+1} - F(s) \cdot \frac{1}{s-1}$$

Solving the last equation for $F(s)$ and using partial fractions give

$$F(s) = \frac{6}{s^3} - \frac{6}{s^4} + \frac{1}{s} - \frac{2}{s+1}$$

The inverse transform is

$$f(t) = 3\mathscr{L}^{-1}\left\{\frac{2!}{s^3}\right\} - \mathscr{L}^{-1}\left\{\frac{3!}{s^4}\right\} + \mathscr{L}^{-1}\left\{\frac{1}{s}\right\} - 2\mathscr{L}^{-1}\left\{\frac{1}{s+1}\right\}$$

$$= 3t^2 - t^3 + 1 - 2e^{-t}$$ ■

Series Circuits In a single loop or series circuit, Kirchhoff's second law states that the sum of the voltage drops across an inductor, resistor, and capacitor is equal to the impressed voltage $E(t)$. Now it is known that the voltage drops across an inductor, resistor, and capacitor are, respectively,

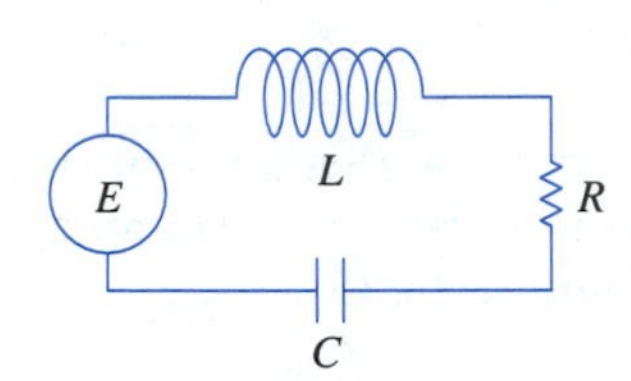

Figure 5.36.

$$L\frac{di}{dt}, \quad R\,i(t), \quad \text{and} \quad \frac{1}{C}\int_0^t i(\tau)\,d\tau$$

where $i(t)$ is the current, and L, R, and C are constants. It follows that the current in a circuit, such as that shown in Figure 5.36, is governed by the **integrodifferential equation**

$$L\frac{di}{dt} + Ri + \frac{1}{C}\int_0^t i(\tau)\,d\tau = E(t) \tag{6}$$

■ **EXAMPLE 8 An Integrodifferential Equation**

Determine the current $i(t)$ in a single loop L-R-C circuit when $L = 0.1$ henry, $R = 20$ ohms, $C = 10^{-3}$ farad, $i(0) = 0$, and the impressed voltage $E(t)$ is as given in Figure 5.37.

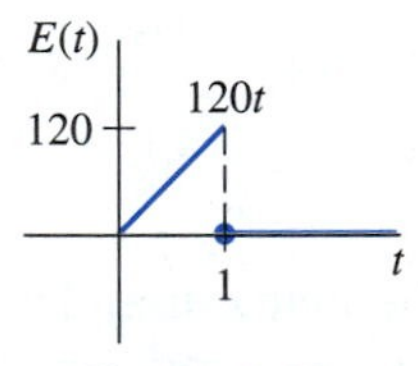

Figure 5.37.

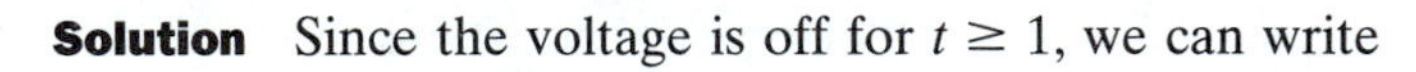

Solution Since the voltage is off for $t \geq 1$, we can write

$$E(t) = 120t - 120t\,\mathscr{U}(t-1)$$

Equation (6) then becomes

$$0.1\frac{di}{dt} + 20i + 10^3\int_0^t i(\tau)\,d\tau = 120t - 120t\,\mathscr{U}(t-1)$$

Now from (4) of this section and (9) of Section 5.2, the transform of the foregoing equation is

$$0.1sI(s) + 20I(s) + 10^3\frac{I(s)}{s} = 120\left[\frac{1}{s^2} - \frac{1}{s^2}e^{-s} - \frac{1}{s}e^{-s}\right]$$

or, after multiplying by $10s$,

$$(s+100)^2 I(s) = 1200\left[\frac{1}{s} - \frac{1}{s}e^{-s} - e^{-s}\right]$$

$$I(s) = 1200\left[\frac{1}{s(s+100)^2} - \frac{1}{s(s+100)^2}e^{-s} - \frac{1}{(s+100)^2}e^{-s}\right]$$

By partial fractions we can write

$$I(s) = 1200\left[\frac{1/10{,}000}{s} - \frac{1/10{,}000}{s+100} - \frac{1/100}{(s+100)^2} - \frac{1/10{,}000}{s}e^{-s}\right.$$
$$\left. + \frac{1/10{,}000}{s+100}e^{-s} + \frac{1/100}{(s+100)^2}e^{-s} - \frac{1}{(s+100)^2}e^{-s}\right]$$

Employing the inverse form of the second translation theorem, we obtain

$$i(t) = \frac{3}{25}[1 - \mathcal{U}(t-1)] - \frac{3}{25}[e^{-100t} - e^{-100(t-1)}\mathcal{U}(t-1)]$$
$$- 12te^{-100t} - 1188(t-1)e^{-100(t-1)}\mathcal{U}(t-1) \qquad \blacksquare$$

EXERCISES 5.3

Answers to odd-numbered problems begin on page 453.

In Problems 1–22, find either $F(s)$ or $f(t)$ as indicated.

1. $\mathcal{L}\{t \cos 2t\}$
2. $\mathcal{L}\{t \sinh 3t\}$
3. $\mathcal{L}\{t^2 \sinh t\}$
4. $\mathcal{L}\{t^2 \cos t\}$
5. $\mathcal{L}\{te^{2t} \sin 6t\}$
6. $\mathcal{L}\{te^{-3t} \cos 3t\}$
7. $\mathcal{L}^{-1}\left\{\frac{s}{(s^2+1)^2}\right\}$
8. $\mathcal{L}^{-1}\left\{\frac{s+1}{(s^2+2s+2)^2}\right\}$
9. $\mathcal{L}\left\{\int_0^t e^{\tau}\,d\tau\right\}$
10. $\mathcal{L}\left\{\int_0^t \cos\tau\,d\tau\right\}$
11. $\mathcal{L}\left\{\int_0^t e^{-\tau}\cos\tau\,d\tau\right\}$
12. $\mathcal{L}\left\{\int_0^t \tau\sin\tau\,d\tau\right\}$
13. $\mathcal{L}\left\{\int_0^t \tau e^{t-\tau}\,d\tau\right\}$
14. $\mathcal{L}\left\{\int_0^t \sin\tau\cos(t-\tau)\,d\tau\right\}$
15. $\mathcal{L}\left\{t\int_0^t \sin\tau\,d\tau\right\}$
16. $\mathcal{L}\left\{t\int_0^t \tau e^{-\tau}\,d\tau\right\}$
17. $\mathcal{L}\{1 * t^3\}$
18. $\mathcal{L}\{1 * e^{-2t}\}$
19. $\mathcal{L}\{t^2 * t^4\}$
20. $\mathcal{L}\{t^2 * te^t\}$
21. $\mathcal{L}\{e^{-t} * e^t \cos t\}$
22. $\mathcal{L}\{e^{2t} * \sin t\}$

In Problems 23 and 24, use Theorem 5.7 in the form ($n = 1$),

$$f(t) = -\frac{1}{t}\mathcal{L}^{-1}\left\{\frac{d}{ds}F(s)\right\}$$

to evaluate the given inverse Laplace transform.

23. $\mathcal{L}^{-1}\left\{\ln\frac{s-3}{s+1}\right\}$
24. $\mathcal{L}^{-1}\left\{\ln\frac{s^2+1}{s^2+4}\right\}$

In Problems 25 and 26, suppose $\mathcal{L}^{-1}\{F(s)\} = f(t)$. Find the inverse Laplace transform of the given function.

25. $\frac{1}{s+5}F(s)$
26. $\frac{s}{s^2+4}F(s)$

In Problems 27–32, use (2) or (5) to find $f(t)$.

27. $\mathcal{L}^{-1}\left\{\frac{1}{s(s+1)}\right\}$
28. $\mathcal{L}^{-1}\left\{\frac{1}{s^3(s-1)}\right\}$
29. $\mathcal{L}^{-1}\left\{\frac{1}{(s+1)(s-2)}\right\}$
30. $\mathcal{L}^{-1}\left\{\frac{1}{(s+1)^2}\right\}$
31. $\mathcal{L}^{-1}\left\{\frac{s}{(s^2+4)^2}\right\}$
32. $\mathcal{L}^{-1}\left\{\frac{1}{(s^2+4s+5)^2}\right\}$
33. Prove the commutative property of the convolution integral $f * g = g * f$.
34. Prove the distributive property of the convolution integral $f * (g + h) = f * g + f * h$.

In Problems 35–38, use the Laplace transform to solve the given differential equation subject to the indicated initial conditions.

35. $y'' + 9y = \cos 3t, \quad y(0) = 2, \quad y'(0) = 5$
36. $y'' + y = \sin t, \quad y(0) = 1, \quad y'(0) = -1$
37. $y'' + 9y = f(t), \quad y(0) = 0, \quad y'(0) = 0$
    where $f(t) = \begin{cases} 0, & 0 \le t < \pi, \\ \cos 3t, & t \ge \pi \end{cases}$
38. $y'' + y = f(t), \quad y(0) = 1, \quad y'(0) = 0$
    where $f(t) = \begin{cases} 1, & 0 \le t < \pi/2, \\ \sin t, & t \ge \pi/2 \end{cases}$

In Problems 39–48, use the Laplace transform to solve the given integral equation or integrodifferential equation.

39. $f(t) + \int_0^t (t - \tau)f(\tau)\, d\tau = t$

40. $f(t) = 2t - 4\int_0^t \sin \tau f(t - \tau)\, d\tau$

41. $f(t) = te^t + \int_0^t \tau f(t - \tau)\, d\tau$

42. $f(t) + 2\int_0^t f(\tau)\cos(t - \tau)\, d\tau = 4e^{-t} + \sin t$

43. $f(t) + \int_0^t f(\tau)\, d\tau = 1$

44. $f(t) = \cos t + \int_0^t e^{-\tau} f(t - \tau)\, d\tau$

45. $f(t) = 1 + t - \frac{8}{3}\int_0^t (\tau - t)^3 f(\tau)\, d\tau$

46. $t - 2f(t) = \int_0^t (e^{\tau} - e^{-\tau})f(t - \tau)\, d\tau$

47. $y'(t) = 1 - \sin t - \int_0^t y(\tau)\, d\tau, \quad y(0) = 0$

48. $\frac{dy}{dt} + 6y(t) + 9\int_0^t y(\tau)\, d\tau = 1, \quad y(0) = 0$

49. Use equation (6) to determine the current $i(t)$ in a single-loop L-R-C circuit when $L = 0.005$ henry, $R = 1$ ohm, $C = 0.02$ farad, $E(t) = 100[1 - \mathcal{U}(t - 1)]$ volts, and $i(0) = 0$.

50. Solve Problem 49 when $E(t) = 100[t - (t - 1)\mathcal{U}(t - 1)]$.

51. A 16-lb weight is attached to a spring whose constant is $k = 4.5$ lb/ft. Beginning at $t = 0$, a force equal to $f(t) = 4 \sin 3t + 2 \cos 3t$ acts on the system. Assuming that no damping forces are present, use the Laplace transform to find the equation of motion if the weight is released from rest from the equilibrium position.

Discussion Problems

52. In this problem, you are asked to show how the convolution integral can be used to find a solution of a certain kind of initial-value problem.

(a) Use the Laplace transform to show that a solution of

$$ay'' + by' + cy = g(t), \quad y(0) = 0, \quad y'(0) = 0$$

is given by $y_2(t) = \frac{1}{a} g * y_1$, where y_1 is a solution of

$$ay'' + by' + cy = 0, \quad y(0) = 0, \quad y'(0) = 1$$

(b) The interesting thing about the solution y_2 in part (a) is that it remains valid even though the Laplace transform of $g(t)$ *may not exist*. Use part (a) to find a solution of the initial-value problem

$$y'' + y = \sec t, \quad y(0) = 0, \quad y'(0) = 0$$

Note that $\sec t$ does not possess a Laplace transform.

53. Although the Laplace transform is not well-suited to solving linear differential equations with variable coefficients, it can nevertheless be used in some instances. Investigate how the Laplace transform and appropriate operational properties can be put to use to find a solution of $ty'' + 2ty' + 2y = t$, subject to $y(0) = 0$.

54. Discuss: Show how the property $\mathcal{L}\{f'(t)\} = sF(s) - f(0)$ can be used to prove that $\mathcal{L}\left\{\int_0^t f(\tau)\, d\tau\right\} = \frac{F(s)}{s}$.

5.4 Periodic Functions

At the mention of the words "periodic function" the trigonometric functions are probably brought to mind. All the trigonometric functions are periodic; $\sin t$ and $\cos t$ are periodic with period 2π. Graphs of periodic functions are repetitive; for example, the function $f(t) = \sin t$ completes one cycle of its graph in an interval of length equal to 2π.

But there are different kinds of periodic functions. In this section, we shall see how to obtain their Laplace transforms.

Transform of a Periodic Function In general, a function f is said to be **periodic** if there exists some number $T > 0$ such that $f(t + T) = f(t)$. The number T is called the **period** of f. A periodic function is often constructed by specifying a rule on an interval such as $[0, T]$ or $[0, T)$ and then *defining* it to be periodic with period T. For example, on the interval $0 \le t < 2$, suppose f is given by

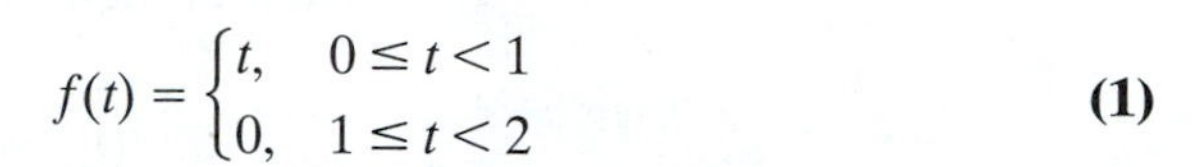

$$f(t) = \begin{cases} t, & 0 \le t < 1 \\ 0, & 1 \le t < 2 \end{cases} \qquad \textbf{(1)}$$

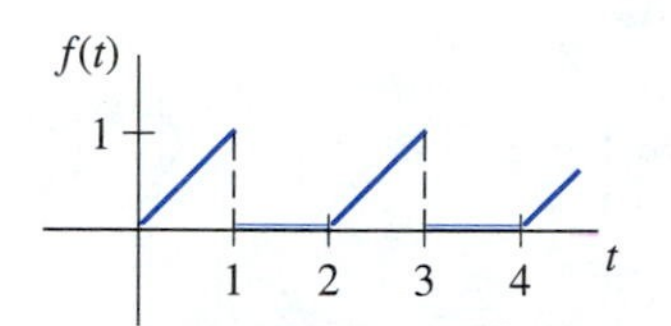

Figure 5.38.

and outside the interval by $f(t + 2) = f(t)$. Here the period of the function is 2; the graph of f on the initial interval is repeated as shown in Figure 5.38. The periodic function is piecewise continuous on $[0, \infty)$. Finding the Laplace transform of a piecewise continuous function such as that given in (1) is not as straightforward a task as finding the Laplace transform of a continuous periodic function such as $f(t) = \sin t$. For this last function, we can simply use (1) in Definition 5.1 and integrate (by parts) over $[0, \infty)$. The next theorem shows that we can find the Laplace transform of (1) by integrating over $[0, T)$.

Theorem 5.9 Transform of a Periodic Function

If f is piecewise continuous on $[0, \infty)$, of exponential order, and periodic with period T, then

$$\mathscr{L}\{f(t)\} = \frac{1}{1 - e^{-sT}} \int_0^T e^{-st} f(t)\, dt \qquad \textbf{(2)}$$

Proof Write the Laplace transform as two integrals:

$$\mathscr{L}\{f(t)\} = \int_0^T e^{-st} f(t)\, dt + \int_T^\infty e^{-st} f(t)\, dt \qquad \textbf{(3)}$$

When we let $t = u + T$, the last integral in (3) becomes

$$\int_T^\infty e^{-st} f(t)\, dt = \int_0^\infty e^{-s(u+T)} f(u + T)\, du = e^{-sT} \int_0^\infty e^{-su} f(u)\, du = e^{-sT} \mathscr{L}\{f(t)\}$$

Hence (3) is $$\mathscr{L}\{f(t)\} = \int_0^T e^{-st} f(t)\, dt + e^{-sT} \mathscr{L}\{f(t)\}$$

Solving for $\mathscr{L}\{f(t)\}$ yields the result given in (2). □

■ **EXAMPLE 1 Transform of a Periodic Function**

Find the Laplace transform of the periodic function shown in Figure 5.38.

Solution On the interval $0 \le t < 2$, the function can be defined by

$$f(t) = \begin{cases} t, & 0 \le t < 1 \\ 0, & 1 \le t < 2 \end{cases}$$

and outside the interval by $f(t + 2) = f(t)$. Identifying $T = 2$, we use (2) and integration by parts to obtain

$$\begin{aligned} \mathscr{L}\{f(t)\} &= \frac{1}{1 - e^{-2s}} \int_0^2 e^{-st} f(t)\, dt = \frac{1}{1 - e^{-2s}} \left[\int_0^1 e^{-st} t\, dt + \int_1^2 e^{-st} 0\, dt \right] \\ &= \frac{1}{1 - e^{-2s}} \left[-\frac{e^{-s}}{s} + \frac{1 - e^{-s}}{s^2} \right] \qquad (4) \\ &= \frac{1 - (s + 1)e^{-s}}{s^2(1 - e^{-2s})} \end{aligned}$$

■

The result in (4) of Example 1 can be obtained without actually integrating by making use of the second translation theorem. If we define

$$g(t) = \begin{cases} t, & 0 \le t < 1 \\ 0, & t \ge 1 \end{cases}$$

then $f(t) = g(t)$ on the interval $[0, T]$, where $T = 2$. But we can express g in terms of unit step functions as

$$g(t) = t - t\mathscr{U}(t - 1) = t - (t - 1)\mathscr{U}(t - 1) - \mathscr{U}(t - 1)$$

Thus

$$\begin{aligned} \mathscr{L}\{f(t)\} &= \frac{1}{1 - e^{-2s}} \mathscr{L}\{g(t)\} \\ &= \frac{1}{1 - e^{-2s}} \mathscr{L}\{t - (t - 1)\mathscr{U}(t - 1) - \mathscr{U}(t - 1)\} \\ &= \frac{1}{1 - e^{-2s}} \left[\frac{1}{s^2} - \frac{1}{s^2} e^{-s} - \frac{1}{s} e^{-s} \right] \end{aligned}$$

Inspection of the expression inside the brackets reveals that it is identical to (4).

■ EXAMPLE 2 A Periodic Impressed Voltage

The differential equation for the current $i(t)$ in a single loop L-R series circuit is

$$L\frac{di}{dt} + Ri = E(t) \qquad (5)$$

Determine the current $i(t)$ when $i(0) = 0$ and $E(t)$ is the square wave function with amplitude 1 shown in Figure 5.39.

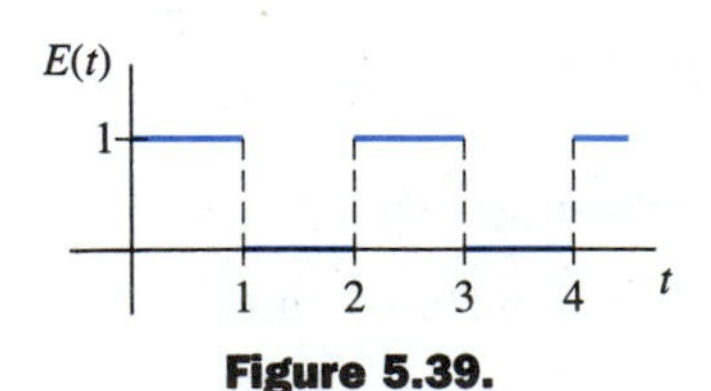

Figure 5.39.

Solution The Laplace transform of the equation is

$$LsI(s) + RI(s) = \mathcal{L}\{E(t)\} \tag{6}$$

Since $E(t)$ is periodic with period $T = 2$, (2) yields

$$\mathcal{L}\{E(t)\} = \frac{1}{1-e^{-2s}}\left(\int_0^1 1 \cdot e^{-st}\,dt + \int_1^2 0 \cdot e^{-st}\,dt\right)$$

$1 - e^{-2s} = (1 + e^{-s})(1 - e^{-s}) \rightarrow$
$$= \frac{1}{1-e^{-2s}}\frac{1-e^{-s}}{s} = \frac{1}{s(1+e^{-s})}$$

Hence from (6) we find

$$I(s) = \frac{1/L}{s(s + R/L)(1 + e^{-s})} \tag{7}$$

To find the inverse Laplace transform of this function, we first make use of a geometric series. With the identification $x = e^{-s}$, $s > 0$, the geometric series

$$\frac{1}{1+x} = 1 - x + x^2 - x^3 + \cdots \quad \text{becomes} \quad \frac{1}{1+e^{-s}} = 1 - e^{-s} + e^{-2s} - e^{-3s} + \cdots$$

From
$$\frac{1}{s(s+R/L)} = \frac{L/R}{s} - \frac{L/R}{s+R/L}$$

we can then rewrite (7) as

$$I(s) = \frac{1}{R}\left(\frac{1}{s} - \frac{1}{s+R/L}\right)(1 - e^{-s} + e^{-2s} - e^{-3s} + \cdots)$$

$$= \frac{1}{R}\left(\frac{1}{s} - \frac{e^{-s}}{s} + \frac{e^{-2s}}{s} - \frac{e^{-3s}}{s} + \cdots\right) - \frac{1}{R}\left(\frac{1}{s+R/L} - \frac{e^{-s}}{s+R/L} + \frac{e^{-2s}}{s+R/L} - \frac{e^{-3s}}{s+R/L} + \cdots\right)$$

By applying the inverse form of the second translation theorem to each term of both series, we obtain

$$i(t) = \frac{1}{R}(1 - \mathcal{U}(t-1) + \mathcal{U}(t-2) - \mathcal{U}(t-3) + \cdots)$$

$$- \frac{1}{R}(e^{-Rt/L} - e^{-R(t-1)/L}\mathcal{U}(t-1) + e^{-R(t-2)/L}\mathcal{U}(t-2) - e^{-R(t-3)/L}\mathcal{U}(t-3) + \cdots)$$

or equivalently,

$$i(t) = \frac{1}{R}(1 - e^{-Rt/L}) + \frac{1}{R}\sum_{n=1}^{\infty}(-1)^n(1 - e^{-R(t-n)/L})\mathcal{U}(t-n)$$ ■

To interpret the solution in Example 2, let us assume for the sake of illustration that $R = 1$, $L = 1$, and $0 \le t < 4$. In this case

$$i(t) = 1 - e^{-t} - (1 - e^{t-1})\mathcal{U}(t-1) + (1 - e^{-(t-2)})\mathcal{U}(t-2) - (1 - e^{-(t-3)})\mathcal{U}(t-3)$$

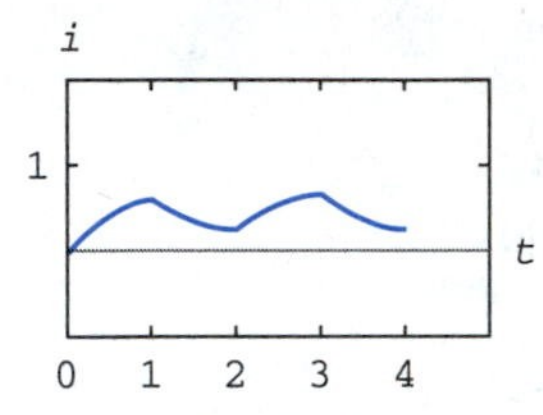

Figure 5.40.

In other words,

$$i(t) = \begin{cases} 1 - e^{-t}, & 0 \le t < 1 \\ -e^{-t} + e^{-(t-1)}, & 1 \le t < 2 \\ 1 - e^{-t} + e^{-(t-1)} - e^{-(t-2)}, & 2 \le t < 3 \\ -e^{-t} + e^{-(t-1)} - e^{-(t-2)} + e^{-(t-3)}, & 3 \le t < 4 \end{cases}$$

The graph of $i(t)$ on the interval $0 \le t < 4$, given in Figure 5.40, was obtained with the help of a graphing utility. Notice that even though the input $E(t)$ is discontinuous, the response $i(t)$ is a continuous function.

EXERCISES 5.4

Answers to odd-numbered problems begin on page 453.

In Problems 1–8, use Theorem 5.9 to find the Laplace transform of the given periodic function.

1. $f(t) = \sin t$, $f(t + 2\pi) = f(t)$

2. $f(t) = \cos 2t$, $f(t + \pi) = f(t)$

3.

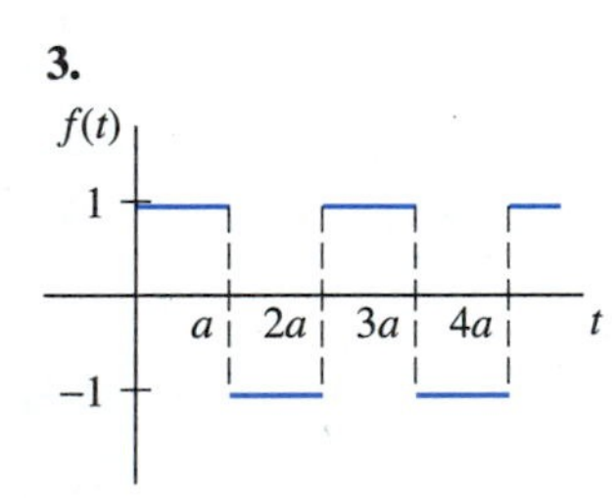

Meander function

Figure 5.41.

4.

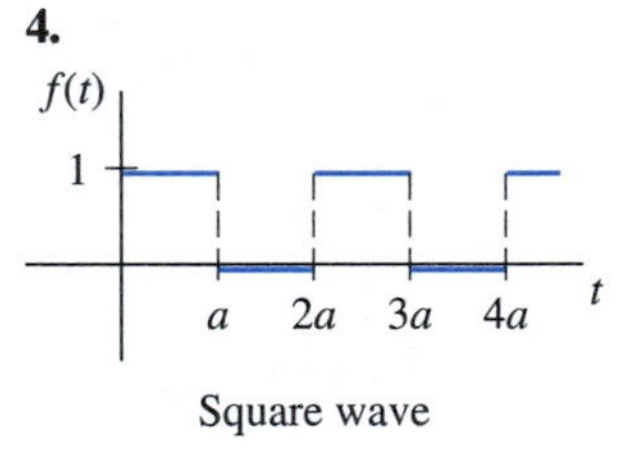

Square wave

Figure 5.42.

5. **6.**

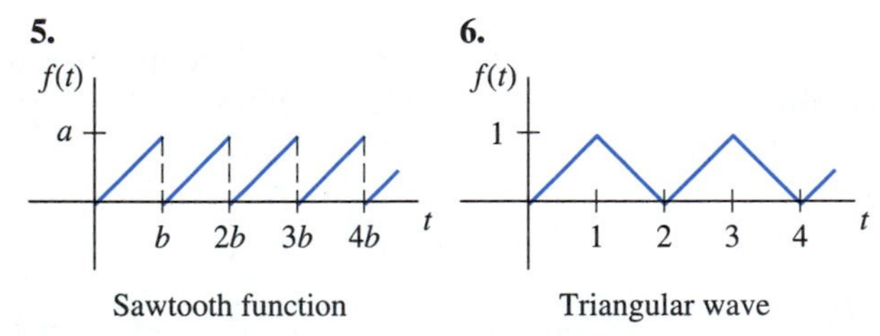

Sawtooth function

Figure 5.43.

Triangular wave

Figure 5.44.

7.

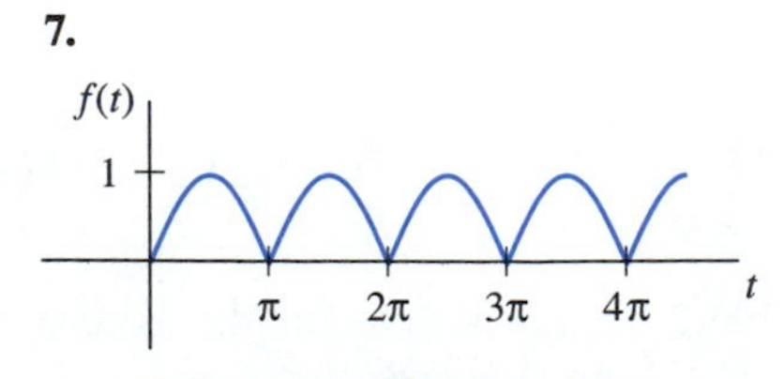

Full-wave rectification of sin t

Figure 5.45.

8.

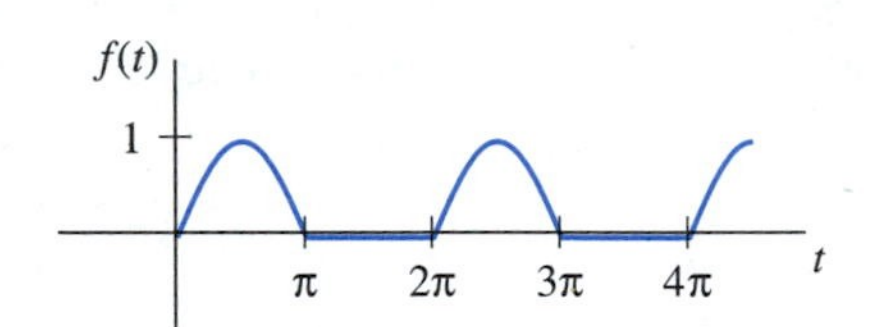

Half-wave rectification of sin t

Figure 5.46.

9. Solve equation (5) subject to $i(0) = 0$, where $E(t)$ is the sawtooth function given in Problem 5 with $a = b = 1$. Specify the solution on the interval $0 \le t < 2$.

10. Solve equation (5) subject to $i(0) = 0$, where $E(t)$ is the meander function given in Problem 3 with $a = 1$. Specify the solution on the interval $0 \le t < 3$.

In Section 3.6, we saw that

$$m\frac{d^2x}{dt^2} + \beta\frac{dx}{dt} + kx = f(t), \quad x(0) = 0, \quad x'(0) = 0 \tag{8}$$

is a model for a driven spring/mass system with damping. In Problems 11 and 12, solve (8) with the driving function f as specified. Using a graphing utility to obtain the graph of $x(t)$ on the indicated interval.

11. $m = \frac{1}{2}$, $\beta = 1$, $k = 5$, f is the meander function in Problem 3 with amplitude 10, and $a = \pi$, $0 \le t < 2\pi$.

12. $m = 1$, $\beta = 2$, $k = 1$, f is the square wave in Problem 4 with amplitude 5, and $a = \pi$, $0 \le t < 4\pi$.

5.5 Dirac Delta Function

The *delta function* was the invention of the contemporary British physicist and Nobel laureate Paul Dirac (1902–1984). Dirac was one of the founding fathers, in the era 1900–1930, of a new way of describing the behavior of atoms and elementary particles, called *quantum mechanics*. Used extensively throughout his 1932 classic treatise on this subject, Dirac's delta function did not ''behave'' like ordinary functions and so mathematicians initially treated it with scorn. But an entirely new branch of mathematics, the *theory of distributions*, grew out of the process of putting the delta function on a rigorous footing.

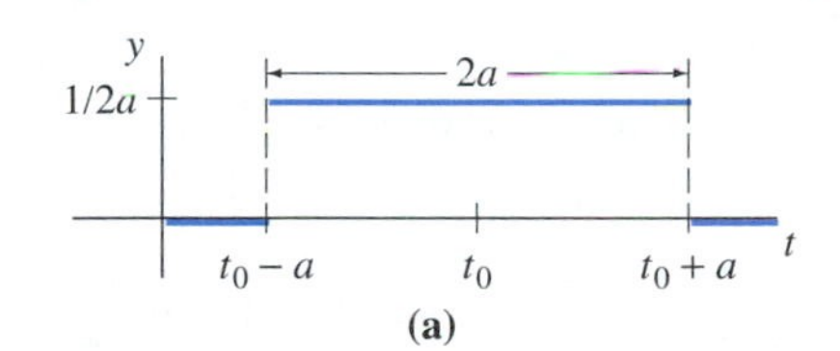

(a)

(b) Behavior of δ_a as $a \to 0$

Figure 5.47.

Unit Impulse Mechanical systems are often acted upon by an external force (or emf in an electrical circuit) of large magnitude that acts only for a very short period of time. For example, a vibrating airplane wing could be struck by a bolt of lightning, a mass on a spring could be given a sharp blow by a ball peen hammer, a ball (baseball, golf ball, tennis ball) could be sent soaring when struck violently by some kind of club (baseball bat, golf club, tennis racket). If we assume that the magnitude of such a force is constant over a short interval of time, then the following function could serve as its mathematical model

$$\delta_a(t - t_0) = \begin{cases} 0, & 0 \le t < t_0, \\ \dfrac{1}{2a}, & t_0 - a \le t < t_0 + a, \\ 0, & t \ge t_0 + a \end{cases} \tag{1}$$

where $a > 0$ and $t_0 > 0$. As illustrated in Figure 5.47(a), $\delta_a(t - t_0)$ is a piecewise-defined function with the constant value $1/2a$ over the interval $(t_0 - a, t_0 + a)$ and zero elsewhere. As we take the values of a to be smaller and smaller (that is, $a \to 0$), we see in Figure 5.47(b) that $\delta_a(t - t_0)$ is a function of large magnitude that is "on" for only a short period of time around t_0. The function $\delta_a(t - t_0)$ is called an **impulse**, or more precisely in this case, a **unit impulse** since it possesses the integration property $\int_0^\infty \delta_a(t - t_0)\, dt = 1$.

Dirac Delta Function In practice, it is convenient to work with another type of unit impulse—a "function" that approximates $\delta_a(t - t_0)$ and is defined by the limit

$$\delta(t - t_0) = \lim_{a \to 0} \delta_a(t - t_0) \tag{2}$$

The expression in (2), which is not a function in the usual sense of the word, can be characterized by the two properties

$$\textbf{(i)}\ \delta(t-t_0)=\begin{cases}\infty, & t=t_0\\ 0, & t\neq t_0\end{cases},\quad\text{and}\quad \textbf{(ii)}\ \int_0^\infty \delta(t-t_0)\,dt=1$$

The expression $\delta(t-t_0)$ is called the **Dirac delta function**.

It is possible to obtain the Laplace transform of the Dirac delta function by the formal assumption that $\mathscr{L}\{\delta(t-t_0)\}=\lim_{a\to 0}\mathscr{L}\{\delta_a(t-t_0)\}$.

Theorem 5.10 Transform of the Dirac Delta Function

For $t_0>0$,

$$\mathscr{L}\{\delta(t-t_0)\}=e^{-st_0} \tag{3}$$

Proof To begin, we can write $\delta_a(t-t_0)$ in terms of the unit step function by virtue of (5) and (6) of Section 5.2:

$$\delta_a(t-t_0)=\frac{1}{2a}[\mathscr{U}(t-(t_0-a))-\mathscr{U}(t-(t_0+a))]$$

By linearity and (7) of Section 5.2, the Laplace transform of this last expression is

$$\mathscr{L}\{\delta_a(t-t_0)\}=\frac{1}{2a}\left[\frac{e^{-s(t_0-a)}}{s}-\frac{e^{-s(t_0+a)}}{s}\right]=e^{-st_0}\left(\frac{e^{sa}-e^{-sa}}{2sa}\right) \tag{4}$$

Since (4) has the indeterminate form 0/0 as $a\to 0$, we apply L'Hôpital's rule:

$$\mathscr{L}\{\delta(t-t_0)\}=\lim_{a\to 0}\mathscr{L}\{\delta_a(t-t_0)\}=e^{-st_0}\lim_{a\to 0}\left(\frac{e^{sa}-e^{-sa}}{2sa}\right)=e^{-st_0} \qquad \square$$

Now when $t_0=0$, it seems plausible to conclude from (3) that

$$\mathscr{L}\{\delta(t)\}=1$$

The last result emphasizes the fact that $\delta(t)$ is not the usual type of function that we have been considering since we expect from Theorem 5.4 that $\mathscr{L}\{f(t)\}\to 0$ as $s\to\infty$.

EXAMPLE 1 Two Initial-Value Problems

Solve $y''+y=4\delta(t-2\pi)$ subject to **(a)** $y(0)=1$, $y'(0)=0$, and **(b)** $y(0)=0$, $y'(0)=0$. The two initial-value problems could serve as models for describing the motion of a mass on a spring moving in a medium in which damping is negligible. At $t=2\pi$ s, the mass is given a sharp blow. In (a) the mass is released from rest 1 unit below the equilibrium position. In (b) the mass is at rest in the equilibrium position.

Solution **(a)** From (3), the Laplace transform of the differential equation is

$$s^2Y(s) - s + Y(s) = 4e^{-2\pi s} \quad \text{or} \quad Y(s) = \frac{s}{s^2+1} + \frac{4e^{-2\pi s}}{s^2+1}$$

Using the inverse form of the second translation theorem, we find

$$y(t) = \cos t + 4\sin(t - 2\pi)\mathscr{U}(t - 2\pi)$$

Since $\sin(t - 2\pi) = \sin t$, the foregoing solution can be written as

$$y(t) = \begin{cases} \cos t, & 0 \le t < 2\pi \\ \cos t + 4\sin t, & t \ge 2\pi \end{cases} \qquad \textbf{(5)}$$

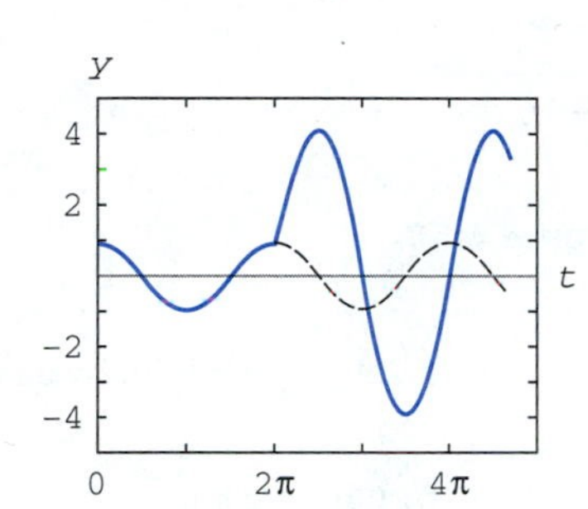

Figure 5.48.

In Figure 5.48, we see from the graph of (5) that the mass is exhibiting simple harmonic motion until it is struck at $t = 2\pi$. The influence of the unit impulse is to increase the amplitude of vibration to $\sqrt{17}$ for $t > 2\pi$.

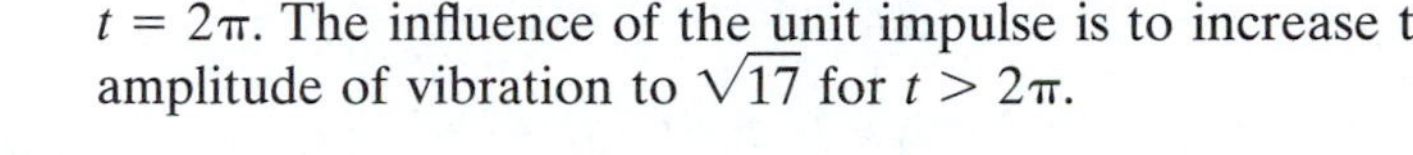

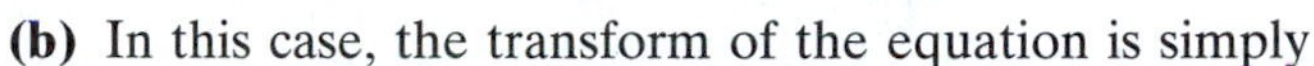

(b) In this case, the transform of the equation is simply

$$Y(s) = \frac{4e^{-2\pi s}}{s^2+1}$$

and so

$$y(t) = 4\sin(t - 2\pi)\mathscr{U}(t - 2\pi)$$

$$= \begin{cases} 0, & 0 \le t < 2\pi \\ 4\sin t, & t \ge 2\pi \end{cases} \qquad \textbf{(6)}$$

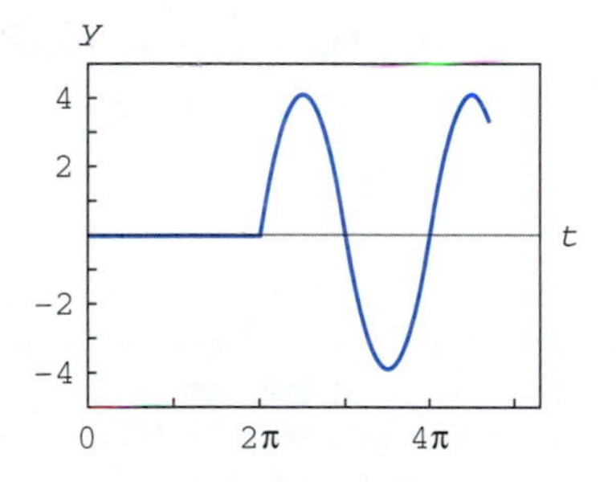

Figure 5.49.

The graph of (6) in Figure 5.49 shows, as we would expect from the initial conditions, that the mass exhibits no motion until it is struck at $t = 2\pi$. ■

Remarks

If $\delta(t - t_0)$ were a function in the strictest interpretation of that word, then property (*i*) on page 354 would imply that $\int_0^\infty \delta(t - t_0)\,dt = 0$ rather than $\int_0^\infty \delta(t - t_0)\,dt = 1$. In the theory of distributions mentioned in the introduction to this section, (2) is not an accepted definition of $\delta(t - t_0)$, nor does one speak of a function whose values are either ∞ or 0. Although we shall not pursue this topic further, suffice it to say that the Dirac delta function is best characterized by its effect on other functions. If f is a continuous function, then

$$\int_0^\infty f(t)\,\delta(t - t_0)\,dt = f(t_0) \qquad \textbf{(7)}$$

can be taken to be the *definition* of $\delta(t - t_0)$. This result is known as the **sifting property** since $\delta(t - t_0)$ has the effect of sifting the value $f(t_0)$ out of the set of values of f on $[0, \infty)$. Note that property (*ii*) (with $f(t) = 1$) and (3) (with $f(t) = e^{-st}$) are consistent with (7).

EXERCISES 5.5

Answers to odd-numbered problems begin on page 453.

In Problems 1–12, use the Laplace transform to solve the given differential equation subject to the indicated initial conditions.

1. $y' - 3y = \delta(t - 2), \quad y(0) = 0$

2. $y' + y = \delta(t - 1), \quad y(0) = 2$

3. $y'' + y = \delta(t - 2\pi), \quad y(0) = 0, y'(0) = 1$

4. $y'' + 16y = \delta(t - 2\pi), \quad y(0) = 0, y'(0) = 0$

5. $y'' + y = \delta\left(t - \frac{\pi}{2}\right) + \delta\left(t - \frac{3\pi}{2}\right), \quad y(0) = 0, y'(0) = 0$

6. $y'' + y = \delta(t - 2\pi) + \delta(t - 4\pi), \quad y(0) = 1, y'(0) = 0$

7. $y'' + 2y' = \delta(t - 1), \quad y(0) = 0, y'(0) = 1$

8. $y'' - 2y' = 1 + \delta(t - 2), \quad y(0) = 0, y'(0) = 1$

9. $y'' + 4y' + 5y = \delta(t - 2\pi), \quad y(0) = 0, y'(0) = 0$

10. $y'' + 2y' + y = \delta(t - 1), \quad y(0) = 0, y'(0) = 0$

11. $y'' + 4y' + 13y = \delta(t - \pi) + \delta(t - 3\pi)$,
$y(0) = 1, y'(0) = 0$

12. $y'' - 7y' + 6y = e^t + \delta(t - 2) + \delta(t - 4)$,
$y(0) = 0, y'(0) = 0$

13. A uniform beam of length L carries a concentrated load w_0 at $x = L/2$. The beam is embedded at its left end and is free at its right end. Use the Laplace transform to determine the deflection $y(x)$ from

$$EI\frac{d^4y}{dx^4} = w_0\delta\left(x - \frac{L}{2}\right)$$

where $y(0) = 0$, $y'(0) = 0$, $y''(L) = 0$, and $y'''(L) = 0$.

14. Solve the differential equation in Problem 13 subject to $y(0) = 0$, $y'(0) = 0$, $y(L) = 0$, $y'(L) = 0$. In this case, the beam is embedded at both ends. See Figure 5.50.

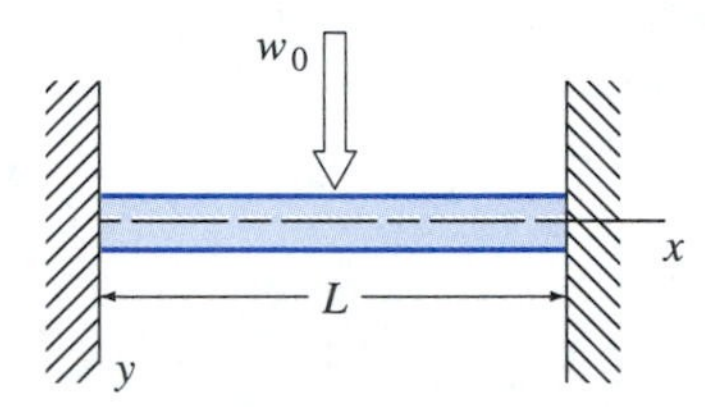

Figure 5.50.

15. Use the sifting property (7) in solving the initial-value problem

$$y'' + 2y' + 2y = \cos t\, \delta(t - 3\pi),$$
$$y(0) = 1, y'(0) = -1$$

Use a graphing utility to obtain a graph of the solution.

16. When a uniform beam is supported by an elastic foundation, the differential equation for its deflection $y(x)$ is

$$\frac{d^4y}{dx^4} + 4a^4y = \frac{w(x)}{EI}$$

where a is a constant. In the case when $a = 1$, find the deflection $y(x)$ of an elastically supported beam of length π that is embedded in concrete at both ends when a concentrated load w_0 is applied at $x = \pi/2$. [*Hint*: Use the table of Laplace transforms in Appendix II.]

Discussion Problem

17. Discuss the unusual aspect of the initial-value problem $y'' + \omega^2y = \delta(t)$, $y(0) = 0$, $y'(0) = 0$, and its solution.

5.6 Systems of Differential Equations

When initial conditions are specified, the Laplace transform reduces a system of linear DEs with constant coefficients to a set of simultaneous algebraic equations. We solve these algebraic equations for the transformed functions; the inverse Laplace transforms of these functions are solutions of the linear system.

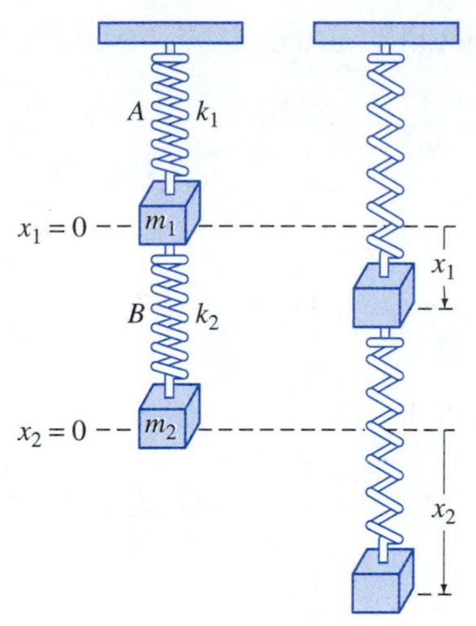

Figure 5.51.

Coupled Springs Suppose two masses m_1 and m_2 are connected to two springs A and B of negligible mass and with spring constants k_1 and k_2, respectively. In turn, the two springs are attached as shown in Figure 5.51. Let $x_1(t)$ and $x_2(t)$ denote the vertical displacements of the masses from their equilibrium positions. When the system is in motion, spring B is subject to both an elongation and a compression; hence its net elongation is $x_2 - x_1$. Therefore it follows from Hooke's law that spring A and B exert forces $-k_1x_1$ and $k_2(x_2 - x_1)$, respectively, on m_1. If no external force is impressed on the system and if no damping force is present, then the net force on m_1 is $-k_1x_1 + k_2(x_2 - x_1)$. By Newton's second law we can write

$$m_1\frac{d^2x_1}{dt^2} = -k_1x_1 + k_2(x_2 - x_1)$$

Similarly, the net force exerted on mass m_2 is due solely to the net elongation of B; that is, $-k_2(x_2 - x_1)$. Hence we have

$$m_2\frac{d^2x_2}{dt^2} = -k_2(x_2 - x_1)$$

In other words, the motion of the system of coupled springs is represented by the simultaneous equations

$$\begin{aligned} m_1x_1'' &= -k_1x_1 + k_2(x_2 - x_1) \\ m_2x_2'' &= -k_2(x_2 - x_1) \end{aligned} \tag{1}$$

In the example that follows, we solve the system (1) under the assumption that $k_1 = 6$, $k_2 = 4$, $m_1 = 1$, $m_2 = 1$, and that the masses start from the equilibrium position with opposite unit velocities.

EXAMPLE 1 Coupled Springs

Solve

$$\begin{aligned} x_1'' + 10x_1 \quad - 4x_2 &= 0 \\ -4x_1 + x_2'' + 4x_2 &= 0 \end{aligned} \tag{2}$$

subject to $x_1(0) = 0$, $x_1'(0) = 1$, $x_2(0) = 0$, $x_2'(0) = -1$.

Solution The Laplace transform of each equation is

$$s^2X_1(s) - sx_1(0) - x_1'(0) + 10X_1(s) - 4X_2(s) = 0$$

$$-4X_1(s) + s^2X_2(s) - sx_2(0) - x_2'(0) + 4X_2(s) = 0$$

where $X_1(s) = \mathscr{L}\{x_1(t)\}$ and $X_2(s) = \mathscr{L}\{x_2(t)\}$. The preceding system is the same as

$$\begin{aligned} (s^2 + 10)X_1(s) - \qquad 4X_2(s) &= 1 \\ -4X_1(s) + (s^2 + 4)X_2(s) &= -1 \end{aligned} \tag{3}$$

Solving (3) for $X_1(s)$, and using partial fractions on the result, yield

$$X_1(s) = \frac{s^2}{(s^2+2)(s^2+12)} = -\frac{1/5}{s^2+2} + \frac{6/5}{s^2+12}$$

and therefore

$$x_1(t) = -\frac{1}{5\sqrt{2}}\mathcal{L}^{-1}\left\{\frac{\sqrt{2}}{s^2+2}\right\} + \frac{6}{5\sqrt{12}}\mathcal{L}^{-1}\left\{\frac{\sqrt{12}}{s^2+12}\right\}$$
$$= -\frac{\sqrt{2}}{10}\sin\sqrt{2}t + \frac{\sqrt{3}}{5}\sin 2\sqrt{3}t$$

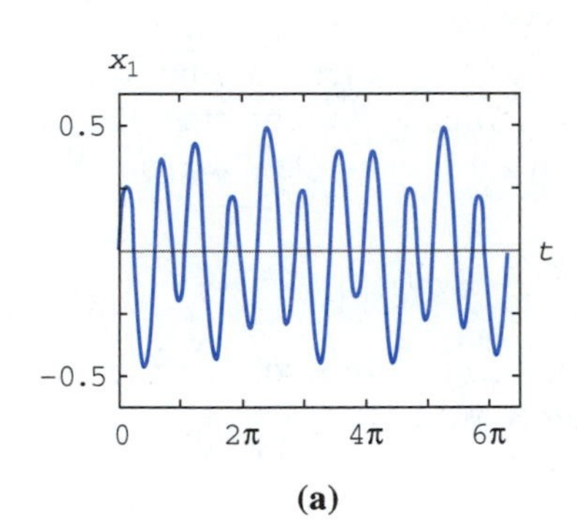

(a)

Substituting the expression for $X_1(s)$ into the first equation of (3) gives us

$$X_2(s) = -\frac{s^2+6}{(s^2+2)(s^2+12)} = -\frac{2/5}{s^2+2} - \frac{3/5}{s^2+12}$$

and
$$x_2(t) = -\frac{2}{5\sqrt{2}}\mathcal{L}^{-1}\left\{\frac{\sqrt{2}}{s^2+2}\right\} - \frac{3}{5\sqrt{12}}\mathcal{L}^{-1}\left\{\frac{\sqrt{12}}{s^2+12}\right\}$$
$$= -\frac{\sqrt{2}}{5}\sin\sqrt{2}t - \frac{\sqrt{3}}{10}\sin 2\sqrt{3}t$$

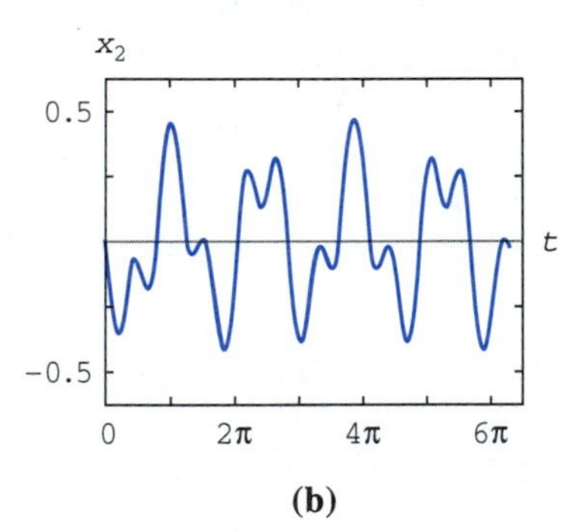

(b)

Figure 5.52.

Finally the solution to the given system (2) is

$$\begin{aligned} x_1(t) &= -\frac{\sqrt{2}}{10}\sin\sqrt{2}t + \frac{\sqrt{3}}{5}\sin 2\sqrt{3}t \\ x_2(t) &= -\frac{\sqrt{2}}{5}\sin\sqrt{2}t - \frac{\sqrt{3}}{10}\sin 2\sqrt{3}t \end{aligned} \qquad \textbf{(4)}$$

The graphs of x_1 and x_2 given in Figure 5.52 reveal the complicated oscillatory motion of each mass. ■

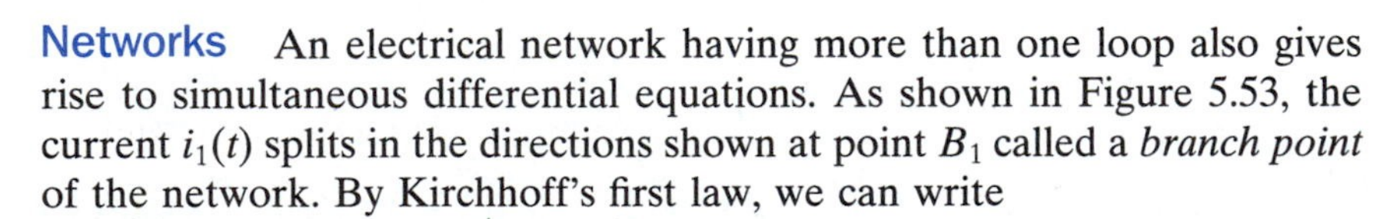

Networks An electrical network having more than one loop also gives rise to simultaneous differential equations. As shown in Figure 5.53, the current $i_1(t)$ splits in the directions shown at point B_1 called a *branch point* of the network. By Kirchhoff's first law, we can write

$$i_1(t) = i_2(t) + i_3(t) \qquad \textbf{(5)}$$

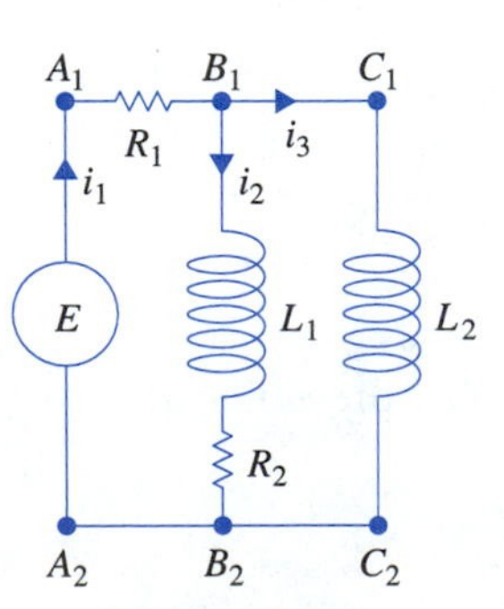

Figure 5.53.

In addition, we can also apply **Kirchhoff's second law** to each loop. For loop $A_1B_1B_2A_2A_1$, summing the voltage drops across each part of the loop gives

$$E(t) = i_1R_1 + L_1\frac{di_2}{dt} + i_2R_2 \qquad \textbf{(6)}$$

Similarly, for loop $A_1B_1C_1C_2B_2A_2A_1$, we find

$$E(t) = i_1R_1 + L_2\frac{di_3}{dt} \qquad \textbf{(7)}$$

Using (5) to eliminate i_1 in (6) and (7) yields two first-order equations for the currents $i_2(t)$ and $i_3(t)$:

$$\begin{aligned} L_1\frac{di_2}{dt} + (R_1 + R_2)i_2 + R_1 i_3 &= E(t) \\ L_2\frac{di_3}{dt} + \qquad R_1 i_2 + R_1 i_3 &= E(t) \end{aligned} \tag{8}$$

Given the natural initial conditions $i_2(0) = 0$, $i_3(0) = 0$, the system (8) is amenable to solution by the Laplace transform.

We leave it as an exercise (see Problem 18) to show that the system of differential equations describing the currents $i_1(t)$ and $i_2(t)$ in the network containing a resistor, an inductor, and a capacitor, shown in Figure 5.54 is

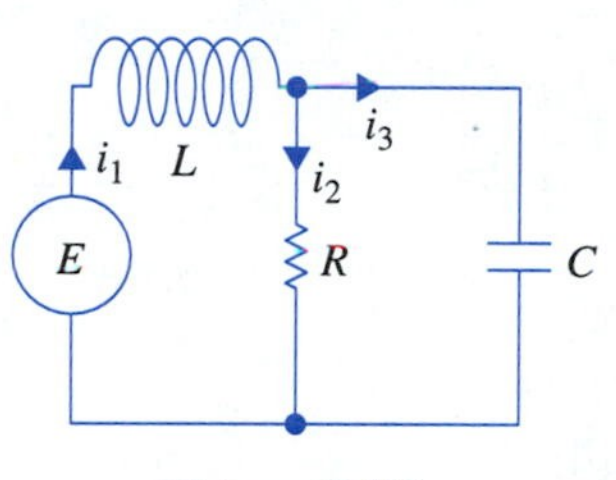

Figure 5.54.

$$\begin{aligned} L\frac{di_1}{dt} + Ri_2 \quad &= E(t) \\ RC\frac{di_2}{dt} + i_2 - i_1 &= 0 \end{aligned} \tag{9}$$

EXAMPLE 2 An Electrical Network

Solve the system (9) under the conditions $E = 60$ volts, $L = 1$ henry, $R = 50$ ohms, $C = 10^{-4}$ farad, and i_1 and i_2 are initially zero.

Solution We must solve

$$\begin{aligned} \frac{di_1}{dt} + 50i_2 \quad &= 60 \\ 50(10^{-4})\frac{di_2}{dt} + i_2 - i_1 &= 0 \end{aligned}$$

subject to $i_1(0) = 0$, $i_2(0) = 0$.

Applying the Laplace transform to each equation of the system, and simplifying, give

$$\begin{aligned} sI_1(s) + \qquad 50I_2(s) &= \frac{60}{s} \\ -200I_1(s) + (s + 200)I_2(s) &= 0 \end{aligned}$$

where $I_1(s) = \mathscr{L}\{i_1(t)\}$ and $I_2(s) = \mathscr{L}\{i_2(t)\}$. Solving the system for I_1 and I_2, and decomposing the results into partial fractions, give

$$\begin{aligned} I_1(s) &= \frac{60s + 12{,}000}{s(s + 100)^2} = \frac{6/5}{s} - \frac{6/5}{s + 100} - \frac{60}{(s + 100)^2} \\ I_2(s) &= \frac{12{,}000}{s(s + 100)^2} = \frac{6/5}{s} - \frac{6/5}{s + 100} - \frac{120}{(s + 100)^2} \end{aligned}$$

Taking the inverse Laplace transform, we find the currents to be

$$i_1(t) = \frac{6}{5} - \frac{6}{5}e^{-100t} - 60te^{-100t}, \quad i_2(t) = \frac{6}{5} - \frac{6}{5}e^{-100t} - 120te^{-100t}$$ ■

Note that both $i_1(t)$ and $i_2(t)$ in Example 2 tend toward the value $E/R = 6/5$ as $t \to \infty$. Furthermore, since the current through the capacitor is $i_3(t) = i_1(t) - i_2(t) = 60te^{-100t}$, we observe that $i_3(t) \to 0$ as $t \to \infty$.

EXERCISES 5.6

Answers to odd-numbered problems begin on page 454.

In Problems 1–12, use the Laplace transform to solve the given system of differential equations.

1. $\frac{dx}{dt} = -x + y$

$\frac{dy}{dt} = 2x$

$x(0) = 0,\ y(0) = 1$

2. $\frac{dx}{dt} = 2y + e^t$

$\frac{dy}{dt} = 8x - t$

$x(0) = 1,\ y(0) = 1$

3. $\frac{dx}{dt} = x - 2y$

$\frac{dy}{dt} = 5x - y$

$x(0) = -1,\ y(0) = 2$

4. $\frac{dx}{dt} + 3x + \frac{dy}{dt} = 1$

$\frac{dx}{dt} - x + \frac{dy}{dt} - y = e^t$

$x(0) = 0,\ y(0) = 0$

5. $2\frac{dx}{dt} + \frac{dy}{dt} - 2x = 1$

$\frac{dx}{dt} + \frac{dy}{dt} - 3x - 3y = 2$

$x(0) = 0,\ y(0) = 0$

6. $\frac{dx}{dt} + x - \frac{dy}{dt} + y = 0$

$\frac{dx}{dt} + \frac{dy}{dt} + 2y = 0$

$x(0) = 0,\ y(0) = 1$

7. $\frac{d^2x}{dt^2} + x - y = 0$

$\frac{d^2y}{dt^2} + y - x = 0$

$x(0) = 0,\ x'(0) = -2$

$y(0) = 0,\ y'(0) = 1$

8. $\frac{d^2x}{dt^2} + \frac{dx}{dt} + \frac{dy}{dt} = 0$

$\frac{d^2y}{dt^2} + \frac{dy}{dt} - 4\frac{dx}{dt} = 0$

$x(0) = 1,\ x'(0) = 0,$

$y(0) = -1,\ y'(0) = 5$

9. $\frac{d^2x}{dt^2} + \frac{d^2y}{dt^2} = t^2$

$\frac{d^2x}{dt^2} - \frac{d^2y}{dt^2} = 4t$

$x(0) = 8,\ x'(0) = 0,$

$y(0) = 0,\ y'(0) = 0$

10. $\frac{dx}{dt} - 4x + \frac{d^3y}{dt^3} = 6\sin t$

$\frac{dx}{dt} + 2x - 2\frac{d^3y}{dt^3} = 0$

$x(0) = 0,\ y(0) = 0,$

$y'(0) = 0,\ y''(0) = 0$

11. $\frac{d^2x}{dt^2} + 3\frac{dy}{dt} + 3y = 0$

$\frac{d^2x}{dt^2} + 3y = te^{-t}$

$x(0) = 0,\ x'(0) = 2,\ y(0) = 0$

12. $\frac{dx}{dt} = 4x - 2y + 2\mathscr{U}(t-1)$

$\frac{dy}{dt} = 3x - y + \mathscr{U}(t-1)$

$x(0) = 0,\ y(0) = \frac{1}{2}$

13. Solve system (1) when $k_1 = 3$, $k_2 = 2$, $m_1 = 1$, $m_2 = 1$ and $x_1(0) = 0$, $x_1'(0) = 1$, $x_2(0) = 1$, $x_2'(0) = 0$.

14. Derive the system of differential equations describing the straight-line vertical motion of the coupled springs shown in Figure 5.55. Use the Laplace transform to solve the system when $k_1 = 1$, $k_2 = 1$, $k_3 = 1$, $m_1 = 1$, $m_2 = 1$, and $x_1(0) = 0$, $x_1'(0) = -1$, $x_2(0) = 0$, $x_2'(0) = 1$.

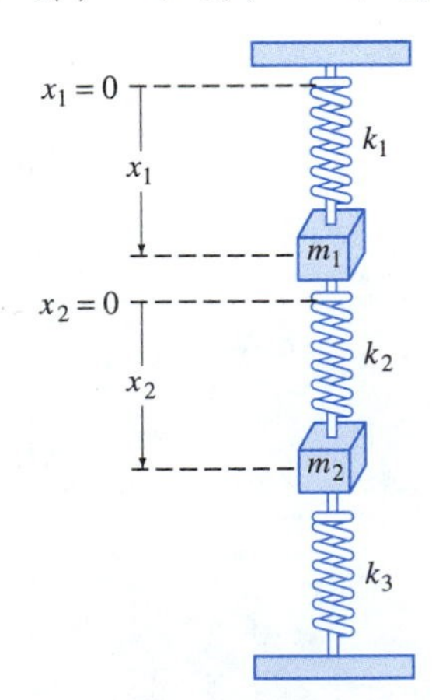

Figure 5.55.

15. **(a)** Show that the system of differential equations for the currents $i_2(t)$ and $i_3(t)$ in the electrical network shown in Figure 5.56 is

$$L_1 \frac{di_2}{dt} + Ri_2 + Ri_3 = E(t)$$

$$L_2 \frac{di_3}{dt} + Ri_2 + Ri_3 = E(t)$$

(b) Solve the system in part (a) if $R = 5$ ohms, $L_1 = 0.01$ henry, $L_2 = 0.0125$ henry, $E = 100$ volts, $i_2(0) = 0$, and $i_3(0) = 0$.

(c) Determine the current $i_1(t)$.

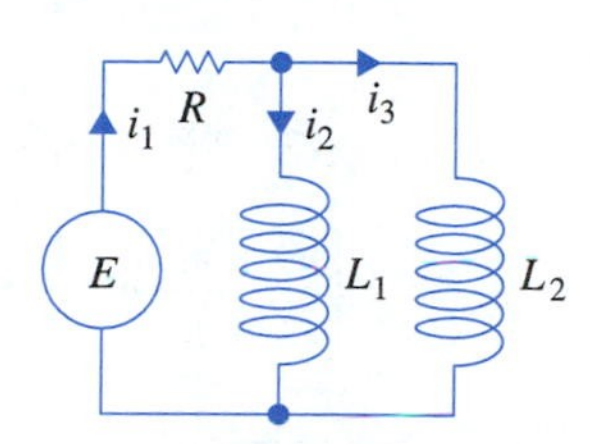

Figure 5.56.

16. **(a)** Show that the system of differential equations for the currents $i_2(t)$ and $i_3(t)$ in the electrical network shown in Figure 5.57 is

$$L \frac{di_2}{dt} + L \frac{di_3}{dt} + R_1 i_2 = E(t)$$

$$-R_1 \frac{di_2}{dt} + R_2 \frac{di_3}{dt} + \frac{1}{C} i_3 = 0$$

(b) Solve the system in part (a) if $R_1 = 10$ ohms, $R_2 = 5$ ohms, $L = 1$ henry, $C = 0.2$ farad,

$$E(t) = \begin{cases} 120, & 0 \le t < 2 \\ 0, & t \ge 2 \end{cases}$$

$i_2(0) = 0$, and $i_3(0) = 0$.

(c) Determine the current $i_1(t)$.

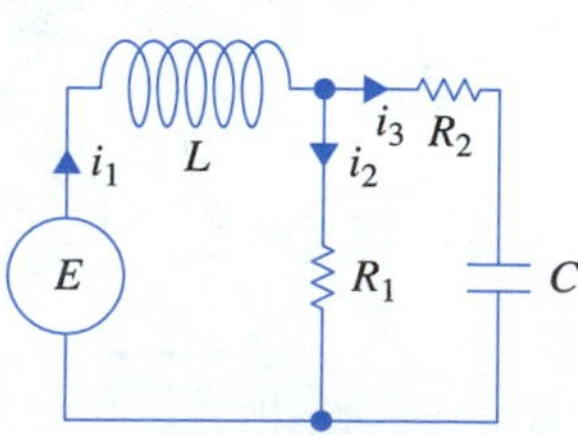

Figure 5.57.

17. Solve the system given in (8) when $R_1 = 6$ ohms, $R_2 = 5$ ohms, $L_1 = 1$ henry, $L_2 = 1$ henry, $E(t) = 50 \sin t$ volts.

18. Derive the system of equations in (9).

19. Solve (9) when $E = 60$ volts, $L = \frac{1}{2}$ henry, $R = 50$ ohms, $C = 10^{-4}$ farad, $i_1(0) = 0$, $i_2(0) = 0$.

20. Solve (9) when $E = 60$ volts, $L = 2$ henrys, $R = 50$ ohms, $C = 10^{-4}$ farad, $i_1(0) = 0$, $i_2(0) = 0$.

21. **(a)** Show that the system of differential equations for the charge on the capacitor $q(t)$ and the current $i_3(t)$ in the electrical network shown in Figure 5.58 is

$$R_1 \frac{dq}{dt} + \frac{1}{C} q + R_1 i_3 = E(t)$$

$$L \frac{di_3}{dt} + R_2 i_3 - \frac{1}{C} q = 0$$

(b) Find the charge on the capacitor when $L = 1$ henry, $R_1 = 1$ ohm, $R_2 = 1$ ohm, $C = 1$ farad,

$$E(t) = \begin{cases} 0, & 0 < t < 1 \\ 50e^{-t} & t \ge 1 \end{cases}$$

$i_3(0) = 0$, and $q(0) = 0$.

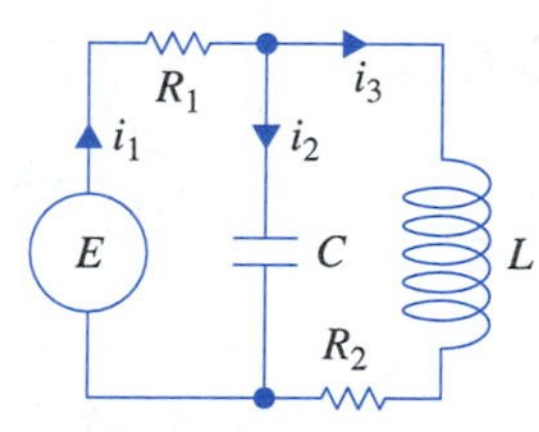

Figure 5.58.

Discussion Problems

22. **(a)** Use a graphing utility to reproduce the graphs of the solution in (4), but use a larger interval on the t-axis than is shown in Figure 5.52.

(b) Discuss: Which mass has extreme displacements of greater magnitude? Why?

(c) Discuss: Certainly the graphs of $x_1(t)$ and $x_2(t)$ are oscillatory, but are they periodic? Back your viewpoint with sound mathematics.

Chapter 5 in Review

Answers to this review are posted at our Web site diffeq.brookscole.com/books.html#zill 98

In Problems 1–4, answer true or false.

1. If f is not piecewise continuous on $[0, \infty)$, then $\mathscr{L}\{f(t)\}$ will not exist.
2. The function $f(t) = (e^t)^{1000}$ is not of exponential order.
3. If f is bounded for $t \geq 0$, that is, $-B \leq f(t) \leq B$ for some $B \geq 0$, then f is of exponential order.
4. If $\mathscr{L}\{f(t)\} = F(s)$ and $\mathscr{L}\{g(t)\} = G(s)$, then $\mathscr{L}^{-1}\{F(s)G(s)\} = f(t)g(t)$.

In Problems 5–12, fill in the blanks.

5. $\mathscr{L}\{e^{-5t} \sin 2t\} =$ ________.
6. $\mathscr{L}\{\sin 2t\, \mathscr{U}(t - \pi)\} =$ ________.
7. $\mathscr{L}^{-1}\left\{\dfrac{1}{3s - 1}\right\} =$ ________.
8. $\mathscr{L}^{-1}\left\{\dfrac{s + \pi}{s^2 + \pi^2} e^{-s}\right\} =$ ________.
9. If $\mathscr{L}\{f(t)\} = F(s)$, then $\mathscr{L}\{te^{8t}f(t)\} =$ ________.
10. If $\mathscr{L}\{f(t)\} = F(s)$ and $k > 0$, then $\mathscr{L}\{e^{at} f(t - k)\mathscr{U}(t - k)\} =$ ________.
11. $\mathscr{L}\{e^{5t}\}$ exists for $s >$ ________.
12. $\mathscr{L}\left\{\int_0^t e^{a\tau} f(\tau)\, d\tau\right\} =$ ________, whereas $\mathscr{L}\left\{e^{at} \int_0^t f(\tau)\, d\tau\right\} =$ ________.

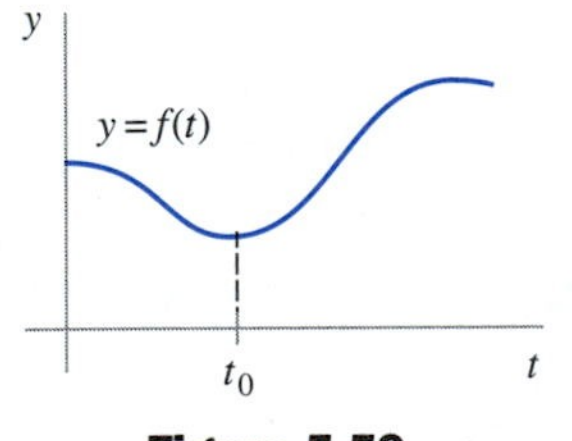

Figure 5.59.

In Problems 13–16, use the unit step function to write down an equation for each graph in terms of the function $y = f(t)$ whose graph is given in Figure 5.59.

13.

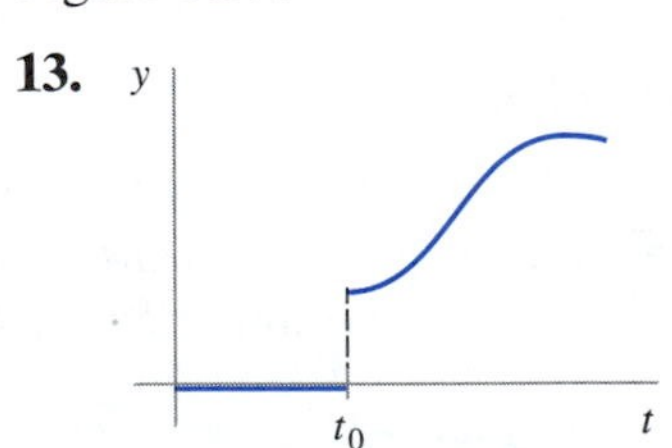

Figure 5.60.

14.

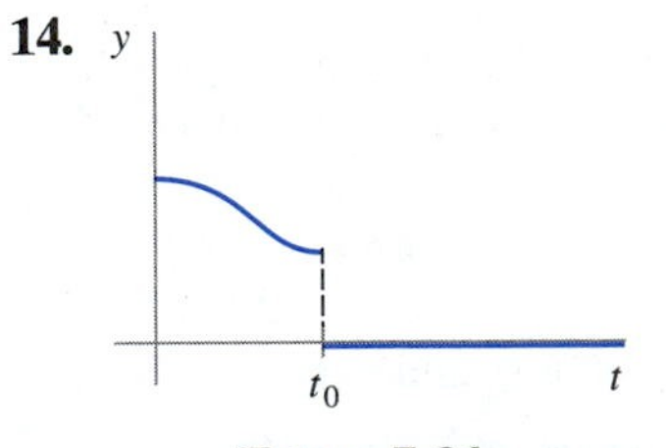

Figure 5.61.

15.

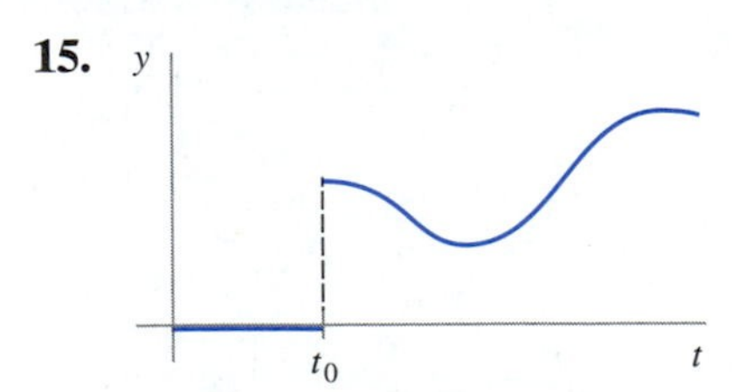

Figure 5.62.

16.

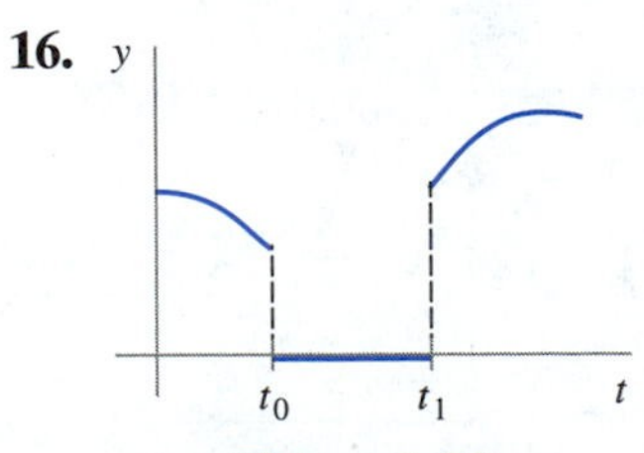

Figure 5.63.

17. Let $\mathscr{L}\{e^{-t^2}\} = F(s)$. Use the Laplace transform to solve

$$y'' + y = e^{-t^2}, \quad y(0) = 0, y'(0) = 0$$

18. Use the Laplace transform to solve

$$y'' + 2y' + 10y = 0, \quad y(0) = 3, y'(0) = -5$$

19. The current $i(t)$ in an RC-series circuit can be determined from the integral equation

$$Ri + \frac{1}{C}\int_0^t i(\tau)\, d\tau = E(t)$$

where $E(t)$ is the impressed voltage. Determine $i(t)$ when $R = 10$ ohms, $C = 0.5$ farad, and $E(t) = 2(t^2 + t)$.

20. A uniform cantilever beam of length L is embedded at its left end ($x = 0$) and free at its right end. Find the deflection $y(x)$ if the load per unit length is given by

$$w(x) = \frac{2w_0}{L}\left[\frac{L}{2} - x + \left(x - \frac{L}{2}\right)\mathscr{U}\left(x - \frac{L}{2}\right)\right]$$

Projects and Computer Lab Experiments for Chapter 5

Section 5.1

1. Using Complex Numbers

(a) Look up the definition of equality of two complex numbers.

(b) Assume (or prove it at your instructor's direction) that the result in part (c) of Theorem 5.1 holds for the pure imaginary number $a = ikt$, where $i^2 = -1$ and k is real. Show how $\mathscr{L}\{e^{ikt}\}$ can be used to find $\mathscr{L}\{\cos kt\}$ and $\mathscr{L}\{\sin kt\}$.

2. Laplace Transform Doesn't Exist Show that the function $f(t) = 1/t^2$ does not possess a Laplace transform.

3. Transform of the Logarithm Because $f(t) = \ln t$ has a bad discontinuity at $t = 0$, it might be assumed that $\mathscr{L}\{\ln t\}$ does not exist; however, this is incorrect. The point of this project then is to find the Laplace transform of $f(t) = \ln t$, $t > 0$. Because of the discontinuity at $t = 0$, we define

$$\mathscr{L}\{\ln t\} = \lim_{\substack{b\to\infty \\ a\to 0}} \int_a^b e^{-st} \ln t\, dt$$

(a) Find the definition of **Euler's constant** $\gamma = 0.57721566\ldots$. To solve part (b), you will also need to find a representation of γ as an improper integral.

(b) Show that

$$\mathcal{L}\{\ln t\} = -\frac{\gamma}{s} - \frac{\ln s}{s}, \quad s > 0$$

Section 5.3

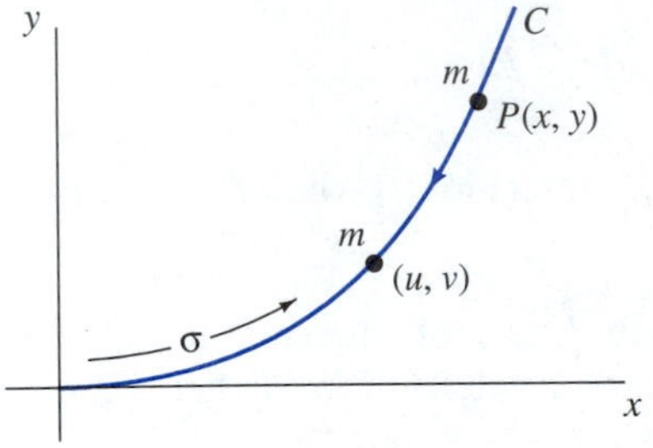

Figure 5.64.

4. Tautochrone Consider a frictionless wire bent in the form of a smooth curve C. As shown in Figure 5.64, a bead with mass m starts from rest at the point $P(x, y)$ and slides down the wire under the influence of its own weight. We want to find the shape of the wire so that the time of descent to its lowest end (the origin in the figure) is the same constant T, regardless of the position of the starting point P. Let σ denote arc length along C.

(a) By considering the sums of the potential and kinetic energies at (x, y) and (u, v), show how the integral equation

$$T = \frac{1}{\sqrt{2g}} \int_0^y \frac{\sigma'(v)}{\sqrt{y - v}}\, dv$$

arises in this problem.

(b) Show how the Laplace transform is useful in the eventual solution of the problem. The curve C is called a **tautochrone** (from Greek: *tautos* meaning *same* and *chronos* meaning *time*.)

Section 5.5

5. Sliding Friction Consider a spring/mass system in which the mass m slides over a dry horizontal surface whose coefficient of kinetic friction is μ. The frictional force of constant magnitude $f_k = \mu mg$ acts opposite to the direction of motion. Suppose that the mass is released from rest from a point $x(0) = a > 0$ and that there are no other external forces. Then the differential equations describing the motion of the mass are

$$x'' + \omega^2 x = F, \qquad 0 < x < T/2$$

$$x'' + \omega^2 x = -F, \quad T/2 < x < 3T/2$$

$$x'' + \omega^2 x = F, \quad 3T/2 < x < 2T$$

and so on, where $\omega^2 = k/m$, $F = f_k/m = \mu g$, and $T = 2\pi/\omega$. (See Problem 26, *Projects and Computer Lab Experiments for Chapter 3*.)

(a) Explain why, in general, the initial displacement must satisfy $\omega^2|a| > F$.

(b) Explain why the interval $-F/\omega^2 \le x \le F/\omega^2$ is appropriately called the "dead zone" of the system.

(c) Use the Laplace transform and the concept of the meander function to solve for the displacement $x(t)$ for $t \ge 0$.

(d) Show that each successive oscillation is $2F/\omega^2$ shorter than the preceding one.

(e) Predict the long-term behavior of the system.

6. Pure Resonance In Section 3.6, we saw that an undamped spring/mass system, whose model is $x'' + \omega^2 x = f(t)$, $x(0) = 0$, $x'(0) = 0$, is in a state of pure resonance when the driving force is $f(t) = F_0 \sin \omega t$ (or $f(t) = F_0 \cos \omega t$). That is, f is a continuous function with period $2\pi/\omega$, which is equal to the period of free vibrations. Assume for simplicity that $\omega = 1$ so that the period of free vibrations is 2π. Investigate whether pure resonance exists when f is a discontinuous 2π-periodic function defined by the square wave in Problem 4 in Exercises 5.4.

Section 5.6

7. System of DEs Use the Laplace transform to solve the system of differential equations

$$\frac{dx}{dt} + x + \frac{d^2y}{dt^2} = e^t$$

$$\frac{dy}{dt} + x = -e^t$$

$$x(0) = -1,\ y(0) = 1,\ y'(0) = 0$$

Examine your solution in careful detail. State, and try to explain, any peculiarities.

8. Double Pendulum As shown in Figure 5.65(a), a double pendulum oscillates in a vertical plane under the influence of gravity. For small displacements $\theta_1(t)$ and $\theta_2(t)$, it can be shown that the system of differential equations describing the motion are

$$(m_1 + m_2)l_1^2\theta_1'' + m_2 l_1 l_2 \theta_2'' + (m_1 + m_2)l_1 g\theta_1 = 0$$

$$m_2 l_2^2 \theta_2'' + m_2 l_1 l_2 \theta_1'' + m_2 l_2 g\theta_2 = 0$$

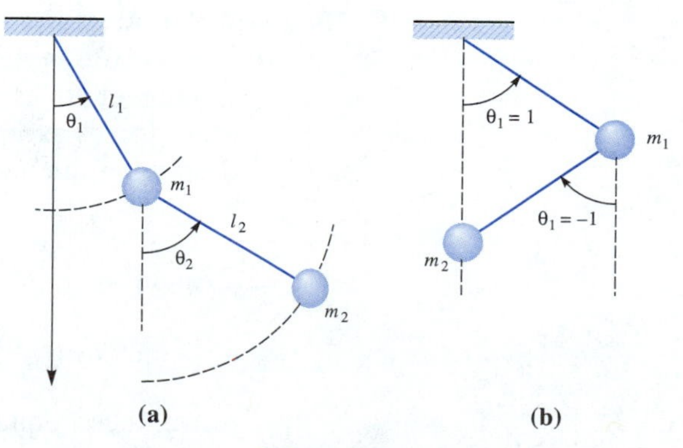

Figure 5.65.

As indicated in Figure 5.65(a), θ_1 is measured (in radians) from a vertical line extending downward from the pivot of the system and θ_2 is measured from a vertical line extending downward from the center of mass m_1. The positive direction is to the right, and the negative direction is to the left.

(a) Use the Laplace transform to solve the system when $m_1 = 3$, $m_2 = 1$, $l_1 = l_2 = 16$, $\theta_1(0) = 1$, $\theta_2(0) = -1$, $\theta_1'(0) = 0$, and $\theta_2'(0) = 0$.

(b) Use a graphing utility to obtain the graphs of $\theta_1(t)$ and $\theta_2(t)$ in the $t\theta$-plane. Which mass has extreme displacements of greater magnitude? Use the graphs to estimate the first time that each mass passes through its equilibrium position. Discuss whether the motion of the pendulums is periodic.

(c) As parametric equations, graph $\theta_1(t)$ and $\theta_2(t)$ in the $\theta_1\theta_2$-plane. The curve defined by these parametric equations is called a Lissajous curve.

(d) The position of the masses at $t = 0$ is given in Figure 5.65(b). Note that we have used 1 radian $= 57.3°$. Use a calculator or a table application in a CAS to construct a table of values of the angles θ_1 and θ_2 for $t = 1, 2, \ldots, 10$ s. Then carefully plot the positions of the two masses at these times.

(e) Use a CAS to find the first time that $\theta_1(t) = \theta_2(t)$ and compute the corresponding angular value. Plot the positions of the two masses at these times.

9. Movie of the Double Pendulum Some computer algebra systems have an animation capability. That is, it is possible to make a "movie" by selecting a sequence of graphic cells (the frames of the movie) and displaying them in rapid succession. Check the CAS you have on hand to see whether it has this capability. If so, use the solution of Problem 8 to plot the positions of each mass for a chosen sequence of time values. Utilize the CAS also to draw appropriate lines to simulate the pendulum rods as in Figure 5.65(b). Make a "movie" of the motion of the double pendulum.

10. Projectile Motion

(a) A projectile shot from a gun has weight $w = mg$ and initial velocity $\mathbf{v}_0$ tangent to its path of motion. Ignoring air resistance and all other forces except its weight, the system of differential equations describing the motion of the projectile is

$$m\frac{d^2x}{dt^2} = 0$$

$$m\frac{d^2y}{dt^2} = -mg$$

Solve these equations subject to $x(0) = 0$, $y(0) = 0$, $x'(0) = v_0 \cos\theta$, $y'(0) = v_0 \sin\theta$, where $v_0 = |\mathbf{v}_0|$ is constant and θ is the constant angle of elevation. See Figure 5.66.

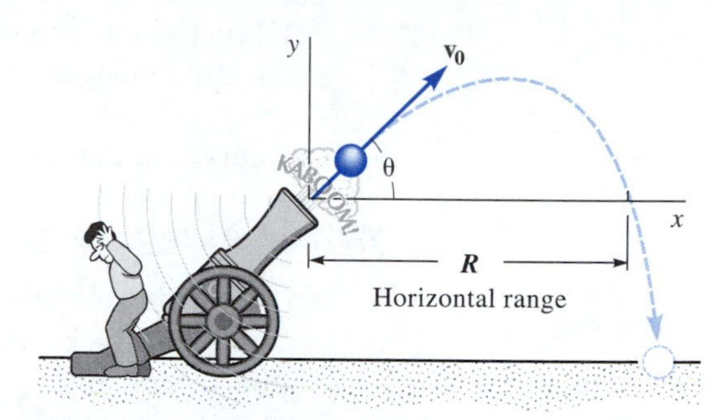

Figure 5.66.

(b) Use $x(t)$ obtained in part (a) to eliminate the variable t from $y(t)$. Use the expression y to show that the horizontal range R of the projectile is given by

$$R = \frac{v_0^2}{g} \sin 2\theta$$

From this formula, we see that R is a maximum when $\sin 2\theta = 1$ or $\theta = \pi/4$. A range less than the maximum—a submaximum range—can be attained by firing the gun at either of two complementary angles. For example, the projectile has the same range for $\theta = \pi/6$ (30°) as it does when $\theta = \pi/3$ (60°), the only difference being that the smaller angle results in a low trajectory whereas the larger gives a high trajectory.

(c) Now suppose that air resistance is a retarding force tangent to the path but acts opposite to the motion. Since air resistance is then a multiple of the velocity, the system of differential equations describing the motion of the projectile is

$$m \frac{d^2x}{dt^2} = -\beta \frac{dx}{dt}$$

$$m \frac{d^2y}{dt^2} = -mg - \beta \frac{dy}{dt}$$

where $\beta > 0$. Use the Laplace transform to solve these equations subject to $x(0) = 0$, $y(0) = 0$, $x'(0) = v_0 \cos \theta$, $y'(0) = v_0 \sin \theta$, where $v_0 = |\mathbf{v}_0|$ and θ are constant.

(d) Suppose $m = \frac{1}{4}$ slug, $g = 32$ ft/s^2, $\beta = 0.02$, $\theta = 38°$, and $v_0 = 300$ ft/s. Use a CAS to find the time when the projectile hits the ground and then compute the corresponding horizontal range.

(e) Use the equations $x(t)$ and $y(t)$ in part (a) with $\theta = 38°$, and $v_0 = 300$ ft/s and the analogous equations in part (d) as parametric equations to plot the path of the projectile—the ballistic curve—on the same coordinate system. Give a time interval $0 \le t \le T$, where T is the time the projectile hits the ground.

(f) For the equation in part (a), also plot the ballistic curve for the complementary angle $\theta = 52°$. In part (d), show that the complementary angle does not give the same range. Also plot this ballistic curve on the same coordinate system. See Problem 12.

Writing Assignments

11. No One Said Life Was Fair Write a report on the life, times, discoveries, and hard-luck career of Oliver Heaviside. Stress his contributions to the theory and applications of the Laplace transform. You might start with "Heaviside and the Operational Calculus" by J. L. B. Cooper, *Mathematics Gazette* 36 (1952) 5–19.

12. There Really Are Two Angles In Problem 10, we saw that in the absence of air resistance a projectile will attain the same submaximum range at two angles of elevation in the interval $[0, \pi/2]$. This is also true when air resistance is taken to be proportional to the velocity of the projectile, but as part (f) of Problem 10 shows, the angles are not complements. Write or give an oral report on "Tartaglia's Inverse Problem in a Resistive Medium" by C. W. Groetsch, *The American Mathematical Monthly* 103 (1996), 546–551.

Chapter 6

Series Solutions

6.1 Solutions About Ordinary Points

The Cauchy-Euler equation that we discussed in Section 3.5 notwithstanding, *most* linear higher-order differential equations with variable coefficients cannot be solved in terms of elementary functions. A standard technique for solving such DEs is to try to find a solution in the form of an *infinite series*. In this section we are going to find power series solutions. The center of the series x_0 will be a special point called an *ordinary point* of the DE.

Note Throughout this chapter we are going to use the traditional symbol x to denote the independent variable.

Power Series Solutions Suppose the linear second-order differential equation

$$a_2(x)y'' + a_1(x)y' + a_0(x)y = 0 \tag{1}$$

is put into the standard form

$$y'' + P(x)y' + Q(x)y = 0 \tag{2}$$

by dividing by the leading coefficient $a_2(x)$. Our ability to find a solution of (1) in the form of a **power series**,

$$y = c_0 + c_1(x - x_0) + c_2(x - x_0)^2 + \cdots$$

depends in an essential way on how the functions P and Q, defined from the standard form (2), behave near the point x_0. We begin with an important definition.

Definition 6.1 Ordinary and Singular Points

A point x_0 is said to be an **ordinary point** of the differential equation (1) if P and Q in the standard form (2) are analytic at x_0. A point that is not an ordinary point is said to be a **singular point** of the equation.

A function f is said to be analytic at a point x_0 if it can be expanded in a power series, centered at x_0, which converges to $f(x)$ for every x in an open interval containing x_0. Such an expansion is called a Taylor series for f. For example, every finite value of x is an ordinary point of the differential equation $y'' + e^x y' + (\sin x)y = 0$. In particular, we see that 0 is an ordinary point of the equation since the coefficients $P(x) = e^x$ and $Q(x) = \sin x$ are analytic at this point. Recall from calculus that the

Taylor series

$$e^x = 1 + \frac{x}{1!} + \frac{x^2}{2!} + \cdots \quad \text{and} \quad \sin x = x - \frac{x^3}{3!} + \frac{x^5}{5!} + \cdots$$

converge for all x. A Taylor series centered at 0 is also called a Maclaurin series.

In the example that follows, we use the fact that the function $Q(x) = (\sin x)/x$ has a removable discontinuity at $x = 0$.

EXAMPLE 1 Ordinary/Singular Points

(a) The differential equation $xy'' + (\sin x)y = 0$ has an ordinary point at $x = 0$ since $Q(x) = (\sin x)/x$ possesses the power series expansion

$$Q(x) = 1 - \frac{x^2}{3!} + \frac{x^4}{5!} - \frac{x^6}{7!} + \cdots$$

that converges for all finite values of x.

(b) The differential equation $y'' + (\ln x)y = 0$ has a singular point at $x = 0$ because $Q(x) = \ln x$ possesses no power series in x. ■

Polynomial Coefficients Primarily we shall be concerned with the case when (1) has *polynomial* coefficients. As a consequence of Definition 6.1, we note that when $a_2(x)$, $a_1(x)$, and $a_0(x)$ are polynomials with *no common factors*, a point $x = x_0$ is

(i) an ordinary point if $a_2(x_0) \neq 0$, or **(ii)** a singular point if $a_2(x_0) = 0$

EXAMPLE 2 Singular/Ordinary Points

(a) The singular points of the equation $(x^2 - 1)y'' + 2xy' + 6y = 0$ are the solutions of $x^2 - 1 = 0$ or $x = \pm 1$. All other finite values of x are ordinary points.

(b) Singular points need not be real numbers. The equation $(x^2 + 1)y'' + xy' - y = 0$ has singular points at the solutions of $x^2 + 1 = 0$, namely, $x = \pm i$. All other finite values of x, real or complex, are ordinary points.

(c) The Cauchy-Euler equation $ax^2y'' + bxy' + cy = 0$, where a, b, and c are constants, has a singular point at $x = 0$. All other finite values of x, real or complex, are ordinary points. ■

For our purpose, ordinary points and singular points will always be finite points. It is possible for a differential equation to have, say, a singular point at infinity.

We state the following theorem about the existence of power series solutions without proof.

Theorem 6.1 Existence of Power Series Solutions

If $x = x_0$ is an ordinary point of the differential equation (1), we can always find two linearly independent solutions in the form of power series centered at x_0:

$$y = \sum_{n=0}^{\infty} c_n(x - x_0)^n \tag{3}$$

A series solution converges at least for $|x - x_0| < R$, where R is the distance from x_0 to the closest singular point (real or complex).

A solution of a differential equation of the form given in (3) is said to be a solution *about* the ordinary point x_0. The distance R given in Theorem 6.1 is the minimum value for the radius of convergence. A differential equation could have a finite singular point and yet a solution could be valid for all x; for example, the differential equation may possess a polynomial solution.

To solve a linear second-order equation such as (1), we find two sets of coefficients c_n so that we have two distinct power series $y_1(x)$ and $y_2(x)$, both expanded about the same ordinary point x_0. The procedure used to solve a second-order equation is to assume a solution $y = \sum_{n=0}^{\infty} c_n(x - x_0)^n$ and then determine the c_n. The general solution of the differential equation is $y = C_1y_1(x) + C_2y_2(x)$; in fact, it can be shown that $C_1 = c_0$ and $C_2 = c_1$, where c_0 and c_1 are arbitrary.

Note For the sake of simplicity, we assume an ordinary point is always located at $x = 0$, since, if not, the substitution $t = x - x_0$ translates the value $x = x_0$ to $t = 0$.

EXAMPLE 3 Power Series Solution About an Ordinary Point

Solve $y'' + xy = 0$.

Solution We see that $x = 0$ is an ordinary point of the equation. Since there are no finite singular points, Theorem 6.1 guarantees two power series solutions, centered at 0, convergent for $|x| < \infty$. Substituting

$$y = \sum_{n=0}^{\infty} c_n x^n \quad \text{and} \quad y'' = \sum_{n=2}^{\infty} n(n-1)c_n x^{n-2}$$

into the differential equation gives

$$
\begin{aligned}
y'' + xy &= \sum_{n=2}^{\infty} n(n-1)c_n x^{n-2} + \sum_{n=0}^{\infty} c_n x^{n+1} \\
&= 2 \cdot 1 c_2 x^0 + \underbrace{\sum_{n=3}^{\infty} n(n-1)c_n x^{n-2} + \sum_{n=0}^{\infty} c_n x^{n+1}}_{\text{both series start with } x \text{ to the first power}}
\end{aligned}
$$

Letting $k = n - 2$ in the first series and $k = n + 1$ in the second, we have

$$
\begin{aligned}
y'' + xy &= 2c_2 + \sum_{k=1}^{\infty} (k+2)(k+1)c_{k+2}x^k + \sum_{k=1}^{\infty} c_{k-1}x^k \\
&= 2c_2 + \sum_{k=1}^{\infty} [(k+2)(k+1)c_{k+2} + c_{k-1}]x^k = 0
\end{aligned}
$$

We must then have $2c_2 = 0$, which implies $c_2 = 0$, and

$$(k + 2)(k + 1)c_{k+2} + c_{k-1} = 0$$

The last expression, called a **recurrence relation**, is the same as

$$c_{k+2} = -\frac{c_{k-1}}{(k+2)(k+1)}, \qquad k = 1, 2, 3, \ldots$$

Iteration gives

$$
\begin{aligned}
c_3 &= -\frac{c_0}{3 \cdot 2} \\
c_4 &= -\frac{c_1}{4 \cdot 3} \\
c_5 &= -\frac{c_2}{5 \cdot 4} = 0 && \leftarrow c_2 \text{ is zero} \\
c_6 &= -\frac{c_3}{6 \cdot 5} = \frac{1}{6 \cdot 5 \cdot 3 \cdot 2} c_0 \\
c_7 &= -\frac{c_4}{7 \cdot 6} = \frac{1}{7 \cdot 6 \cdot 4 \cdot 3} c_1 \\
c_8 &= -\frac{c_5}{8 \cdot 7} = 0 && \leftarrow c_5 \text{ is zero} \\
c_9 &= -\frac{c_6}{9 \cdot 8} = -\frac{1}{9 \cdot 8 \cdot 6 \cdot 5 \cdot 3 \cdot 2} c_0 \\
c_{10} &= -\frac{c_7}{10 \cdot 9} = -\frac{1}{10 \cdot 9 \cdot 7 \cdot 6 \cdot 4 \cdot 3} c_1 \\
c_{11} &= -\frac{c_8}{11 \cdot 10} = 0 && \leftarrow c_8 \text{ is zero}
\end{aligned}
$$

and so on. It should be apparent that both c_0 and c_1 are arbitrary. Now

$$\begin{aligned} y &= c_0 + c_1x + c_2x^2 + c_3x^3 + c_4x^4 + c_5x^5 + c_6x^6 + c_7x^7 + c_8x^8 \\ &\quad + c_9x^9 + c_{10}x^{10} + c_{11}x^{11} + \cdots \\ &= c_0 + c_1x + 0 - \frac{1}{3\cdot 2}c_0x^3 - \frac{1}{4\cdot 3}c_1x^4 + 0 + \frac{1}{6\cdot 5\cdot 3\cdot 2}c_0x^6 \\ &\quad + \frac{1}{7\cdot 6\cdot 4\cdot 3}c_1x^7 + 0 - \frac{1}{9\cdot 8\cdot 6\cdot 5\cdot 3\cdot 2}c_0x^9 \\ &\quad - \frac{1}{10\cdot 9\cdot 7\cdot 6\cdot 4\cdot 3}c_1x^{10} + 0 + \cdots \\ &= c_0\left[1 - \frac{1}{3\cdot 2}x^3 + \frac{1}{6\cdot 5\cdot 3\cdot 2}x^6 - \frac{1}{9\cdot 8\cdot 6\cdot 5\cdot 3\cdot 2}x^9 + \cdots\right] \\ &\quad + c_1\left[x - \frac{1}{4\cdot 3}x^4 + \frac{1}{7\cdot 6\cdot 4\cdot 3}x^7 - \frac{1}{10\cdot 9\cdot 7\cdot 6\cdot 4\cdot 3}x^{10} + \cdots\right] \end{aligned}$$ ■

Although the pattern of the coefficients in Example 3 should be clear, it is sometimes useful to write the solutions in terms of summation notation. By using the properties of the factorial, we can write

$$y_1(x) = c_0\left[1 + \sum_{k=1}^{\infty} \frac{(-1)^k[1\cdot 4\cdot 7\cdots(3k-2)]}{(3k)!}x^{3k}\right]$$

and $$y_2(x) = c_1\left[x + \sum_{k=1}^{\infty} \frac{(-1)^k[2\cdot 5\cdot 8\cdots(3k-1)]}{(3k+1)!}x^{3k+1}\right]$$

In this form, the ratio test can be used to show that each series converges for $|x| < \infty$.

The differential equation in Example 3 is called **Airy's equation** and is encountered in the study of diffraction of light, diffraction of radio waves around the surface of the Earth, aerodynamics, and the deflection of a uniform thin vertical column that bends under its own weight. Other common forms of Airy's equation are $y'' - xy = 0$ and $y'' + \alpha^2xy = 0$. See Exercises 6.3 for applications of the last equation.

■ EXAMPLE 4 Power Series Solution About an Ordinary Point

Solve $(x^2 + 1)y'' + xy' - y = 0$.

Solution The assumption $y = \sum_{n=0}^{\infty} c_n x^n$ leads to

$$(x^2 + 1) \sum_{n=2}^{\infty} n(n-1)c_n x^{n-2} + x \sum_{n=1}^{\infty} nc_n x^{n-1} - \sum_{n=0}^{\infty} c_n x^n$$

$$= \sum_{n=2}^{\infty} n(n-1)c_n x^n + \sum_{n=2}^{\infty} n(n-1)c_n x^{n-2} + \sum_{n=1}^{\infty} nc_n x^n - \sum_{n=0}^{\infty} c_n x^n$$

$$= 2c_2 x^0 - c_0 x^0 + 6c_3 x + c_1 x - c_1 x + \underbrace{\sum_{n=2}^{\infty} n(n-1)c_n x^n}_{k=n}$$

$$+ \underbrace{\sum_{n=4}^{\infty} n(n-1)c_n x^{n-2}}_{k=n-2} + \underbrace{\sum_{n=2}^{\infty} nc_n x^n}_{k=n} - \underbrace{\sum_{n=2}^{\infty} c_n x^n}_{k=n}$$

$$= 2c_2 - c_0 + 6c_3 x + \sum_{k=2}^{\infty} [k(k-1)c_k + (k+2)(k+1)c_{k+2} + kc_k - c_k]x^k$$

$$= 2c_2 - c_0 + 6c_3 x + \sum_{k=2}^{\infty} [(k+1)(k-1)c_k + (k+2)(k+1)c_{k+2}]x^k = 0$$

Thus

$$2c_2 - c_0 = 0, \quad c_3 = 0$$

$$(k+1)(k-1)c_k + (k+2)(k+1)c_{k+2} = 0$$

or

$$c_2 = \frac{1}{2}c_0, \quad c_3 = 0,$$

and

$$c_{k+2} = \frac{1-k}{k+2}c_k, \quad k = 2, 3, 4, \ldots$$

Iteration of the last formula gives

$$c_4 = -\frac{1}{4}c_2 = -\frac{1}{2 \cdot 4}c_0 = -\frac{1}{2^2 2!}c_0$$

$$c_5 = -\frac{2}{5}c_3 = 0 \qquad \leftarrow c_3 \text{ is zero}$$

$$c_6 = -\frac{3}{6}c_4 = \frac{3}{2 \cdot 4 \cdot 6}c_0 = \frac{1 \cdot 3}{2^3 3!}c_0$$

$$c_7 = -\frac{4}{7}c_5 = 0$$

$$c_8 = -\frac{5}{8}c_6 = -\frac{3 \cdot 5}{2 \cdot 4 \cdot 6 \cdot 8}c_0 = -\frac{1 \cdot 3 \cdot 5}{2^4 4!}c_0$$

$$c_9 = -\frac{6}{9}c_7 = 0$$

$$c_{10} = -\frac{7}{10}c_8 = \frac{3 \cdot 5 \cdot 7}{2 \cdot 4 \cdot 6 \cdot 8 \cdot 10}c_0 = \frac{1 \cdot 3 \cdot 5 \cdot 7}{2^5 5!}c_0$$

and so on. Therefore

$$y = c_0 + c_1x + c_2x^2 + c_3x^3 + c_4x^4 + c_5x^5 + c_6x^6 + c_7x^7 + c_8x^8 + c_9x^9 + c_{10}x^{10} + \cdots$$

$$= c_1x + c_0\left[1 + \frac{1}{2}x^2 - \frac{1}{2^2 2!}x^4 + \frac{1 \cdot 3}{2^3 3!}x^6 - \frac{1 \cdot 3 \cdot 5}{2^4 4!}x^8 + \frac{1 \cdot 3 \cdot 5 \cdot 7}{2^5 5!}x^{10} - \cdots\right]$$

The solutions are the polynomial $y_2(x) = c_1x$ and the series

$$y_1(x) = c_0\left[1 + \frac{1}{2}x^2 + \sum_{n=2}^{\infty}(-1)^{n-1}\frac{1 \cdot 3 \cdot 5 \cdots (2n-3)}{2^n n!}x^{2n}\right], \quad |x| < 1$$ ■

Note In the complex plane, the distance between two complex numbers $a + bi$ and $c + di$ is simply the distance between the points (a, b) and (c, d), that is, $\sqrt{(a-c)^2 + (b-d)^2}$. Thus in Example 4, since the only singular points of the differential equation are the complex numbers i and $-i$, and because the distance between 0 and $\pm i$ is 1, we are guaranteed that the power series solution $y_1(x)$ converges at least for $|x| < 1$. Of course, the polynomial solution $y_2(x) = x$ is valid for $|x| < \infty$.

Nonpolynomial Coefficients The next example illustrates how to find a power series solution about an ordinary point of a differential equation when its coefficients are not polynomials. In this example, we see an application of multiplication of two power series that is discussed in Example 2 of Appendix III.

■ EXAMPLE 5 DE with Nonpolynomial Coefficients

Solve $y'' + (\cos x)y = 0$.

Solution Since $\cos x = 1 - \frac{x^2}{2!} + \frac{x^4}{4!} - \frac{x^6}{6!} + \cdots$, it is seen that $x = 0$ is an ordinary point. Thus the assumption $y = \sum_{n=0}^{\infty} c_n x^n$ leads to

$$y'' + (\cos x)y = \sum_{n=2}^{\infty} n(n-1)c_n x^{n-2} + \left(1 - \frac{x^2}{2!} + \frac{x^4}{4!} - \cdots\right)\sum_{n=0}^{\infty} c_n x^n$$

$$= (2c_2 + 6c_3x + 12c_4x^2 + 20c_5x^3 + \cdots)$$

$$+ \left(1 - \frac{x^2}{2} + \frac{x^4}{24} - \cdots\right)(c_0 + c_1x + c_2x^2 + c_3x^3 + \cdots)$$

$$= 2c_2 + c_0 + (6c_3 + c_1)x + \left(12c_4 + c_2 - \frac{1}{2}c_0\right)x^2 + \left(20c_5 + c_3 - \frac{1}{2}c_1\right)x^3 + \cdots$$

Since the last line is to be identically zero, we must have

$$2c_2 + c_0 = 0$$
$$6c_3 + c_1 = 0$$
$$12c_4 + c_2 - \tfrac{1}{2}c_0 = 0$$
$$20c_5 + c_3 - \tfrac{1}{2}c_1 = 0$$

and so on. Since c_0 and c_1 are arbitrary, we find

$$y_1(x) = c_0\left[1 - \frac{1}{2}x^2 + \frac{1}{12}x^4 - \cdots\right] \quad \text{and} \quad y_2(x) = c_1\left[x - \frac{1}{6}x^3 + \frac{1}{30}x^5 - \cdots\right]$$

Since the differential equation has no singular points, both series converge for all finite values of x. ■

Remarks

In several of the problems in Exercises 6.1, the power series method leads to recurrence relations that are more complicated than the ones in Examples 3 and 4. In Problem 8, for example, substituting $y = \sum_{n=0}^{\infty} c_n x^n$ into the equation yields

$$c_{k+2} = \frac{c_k + c_{k-1}}{(k+1)(k+2)}, \quad k = 1, 2, 3, \ldots$$

To simplify the iteration of this *three-term recurrence relation*, we can first choose $c_0 \neq 0$, $c_1 = 0$ to obtain one solution and then choose $c_0 = 0$, $c_1 \neq 0$ to obtain the other.

EXERCISES 6.1

Answers to odd-numbered problems begin on page 454.

In Problems 1–14, for each differential equation find two linearly independent power series solutions about the ordinary point $x = 0$.

1. $y'' - xy = 0$

2. $y'' + x^2y = 0$

3. $y'' - 2xy' + y = 0$

4. $y'' - xy' + 2y = 0$

5. $y'' + x^2y' + xy = 0$

6. $y'' + 2xy' + 2y = 0$

7. $(x - 1)y'' + y' = 0$

8. $y'' - (1 + x)y = 0$

9. $(x^2 - 1)y'' + 4xy' + 2y = 0$

10. $(x^2 + 1)y'' - 6y = 0$

11. $(x^2 + 2)y'' + 3xy' - y = 0$

12. $(x^2 - 1)y'' + xy' - y = 0$

13. $y'' - (x + 1)y' - y = 0$

14. $y'' - xy' - (x + 2)y = 0$

In Problems 15–18, use the power series method to solve the given differential equation subject to the indicated initial conditions.

15. $(x - 1)y'' - xy' + y = 0, \quad y(0) = -2, y'(0) = 6$

16. $(x + 1)y'' - (2 - x)y' + y = 0, \quad y(0) = 2, y'(0) = -1$

17. $y'' - 2xy' + 8y = 0, \quad y(0) = 3, y'(0) = 0$

18. $(x^2 + 1)y'' + 2xy' = 0, \quad y(0) = 0, y'(0) = 1$

In Problems 19–22, use the procedure illustrated in Example 5 to find two power series solutions of the given differential equation about the ordinary point $x = 0$.

19. $y'' + (\sin x)y = 0$

20. $xy'' + (\sin x)y = 0$ [*Hint*: See Example 1.]

21. $y'' + e^{-x}y = 0$

22. $y'' + e^{x}y' - y = 0$

Discussion Problems

23. Note that $x = 0$ is an ordinary point of the differential equation $y'' + xy' + y = 0$. Discuss: Why isn't it a good idea to try to solve the initial-value problem

$$y'' + xy' + y = 0, \quad y(1) = -5, \quad y'(1) = 2$$

using power series centered at $x = 0$? Using power series, find a more convenient way to solve the problem.

24. Without actually solving again, use the results obtained in Example 3 to write down the general solution of Airy's equation in the form $y'' - xy = 0$.

25. All of the differential equations considered in this section have been homogeneous. Discuss how the power series method can be used to solve a nonhomogeneous equation such as $y'' - xy = 1$. Carry out your ideas and write the solution of the equation in the form $y = y_c + y_p$. Then solve $y'' - 4xy' - 4y = e^x$.

26. **(a)** Note that $x = 0$ is an ordinary point of each of the differential equations

$$(x^2 + 7x + 12)y'' + xy' + y = 0 \quad \text{and}$$
$$(x^2 - 2x + 2)y'' - 5xy' + y = 0$$

Discuss: Without actually solving either equation, what is the minimum value for R for a power series solution $y = \sum_{n=0}^{\infty} c_n x^n$, $|x| < R$?

(b) Discuss: Without actually solving either equation, what is the minimum value for R for a power series solution $y = \sum_{n=0}^{\infty} c_n (x - 1)^n$, $|x| < R$?

6.2 Solutions About Singular Points

The two differential equations $y'' + xy = 0$ and $xy'' + y = 0$ are similar only in that they are both examples of simple linear second-order equations with variable coefficients. That is all they have in common. We saw in the preceding section that since $x = 0$ is an *ordinary point* of the first equation, there was no problem in finding two linearly independent power series solutions centered at that point. In contrast, because $x = 0$ is a *singular point* of the second DE, finding two infinite series solutions of the equation about that point becomes a more difficult task.

Series Solutions When $x = x_0$ is a singular point of a linear second-order differential equation, it is generally not possible to find a solution in the form of a power series $y = \sum_{n=0}^{\infty} c_n(x - x_0)^n$; but we may be able to find an infinite series solution of the form $y = \sum_{n=0}^{\infty} c_n(x - x_0)^{n+r}$, where r is a constant that has to be determined. If r turns out to be a number that is not a nonnegative integer, then this second kind of series is not a power series.

Regular and Irregular Singular Points We saw in Section 6.1 that $x = x_0$ is a singular point of the differential equation

$$a_2(x)y'' + a_1(x)y' + a_0(x)y = 0 \tag{1}$$

if at least one of the coefficients P and Q in its standard form

$$y'' + P(x)y' + Q(x)y = 0 \tag{2}$$

fails to be analytic at x_0. Singular points are further classified as either regular or irregular. The classification again depends on the functions P and Q in (2).

Definition 6.2 Regular/Irregular Singular Points

A singular point x_0 is said to be a **regular singular point** of the differential equation (1) if both $(x - x_0)P(x)$ and $(x - x_0)^2Q(x)$ are analytic at x_0. A singular point that is not regular is said to be an **irregular singular point** of the equation.

Polynomial Coefficients In the case in which the coefficients in (1) are polynomials with no common factors, Definition 6.2 is equivalent to the following:

Let $a_2(x_0) = 0$. Form $P(x)$ and $Q(x)$ by reducing $a_1(x)/a_2(x)$ and $a_0(x)/a_2(x)$ to lowest terms, respectively. If the factor $(x - x_0)$ appears *at most* to the first power in the denominator of $P(x)$ and *at most* to the second power in the denominator of $Q(x)$, then $x = x_0$ is a regular singular point.

■ EXAMPLE 1 Classification of Singular Points

It should be clear that $x = -2$ and $x = 2$ are singular points of the equation

$$(x^2 - 4)^2y'' + (x - 2)y' + y = 0$$

Dividing the equation by $(x^2 - 4)^2 = (x - 2)^2(x + 2)^2$, we find that

$$P(x) = \frac{1}{(x-2)(x+2)^2} \quad \text{and} \quad Q(x) = \frac{1}{(x-2)^2(x+2)^2}$$

We now test $P(x)$ and $Q(x)$ at each singular point.

In order that $x = -2$ be a regular singular point, the factor $x + 2$ can appear at most to the first power in the denominator of $P(x)$, and can appear at most to the second power in the denominator of $Q(x)$. Inspection of $P(x)$ and $Q(x)$ shows that the first condition does not obtain, and so we conclude that $x = -2$ is an irregular singular point.

In order that $x = 2$ be a regular singular point, the factor $x - 2$ can appear at most to the first power in the denominator of $P(x)$, and can appear at most to the second power in the denominator of $Q(x)$. Further inspection of $P(x)$ and $Q(x)$ shows that both these conditions are satisfied, so $x = 2$ is a regular singular point. ■

■ EXAMPLE 2 Classification of Singular Points

(a) $x = 1$ and $x = -1$ are regular singular points of

$$(1 - x^2)y'' - 2xy' + 30y = 0$$

(b) $x = 0$ is an irregular singular point of $x^3y'' - 2xy' + 5y = 0$ since

$$P(x) = -\frac{2}{x^2} \quad \text{and} \quad Q(x) = \frac{5}{x^3}$$

(c) $x = 0$ is a regular singular point of $xy'' - 2xy' + 5y = 0$ since

$$P(x) = -2 \quad \text{and} \quad Q(x) = \frac{5}{x}$$

■

In part (c) of Example 2, notice that $(x - 0)$ and $(x - 0)^2$ do not even appear in the denominators of $P(x)$ and $Q(x)$, respectively. Remember, these factors can appear at most in this fashion. For a singular point $x = x_0$, any nonnegative power of $(x - x_0)$ less than one (namely, zero) and any nonnegative power less than two (namely, zero and one) in the denominators of $P(x)$ and $Q(x)$, respectively, imply x_0 is a regular singular point.

A Cauchy-Euler equation $ax^2y'' + bxy' + cy = 0$ has a regular singular point at $x = 0$. The two solutions of $x^2y'' - 3xy' + 4y = 0$ on the interval $(0, \infty)$ are $y_1 = x^2$ and $y_2 = x^2 \ln x$. If the procedure of Theorem 6.1 is attempted at $x = 0$ (that is, assume a solution of the form $y = \sum_{n=0}^{\infty} c_n x^n$), we would succeed in obtaining only the polynomial solution $y_1 = x^2$. The fact that we would not obtain the second solution is not really surprising since $\ln x$ does not possess a Taylor series expansion about $x = 0$. It follows that $y_2 = x^2 \ln x$ does not have a power series in x.

Method of Frobenius To solve a differential equation (1) about a regular singular point, we employ the following theorem due to Frobenius.

Theorem 6.2 Frobenius' Theorem

If $x = x_0$ is a regular singular point of the differential equation (1), then there exists at least one solution of the form

$$y = (x - x_0)^r \sum_{n=0}^{\infty} c_n(x - x_0)^n = \sum_{n=0}^{\infty} c_n(x - x_0)^{n+r} \tag{3}$$

where the number r is a constant to be determined. The series will converge at least on some interval $0 < x - x_0 < R$.

Notice the words *at least* in the first sentence of Theorem 6.2. This means that, as opposed to Theorem 6.1, we are not guaranteed two solutions of the form indicated in (3). The **method of Frobenius** consists of identifying a regular singular point x_0, substituting $y = \sum_{n=0}^{\infty} c_n(x - x_0)^{n+r}$ into the given differential equation, and determining the unknown exponent r and the coefficients c_n.

As we did in the discussion of solutions about ordinary points, we shall always assume, for the sake of simplicity, that $x_0 = 0$.

EXAMPLE 3 Series Solution About a Regular Singular Point

Since $x = 0$ is a regular singular point of the differential equation

$$3xy'' + y' - y = 0 \tag{4}$$

we try a solution of the form $y = \sum_{n=0}^{\infty} c_n x^{n+r}$. Now

$$y' = \sum_{n=0}^{\infty} (n + r)c_n x^{n+r-1} \quad \text{and} \quad y'' = \sum_{n=0}^{\infty} (n + r)(n + r - 1)c_n x^{n+r-2}$$

so that

$$3xy'' + y' - y = 3\sum_{n=0}^{\infty} (n + r)(n + r - 1)c_n x^{n+r-1} + \sum_{n=0}^{\infty} (n + r)c_n x^{n+r-1} - \sum_{n=0}^{\infty} c_n x^{n+r}$$

$$= \sum_{n=0}^{\infty} (n + r)(3n + 3r - 2)c_n x^{n+r-1} - \sum_{n=0}^{\infty} c_n x^{n+r}$$

$$= x^r\left[r(3r - 2)c_0 x^{-1} + \underbrace{\sum_{n=1}^{\infty} (n + r)(3n + 3r - 2)c_n x^{n-1}}_{k = n - 1} - \underbrace{\sum_{n=0}^{\infty} c_n x^n}_{k = n}\right]$$

$$= x^r\left[r(3r - 2)c_0 x^{-1} + \sum_{k=0}^{\infty} [(k + r + 1)(3k + 3r + 1)c_{k+1} - c_k]x^k\right] = 0$$

which implies

$$r(3r - 2)c_0 = 0$$

$$(k + r + 1)(3k + 3r + 1)c_{k+1} - c_k = 0, \qquad k = 0, 1, 2, \ldots$$

Since nothing is gained by taking $c_0 = 0$, we must then have

$$r(3r - 2) = 0 \tag{5}$$

and

$$c_{k+1} = \frac{c_k}{(k + r + 1)(3k + 3r + 1)}, \qquad k = 0, 1, 2, \ldots \tag{6}$$

The two values of r that satisfy (5), $r_1 = \frac{2}{3}$ and $r_2 = 0$, when substituted in (6), give two different recurrence relations:

$$r_1 = \frac{2}{3}, \qquad c_{k+1} = \frac{c_k}{(3k + 5)(k + 1)}, \quad k = 0, 1, 2, \ldots \tag{7}$$

and

$$r_2 = 0, \qquad c_{k+1} = \frac{c_k}{(k + 1)(3k + 1)}, \quad k = 0, 1, 2, \ldots \tag{8}$$

Iteration of (7):	Iteration of (8):
$c_1 = \frac{c_0}{5 \cdot 1}$	$c_1 = \frac{c_0}{1 \cdot 1}$
$c_2 = \frac{c_1}{8 \cdot 2} = \frac{c_0}{2!5 \cdot 8}$	$c_2 = \frac{c_1}{2 \cdot 4} = \frac{c_0}{2!1 \cdot 4}$
$c_3 = \frac{c_2}{11 \cdot 3} = \frac{c_0}{3!5 \cdot 8 \cdot 11}$	$c_3 = \frac{c_2}{3 \cdot 7} = \frac{c_0}{3!1 \cdot 4 \cdot 7}$
$c_4 = \frac{c_3}{14 \cdot 4} = \frac{c_0}{4!5 \cdot 8 \cdot 11 \cdot 14}$	$c_4 = \frac{c_3}{4 \cdot 10} = \frac{c_0}{4!1 \cdot 4 \cdot 7 \cdot 10}$
$\vdots$	$\vdots$
$c_n = \frac{c_0}{n!5 \cdot 8 \cdot 11 \cdots (3n+2)}$,	$c_n = \frac{c_0}{n!1 \cdot 4 \cdot 7 \cdots (3n-2)}$,

where $n = 1, 2, 3, \ldots$. Thus we obtain two series solutions

$$y_1 = c_0 x^{2/3}\left[1 + \sum_{n=1}^{\infty} \frac{1}{n!5 \cdot 8 \cdot 11 \cdots (3n+2)} x^n\right] \tag{9}$$

and

$$y_2 = c_0 x^0\left[1 + \sum_{n=1}^{\infty} \frac{1}{n!1 \cdot 4 \cdot 7 \cdots (3n-2)} x^n\right]. \tag{10}$$

By the ratio test it can be demonstrated that both (9) and (10) converge for all finite values of x, that is, $|x| < \infty$. Also, it should be clear from the form of these solutions that neither series is a constant multiple of the other and, therefore, $y_1(x)$ and $y_2(x)$ are linearly independent on the entire x-axis. Hence by the superposition principle, $y = C_1y_1(x) + C_2y_2(x)$ is another solution of (4). On any interval not containing the origin, such as $(0, \infty)$, this linear combination represents the general solution of the differential equation. ■

Indicial Equation Equation (5) is called the **indicial equation** of the problem, and the values $r_1 = \frac{2}{3}$ and $r_2 = 0$ are called the **indicial roots**, or **exponents**, of the singularity $x = 0$. In general, after substituting $y = \sum_{n=0}^{\infty} c_n x^{n+r}$ into the given differential equation and simplifying, the indicial equation is a quadratic equation in r that results from equating the *total coefficient of the lowest power of x to zero*. We solve the indicial equation for the two values of the exponents and substitute these values into a recurrence relation such as (6). Theorem 6.2 guarantees that at least one solution of the assumed series form can be found.

■ EXAMPLE 4 **Series Solution About a Regular Singular Point**

Solve $2xy'' + (1 + x)y' + y = 0$.

Solution If $y = \sum_{n=0}^{\infty} c_n x^{n+r}$, then

$$2xy'' + (1+x)y' + y = 2\sum_{n=0}^{\infty}(n+r)(n+r-1)c_n x^{n+r-1} + \sum_{n=0}^{\infty}(n+r)c_n x^{n+r-1}$$

$$+ \sum_{n=0}^{\infty}(n+r)c_n x^{n+r} + \sum_{n=0}^{\infty} c_n x^{n+r}$$

$$= \sum_{n=0}^{\infty}(n+r)(2n+2r-1)c_n x^{n+r-1} + \sum_{n=0}^{\infty}(n+r+1)c_n x^{n+r}$$

$$= x^r\Big[r(2r-1)c_0x^{-1} + \underbrace{\sum_{n=1}^{\infty}(n+r)(2n+2r-1)c_n x^{n-1}}_{k=n-1} + \underbrace{\sum_{n=0}^{\infty}(n+r+1)c_n x^n}_{k=n}\Big]$$

$$= x^r\Big[r(2r-1)c_0x^{-1} + \sum_{k=0}^{\infty}[(k+r+1)(2k+2r+1)c_{k+1} + (k+r+1)c_k]x^k\Big] = 0$$

which implies

$$r(2r-1) = 0 \tag{11}$$

$$(k+r+1)(2k+2r+1)c_{k+1} + (k+r+1)c_k = 0, \qquad k = 0, 1, 2, \ldots \tag{12}$$

From (11), see that the indicial roots are $r_1 = \frac{1}{2}$ and $r_2 = 0$.

For $r_1 = \frac{1}{2}$, we can divide by $k + \frac{3}{2}$ in (12) to obtain

$$c_{k+1} = \frac{-c_k}{2(k+1)} \tag{13}$$

whereas for $r_2 = 0$, (12) becomes

$$c_{k+1} = \frac{-c_k}{2k+1} \tag{14}$$

Iteration of (13):	Iteration of (14):
$c_1 = \frac{-c_0}{2 \cdot 1}$	$c_1 = \frac{-c_0}{1}$
$c_2 = \frac{-c_1}{2 \cdot 2} = \frac{c_0}{2^2 \cdot 2!}$	$c_2 = \frac{-c_1}{3} = \frac{c_0}{1 \cdot 3}$
$c_3 = \frac{-c_2}{2 \cdot 3} = \frac{-c_0}{2^3 \cdot 3!}$	$c_3 = \frac{-c_2}{5} = \frac{-c_0}{1 \cdot 3 \cdot 5}$
$c_4 = \frac{-c_3}{2 \cdot 4} = \frac{c_0}{2^4 \cdot 4!}$	$c_4 = \frac{-c_3}{7} = \frac{c_0}{1 \cdot 3 \cdot 5 \cdot 7}$
$\vdots$	$\vdots$
$c_n = \frac{(-1)^n c_0}{2^n n!}$,	$c_n = \frac{(-1)^n c_0}{1 \cdot 3 \cdot 5 \cdot 7 \cdot (2n-1)}$,

where $n = 1, 2, 3, \ldots$. Thus for the indicial root $r_1 = \frac{1}{2}$ we obtain the solution

$$y_1 = c_0 x^{1/2}\left[1 + \sum_{n=1}^{\infty} \frac{(-1)^n}{2^n n!} x^n\right] = c_0 \sum_{n=0}^{\infty} \frac{(-1)^n}{2^n n!} x^{n+1/2}$$

which converges for $x \geq 0$. As given, this series is not defined for negative values of x because of the presence of $x^{1/2}$. For $r_2 = 0$, a second solution is

$$y_2 = c_0\left[1 + \sum_{n=1}^{\infty} \frac{(-1)^n}{1 \cdot 3 \cdot 5 \cdot 7 \cdots (2n-1)} x^n\right], \quad |x| < \infty$$

On the interval $(0, \infty)$, the general solution of the given differential equation is $y = C_1 y_1(x) + C_2 y_2(x)$. ■

■ **EXAMPLE 5** **Method Gives Only One Solution**

Solve $xy'' + 3y' - y = 0$.

Solution You should verify that the indicial roots are $r_1 = 0$ and $r_2 = -2$ and that the two corresponding recurrence relations yield exactly the same set of coefficients. In other words, in this case, the method of Frobenius produces only a single series solution

$$y_1 = \sum_{n=0}^{\infty} \frac{2}{n!(n+2)!} x^n = 1 + \frac{1}{3}x + \frac{1}{24}x^2 + \frac{1}{360}x^3 + \cdots$$ ■

Three Cases When using the method of Frobenius, we usually distinguish three cases corresponding to the nature of the indicial roots r_1 and r_2. For the sake of discussion let us suppose that r_1 and r_2 are real solutions of the indicial equation and that r_1 denotes the largest root.

CASE I If r_1 and r_2 are distinct and do not differ by an integer, there exist two linearly independent solutions y_1 and y_2 of (1) of the form given in (3).

This is the case illustrated in Examples 3 and 4. In the remaining two cases, we see that when the indicial roots differ by an integer, we *may or may not* be able to find two solutions having the form (3).

CASE II If $r_1 - r_2 = N$, where N is a positive integer, then there exist two linearly independent solutions of equation (1) of the form

$$y_1 = \sum_{n=0}^{\infty} c_n x^{n+r_1} \tag{15}$$

$$y_2 = C y_1(x) \ln x + \sum_{n=0}^{\infty} b_n x^{n+r_2} \tag{16}$$

where C is a constant that could be zero.

This is the situation in Example 5; notice that $r_1 - r_2 = 2$ is an integer. When the method fails to give a second solution, as it did in Example 5, then $C \neq 0$. One way of obtaining the solution (16) with the logarithmic term is to use the general formula developed in Problem 52 in Exercises 3.2:

$$y_2(x) = y_1(x) \int \frac{e^{-\int P(x)\,dx}}{y_1^2(x)}\,dx \tag{17}$$

where y_1 is the known solution and P is the coefficient of y' in the standard form of the differential equation (2). We will illustrate how to use this formula in Example 6. Finally in the last case—the case when the indicial roots are equal—a second solution will *always* contain a logarithm. This situation is analogous to the solution of a Cauchy-Euler equation when the roots of the auxiliary equation are equal.

CASE III If $r_1 = r_2$, then there always exist two linearly independent solutions of the equation (1) of the form

$$y_1 = \sum_{n=0}^{\infty} c_n x^{n+r_1} \tag{18}$$

$$y_2 = y_1(x) \ln x + \sum_{n=1}^{\infty} b_n x^{n+r_1} \tag{19}$$

The solution y_2 whose form is given in (19) can also be obtained using formula (17).

EXAMPLE 6 Example 5 Revisited

Find the general solution of $xy'' + 3y' - y = 0$.

Solution From the known solution given in Example 5,

$$y_1 = 1 + \frac{1}{3}x + \frac{1}{24}x^2 + \frac{1}{360}x^3 + \cdots$$

we can construct a second solution y_2 using (17). For those with the time, energy, and patience, it is possible to carry out by hand the drudgery of squaring a series, long division, and integration of the quotient. But life can be simplified since all these operations can be done with relative ease with the help of a CAS. We give the results:

$$y_2 = y_1(x) \int \frac{e^{-\int (3/x)\,dx}}{y_1^2(x)}\,dx = y_1(x) \int \frac{dx}{x^3\left[1 + \frac{1}{3}x + \frac{1}{24}x^2 + \frac{1}{360}x^3 + \cdots\right]^2}$$

$$= y_1(x) \int \frac{dx}{x^3\left[1 + \frac{2}{3}x + \frac{7}{36}x^2 + \frac{1}{30}x^3 + \cdots\right]} \qquad \leftarrow \text{squaring}$$

$$= y_1(x) \int \frac{1}{x^3}\left[1 - \frac{2}{3}x + \frac{1}{4}x^2 - \frac{19}{270}x^3 + \cdots\right] dx \quad \leftarrow \text{long division}$$

$$= y_1(x) \int \left[\frac{1}{x^3} - \frac{2}{3x^2} + \frac{1}{4x} - \frac{19}{270} + \cdots\right] dx$$

$$= y_1(x) \left[-\frac{1}{2x^2} + \frac{2}{3x} + \frac{1}{4}\ln x - \frac{19}{270}x + \cdots\right]$$

or $$y_2 = \frac{1}{4}y_1(x)\ln x + y_1(x)\left[-\frac{1}{2x^2} + \frac{2}{3x} - \frac{19}{270}x + \cdots\right]$$

On the interval $(0, \infty)$, the general solution is $y = C_1y_1(x) + C_2y_2(x)$. ■

Remarks

(*i*) Since r is a root of a quadratic equation, it could be complex. We shall not, however, investigate this case.

(*ii*) If $x_0 = 0$ is an irregular singular point, we may not be able to find any solution of the form $y = \sum_{n=0}^{\infty} c_n x^{n+r}$.

EXERCISES 6.2

Answers to odd-numbered problems begin on page 455.

In Problems 1–10, determine the singular points of each differential equation. Classify each singular point as regular or irregular.

1. $x^3y'' + 4x^2y' + 3y = 0$
2. $xy'' - (x + 3)^{-2}y = 0$
3. $(x^2 - 9)^2y'' + (x + 3)y' + 2y = 0$
4. $y'' - \frac{1}{x}y' + \frac{1}{(x-1)^3}y = 0$
5. $(x^3 + 4x)y'' - 2xy' + 6y = 0$
6. $x^2(x - 5)^2y'' + 4xy' + (x^2 - 25)y = 0$
7. $(x^2 + x - 6)y'' + (x + 3)y' + (x - 2)y = 0$
8. $x(x^2 + 1)^2y'' + y = 0$
9. $x^3(x^2 - 25)(x - 2)^2y'' + 3x(x - 2)y' + 7(x + 5)y = 0$
10. $(x^3 - 2x^2 - 3x)^2y'' + x(x - 3)^2y' - (x + 1)y = 0$

In Problems 11–22, show that the indicial roots do not differ by an integer. Use the method of Frobenius to obtain two linearly independent series solutions about the regular singular point $x = 0$. Form the general solution on $(0, \infty)$.

11. $2xy'' - y' + 2y = 0$
12. $2xy'' + 5y' + xy = 0$
13. $4xy'' + \frac{1}{2}y' + y = 0$
14. $2x^2y'' - xy' + (x^2 + 1)y = 0$
15. $3xy'' + (2 - x)y' - y = 0$
16. $x^2y'' - (x - \frac{2}{9})y = 0$
17. $2xy'' - (3 + 2x)y' + y = 0$
18. $x^2y'' + xy' + (x^2 - \frac{4}{9})y = 0$
19. $9x^2y'' + 9x^2y' + 2y = 0$
20. $2x^2y'' + 3xy' + (2x - 1)y = 0$
21. $2x^2y'' - x(x - 1)y' - y = 0$
22. $x(x - 2)y'' + y' - 2y = 0$

In Problems 23–30, show that the indicial roots differ by an integer. Use the method of Frobenius to obtain at least one series solution about the regular singular point $x = 0$. If necessary, use (17) to find a second solution. Form the general solution on $(0, \infty)$.

23. $xy'' + 2y' - xy = 0$
24. $x^2y'' + xy' + (x^2 - \frac{1}{4})y = 0$
25. $x(x - 1)y'' + 3y' - 2y = 0$

26. $y'' + \frac{3}{x}y' - 2y = 0$

27. $xy'' + (1 - x)y' - y = 0$

28. $xy'' + y = 0$

29. $xy'' + y' + y = 0$

30. $xy'' - xy' + y = 0$

31. **(a)** The differential equation $x^4y'' + \lambda y = 0$ has an irregular singular point at $x = 0$. Show that the substitution $t = 1/x$ yields the differential equation

$$\frac{d^2y}{dt^2} + \frac{2}{t}\frac{dy}{dt} + \lambda y = 0$$

which now has a regular singular point at $t = 0$.

(b) Use the method of this section to find two series solutions of the second equation in part (a) about the singular point $t = 0$.

(c) Express each series solution of the original equation in terms of elementary functions.

32. In Example 3 of Section 3.7, we saw that when a constant vertical compressive force or load P was applied to a thin column of uniform cross section, the deflection $y(x)$ satisfied the boundary-value problem

$$EI\frac{d^2y}{dx^2} + Py = 0, \quad y(0) = 0, \quad y(L) = 0 \qquad \textbf{(20)}$$

The column will buckle or deflect only when the compressive force is a critical load P_n.

(a) Now suppose the column is again of length L and has circular cross sections but is tapered, as shown in Figure 6.1(a). If the column, a truncated cone, has a linear taper $y = cx$ as shown in cross section in Figure 6.1(b), the moment of inertia of a cross section with respect to an axis perpendicular to the xy-plane is $I = \frac{1}{4}\pi r^4$, where $r = y$ and $y = cx$. Hence we can write $I(x) = I_0(x/b)^4$, where $I_0 = I(b) = \frac{1}{4}\pi(cb)^4$. Substituting $I(x)$ into the differential equation in (20), we see that the deflection in this case is determined from the boundary-value problem

$$x^4\frac{d^2y}{dx^2} + \lambda y = 0, \quad y(a) = 0, \quad y(b) = 0$$

where $\lambda = Pb^4/EI_0$. Use the results of Problem 31 to find the critical loads P_n for the tapered column. Use an appropriate identity to express the buckling modes $y_n(x)$ as a single function.

(b) Use a CAS to obtain the graph of the first buckling mode $y_1(x)$ corresponding to the Euler load P_1 when $b = 11$ and $a = 1$.

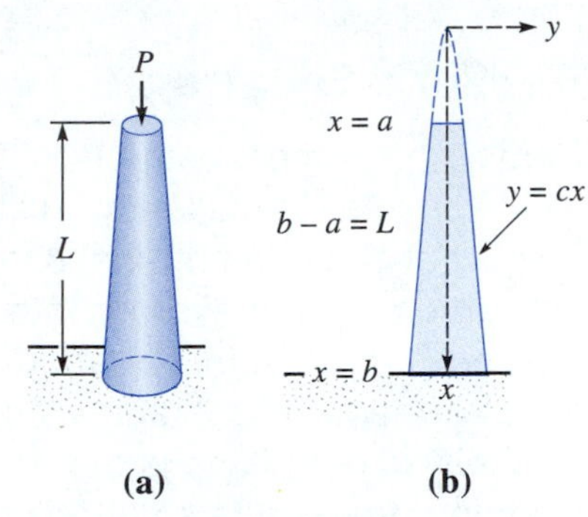

Figure 6.1.

Discussion Problems

33. Discuss: How would you define a regular singular point for a linear third-order differential equation

$$a_3(x)y''' + a_2(x)y'' + a_1(x)y' + a_0(x)y = 0$$

34. Each of the differential equations $x^3y'' + y = 0$ and $x^2y'' + (3x - 1)y' + y = 0$ has an irregular singular point at $x = 0$. Determine whether the method of Frobenius yields a series solution of each differential equation about $x = 0$. Discuss and explain your findings.

6.3 Bessel's Equation

Some of the most famous differential equations are linear second-order equations with variable coefficients. Historically, these equations often arose in someone's investigations of a physical phenomenon. The study of these DEs, their solutions, and properties, in turn evolved into a branch of mathematics known as *special functions*. One such differential equation occurred in the analysis of oscillating chains, vibrating circular membranes, and planetary motion. It is the German astronomer Friedrich Bessel* whose name is permanently associated with the differential equation we now examine; the series solutions of the DE—the special functions—not surprisingly are called *Bessel functions*.

Over the last two hundred years, much has been written about the solutions of the linear differential equation

$$x^2y'' + xy' + (x^2 - \nu^2)y = 0 \tag{1}$$

known as **Bessel's equation**. Since $x = 0$ is a regular singular point, we can find an infinite series solution of (1) using the method of the preceding section. We assume that ν is a constant and that $\nu \geq 0$.

Solution of Bessel's Equation If we assume a solution $y = \sum_{n=0}^{\infty} c_n x^{n+r}$, then

$$\begin{aligned} x^2y'' + xy' + (x^2 - \nu^2)y &= \sum_{n=0}^{\infty} c_n(n+r)(n+r-1)x^{n+r} + \sum_{n=0}^{\infty} c_n(n+r)x^{n+r} + \sum_{n=0}^{\infty} c_n x^{n+r+2} \\ &\quad - \nu^2 \sum_{n=0}^{\infty} c_n x^{n+r} \\ &= c_0(r^2 - r + r - \nu^2)x^r + x^r \sum_{n=1}^{\infty} c_n[(n+r)(n+r-1) + (n+r) - \nu^2]x^n \\ &\quad + x^r \sum_{n=0}^{\infty} c_n x^{n+2} \\ &= c_0(r^2 - \nu^2)x^r + x^r \sum_{n=1}^{\infty} c_n[(n+r)^2 - \nu^2]x^n + x^r \sum_{n=0}^{\infty} c_n x^{n+2} \end{aligned} \tag{2}$$

*Friedrich Wilhelm Bessel (1784–1846) was the first person to measure the distance to a star.

From (2) we see that the indicial equation is $r^2 - \nu^2 = 0$, so that the indicial roots are $r_1 = \nu$ and $r_2 = -\nu$. When $r_1 = \nu$, (2) becomes

$$x^\nu \sum_{n=1}^{\infty} c_n n(n + 2\nu)x^n + x^\nu \sum_{n=0}^{\infty} c_n x^{n+2}$$

$$= x^\nu \left[(1 + 2\nu)c_1 x + \underbrace{\sum_{n=2}^{\infty} c_n n(n + 2\nu)x^n}_{k = n - 2} + \underbrace{\sum_{n=0}^{\infty} c_n x^{n+2}}_{k = n} \right]$$

$$= x^\nu \left[(1 + 2\nu)c_1 x + \sum_{k=0}^{\infty} [(k + 2)(k + 2 + 2\nu)c_{k+2} + c_k]x^{k+2} \right] = 0$$

Therefore, by the usual argument we can write $(1 + 2\nu)c_1 = 0$ and

$$c_{k+2} = \frac{-c_k}{(k + 2)(k + 2 + 2\nu)}, \qquad k = 0, 1, 2, \ldots \tag{3}$$

The choice $c_1 = 0$ in (3) implies that $c_3 = c_5 = c_7 = \cdots = 0$, so for $k = 0, 2, 4, \ldots$, we find, after letting $k + 2 = 2n$, $n = 1, 2, 3, \ldots$, that

$$c_{2n} = -\frac{c_{2n-2}}{2^2 n(n + \nu)} \tag{4}$$

Thus $\quad c_2 = -\dfrac{c_0}{2^2 \cdot 1 \cdot (1 + \nu)}$

$$c_4 = -\frac{c_2}{2^2 \cdot 2(2 + \nu)} = \frac{c_0}{2^4 \cdot 1 \cdot 2(1 + \nu)(2 + \nu)}$$

$$c_6 = -\frac{c_4}{2^2 \cdot 3(3 + \nu)} = -\frac{c_0}{2^6 \cdot 1 \cdot 2 \cdot 3(1 + \nu)(2 + \nu)(3 + \nu)}$$

$$\vdots$$

$$c_{2n} = \frac{(-1)^n c_0}{2^{2n} n!(1 + \nu)(2 + \nu) \cdots (n + \nu)}, \qquad n = 1, 2, 3, \ldots \tag{5}$$

It is standard practice to choose c_0 to be a specific value, namely,

$$c_0 = \frac{1}{2^\nu \Gamma(1 + \nu)}$$

where $\Gamma(1 + \nu)$ is the Gamma function. See Problem 31 in Exercises 5.1. Since this latter function possesses the convenient property $\Gamma(1 + \alpha) = \alpha\Gamma(\alpha)$, we can reduce the indicated product in the denominator of (5) to one term. For example,

$$\Gamma(1 + \nu + 1) = (1 + \nu)\Gamma(1 + \nu)$$

$$\Gamma(1 + \nu + 2) = (2 + \nu)\Gamma(2 + \nu) = (2 + \nu)(1 + \nu)\Gamma(1 + \nu)$$

Hence for $n = 0, 1, 2, \ldots$ we can write (5) as

$$c_{2n} = \frac{(-1)^n}{2^{2n+\nu} n!(1+\nu)(2+\nu)\cdots(n+\nu)\Gamma(1+\nu)} = \frac{(-1)^n}{2^{2n+\nu} n!\Gamma(1+\nu+n)}$$

Note for $\nu = 0$, $\Gamma(1 + n) = n!$. See Problem 35 in Exercises 5.1.

Bessel Functions of the First Kind The series solution $y = \sum_{n=0}^{\infty} c_{2n} x^{2n+\nu}$ is usually denoted by $J_\nu(x)$:

$$J_\nu(x) = \sum_{n=0}^{\infty} \frac{(-1)^n}{n!\Gamma(1+\nu+n)} \left(\frac{x}{2}\right)^{2n+\nu} \tag{6}$$

If $\nu \geq 0$, the series converges at least on the interval $[0, \infty)$. Also, for the second exponent $r_2 = -\nu$, we obtain, in exactly the same manner,

$$J_{-\nu}(x) = \sum_{n=0}^{\infty} \frac{(-1)^n}{n!\Gamma(1-\nu+n)} \left(\frac{x}{2}\right)^{2n-\nu} \tag{7}$$

The functions $J_\nu(x)$ and $J_{-\nu}(x)$ are called **Bessel functions of the first kind** of order ν and $-\nu$, respectively. Depending on the value of ν, (7) may contain negative powers of x and hence converges on $(0, \infty)$.*

Now some care must be taken in writing the general solution of (1). When $\nu = 0$, it is apparent that (6) and (7) are the same. If $\nu > 0$ and $r_1 - r_2 = \nu - (-\nu) = 2\nu$ is not a positive integer, it follows from Case I of Section 6.2 that $J_\nu(x)$ and $J_{-\nu}(x)$ are linearly independent solutions of (1) on $(0, \infty)$, and so the general solution on the interval would be $y = c_1 J_\nu(x) + c_2 J_{-\nu}(x)$. But we also know that when $r_1 - r_2 = 2\nu$ is a positive integer, a second series solution of (1) *may* exist. In this second case, we distinguish two possibilities. When $\nu = m$ = positive integer, $J_{-m}(x)$ defined by (7) and $J_m(x)$ are not linearly independent solutions. It can be shown that J_{-m} is a constant multiple of J_m (see Property (i) on page 391). In addition, $r_1 - r_2 = 2\nu$ can be a positive integer when ν is half an odd positive integer. It can be shown in this latter event that $J_\nu(x)$ and $J_{-\nu}(x)$ are linearly independent. In other words, the general solution of (1) on $(0, \infty)$ is

$$y = c_1 J_\nu(x) + c_2 J_{-\nu}(x), \qquad \nu \neq \text{integer} \tag{8}$$

Because of their importance, Bessel functions are built-in functions in most computer algebra systems. With the aid of a CAS, we get the graphs of $y = J_0(x)$ and $y = J_1(x)$ shown in Figure 6.2.

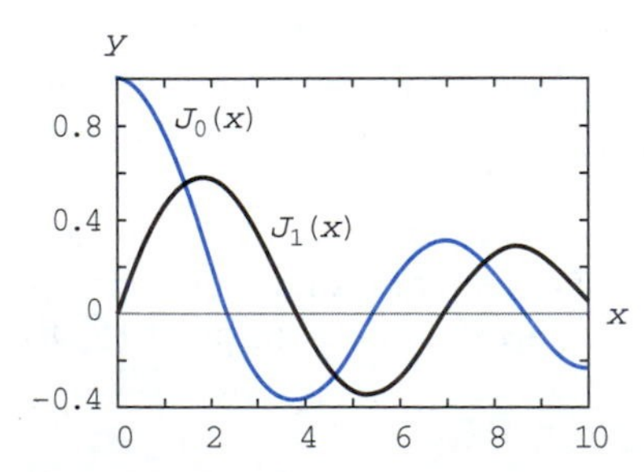

Figure 6.2. Bessel functions of the first kind.

*When we replace x by $|x|$, the series given in (7) and (8) converge for $0 < |x| < \infty$.

■ **EXAMPLE 1 General Solution: ν Not an Integer**

By identifying $\nu^2 = \frac{1}{4}$ and $\nu = \frac{1}{2}$, we see from (8) that the general solution of the equation $x^2y'' + xy' + (x^2 - \frac{1}{4})y = 0$ on $(0, \infty)$ is

$$y = c_1J_{1/2}(x) + c_2J_{-1/2}(x) \qquad ■$$

Bessel Functions of the Second Kind If $\nu \neq$ integer, the function defined by the linear combination

$$Y_\nu(x) = \frac{\cos \nu\pi J_\nu(x) - J_{-\nu}(x)}{\sin \nu\pi} \tag{9}$$

and the function $J_\nu(x)$ are linearly independent solutions of (1). Thus another form of the general solution of (1) is $y = c_1J_\nu(x) + c_2Y_\nu(x)$, provided $\nu \neq$ integer. As $\nu \to m$, m an integer, (9) has the indeterminate form 0/0. However, it can be shown by L'Hôpital's rule that $\lim_{\nu \to m} Y_\nu(x)$ exists. Moreover, the function

$$Y_m(x) = \lim_{\nu \to m} Y_\nu(x)$$

and $J_m(x)$ are linearly independent solutions of $x^2y'' + xy' + (x^2 - m^2)y = 0$. Hence for *any* value of ν, the general solution of (1) on $(0, \infty)$ can be written as

$$y = c_1J_\nu(x) + c_2Y_\nu(x) \tag{10}$$

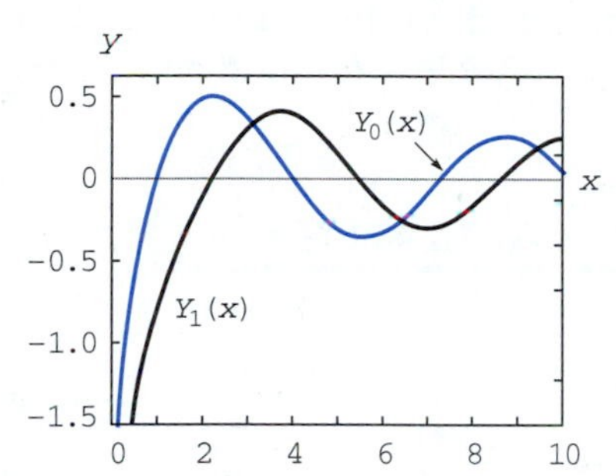

Figure 6.3. Bessel functions of the second kind.

$Y_\nu(x)$ is called the **Bessel function of the second kind** of order ν. Figure 6.3 shows the graphs of $Y_0(x)$ and $Y_1(x)$.

■ **EXAMPLE 2 General Solution: ν an Integer**

By identifying $\nu^2 = 9$ and $\nu = 3$, we see from (10) that the general solution of the equation $x^2y'' + xy' + (x^2 - 9)y = 0$ on $(0, \infty)$ is $y = c_1J_3(x) + c_2Y_3(x)$. ■

Parametric Bessel Equation By replacing x by λx in (1) and using the chain rule, we obtain an alternative form of Bessel's equation known as the **parametric Bessel equation**:

$$x^2y'' + xy' + (\lambda^2x^2 - \nu^2)y = 0 \tag{11}$$

The general solution of (11) is

$$y = c_1J_\nu(\lambda x) + c_2Y_\nu(\lambda x) \tag{12}$$

Properties We list below a few of the more useful properties of Bessel functions of order m, $m = 0, 1, 2, \ldots$:

(i) $J_{-m}(x) = (-1)^mJ_m(x)$ **(ii)** $J_m(-x) = (-1)^mJ_m(x)$

(iii) $J_m(0) = \begin{cases} 0, & m > 0 \\ 1, & m = 0 \end{cases}$ **(iv)** $\lim_{x \to 0^+} Y_m(x) = -\infty$

Note that Property (ii) indicates that $J_m(x)$ is an even function if m is an even integer and an odd function if m is an odd integer. The graphs of $Y_0(x)$ and $Y_1(x)$ in Figure 6.3 illustrate Property (iv): $Y_m(x)$ is unbounded at the origin. This last fact is not obvious from (9). It can be shown that for $x > 0$,

$$Y_0(x) = \frac{2}{\pi} J_0(x) \left[\gamma + \ln \frac{x}{2} \right] - \frac{2}{\pi} \sum_{k=1}^{\infty} \frac{(-1)^k}{(k!)^2} \left(1 + \frac{1}{2} + \cdots + \frac{1}{k}\right) \left(\frac{x}{2}\right)^{2k}$$

where $\gamma = 0.57721566 \ldots$ is **Euler's constant**.

Numerical Values Some functional values of $J_0(x)$, $J_1(x)$, $Y_0(x)$, and $Y_1(x)$ for selected values of x are given in Table 6.1. The first five nonnegative zeros of $J_0(x)$, $J_1(x)$, $Y_0(x)$, and $Y_1(x)$ are given in Table 6.2.

Sometimes it is possible to transform a given differential equation into equation (1) by means of a change of variable. We can then express

Table 6.1 Numerical Values of J_0, J_1, Y_0, and Y_1

x	$J_0(x)$	$J_1(x)$	$Y_0(x)$	$Y_1(x)$
0	1.0000	0.0000	—	—
1	0.7652	0.4401	0.0883	−0.7812
2	0.2239	0.5767	0.5104	−0.1070
3	−0.2601	0.3391	0.3769	0.3247
4	−0.3971	−0.0660	−0.0169	0.3979
5	−0.1776	−0.3276	−0.3085	0.1479
6	0.1506	−0.2767	−0.2882	−0.1750
7	0.3001	−0.0047	−0.0259	−0.3027
8	0.1717	0.2346	0.2235	−0.1581
9	−0.0903	0.2453	0.2499	0.1043
10	−0.2459	0.0435	0.0557	0.2490
11	−0.1712	−0.1768	−0.1688	0.1637
12	0.0477	−0.2234	−0.2252	−0.0571
13	0.2069	−0.0703	−0.0782	−0.2101
14	0.1711	0.1334	0.1272	−0.1666
15	−0.0142	0.2051	0.2055	0.0211

Table 6.2 Zeros of J_0, J_1, Y_0, and Y_1

$J_0(x)$	$J_1(x)$	$Y_0(x)$	$Y_1(x)$
2.4048	0.0000	0.8936	2.1971
5.5201	3.8317	3.9577	5.4297
8.6537	7.0156	7.0861	8.5960
11.7915	10.1735	10.2223	11.7492
14.9309	13.3237	13.3611	14.8974

the general solution of the original equation in terms of Bessel functions. The next example illustrates this technique.

■ EXAMPLE 3 **The Aging Spring Revisited**

In Section 3.6, we saw that one mathematical model for the free undamped motion of a mass on an aging spring is given by $mx'' + ke^{-\alpha t}x = 0$, $\alpha > 0$. We are now in a position to find the general solution of this equation. It is left as a problem to show that the change of variables $s = \dfrac{2}{\alpha}\sqrt{\dfrac{k}{m}}\,e^{-\alpha t/2}$ transforms the differential equation of the aging spring into

$$s^2\frac{d^2x}{ds^2} + s\frac{dx}{ds} + s^2x = 0$$

The last equation is recognized as (1) with $\nu = 0$, where the symbols x and s play the roles of y and x, respectively. The general solution of the new equation is $x = c_1J_0(s) + c_2Y_0(s)$. By resubstituting s in the last expression, we see that the general solution of $mx'' + ke^{-\alpha t}x = 0$ is

$$x(t) = c_1J_0\left(\frac{2}{\alpha}\sqrt{\frac{k}{m}}\,e^{-\alpha t/2}\right) + c_2Y_0\left(\frac{2}{\alpha}\sqrt{\frac{k}{m}}\,e^{-\alpha t/2}\right)$$

See Problems 23 and 24 in Exercises 6.3. ■

The other model discussed in Section 3.6 of a spring whose characteristics change with time was $mx'' + ktx = 0$. By dividing through by m, we see that this equation is Airy's equation $y'' + \alpha^2xy = 0$, where the symbols x and t play, in turn, the roles of the symbols y and x. See Example 3 in Section 6.1. The general solution of Airy's differential equation can also be written in terms of Bessel functions. See Problems 25–27 in Exercises 6.3.

EXERCISES 6.3

Answers to odd-numbered problems begin on page 455.

In Problems 1–8, find the general solution of the given differential equation on $(0, \infty)$.

1. $x^2y'' + xy' + (x^2 - \frac{1}{9})y = 0$

2. $x^2y'' + xy' + (x^2 - 1)y = 0$

3. $4x^2y'' + 4xy' + (4x^2 - 25)y = 0$

4. $16x^2y'' + 16xy' + (16x^2 - 1)y = 0$

5. $xy'' + y' + xy = 0$

6. $\dfrac{d}{dx}[xy'] + \left(x - \dfrac{4}{x}\right)y = 0$

7. $x^2y'' + xy' + (9x^2 - 4)y = 0$

8. $x^2y'' + xy' + (36x^2 - \frac{1}{4})y = 0$

9. Use the change of variables $y = x^{-1/2}w(x)$ to find the general solution of the equation

$$x^2y'' + 2xy' + \lambda^2x^2y = 0, \qquad x > 0$$

10. Verify that the differential equation

$$xy'' + (1 - 2n)y' + xy = 0, \qquad x > 0$$

possesses the particular solution $y = x^nJ_n(x)$.

11. Verify that the differential equation

$$xy'' + (1 + 2n)y' + xy = 0, \qquad x > 0$$

possesses the particular solution $y = x^{-n}J_n(x)$.

12. Verify that the differential equation

$$x^2y'' + (\lambda^2x^2 - \nu^2 + \tfrac{1}{4})y = 0, \qquad x > 0$$

possesses the particular solution $y = \sqrt{x}J_\nu(\lambda x)$, where $\lambda > 0$.

In Problems 13–18, use the results of Problems 10, 11, and 12 to find a particular solution of the given differential equation on $(0, \infty)$.

13. $y'' + y = 0$

14. $xy'' - y' + xy = 0$

15. $xy'' + 3y' + xy = 0$

16. $4x^2y'' + (16x^2 + 1)y = 0$

17. $x^2y'' + (x^2 - 2)y = 0$

18. $xy'' - 5y' + xy = 0$

In Problems 19 and 20, use (6) to derive the given differential recurrence relation.

19. $xJ'_\nu(x) = \nu J_\nu(x) - xJ_{\nu+1}(x)$ [*Hint:* Write $xJ'_\nu(x)$ as two series.]

20. $xJ'_\nu(x) = -\nu J_\nu(x) + xJ_{\nu-1}(x)$ [*Hint:* Write $2n + \nu = 2(n + \nu) - \nu$.]

In Problems 21 and 22, use the notion of an integrating factor for a linear first-order differential equation along with Problems 19 and 20 to derive the given differential recurrence relation.

21. $\dfrac{d}{dx}[x^{-\nu}J_\nu(x)] = -x^{-\nu}J_{\nu+1}(x)$

22. $\dfrac{d}{dx}[x^{\nu}J_\nu(x)] = x^{\nu}J_{\nu-1}(x)$

23. Use the change of variables $s = \dfrac{2}{\alpha}\sqrt{\dfrac{k}{m}}e^{-\alpha t/2}$ to show that the differential equation of an aging spring $mx'' + ke^{-\alpha t}x = 0$, $\alpha > 0$, becomes

$$s^2\frac{d^2x}{ds^2} + s\frac{dx}{ds} + s^2x = 0$$

24. **(a)** Use the general solution given in Example 3 to solve the initial-value problem

$$4x'' + e^{-0.1t}x = 0, \qquad x(0) = 1, \quad x'(0) = -\frac{1}{2}$$

When $\nu = 0$, the relation in Problem 21 implies $J'_0(x) = -J_1(x)$. Likewise it can be shown that $Y'_0(x) = -Y_1(x)$. Use these facts along with Table 6.1, or a CAS, to evaluate coefficients.

(b) Use a CAS to plot the graph of the solution obtained in part (a) over the interval $0 \le t \le 200$. Does the graph corroborate your conjecture in Problem 43(a) in Exercises 3.6?

25. Show that $y = x^{1/2}w(\tfrac{2}{3}\alpha x^{3/2})$ is a solution of **Airy's differential equation** $y'' + \alpha^2xy = 0$, $x > 0$ whenever w is a solution of Bessel's equation $t^2w'' + tw' + (t^2 - \tfrac{1}{9})w = 0$, $t > 0$. [*Hint:* After differentiating, substituting, and simplifying, then let $t = \tfrac{2}{3}\alpha x^{3/2}$.]

26. Use the result of Problem 25 to express the general solution of Airy's equation for $x > 0$ in terms of Bessel functions.

27. **(a)** Use the general solution obtained in Problem 26 to solve the initial-value problem

$$4x'' + tx = 0, \quad x(0.1) = 1, \quad x'(0.1) = -\frac{1}{2}$$

Use a CAS to evaluate coefficients.

(b) Use a CAS to plot the graph of the solution obtained in part (a) over the interval $0 \le t \le 200$. Does the graph corroborate your conjecture in Problem 43(b) in Exercises 3.6?

28. A uniform thin column, positioned vertically with one end embedded in the ground, will deflect or bend away from the vertical under the influence of its own weight when its length is greater than a certain critical height. It can be shown that the angular deflection $\theta(x)$ of the column from the vertical at a point $P(x)$ is a solution of the boundary-value problem:

$$EI\frac{d^2\theta}{dx^2} + \delta g(L - x)\theta = 0, \qquad \theta(0) = 0, \quad \theta'(L) = 0$$

where E is Young's modulus, I is the cross-sectional moment of inertia, δ is the constant linear density, and x is distance along the column measured from its base. See Figure 6.4. The column will bend only if this boundary-value problem has a nontrivial solution.

(a) First make the change of variables $t = L - x$ and

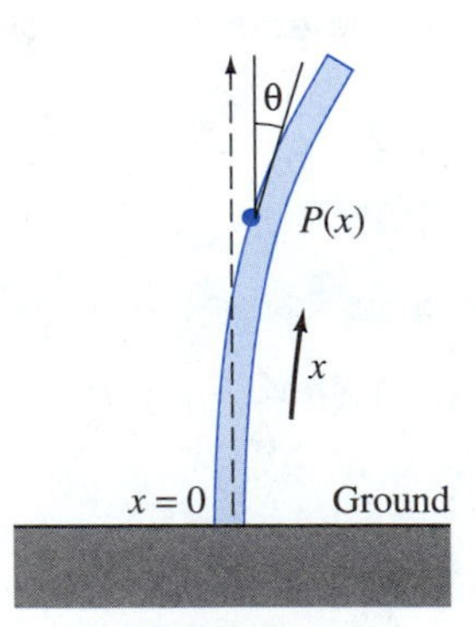

Figure 6.4.

state the resulting boundary-value problem. Then use the result of Problem 26 to express the general solution of the differential equation in terms of Bessel functions.

(b) With the aid of a CAS, find the critical length of a solid steel rod of radius $r = 0.05$ in., $\delta g = 0.28\ A$ lb/in., $E = 2.6 \times 10^7$ lb/in.^{2}, $A = \pi r^2$, and $I = \frac{1}{4}\pi r^4$.

29. It can be shown by a change of variable that the general solution of the differential equation $xy'' + \lambda y = 0$ on the interval $(0, \infty)$ is

$$y = c_1\sqrt{x}\,J_1(2\sqrt{\lambda x}) + c_2\sqrt{x}\,Y_1(2\sqrt{\lambda x})$$

Verify by direct substitution that $y = \sqrt{x}\,J_1(2\sqrt{x})$ is a particular solution of the equation in the case $\lambda = 1$.

30. In Example 3 of Section 3.7, we saw that when a constant vertical compressive force or load P was applied to a thin column of uniform cross section, the deflection $y(x)$ satisfied the boundary-value problem

$$EI\frac{d^2y}{dx^2} + Py = 0, \qquad y(0) = 0, \qquad y(L) = 0$$

(a) If the bending stiffness factor EI is proportional to x, then $EI(x) = kx$, where k is a constant of proportionality. If $EI(L) = kL = M$ is the maximum stiffness factor, then $k = M/L$ and so $EI(x) = Mx/L$. Use the information in Problem 29 to find the solution of

$$M\frac{x}{L}\frac{d^2y}{dx^2} + Py = 0, \qquad y(0) = 0, \qquad y(L) = 0$$

if it is known that $\sqrt{x}\,Y_1(2\sqrt{\lambda x})$ is *not* zero at $x = 0$.

(b) Use Table 6.2 to find the Euler load P_1 for the column.

(c) Use a CAS to obtain the graph of the first buckling mode $y_1(x)$ corresponding to the Euler load P_1. For simplicity, assume that $c_1 = 1$ and $L = 1$.

Discussion Problems

31. For the purposes of this problem, ignore the graphs given in Figure 6.2. Use the substitution $y = u/\sqrt{x}$ to show that Bessel's equation (1) has the alternative form

$$\frac{d^2u}{dx^2} + \left(1 - \frac{\nu^2 - \frac{1}{4}}{x^2}\right)u = 0$$

This is a form of the differential equation in Problem 12. For a fixed value of ν, discuss how this last equation enables us to discern the qualitative behavior of a solution of (1) as $x \to \infty$.

32. Use Table 6.2 or a CAS and take the difference between successive zeros of J_0. Do the same for J_1. Use tables or a CAS and repeat for J_2. If possible, formulate a conjecture. Discuss how these data tie in with Problem 31.

33. Discuss how Problem 21 and Rolle's theorem from calculus can be used to demonstrate that the zeros of J_0 and J_1 interlace, that is, between any two consecutive zeros of J_0 there exists a zero of J_1. See Figure 6.2 and Table 6.2.

34. When $\nu = n$, n a nonnegative integer, the Bessel function of the first kind $J_n(x)$ can be represented by a definite integral. Verify by direct substitution that, in the case $n = 0$,

$$J_0(x) = \frac{1}{\pi}\int_0^{\pi} \cos(x \sin t)\,dt$$

is a solution of (1). [*Hint:* After computing $J_0'(x)$, use integration by parts.]

35. The general form of the integral representation given in Problem 34 is

$$J_n(x) = \frac{1}{\pi}\int_0^{\pi} \cos(x \sin t - nt)\,dt$$

Discuss: For n a nonnegative integer, how does it follow from this integral that $|J_n(x)| \le 1$ for all x?

Chapter 6 in Review

Answers to this review are posted at our Web site diffeq.brookscole.com/books.html#zill_98

In Problems 1–4, answer true or false.

1. The differential equation $xy'' + (1 - \cos x)y' + x^2y = 0$ has a singular point at $x = 0$.

2. Since the differential equation $(x^2 - 2x + 10)y'' + y = 0$ has no real singular points, a power series solution, centered at 0, will converge for all x.

3. Since $x^3y'' + xy' - y = 0$ has an irregular singular point at $x = 0$, the differential equation possesses no solution that is analytic at $x = 0$.

4. The general solution of the Bessel equation $x^2y'' + xy' + (x^2 - 1)y = 0$ for $x > 0$ is $y = c_2J_1(x) + c_2J_{-1}(x)$.

5. Construct a linear second-order differential equation that has a regular singular point at $x = 1$ but an irregular singular point at $x = 0$.

6. Note that $x = 0$ is an ordinary point of the differential equation $y'' + x^2y' + 2xy = 5 - 2x + 10x^3$. Use the assumption $y = \sum_{n=0}^{\infty} c_nx^n$ to find the general solution $y = y_c + y_p$ that consists of three power series centered at $x = 0$.

In Problems 7 and 8, use an appropriate infinite series method to find two linearly independent solutions of the given differential equation.

7. $2xy'' + y' + y = 0$

8. $y'' - xy' - y = 0$

In Problems 9 and 10, solve the given initial-value problem.

9. $y'' + xy' + 2y = 0, \quad y(0) = 3, \quad y'(0) = -2$

10. $(x + 2)y'' + 3y = 0, \quad y(0) = 0, \quad y'(0) = 1$

Projects and Computer Lab Experiments for Chapter 6

Section 6.1

1. **Special Functions** Historically many interesting linear second-order differential equations with variable coefficients arose from someone's study of a physical problem. The study of the solutions of these equations, and other functions related to these solutions, eventually gave rise to a separate branch of mathematics known as **special functions**. Moreover, some of these differential equations possess polynomial solutions obtained, of course, by trying to find power series solutions centered at the ordinary point $x = 0$.

   (a) Let $n \geq 0$ be an integer. Find polynomial solutions of the following equations corresponding to $n = 0, 1, 2, 3$. For convenience, choose $c_0 = 1$, $c_1 = 1$ when iterating your recurrence relation.

   $y'' - 2xy' + 2ny = 0$ **Hermite's equation**

   $(1 - x^2)y'' - 2xy' + n(n + 1)y = 0$ **Legendre's equation**

   $(1 - x^2)y'' - xy' + n^2y = 0$ **Chebyshev's equation**

(b) In the case of Legendre's equation, find the four polynomial solutions that satisfy $y(1) = 1$. These solutions, called Legendre polynomials, are denoted by $P_0(x)$, $P_1(x)$, $P_2(x)$, and $P_3(x)$. Use a graphing utility to graph the Legendre polynomials on the interval $[-1, 1]$. Use these graphs to conjecture a property of the polynomials $P_n(x)$ for n even and for n odd.

2. Hermite Polynomials In this problem, we concentrate on the polynomials obtained from Hermite's equation in Problem 1. Denote your solutions by $h_0(x)$, $h_1(x)$, $h_2(x)$, and $h_3(x)$. The polynomial solutions $h_n(x)$ are not what is commonly accepted as *the* **Hermite polynomials** $H_0(x)$, $H_1(x)$, $H_2(x)$, and $H_3(x)$. In general, the H_n are constant multiples of the h_n, that is, $H_n(x) = C_n h_n(x)$, where C_n depends on the value of $n \geq 0$.

(a) There are several ways of defining the H_n—one is by means of the **generating function** $e^{2xt - t^2}$:

$$e^{2xt - t^2} = \sum_{n=0}^{\infty} \frac{H_n(x)}{n!} t^n$$

Obtain the first four terms of a power series expansion of $e^{2xt - t^2}$ centered at $t = 0$ (Maclaurin series) to obtain $H_0(x)$, $H_1(x)$, $H_2(x)$, and $H_3(x)$. Use this result to find C_0, C_1, C_2, and C_3. Can you discern a rule for choosing C_n?

(b) The Hermite polynomials $H_n(x)$ can also be obtained from a differential relation known as **Rodrigues' formula**:

$$H_n(x) = (-1)^n e^{x^2} \frac{d^n}{dx^n} e^{-x^2}$$

Use this formula to obtain $H_0(x)$, $H_1(x)$, $H_2(x)$, and $H_3(x)$.

(c) Derive the Rodrigues formula in part (b) from the generating function in part (a). Start by observing that $e^{2xt - t^2} = e^{-x^2} e^{-(x - t)^2}$ and then use the general form of a Maclaurin series.

(d) Use a graphing utility to obtain the graphs of $h_0(x)$, $h_1(x)$, $h_2(x)$, and $h_3(x)$ on one set of coordinate axes and $H_0(x)$, $H_1(x)$, $H_2(x)$, and $H_3(x)$ on a different set of axes.

Section 6.2

3. A Very Special DE The **hypergeometric equation**

$$x(1 - x)y'' + [\gamma - (\alpha + \beta + 1)x]y' - \alpha\beta y = 0$$

where α, β, γ are real constants, has three regular singular points: 0, 1, and ∞. This equation has been extensively studied and is important because any linear second-order differential equation with three distinct regular singular points can be transformed into the hypergeometric equation.

(a) Show that 0 is a regular singular point and that the indicial roots are 0 and $1 - \gamma$.

(b) Show that a solution corresponding to the indicial root 0 is given by

$$y_1(x) = 1 + \frac{\alpha\beta}{1!\gamma}x + \frac{\alpha(\alpha+1)\beta(\beta+1)}{2!\gamma(\gamma+1)}x^2 + \frac{\alpha(\alpha+1)(\alpha+2)\beta(\beta+1)(\beta+2)}{3!\gamma(\gamma+1)(\gamma+2)}x^3 + \cdots$$

provided γ is not zero or a negative integer. This solution is known as the **hypergeometric series** and is denoted by $F(\alpha, \beta, \gamma, x)$. The series converges for at least $|x| < 1$.

(c) For γ not a positive integer, express the solution $y_2(x)$ corresponding to the indicial root $1 - \gamma$ in terms of F. For γ not an integer, express the general solution of the differential equation in terms of F.

(d) Identify the functions defined by $F(1, \beta, \beta, x)$ and $F(-k, \beta, \beta, -x)$.

(e) Use the substitution $\frac{1}{2}(1 - x) = t$ to show that a particular solution of Legendre's differential equation

$$(1 - x^2)y'' - 2xy' + n(n + 1)y = 0$$

is given by $y = F(n + 1, -n, 1, \frac{1}{2}(1 - x))$. Find particular solutions of Legendre's equation for $n = 1$ and for $n = 2$. (See Problem 1.)

Section 6.3

4. **Spherical Bessel Functions**

(a) Use (6) of Section 6.3 along with the properties $\Gamma(\frac{1}{2}) = \sqrt{\pi}$ and $\Gamma(1 + \alpha) = \alpha\Gamma(\alpha)$ to show that

$$J_{1/2}(x) = \sqrt{\frac{2}{\pi x}}\sin x$$

(b) Use (7) of Section 6.3 to show that

$$J_{-1/2}(x) = \sqrt{\frac{2}{\pi x}}\cos x$$

(c) Use Problems 19 and 20 in Exercises 6.3 to derive the recurrence relation

$$2\nu J_\nu(x) = xJ_{\nu+1}(x) + xJ_{\nu-1}(x)$$

It follows from (a) and (b) and this recurrence relation that when ν = half an odd integer, $J_\nu(x)$ can be expressed in terms of $\sin x$, $\cos x$, and powers of x. Such Bessel functions are called **spherical Bessel functions**. Find the spherical Bessel functions $J_{3/2}(x)$ and $J_{-3/2}(x)$.

(d) Use a graphing utility to obtain the graphs of $J_{1/2}(x)$, $J_{-1/2}(x)$, $J_{3/2}(x)$ and $J_{-3/2}(x)$ for $x \geq 0$.

5. **A Solution in Terms of Bessel Functions** The differential equation $dy/dx = x^2 + y^2$ cannot be solved in terms of elementary functions.

However, a solution of this first-order equation can be expressed in terms of Bessel functions.

(a) Show that the substitution $y = -\dfrac{1}{u}\dfrac{du}{dx}$ leads to the equation $u'' + x^2u = 0$.

(b) Show that $u = x^{1/2}\,w(\tfrac{1}{2}x^2)$ is a solution of $u'' + x^2u = 0$ whenever w is a solution of Bessel's equation $x^2w'' + xw' + (x^2 - \tfrac{1}{16})\,w = 0$.

(c) Use Problems 19 and 20 in Exercises 6.3 in the forms

$$J_\nu'(x) = \frac{\nu}{x}J_\nu(x) - J_{\nu+1}(x) \quad \text{and} \quad J_\nu'(x) = -\frac{\nu}{x}J_\nu(x) + J_{\nu-1}(x)$$

as an aid to show that a one-parameter family of solutions of $dy/dx = x^2 + y^2$ is given by

$$y = x\frac{J_{3/4}(\tfrac{1}{2}x^2) - cJ_{-3/4}(\tfrac{1}{2}x^2)}{cJ_{1/4}(\tfrac{1}{2}x^2) + J_{-1/4}(\tfrac{1}{2}x^2)}$$

(d) The initial-value problem $dy/dx = x^2 + y^2$, $y(0) = 1$ possesses a solution. Use the family of solutions in part (c) along with (6) and (7) of this section to show that the initial condition $y(0) = 1$ yields $c = -\Gamma(\tfrac{1}{4})/2\Gamma(\tfrac{3}{4})$.

6. **Graphs** Most computer algebra systems have built-in special functions. Among these are the Bessel functions. Use a CAS to plot the graphs of $J_2(x)$, $J_3(x)$, $Y_{1/2}(x)$, $Y_2(x)$, and $Y_3(x)$. Find the first four positive x-intercepts and the location of the relative maximum nearest the y-axis.

7. **Bessel Function of Order Zero** From (6) of Section 6.3, the Bessel function of order 0 is defined by $J_0(x) = \sum_{n=0}^{\infty} \dfrac{(-1)^n}{(n!)^2 2^{2n}} x^{2n}$. Each partial sum of this series is a polynomial of an even degree and can be used to approximate $J_0(x)$ on certain intervals.

   (a) Write down the twelve approximating polynomials of degrees 0, 2, 4, . . . , 22.

   (b) Use a CAS in which $J_0(x)$ is a built-in function. On the same set of coordinate axes, graph $J_0(x)$ and the approximating polynomial of degree 0; on a different set of coordinate axes, graph $J_0(x)$ and the approximating polynomial of degree 2, and so on.

   (c) In each case in part (b), determine an interval on which you think the polynomial best approximates $J_0(x)$. Compare roots of the approximating polynomial and of $J_0(x)$.

8. **Bessel Function of the Second Kind** On page 391 we saw that the Bessel function of a second kind $Y_m(x)$, m a nonnegative integer, has the unusual definition $Y_m(x) = \lim_{\nu \to m} Y_\nu(x)$. To get a feeling for the existence of this limit, as $\nu \to 2$ for example, substitute $\nu = 2.8$, 2.7, . . . , 2.1 in the definition of $Y_\nu(x)$ given in (9) of Section 6.3. Then

use a CAS to plot the graphs of the functions $Y_\nu(x)$ for $\nu = 2.8, 2.7, \ldots, 2.1$ defined in this manner. Then obtain the graph of $Y_2(x)$ directly from the CAS and compare with $Y_{2.1}(x)$. Finally, if your CAS has an animation capability, make a "movie" of the eight graphs "morphing" into $Y_2(x)$. Describe what you see.

9. **Eigenvalues*** In some applications involving Bessel functions it is necessary to find eigenvalues λ that are defined in terms of the positive roots of equations such as $J_2(3\lambda) = 0$ and $3J_0(4\lambda) + 4\lambda J_0'(4\lambda) = 0$.
   (a) Use a CAS to plot the graphs of the functions $y = J_2(x)$ and $y = 3J_0(x) + 4J_0'(x)$ on intervals large enough to show the first four positive x-intercepts of each graph.
   (b) Use the root finding capability of your CAS to approximate the first four positive roots x_i of the equations $J_2(x) = 0$ and $3J_0(x) + 4J_0'(x) = 0$.
   (c) Use the data obtained in part (b) to find the first four positive values of λ_i that satisfy $J_2(3\lambda) = 0$ and the first four positive values of λ_i that satisfy $3J_0(4\lambda) + 4\lambda J_0'(4\lambda) = 0$.

Writing Assignments

10. **Almost Periodic** Write a short report on why $J_0(x)$ and $J_1(x)$ could be called "almost periodic functions." Discuss the connection between the nonnegative zeros of $J_0(x)$ and $J_1(x)$ and the concept of "almost periodic."

*See Section 3.7.

Appendices

Appendix I

Introduction to Matrices

A.1 Basic Definitions and Theory

Definition A.1 Matrix

A **matrix A** is any rectangular array of numbers or functions:

$$\mathbf{A} = \begin{pmatrix} a_{11} & a_{12} & \cdots & a_{1n} \\ a_{21} & a_{22} & \cdots & a_{2n} \\ \vdots & \vdots & & \vdots \\ a_{m1} & a_{m2} & \cdots & a_{mn} \end{pmatrix} \tag{1}$$

If a matrix has m rows and n columns, we say that its **size** is m by n (written $m \times n$). An $n \times n$ matrix is called a **square** matrix of order n.

The element, or entry, in the ith row and jth column of an $m \times n$ matrix $\mathbf{A}$ is written a_{ij}. An $m \times n$ matrix $\mathbf{A}$ is then abbreviated as $\mathbf{A} = (a_{ij})_{m \times n}$, or simply $\mathbf{A} = (a_{ij})$. A 1×1 matrix is simply one constant or function.

Definition A.2 Equality of Matrices

Two $m \times n$ matrices $\mathbf{A}$ and $\mathbf{B}$ are **equal** if $a_{ij} = b_{ij}$ for each i and j.

Definition A.3 Column Matrix

A **column matrix X** is any matrix having n rows and one column:

$$\mathbf{X} = \begin{pmatrix} b_{11} \\ b_{21} \\ \vdots \\ b_{n1} \end{pmatrix} = (b_{i1})_{n \times 1}$$

A column matrix is also called a **column vector** or simply a **vector**.

Definition A.4 Multiples of Matrices

A **multiple** of a matrix **A** is defined to be

$$k\mathbf{A} = \begin{pmatrix} ka_{11} & ka_{12} & \cdots & ka_{1n} \\ ka_{21} & ka_{22} & \cdots & ka_{2n} \\ \vdots & \vdots & & \vdots \\ ka_{m1} & ka_{m2} & \cdots & ka_{mn} \end{pmatrix} = (ka_{ij})_{m \times n}$$

where k is a constant or a function.

EXAMPLE 1 Multiples of Matrices

$$\textbf{(a)}\ 5\begin{pmatrix} 2 & -3 \\ 4 & -1 \\ \frac{1}{5} & 6 \end{pmatrix} = \begin{pmatrix} 10 & -15 \\ 20 & -5 \\ 1 & 30 \end{pmatrix} \qquad \textbf{(b)}\ e^t\begin{pmatrix} 1 \\ -2 \\ 4 \end{pmatrix} = \begin{pmatrix} e^t \\ -2e^t \\ 4e^t \end{pmatrix}$$

We note in passing that for any matrix **A**, the product $k\mathbf{A}$ is the same as $\mathbf{A}k$. For example,

$$e^{-3t}\begin{pmatrix} 2 \\ 5 \end{pmatrix} = \begin{pmatrix} 2e^{-3t} \\ 5e^{-3t} \end{pmatrix} = \begin{pmatrix} 2 \\ 5 \end{pmatrix}e^{-3t}$$

Definition A.5 Addition of Matrices

The **sum** of two $m \times n$ matrices **A** and **B** is defined to be the matrix

$$\mathbf{A} + \mathbf{B} = (a_{ij} + b_{ij})_{m \times n}$$

In other words, when adding two matrices of the same size, we add the corresponding elements.

EXAMPLE 2 Matrix Addition

The sum of $\mathbf{A} = \begin{pmatrix} 2 & -1 & 3 \\ 0 & 4 & 6 \\ -6 & 10 & -5 \end{pmatrix}$ and $\mathbf{B} = \begin{pmatrix} 4 & 7 & -8 \\ 9 & 3 & 5 \\ 1 & -1 & 2 \end{pmatrix}$ is

$$\mathbf{A} + \mathbf{B} = \begin{pmatrix} 2+4 & -1+7 & 3+(-8) \\ 0+9 & 4+3 & 6+5 \\ -6+1 & 10+(-1) & -5+2 \end{pmatrix} = \begin{pmatrix} 6 & 6 & -5 \\ 9 & 7 & 11 \\ -5 & 9 & -3 \end{pmatrix}$$

EXAMPLE 3 **A Matrix Written as a Sum of Column Matrices**

The single matrix $\begin{pmatrix} 3t^2 - 2e^t \\ t^2 + 7t \\ 5t \end{pmatrix}$ can be written as the sum of three column vectors:

$$\begin{pmatrix} 3t^2 - 2e^t \\ t^2 + 7t \\ 5t \end{pmatrix} = \begin{pmatrix} 3t^2 \\ t^2 \\ 0 \end{pmatrix} + \begin{pmatrix} 0 \\ 7t \\ 5t \end{pmatrix} + \begin{pmatrix} -2e^t \\ 0 \\ 0 \end{pmatrix} = \begin{pmatrix} 3 \\ 1 \\ 0 \end{pmatrix} t^2 + \begin{pmatrix} 0 \\ 7 \\ 5 \end{pmatrix} t + \begin{pmatrix} -2 \\ 0 \\ 0 \end{pmatrix} e^t$$

The **difference** of two $m \times n$ matrices is defined in the usual manner: $\mathbf{A} - \mathbf{B} = \mathbf{A} + (-\mathbf{B})$, where $-\mathbf{B} = (-1)\mathbf{B}$.

Definition A.6 **Multiplication of Matrices**

Let $\mathbf{A}$ be a matrix having m rows and n columns and $\mathbf{B}$ be a matrix having n rows and p columns. We define the **product AB** to be the $m \times p$ matrix

$$\mathbf{AB} = \begin{pmatrix} a_{11} & a_{12} & \cdots & a_{1n} \\ a_{21} & a_{22} & \cdots & a_{2n} \\ \vdots & \vdots & & \vdots \\ a_{m1} & a_{m2} & \cdots & a_{mn} \end{pmatrix} \begin{pmatrix} b_{11} & b_{12} & \cdots & b_{1p} \\ b_{21} & b_{22} & \cdots & b_{2p} \\ \vdots & \vdots & & \vdots \\ b_{n1} & b_{n2} & \cdots & b_{np} \end{pmatrix}$$

$$= \begin{pmatrix} a_{11}b_{11} + a_{12}b_{21} + \cdots + a_{1n}b_{n1} & \cdots & a_{11}b_{1p} + a_{12}b_{2p} + \cdots + a_{1n}b_{np} \\ a_{21}b_{11} + a_{22}b_{21} + \cdots + a_{2n}b_{n1} & \cdots & a_{21}b_{1p} + a_{22}b_{2p} + \cdots + a_{2n}b_{np} \\ \vdots & & \vdots \\ a_{m1}b_{11} + a_{m2}b_{21} + \cdots + a_{mn}b_{n1} & \cdots & a_{m1}b_{1p} + a_{m2}b_{2p} + \cdots + a_{mn}b_{np} \end{pmatrix}$$

$$= \left(\sum_{k=1}^{n} a_{ik}b_{kj} \right)_{m \times p}$$

Be careful to note in Definition A.6 that the product $\mathbf{AB} = \mathbf{C}$ is defined only when the number of columns in the matrix $\mathbf{A}$ is the same as the number of rows in $\mathbf{B}$. The size of the product can be determined from

$$\mathbf{A}_{m \times n}\mathbf{B}_{n \times p} = \mathbf{C}_{m \times p}$$

Also, you might recognize that the entries in, say, the ith row of the final matrix $\mathbf{AB}$ are formed by using the component definition of the inner or dot product of the ith row of $\mathbf{A}$ with each of the columns of $\mathbf{B}$.

■ EXAMPLE 4 **Multiplication of Matrices**

(a) For $\mathbf{A} = \begin{pmatrix} 4 & 7 \\ 3 & 5 \end{pmatrix}$ and $\mathbf{B} = \begin{pmatrix} 9 & -2 \\ 6 & 8 \end{pmatrix}$,

$$\mathbf{AB} = \begin{pmatrix} 4 \cdot 9 + 7 \cdot 6 & 4 \cdot (-2) + 7 \cdot 8 \\ 3 \cdot 9 + 5 \cdot 6 & 3 \cdot (-2) + 5 \cdot 8 \end{pmatrix} = \begin{pmatrix} 78 & 48 \\ 57 & 34 \end{pmatrix}$$

(b) For $\mathbf{A} = \begin{pmatrix} 5 & 8 \\ 1 & 0 \\ 2 & 7 \end{pmatrix}$ and $\mathbf{B} = \begin{pmatrix} -4 & -3 \\ 2 & 0 \end{pmatrix}$,

$$\mathbf{AB} = \begin{pmatrix} 5 \cdot (-4) + 8 \cdot 2 & 5 \cdot (-3) + 8 \cdot 0 \\ 1 \cdot (-4) + 0 \cdot 2 & 1 \cdot (-3) + 0 \cdot 0 \\ 2 \cdot (-4) + 7 \cdot 2 & 2 \cdot (-3) + 7 \cdot 0 \end{pmatrix} = \begin{pmatrix} -4 & -15 \\ -4 & -3 \\ 6 & -6 \end{pmatrix}$$

■

In general, *matrix multiplication is not commutative*; that is, $\mathbf{AB} \neq \mathbf{BA}$. Observe in part (a) of Example 4 that $\mathbf{BA} = \begin{pmatrix} 30 & 53 \\ 48 & 82 \end{pmatrix}$, whereas in part (b), the product **BA** is not defined since Definition A.6 requires that the first matrix, (in this case **B**) have the same number of columns as the second matrix has rows.

We are particularly interested in the product of a square matrix and a column vector.

■ EXAMPLE 5 **Multiplication of Matrices**

(a) $$\begin{pmatrix} 2 & -1 & 3 \\ 0 & 4 & 5 \\ 1 & -7 & 9 \end{pmatrix}\begin{pmatrix} -3 \\ 6 \\ 4 \end{pmatrix} = \begin{pmatrix} 2 \cdot (-3) + (-1) \cdot 6 + 3 \cdot 4 \\ 0 \cdot (-3) + \quad 4 \cdot 6 + 5 \cdot 4 \\ 1 \cdot (-3) + (-7) \cdot 6 + 9 \cdot 4 \end{pmatrix} = \begin{pmatrix} 0 \\ 44 \\ -9 \end{pmatrix}$$

(b) $$\begin{pmatrix} -4 & 2 \\ 3 & 8 \end{pmatrix}\begin{pmatrix} x \\ y \end{pmatrix} = \begin{pmatrix} -4x + 2y \\ 3x + 8y \end{pmatrix}$$

■

Multiplicative Identity For a given positive integer n, the $n \times n$ matrix

$$\mathbf{I} = \begin{pmatrix} 1 & 0 & 0 & \cdots & 0 \\ 0 & 1 & 0 & \cdots & 0 \\ \vdots & \vdots & & & \vdots \\ 0 & 0 & 0 & \cdots & 1 \end{pmatrix}$$

is called the **multiplicative identity matrix**. It follows from Definition A.6 that for any $n \times n$ matrix $\mathbf{A}$,

$$\mathbf{AI} = \mathbf{IA} = \mathbf{A}$$

Also, it is readily verified that if $\mathbf{X}$ is an $n \times 1$ column matrix, then $\mathbf{IX} = \mathbf{X}$.

Zero Matrix A matrix consisting of all zero entries is called a **zero matrix** and is denoted by $\mathbf{0}$. For example,

$$\mathbf{0} = \begin{pmatrix} 0 \\ 0 \end{pmatrix}, \qquad \mathbf{0} = \begin{pmatrix} 0 & 0 \\ 0 & 0 \end{pmatrix}, \qquad \mathbf{0} = \begin{pmatrix} 0 & 0 \\ 0 & 0 \\ 0 & 0 \end{pmatrix}$$

and so on. If $\mathbf{A}$ and $\mathbf{0}$ are $m \times n$ matrices, then

$$\mathbf{A} + \mathbf{0} = \mathbf{0} + \mathbf{A} = \mathbf{A}$$

Associative Law Although we shall not prove it, matrix multiplication is **associative**. If $\mathbf{A}$ is an $m \times p$ matrix, $\mathbf{B}$ a $p \times r$ matrix, and $\mathbf{C}$ an $r \times n$ matrix, then

$$\mathbf{A}(\mathbf{BC}) = (\mathbf{AB})\mathbf{C}$$

is an $m \times n$ matrix.

Distributive Law If all products are defined, multiplication is **distributive** over addition,

$$\mathbf{A}(\mathbf{B} + \mathbf{C}) = \mathbf{AB} + \mathbf{AC} \quad \text{and} \quad (\mathbf{B} + \mathbf{C})\mathbf{A} = \mathbf{BA} + \mathbf{CA}$$

Determinant of a Matrix Associated with every *square* matrix $\mathbf{A}$ of constants, there is a number called the **determinant of the matrix**, which is denoted by det $\mathbf{A}$.

EXAMPLE 6 Determinant of a Square Matrix

For $\mathbf{A} = \begin{pmatrix} 3 & 6 & 2 \\ 2 & 5 & 1 \\ -1 & 2 & 4 \end{pmatrix}$ we expand det $\mathbf{A}$ by cofactors of the first row:

$$\det \mathbf{A} = \begin{vmatrix} 3 & 6 & 2 \\ 2 & 5 & 1 \\ -1 & 2 & 4 \end{vmatrix} = 3\begin{vmatrix} 5 & 1 \\ 2 & 4 \end{vmatrix} - 6\begin{vmatrix} 2 & 1 \\ -1 & 4 \end{vmatrix} + 2\begin{vmatrix} 2 & 5 \\ -1 & 2 \end{vmatrix}$$

$$= 3(20 - 2) - 6(8 + 1) + 2(4 + 5) = 18 \quad ■$$

It can be proved that a determinant det **A** can be expanded by cofactors using any row or column. If det **A** has a row (or a column) containing many zero entries, then wisdom dictates that we expand the determinant by that row (or column).

Definition A.7 Transpose of a Matrix

The **transpose** of the $m \times n$ matrix (1) is the $n \times m$ matrix $\mathbf{A}^T$ given by

$$\mathbf{A}^T = \begin{pmatrix} a_{11} & a_{21} & \cdots & a_{m1} \\ a_{12} & a_{22} & \cdots & a_{m2} \\ \vdots & \vdots & & \vdots \\ a_{1n} & a_{2n} & \cdots & a_{mn} \end{pmatrix}$$

In other words, the rows of a matrix **A** become the columns of its transpose $\mathbf{A}^T$.

■ EXAMPLE 7 **Transpose of a Matrix**

(a) The transpose of $\mathbf{A} = \begin{pmatrix} 3 & 6 & 2 \\ 2 & 5 & 1 \\ -1 & 2 & 4 \end{pmatrix}$ is $\mathbf{A}^T = \begin{pmatrix} 3 & 2 & -1 \\ 6 & 5 & 2 \\ 2 & 1 & 4 \end{pmatrix}$.

(b) If $\mathbf{X} = \begin{pmatrix} 5 \\ 0 \\ 3 \end{pmatrix}$, then $\mathbf{X}^T = (5 \quad 0 \quad 3)$. ■

Definition A.8 Multiplicative Inverse of a Matrix

Let **A** be an $n \times n$ matrix. If there exists an $n \times n$ matrix **B** such that

$$\mathbf{AB} = \mathbf{BA} = \mathbf{I}$$

where **I** is the multiplicative identity, then **B** is said to be the **multiplicative inverse of A** and is denoted by $\mathbf{B} = \mathbf{A}^{-1}$.

Definition A.9 Nonsingular/Singular Matrices

Let **A** be an $n \times n$ matrix. If det $\mathbf{A} \neq 0$, then **A** is said to be **nonsingular**. If det $\mathbf{A} = 0$, then **A** is said to be **singular**.

The following theorem gives a necessary and sufficient condition for a square matrix to have a multiplicative inverse.

Theorem A.1 Nonsingularity Implies A Has an Inverse

An $n \times n$ matrix $\mathbf{A}$ has a multiplicative inverse $\mathbf{A}^{-1}$ if and only if $\mathbf{A}$ is nonsingular.

The following theorem gives one way of finding the multiplicative inverse for a nonsingular matrix.

Theorem A.2 A Formula for the Inverse of a Matrix

Let $\mathbf{A}$ be an $n \times n$ nonsingular matrix, and let $C_{ij} = (-1)^{i+j}M_{ij}$, where M_{ij} is the determinant of the $(n-1) \times (n-1)$ matrix obtained by deleting the ith row and jth column from $\mathbf{A}$. Then

$$\mathbf{A}^{-1} = \frac{1}{\det \mathbf{A}}(C_{ij})^T \tag{2}$$

Each C_{ij} in Theorem A.2 is simply the **cofactor** (signed minor) of the corresponding entry a_{ij} in $\mathbf{A}$. Note that the transpose is utilized in formula (2).

For future reference, we observe in the case of a 2×2 nonsingular matrix

$$\mathbf{A} = \begin{pmatrix} a_{11} & a_{12} \\ a_{21} & a_{22} \end{pmatrix}$$

that $C_{11} = a_{22}$, $C_{12} = -a_{21}$, $C_{21} = -a_{12}$, and $C_{22} = a_{11}$. Thus

$$\mathbf{A}^{-1} = \frac{1}{\det \mathbf{A}}\begin{pmatrix} a_{22} & -a_{21} \\ -a_{12} & a_{11} \end{pmatrix}^T = \frac{1}{\det \mathbf{A}}\begin{pmatrix} a_{22} & -a_{12} \\ -a_{21} & a_{11} \end{pmatrix} \tag{3}$$

For a 3×3 nonsingular matrix

$$\mathbf{A} = \begin{pmatrix} a_{11} & a_{12} & a_{13} \\ a_{21} & a_{22} & a_{23} \\ a_{31} & a_{32} & a_{33} \end{pmatrix}$$

$$C_{11} = \begin{vmatrix} a_{22} & a_{23} \\ a_{32} & a_{33} \end{vmatrix}, \qquad C_{12} = -\begin{vmatrix} a_{21} & a_{23} \\ a_{31} & a_{33} \end{vmatrix}, \qquad C_{13} = \begin{vmatrix} a_{21} & a_{22} \\ a_{31} & a_{32} \end{vmatrix}$$

and so on. Carrying out the transposition gives

$$\mathbf{A}^{-1} = \frac{1}{\det \mathbf{A}} \begin{pmatrix} C_{11} & C_{21} & C_{31} \\ C_{12} & C_{22} & C_{32} \\ C_{13} & C_{23} & C_{33} \end{pmatrix} \tag{4}$$

EXAMPLE 8 Inverse of a 2 × 2 Matrix

Find the multiplicative inverse for $\mathbf{A} = \begin{pmatrix} 1 & 4 \\ 2 & 10 \end{pmatrix}$.

Solution Since $\det \mathbf{A} = 10 - 8 = 2 \neq 0$, $\mathbf{A}$ is nonsingular. It follows from Theorem A.1 that $\mathbf{A}^{-1}$ exists. From (3) we find

$$\mathbf{A}^{-1} = \frac{1}{2}\begin{pmatrix} 10 & -4 \\ -2 & 1 \end{pmatrix} = \begin{pmatrix} 5 & -2 \\ -1 & \frac{1}{2} \end{pmatrix}$$

■

Not every square matrix has a multiplicative inverse. The matrix $\mathbf{A} = \begin{pmatrix} 2 & 2 \\ 3 & 3 \end{pmatrix}$ is singular since $\det \mathbf{A} = 0$. Hence $\mathbf{A}^{-1}$ does not exist.

EXAMPLE 9 Inverse of a 3 × 3 Matrix

Find the multiplicative inverse for $\mathbf{A} = \begin{pmatrix} 2 & 2 & 0 \\ -2 & 1 & 1 \\ 3 & 0 & 1 \end{pmatrix}$.

Solution Since $\det \mathbf{A} = 12 \neq 0$, the given matrix is nonsingular. The cofactors corresponding to the entries in each row of $\det \mathbf{A}$ are

$$C_{11} = \begin{vmatrix} 1 & 1 \\ 0 & 1 \end{vmatrix} = 1 \qquad C_{12} = -\begin{vmatrix} -2 & 1 \\ 3 & 1 \end{vmatrix} = 5 \qquad C_{13} = \begin{vmatrix} -2 & 1 \\ 3 & 0 \end{vmatrix} = -3$$

$$C_{21} = -\begin{vmatrix} 2 & 0 \\ 0 & 1 \end{vmatrix} = -2 \qquad C_{22} = \begin{vmatrix} 2 & 0 \\ 3 & 1 \end{vmatrix} = 2 \qquad C_{23} = -\begin{vmatrix} 2 & 2 \\ 3 & 0 \end{vmatrix} = 6$$

$$C_{31} = \begin{vmatrix} 2 & 0 \\ 1 & 1 \end{vmatrix} = 2 \qquad C_{32} = -\begin{vmatrix} 2 & 0 \\ -2 & 1 \end{vmatrix} = -2 \qquad C_{33} = \begin{vmatrix} 2 & 2 \\ -2 & 1 \end{vmatrix} = 6$$

It follows from (4) that

$$\mathbf{A}^{-1} = \frac{1}{12}\begin{pmatrix} 1 & -2 & 2 \\ 5 & 2 & -2 \\ -3 & 6 & 6 \end{pmatrix} = \begin{pmatrix} \frac{1}{12} & -\frac{1}{6} & \frac{1}{6} \\ \frac{5}{12} & \frac{1}{6} & -\frac{1}{6} \\ -\frac{1}{4} & \frac{1}{2} & \frac{1}{2} \end{pmatrix}$$

You are urged to verify that $\mathbf{A}^{-1}\mathbf{A} = \mathbf{A}\mathbf{A}^{-1} = \mathbf{I}$. ■

Formula (2) presents obvious difficulties for nonsingular matrices larger than 3×3. For example, to apply (2) to a 4×4 matrix, we would have to calculate *sixteen* 3×3 determinants.* In the case of a large matrix, there are more efficient ways of finding $\mathbf{A}^{-1}$. The curious reader is referred to any text in linear algebra.

Since our goal is to apply the concept of a matrix to systems of linear differential equations in normal form, we need the following definitions.

Definition A.10 Derivative of a Matrix of Functions

If $\mathbf{A}(t) = (a_{ij}(t))_{m \times n}$ is a matrix whose entries are functions differentiable on a common interval, then

$$\frac{d\mathbf{A}}{dt} = \left(\frac{d}{dt} a_{ij}\right)_{m \times n}$$

Definition A.11 Integral of a Matrix of Functions

If $\mathbf{A}(t) = (a_{ij}(t))_{m \times n}$ is a matrix whose entries are functions continuous on a common interval containing t and t_0, then

$$\int_{t_0}^{t} \mathbf{A}(s)\, ds = \left(\int_{t_0}^{t} a_{ij}(s)\, ds\right)_{m \times n}$$

To differentiate (integrate) a matrix of functions, we simply differentiate (integrate) each entry. The derivative of a matrix is also denoted by $\mathbf{A}'(t)$.

EXAMPLE 10 Derivative/Integral of a Matrix

$$\text{If} \quad \mathbf{X}(t) = \begin{pmatrix} \sin 2t \\ e^{3t} \\ 8t - 1 \end{pmatrix}, \quad \text{then} \quad \mathbf{X}'(t) = \begin{pmatrix} \dfrac{d}{dt}\sin 2t \\ \dfrac{d}{dt} e^{3t} \\ \dfrac{d}{dt}(8t - 1) \end{pmatrix} = \begin{pmatrix} 2\cos 2t \\ 3e^{3t} \\ 8 \end{pmatrix}$$

$$\text{and} \qquad \int_0^t \mathbf{X}(s)\, ds = \begin{pmatrix} \int_0^t \sin 2s\, ds \\ \int_0^t e^{3s}\, ds \\ \int_0^t (8s - 1)\, ds \end{pmatrix} = \begin{pmatrix} -\frac{1}{2}\cos 2t + \frac{1}{2} \\ \frac{1}{3}e^{3t} - \frac{1}{3} \\ 4t^2 - t \end{pmatrix}$$

* Strictly speaking, a determinant is a number, but it is sometimes convenient to refer to a determinant as if it were an array.

A.2 Gaussian and Gauss-Jordan Elimination

Matrices are an invaluable aid in solving algebraic systems of n linear equations in n unknowns

$$\begin{aligned} a_{11}x_1 + a_{12}x_2 + \cdots + a_{1n}x_n &= b_1 \\ a_{21}x_1 + a_{22}x_2 + \cdots + a_{2n}x_n &= b_2 \\ \vdots \qquad\qquad\qquad\qquad &\ \ \vdots \\ a_{n1}x_1 + a_{n2}x_2 + \cdots + a_{nn}x_n &= b_n \end{aligned} \tag{5}$$

If $\mathbf{A}$ denotes the matrix of coefficients in (5), we know that Cramer's rule could be used to solve the system whenever $\det \mathbf{A} \neq 0$. However, that rule requires a herculean effort if $\mathbf{A}$ is larger than 3×3. The procedure that we shall now consider has the distinct advantage of being not only an efficient way of handling large systems but also a means of solving consistent systems (5) in which $\det \mathbf{A} = 0$, and a means of solving m linear equations in n unknowns.

Definition A.12 Augmented Matrix

The **augmented matrix** of the system (5) is the $n \times (n + 1)$ matrix

$$\left(\begin{array}{cccc|c} a_{11} & a_{12} & \cdots & a_{1n} & b_1 \\ a_{21} & a_{22} & \cdots & a_{2n} & b_2 \\ \vdots & \vdots & & & \vdots \\ a_{n1} & a_{n2} & \cdots & a_{nn} & b_n \end{array}\right)$$

If $\mathbf{B}$ is the column matrix of the b_i, $i = 1, 2, \ldots, n$, the augmented matrix of (5) is denoted by $(\mathbf{A} \mid \mathbf{B})$.

Elementary Row Operations Recall from algebra that we can transform an algebraic system of equations into an equivalent system (that is, one having the same solution) by multiplying an equation by a nonzero constant, interchanging the positions of any two equations in a system, and adding a nonzero constant multiple of an equation to another equation. These operations on equations in a system are, in turn, equivalent to **elementary row operations** on an augmented matrix:

(i) Multiply a row by a nonzero constant.

(ii) Interchange any two rows.

(iii) Add a nonzero constant multiple of one row to any other row.

Elimination Methods To solve a system such as (5) using an augmented matrix, we use either **Gaussian elimination** or the **Gauss-Jordan elimination method**. In the former method we carry out a succession of elementary row operations until we arrive at an augmented matrix in **row-echelon form**:

(i) The first nonzero entry in a nonzero row is 1.

(ii) In consecutive nonzero rows, the first entry 1 in the lower row appears to the right of the first 1 in the higher row.

(iii) Rows consisting of all 0's are at the bottom of the matrix.

In the Gauss-Jordan method, the row operations are continued until we obtain an augmented matrix that is in **reduced row-echelon form**. A reduced row-echelon matrix has the same three properties just listed in addition to

(iv) A column containing a first entry 1 has 0's everywhere else.

■ EXAMPLE 11 **Row-Echelon/Reduced Row-Echelon Form**

(a) The augmented matrices

$$\left(\begin{array}{ccc|c} 1 & 5 & 0 & 2 \\ 0 & 1 & 0 & -1 \\ 0 & 0 & 0 & 0 \end{array}\right) \quad \text{and} \quad \left(\begin{array}{ccccc|c} 0 & 0 & 1 & -6 & 2 & 2 \\ 0 & 0 & 0 & 0 & 1 & 4 \end{array}\right)$$

are in row-echelon form. You should verify that the three criteria are satisfied.

(b) The augmented matrices

$$\left(\begin{array}{ccc|c} 1 & 0 & 0 & 7 \\ 0 & 1 & 0 & -1 \\ 0 & 0 & 0 & 0 \end{array}\right) \quad \text{and} \quad \left(\begin{array}{ccccc|c} 0 & 0 & 1 & -6 & 0 & -6 \\ 0 & 0 & 0 & 0 & 1 & 4 \end{array}\right)$$

are in reduced row-echelon form. Note that the remaining entries in the columns containing a leading entry 1 are all 0's. ■

Note that in Gaussian elimination, we stop after we have obtained *an* augmented matrix in row-echelon form. In other words, by using different sequences of row operations we may arrive at different row-echelon forms. This method then requires the use of back-substitution. In Gauss-Jordan elimination, we stop when we have obtained *the* augmented matrix in reduced row-echelon form. Any sequence of row operations will lead to the same augmented matrix in reduced row-echelon form. This method does not require back-substitution; the solution of the system will be apparent by inspection of the final matrix. In terms of the equations of the original system, our goal in both methods is simply to make the coefficient of x_1 in the first equation* equal to 1 and then use multiples of that equation to eliminate x_1 from other equations. The process is repeated on the other variables.

To keep track of the row operations on an augmented matrix, we utilize the following notation:

Symbol	Meaning
R_{ij}	Interchange rows i and j
cR_i	Multiply the ith row by the nonzero constant c
$cR_i + R_j$	Multiply the ith row by c and add to the jth row

■ EXAMPLE 12 Solution by Elimination

Solve

$$\begin{aligned} 2x_1 + 6x_2 + x_3 &= 7 \\ x_1 + 2x_2 - x_3 &= -1 \\ 5x_1 + 7x_2 - 4x_3 &= 9 \end{aligned}$$

using **(a)** Gaussian elimination and **(b)** Gauss-Jordan elimination.

Solution

(a) Using row operations on the augmented matrix of the system, we obtain

$$\left(\begin{array}{ccc|c} 2 & 6 & 1 & 7 \\ 1 & 2 & -1 & -1 \\ 5 & 7 & -4 & 9 \end{array}\right) \xrightarrow{R_{12}} \left(\begin{array}{ccc|c} 1 & 2 & -1 & -1 \\ 2 & 6 & 1 & 7 \\ 5 & 7 & -4 & 9 \end{array}\right) \xrightarrow[-5R_1 + R_3]{-2R_1 + R_2} \left(\begin{array}{ccc|c} 1 & 2 & -1 & -1 \\ 0 & 2 & 3 & 9 \\ 0 & -3 & 1 & 14 \end{array}\right)$$

$$\xrightarrow{\frac{1}{2}R_2} \left(\begin{array}{ccc|c} 1 & 2 & -1 & -1 \\ 0 & 1 & \frac{3}{2} & \frac{9}{2} \\ 0 & -3 & 1 & 14 \end{array}\right) \xrightarrow{3R_2 + R_3} \left(\begin{array}{ccc|c} 1 & 2 & -1 & -1 \\ 0 & 1 & \frac{3}{2} & \frac{9}{2} \\ 0 & 0 & \frac{11}{2} & \frac{55}{2} \end{array}\right) \xrightarrow{\frac{2}{11}R_3} \left(\begin{array}{ccc|c} 1 & 2 & -1 & -1 \\ 0 & 1 & \frac{3}{2} & \frac{9}{2} \\ 0 & 0 & 1 & 5 \end{array}\right)$$

The last matrix is in row-echelon form and represents the system

$$\begin{aligned} x_1 + 2x_2 - x_3 &= -1 \\ x_2 + \frac{3}{2}x_3 &= \frac{9}{2} \\ x_3 &= 5 \end{aligned}$$

Substituting $x_3 = 5$ into the second equation then gives $x_2 = -3$. Substituting both these values back into the first equation finally yields $x_1 = 10$.

---

* We can always interchange equations so that the first equation contains the variable x_1.

(b) We start with the last augmented matrix in part (a). Since the first entries in the second and third rows are 1's, we must, in turn, make the remaining entries in the second and third columns 0's:

$$\left(\begin{array}{ccc|c} 1 & 2 & -1 & -1 \\ 0 & 1 & \frac{3}{2} & \frac{9}{2} \\ 0 & 0 & 1 & 5 \end{array}\right) \xrightarrow{-2R_2 + R_1} \left(\begin{array}{ccc|c} 1 & 0 & -4 & -10 \\ 0 & 1 & \frac{3}{2} & \frac{9}{2} \\ 0 & 0 & 1 & 5 \end{array}\right) \xrightarrow[-\frac{3}{2}R_3 + R_2]{4R_3 + R_1} \left(\begin{array}{ccc|c} 1 & 0 & 0 & 10 \\ 0 & 1 & 0 & -3 \\ 0 & 0 & 1 & 5 \end{array}\right)$$

The last matrix is now in reduced row-echelon form. Because of what the matrix means in terms of equations, it is evident that the solution of the system is $x_1 = 10$, $x_2 = -3$, $x_3 = 5$. ■

■ EXAMPLE 13 Gauss-Jordan Elimination

Solve

$$\begin{aligned} x + 3y - 2z &= -7 \\ 4x + y + 3z &= 5 \\ 2x - 5y + 7z &= 19 \end{aligned}$$

Solution We solve the system using Gauss-Jordan elimination:

$$\left(\begin{array}{ccc|c} 1 & 3 & -2 & -7 \\ 4 & 1 & 3 & 5 \\ 2 & -5 & 7 & 19 \end{array}\right) \xrightarrow[-2R_1 + R_3]{-4R_1 + R_2} \left(\begin{array}{ccc|c} 1 & 3 & -2 & -7 \\ 0 & -11 & 11 & 33 \\ 0 & -11 & 11 & 33 \end{array}\right)$$

$$\xrightarrow[-\frac{1}{11}R_3]{-\frac{1}{11}R_2} \left(\begin{array}{ccc|c} 1 & 3 & -2 & -7 \\ 0 & 1 & -1 & -3 \\ 0 & 1 & -1 & -3 \end{array}\right) \xrightarrow[-R_2 + R_3]{-3R_2 + R_1} \left(\begin{array}{ccc|c} 1 & 0 & 1 & 2 \\ 0 & 1 & -1 & -3 \\ 0 & 0 & 0 & 0 \end{array}\right)$$

In this case, the last matrix in reduced row-echelon form implies that the original system of three equations in three unknowns is really equivalent to two equations in three unknowns. Since only z is common to both equations (the nonzero rows), we can assign its values arbitrarily. If we let $z = t$, where t represents any real number, then we see that the system has infinitely many solutions: $x = 2 - t$, $y = -3 + t$, $z = t$. Geometrically, these equations are the parametric equations for the line of intersection of the planes $x + 0y + z = 2$ and $0x + y - z = -3$. ■

A.3 The Eigenvalue Problem

Gauss-Jordan elimination can be used to find the **eigenvectors** of a square matrix.

Definition A.13 Eigenvalues and Eigenvectors

Let $\mathbf{A}$ be an $n \times n$ matrix. A number λ is said to be an **eigenvalue** of $\mathbf{A}$ if there exists a *nonzero* solution vector $\mathbf{K}$ of the linear system

$$\mathbf{AK} = \lambda\mathbf{K} \tag{6}$$

The solution vector $\mathbf{K}$ is said to be an **eigenvector** corresponding to the eigenvalue λ.

The word *eigenvalue* is a combination of German and English terms adapted from the German word *eigenwert*, which, translated literally, is "proper value." Eigenvalues and eigenvectors are also called **characteristic values** and **characteristic vectors**, respectively.

■ EXAMPLE 14 Eigenvector of a Matrix

Verify that $\mathbf{K} = \begin{pmatrix} 1 \\ -1 \\ 1 \end{pmatrix}$ is an eigenvector of the matrix

$$\mathbf{A} = \begin{pmatrix} 0 & -1 & -3 \\ 2 & 3 & 3 \\ -2 & 1 & 1 \end{pmatrix}.$$

Solution By carrying out the multiplication $\mathbf{AK}$, we see that

$$\mathbf{AK} = \begin{pmatrix} 0 & -1 & -3 \\ 2 & 3 & 3 \\ -2 & 1 & 1 \end{pmatrix}\begin{pmatrix} 1 \\ -1 \\ 1 \end{pmatrix} = \begin{pmatrix} -2 \\ 2 \\ -2 \end{pmatrix} = (-2)\begin{pmatrix} 1 \\ -1 \\ 1 \end{pmatrix} = (\underset{\text{eigenvalue}}{\underbrace{-2}})\mathbf{K}$$

We see from the preceding line and Definition A.13 that $\lambda = -2$ is an eigenvalue of $\mathbf{A}$. ■

Using properties of matrix algebra, we can write (6) in the alternative form

$$(\mathbf{A} - \lambda\mathbf{I})\mathbf{K} = \mathbf{0} \tag{7}$$

where $\mathbf{I}$ is the multiplicative identity. If we let

$$\mathbf{K} = \begin{pmatrix} k_1 \\ k_2 \\ \vdots \\ k_n \end{pmatrix}$$

then (7) is the same as

$$\begin{aligned}
(a_{11}-\lambda)k_1 + \quad a_{12}k_2 + \cdots + \quad a_{1n}k_n &= 0\\
a_{21}k_1 + (a_{22}-\lambda)k_2 + \cdots + \quad a_{2n}k_n &= 0\\
\vdots \qquad\qquad\qquad\qquad & \ \ \vdots\\
a_{n1}k_1 + \quad a_{n2}k_2 + \cdots + (a_{nn}-\lambda)k_n &= 0
\end{aligned} \tag{8}$$

Although an obvious solution of (8) is $k_1 = 0$, $k_2 = 0, \ldots, k_n = 0$, we are seeking only nontrivial solutions. Now it is known that a homogeneous system of n linear equations in n unknowns (that is, $b_i = 0$, $i = 1, 2, \ldots, n$, in (5)) has a nontrivial solution if and only if the determinant of the coefficient matrix is equal to zero. Thus to find a nonzero solution $\mathbf{K}$ for (7), we must have

$$\det(\mathbf{A} - \lambda\mathbf{I}) = 0 \tag{9}$$

Inspection of (8) shows that the expansion of $\det(\mathbf{A} - \lambda\mathbf{I})$ by cofactors results in an nth-degree polynomial in λ. The equation (9) is called the **characteristic equation** of $\mathbf{A}$. Thus *the eigenvalues of* $\mathbf{A}$ *are the roots of the characteristic equation.* To find an eigenvector corresponding to an eigenvalue λ, we simply solve the system of equations $(\mathbf{A} - \lambda\mathbf{I})\mathbf{K} = \mathbf{0}$ by applying Gauss-Jordan elimination to the augmented matrix $(\mathbf{A} - \lambda\mathbf{I}|\mathbf{0})$.

■ EXAMPLE 15 Eigenvalues/Eigenvectors

Find the eigenvalues and eigenvectors of $\mathbf{A} = \begin{pmatrix} 1 & 2 & 1 \\ 6 & -1 & 0 \\ -1 & -2 & -1 \end{pmatrix}$.

Solution To expand the determinant in the characteristic equation, we use the cofactors of the second row:

$$\det(\mathbf{A} - \lambda\mathbf{I}) = \begin{vmatrix} 1-\lambda & 2 & 1 \\ 6 & -1-\lambda & 0 \\ -1 & -2 & -1-\lambda \end{vmatrix} = -\lambda^3 - \lambda^2 + 12\lambda = 0$$

From $-\lambda^3 - \lambda^2 + 12\lambda = -\lambda(\lambda + 4)(\lambda - 3) = 0$, we see that the eigenvalues are $\lambda_1 = 0$, $\lambda_2 = -4$, and $\lambda_3 = 3$. To find the eigenvectors, we must now reduce $(\mathbf{A} - \lambda\mathbf{I}|\mathbf{0})$ three times corresponding to the three distinct eigenvalues.

For $\lambda_1 = 0$, we have

$$(\mathbf{A} - 0\mathbf{I}|\mathbf{0}) = \left(\begin{array}{ccc|c} 1 & 2 & 1 & 0 \\ 6 & -1 & 0 & 0 \\ -1 & -2 & -1 & 0 \end{array}\right) \xrightarrow[R_1 + R_3]{-6R_1 + R_2} \left(\begin{array}{ccc|c} 1 & 2 & 1 & 0 \\ 0 & -13 & -6 & 0 \\ 0 & 0 & 0 & 0 \end{array}\right)$$

$$\xrightarrow{-\frac{1}{13}R_2} \left(\begin{array}{ccc|c} 1 & 2 & 1 & 0 \\ 0 & 1 & \frac{6}{13} & 0 \\ 0 & 0 & 0 & 0 \end{array}\right) \xrightarrow{-2R_2 + R_1} \left(\begin{array}{ccc|c} 1 & 0 & \frac{1}{13} & 0 \\ 0 & 1 & \frac{6}{13} & 0 \\ 0 & 0 & 0 & 0 \end{array}\right)$$

Thus we see that $k_1 = -\frac{1}{13}k_3$ and $k_2 = -\frac{6}{13}k_3$. Choosing $k_3 = -13$, we get the eigenvector*

$$\mathbf{K}_1 = \begin{pmatrix} 1 \\ 6 \\ -13 \end{pmatrix}$$

For $\lambda_2 = -4$,

$$(\mathbf{A} + 4\mathbf{I}|\mathbf{0}) = \left(\begin{array}{ccc|c} 5 & 2 & 1 & 0 \\ 6 & 3 & 0 & 0 \\ -1 & -2 & 3 & 0 \end{array}\right) \xrightarrow[R_{31}]{-R_3} \left(\begin{array}{ccc|c} 1 & 2 & -3 & 0 \\ 6 & 3 & 0 & 0 \\ 5 & 2 & 1 & 0 \end{array}\right)$$

$$\xrightarrow[-5R_1 + R_3]{-6R_1 + R_2} \left(\begin{array}{ccc|c} 1 & 2 & -3 & 0 \\ 0 & -9 & 18 & 0 \\ 0 & -8 & 16 & 0 \end{array}\right) \xrightarrow[-\frac{1}{8}R_3]{-\frac{1}{9}R_2} \left(\begin{array}{ccc|c} 1 & 2 & -3 & 0 \\ 0 & 1 & -2 & 0 \\ 0 & 1 & -2 & 0 \end{array}\right) \xrightarrow[-R_2 + R_3]{-2R_2 + R_1} \left(\begin{array}{ccc|c} 1 & 0 & 1 & 0 \\ 0 & 1 & -2 & 0 \\ 0 & 0 & 0 & 0 \end{array}\right)$$

implies $k_1 = -k_3$ and $k_2 = 2k_3$. Choosing $k_3 = 1$ then yields the second eigenvector:

$$\mathbf{K}_2 = \begin{pmatrix} -1 \\ 2 \\ 1 \end{pmatrix}$$

Finally, for $\lambda_3 = 3$, Gauss-Jordan elimination gives

$$(\mathbf{A} - 3\mathbf{I}|\mathbf{0}) = \left(\begin{array}{ccc|c} -2 & 2 & 1 & 0 \\ 6 & -4 & 0 & 0 \\ -1 & -2 & -4 & 0 \end{array}\right) \xrightarrow{\text{row operations}} \left(\begin{array}{ccc|c} 1 & 0 & 1 & 0 \\ 0 & 1 & \frac{3}{2} & 0 \\ 0 & 0 & 0 & 0 \end{array}\right)$$

and so $k_1 = -k_3$ and $k_2 = -\frac{3}{2}k_3$. The choice of $k_3 = -2$ leads to the third eigenvector:

$$\mathbf{K}_3 = \begin{pmatrix} 2 \\ 3 \\ -2 \end{pmatrix}$$

■

* Of course k_3 could be chosen as any nonzero number. In other words, a nonzero constant multiple of an eigenvector is also an eigenvector.

When an $n \times n$ matrix $\mathbf{A}$ possesses n distinct eigenvalues $\lambda_1, \lambda_2, \ldots, \lambda_n$, it can be proved that a set of n linearly independent* eigenvectors $\mathbf{K}_1, \mathbf{K}_2, \ldots, \mathbf{K}_n$ can be found. However, when the characteristic equation has repeated roots, it may not be possible to find n linearly independent eigenvectors for $\mathbf{A}$.

EXAMPLE 16 **Eigenvalues/Eigenvectors**

Find the eigenvalues and eigenvectors of $\mathbf{A} = \begin{pmatrix} 3 & 4 \\ -1 & 7 \end{pmatrix}$.

Solution From the characteristic equation

$$\det(\mathbf{A} - \lambda\mathbf{I}) = \begin{vmatrix} 3-\lambda & 4 \\ -1 & 7-\lambda \end{vmatrix} = (\lambda - 5)^2 = 0$$

we see that $\lambda_1 = \lambda_2 = 5$ is an eigenvalue of multiplicity two. In the case of a 2×2 matrix, there is no need to use Gauss-Jordan elimination. To find the eigenvector(s) corresponding to $\lambda_1 = 5$, we resort to the system $(\mathbf{A} - 5\mathbf{I}|\mathbf{0})$ in its equivalent form

$$\begin{aligned} -2k_1 + 4k_2 &= 0 \\ -k_1 + 2k_2 &= 0 \end{aligned}$$

It is apparent from this system that $k_1 = 2k_2$. Thus if we choose $k_2 = 1$, we find the single eigenvector

$$\mathbf{K}_1 = \begin{pmatrix} 2 \\ 1 \end{pmatrix}$$

■

EXAMPLE 17 **Eigenvalues/Eigenvectors**

Find the eigenvalues and eigenvectors of $\mathbf{A} = \begin{pmatrix} 9 & 1 & 1 \\ 1 & 9 & 1 \\ 1 & 1 & 9 \end{pmatrix}$.

Solution The characteristic equation

$$\det(\mathbf{A} - \lambda\mathbf{I}) = \begin{vmatrix} 9-\lambda & 1 & 1 \\ 1 & 9-\lambda & 1 \\ 1 & 1 & 9-\lambda \end{vmatrix} = -(\lambda - 11)(\lambda - 8)^2 = 0$$

shows that $\lambda_1 = 11$ and that $\lambda_2 = \lambda_3 = 8$ is an eigenvalue of multiplicity two.

---

* Linear independence of column vectors is defined in exactly the same manner as for functions.

For $\lambda_1 = 11$, Gauss-Jordan elimination gives

$$(\mathbf{A} - 11\mathbf{I}|\mathbf{0}) = \left(\begin{array}{rrr|r} -2 & 1 & 1 & 0 \\ 1 & -2 & 1 & 0 \\ 1 & 1 & -2 & 0 \end{array}\right) \xrightarrow{\text{row operations}} \left(\begin{array}{rrr|r} 1 & 0 & -1 & 0 \\ 0 & 1 & -1 & 0 \\ 0 & 0 & 0 & 0 \end{array}\right)$$

Hence $k_1 = k_3$ and $k_2 = k_3$. If $k_3 = 1$, then

$$\mathbf{K}_1 = \begin{pmatrix} 1 \\ 1 \\ 1 \end{pmatrix}$$

Now for $\lambda_2 = 8$ we have

$$(\mathbf{A} - 8\mathbf{I}|\mathbf{0}) = \left(\begin{array}{rrr|r} 1 & 1 & 1 & 0 \\ 1 & 1 & 1 & 0 \\ 1 & 1 & 1 & 0 \end{array}\right) \xrightarrow{\text{row operations}} \left(\begin{array}{rrr|r} 1 & 1 & 1 & 0 \\ 0 & 0 & 0 & 0 \\ 0 & 0 & 0 & 0 \end{array}\right)$$

In the equation $k_1 + k_2 + k_3 = 0$, we are free to select two of the variables arbitrarily. Choosing, on the one hand, $k_2 = 1$, $k_3 = 0$ and, on the other, $k_2 = 0$, $k_3 = 1$, we obtain two linearly independent eigenvectors

$$\mathbf{K}_2 = \begin{pmatrix} -1 \\ 1 \\ 0 \end{pmatrix} \quad \text{and} \quad \mathbf{K}_3 = \begin{pmatrix} -1 \\ 0 \\ 1 \end{pmatrix}$$

■

APPENDIX I EXERCISES

Answers to odd-numbered problems begin on page 456.

A.1

1. If $\mathbf{A} = \begin{pmatrix} 4 & 5 \\ -6 & 9 \end{pmatrix}$ and $\mathbf{B} = \begin{pmatrix} -2 & 6 \\ 8 & -10 \end{pmatrix}$, find **(a)** $\mathbf{A} + \mathbf{B}$, **(b)** $\mathbf{B} - \mathbf{A}$, **(c)** $2\mathbf{A} + 3\mathbf{B}$.

2. If $\mathbf{A} = \begin{pmatrix} -2 & 0 \\ 4 & 1 \\ 7 & 3 \end{pmatrix}$ and $\mathbf{B} = \begin{pmatrix} 3 & -1 \\ 0 & 2 \\ -4 & -2 \end{pmatrix}$, find **(a)** $\mathbf{A} - \mathbf{B}$, **(b)** $\mathbf{B} - \mathbf{A}$, **(c)** $2(\mathbf{A} + \mathbf{B})$.

3. If $\mathbf{A} = \begin{pmatrix} 2 & -3 \\ -5 & 4 \end{pmatrix}$ and $\mathbf{B} = \begin{pmatrix} -1 & 6 \\ 3 & 2 \end{pmatrix}$, find **(a)** $\mathbf{AB}$, **(b)** $\mathbf{BA}$, **(c)** $\mathbf{A}^2 = \mathbf{AA}$, **(d)** $\mathbf{B}^2 = \mathbf{BB}$.

4. If $\mathbf{A} = \begin{pmatrix} 1 & 4 \\ 5 & 10 \\ 8 & 12 \end{pmatrix}$ and $\mathbf{B} = \begin{pmatrix} -4 & 6 & -3 \\ 1 & -3 & 2 \end{pmatrix}$, find **(a)** $\mathbf{AB}$, **(b)** $\mathbf{BA}$.

5. If $\mathbf{A} = \begin{pmatrix} 1 & -2 \\ -2 & 4 \end{pmatrix}$, $\mathbf{B} = \begin{pmatrix} 6 & 3 \\ 2 & 1 \end{pmatrix}$, and $\mathbf{C} = \begin{pmatrix} 0 & 2 \\ 3 & 4 \end{pmatrix}$, find **(a)** $\mathbf{BC}$, **(b)** $\mathbf{A}(\mathbf{BC})$, **(c)** $\mathbf{C}(\mathbf{BA})$, **(d)** $\mathbf{A}(\mathbf{B} + \mathbf{C})$.

6. If $\mathbf{A} = (5 \quad -6 \quad 7)$, $\mathbf{B} = \begin{pmatrix} 3 \\ 4 \\ -1 \end{pmatrix}$, and $\mathbf{C} = \begin{pmatrix} 1 & 2 & 4 \\ 0 & 1 & -1 \\ 3 & 2 & 1 \end{pmatrix}$, find **(a)** $\mathbf{AB}$, **(b)** $\mathbf{BA}$, **(c)** $(\mathbf{BA})\mathbf{C}$, **(d)** $(\mathbf{AB})\mathbf{C}$.

7. If $\mathbf{A} = \begin{pmatrix} 4 \\ 8 \\ -10 \end{pmatrix}$ and $\mathbf{B} = (2 \quad 4 \quad 5)$, find **(a)** $\mathbf{A}^T\mathbf{A}$, **(b)** $\mathbf{B}^T\mathbf{B}$, **(c)** $\mathbf{A} + \mathbf{B}^T$.

8. If $\mathbf{A} = \begin{pmatrix} 1 & 2 \\ 2 & 4 \end{pmatrix}$ and $\mathbf{B} = \begin{pmatrix} -2 & 3 \\ 5 & 7 \end{pmatrix}$, find
(a) $\mathbf{A} + \mathbf{B}^T$, **(b)** $2\mathbf{A}^T - \mathbf{B}^T$, **(c)** $\mathbf{A}^T(\mathbf{A} - \mathbf{B})$.

9. If $\mathbf{A} = \begin{pmatrix} 3 & 4 \\ 8 & 1 \end{pmatrix}$ and $\mathbf{B} = \begin{pmatrix} 5 & 10 \\ -2 & -5 \end{pmatrix}$, find
(a) $(\mathbf{AB})^T$, **(b)** $\mathbf{B}^T\mathbf{A}^T$.

10. If $\mathbf{A} = \begin{pmatrix} 5 & 9 \\ -4 & 6 \end{pmatrix}$ and $\mathbf{B} = \begin{pmatrix} -3 & 11 \\ -7 & 2 \end{pmatrix}$, find
(a) $\mathbf{A}^T + \mathbf{B}^T$, **(b)** $(\mathbf{A} + \mathbf{B})^T$.

In Problems 11–14, write the given sum as a single column matrix.

11. $4\begin{pmatrix} -1 \\ 2 \end{pmatrix} - 2\begin{pmatrix} 2 \\ 8 \end{pmatrix} + 3\begin{pmatrix} -2 \\ 3 \end{pmatrix}$

12. $3t\begin{pmatrix} 2 \\ t \\ -1 \end{pmatrix} + (t-1)\begin{pmatrix} -1 \\ -t \\ 3 \end{pmatrix} - 2\begin{pmatrix} 3t \\ 4 \\ -5t \end{pmatrix}$

13. $\begin{pmatrix} 2 & -3 \\ 1 & 4 \end{pmatrix}\begin{pmatrix} -2 \\ 5 \end{pmatrix} - \begin{pmatrix} -1 & 6 \\ -2 & 3 \end{pmatrix}\begin{pmatrix} -7 \\ 2 \end{pmatrix}$

14. $\begin{pmatrix} 1 & -3 & 4 \\ 2 & 5 & -1 \\ 0 & -4 & -2 \end{pmatrix}\begin{pmatrix} t \\ 2t-1 \\ -t \end{pmatrix} + \begin{pmatrix} -t \\ 1 \\ 4 \end{pmatrix} - \begin{pmatrix} 2 \\ 8 \\ -6 \end{pmatrix}$

In Problems 15–22, determine whether the given matrix is singular or nonsingular. If nonsingular, find $\mathbf{A}^{-1}$.

15. $\mathbf{A} = \begin{pmatrix} -3 & 6 \\ -2 & 4 \end{pmatrix}$

16. $\mathbf{A} = \begin{pmatrix} 2 & 5 \\ 1 & 4 \end{pmatrix}$

17. $\mathbf{A} = \begin{pmatrix} 4 & 8 \\ -3 & -5 \end{pmatrix}$

18. $\mathbf{A} = \begin{pmatrix} 7 & 10 \\ 2 & 2 \end{pmatrix}$

19. $\mathbf{A} = \begin{pmatrix} 2 & 1 & 0 \\ -1 & 2 & 1 \\ 1 & 2 & 1 \end{pmatrix}$

20. $\mathbf{A} = \begin{pmatrix} 3 & 2 & 1 \\ 4 & 1 & 0 \\ -2 & 5 & -1 \end{pmatrix}$

21. $\mathbf{A} = \begin{pmatrix} 2 & 1 & 1 \\ 1 & -2 & -3 \\ 3 & 2 & 4 \end{pmatrix}$

22. $\mathbf{A} = \begin{pmatrix} 4 & 1 & -1 \\ 6 & 2 & -3 \\ -2 & -1 & 2 \end{pmatrix}$

In Problems 23 and 24, show that the given matrix is nonsingular for every real value of t. Find $\mathbf{A}^{-1}(t)$.

23. $\mathbf{A}(t) = \begin{pmatrix} 2e^{-t} & e^{4t} \\ 4e^{-t} & 3e^{4t} \end{pmatrix}$

24. $\mathbf{A}(t) = \begin{pmatrix} 2e^t \sin t & -2e^t \cos t \\ e^t \cos t & e^t \sin t \end{pmatrix}$

In Problems 25–28, find $d\mathbf{X}/dt$.

25. $\mathbf{X} = \begin{pmatrix} 5e^{-t} \\ 2e^{-t} \\ -7e^{-t} \end{pmatrix}$

26. $\mathbf{X} = \begin{pmatrix} \frac{1}{2}\sin 2t - 4\cos 2t \\ -3\sin 2t + 5\cos 2t \end{pmatrix}$

27. $\mathbf{X} = 2\begin{pmatrix} 1 \\ -1 \end{pmatrix} e^{2t} + 4\begin{pmatrix} 2 \\ 1 \end{pmatrix} e^{-3t}$

28. $\mathbf{X} = \begin{pmatrix} 5te^{2t} \\ t \sin 3t \end{pmatrix}$

29. Let $\mathbf{A}(t) = \begin{pmatrix} e^{4t} & \cos \pi t \\ 2t & 3t^2 - 1 \end{pmatrix}$.

Find **(a)** $\dfrac{d\mathbf{A}}{dt}$, **(b)** $\int_0^2 \mathbf{A}(t)\,dt$, **(c)** $\int_0^t \mathbf{A}(s)\,ds$.

30. Let $\mathbf{A}(t) = \begin{pmatrix} \dfrac{1}{t^2+1} & 3t \\ t^2 & t \end{pmatrix}$ and $\mathbf{B}(t) = \begin{pmatrix} 6t & 2 \\ 1/t & 4t \end{pmatrix}$.

Find **(a)** $\dfrac{d\mathbf{A}}{dt}$, **(b)** $\dfrac{d\mathbf{B}}{dt}$, **(c)** $\int_0^1 \mathbf{A}(t)\,dt$, **(d)** $\int_1^2 \mathbf{B}(t)\,dt$,
(e) $\mathbf{A}(t)\mathbf{B}(t)$, **(f)** $\dfrac{d}{dt}\mathbf{A}(t)\mathbf{B}(t)$, **(g)** $\int_1^t \mathbf{A}(s)\mathbf{B}(s)\,ds$.

A.2

In Problems 31–38, solve the given system of equations by either Gaussian elimination or Gauss-Jordan elimination.

31.
$$\begin{aligned} x + y - 2z &= 14 \\ 2x - y + z &= 0 \\ 6x + 3y + 4z &= 1 \end{aligned}$$

32.
$$\begin{aligned} 5x - 2y + 4z &= 10 \\ x + y + z &= 9 \\ 4x - 3y + 3z &= 1 \end{aligned}$$

33.
$$\begin{aligned} y + z &= -5 \\ 5x + 4y - 16z &= -10 \\ x - y - 5z &= 7 \end{aligned}$$

34.
$$\begin{aligned} 3x + y + z &= 4 \\ 4x + 2y - z &= 7 \\ x + y - 3z &= 6 \end{aligned}$$

35.
$$\begin{aligned} 2x + y + z &= 4 \\ 10x - 2y + 2z &= -1 \\ 6x - 2y + 4z &= 8 \end{aligned}$$

36.
$$\begin{aligned} x + 2z &= 8 \\ x + 2y - 2z &= 4 \\ 2x + 5y - 6z &= 6 \end{aligned}$$

37.
$$\begin{aligned} x_1 + x_2 - x_3 - x_4 &= -1 \\ x_1 + x_2 + x_3 + x_4 &= 3 \\ x_1 - x_2 + x_3 - x_4 &= 3 \\ 4x_1 + x_2 - 2x_3 + x_4 &= 0 \end{aligned}$$

38.
$$\begin{aligned} 2x_1 + x_2 + x_3 &= 0 \\ x_1 + 3x_2 + x_3 &= 0 \\ 7x_1 + x_2 + 3x_3 &= 0 \end{aligned}$$

In Problems 39 and 40, use Gauss-Jordan elimination to demonstrate that the given system of equations has no solution.

39. $$\begin{aligned} x + 2y + 4z &= 2 \\ 2x + 4y + 3z &= 1 \\ x + 2y - z &= 7 \end{aligned}$$

40. $$\begin{aligned} x_1 + x_2 - x_3 + 3x_4 &= 1 \\ x_2 - x_3 - 4x_4 &= 0 \\ x_1 + 2x_2 - 2x_3 - x_4 &= 6 \\ 4x_1 + 7x_2 - 7x_3 \qquad &= 9 \end{aligned}$$

A.3

In Problems 41–48, find the eigenvalues and eigenvectors of the given matrix.

41. $\begin{pmatrix} -1 & 2 \\ -7 & 8 \end{pmatrix}$

42. $\begin{pmatrix} 2 & 1 \\ 2 & 1 \end{pmatrix}$

43. $\begin{pmatrix} -8 & -1 \\ 16 & 0 \end{pmatrix}$

44. $\begin{pmatrix} 1 & 1 \\ \frac{1}{4} & 1 \end{pmatrix}$

45. $\begin{pmatrix} 5 & -1 & 0 \\ 0 & -5 & 9 \\ 5 & -1 & 0 \end{pmatrix}$

46. $\begin{pmatrix} 3 & 0 & 0 \\ 0 & 2 & 0 \\ 4 & 0 & 1 \end{pmatrix}$

47. $\begin{pmatrix} 0 & 4 & 0 \\ -1 & -4 & 0 \\ 0 & 0 & -2 \end{pmatrix}$

48. $\begin{pmatrix} 1 & 6 & 0 \\ 0 & 2 & 1 \\ 0 & 1 & 2 \end{pmatrix}$

In Problems 49 and 50, show that the given matrix has complex eigenvalues. Find the eigenvectors of the matrix.

49. $\begin{pmatrix} -1 & 2 \\ -5 & 1 \end{pmatrix}$

50. $\begin{pmatrix} 2 & -1 & 0 \\ 5 & 2 & 4 \\ 0 & 1 & 2 \end{pmatrix}$

51. If $\mathbf{A}(t)$ is a 2×2 matrix of differentiable functions and $\mathbf{X}(t)$ is a 2×1 column matrix of differentiable functions, prove the product rule

$$\frac{d}{dt}[\mathbf{A}(t)\mathbf{X}(t)] = \mathbf{A}(t)\mathbf{X}'(t) + \mathbf{A}'(t)\mathbf{X}(t)$$

52. Derive formula (3). [*Hint*: Find a matrix $\mathbf{B} = \begin{pmatrix} b_{11} & b_{12} \\ b_{21} & b_{22} \end{pmatrix}$ for which $\mathbf{AB} = \mathbf{I}$. Solve for b_{11}, b_{12}, b_{21}, and b_{22}. Then show that $\mathbf{BA} = \mathbf{I}$.]

53. If $\mathbf{A}$ is nonsingular and $\mathbf{AB} = \mathbf{AC}$, show that $\mathbf{B} = \mathbf{C}$.

54. If $\mathbf{A}$ and $\mathbf{B}$ are nonsingular, show that $(\mathbf{AB})^{-1} = \mathbf{B}^{-1}\mathbf{A}^{-1}$.

55. Let $\mathbf{A}$ and $\mathbf{B}$ be $n \times n$ matrices. In general, is $(\mathbf{A} + \mathbf{B})^2 = \mathbf{A}^2 + 2\mathbf{AB} + \mathbf{B}^2$?

Appendix II

Laplace Transforms

$f(t)$	$\mathscr{L}\{f(t)\} = F(s)$
1. 1	$\dfrac{1}{s}$
2. t	$\dfrac{1}{s^2}$
3. t^n	$\dfrac{n!}{s^{n+1}}$, n a positive integer
4. $t^{-1/2}$	$\sqrt{\dfrac{\pi}{s}}$
5. $t^{1/2}$	$\dfrac{\sqrt{\pi}}{2s^{3/2}}$
6. t^{α}	$\dfrac{\Gamma(\alpha+1)}{s^{\alpha+1}}$, $\alpha > -1$
7. $\sin kt$	$\dfrac{k}{s^2+k^2}$
8. $\cos kt$	$\dfrac{s}{s^2+k^2}$
9. $\sin^2 kt$	$\dfrac{2k^2}{s(s^2+4k^2)}$
10. $\cos^2 kt$	$\dfrac{s^2+2k^2}{s(s^2+4k^2)}$
11. e^{at}	$\dfrac{1}{s-a}$
12. $\sinh kt$	$\dfrac{k}{s^2-k^2}$
13. $\cosh kt$	$\dfrac{s}{s^2-k^2}$

$f(t)$	$\mathscr{L}\{f(t)\} = F(s)$
14. $\sinh^2 kt$	$\dfrac{2k^2}{s(s^2 - 4k^2)}$
15. $\cosh^2 kt$	$\dfrac{s^2 - 2k^2}{s(s^2 - 4k^2)}$
16. te^{at}	$\dfrac{1}{(s-a)^2}$
17. $t^n e^{at}$	$\dfrac{n!}{(s-a)^{n+1}}$, n a positive integer
18. $e^{at} \sin kt$	$\dfrac{k}{(s-a)^2 + k^2}$
19. $e^{at} \cos kt$	$\dfrac{s-a}{(s-a)^2 + k^2}$
20. $e^{at} \sinh kt$	$\dfrac{k}{(s-a)^2 - k^2}$
21. $e^{at} \cosh kt$	$\dfrac{s-a}{(s-a)^2 - k^2}$
22. $t \sin kt$	$\dfrac{2ks}{(s^2 + k^2)^2}$
23. $t \cos kt$	$\dfrac{s^2 - k^2}{(s^2 + k^2)^2}$
24. $\sin kt + kt \cos kt$	$\dfrac{2ks^2}{(s^2 + k^2)^2}$
25. $\sin kt - kt \cos kt$	$\dfrac{2k^3}{(s^2 + k^2)^2}$
26. $t \sinh kt$	$\dfrac{2ks}{(s^2 - k^2)^2}$
27. $t \cosh kt$	$\dfrac{s^2 + k^2}{(s^2 - k^2)^2}$
28. $\dfrac{e^{at} - e^{bt}}{a - b}$	$\dfrac{1}{(s-a)(s-b)}$
29. $\dfrac{ae^{at} - be^{bt}}{a - b}$	$\dfrac{s}{(s-a)(s-b)}$
30. $1 - \cos kt$	$\dfrac{k^2}{s(s^2 + k^2)}$

$f(t)$	$\mathscr{L}\{f(t)\} = F(s)$
31. $kt - \sin kt$	$\dfrac{k^3}{s^2(s^2 + k^2)}$
32. $\dfrac{a \sin bt - b \sin at}{ab(a^2 - b^2)}$	$\dfrac{1}{(s^2 + a^2)(s^2 + b^2)}$
33. $\dfrac{\cos bt - \cos at}{a^2 - b^2}$	$\dfrac{s}{(s^2 + a^2)(s^2 + b^2)}$
34. $\sin kt \sinh kt$	$\dfrac{2k^2 s}{s^4 + 4k^4}$
35. $\sin kt \cosh kt$	$\dfrac{k(s^2 + 2k^2)}{s^4 + 4k^4}$
36. $\cos kt \sinh kt$	$\dfrac{k(s^2 - 2k^2)}{s^4 + 4k^4}$
37. $\cos kt \cosh kt$	$\dfrac{s^3}{s^4 + 4k^4}$
38. $J_0(kt)$	$\dfrac{1}{\sqrt{s^2 + k^2}}$
39. $\dfrac{e^{bt} - e^{at}}{t}$	$\ln \dfrac{s - a}{s - b}$
40. $\dfrac{2(1 - \cos kt)}{t}$	$\ln \dfrac{s^2 + k^2}{s^2}$
41. $\dfrac{2(1 - \cosh kt)}{t}$	$\ln \dfrac{s^2 - k^2}{s^2}$
42. $\dfrac{\sin at}{t}$	$\arctan \left(\dfrac{a}{s}\right)$
43. $\dfrac{\sin at \cos bt}{t}$	$\dfrac{1}{2} \arctan \dfrac{a + b}{s} + \dfrac{1}{2} \arctan \dfrac{a - b}{s}$
44. $\delta(t)$	1
45. $\delta(t - t_0)$	e^{-st_0}
46. $e^{at} f(t)$	$F(s - a)$
47. $f(t - a)\mathscr{U}(t - a)$	$e^{-as} F(s)$
48. $g(t)\mathscr{U}(t - a)$	$e^{-as} \mathscr{L}\{g(t + a)\}$

$f(t)$	$\mathscr{L}\{f(t)\} = F(s)$
49. $\mathscr{U}(t - a)$	$\frac{e^{-as}}{s}$
50. $f^{(n)}(t)$	$s^n F(s) - s^{(n-1)} f(0) - \cdots - f^{(n-1)}(0)$
51. $t^n f(t)$	$(-1)^n \frac{d^n}{ds^n} F(s)$
52. $\int_0^t f(\tau) g(t - \tau)\, d\tau$	$F(s)G(s)$

Appendix III

Review of Power Series

In Section 6.1, power series are used to find solutions of certain kinds of linear differential equations with variable coefficients. Because of this, it is appropriate to list some of the more important facts about power series. For a more in-depth review of the infinite series concept, you should consult a calculus text.

- **Definition of a Power Series** A **power series** in $x - a$ is an infinite series of the form $\sum_{n=0}^{\infty} c_n(x - a)^n$. A series such as this is also said to be a **power series centered at a**. For example, $\sum_{n=1}^{\infty} \frac{(-1)^{n+1}}{n^2} x^n$ is a power series in x; the series is centered at zero.

- **Convergence** For a specified value of x, a power series is a series of constants. If the series equals a finite real constant for the given x, then the series is said to **converge** at x. If the series does not converge at x, it is said to **diverge** at x.

- **Interval of Convergence** Every power series has an **interval of convergence**. The interval of convergence is the set of all numbers for which the series converges.

- **Radius of Convergence** Every interval of convergence has a **radius of convergence** R. For a power series $\sum_{n=0}^{\infty} c_n(x - a)^n$, we have just three possibilities:

  (i) The series converges only at its center a. In this case $R = 0$.

  (ii) The series converges for all x satisfying $|x - a| < R$, where $R > 0$. The series diverges for $|x - a| > R$.

  (iii) The series converges for all x. In this case we write $R = \infty$.

- **Convergence at an Endpoint** Recall that the absolute-value inequality $|x - a| < R$ is equivalent to $-R < x - a < R$, or $a - R < x < a + R$. If a power series converges for $|x - a| < R$, where $R > 0$, it may or may not converge at the endpoints of the interval $a - R < x < a + R$. Figure III.1 shows four possible intervals of convergence.

- **Absolute Convergence** Within its interval of convergence, a power series converges absolutely. In other words, for x in the interval of convergence the series of absolute values $\sum_{n=0}^{\infty} |c_n||(x - a)^n|$ converges.

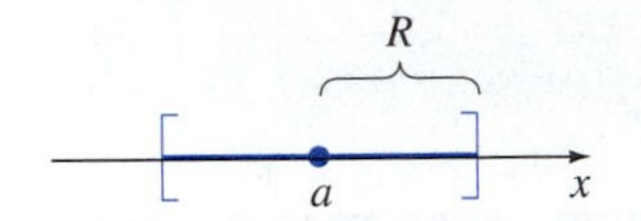

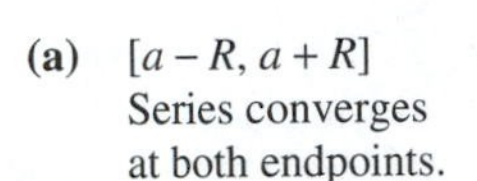

(a) $[a-R, a+R]$
Series converges at both endpoints.

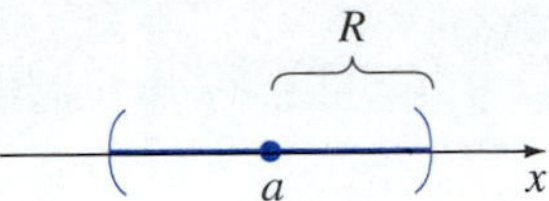

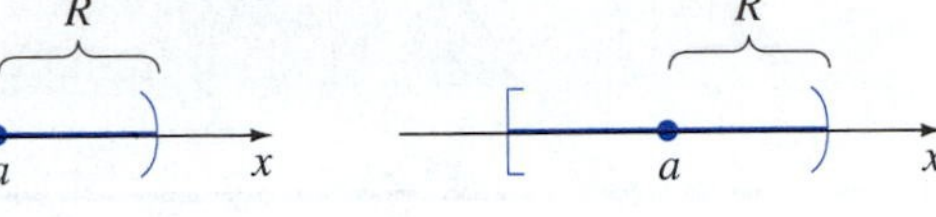

(b) $(a-R, a+R)$
Series diverges at both endpoints.

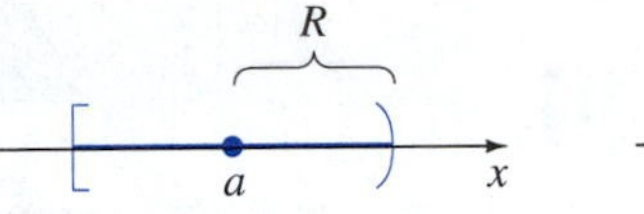

(c) $[a-R, a+R)$
Series converges at $a-R$, diverges at $a+R$.

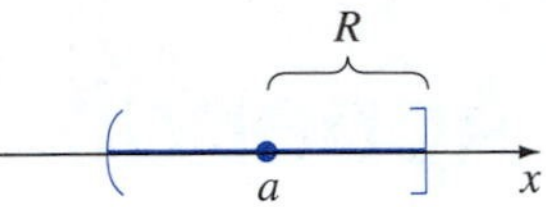

(d) $(a-R, a+R]$
Series diverges at $a-R$, converges at $a+R$.

Figure III.1.

- **Finding the Interval of Convergence** Convergence of a power series can often be determined by the **ratio test**:

$$\lim_{n\to\infty}\left|\frac{c_{n+1}}{c_n}\right| |x-a| = L$$

The series will converge absolutely for those values of x for which $L < 1$. From this test, we see that the radius of convergence is given by

$$R = \lim_{n\to\infty}\left|\frac{c_n}{c_{n+1}}\right| \tag{1}$$

provided the limit exists.

- **A Power Series Represents a Function** A power series represents a function

$$f(x) = \sum_{n=0}^{\infty} c_n(x-a)^n = c_0 + c_1(x-a) + c_2(x-a)^2 + c_3(x-a)^3 + \cdots$$

whose domain is the interval of convergence of the series. If the series has a radius of convergence $R > 0$, then f is continuous, differentiable, and integrable on the interval $(a - R, a + R)$. Moreover, $f'(x)$ and $\int f(x)\,dx$ can be found from term-by-term differentiation and integration:

$$f'(x) = c_1 + 2c_2(x-a) + 3c_3(x-a)^2 + \cdots = \sum_{n=1}^{\infty} nc_n(x-a)^{n-1}$$

$$\int f(x)\,dx = C + c_0(x-a) + c_1\frac{(x-a)^2}{2} + c_2\frac{(x-a)^3}{3} + \cdots = C + \sum_{n=0}^{\infty} c_n\frac{(x-a)^{n+1}}{n+1}$$

Although the radius of convergence for both these series is R, the interval of convergence may differ from the original series in that convergence at an endpoint may be either lost by differentiation or gained through integration.

- **Series That Are Identically Zero** If $\sum_{n=0}^{\infty} c_n(x-a)^n = 0$, $R > 0$, for all real numbers x in the interval of convergence, then $c_n = 0$ for all n.

- **Analytic at a Point** In calculus, it is seen that functions such as e^x, $\cos x$, and $\ln(x-1)$ can be represented by power series expansions in either Maclaurin or Taylor series. We say that a function f is **analytic at point *a*** if it can be represented by a power series in $x - a$ with a positive radius of convergence. The notion of analyticity at a point is important in Sections 6.2 and 6.3.

- **Arithmetic of Power Series** Power series can be combined through the operations of addition, multiplication, and division. The procedures for power series are similar to the way in which two polynomials are added, multiplied, and divided; that is, we add coefficients of like powers of x, use the distributive law and collect like terms, and perform long division. For example, if the power series $f(x) = \sum_{n=0}^{\infty} c_n x^n$ and $g(x) = \sum_{n=0}^{\infty} b_n x^n$ both converge for $|x| < R$, then

$$f(x) + g(x) = (c_0 + b_0) + (c_1 + b_1)x + (c_2 + b_2)x^2 + \cdots$$

$$f(x)g(x) = c_0b_0 + (c_0b_1 + c_1b_0)x + (c_0b_2 + c_1b_1 + c_2b_0)x^2 + \cdots$$

EXAMPLE 1 Interval of Convergence

Find the interval of convergence of the power series $\displaystyle\sum_{n=1}^{\infty} \frac{(x-3)^n}{2^n n}$.

Solution The power series is centered at 3. From (1), the radius of convergence is

$$R = \lim_{n\to\infty} \frac{2^{n+1}(n+1)}{2^n n} = 2$$

The series converges absolutely for $|x - 3| < 2$, or $1 < x < 5$. At the left endpoint $x = 1$, we find that the series of constants $\sum_{n=1}^{\infty}((-1)^n/n)$ is convergent by the alternating series test. At the right endpoint $x = 5$, we find that the series is the divergent harmonic series $\sum_{n=1}^{\infty}(1/n)$. Thus the interval of convergence is $[1,5)$. ■

EXAMPLE 2 Multiplication of Two Power Series

Find the first four terms of a power series in x for $e^x \cos x$.

Solution From calculus, the Maclaurin series for e^x and $\cos x$ are, respectively,

$$e^x = 1 + x + \frac{x^2}{2} + \frac{x^3}{6} + \frac{x^4}{24} + \cdots \quad \text{and} \quad \cos x = 1 - \frac{x^2}{2} + \frac{x^4}{24} - \cdots$$

Multiplying out and collecting like terms yield

$$e^x \cos x = \left(1 + x + \frac{x^2}{2} + \frac{x^3}{6} + \frac{x^4}{24} + \cdots\right)\left(1 - \frac{x^2}{2} + \frac{x^4}{24} - \cdots\right)$$

$$= 1 + (1)x + \left(-\frac{1}{2} + \frac{1}{2}\right)x^2 + \left(-\frac{1}{2} + \frac{1}{6}\right)x^3 + \left(\frac{1}{24} - \frac{1}{4} + \frac{1}{24}\right)x^4 + \cdots$$

$$= 1 + x - \frac{x^3}{3} - \frac{x^4}{6} + \cdots$$ ■

In Example 2, the interval of convergence for the Maclaurin series for both e^x and $\cos x$ is $(-\infty, \infty)$. Consequently, the interval of convergence for the power series for $e^x \cos x$ is also $(-\infty, \infty)$.

■ EXAMPLE 3 **Division by a Power Series**

Find the first four terms of a power series in x for $\sec x$.

Solution One way of proceeding is to use the Maclaurin series for $\cos x$ given in Example 2 and then to use long division. Since $\sec x = 1/\cos x$, we have

$$\begin{array}{r|l}
 & 1 + \frac{x^2}{2} + \frac{5x^4}{24} + \frac{61x^6}{720} + \cdots \\
\hline
\cos x = 1 - \frac{x^2}{2} + \frac{x^4}{24} - \frac{x^6}{720} + \cdots & 1 \\
 & 1 - \frac{x^2}{2} + \frac{x^4}{24} - \frac{x^6}{720} + \cdots \\
\cline{2-2}
 & \quad \frac{x^2}{2} - \frac{x^4}{24} + \frac{x^6}{720} - \cdots \\
 & \quad \frac{x^2}{2} - \frac{x^4}{4} + \frac{x^6}{48} - \cdots \\
\cline{2-2}
 & \qquad \frac{5x^4}{24} - \frac{7x^6}{360} + \cdots \\
 & \qquad \frac{5x^4}{24} - \frac{5x^6}{48} + \cdots \\
\cline{2-2}
 & \qquad\qquad \frac{61x^6}{720} - \cdots
\end{array}$$

Thus

$$\sec x = 1 + \frac{x^2}{2} + \frac{5x^4}{24} + \frac{61x^6}{720} + \cdots \tag{2}$$

The interval of convergence of this series is $(-\pi/2, \pi/2)$. (Why?) ■

The procedures illustrated in Examples 2 and 3 are obviously tedious to do by hand. Problems of this sort can be done with minimal fuss using a computer algebra system such as *Mathematica* or *Maple*. In *Mathematica*, the division in Example 3 is avoided by using the command **Series[Sec[x], {x, 0, 8}]**.

It is important that you become adept at simplifying the sum of two or more power series, each series expressed in summation (sigma) notation, to an expression with a single Σ. This often requires a shift of the summation indices.

EXAMPLE 4 Adding Two Power Series

Write $\sum_{n=1}^{\infty} 2nc_n x^{n-1} + \sum_{n=0}^{\infty} 6c_n x^{n+1}$ as one series.

Solution In order to add the series, we require that both summation indices start with the same number and that the powers of x in each series be "in phase"; that is, if one series starts with a multiple of, say, x to the first power, then we want the other series to start with the same power. By writing

series starts with x for $n = 2$ ↓ series starts with x for $n = 0$ ↓

$$\sum_{n=1}^{\infty} 2nc_n x^{n-1} + \sum_{n=0}^{\infty} 6c_n x^{n+1} = 2 \cdot 1 \cdot c_1 x^0 + \sum_{n=2}^{\infty} 2nc_n x^{n-1} + \sum_{n=0}^{\infty} 6c_n x^{n+1} \tag{3}$$

we have both series on the right side start with x^1. To get the same summation index, we are inspired by the exponents of x; we let $k = n - 1$ in the first series and at the same time let $k = n + 1$ in the second series. Thus the right side of (3) becomes

$$2c_1 + \sum_{k=1}^{\infty} 2(k+1)c_{k+1}x^k + \sum_{k=1}^{\infty} 6c_{k-1}x^k \tag{4}$$

Recall that the summation index is a "dummy" variable. The fact that $k = n - 1$ in one case and $k = n + 1$ in the other should cause no confusion if you keep in mind that it is the *value* of the summation index that is important. In both cases k takes on the same successive values 1, 2, 3, . . . for $n = 2, 3, 4, \ldots$ (for $k = n - 1$) and $n = 0, 1, 2, \ldots$ (for $k = n + 1$).

We are now in a position to add the series in (4) term by term:

$$\sum_{n=1}^{\infty} 2nc_n x^{n-1} + \sum_{n=0}^{\infty} 6c_n x^{n+1} = 2c_1 + \sum_{k=1}^{\infty} [2(k+1)c_{k+1} + 6c_{k-1}]x^k \tag{5}$$

If you are not convinced, then write out a few terms on both sides of (5). ■

Answers to Selected Odd-Numbered Problems

Exercises 1.1, Page 9

1. linear, second-order **3.** nonlinear, first-order **5.** linear, fourth-order **7.** nonlinear, second-order **9.** linear, third-order **25.** one interval is $-\pi/10 < t < \pi/10$

Exercises 1.2, Page 17

1. $y = \dfrac{1}{1 - 4e^{-t}}$ **3.** $x = -\cos t + 8 \sin t$ **5.** $x = \dfrac{\sqrt{3}}{4}\cos t + \dfrac{1}{4}\sin t$ **7.** $y = \dfrac{3}{2}e^{t} - \dfrac{1}{2}e^{-t}$ **9.** $y = 5e^{-(t+1)}$

11. $y = 0$, $y = t^3$ **13.** half-planes defined by either $y > 0$ or $y < 0$ **15.** half-planes defined by either $t > 0$ or $t < 0$ **17.** the regions defined by $y > 2$, or $y < -2$, or $-2 < y < 2$ **19.** any region not containing $(0, 0)$ **21.** yes **23.** no **25. (b)** $y = \tan t$; $y = \tan t$ is not differentiable on $(-2, 2)$ **(c)** $-\pi/2 < t < \pi/2$

Exercises 2.1, Page 34

1.

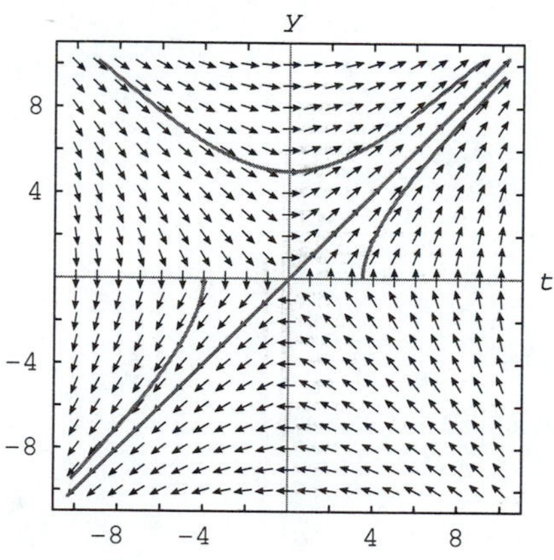

3.

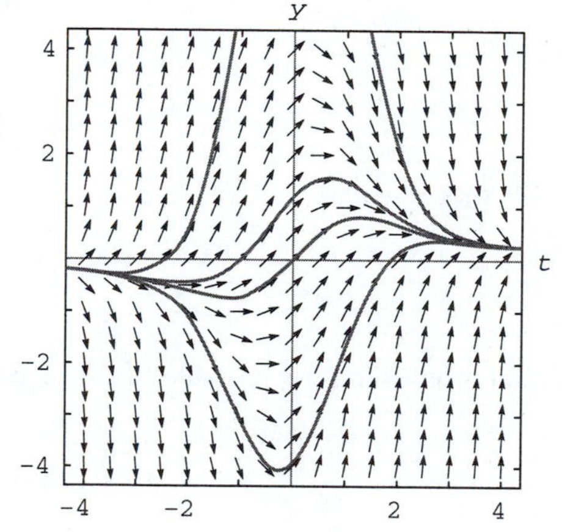

5.

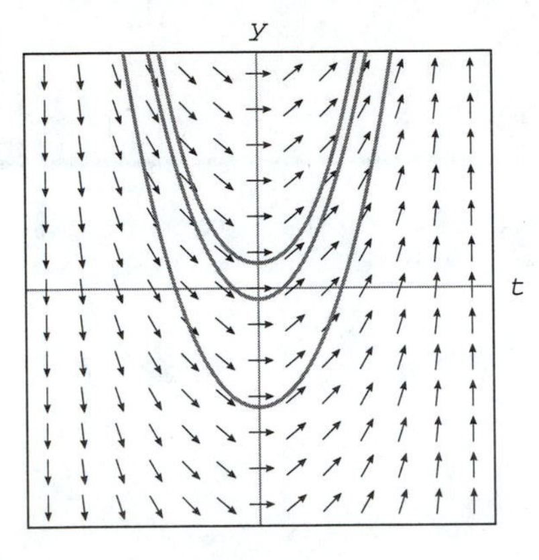

7.

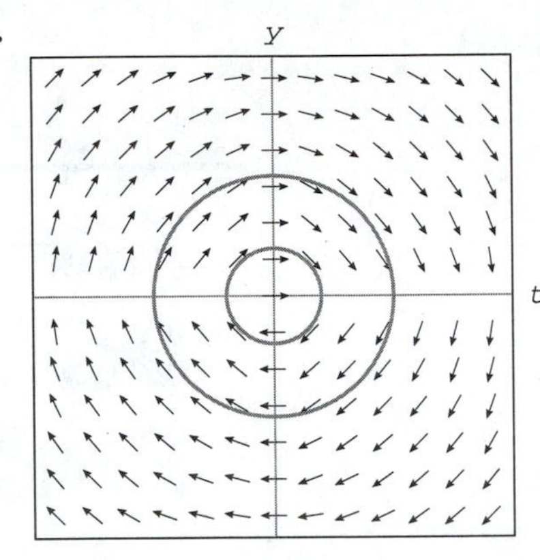

9.

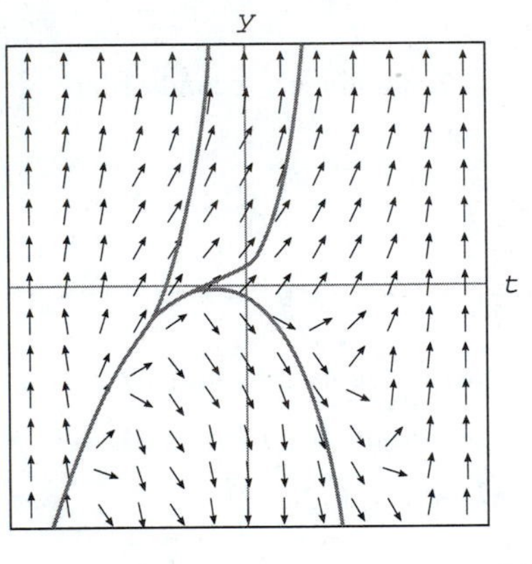

11.

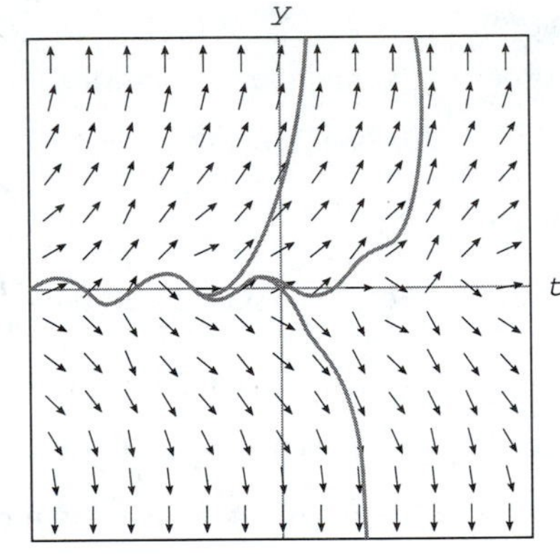

13. 0 is asymptotically stable (attractor); 3 is unstable (repeller)

3
0

15. 2 is unstable (not a repeller)

2

17. -2 is unstable (repeller); 0 is unstable (not a repeller); 2 is asymptotically stable (attractor)

2
0
-2

19. -1 is asymptotically stable (attractor); 0 is unstable (repeller)

0
-1

21.

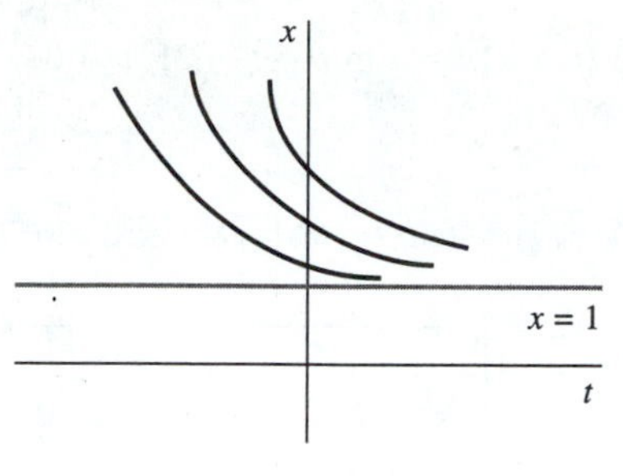

(a)

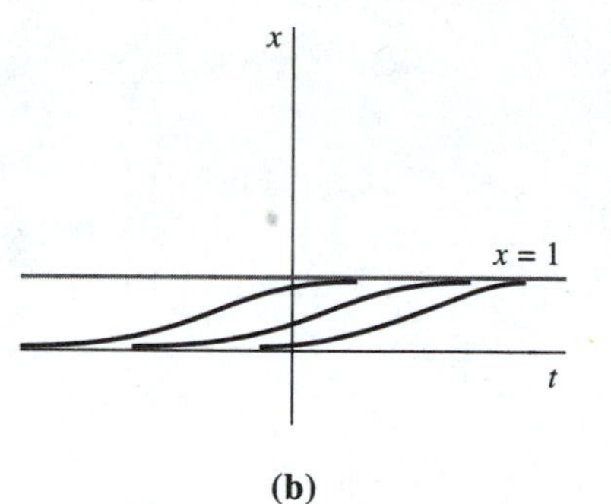

(b)

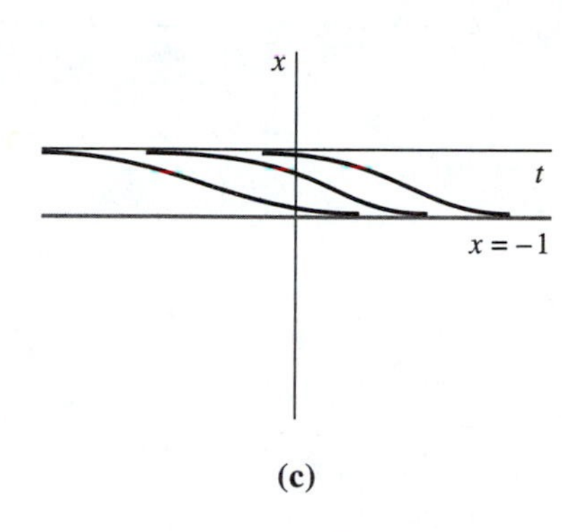

(c)

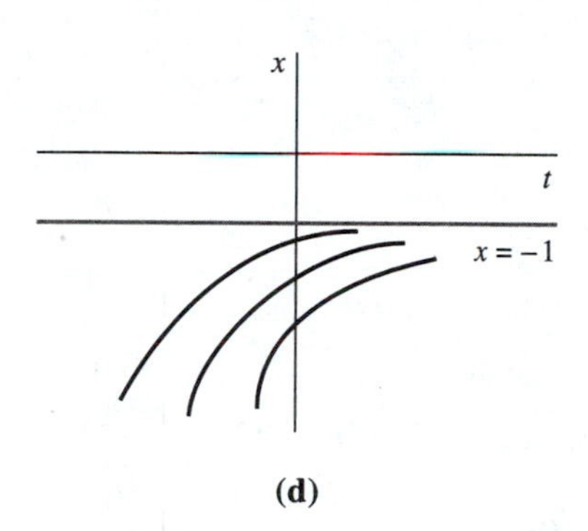

(d)

23. **(a)** mg/k **(b)** $\sqrt{mg/k}$

25. unstable for any positive integer n; unstable for n even and stable for n odd

Exercises 2.2, Page 45

1. $y = -\frac{1}{5}\cos 5t + c$ **3.** $y = \frac{1}{3}e^{-3t} + c$ **5.** $y = ct^4$ **7.** $-3e^{-2y} = 2e^{3x} + c$

9. $\frac{x^3}{3}\ln|x| - \frac{1}{9}x^3 = \frac{y^2}{2} + 2y + \ln|y| + c$ **11.** $4\cos y = 2t + \sin 2t + c$ **13.** $(e^t + 1)^{-2} + 2(e^y + 1)^{-1} = c$

15. $S = ce^{kr}$ **17.** $P = \frac{ce^t}{1 + ce^t}$ **19.** $(y + 3)^5 = c_1(x + 4)^5 e^{y - x}$ **21.** $x = \tan(4t - 3\pi/4)$

23. $xy = e^{-(1 + 1/x)}$ **25.** **(a)** $y = 3\frac{1 - e^{6t}}{1 + e^{6t}}$ **(b)** $y = 3$ **(c)** $y = 3\frac{2 - e^{6t-2}}{2 + e^{6t-2}}$

27. $t\ln|t| + y = ct$ **29.** $t + y\ln|t| = cy$ **31.** $\ln(x^2 + y^2) + 2\tan^{-1}(y/x) = c$

Exercises 2.3, Page 56

1. $y = ce^{5t}, -\infty < t < \infty$ **3.** $y = \frac{1}{4}e^{3t} + ce^{-t}, -\infty < t < \infty$ **5.** $y = \frac{1}{3} + ce^{-t^3}, -\infty < t < \infty$

7. $y = t^{-1}\ln t + ct^{-1}, t > 0$ **9.** $y = ct - t\cos t, t > 0$ **11.** $y = \frac{1}{7}t^3 - \frac{1}{5}t + ct^{-4}, t > 0$

13. $y = \frac{1}{2t^2}e^t + \frac{c}{t^2}e^{-t}, t > 0$ **15.** $x = 2y^6 + cy^4, y > 0$ **17.** $y = \sin t + c\cos t, -\pi/2 < t < \pi/2$

19. $(t + 1)e^t y = t^2 + c, t > -1$ **21.** $(\sec\theta + \tan\theta)r = \theta - \cos\theta + c, -\pi/2 < \theta < \pi/2$

23. $y = \dfrac{e^t}{t} + \dfrac{2-e}{t}, t > 0$ **25.** $i = E/R + (i_0 - E/R)e^{-Rt/L}, -\infty < t < \infty$ **27.** $(t+1)x = t \ln t - t + 21, t > 0$

29. $y = \begin{cases} \dfrac{1}{2}(1-e^{-2t}), 0 \le t \le 3 \\ \dfrac{1}{2}(e^6 - 1)e^{-2t}, t > 3 \end{cases}$ **31.** $y = \begin{cases} \dfrac{1}{2} + \dfrac{3}{2}e^{-t^2}, 0 \le t \le 1 \\ \left(\dfrac{1}{2}e + \dfrac{3}{2}\right)e^{-t^2}, t > 1 \end{cases}$ **33.** **(a)** $y = (\sqrt{\pi}/2)e^{t^2}\operatorname{erfc}(t)$ **(b)** $y(2) \approx 0.226339$

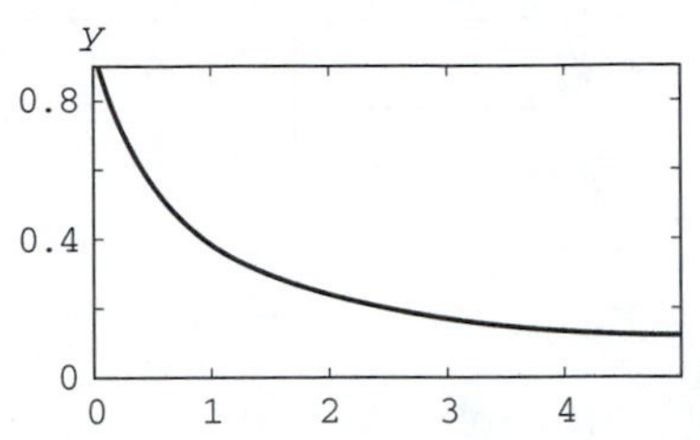

35. $y^3 = 1 + ct^{-3}$ **37.** $y^{-3} = -\dfrac{9}{5}t^{-1} + \dfrac{49}{5}t^{-6}$

Exercises 2.4, Page 72

1. 7.9 years; 10 years **3.** 760 **5.** 11 hr **7.** 136.5 hr

9. $I(15) = 0.00098I_0$, or $I(15)$ is approximately 0.1% of I_0 **11.** 15,600 years

13. $T(1) = 36.67$ degrees; approximately 3.06 min **15.** $i(t) = \frac{3}{5} - \frac{3}{5}e^{-500t}$; $i \to \frac{3}{5}$ as $t \to \infty$

17. $q(t) = \frac{1}{100} - \frac{1}{100}e^{-50t}$; $i(t) = \frac{1}{2}e^{-50t}$ **19.** $i(t) = \begin{cases} 60 - 60e^{-t/10}, & 0 \le t \le 20 \\ 60(e^2 - 1)e^{-t/10}, & t > 20 \end{cases}$ **21.** $A(t) = 200 - 170e^{-t/50}$

23. $A(t) = 1000 - 1000e^{-t/100}$ **25.** 64.38 lb

27. **(a)** $\dfrac{dA}{dt} = k(M - A), k > 0$ **(b)** $\dfrac{dA}{dt} = k_1(M - A) - k_2A, k_1 > 0, k_2 > 0$

(c) For part (a), $A = M(1 - e^{-kt})$. As $t \to \infty$, $A \to M$. Over a long period of time practically all the material will be memorized.

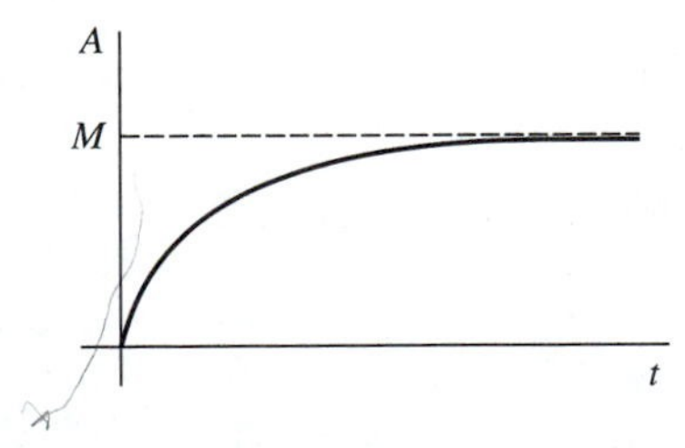

For part (b), $A = \dfrac{k_1M}{k_1 + k_2}(1 - e^{-(k_1+k_2)t})$. As $t \to \infty$, $A \to \dfrac{k_1M}{k_1 + k_2}$. If $k_1 > 0$ and $k_2 > 0$, $\dfrac{k_1M}{k_1 + k_2} < M$, and so the material will never be completely memorized.

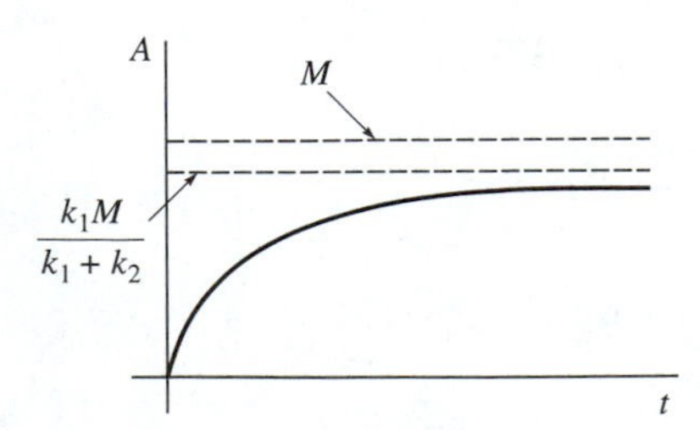

29. 1834; 2000 **31.** 1,000,000; 52.9 months **33.** For $\alpha \neq \beta$, $X = \dfrac{\alpha - c\beta e^{(\alpha-\beta)kt}}{1 - ce^{(\alpha-\beta)kt}}$; for $\alpha = \beta$, $X = \alpha - \dfrac{1}{kt + c}$

35. (a) $v(t) = \sqrt{\dfrac{mg}{k}} \tanh\left(\sqrt{\dfrac{kg}{m}}\, t + c_1\right)$ where $c_1 = \tanh^{-1}\left(\sqrt{\dfrac{k}{mg}}\, v_0\right)$ **(b)** $\sqrt{\dfrac{mg}{k}}$

37. (a) $\dfrac{dy}{dx} = -\dfrac{y}{\sqrt{s^2 - y^2}}$ **(b)** $x = 10 \ln\left(\dfrac{10 + \sqrt{100 - y^2}}{y}\right) - \sqrt{100 - y^2}$

Exercises 2.5, Page 88

1. (a) $y = 1 - t + \tan(t + \pi/4)$

(b) $h = 0.1$

t_n	y_n	True Value
0.00	2.0000	2.0000
0.10	2.1000	2.1230
0.20	2.2440	2.3085
0.30	2.4525	2.5958
0.40	2.7596	3.0650
0.50	3.2261	3.9082

$h = 0.05$

t_n	y_n	True Value
0.00	2.0000	2.0000
0.05	2.0500	2.0554
0.10	2.1105	2.1230
0.15	2.1838	2.2061
0.20	2.2727	2.3085
0.25	2.3812	2.4358
0.30	2.5142	2.5958
0.35	2.6788	2.7997
0.40	2.8845	3.0650
0.45	3.1455	3.4189
0.50	3.4823	3.9082

3. (a)

t_n	y_n
1.00	5.0000
1.10	3.8000
1.20	2.9800
1.30	2.4260
1.40	2.0582
1.50	1.8207

(b)

t_n	y_n
1.00	5.0000
1.05	4.4000
1.10	3.8950
1.15	3.4707
1.20	3.1151
1.25	2.8179
1.30	2.5702
1.35	2.3647
1.40	2.1950
1.45	2.0557
1.50	1.9424

5. (a)

t_n	y_n
0.00	0.0000
0.10	0.1000
0.20	0.2010
0.30	0.3050
0.40	0.4143
0.50	0.5315

(b)

t_n	y_n
0.00	0.0000
0.05	0.0500
0.10	0.1001
0.15	0.1506
0.20	0.2018
0.25	0.2538
0.30	0.3070
0.35	0.3617
0.40	0.4183
0.45	0.4770
0.50	0.5384

7. (a)

t_n	y_n
0.00	0.0000
0.10	0.1000
0.20	0.1905
0.30	0.2731
0.40	0.3492
0.50	0.4198

(b)

t_n	y_n
0.00	0.0000
0.05	0.0500
0.10	0.0976
0.15	0.1429
0.20	0.1863
0.25	0.2278
0.30	0.2676
0.35	0.3058
0.40	0.3427
0.45	0.3782
0.50	0.4124

9. (a)

t_n	y_n
0.00	0.5000
0.10	0.5250
0.20	0.5431
0.30	0.5548
0.40	0.5613
0.50	0.5639

(b)

t_n	y_n
0.00	0.5000
0.05	0.5125
0.10	0.5232
0.15	0.5322
0.20	0.5395
0.25	0.5452
0.30	0.5496
0.35	0.5527
0.40	0.5547
0.45	0.5559
0.50	0.5565

11. (a)

t_n	y_n
1.00	1.0000
1.10	1.0000
1.20	1.0191
1.30	1.0588
1.40	1.1231
1.50	1.2194

(b)

t_n	y_n
1.00	1.0000
1.05	1.0000
1.10	1.0049
1.15	1.0147
1.20	1.0298
1.25	1.0506
1.30	1.0775
1.35	1.1115
1.40	1.1538
1.45	1.2057
1.50	1.2696

13. (a) $h = 0.1$

t_n	y_n
1.00	5.0000
1.10	3.9900
1.20	3.2545
1.30	2.7236
1.40	2.3451
1.50	2.0801

$h = 0.05$

t_n	y_n
1.00	5.0000
1.05	4.4475
1.10	3.9763
1.15	3.5751
1.20	3.2342
1.25	2.9452
1.30	2.7009
1.35	2.4952
1.40	2.3226
1.45	2.1786
1.50	2.0592

(b) $h = 0.1$

t_n	y_n
0.00	0.0000
0.10	0.1005
0.20	0.2030
0.30	0.3098
0.40	0.4234
0.50	0.5470

$h = 0.05$

t_n	y_n
0.00	0.0000
0.05	0.0501
0.10	0.1004
0.15	0.1512
0.20	0.2028
0.25	0.2554
0.30	0.3095
0.35	0.3652
0.40	0.4230
0.45	0.4832
0.50	0.5465

(c) $h = 0.1$

t_n	y_n
0.00	0.0000
0.10	0.0952
0.20	0.1822
0.30	0.2622
0.40	0.3363
0.50	0.4053

$h = 0.05$

t_n	y_n
0.00	0.0000
0.05	0.0488
0.10	0.0953
0.15	0.1397
0.20	0.1823
0.25	0.2231
0.30	0.2623
0.35	0.3001
0.40	0.3364
0.45	0.3715
0.50	0.4054

(d) $h = 0.1$

t_n	y_n
0.00	0.5000
0.10	0.5215
0.20	0.5362
0.30	0.5449
0.40	0.5490
0.50	0.5503

$h = 0.05$

t_n	y_n
0.00	0.5000
0.05	0.5116
0.10	0.5214
0.15	0.5294
0.20	0.5359
0.25	0.5408
0.30	0.5444
0.35	0.5469
0.40	0.5484
0.45	0.5492
0.50	0.5495

(e) $h = 0.1$

t_n	y_n
1.00	1.0000
1.10	1.0095
1.20	1.0404
1.30	1.0967
1.40	1.1866
1.50	1.3260

$h = 0.05$

t_n	y_n
1.00	1.0000
1.05	1.0024
1.10	1.0100
1.15	1.0228
1.20	1.0414
1.25	1.0663
1.30	1.0984
1.35	1.1389
1.40	1.1895
1.45	1.2526
1.50	1.3315

15. (a) The appearance of the graph will depend on the ODE solver used. The following graph was obtained using *Mathematica* on the interval [1, 1.3556].

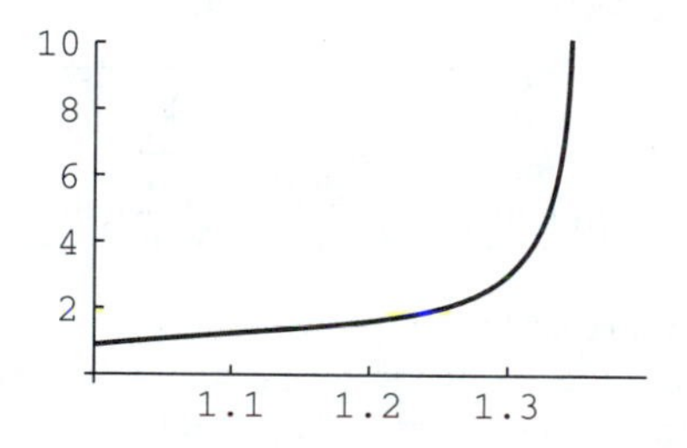

(b)

t_n	Euler	Improved Euler
1.0	1.0000	1.0000
1.1	1.2000	1.2469
1.2	1.4938	1.6668
1.3	1.9711	2.6427
1.4	2.9060	8.7989

17. (a) $y_1 = 1.2$ **(b)** $y''(c)\dfrac{h^2}{2} = 4e^{2c}\dfrac{(0.1)^2}{2} = 0.02\,e^{2c} \leq 0.02\,e^{0.2} = 0.0244$

(c) Actual value is $y(0.1) = 1.2214$. Error is 0.0214. **(d)** If $h = 0.05$, $y_2 = 1.21$. **(e)** Error with $h = 0.1$ is 0.0214. Error with $h = 0.05$ is 0.0114.

19. (a) $y_1 = 0.8$ **(b)** $y''(c)\dfrac{h^2}{2} = 5e^{-2c}\dfrac{(0.1)^2}{2} = 0.025\,e^{-2c} \leq 0.025$ for $0 \leq c \leq 0.1$ **(c)** Actual value is $y(0.1) = 0.8234$. Error is 0.0234. **(d)** If $h = 0.05$, $y_2 = 0.8125$. **(e)** Error with $h = 0.1$ is 0.0234. Error with $h = 0.05$ is 0.0109.

21. (a) Error is $19h^2e^{-3(c-1)}$. **(b)** $y''(c)\dfrac{h^2}{2} \leq 19(0.1)^2(1) = 0.19$ **(c)** If $h = 0.1$, $y_5 = 1.8207$. If $h = 0.05$, $y_{10} = 1.9424$. **(d)** Error with $h = 0.1$ is 0.2325. Error with $h = 0.05$ is 0.1109.

23. (a) Error is $\dfrac{1}{(c+1)^2}\dfrac{h^2}{2}$. **(b)** $\left|y''(c)\dfrac{h^2}{2}\right| \leq (1)\dfrac{(0.1)^2}{2} = 0.005$ **(c)** If $h = 0.1$, $y_5 = 0.4198$. If $h = 0.05$, $y_{10} = 0.4124$. **(d)** Error with $h = 0.1$ is 0.0143. Error with $h = 0.05$ is 0.0069.

25.

t_n	y_n	True Value
0.00	2.0000	2.0000
0.10	2.1230	2.1230
0.20	2.3085	2.3085
0.30	2.5958	2.5958
0.40	3.0649	3.0650
0.50	3.9078	3.9082

27.

t_n	y_n
1.00	5.0000
1.10	3.9724
1.20	3.2284
1.30	2.6945
1.40	2.3163
1.50	2.0533

29.

t_n	y_n
0.00	0.0000
0.10	0.1003
0.20	0.2027
0.30	0.3093
0.40	0.4228
0.50	0.5463

31.

t_n	y_n
0.00	0.0000
0.10	0.0953
0.20	0.1823
0.30	0.2624
0.40	0.3365
0.50	0.4055

33.

t_n	y_n
0.00	0.5000
0.10	0.5213
0.20	0.5358
0.30	0.5443
0.40	0.5482
0.50	0.5493

35.

t_n	y_n
1.00	1.0000
1.10	1.0101
1.20	1.0417
1.30	1.0989
1.40	1.1905
1.50	1.3333

37. (a) $v(5) = 35.7678$

(b)

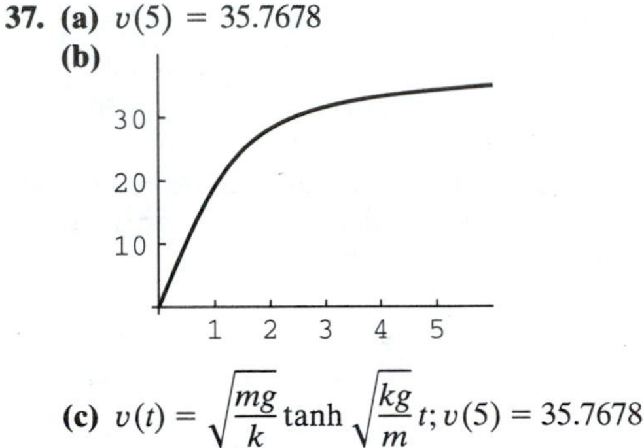

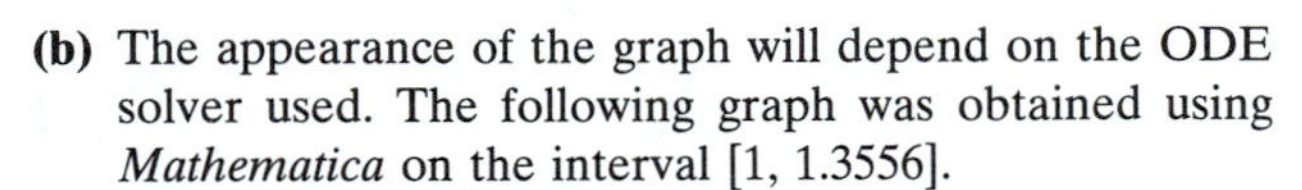
(c) $v(t) = \sqrt{\dfrac{mg}{k}}\tanh\sqrt{\dfrac{kg}{m}}\,t;\, v(5) = 35.7678$

39. (a) $h = 0.1$

t_n	y_n
1.00	1.0000
1.10	1.2511
1.20	1.6934
1.30	2.9425
1.40	903.0282

$h = 0.05$

t_n	y_n
1.00	1.0000
1.05	1.1112
1.10	1.2511
1.15	1.4348
1.20	1.6934
1.25	2.1047
1.30	2.9560
1.35	7.8981
1.40	1.1E + 15

(b) The appearance of the graph will depend on the ODE solver used. The following graph was obtained using *Mathematica* on the interval [1, 1.3556].

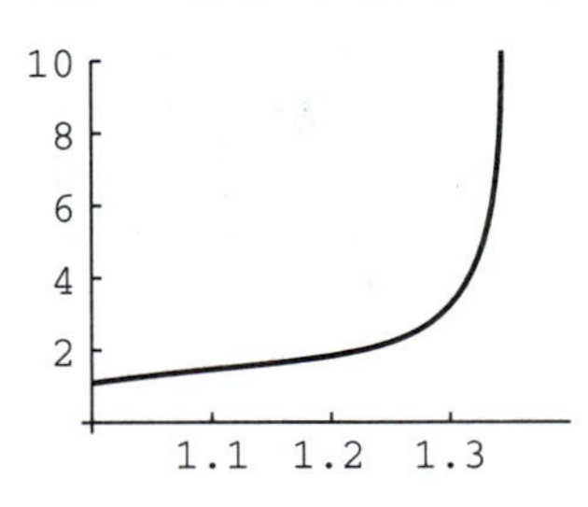

41. (a) $y_1 = 0.82341667$ **(b)** $y^{(5)}(c)\dfrac{h^5}{5!} = 40e^{-2c}\dfrac{h^5}{5!} \le 40e^{2(0)}\dfrac{(0.1)^5}{5!} = 3.333 \times 10^{-6}$

(c) Actual value is $y(0.1) = 0.8234134413$. Error is $3.225 \times 10^{-6} \le 3.333 \times 10^{-6}$. **(d)** If $h = 0.05$, $y_2 = 0.82341363$. **(e)** Error with $h = 0.1$ is 3.225×10^{-6}. Error with $h = 0.05$ is 1.854×10^{-7}.

43. (a) $y^{(5)}(c)\dfrac{h^5}{5!} = \dfrac{24}{(c+1)^5}\dfrac{h^5}{5!}$ **(b)** $\dfrac{24}{(c+1)^5}\dfrac{h^5}{5!} \le 24\dfrac{(0.1)^5}{5!} = 2.0000 \times 10^{-6}$ **(c)** From calculation with $h = 0.1$, $y_5 = 0.40546517$. From calculation with $h = 0.05$, $y_{10} = 0.40546511$.

Exercises 3.1, Page 129

1. $y = \frac{1}{2}e^t - \frac{1}{2}e^{-t}$ **3.** $y = \frac{3}{5}e^{4t} + \frac{2}{5}e^{-t}$ **5.** $y = 3t - 4t \ln t$ **7.** $y = 0, y = t^2$

9. (a) $y = e^t \cos t - e^t \sin t$ **(b)** no solution **(c)** $y = e^t \cos t + e^{-\pi/2}e^t \sin t$ **(d)** $y = c_2 e^t \sin t$, where c_2 is arbitrary

11. $(-\infty, 2)$ **13.** dependent **15.** dependent **17.** dependent **19.** independent

33. $y_p = t^2 + 3t + 3e^{2t}$; $y_p = -2t^2 - 6t - \frac{1}{3}e^{2t}$

Exercises 3.2, Page 137

1. $y = c_1 + c_2 e^{-t/4}$ **3.** $y = c_1 \cos 3t + c_2 \sin 3t$ **5.** $y = c_1 e^{3t} + c_2 e^{-2t}$ **7.** $y = c_1 e^{-4t} + c_2 t e^{-4t}$

9. $y = c_1 e^{(-3+\sqrt{29})t/2} + c_2 e^{(-3-\sqrt{29})t/2}$ **11.** $y = c_1 e^{2t/3} + c_2 e^{-t/4}$ **13.** $y = e^{2t}(c_1 \cos t + c_2 \sin t)$

15. $y = c_1 + c_2 e^{-t} + c_3 e^{5t}$ **17.** $y = c_1 e^t + e^{-t/2}\left(c_2 \cos \frac{\sqrt{3}}{2}t + c_3 \sin \frac{\sqrt{3}}{2}t\right)$ **19.** $y = c_1 e^{-t} + c_2 e^{3t} + c_3 t e^{3t}$

21. $y = c_1 e^t + e^{-t}(c_2 \cos t + c_3 \sin t)$ **23.** $y = c_1 e^{-t} + c_2 t e^{-t} + c_3 t^2 e^{-t}$

25. $y = c_1 + c_2 t + e^{-t/2}\left(c_3 \cos \frac{\sqrt{3}}{2}t + c_4 \sin \frac{\sqrt{3}}{2}t\right)$ **27.** $y = c_1 \cos \frac{\sqrt{3}}{2}x + c_2 \sin \frac{\sqrt{3}}{2}x + c_3 x \cos \frac{\sqrt{3}}{2}x + c_4 x \sin \frac{\sqrt{3}}{2}x$

29. $y = c_1 + c_2 e^{-2x} + c_3 e^{2x} + c_4 \cos 2x + c_5 \sin 2x$ **31.** $y = c_1 e^r + c_2 r e^r + c_3 e^{-r} + c_4 r e^{-r} + c_5 e^{-5r}$

33. $y = -e^{t/2} \cos(t/2) + e^{t/2} \sin(t/2)$ **35.** $y = 0$ **37.** $y = e^{2(t-1)} - e^{t-1}$ **39.** $y = \frac{5}{36} - \frac{5}{36}e^{-6t} + \frac{1}{6}te^{-6t}$

41. $y = 2 - 2e^t + 2te^t - \frac{1}{2}t^2 e^t$ **43.** $y = e^{5t} - te^{5t}$ **45.** $y = -2 \cos t$

Exercises 3.3, Page 148

1.

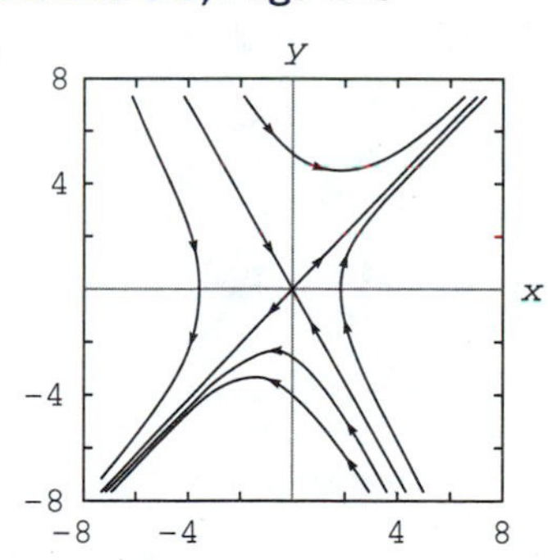

3. $x(t) = e^t$ and $x(t) = 10e^t$ correspond to the same trajectory, $y = x$, $x > 0$.

5.

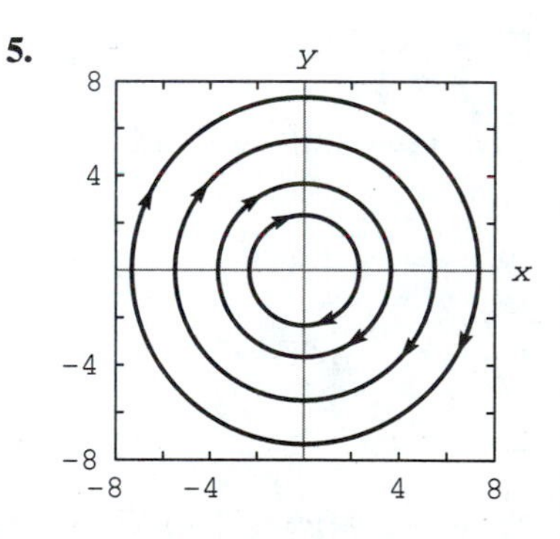

7. saddle point; unstable **9.** node; unstable **11.** spiral point; unstable **13.** node; asymptotically stable

15. For example, in part (a)

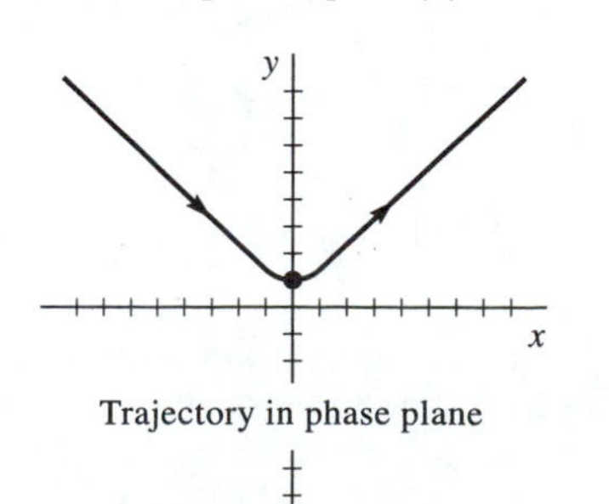

Trajectory in phase plane

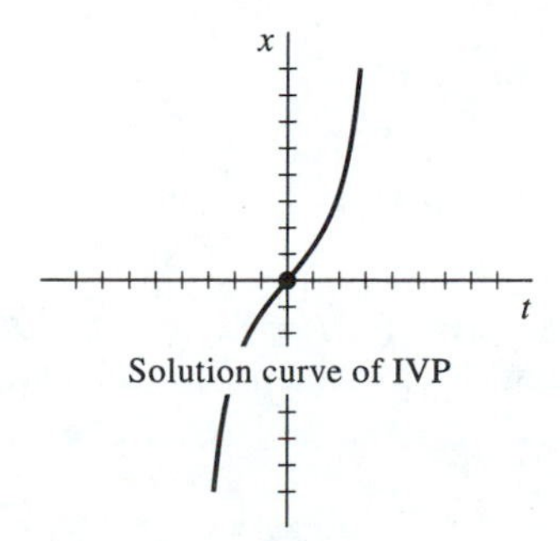

Solution curve of IVP

Exercises 3.4, Page 163

1. $y = c_1e^{-t} + c_2e^{-2t} + 3$ **3.** $y = c_1e^{5t} + c_2te^{5t} + \frac{6}{5}t + \frac{3}{5}$ **5.** $y = c_1e^{-2t} + c_2te^{-2t} + t^2 - 4t + \frac{7}{2}$

7. $y = c_1 \cos \sqrt{3}\,t + c_2 \sin \sqrt{3}\,t + (-4t^2 + 4t - \frac{4}{3})e^{3t}$ **9.** $y = c_1 + c_2e^t + 3t$ **11.** $y = c_1e^{t/2} + c_2te^{t/2} + 12 + \frac{1}{2}t^2e^{t/2}$

13. $y = c_1 \cos 2t + c_2 \sin 2t - \frac{3}{4}t \cos 2t$ **15.** $y = c_1 \cos t + c_2 \sin t - \frac{1}{2}t^2 \cos t + \frac{1}{2}t \sin t$

17. $y = c_1e^t \cos 2t + c_2e^t \sin 2t + \frac{1}{4}te^t \sin 2t$ **19.** $y = c_1e^{-t} + c_2te^{-t} - \frac{1}{2}\cos t + \frac{12}{25}\sin 2t - \frac{9}{25}\cos 2t$

21. $y = c_1 + c_2t + c_3e^{6t} - \frac{1}{4}t^2 - \frac{6}{37}\cos t + \frac{1}{37}\sin t$ **23.** $y = c_1e^t + c_2te^t + c_3t^2e^t - t - 3 - \frac{2}{3}t^3e^t$

25. $y = \sqrt{2}\sin 2t - \frac{1}{2}$ **27.** $y = -200 + 200e^{-t/5} - 3t^2 + 30t$

29. $y = c_1 \cos t + c_2 \sin t + t \sin t + \cos t \ln|\cos t|$; $(-\pi/2, \pi/2)$

31. $y = c_1 \cos t + c_2 \sin t + \frac{1}{2}\sin t - \frac{1}{2}t \cos t = c_1 \cos t + c_3 \sin t - \frac{1}{2}t \cos t$; $(-\infty, \infty)$

33. $y = c_1 \cos t + c_2 \sin t + \frac{1}{2} - \frac{1}{6}\cos 2t$; $(-\infty, \infty)$

35. $y = c_1e^t + c_2e^{-t} + \frac{1}{4}te^t - \frac{1}{4}te^{-t} = c_1e^t + c_2e^{-t} + \frac{1}{2}t \sinh t$; $(-\infty, \infty)$

37. $y = c_1e^{2t} + c_2e^{-2t} + \dfrac{1}{4}\left(e^{2t}\ln|t| - e^{-2t}\displaystyle\int_{t_0}^{t}\frac{e^{4v}}{v}\,dv\right)$, $t_0 > 0$; $(0, \infty)$

39. $y = c_1e^{-t} + c_2e^{-2t} + (e^{-t} + e^{-2t})\ln(1 + e^t)$; $(-\infty, \infty)$ **41.** $y = c_1e^{-2t} + c_2e^{-t} - e^{-2t}\sin e^t$; $(-\infty, \infty)$

43. $y = c_1e^t + c_2te^t - \frac{1}{2}e^t \ln(1 + t^2) + te^t \tan^{-1} t$; $(-\infty, \infty)$ **45.** $y = c_1e^{-t} + c_2te^{-t} + \frac{1}{2}t^2e^{-t}\ln t - \frac{3}{4}t^2e^{-t}$; $(0, \infty)$

47. $y = c_1e^t \cos 3t + c_2e^t \sin 3t - \frac{1}{27}e^t \cos 3t \ln|\sec 3t + \tan 3t|$; $(-\pi/6, \pi/6)$

49. $y = c_1 + c_2 \cos t + c_3 \sin t - \ln|\cos t| - \sin t \ln|\sec t + \tan t|$; $(-\pi/2, \pi/2)$

51. $y = c_1e^t + c_2e^{2t} + c_3e^{-t} + \frac{1}{8}e^{3t}$; $(-\infty, \infty)$ **53.** $y = \frac{1}{4}e^{-t/2} + \frac{3}{4}e^{t/2} + \frac{1}{8}t^2e^{t/2} - \frac{1}{4}te^{t/2}$ **55.** $y = \frac{4}{9}e^{-4t} + \frac{25}{36}e^{2t} - \frac{1}{4}e^{-2t} + \frac{1}{9}e^{-t}$

Exercises 3.5, Page 171

1. $y = c_1t^{-1} + c_2t^2$ **3.** $y = c_1 + c_2 \ln t$ **5.** $y = c_1 \cos(2 \ln t) + c_2 \sin(2 \ln t)$ **7.** $y = c_1t^{(2-\sqrt{6})} + c_2t^{(2+\sqrt{6})}$

9. $y_1 = c_1 \cos(\frac{1}{5}\ln t) + c_2 \sin(\frac{1}{5}\ln t)$ **11.** $y = c_1t^{-2} + c_2t^{-2}\ln t$ **13.** $y = t[c_1 \cos(\ln t) + c_2 \sin(\ln t)]$

15. $y = t^{-1/2}\left[c_1 \cos\left(\dfrac{\sqrt{3}}{6}\ln t\right) + c_2 \sin\left(\dfrac{\sqrt{3}}{6}\ln t\right)\right]$ **17.** $y = c_1t^3 + c_2 \cos(\sqrt{2}\ln t) + c_3 \sin(\sqrt{2}\ln t)$

19. $y = c_1t^{-1} + c_2t^2 + c_3t^4$ **21.** $y = c_1 + c_2t + c_3t^2 + c_4t^{-3}$ **23.** $y = 2 - 2t^{-2}$ **25.** $y = \cos(\ln t) + 2 \sin(\ln t)$

27. $y = 2(-t)^{1/2} - 5(-t)^{1/2}\ln(-t)$ **29.** $y = c_1 + c_2 \ln t + \dfrac{t^2}{4}$ **31.** $y = c_1t^{-1/2} + c_2t^{-1} + \frac{1}{15}t^2 - \frac{1}{6}t$

33. $y = c_1t + c_2t \ln t + t(\ln t)^2$ **35.** $y = c_1t^{-1} + c_2t^{-8} + \frac{1}{30}t^2$ **37.** $y = t^2[c_1 \cos(3 \ln t) + c_2 \sin(3 \ln t)] + \frac{4}{13} + \frac{3}{10}t$

39. $y = c_1t^2 + c_2t^{-10} - \frac{1}{7}t^{-3}$

Exercises 3.6, Page 185

1. $\sqrt{2}\,\pi/8$ **3.** $x(t) = -\frac{1}{4}\cos 4\sqrt{6}\,t$

5. **(a)** $x(\pi/12) = -1/4$; $x(\pi/8) = -1/2$; $x(\pi/6) = -1/4$; $x(\pi/4) = 1/2$; $x(9\pi/32) = \sqrt{2}/4$ **(b)** 4 ft/s; downward **(c)** $t = (2n + 1)\pi/16$, $n = 0, 1, 2, \ldots$

7. **(a)** the 20-kg mass **(b)** the 20-kg mass; the 50-kg mass **(c)** $t = n\pi$, $n = 0, 1, 2, \ldots$; at the equilibrium position; the 50-kg mass is moving upward whereas the 20-kg mass is moving upward when n is even and downward when n is odd.

9. $x(t) = \frac{1}{2}\cos 2t + \frac{3}{4}\sin 2t$

$$= \frac{\sqrt{13}}{4}\sin(2t + 0.5880)$$

11. **(a)** $x(t) = -\frac{2}{3}\cos 10t + \frac{1}{2}\sin 10t = \frac{5}{6}\sin(10t - 0.927)$ **(b)** $\frac{5}{6}$ ft; $\pi/5$ **(c)** 15 cycles **(d)** 0.721 s **(e)** $(2n + 1)\pi/20 + 0.0927$, $n = 0, 1, 2, \ldots$ **(f)** $x(3) = -0.597$ ft **(g)** $x'(3) = -5.814$ ft/s **(h)** $x''(3) = 59.702$ ft/s^2 **(i)** $\pm 8\frac{1}{3}$ ft/s **(j)** $0.1451 + n\pi/5$; $0.3545 + n\pi/5$, $n = 0, 1, 2, \ldots$ **(k)** $0.3545 + n\pi/5$, $n = 0, 1, 2, \ldots$

13. 120 lb/ft; $x(t) = \dfrac{\sqrt{3}}{12}\sin 8\sqrt{3}\,t$

15. $\frac{1}{4}$ s; $\frac{1}{2}$ s, $x(\frac{1}{2}) = e^{-2}$; that is, the weight is approximately 0.14 ft below the equilibrium position.

17. **(a)** $x(t) = \frac{4}{3}e^{-2t} - \frac{1}{3}e^{-8t}$ **(b)** $x(t) = -\frac{2}{3}e^{-2t} + \frac{5}{3}e^{-8t}$

19. **(a)** $x(t) = e^{-2t}[-\cos 4t - \frac{1}{2}\sin 4t]$ **(b)** $x(t) = \dfrac{\sqrt{5}}{2}e^{-2t}\sin(4t + 4.249)$ **(c)** $t = 1.294$ s

21. **(a)** $\beta > \frac{5}{2}$ **(b)** $\beta = \frac{5}{2}$ **(c)** $0 < \beta < \frac{5}{2}$

23. $x(t) = e^{-t/2}\left(-\dfrac{4}{3}\cos\dfrac{\sqrt{47}}{2}t - \dfrac{64}{3\sqrt{47}}\sin\dfrac{\sqrt{47}}{2}t\right) + \dfrac{10}{3}(\cos 3t + \sin 3t)$

25. $x(t) = \frac{1}{4}e^{-4t} + te^{-4t} - \frac{1}{4}\cos 4t$

27. $x(t) = -\frac{1}{2}\cos 4t + \frac{9}{4}\sin 4t + \frac{1}{2}e^{-2t}\cos 4t - 2e^{-2t}\sin 4t$

29. **(a)** $m\dfrac{d^2x}{dt^2} = -k(x - h) - \beta\dfrac{dx}{dt}$ or $\dfrac{d^2x}{dt^2} + 2\lambda\dfrac{dx}{dt} + \omega^2 x = \omega^2 h(t)$ where $2\lambda = \beta/m$ and $\omega^2 = k/m$ **(b)** $x(t) = e^{-2t}\left(-\frac{56}{13}\cos 2t - \frac{72}{13}\sin 2t\right) + \frac{56}{13}\cos t + \frac{32}{13}\sin t$

31. 4.568 coulombs; 0.0509 s

33. $q(t) = 10 - 10e^{-3t}(\cos 3t + \sin 3t)$; $i(t) = 60e^{-3t}\sin 3t$; 10.432 coulombs

35. $q_p = \frac{100}{13}\sin t + \frac{150}{13}\cos t$; $i_p = \frac{100}{13}\cos t - \frac{150}{13}\sin t$

Exercises 3.7, Page 197

1. **(a)** $y(x) = \dfrac{w_0}{24EI}(6L^2x^2 - 4Lx^3 + x^4)$ **(b)**

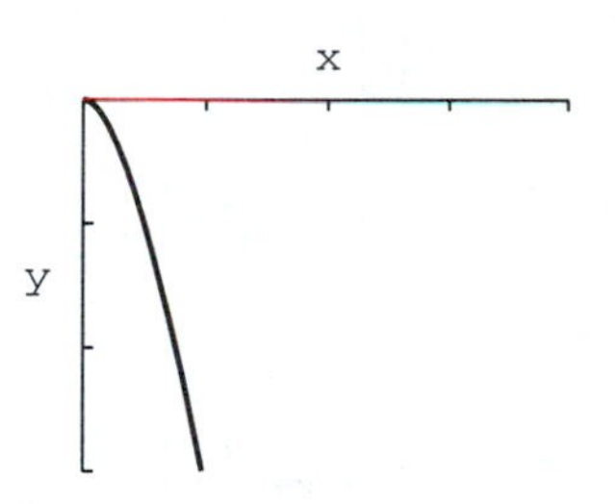

3. **(a)** $y(x) = \dfrac{w_0}{48EI}(3L^2x^2 - 5Lx^3 + 2x^4)$ **(b)**

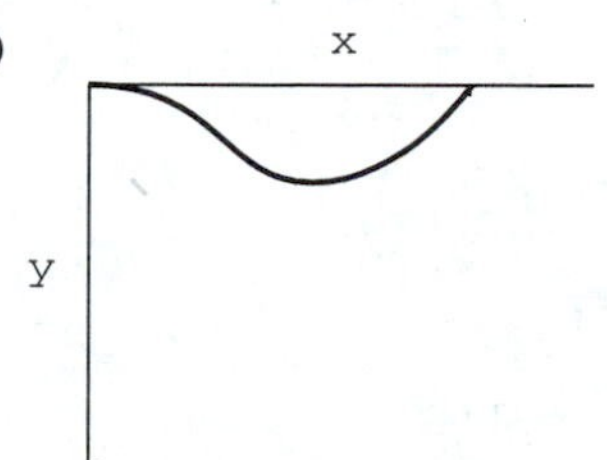

5. **(a)** $y_{\max} = \dfrac{w_0L^4}{8EI}$ **(b)** $\frac{1}{16}$th of the maximum deflection in part (a)

7. $y(x) = -\dfrac{w_0 EI}{P^2}\cosh\sqrt{\dfrac{P}{EI}}x + \left(\dfrac{w_0 EI}{P^2}\sinh\sqrt{\dfrac{P}{EI}}L - \dfrac{w_0 L\sqrt{EI}}{P\sqrt{P}}\right)\dfrac{\sinh\sqrt{\dfrac{P}{EI}}x}{\cosh\sqrt{\dfrac{P}{EI}}L} + \dfrac{w_0}{2P}x^2 + \dfrac{w_0 EI}{P^2}$

9. $\lambda = n^2$, $n = 1, 2, 3, \ldots$, $y = \sin nx$ **11.** $\lambda = (2n-1)^2\pi^2/4L^2$, $n = 1, 2, 3, \ldots$, $y = \cos((2n-1)\pi x/2L)$

13. $\lambda = n^2$, $n = 0, 1, 2, \ldots$, $y = \cos nx$ **15.** $\lambda = n^2\pi^2/25$, $n = 1, 2, 3, \ldots$, $y = e^{-x}\sin(n\pi x/5)$

17. $\lambda = n\pi/L$, $n = 1, 2, 3, \ldots$, $y = \sin(n\pi x/L)$ **19.** $\lambda = n^2$, $n = 1, 2, 3, \ldots$, $y = \sin(n\ln x)$

21. for $\lambda = 0$, $y = 1$; for $\lambda = n^2\pi^2/4$, $n = 1, 2, 3, \ldots$, $y = \cos((n\pi/2)\ln x)$

23. **(b)** From $\lambda_n = x_n^2$, we see no new eigenvalues result when x_n is negative. For $\lambda = 0$, the family of solutions for $y'' = 0$ is $y = c_1x + c_2$. The only member of this family that satisfies the given boundary conditions is $y = 0$.
(c) $\lambda_1 = 4.1159$, $\lambda_2 = 24.1393$, $\lambda_3 = 63.9652$, $\lambda_4 = 122.8883$

25. $\omega_n = n\pi\sqrt{T}/L\sqrt{\rho}$, $n = 1, 2, 3, \ldots$, $y = \sin(n\pi x/L)$ **27.** $u(r) = \left(\dfrac{u_0 - u_1}{b-a}\right)\dfrac{ab}{r} + \dfrac{u_1 b - u_0 a}{b-a}$

Exercises 3.8, Page 203

1. $x = c_1e^t + c_2te^t$
$y = (c_1 - c_2)e^t + c_2te^t$

3. $x = c_1\cos t + c_2\sin t + t + 1$
$y = c_1\sin t - c_2\cos t + t - 1$

5. $x = \frac{1}{2}c_1\sin t + \frac{1}{2}c_2\cos t - 2c_3\sin\sqrt{6}t - 2c_4\cos\sqrt{6}t$
$y = c_1\sin t + c_2\cos t + c_3\sin\sqrt{6}t + c_4\cos\sqrt{6}t$

7. $x = c_1e^{2t} + c_2e^{-2t} + c_3\sin 2t + c_4\cos 2t + \frac{1}{5}e^t$
$y = c_1e^{2t} + c_2e^{-2t} - c_3\sin 2t - c_4\cos 2t - \frac{1}{5}e^t$

9. $x = c_1 - c_2\cos t + c_3\sin t + \frac{17}{15}e^{3t}$
$y = c_1 + c_2\sin t + c_3\cos t - \frac{4}{15}e^{3t}$

11. $x = c_1e^t + c_2e^{-t/2}\cos\dfrac{\sqrt{3}}{2}t + c_3e^{-t/2}\sin\dfrac{\sqrt{3}}{2}t$

$y = \left(-\dfrac{3}{2}c_2 - \dfrac{\sqrt{3}}{2}c_3\right)e^{-t/2}\cos\dfrac{\sqrt{3}}{2}t + \left(\dfrac{\sqrt{3}}{2}c_2 - \dfrac{3}{2}c_3\right)e^{-t/2}\sin\dfrac{\sqrt{3}}{2}t$

13. $x = c_1e^{4t} + \frac{4}{3}e^t$
$y = -\frac{3}{4}c_1e^{4t} + c_2 + 5e^t$

15. $x = c_1 + c_2t + c_3e^t + c_4e^{-t} - \frac{1}{2}t^2$
$y = (c_1 - c_2 + 2) + (c_2 + 1)t + c_4e^{-t} - \frac{1}{2}t^2$

17. $x = c_1e^t + c_2e^{-t/2}\sin\dfrac{\sqrt{3}}{2}t + c_3e^{-t/2}\cos\dfrac{\sqrt{3}}{2}t$

$y = c_1e^t + \left(-\dfrac{1}{2}c_2 - \dfrac{\sqrt{3}}{2}c_3\right)e^{-t/2}\sin\dfrac{\sqrt{3}}{2}t + \left(\dfrac{\sqrt{3}}{2}c_2 - \dfrac{1}{2}c_3\right)e^{-t/2}\cos\dfrac{\sqrt{3}}{2}t$

$z = c_1e^t + \left(-\dfrac{1}{2}c_2 + \dfrac{\sqrt{3}}{2}c_3\right)e^{-t/2}\sin\dfrac{\sqrt{3}}{2}t + \left(-\dfrac{\sqrt{3}}{2}c_2 - \dfrac{1}{2}c_3\right)e^{-t/2}\cos\dfrac{\sqrt{3}}{2}t$

19. $x = -6c_1e^{-t} - 3c_2e^{-2t} + 2c_3e^{3t}$
$y = c_1e^{-t} + c_2e^{-2t} + c_3e^{3t}$
$z = 5c_1e^{-t} + c_2e^{-2t} + c_3e^{3t}$

21. $x = e^{-3t+3} - te^{-3t+3}$
$y = -e^{-3t+3} + 2te^{-3t+3}$

Exercises 3.9, Page 208

1. $y(t) = -2e^{2t} + 5te^{2t}$; $y(0.2) = -1.4918$, $y_2 = -1.6800$ **3.** $y_1 = -1.4928$, $y_2 = -1.4919$ **5.** $y_1 = 1.4640$, $y_2 = 1.4640$

7. $x_1 = 8.3055$, $y_1 = 3.4199$
$x_2 = 8.3055$, $y_2 = 3.4199$

9. $x_1 = -3.9123$, $y_1 = 4.2857$
$x_2 = -3.9123$, $y_2 = 4.2857$

11. $x_1 = 0.4179$, $y_1 = -2.1824$
$x_2 = 0.4173$, $y_2 = -2.1821$

Exercises 3.10, Page 216

1. (a)

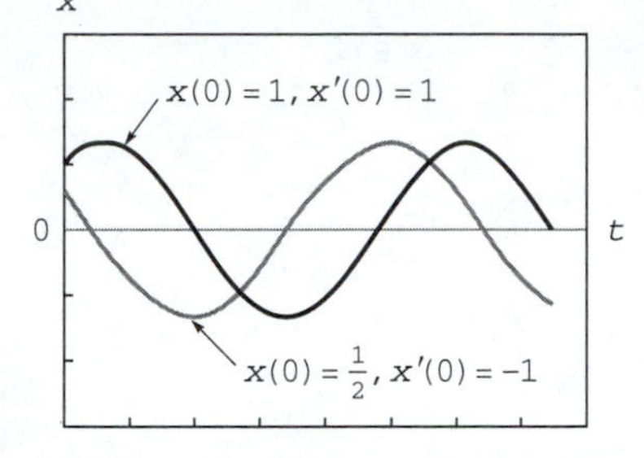

For the first IVP the period T is approximately 6; for the second IVP the period T is approximately 6.3.

(b) (0, 0)

(c)

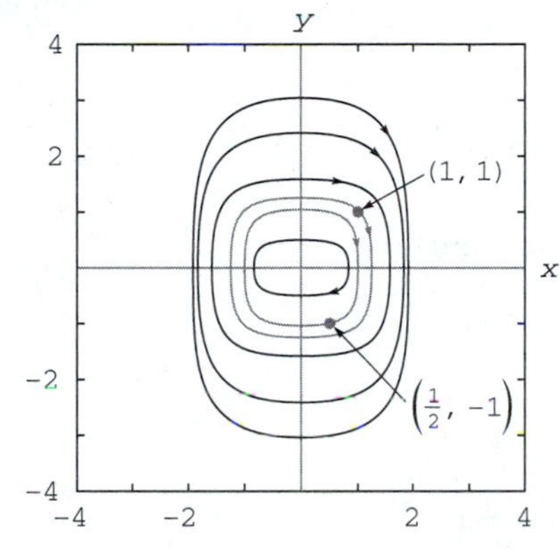

(0, 0) is a (stable) center

3. (a)

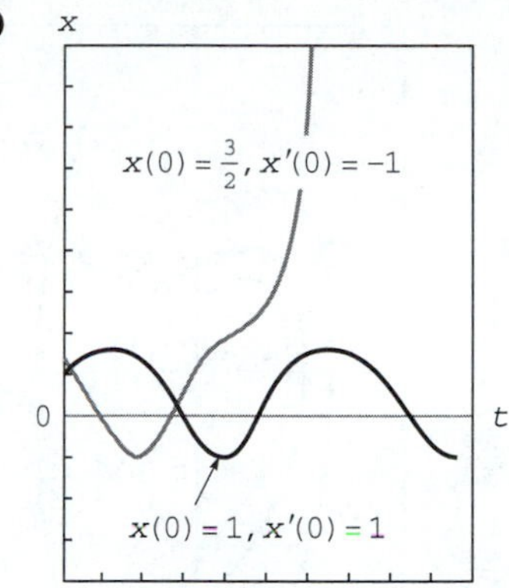

For the first IVP the period T is approximately 6; the solution of the second IVP appears not to be periodic.

(b) (0, 0), (2, 0)

(c)

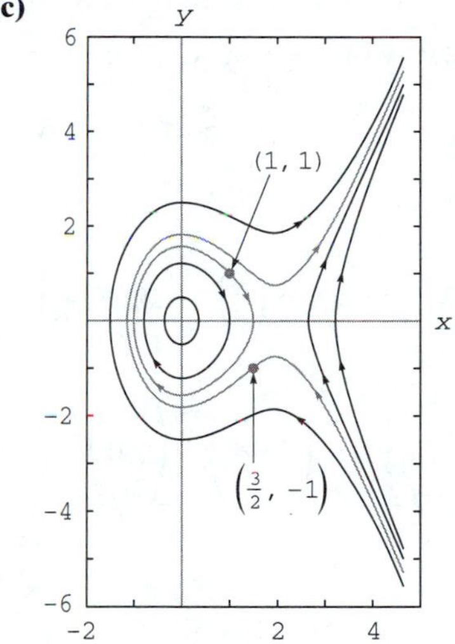

(0, 0) is a (stable) center; (2, 0) is an (unstable) saddle point.

5. (a) $|x_1| \approx 1.2$
(b) $-0.8 \le x \le 1.1$

7. $\dfrac{d^2x}{dt^2} + x = 0$

9. (a) expect $x \to 0$ as $t \to \infty$ **(b)**

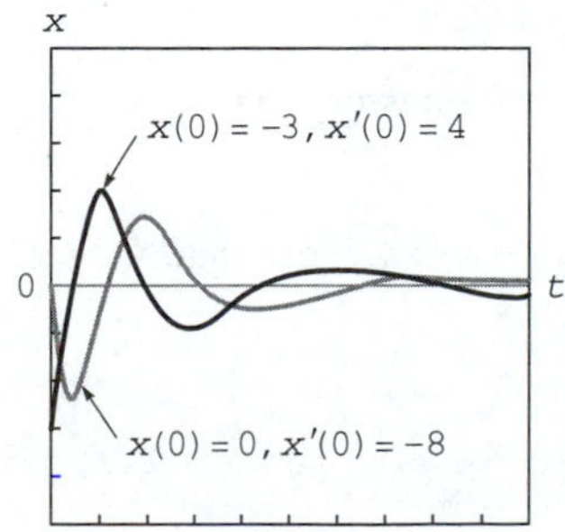

(c) (0, 0) **(d)**

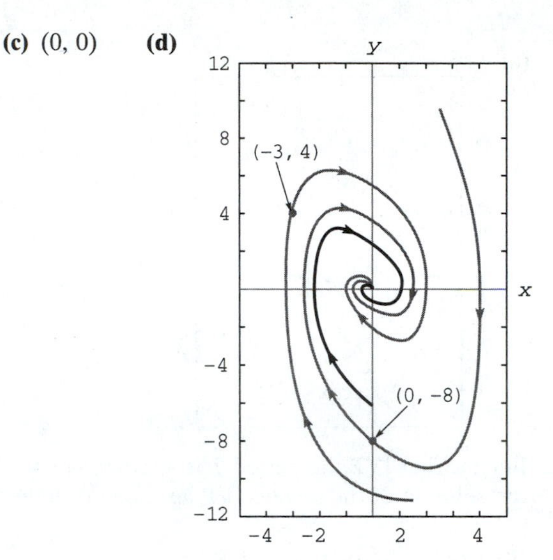

(0, 0) is an asymptotically stable spiral point.

11. The system is always underdamped. (0, 0) is an asymptotically stable spiral point.

13. $y = -\dfrac{1}{c_1}\tan^{-1}\dfrac{t}{c_1} + c_2$ **15.** $y = c_2 e^{c_1 t}$ **17.** $y = \dfrac{T}{w}\cosh\dfrac{w}{T}x$

Exercises 4.1 Page 248

1. $\mathbf{X}' = \begin{pmatrix} 3 & -5 \\ 4 & 8 \end{pmatrix}\mathbf{X}$, where $\mathbf{X} = \begin{pmatrix} x \\ y \end{pmatrix}$ **3.** $\mathbf{X}' = \begin{pmatrix} -3 & 4 & -9 \\ 6 & -1 & 0 \\ 10 & 4 & 3 \end{pmatrix}\mathbf{X}$, where $\mathbf{X} = \begin{pmatrix} x \\ y \\ z \end{pmatrix}$

5. $\mathbf{X}' = \begin{pmatrix} 1 & -1 & 1 \\ 2 & 1 & -1 \\ 1 & 1 & 1 \end{pmatrix}\mathbf{X} + \begin{pmatrix} 0 \\ -3t^2 \\ t^2 \end{pmatrix} + \begin{pmatrix} t \\ 0 \\ -t \end{pmatrix} + \begin{pmatrix} -1 \\ 0 \\ 2 \end{pmatrix}$, where $\mathbf{X} = \begin{pmatrix} x \\ y \\ z \end{pmatrix}$ **7.** $\dfrac{dx}{dt} = 4x + 2y + e^t$

$\dfrac{dy}{dt} = -x + 3y - e^t$

9. $\dfrac{dx}{dt} = x - y + 2z + e^{-t} - 3t$

$\dfrac{dy}{dt} = 3x - 4y + z + 2e^{-t} + t$

$\dfrac{dz}{dt} = -2x + 5y + 6z + 2e^{-t} - t$

17. Yes; $W(\mathbf{X}_1, \mathbf{X}_2) = -2e^{-8t} \neq 0$ implies $\mathbf{X}_1$ and $\mathbf{X}_2$ are linearly independent on $(-\infty, \infty)$.

19. No; $W(\mathbf{X}_1, \mathbf{X}_2, \mathbf{X}_3) = \begin{vmatrix} 1+t & 1 & 3+2t \\ -2+2t & -2 & -6+4t \\ 4+2t & 4 & 12+4t \end{vmatrix} = 0$ for every t.

The solution vectors are linearly dependent on $(-\infty, \infty)$. Note that $\mathbf{X}_3 = 2\mathbf{X}_1 + \mathbf{X}_2$.

Exercises 4.2, Page 262

1. $\mathbf{X} = c_1\begin{pmatrix} 1 \\ 2 \end{pmatrix}e^{5t} + c_2\begin{pmatrix} 1 \\ -1 \end{pmatrix}e^{-t}$ **3.** $\mathbf{X} = c_1\begin{pmatrix} 2 \\ 1 \end{pmatrix}e^{-3t} + c_2\begin{pmatrix} 2 \\ 5 \end{pmatrix}e^{t}$ **5.** $\mathbf{X} = c_1\begin{pmatrix} 5 \\ 2 \end{pmatrix}e^{8t} + c_2\begin{pmatrix} 1 \\ 4 \end{pmatrix}e^{-10t}$

7. $\mathbf{X} = c_1\begin{pmatrix}1\\0\\0\end{pmatrix}e^{t} + c_2\begin{pmatrix}2\\3\\1\end{pmatrix}e^{2t} + c_3\begin{pmatrix}1\\0\\2\end{pmatrix}e^{-t}$ **9.** $\mathbf{X} = c_1\begin{pmatrix}-1\\0\\1\end{pmatrix}e^{-t} + c_2\begin{pmatrix}1\\4\\3\end{pmatrix}e^{3t} + c_3\begin{pmatrix}1\\-1\\3\end{pmatrix}e^{-2t}$

11. $\mathbf{X} = c_1\begin{pmatrix}4\\0\\-1\end{pmatrix}e^{-t} + c_2\begin{pmatrix}-12\\6\\5\end{pmatrix}e^{-t/2} + c_3\begin{pmatrix}4\\2\\-1\end{pmatrix}e^{-3t/2}$ **13.** $\mathbf{X} = 3\begin{pmatrix}1\\1\end{pmatrix}e^{t/2} + 2\begin{pmatrix}0\\1\end{pmatrix}e^{-t/2}$

15. $\mathbf{X} = c_1\begin{pmatrix}0.382175\\0.851161\\0.359815\end{pmatrix}e^{8.58979t} + c_2\begin{pmatrix}0.405188\\-0.676043\\0.615458\end{pmatrix}e^{2.25684t} + c_3\begin{pmatrix}-0.923562\\-0.132174\\0.35995\end{pmatrix}e^{-0.0466321t}$

17. $\mathbf{X} = c_1\begin{pmatrix}1\\3\end{pmatrix} + c_2\left[\begin{pmatrix}1\\3\end{pmatrix}t + \begin{pmatrix}\frac{1}{4}\\-\frac{1}{4}\end{pmatrix}\right]$ **19.** $\mathbf{X} = c_1\begin{pmatrix}1\\1\end{pmatrix}e^{2t} + c_2\left[\begin{pmatrix}1\\1\end{pmatrix}te^{2t} + \begin{pmatrix}-\frac{1}{3}\\0\end{pmatrix}e^{2t}\right]$

21. $\mathbf{X} = c_1\begin{pmatrix}1\\1\\1\end{pmatrix}e^{t} + c_2\begin{pmatrix}1\\1\\0\end{pmatrix}e^{2t} + c_3\begin{pmatrix}1\\0\\1\end{pmatrix}e^{2t}$ **23.** $\mathbf{X} = c_1\begin{pmatrix}-4\\-5\\2\end{pmatrix} + c_2\begin{pmatrix}2\\0\\-1\end{pmatrix}e^{5t} + c_3\left[\begin{pmatrix}2\\0\\-1\end{pmatrix}te^{5t} + \begin{pmatrix}-\frac{1}{2}\\-\frac{1}{2}\\-1\end{pmatrix}e^{5t}\right]$

25. $\mathbf{X} = c_1\begin{pmatrix}0\\1\\1\end{pmatrix}e^{t} + c_2\left[\begin{pmatrix}0\\1\\1\end{pmatrix}te^{t} + \begin{pmatrix}0\\1\\0\end{pmatrix}e^{t}\right] + c_3\left[\begin{pmatrix}0\\1\\1\end{pmatrix}\frac{t^2}{2}e^{t} + \begin{pmatrix}0\\1\\0\end{pmatrix}te^{t} + \begin{pmatrix}\frac{1}{2}\\0\\0\end{pmatrix}e^{t}\right]$ **27.** $\mathbf{X} = -7\begin{pmatrix}2\\1\end{pmatrix}e^{4t} + 13\begin{pmatrix}2t+1\\t+1\end{pmatrix}e^{4t}$

29. Corresponding to the eigenvalue $\lambda_1 = 2$ of multiplicity five, eigenvectors are $\mathbf{K}_1 = \begin{pmatrix}1\\0\\0\\0\\0\end{pmatrix}, \mathbf{K}_2 = \begin{pmatrix}0\\0\\1\\0\\0\end{pmatrix}, \mathbf{K}_3 = \begin{pmatrix}0\\0\\0\\1\\0\end{pmatrix}$

31. $\mathbf{X} = c_1\begin{pmatrix}\cos t\\2\cos t + \sin t\end{pmatrix}e^{4t} + c_2\begin{pmatrix}\sin t\\2\sin t - \cos t\end{pmatrix}e^{4t}$ **33.** $\mathbf{X} = c_1\begin{pmatrix}\cos t\\-\cos t - \sin t\end{pmatrix}e^{4t} + c_2\begin{pmatrix}\sin t\\-\sin t + \cos t\end{pmatrix}e^{4t}$

35. $\mathbf{X} = c_1\begin{pmatrix}5\cos 3t\\4\cos 3t + 3\sin 3t\end{pmatrix} + c_2\begin{pmatrix}5\sin 3t\\4\sin 3t - 3\cos 3t\end{pmatrix}$ **37.** $\mathbf{X} = c_1\begin{pmatrix}1\\0\\0\end{pmatrix} + c_2\begin{pmatrix}-\cos t\\\cos t\\\sin t\end{pmatrix} + c_3\begin{pmatrix}\sin t\\-\sin t\\\cos t\end{pmatrix}$

39. $\mathbf{X} = c_1\begin{pmatrix}0\\2\\1\end{pmatrix}e^{t} + c_2\begin{pmatrix}\sin t\\\cos t\\\cos t\end{pmatrix}e^{t} + c_3\begin{pmatrix}\cos t\\-\sin t\\-\sin t\end{pmatrix}e^{t}$

41. $\mathbf{X} = \begin{pmatrix}28\\-5\\25\end{pmatrix}e^{2t} + c_2\begin{pmatrix}5\cos 3t\\-4\cos 3t - 3\sin 3t\\0\end{pmatrix}e^{-2t} + c_3\begin{pmatrix}5\sin 3t\\-4\sin 3t + 3\cos 3t\\0\end{pmatrix}e^{-2t}$

43. $\mathbf{X} = -\begin{pmatrix}25\\-7\\6\end{pmatrix}e^{t} - \begin{pmatrix}\cos 5t - 5\sin 5t\\\cos 5t\\\cos 5t\end{pmatrix} + 6\begin{pmatrix}5\cos 5t + \sin 5t\\\sin 5t\\\sin 5t\end{pmatrix}$

Exercises 4.3, Page 268

1. $\mathbf{X} = c_1\begin{pmatrix}1\\1\end{pmatrix} + c_2\begin{pmatrix}3\\2\end{pmatrix}e^{t} - \begin{pmatrix}11\\11\end{pmatrix}t - \begin{pmatrix}15\\10\end{pmatrix}$ **3.** $\mathbf{X} = c_1\begin{pmatrix}2\\1\end{pmatrix}e^{t/2} + c_2\begin{pmatrix}10\\3\end{pmatrix}e^{3t/2} - \begin{pmatrix}\frac{13}{2}\\\frac{13}{4}\end{pmatrix}te^{t/2} - \begin{pmatrix}\frac{15}{2}\\\frac{9}{4}\end{pmatrix}e^{t/2}$

5. $\mathbf{X} = c_1\begin{pmatrix}2\\1\end{pmatrix}e^t + c_2\begin{pmatrix}1\\1\end{pmatrix}e^{2t} + \begin{pmatrix}3\\3\end{pmatrix}e^t + \begin{pmatrix}4\\2\end{pmatrix}te^t$ **7.** $\mathbf{X} = c_1\begin{pmatrix}4\\1\end{pmatrix}e^{3t} + c_2\begin{pmatrix}-2\\1\end{pmatrix}e^{-3t} + \begin{pmatrix}-12\\0\end{pmatrix}t - \begin{pmatrix}\frac{4}{3}\\\frac{4}{3}\end{pmatrix}$

9. $\mathbf{X} = c_1\begin{pmatrix}1\\-1\end{pmatrix}e^t + c_2\begin{pmatrix}t\\\frac{1}{2}-t\end{pmatrix}e^t + \begin{pmatrix}\frac{1}{2}\\-2\end{pmatrix}e^{-t}$ **11.** $\mathbf{X} = c_1\begin{pmatrix}\cos t\\\sin t\end{pmatrix} + c_2\begin{pmatrix}\sin t\\-\cos t\end{pmatrix} + \begin{pmatrix}\cos t\\\sin t\end{pmatrix}t + \begin{pmatrix}-\sin t\\\cos t\end{pmatrix}\ln|\cos t|$

13. $\mathbf{X} = c_1\begin{pmatrix}\cos t\\\sin t\end{pmatrix}e^t + c_2\begin{pmatrix}\sin t\\-\cos t\end{pmatrix}e^t + \begin{pmatrix}\cos t\\\sin t\end{pmatrix}te^t$

15. $\mathbf{X} = c_1\begin{pmatrix}\cos t\\-\sin t\end{pmatrix} + c_2\begin{pmatrix}\sin t\\\cos t\end{pmatrix} + \begin{pmatrix}\cos t\\-\sin t\end{pmatrix}t + \begin{pmatrix}-\sin t\\\sin t\tan t\end{pmatrix} - \begin{pmatrix}\sin t\\\cos t\end{pmatrix}\ln|\cos t|$

17. $\mathbf{X} = c_1\begin{pmatrix}2\sin t\\\cos t\end{pmatrix}e^t + c_2\begin{pmatrix}2\cos t\\-\sin t\end{pmatrix}e^t + \begin{pmatrix}3\sin t\\\frac{3}{2}\cos t\end{pmatrix}te^t + \begin{pmatrix}\cos t\\-\frac{1}{2}\sin t\end{pmatrix}e^t\ln|\sin t| + \begin{pmatrix}2\cos t\\-\sin t\end{pmatrix}e^t\ln|\cos t|$

19. $\mathbf{X} = c_1\begin{pmatrix}1\\-1\\0\end{pmatrix} + c_2\begin{pmatrix}1\\1\\0\end{pmatrix}e^{2t} + c_3\begin{pmatrix}0\\0\\1\end{pmatrix}e^{3t} + \begin{pmatrix}-\frac{1}{4}e^{2t}+\frac{1}{2}te^{2t}\\-e^t+\frac{1}{4}e^{2t}+\frac{1}{2}te^{2t}\\\frac{1}{2}t^2e^{3t}\end{pmatrix}$ **21.** $\mathbf{X} = \begin{pmatrix}2\\2\end{pmatrix}te^{2t} + \begin{pmatrix}-1\\1\end{pmatrix}e^{2t} + \begin{pmatrix}-2\\2\end{pmatrix}te^{4t} + \begin{pmatrix}2\\0\end{pmatrix}e^{4t}$

Exercises 4.4, Page 274

1. $\mathbf{X} = x_0\begin{pmatrix}0\\0\\1\end{pmatrix} + x_0\begin{pmatrix}1\\-\dfrac{\lambda_1}{\lambda_1-\lambda_2}\\\dfrac{\lambda_2}{\lambda_1-\lambda_2}\end{pmatrix}e^{-\lambda_1 t} + x_0\begin{pmatrix}0\\\dfrac{\lambda_1}{\lambda_1-\lambda_2}\\-\dfrac{\lambda_1}{\lambda_1-\lambda_2}\end{pmatrix}e^{-\lambda_2 t}$ **3.** $\mathbf{X} = -25\begin{pmatrix}-\frac{1}{2}\\1\end{pmatrix}e^{-3t/25} + 25\begin{pmatrix}\frac{1}{2}\\1\end{pmatrix}e^{-t/25}$

5. $\mathbf{X} = 25\begin{pmatrix}-\frac{1}{2}\\1\end{pmatrix}e^{-3t/25} - 125\begin{pmatrix}\frac{1}{2}\\1\end{pmatrix}e^{-t/25} + \begin{pmatrix}100\\100\end{pmatrix}$

7. (a) $\begin{aligned}\frac{dx_1}{dt} &= 3\frac{x_2}{100-t} - 2\frac{x_1}{100+t}\\ \frac{dx_2}{dt} &= 2\frac{x_1}{100+t} - 3\frac{x_2}{100-t}\end{aligned}$, $x_1(0) = 100, x_2(0) = 50$

(b) Since the system is closed, no salt enters or leaves the system and so $x_1(t) + x_2(t) = 100 + 50 = 150$ for all time $t \geq 0$. Use $x_1 = 150 - x_2$ in the first differential equation in part (a) to eliminate x_1. Solve the resulting linear first-order equation for $x_2(t)$. From this solution, we find that $x_2(30) \approx 47.4$ lb.

9.

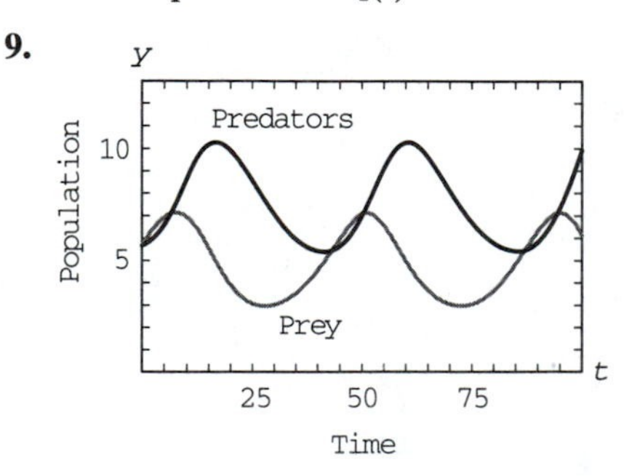

Populations are first equal at about $t = 5.6$. Periods are about 45.

11. In all cases $x(t) \to 6$ and $y(t) \to 8$ as $t \to \infty$.

Exercises 4.5, Page 293

1. (0, 0) is an unstable degenerate node.

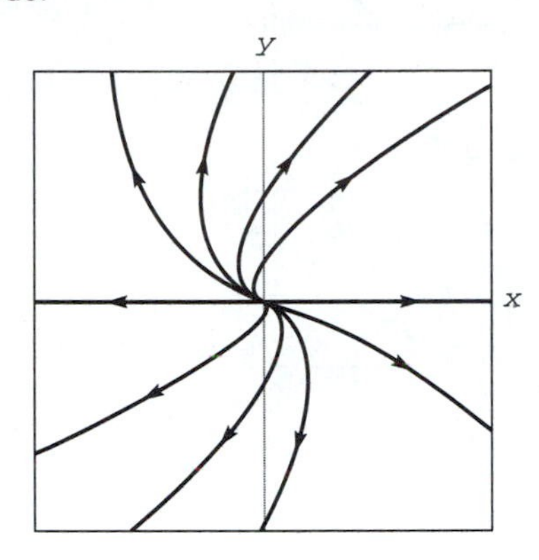

3. (0, 0) is an asymptotically stable node.

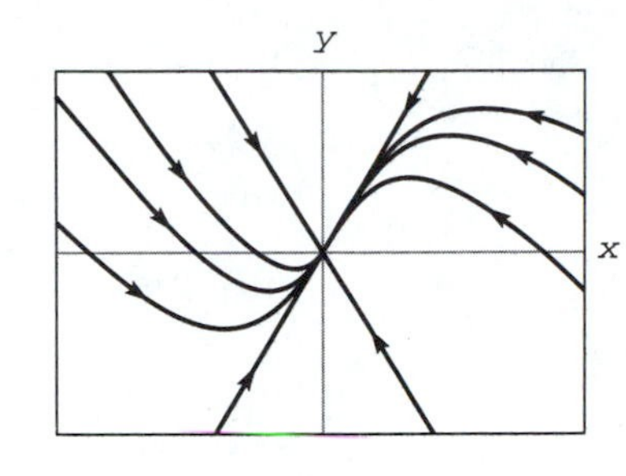

5. (0, 0) is a center.

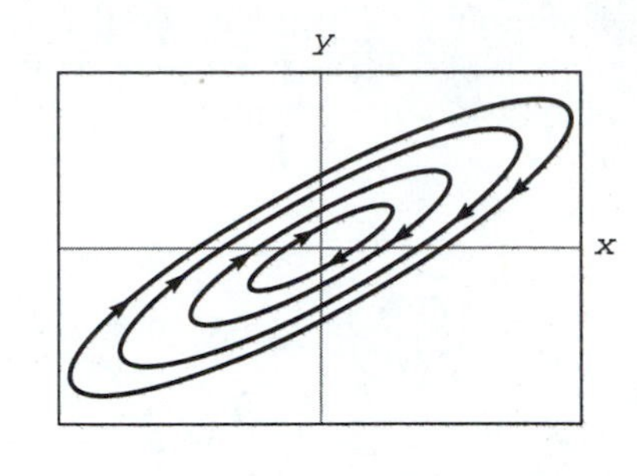

7. (0, 0) is an unstable spiral point.

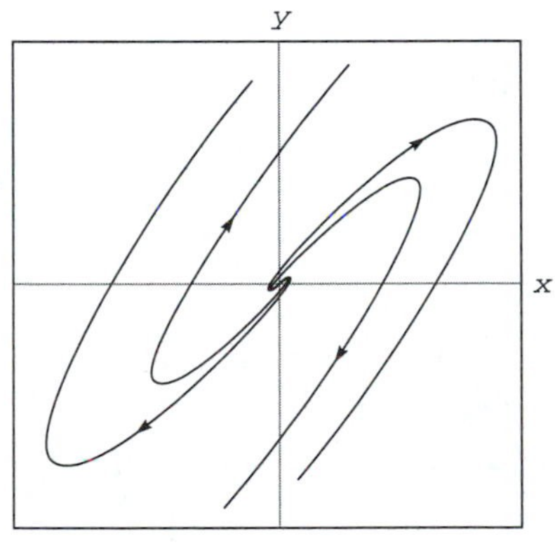

9. (−3, 4) is a saddle point.

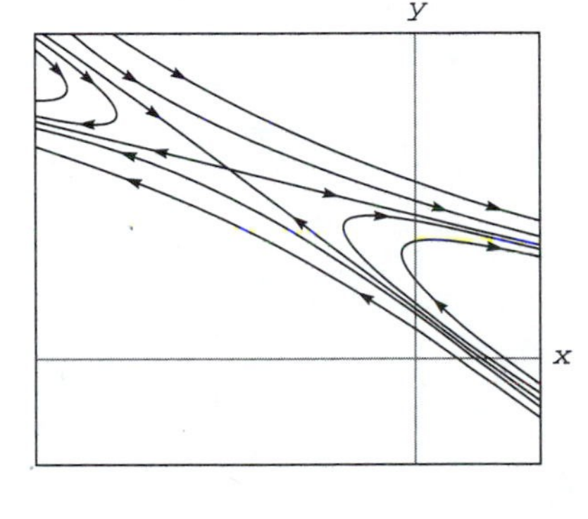

11. $(\frac{1}{2}, 2)$ is an unstable spiral point.

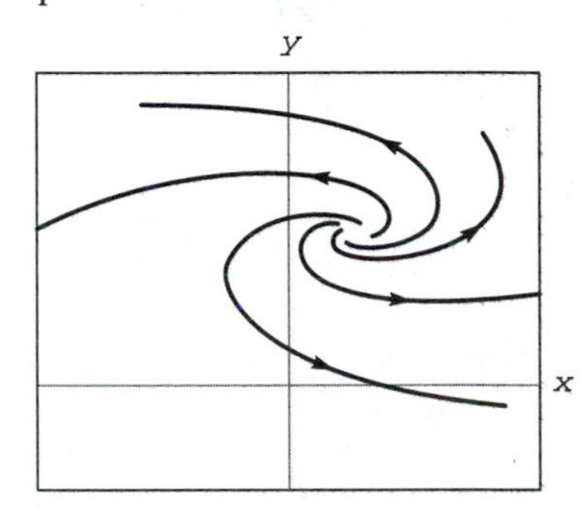

13. **(a)** $c < 0$ **(b)** $0 < c < 1$ **(c)** no values of c **(d)** $c = 0$ **(e)** $c > 1$ **15.** $-3 < a < -1$

17. (0, 0) is an unstable node. **19.** (0, 0) is an unstable spiral point. **21.** (0, 0) is a saddle point.

23. (0, 0) is an unstable node.

25. (1, 1) is an unstable spiral point; (−1, 1) is a saddle point.

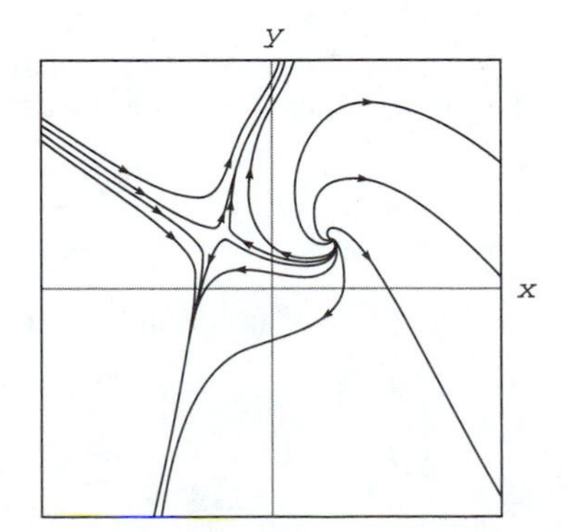

27. (1, 1) is a saddle point; (2, 1) is a saddle point; and $(\frac{3}{2}, \frac{3}{2})$ is an asymptotically stable spiral point.

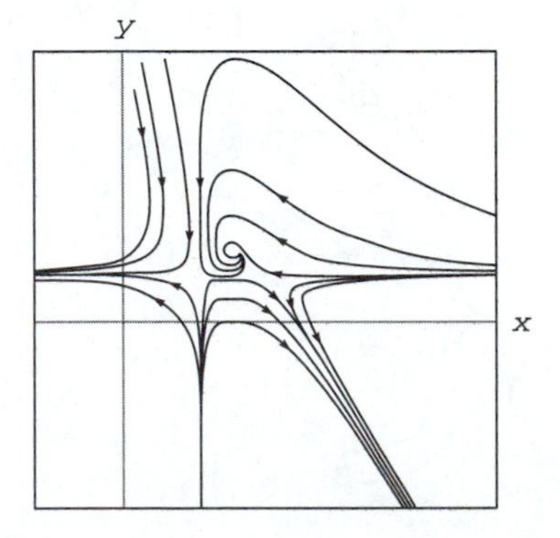

29. Critical points are $((2m + 1)\pi/2, n\pi)$, $m = 0, \pm 1, \pm 2, \ldots$, $n = 0, \pm 1, \pm 2, \ldots$. On a vertical line $x = (2m + 1)\pi/2$, the critical points alternate between centers and saddle points. For example, on $x = \pi/2$ ($m = 0$), $(\pi/2, 2\pi)$ is a center, $(\pi/2, \pi)$ is a saddle point, $(\pi/2, 0)$ is a center, $(\pi/2, -\pi)$ is a saddle point, and so on.

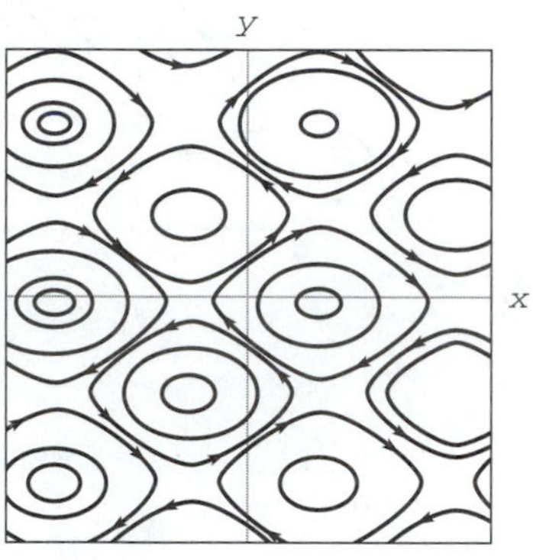

31. The only critical point is (0, 0). The phase portrait suggests that (0, 0) is stable but is not classifiable by type.

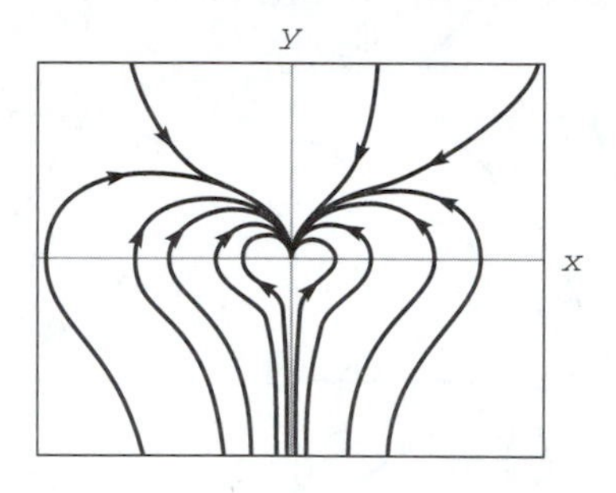

33. **(a)** The linearized system is

$$\frac{dx}{dt} = y$$

$$\frac{dy}{dt} = -x$$

(b) (0, 0) is a center.

(c) From the phase portrait on the left, we see that (0, 0) is a center for the first system. From the phase portrait on the right, we see that (0, 0) is an asymptotically stable spiral point for the second system.

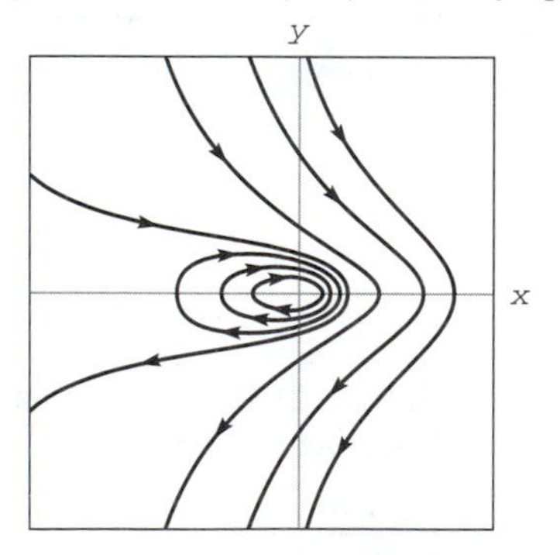

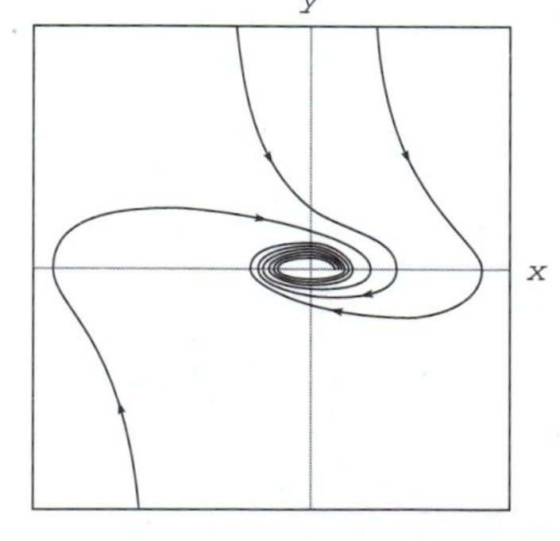

35. The following phase portrait suggests that the differential equation possesses no nonconstant periodic solutions.

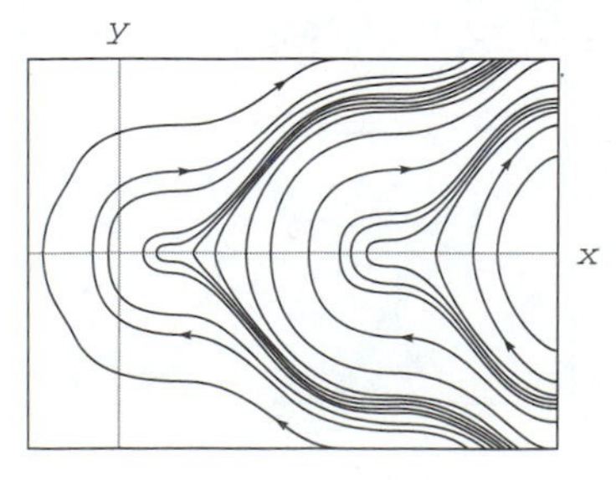

37. If a trajectory passes through (r_0, θ_0) when $t = 0$, then

$$r^2 = \frac{1}{\dfrac{1}{r_0^2} - 2t}, \quad \theta = t + \theta_0 \quad \text{or} \quad r^2 = \frac{1}{\dfrac{1}{r_0^2} - 2\theta + 2\theta_0}$$

describes a spiral curve. (0, 0) is an unstable spiral point.

41. (a) Critical point in the first quadrant is (8, 5).

(b)

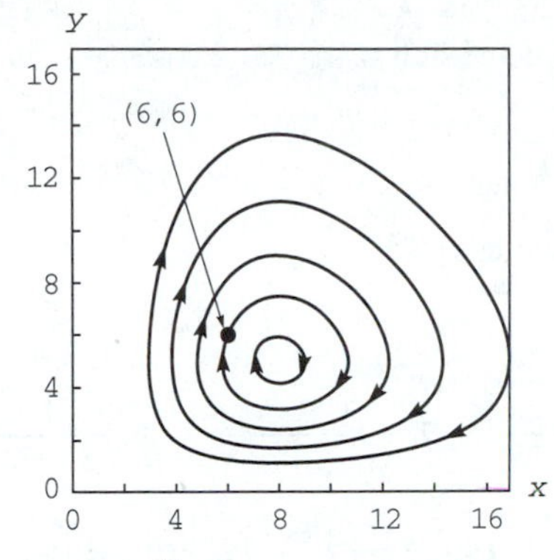

(c) Approximate period of $x(t)$ and $y(t)$ is $10\sqrt{2}\pi \approx 44.4$.

(d) Populations are first equal at (approximately) $t = 6$. Approximate period is 45.

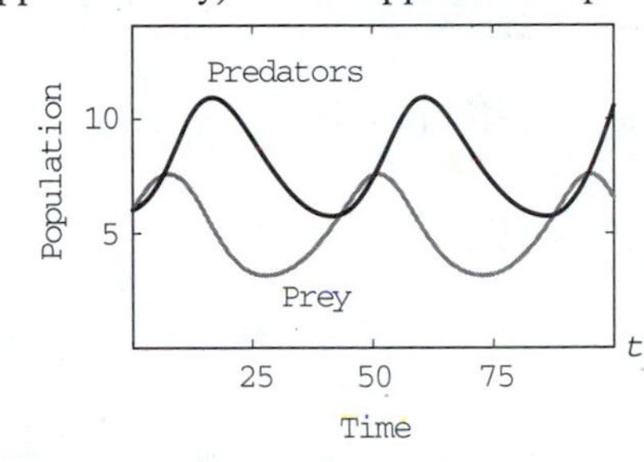

43. (a) Of the critical points (0, 0), (−2, 0), (0, 6), and (2, 4), only (2, 4) is in the first quadrant. By linearization, verified by the phase portrait, we find that (2, 4) is an asymptotically stable spiral point. In other words, the populations $x(t)$ and $y(t)$ are not periodic in this model. After a long time, both populations survive and approach equilibrium, that is, $x(t) \to 2$ and $y(t) \to 4$ as $t \to \infty$. The point (0, 6) is a saddle point and is unstable.

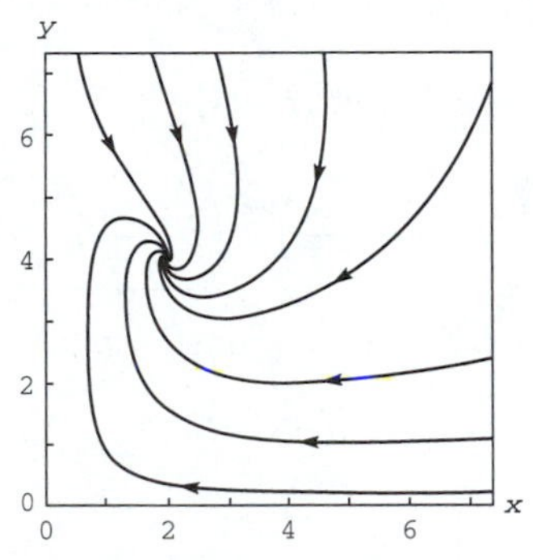

(b) None of the critical points (0, 0), (0, 2), and (−2, 0) are in the first quadrant. Nothing is learned from linearization at (0, 2), but the phase portrait indicates that (0, 2) is an asymptotically stable node. One population (predators) becomes extinct because $x(t) \to 0$ as $t \to \infty$; the other population (prey) survives because $y(t) \to 2$ as $t \to \infty$.

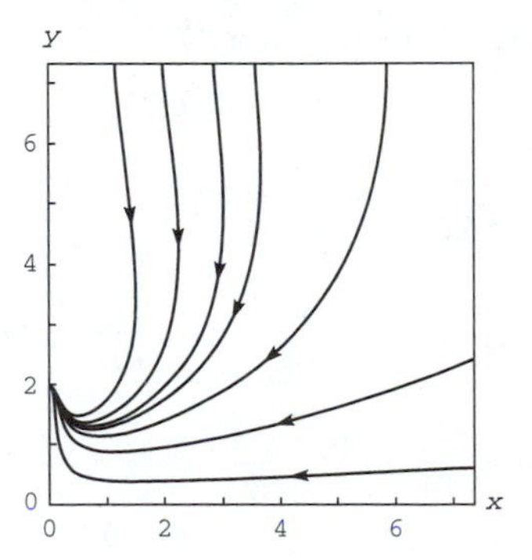

Exercises 5.1, Page 323

1. $\frac{2}{s}e^{-s} - \frac{1}{s}$ **3.** $\frac{1}{s^2} - \frac{1}{s^2}e^{-s}$ **5.** $\frac{1 + e^{-s\pi}}{s^2 + 1}$ **7.** $\frac{e^{-s}}{s} + \frac{e^{-s}}{s^2}$ **9.** $\frac{1}{s} - \frac{1}{s^2} + \frac{e^{-s}}{s^2}$ **11.** $\frac{e^7}{s - 1}$ **13.** $\frac{1}{(s - 4)^2}$

15. $\frac{k}{s^2 + k^2}$ **17.** $\frac{s^2 - 1}{(s^2 + 1)^2}$ **19.** $\frac{48}{s^5} - \frac{10}{s}$ **21.** $\frac{1}{s} + \frac{1}{s - 4}$ **23.** $\frac{6}{s^4} + \frac{6}{s^3} + \frac{3}{s^2} + \frac{1}{s}$ **25.** $\frac{8}{s^3} - \frac{15}{s^2 + 9}$ **27.** $\frac{k}{s^2 - k^2}$

29. $\frac{1}{2s} + \frac{1}{2}\frac{s}{s^2+4}$ **31.** (c) $\sqrt{\frac{\pi}{s}}$; $\frac{\sqrt{\pi}}{2s^{3/2}}$; $\frac{3\sqrt{\pi}}{4s^{5/2}}$ **39.** $\frac{1}{2}t^2$ **41.** $t - 2t^4$ **43.** $1 + 3t + \frac{3}{2}t^2 + \frac{1}{6}t^3$ **45.** $t - 1 + e^{2t}$

47. $\frac{1}{4}e^{-t/4}$ **49.** $\frac{5}{7}\sin 7t$ **51.** $\cos(t/2)$ **53.** $\frac{1}{4}\sinh 4t$ **55.** $2\cos 3t - 2\sin 3t$ **57.** $\frac{1}{3} - \frac{1}{3}e^{-3t}$ **59.** $\frac{3}{4}e^{-3t} + \frac{1}{4}e^{t}$

61. $0.3e^{0.1t} + 0.6e^{-0.2t}$ **63.** $\frac{1}{2}e^{2t} - e^{3t} + \frac{1}{2}e^{6t}$ **65.** $-\frac{1}{3}e^{-t} + \frac{8}{15}e^{2t} - \frac{1}{5}e^{-3t}$ **67.** $\frac{1}{4}t - \frac{1}{8}\sin 2t$

69. $-\frac{1}{4}e^{-2t} + \frac{1}{4}\cos 2t + \frac{1}{4}\sin 2t$ **71.** $y = -1 + e^{t}$ **73.** $y = te^{-4t} + 2e^{-4t}$ **75.** $y = \frac{4}{3}e^{-t} - \frac{1}{3}e^{-4t}$

77. $y = -\frac{8}{9}e^{-t/2} + \frac{1}{9}e^{-2t} + \frac{5}{18}e^{t} + \frac{1}{2}e^{-t}$ **79.** $y = \cos t$

Exercises 5.2, Page 335

1. $\frac{1}{(s-10)^2}$ **3.** $\frac{6}{(s+2)^4}$ **5.** $\frac{3}{(s-1)^2+9}$ **7.** $\frac{3}{(s-5)^2-9}$ **9.** $\frac{1}{(s-2)^2} + \frac{2}{(s-3)^2} + \frac{1}{(s-4)^2}$

11. $\frac{1}{2}\left[\frac{1}{s+1} - \frac{s+1}{(s+1)^2+4}\right]$ **13.** $\frac{1}{2}t^2e^{-2t}$ **15.** $e^{3t}\sin t$ **17.** $e^{-2t}\cos t - 2e^{-2t}\sin t$ **19.** $e^{-t} - te^{-t}$

21. $5 - t - 5e^{-t} - 4te^{-t} - \frac{3}{2}t^2e^{-t}$ **23.** $\frac{e^{-s}}{s^2}$ **25.** $\frac{e^{-2s}}{s^2} + 2\frac{e^{-2s}}{s}$ **27.** $\frac{s}{s^2+4}e^{-\pi s}$ **29.** $\frac{6e^{-s}}{(s-1)^4}$

31. $\frac{1}{2}(t-2)^2\mathscr{U}(t-2)$ **33.** $-\sin t\,\mathscr{U}(t-\pi)$ **35.** $\mathscr{U}(t-1) - e^{-(t-1)}\mathscr{U}(t-1)$ **37.** (c) **39.** (f) **41.** (a)

43. $f(t) = 2 - 4\mathscr{U}(t-3)$; $\mathscr{L}\{f(t)\} = \frac{2}{s} - \frac{4}{s}e^{-3s}$

45. $f(t) = t^2\mathscr{U}(t-1) = (t-1)^2\mathscr{U}(t-1) + 2(t-1)\mathscr{U}(t-1) + \mathscr{U}(t-1)$; $\mathscr{L}\{f(t)\} = 2\frac{e^{-s}}{s^3} + 2\frac{e^{-s}}{s^2} + \frac{e^{-s}}{s}$

47. $f(t) = t - t\mathscr{U}(t-2) = t - (t-2)\mathscr{U}(t-2) - 2\mathscr{U}(t-2)$; $\mathscr{L}\{f(t)\} = \frac{1}{s^2} - \frac{e^{-2s}}{s^2} - 2\frac{e^{-2s}}{s}$

49. $f(t) = \mathscr{U}(t-a) - \mathscr{U}(t-b)$; $\mathscr{L}\{f(t)\} = \frac{e^{-as}}{s} - \frac{e^{-bs}}{s}$ **51.** $y = -\frac{3}{2}e^{3t}\sin 2t$ **53.** $y = \frac{1}{20}t^5e^{2t}$

55. $y = \frac{1}{2} - \frac{1}{2}e^{t}\cos t + \frac{1}{2}e^{t}\sin t$ **57.** $y = [5 - 5e^{-(t-1)}]\mathscr{U}(t-1)$

59. $y = -\frac{1}{4} + \frac{1}{2}t + \frac{1}{4}e^{-2t} - \frac{1}{4}\mathscr{U}(t-1) - \frac{1}{2}(t-1)\mathscr{U}(t-1) + \frac{1}{4}e^{-2(t-1)}\mathscr{U}(t-1)$

61. $y = \cos 2t - \frac{1}{6}\sin 2(t-2\pi)\mathscr{U}(t-2\pi) + \frac{1}{3}\sin(t-2\pi)\mathscr{U}(t-2\pi)$

63. $y = \sin t + [1 - \cos(t-\pi)]\mathscr{U}(t-\pi) - [1 - \cos(t-2\pi)]\mathscr{U}(t-2\pi)$ **65.** $y = (e+1)te^{-t} + (e-1)e^{-t}$

67. $q(t) = \frac{E_0C}{1-kRC}(e^{-kt} - e^{-t/RC})$; $q(t) = \frac{E_0}{R}te^{-t/RC}$ **69.** $q(t) = \frac{2}{5}\mathscr{U}(t-3) - \frac{2}{5}e^{-5(t-3)}\mathscr{U}(t-3)$

71. **(a)** $i(t) = \frac{1}{101}e^{-10t} - \frac{1}{101}\cos t + \frac{10}{101}\sin t - \frac{10}{101}e^{-10(t-3\pi/2)}\mathscr{U}\left(t - \frac{3\pi}{2}\right) + \frac{10}{101}\cos\left(t - \frac{3\pi}{2}\right)\mathscr{U}\left(t - \frac{3\pi}{2}\right)$

$+ \frac{1}{101}\sin\left(t - \frac{3\pi}{2}\right)\mathscr{U}\left(t - \frac{3\pi}{2}\right)$

(b) $i_{max} \approx 0.1$ at $t \approx 1.6$, $i_{min} \approx -0.1$ at $t \approx 4.7$

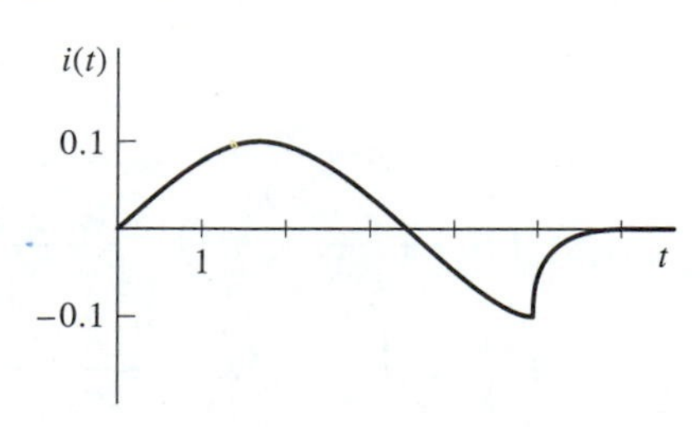

73. $q(t) = \frac{3}{5}e^{-10t} + 6te^{-10t} - \frac{3}{5}\cos 10t$; $i(t) = -60te^{-10t} + 6\sin 10t$; steady-state current is $6\sin 10t$

75. $x(t) = -\frac{3}{2}e^{-7t/2}\cos\frac{\sqrt{15}}{2}t - \frac{7\sqrt{15}}{10}e^{-7t/2}\sin\frac{\sqrt{15}}{2}t$

77. $y(x) = \frac{w_0L^2}{12EI}x^2 - \frac{w_0L}{18EI}x^3 + \frac{w_0}{24EI}\left(x - \frac{L}{3}\right)^4\mathcal{U}\left(x - \frac{L}{3}\right) - \frac{w_0}{24EI}\left(x - \frac{2L}{3}\right)^4\mathcal{U}\left(x - \frac{2L}{3}\right)$

79. $y(x) = \frac{w_0L^2}{48EI}x^2 - \frac{w_0L}{24EI}x^3 + \frac{w_0}{60EIL}\left[\frac{5L}{2}x^4 - x^5 + \left(x - \frac{L}{2}\right)^5\mathcal{U}\left(x - \frac{L}{2}\right)\right]$

Exercises 5.3, Page 347

1. $\frac{s^2-4}{(s^2+4)^2}$ **3.** $\frac{6s^2+2}{(s^2-1)^3}$ **5.** $\frac{12s-24}{[(s-2)^2+36]^2}$ **7.** $\frac{1}{2}t\sin t$ **9.** $\frac{1}{s(s-1)}$ **11.** $\frac{s+1}{s[(s+1)^2+1]}$ **13.** $\frac{1}{s^2(s-1)}$

15. $\frac{3s^2+1}{s^2(s^2+1)^2}$ **17.** $\frac{6}{s^5}$ **19.** $\frac{48}{s^8}$ **21.** $\frac{s-1}{(s+1)[(s-1)^2+1]}$ **23.** $\frac{e^{-t}-e^{3t}}{t}$ **25.** $\int_0^t f(\tau)e^{-5(t-\tau)}\,d\tau$ **27.** $1 - e^{-t}$

29. $-\frac{1}{3}e^{-t} + \frac{1}{3}e^{2t}$ **31.** $\frac{1}{4}t\sin 2t$ **35.** $y = 2\cos 3t + \frac{5}{3}\sin 3t + \frac{1}{6}t\sin 3t$

37. $y = -\frac{1}{6}(t-\pi)\sin 3(t-\pi)\,\mathcal{U}(t-\pi) = \frac{1}{6}(t-\pi)\sin 3t\,\mathcal{U}(t-\pi)$ **39.** $f(t) = \sin t$

41. $f(t) = -\frac{1}{8}e^{-t} + \frac{1}{8}e^{t} + \frac{3}{4}te^{t} + \frac{1}{4}t^2e^{t}$ **43.** $f(t) = e^{-t}$ **45.** $f(t) = \frac{3}{8}e^{2t} + \frac{1}{8}e^{-2t} + \frac{1}{2}\cos 2t + \frac{1}{4}\sin 2t$

47. $y = \sin t - \frac{1}{2}t\sin t$ **49.** $i(t) = 20{,}000[te^{-100t} - (t-1)e^{-100(t-1)}\mathcal{U}(t-1)]$ **51.** $x(t) = \frac{2}{3}t\sin 3t - \frac{4}{3}t\cos 3t + \frac{4}{9}\sin 3t$

Exercises 5.4, Page 352

1. $\frac{1}{s^2+1}$ **3.** $\frac{(1-e^{-as})^2}{s(1-e^{-2as})} = \frac{1-e^{-as}}{s(1+e^{-as})}$ **5.** $\frac{a}{s}\left(\frac{1}{bs} - \frac{1}{e^{bs}-1}\right)$ **7.** $\frac{\coth(\pi s/2)}{s^2+1}$

9. $i(t) = \frac{t}{R} + \frac{L}{R^2}(e^{-Rt/L} - 1) + \frac{1}{R}\sum_{n=1}^{\infty}(e^{-R(t-n)/L} - 1)\mathcal{U}(t-n)$

For $0 \le t < 2$,

$$i(t) = \begin{cases} \frac{t}{R} + \frac{L}{R^2}(e^{-Rt/L} - 1), & 0 \le t < 1 \\ \frac{t}{R} + \frac{L}{R^2}(e^{-Rt/L} - 1) + \frac{1}{R}(e^{-R(t-1)/L} - 1), & 1 \le t < 2 \end{cases}$$

11. $x(t) = 2(1 - e^{-t}\cos 3t - \frac{1}{3}e^{-t}\sin 3t) + 4\sum_{n=1}^{\infty}(-1)^n[1 - e^{-(t-n\pi)}\cos 3(t-n\pi) - \frac{1}{3}e^{-(t-n\pi)}\sin 3(t-n\pi)]\,\mathcal{U}(t-n\pi)$

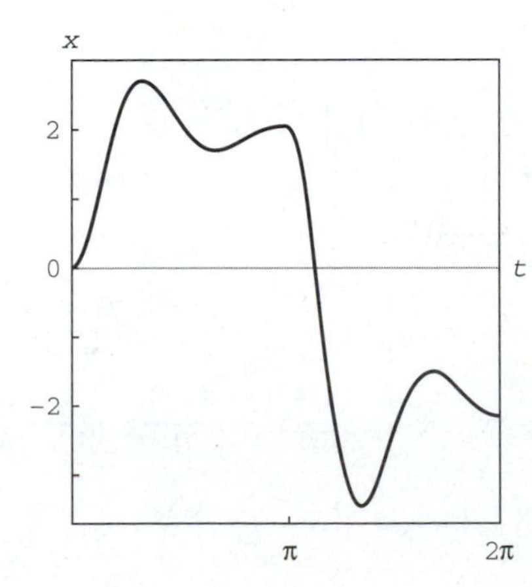

Exercises 5.5, Page 356

1. $y = e^{3(t-2)}\mathcal{U}(t-2)$ **3.** $y = \sin t + \sin t\,\mathcal{U}(t-2\pi)$ **5.** $y = -\cos t\,\mathcal{U}\left(t-\frac{\pi}{2}\right) + \cos t\,\mathcal{U}\left(t-\frac{3\pi}{2}\right)$

7. $y = \frac{1}{2} - \frac{1}{2}e^{-2t} + \left[\frac{1}{2} - \frac{1}{2}e^{-2(t-1)}\right]\mathcal{U}(t-1)$ **9.** $y = e^{-2(t-2\pi)}\sin t\,\mathcal{U}(t-2\pi)$

11. $y = e^{-2t}\cos 3t + \frac{2}{3}e^{-2t}\sin 3t + \frac{1}{3}e^{-2(t-\pi)}\sin 3(t-\pi)\mathcal{U}(t-\pi) + \frac{1}{3}e^{-2(t-3\pi)}\sin 3(t-3\pi)\mathcal{U}(t-3\pi)$

13. $$y(x) = \begin{cases} \dfrac{P_0}{EI}\left(\dfrac{L}{4}x^2 - \dfrac{1}{6}x^3\right), & 0 \le x < L/2 \\ \dfrac{P_0 L^2}{4EI}\left(\dfrac{1}{2}x - \dfrac{L}{12}\right), & L/2 \le x \le L \end{cases}$$

15. $y = e^{-t}\cos t + e^{-(t-3\pi)}\sin t\,\mathcal{U}(t-3\pi)$;

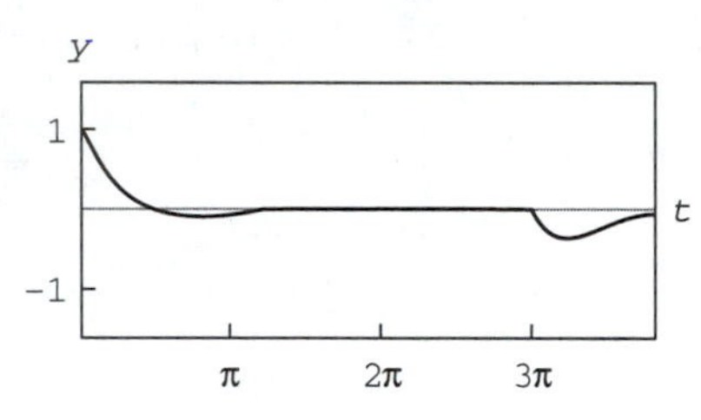

Exercises 5.6, Page 360

1. $x = -\frac{1}{3}e^{-2t} + \frac{1}{3}e^{t}$
$y = \frac{1}{3}e^{-2t} + \frac{2}{3}e^{t}$

3. $x = -\cos 3t - \frac{5}{3}\sin 3t$
$y = 2\cos 3t - \frac{7}{3}\sin 3t$

5. $x = -2e^{3t} + \frac{5}{2}e^{2t} - \frac{1}{2}$
$y = \frac{8}{3}e^{3t} - \frac{5}{2}e^{2t} - \frac{1}{6}$

7. $x = -\frac{1}{2}t - \frac{3}{4}\sqrt{2}\sin\sqrt{2}t$
$y = -\frac{1}{2}t + \frac{3}{4}\sqrt{2}\sin\sqrt{2}t$

9. $x = 8 + \dfrac{2}{3!}t^3 + \dfrac{1}{4!}t^4$
$y = -\dfrac{2}{3!}t^3 + \dfrac{1}{4!}t^4$

11. $x = \frac{1}{2}t^2 + t + 1 - e^{-t}$
$y = -\frac{1}{3} + \frac{1}{3}e^{-t} + \frac{1}{3}te^{-t}$

13. $x_1 = \dfrac{1}{5}\sin t + \dfrac{2\sqrt{6}}{15}\sin\sqrt{6}t + \dfrac{2}{5}\cos t - \dfrac{2}{5}\cos\sqrt{6}t$
$x_2 = \dfrac{2}{5}\sin t - \dfrac{\sqrt{6}}{15}\sin\sqrt{6}t + \dfrac{4}{5}\cos t + \dfrac{1}{5}\cos\sqrt{6}t$

15. (b) $i_2 = \frac{100}{9} - \frac{100}{9}e^{-900t}$
$i_3 = \frac{80}{9} - \frac{80}{9}e^{-900t}$
(c) $i_1 = 20 - 20e^{-900t}$

17. $i_2 = -\frac{20}{13}e^{-2t} + \frac{375}{1469}e^{-15t} + \frac{145}{113}\cos t + \frac{85}{113}\sin t$
$i_3 = \frac{30}{13}e^{-2t} + \frac{250}{1469}e^{-15t} - \frac{280}{113}\cos t + \frac{810}{113}\sin t$

19. $i_1 = \frac{6}{5} - \frac{6}{5}e^{-100t}\cos 100t$
$i_2 = \frac{6}{5} - \frac{6}{5}e^{-100t}\cos 100t - \frac{6}{5}e^{-100t}\sin 100t$

21. (b) $q = 50e^{-t}\sin(t-1)\mathcal{U}(t-1)$

Exercises 6.1, Page 377

1. $y_1(x) = c_0\left[1 + \dfrac{1}{3\cdot 2}x^3 + \dfrac{1}{6\cdot 5\cdot 3\cdot 2}x^6 + \dfrac{1}{9\cdot 8\cdot 6\cdot 5\cdot 3\cdot 2}x^9 + \cdots\right]$
$y_2(x) = c_1\left[x + \dfrac{1}{4\cdot 3}x^4 + \dfrac{1}{7\cdot 6\cdot 4\cdot 3}x^7 + \dfrac{1}{10\cdot 9\cdot 7\cdot 6\cdot 4\cdot 3}x^{10} + \cdots\right]$

3. $y_1(x) = c_0\left[1 - \dfrac{1}{2!}x^2 - \dfrac{3}{4!}x^4 - \dfrac{21}{6!}x^6 - \cdots\right]$
$y_2(x) = c_1\left[x + \dfrac{1}{3!}x^3 + \dfrac{5}{5!}x^5 + \dfrac{45}{7!}x^7 + \cdots\right]$

5. $y_1(x) = c_0\left[1 - \dfrac{1}{3!}x^3 + \dfrac{4^2}{6!}x^6 - \dfrac{7^2\cdot 4^2}{9!}x^9 + \cdots\right]$
$y_2(x) = c_1\left[x - \dfrac{2^2}{4!}x^4 + \dfrac{5^2\cdot 2^2}{7!}x^7 - \dfrac{8^2\cdot 5^2\cdot 2^2}{10!}x^{10} + \cdots\right]$

7. $y_1(x) = c_0$; $y_2(x) = c_1\sum_{n=1}^{\infty}\dfrac{1}{n}x^n$ **9.** $y_1(x) = c_0\sum_{n=0}^{\infty}x^{2n}$; $y_2(x) = c_1\sum_{n=0}^{\infty}x^{2n+1}$

11. $y_1(x) = c_0\left[1 + \frac{1}{4}x^2 - \frac{7}{4\cdot 4!}x^4 + \frac{23\cdot 7}{8\cdot 6!}x^6 - \cdots\right]$

$y_2(x) = c_1\left[x - \frac{1}{6}x^3 + \frac{14}{2\cdot 5!}x^5 - \frac{34\cdot 14}{4\cdot 7!}x^7 - \cdots\right]$

13. $y_1(x) = c_0[1 + \frac{1}{2}x^2 + \frac{1}{6}x^3 + \frac{1}{6}x^4 + \cdots]$

$y_2(x) = c_1[x + \frac{1}{2}x^2 + \frac{1}{2}x^3 + \frac{1}{4}x^4 + \cdots]$

15. $y(x) = -2\left[1 + \frac{1}{2!}x^2 + \frac{1}{3!}x^3 + \frac{1}{4!}x^4 + \cdots\right] + 6x = 8x - 2e^x$

17. $y(x) = 3 - 12x^2 + 4x^4$

19. $y_1(x) = c_0[1 - \frac{1}{6}x^3 + \frac{1}{120}x^5 + \cdots]$

$y_2(x) = c_1[x - \frac{1}{12}x^4 + \frac{1}{180}x^6 + \cdots]$

21. $y_1(x) = c_0[1 - \frac{1}{2}x^2 + \frac{1}{6}x^3 - \frac{1}{40}x^5 + \cdots]$

$y_2(x) = c_1[x - \frac{1}{6}x^3 + \frac{1}{12}x^4 - \frac{1}{60}x^5 + \cdots]$

Exercises 6.2, Page 386

1. $x = 0$, irregular singular point
3. $x = -3$, regular singular point; $x = 3$, irregular singular point
5. $x = 0, 2i, -2i$, regular singular points
7. $x = -3, 2$, regular singular points
9. $x = 0$, irregular singular point; $x = -5, 5, 2$, regular singular points

11. $r_1 = \frac{3}{2}, r_2 = 0;\quad y(x) = C_1x^{3/2}\left[1 - \frac{2}{5}x + \frac{2^2}{7\cdot 5\cdot 2}x^2 - \frac{2^3}{9\cdot 7\cdot 5\cdot 3!}x^3 + \cdots\right] + C_2\left[1 + 2x - 2x^2 + \frac{2^3}{3\cdot 3!}x^3 - \cdots\right]$

13. $r_1 = \frac{7}{8}, r_2 = 0;\quad y(x) = C_1x^{7/8}\left[1 - \frac{2}{15}x + \frac{2^2}{23\cdot 15\cdot 2}x^2 - \frac{2^3}{31\cdot 23\cdot 15\cdot 3!}x^3 + \cdots\right]$

$+ C_2\left[1 - 2x + \frac{2^2}{9\cdot 2}x^2 - \frac{2^3}{17\cdot 9\cdot 3!}x^3 + \cdots\right]$

15. $r_1 = \frac{1}{3}, r_2 = 0;\quad y(x) = C_1x^{1/3}\left[1 + \frac{1}{3}x + \frac{1}{3^2\cdot 2}x^2 + \frac{1}{3^3\cdot 3!}x^3 + \cdots\right] + C_2\left[1 + \frac{1}{2}x + \frac{1}{5\cdot 2}x^2 + \frac{1}{8\cdot 5\cdot 2}x^3 + \cdots\right]$

17. $r_1 = \frac{5}{2}, r_2 = 0;\quad y(x) = C_1x^{5/2}\left[1 + \frac{2\cdot 2}{7}x + \frac{2^2\cdot 3}{9\cdot 7}x^2 + \frac{2^3\cdot 4}{11\cdot 9\cdot 7}x^3 + \cdots\right] + C_2\left[1 + \frac{1}{3}x - \frac{1}{6}x^2 - \frac{1}{6}x^3 - \cdots\right]$

19. $r_1 = \frac{2}{3}, r_2 = \frac{1}{3};\quad y(x) = C_1x^{2/3}[1 - \frac{1}{2}x + \frac{5}{28}x^2 - \frac{1}{21}x^3 + \cdots] + C_2x^{1/3}[1 - \frac{1}{2}x + \frac{1}{5}x^2 - \frac{7}{120}x^3 + \cdots]$

21. $r_1 = 1, r_2 = -\frac{1}{2};\quad y(x) = C_1x\left[1 + \frac{1}{5}x + \frac{1}{5\cdot 7}x^2 + \frac{1}{5\cdot 7\cdot 9}x^3 + \cdots\right] + C_2x^{-1/2}\left[1 + \frac{1}{2}x + \frac{1}{2\cdot 4}x^2 + \frac{1}{2\cdot 4\cdot 6}x^3 + \cdots\right]$

23. $r_1 = 0, r_2 = -1;\quad y(x) = C_1x^{-1}\sum_{n=0}^{\infty}\frac{1}{(2n)!}x^{2n} + C_2x^{-1}\sum_{n=0}^{\infty}\frac{1}{(2n+1)!}x^{2n+1} = \frac{1}{x}[C_1\cosh x + C_2\sinh x]$

25. $r_1 = 4, r_2 = 0;\quad y(x) = C_1\left[1 + \frac{2}{3}x + \frac{1}{3}x^2\right] + C_2\sum_{n=0}^{\infty}(n+1)x^{n+4}$

27. $r_1 = r_2 = 0;\quad y(x) = C_1y_1(x) + C_2\left[y_1(x)\ln x + y_1(x)\left(-x + \frac{1}{4}x^2 - \frac{1}{3\cdot 3!}x^3 + \frac{1}{4\cdot 4!}x^4 - \cdots\right)\right]$

where $y_1(x) = \sum_{n=0}^{\infty}\frac{1}{n!}x^n = e^x$

29. $r_1 = r_2 = 0;\quad y(x) = C_1y_1(x) + C_2[y_1(x)\ln x + y_1(x)(2x + \frac{5}{4}x^2 + \frac{23}{27}x^3 + \cdots)]$

where $y_1(x) = \sum_{n=0}^{\infty}\frac{(-1)^n}{(n!)^{2n}}x^n$

31. **(b)** $y_1 = \sum_{n=0}^{\infty}\frac{(-1)^n}{(2n+1)!}(\sqrt{\lambda}\,t)^{2n} = \frac{\sin\sqrt{\lambda}\,t}{\sqrt{\lambda}\,t}, y_2 = t^{-1}\sum_{n=0}^{\infty}\frac{(-1)^n}{(2n)!}(\sqrt{\lambda}\,t)^{2n} = \frac{\cos\sqrt{\lambda}\,t}{t}$

(c) Since $t = 1/x$, we have the solution $y = c_1x\sin\left(\frac{\sqrt{\lambda}\,x}{x}\right) + c_2x\cos\left(\frac{\sqrt{\lambda}\,x}{x}\right)$

Exercises 6.3, Page 393

1. $y = c_1J_{1/3}(x) + c_2J_{-1/3}(x)$
3. $y = c_1J_{5/2}(x) + c_2J_{-5/2}(x)$
5. $y = c_1J_0(x) + c_2Y_0(x)$
7. $y = c_1J_2(3x) + c_2Y_2(3x)$

9. After we use the change of variables, the differential equation becomes

$$x^2w'' + xw' + (\lambda^2x^2 - \tfrac{1}{4})w = 0$$

Since the solution of the last equation is

$$w = c_1J_{1/2}(\lambda x) + c_2J_{-1/2}(\lambda x)$$

we find

$$y = c_1x^{-1/2}J_{1/2}(\lambda x) + c_2x^{-1/2}J_{-1/2}(\lambda x)$$

13. From Problem 10 with $n = \frac{1}{2}$, we find $y = x^{1/2}J_{1/2}(x)$; from Problem 11 with $n = -\frac{1}{2}$, we find $y = x^{1/2}J_{-1/2}(x)$.

15. From Problem 10 with $n = -1$, we find $y = x^{-1}J_{-1}(x)$; from Problem 11 with $n = 1$, we find $y = x^{-1}J_1(x)$ but since $J_{-1}(x) = -J_1(x)$, no new solution results.

17. From Problem 12 with $\lambda = 1$ and $v = \pm\frac{3}{2}$, we find $y = \sqrt{x}J_{3/2}(x)$ and $y = \sqrt{x}J_{-3/2}(x)$.

27. **(a)** $x(t) = -0.809264t^{1/2}J_{1/3}(\frac{1}{3}t^{3/2}) + 0.782397t^{1/2}J_{-1/3}(\frac{1}{3}t^{3/2})$

Appendix I, Page 420

1. **(a)** $\begin{pmatrix} 2 & 11 \\ 2 & -1 \end{pmatrix}$ **(b)** $\begin{pmatrix} -6 & 1 \\ 14 & -19 \end{pmatrix}$ **(c)** $\begin{pmatrix} 2 & 28 \\ 12 & -12 \end{pmatrix}$

3. **(a)** $\begin{pmatrix} -11 & 6 \\ 17 & -22 \end{pmatrix}$ **(b)** $\begin{pmatrix} -32 & 27 \\ -4 & -1 \end{pmatrix}$ **(c)** $\begin{pmatrix} 19 & -18 \\ -30 & 31 \end{pmatrix}$ **(d)** $\begin{pmatrix} 19 & 6 \\ 3 & 22 \end{pmatrix}$

5. **(a)** $\begin{pmatrix} 9 & 24 \\ 3 & 8 \end{pmatrix}$ **(b)** $\begin{pmatrix} 3 & 8 \\ -6 & -16 \end{pmatrix}$ **(c)** $\begin{pmatrix} 0 & 0 \\ 0 & 0 \end{pmatrix}$ **(d)** $\begin{pmatrix} -4 & -5 \\ 8 & 10 \end{pmatrix}$

7. **(a)** 180 **(b)** $\begin{pmatrix} 4 & 8 & 10 \\ 8 & 16 & 20 \\ 10 & 20 & 25 \end{pmatrix}$ **(c)** $\begin{pmatrix} 6 \\ 12 \\ -5 \end{pmatrix}$ **9.** **(a)** $\begin{pmatrix} 7 & 38 \\ 10 & 75 \end{pmatrix}$ **(b)** $\begin{pmatrix} 7 & 38 \\ 10 & 75 \end{pmatrix}$ **11.** $\begin{pmatrix} -14 \\ 1 \end{pmatrix}$ **13.** $\begin{pmatrix} -38 \\ -2 \end{pmatrix}$

15. singular **17.** nonsingular; $\mathbf{A}^{-1} = \frac{1}{4}\begin{pmatrix} -5 & -8 \\ 3 & 4 \end{pmatrix}$ **19.** nonsingular; $\mathbf{A}^{-1} = \frac{1}{2}\begin{pmatrix} 0 & -1 & 1 \\ 2 & 2 & -2 \\ -4 & -3 & 5 \end{pmatrix}$

21. nonsingular; $\mathbf{A}^{-1} = -\frac{1}{9}\begin{pmatrix} -2 & -2 & -1 \\ -13 & 5 & 7 \\ 8 & -1 & -5 \end{pmatrix}$ **23.** $\det \mathbf{A}(t) = 2e^{3t} \neq 0$ for every value of t; $\mathbf{A}^{-1}(t) = \frac{1}{2e^{3t}}\begin{pmatrix} 3e^{4t} & -e^{4t} \\ -4e^{-t} & 2e^{-t} \end{pmatrix}$

25. $\frac{d\mathbf{X}}{dt} = \begin{pmatrix} -5e^{-t} \\ -2e^{-t} \\ 7e^{-t} \end{pmatrix}$ **27.** $\frac{d\mathbf{X}}{dt} = 4\begin{pmatrix} 1 \\ -1 \end{pmatrix}e^{2t} - 12\begin{pmatrix} 2 \\ 1 \end{pmatrix}e^{-3t}$

29. **(a)** $\begin{pmatrix} 4e^{4t} & -\pi \sin \pi t \\ 2 & 6t \end{pmatrix}$ **(b)** $\begin{pmatrix} \frac{1}{4}e^{8} - \frac{1}{4} & 0 \\ 4 & 6 \end{pmatrix}$ **(c)** $\begin{pmatrix} \frac{1}{4}e^{4t} - \frac{1}{4} & (1/\pi)\sin \pi t \\ t^2 & t^3 - t \end{pmatrix}$ **31.** $x = 3, y = 1, z = -5$

33. $x = 2 + 4t$, $y = -5 - t$, $z = t$ **35.** $x = -\frac{1}{2}$, $y = \frac{3}{2}$, $z = \frac{7}{2}$ **37.** $x_1 = 1$, $x_2 = 0$, $x_3 = 2$, $x_4 = 0$

41. $\lambda_1 = 6, \lambda_2 = 1$, $\mathbf{K}_1 = \begin{pmatrix} 2 \\ 7 \end{pmatrix}, \mathbf{K}_2 = \begin{pmatrix} 1 \\ 1 \end{pmatrix}$ **43.** $\lambda_1 = \lambda_2 = -4$, $\mathbf{K}_1 = \begin{pmatrix} 1 \\ -4 \end{pmatrix}$

45. $\lambda_1 = 0, \lambda_2 = 4, \lambda_3 = -4$, $\mathbf{K}_1 = \begin{pmatrix} 9 \\ 45 \\ 25 \end{pmatrix}, \mathbf{K}_2 = \begin{pmatrix} 1 \\ 1 \\ 1 \end{pmatrix}, \mathbf{K}_3 = \begin{pmatrix} 1 \\ 9 \\ 1 \end{pmatrix}$ **47.** $\lambda_1 = \lambda_2 = \lambda_3 = -2$, $\mathbf{K}_1 = \begin{pmatrix} 2 \\ -1 \\ 0 \end{pmatrix}, \mathbf{K}_2 = \begin{pmatrix} 0 \\ 0 \\ 1 \end{pmatrix}$

49. $\lambda_1 = 3i, \lambda_2 = -3i$, $\mathbf{K}_1 = \begin{pmatrix} 1 - 3i \\ 5 \end{pmatrix}, \mathbf{K}_2 = \begin{pmatrix} 1 + 3i \\ 5 \end{pmatrix}$

53. Since $\mathbf{A}^{-1}$ exists, $\mathbf{AB} = \mathbf{AC}$ implies $\mathbf{A}^{-1}(\mathbf{AB}) = \mathbf{A}^{-1}(\mathbf{AC})$, $(\mathbf{A}^{-1}\mathbf{A})\mathbf{B} = (\mathbf{A}^{-1}\mathbf{A})\mathbf{C}$, $\mathbf{IB} = \mathbf{IC}$, or $\mathbf{B} = \mathbf{C}$.

55. No, since in general $\mathbf{AB} \neq \mathbf{BA}$.

Index

TABLE OF LAPLACE TRANSFORMS

$f(t)$	$\mathscr{L}\{f(t)\} = F(s)$
1. 1	$\dfrac{1}{s}$
2. t	$\dfrac{1}{s^2}$
3. t^n	$\dfrac{n!}{s^{n+1}}$, n a positive integer
4. $t^{-1/2}$	$\sqrt{\dfrac{\pi}{s}}$
5. $t^{1/2}$	$\dfrac{\sqrt{\pi}}{2s^{3/2}}$
6. t^α	$\dfrac{\Gamma(\alpha+1)}{s^{\alpha+1}}, \alpha > -1$
7. $\sin kt$	$\dfrac{k}{s^2+k^2}$
8. $\cos kt$	$\dfrac{s}{s^2+k^2}$
9. $\sin^2 kt$	$\dfrac{2k^2}{s(s^2+4k^2)}$
10. $\cos^2 kt$	$\dfrac{s^2+2k^2}{s(s^2+4k^2)}$
11. e^{at}	$\dfrac{1}{s-a}$
12. $\sinh kt$	$\dfrac{k}{s^2-k^2}$
13. $\cosh kt$	$\dfrac{s}{s^2-k^2}$
14. $\sinh^2 kt$	$\dfrac{2k^2}{s(s^2-4k^2)}$
15. $\cosh^2 kt$	$\dfrac{s^2-2k^2}{s(s^2-4k^2)}$
16. te^{at}	$\dfrac{1}{(s-a)^2}$
17. $t^n e^{at}$	$\dfrac{n!}{(s-a)^{n+1}}$, n a positive integer

$f(t)$	$\mathscr{L}\{f(t)\} = F(s)$
18. $e^{at}\sin kt$	$\dfrac{k}{(s-a)^2+k^2}$
19. $e^{at}\cos kt$	$\dfrac{s-a}{(s-a)^2+k^2}$
20. $e^{at}\sinh kt$	$\dfrac{k}{(s-a)^2-k^2}$
21. $e^{at}\cosh kt$	$\dfrac{s-a}{(s-a)^2-k^2}$
22. $t\sin kt$	$\dfrac{2ks}{(s^2+k^2)^2}$
23. $t\cos kt$	$\dfrac{s^2-k^2}{(s^2+k^2)^2}$
24. $\sin kt + kt\cos kt$	$\dfrac{2ks^2}{(s^2+k^2)^2}$
25. $\sin kt - kt\cos kt$	$\dfrac{2k^3}{(s^2+k^2)^2}$
26. $t\sinh kt$	$\dfrac{2ks}{(s^2-k^2)^2}$
27. $t\cosh kt$	$\dfrac{s^2+k^2}{(s^2-k^2)^2}$
28. $\dfrac{e^{at}-e^{bt}}{a-b}$	$\dfrac{1}{(s-a)(s-b)}$
29. $\dfrac{ae^{at}-be^{bt}}{a-b}$	$\dfrac{s}{(s-a)(s-b)}$
30. $1-\cos kt$	$\dfrac{k^2}{s(s^2+k^2)}$
31. $kt-\sin kt$	$\dfrac{k^3}{s^2(s^2+k^2)}$
32. $\dfrac{a\sin bt - b\sin at}{ab(a^2-b^2)}$	$\dfrac{1}{(s^2+a^2)(s^2+b^2)}$
33. $\dfrac{\cos bt - \cos at}{a^2-b^2}$	$\dfrac{s}{(s^2+a^2)(s^2+b^2)}$
34. $\sin kt\sinh kt$	$\dfrac{2k^2 s}{s^4+4k^4}$